Springer Undergraduate Mathematics Series

The Springer Undergraduate Mathematics Series (SUMS) is a series designed for undergraduates in mathematics and the sciences worldwide. From core foundational material to final year topics, SUMS books take a fresh and modern approach. Textual explanations are supported by a wealth of examples, problems and fully-worked solutions, with particular attention paid to universal areas of difficulty. These practical and concise texts are designed for a one- or two-semester course but the self-study approach makes them ideal for independent use.

Jacek Banasiak

Introduction to Mathematical Methods in Population Theory

Springer

Jacek Banasiak
Department of Mathematics and Applied Mathematics
University of Pretoria
Pretoria, Hatfield, South Africa

Institute of Mathematics
Łódź University of Technology
Łódź, Poland

ISSN 1615-2085 ISSN 2197-4144 (electronic)
Springer Undergraduate Mathematics Series
ISBN 978-3-031-65490-9 ISBN 978-3-031-65491-6 (eBook)
https://doi.org/10.1007/978-3-031-65491-6

This Springer imprint is published by the registered company Springer Nature Switzerland AG
The registered company address is: Gewerbestrasse 11, 6330 Cham, Switzerland

Preface

Why Did I Write This Book?

I see and teach mathematical biology and, in particular, population theory as a part of mathematical modelling, that is, translating the rules of nature into mathematical formulae (most often equations), applying mathematical methods to analyse them and then trying to understand the implications of the obtained results for the original problems. There are many textbooks on mathematical biology or population theory, ranging from excellent to weak. Most of them have, in my opinion, a certain drawback: their main focus is on the biological content, while the mathematics is often ornamental or, at best, sketchy and relegated to appendices or more advanced textbooks. Also, many population dynamics and mathematical biology textbooks use very sophisticated mathematics. Still, it is very difficult to find precise references linking this level of sophistication with standard undergraduate education. My experience as an external examiner, referee and editor dealing with courses, theses and articles on mathematical biology is that postgraduate students and young researchers entering the field make appalling mistakes when trying to work on biologically relevant problems with hopelessly limited mathematical skills. When I started teaching mathematical biology to mathematicians, I found this situation quite frustrating. Thus, I decided to design a course focused more on developing and presenting mathematics inspired by biology than biology itself. Following this, I have tried to write these notes in such a way that a lecturer using them will be able to present mathematical results which are relevant to a biological application but with full mathematical rigour without having to look for proper statements and their proofs elsewhere or to prove them him/herself. Thus, the book is mostly addressed to students and researchers primarily interested in mathematical methods for analysing biological models and developing these methods for their own sake, as a mathematical theory, and not as just a tool for a particular biological application.

In other words, the main topic of this book is mathematics inspired by biology.[1] Such books do exist, but mostly at a much higher level, and thus, I hope that this volume will fill some existing gaps in the literature and find readers outside the mathematical circles.

What Is This Book About?

Population dynamics is a fast-growing discipline spanning biology and mathematics. However, in many existing textbooks, the authors do not provide any mathematically sound introduction, focusing instead on apparently more accessible biological content. Even though its analysis requires sophisticated mathematical tools, the explanation is relegated to appendices or more advanced texts. We hope this book will bridge introductory texts in population dynamics and more advanced texts. Hence, the book is predominantly concerned with the mathematical analysis of known models, which have been drawn from a wide range of very good books in the field, and then with interpreting the obtained mathematical results. Usually, we will not go deeper into the construction of models. However, within the space limitation, we will try to show that the discussed models are not purely phenomenological, that is, constructed just to fit the data, but can be derived from more fundamental principles.

As we said, mathematical population dynamics deals with techniques and methods of solving or analysing the time behaviour of models describing ensembles of living entities, ranging from cells or bacteria (or even genes) through plants and animals to humans. The description can be stochastic or deterministic at the individual, population or community level. In this book, we shall focus on deterministic models, but occasionally, we use basic probabilistic tools. We begin with unstructured single population models in discrete and continuous time and go through exponential, logistic, Ricker, Gompertz and Allee models. Next, we analyse discrete and continuous structured models, discussing, in particular, Leslie and Usher matrices, general projection matrices and Lotka's renewal equation. We also extend some of these results to the nonlinear cases. Mathematically, we will discuss the eigenvalue-eigenvector method of solving discrete and continuous systems of equations, spectral decomposition and the Perron–Frobenius theory leading to the asynchronous exponential growth property. In the nonlinear case, we will focus on stability results beginning from the one-dimensional theory, through the stability by linearization in discrete and continuous cases and the Lyapunov functions method. We provide the basics of phase-plane analysis for classical problems and show its applicability to the analysis of travelling waves. We shall also discuss results based on positivity methods and monotone dynamical systems, adapting the material from [88] and lecture notes [82]. Next, we shall briefly introduce dynamics in

[1] I learnt this phrase from my friend Adam Bobrowski.

infinite dimensional spaces. Here, we discuss the birth-and-death problem and provide its solution using the generating function method. Further, we introduce the McKendrick–von Foerster model, formally deriving it from the Leslie model. We discuss some explicitly solvable cases and provide general well-posedness theory in the linear case. Staying with the linear case, we discuss the asynchronous exponential growth property using the Laplace transform technique (following the method in the lecture notes [55], which are not widely available). Next, we move to nonlinear models, beginning with the so-called linear chain trickery. Further, we introduce the concept of a semigroup of operators, and using the semigroup approach, we provide the theory of solvability of age-structured epidemiology models by casting them into the form of abstract semilinear equations. The book concludes with Appendix A–D that provide some longer proofs of results used in the text and more intricate mathematical considerations.

The book is based on lectures given by the author at the Honours (4th year) level at the University of KwaZulu-Natal in Durban (Mathematical Biology), 4th year level at the Łódź Technical University (Introduction to Mathematical Ecology), for postgraduate students at the SA Centre for Epidemiological Modelling and Analysis in Stellenbosch, at the African Institute for Mathematical Sciences in 2012 (Introduction to Population Theory), the 2011 CIMPA-UNESCO School on Modelling and Simulation in Population Biology in Muizenburg (SA) and recently at the African Mathematical School on Mathematical Modeling in Biology in 2023 at the North West University in Potchefstroom, South Africa. It is written mostly for senior undergraduate and postgraduate students with working knowledge of advanced calculus, basic complex analysis and linear algebra who are interested in applications of mathematics in biology. The main body of it can be used for a course for such an audience. However, some parts of the book touch on some aspects of functional analysis and, while interesting for senior postgraduate students and researchers, do not have to be included in the basic course. I hope the book will equip the students with the mathematical grammar of the language of population dynamics and will allow them to enter the research field without too many disasters.

Acknowledgements As a textbook, the presented material does not contain significant original results. In fact, it would have been impossible without numerous predecessors forming my understanding of mathematical modelling and models' analysis. I gratefully acknowledge the enormous influence of the books such as [25] by M. Braun, [27] by N. Britton, [35] by J. Cushing, [60] by M. Kot, [58] by H. K. Khalil, [67] by D. G. Luenberger, [88] by H.L. Smith and P. Waltman, [93] by H. Thieme or lecture notes [55] by M. Iannelli and [82] by G. Sallet, among others, on my thinking. Their impact is visible in the selection and presentation of the material below.

Finally, I thank Dr A. Błoch and Dr W. Lamb for reading the manuscript and alerting me to its many errors and shortcomings. I did my best to address them, and I take full blame for the ones that inevitably have been left.

I also thank the Springer Editor, Remi Lodh, for his infinite patience, which, inadvertently, allowed the manuscript to evolve into its current form.

Pretoria, Durban (South Africa)
and Łódź (Poland)
May 2024

Jacek Banasiak

Contents

Chapter 1
Mathematical Modelling

1.1 Introductory Remarks

Population dynamics is about building and analysing models that arise in real life. Thus, it is related to a large field of mathematical modelling (that not only applies to biology or population theory). Therefore, we begin by briefly describing it. Mathematical modelling belongs to a broad field of mathematical sciences—a branch of knowledge that aims to understand the world by performing formal symbolic reasoning and computation on abstract structures.

- One aspect of the mathematical sciences involves unearthing and understanding deep relationships among these abstract structures.
- Another one involves capturing certain features of the world by abstract structures through the process of modelling, performing formal reasoning on these abstract structures or using them as a framework for computation, and then reconnecting to make predictions about the world.
- Yet another aspect is using abstract reasoning and structures to make inferences about the world from the data. This is linked to the quest to find ways to turn empirical observations into the means to classify, order and understand reality.

A crucial part is connecting our reasoning gear to the external world. Therefore, the concept of mathematical modelling is central to any application of mathematics.

By a *mathematical model*, we understand a mathematical structure, often an equation, that describes some observed phenomenon or process, can provide its qualitative and quantitative explanation, and, ideally, predicts its properties beyond the known data. A graphical representation of mathematical modelling is presented in Fig. 1.1.

To create a mathematical model, first, we must have a phenomenon of interest that we want to explain and make predictions about. Observing this phenomenon,

J. Banasiak, *Introduction to Mathematical Methods in Population Theory*,
Springer Undergraduate Mathematics Series,
https://doi.org/10.1007/978-3-031-65491-6_1

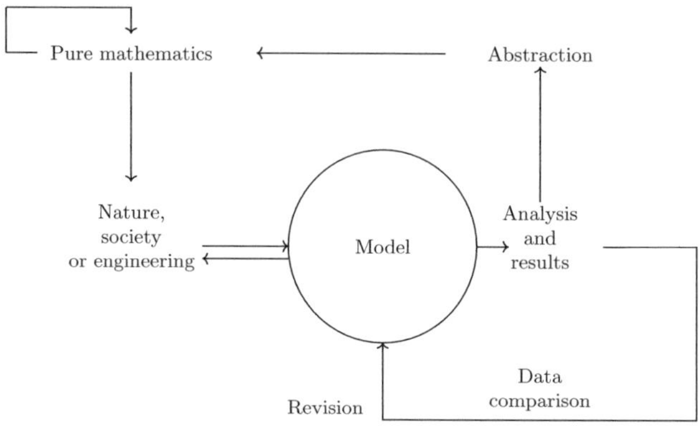

Fig. 1.1 The process of mathematical modelling

we make hypotheses about the quantities most relevant to the problem and the relations between them. In this way, we can devise a mechanism to explain the phenomenon. At this stage, we have to decide whether we want to consider deterministic or stochastic models and how to quantify the observations, that is, for instance, whether time should be continuous or discrete or whether we will consider discrete or continuously distributed attributes of individuals, such as age or position in space. This choice is not always obvious or unique, but one needs to decide on a particular approach before building a model. Sometimes, it is worthwhile to build several models with various configurations and compare their results, see, e.g., [76]. We shall return to these questions a little later. Once a model is built, what we do with it depends on us. If the main purpose of building the model was to quantify a description of the mechanism driving the phenomenon of interest and using it in practice, after a preliminary analysis and interpretation, in Fig. 1.1, we follow the arrow down and, since interesting equations rarely can be solved explicitly, we use numerical methods to obtain results that can be verified against the empirical data. If they agree, we can be satisfied that, at least, our model does not contradict the modelled object, and we can try to probe its area of applicability. If the numerical simulations are too far from observations, we must revise the model. This often has to be repeated many times as it is very difficult to achieve a proper balance between the simplicity and meaningfulness of the model. Indeed, usually, the first model is built using the Ockham razor, or parsimony, principle according to which we should select the simplest of all possible equations agreeing with the model. However, such a model often turns out to be oversimplified, and an insufficient agreement between the actual experiment and the model's prediction indicates that we must have overlooked some important features of the described phenomenon, which requires building a more complex model. On the other hand, sometimes, the built model turns out to be too complicated to yield itself to any meaningful analysis. Then, we have to construct a simpler model that can be analysed but still reflects the main features of the phenomenon under investigation.

It is important to remember that mathematical modelling is not mathematics—proving that a model is correct is impossible. Even if the numerical results are consistent with available empirical data, another experiment can provide data that do not agree with the model. In fact, Box and Draper, the authors of *Empirical Model-Building and Response Surfaces* state, see [37, p. 52],

> Remember that all models are wrong; the practical question is how wrong they have to be not to be useful.

The question of to what extent our models represent real phenomena belongs more to epistemology than to mathematics, and the answer to it depends on how much we believe the world has a rational structure that is synchronized with our cognitive powers. Nevertheless, there are good and bad models. A good model:

- has predictive powers, that is, a model based on available observations gives correct answers in other cases, e.g.,
  - General Theory of Relativity predicted light deflection or gravitational waves;
  - Dirac equations predicted the existence of positrons;
  - Standard Theory predicted the Higgs boson;
- contains earlier working models as subcases, e.g.,
  - Newton's mechanics, which is contained in the Special and General Theory of Relativity for small velocities and away from large masses,
  - Quantum mechanics, which yields the same results as Newton's mechanics for large distances and large energies.

Before we discuss the process of mathematical modelling in detail, it is worthwhile to look closer at other parts of Fig. 1.1 as we will encounter them later in the book. If we wish to have more than a superficial understanding of the model, we turn upward in Fig. 1.1. Once a model is built, we may stop pondering its relation with the real world, begin considering the model as an object of a purely mathematical nature, and build on its foundation a mathematical theory by means of deduction. While the process of model building was not mathematics, the analysis of the model is mathematics—performing logical operations on well-defined objects.

It is important to remember that in this way, we obtain results about the mathematical model and not the modelled process. For instance, if we prove results on the diffusion equation, we infer results about the diffusion equation and not the diffusion process, as the diffusion equation is just a model of the diffusion process and not its exact representation.

Nevertheless, such an analysis does not prohibit us from interpreting the obtained results in the language of the processes the models are concerned with and trying to determine whether they say something relevant about them. It is, however, fair to say that one can follow a different path as one pursues mathematical questions independently of any external feedback and addresses only internally developed problems, as indicated in the left upper corner of Fig. 1.1. It does not mean that abstract results of pure mathematics have nothing to do with the real world—it suffices to mention number theory, considered for centuries to be the queen of pure mathematics, which has found application in coding and cryptography, or linear algebra which provided the key to efficient internet browsers, see, e.g., [28]. Even if we try our best to avoid applications, they can suddenly appear behind the next corner.

As we observed earlier, interesting models usually do not have explicit solutions. Thus, if we are interested in getting results that are relevant to applied sciences, we need to resort to approximate and numerical methods. Here, we need to emphasize that, apart from the intrinsic value of theoretical analysis of models, the lack of a relationship between the construction of a model and its numerical analysis may render the latter incorrect and useless. For instance, if the derived equations have no solutions or are not uniquely solvable, standard numerical methods may not converge at all or may converge to some unspecified combination of multiple solutions. Also, the convergence of numerical algorithms heavily depends on the regularity of solutions, which can only be established through theoretical analysis. If the solutions fail to have the required regularity, the numerical procedure may "overlook" some essential properties of the solutions, such as the gradient catastrophe that appears in solutions of partial differential equations close to the points where the boundary of the domain is not smooth enough. In fact, the gradient catastrophe, manifesting itself as stress concentration at the corners of square windows, was the main reason for several deadly accidents of the first passenger aeroplane de Havilland Comet in the 1950s, see [101]. Another property of differential equations that affects the reliability of numerical methods is the sensitivity of their solutions to small changes in the initial and boundary conditions. This is due to the fact that all the data used in simulations suffer from measurement errors. Thus, for a numerical simulation to have any value, the solutions of equations with close to each other data should also be not too different. The belief that such stability with respect to small changes in the input data is a typical feature of deterministic systems was shattered by Lorenz's discovery of chaotic systems.

Thus, we see that a comprehensive analysis of equations derived in mathematical modelling is important both from the theoretical point of view by contributing to the development of mathematics and from the practical one, helping to validate approximate and numerical methods of applied sciences. Paraphrasing the statement of K. Friedrichs, see [83], we can say that the process of modelling consists of describing nature by approximate equations, but we want to solve the equations exactly.

In other words, once a model is built, it belongs to mathematics and must be considered with full mathematical rigour. The fact that it has been derived in a not necessarily watertight way does not mean that it can be treated in a sloppy manner.

We hope that this book will provide an introduction to such an approach.

1.2 Conservation Principles and Constitutive Relations

Before we begin our discussion of the basic building blocks of a model, we need to distinguish between descriptive and explanatory models. In recent years, we have experienced an unprecedented abundance of information that often results in a statistical fitting of the experimental data to phenomenological formulae. This way, one can get various statistical information such as expectations, medians, variance, and correlations. However, drawing conclusions based on a statistical fit may lead to serious errors. Remember,

do not mistake correlation for causation!

In this book, we focus on models that try to explain the underlying reasons for the observed phenomena. This is not to say that statistical analysis of the data is not important—it serves as a crucial tool for separating the significant part of the data from the noise (though it can overlook black swan events). It cannot, however, replace trying to find causal relations.

We also emphasize that statistical (descriptive) models should not be mixed up with *stochastic models*. Stochastic modelling aims to explain the underlying mechanisms of the observed phenomena, considering the inherent randomness of nature. Such models give probabilities of certain events and are indispensable in modelling small populations. Here, however, we shall focus on *deterministic models* that sometimes can be thought of as stochastic models averaged over many individual trajectories (invoking the law of large numbers) and giving answers in terms of the evolution of the densities of the populations. That is why we often use basic probabilistic arguments to derive a deterministic model.

1.2.1 Conservation Principles

Living matter is a part of the physical universe; thus, all biological processes must obey the laws of physics, particularly all relevant balance laws. To explain this, let Q be a quantity of interest (the number of animals, mass of a pollutant, amount of heat energy, number of infected individuals) in a fixed domain Ω. Over any fixed time interval, in Ω, we can only have

$$\begin{aligned}\text{The change of } Q &= \text{Inflow of } Q - \text{Outflow of } Q \\ &\quad + \text{Creation of } Q - \text{Destruction of } Q,\end{aligned} \tag{1.2.1}$$

see Fig. 1.2. In the probabilistic approach, this is the same as saying that the probability that one of all possible events occurs equals one. To proceed, we must decide whether we model with a *continuous* or *discrete* time.

We use *discrete time models* if we believe that significant changes in the system occur only during evenly spaced time intervals or if we can only observe the system at evenly spaced time instances and have a reason to believe that the essential

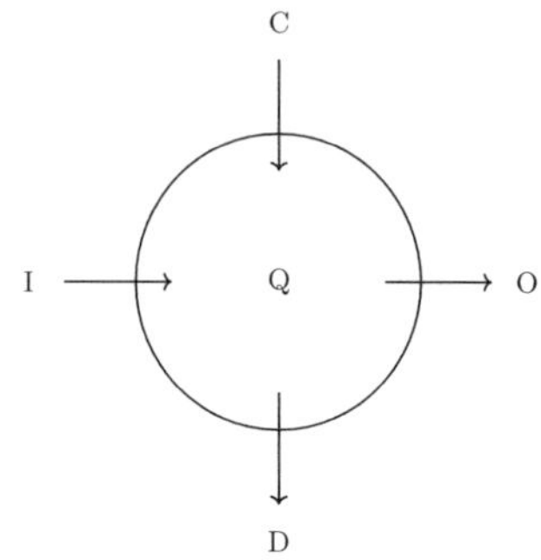

Fig. 1.2 Conservation law for the substance Q

parameters of the system remain unchanged between the successive observations. Then, we use the time between the events/observations as the time unit and count time using the number of elapsed events/observations, and we write (1.2.1) as

$$Q(k+1) - Q(k) = I(k) - O(k) + C(k) - D(k). \tag{1.2.2}$$

The quantities $I(k)$, $O(k)$, $C(k)$, $D(k)$ are the amounts of Q, which, respectively, inflow and outflow, are created and destroyed in the time interval $[k, k+1)$. Some typical examples of populations that can be modelled in discrete time are discussed in more detail in Sect. 2.1.

If the assumptions allowing us to use discrete-time modelling are not satisfied, or we simply cannot identify the basic unit of time, we use *continuous time models*. However, this requires some care as all quantities change continually. Thus, I, O, D and C should be considered as the *rates* of the inflow, outflow, destruction or creation, respectively. In other words, the amount of Q at a given time t will be given by

$$Q(t) = Q(t_0) + \int_{t_0}^{t} I(s)ds - \int_{t_0}^{t} O(s)ds + \int_{t_0}^{t} C(s)ds - \int_{t_0}^{t} D(s)ds,$$

where $Q(t_0)$ is the initial amount of Q.

Hence, assuming that I, O, D and C are continuous functions, so that Q is differentiable, we obtain the conservation law in the differential form,

$$\frac{dQ}{dt} = I - O + C - D. \tag{1.2.3}$$

Note 1 The dimensions of I, O, C and D in (1.2.3) are different than in (1.2.2).

Note 2 If we consider populations, then the value of Q is always a nonnegative integer, and hence, such a function cannot be continuous. Thus (1.2.3) is already an approximation, the validity of which requires Q to be sufficiently large to be considered a continuum.

1.2.2 Constitutive Relations

Conservation laws form a non-negotiable framework on which we build a fully-fledged model by trying to determine the form of Q, I, O, C and D and the relations between them. These are known as *constitutive relations*.

In other words, while the conservation laws are supposed to be universally valid, the constitutive relations reflect our (limited) knowledge about the particular process we model. Thus, they should be viewed as working hypotheses about it. The fact that we rarely have the full information about the process of interest at our disposal was succinctly summarised (though in a different context) by a former US Secretary of State, D. Rumsfeld, [80],

> There are known knowns. These are things we know that we know. There are known unknowns. That is to say, there are things that we know we don't know. But there are also unknown unknowns. There are things we don't know we don't know.

We can model things which we know with good accuracy. We can also try different hypotheses to describe what we know we do not know and reveal the truth by ruling out what is impossible. Clearly, we cannot model unknown unknowns, and we have to wait until the progress of science moves unknown unknowns to the category of known unknowns. A good example of such phenomena is dark matter and dark energy, which were unknown unknowns just a few decades ago but can be considered known unknowns at the time of writing this book.

Before we begin discussing basic models, we briefly list the basic forms of the functions I, O, D and C. They may depend

- on the space variables or other independent quantities, in which case we have partial differential or difference equations that will be discussed in further parts of the book,
- explicitly on time, which leads to non-autonomous equations which will briefly appear later,
- on the unknown Q

  (a) in a nonlinear way such as $I = I(Q) - Q^2$, or
  (b) in a linear way such as $I = I(Q) = 2Q$,

- on other unknown quantities, which leads to systems of equations, presented in more detail below, as they are the main topic of this book.

1.2.3 Modelling Interacting Populations

In some cases, we assume that the quantity Q evolves independently in an environment unaffected by it. In such cases, we describe Q with a single equation in which the environment appears through known parameters. Situations in which such a description is realistic are rare—one would have to imagine a species evolving

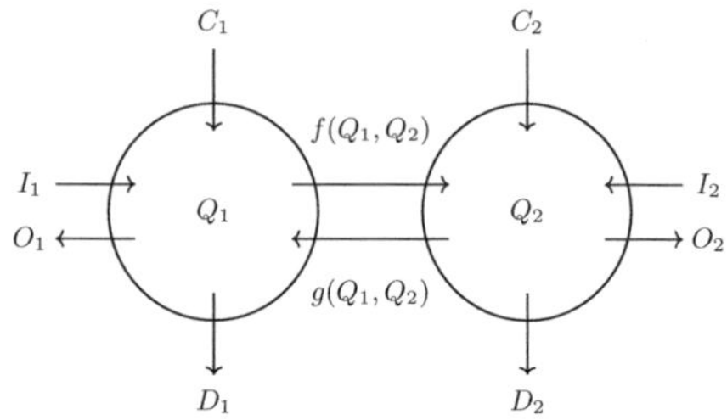

Fig. 1.3 Interactions between populations Q_1 and Q_2, with f and g describing the respective influence of one species on the other

alone in an environment that provides for it but is not affected by it, something like a colony of bacteria on a Petri dish with an infinite nutrient supply. A single equation model can work for a short time when the changes in the nutrient levels are negligible, but, in reality, the species and the environment affect each other and in the case of two interacting objects, we should modify Fig. 1.2, getting Fig. 1.3 with an accompanying modification of (1.2.1) in either (1.2.2) or (1.2.3) version. It is often convenient and possible to separate terms that describe each population's evolution without the influence of the other and terms that describe the effect of the interaction. Then, in the language of differential equations, (1.2.3) will take the form

$$\begin{aligned} \frac{dQ_1}{dt} &= \text{Intrinsic rate of change of } Q_1 + \text{Impact of } Q_2 \text{ on } Q_1, \\ \frac{dQ_2}{dt} &= \text{Intrinsic rate of change of } Q_2 + \text{Impact of } Q_1 \text{ on } Q_2, \end{aligned} \tag{1.2.4}$$

where the first term in each equation corresponds to the right-hand side of (1.2.3).

Part I
Unstructured Models

Chapter 2
Models with Discrete Time

We begin our discussion with models of a single population in a non-interactive environment, that is, as explained in the introduction, when the impact of the environment on the observed population is modelled by parameters appearing in the equations. For instance, in many models described below, we recognize that there is a reaction of the environment to overpopulation. However, we do not model this reaction explicitly; rather, we modify the vital parameters of the populations, such as the birth and death rates, so that a certain parameter represents the environment, for instance, the carrying capacity in the logistic equation.

2.1 Discrete Models

2.1.1 Background

Many plants and animals breed only during a short, well-defined breeding season. Also, often, the adult population dies soon after breeding. Such populations are ideal for modelling using discrete time. Let us consider a few typical examples, [27].

(i) So called monocarpic plants flower once and then die. Such plants may be annual, but, for instance, bamboos grow vegetatively for 20 years and then flower and die.

(ii) Animals with such a life cycle are called semelparous.

   (a) Insects typically die after laying eggs, but their life cycle may range from several days (e.g. house flies) to 13–17 years (cicadas).

   (b) A similar life cycle is observed in some fish species, such as the Pacific salmon or European eel. For instance, the latter live 10–15 years in freshwater European lakes, migrate to the Sargasso Sea, spawn and die.

J. Banasiak, *Introduction to Mathematical Methods in Population Theory*, Springer Undergraduate Mathematics Series,
https://doi.org/10.1007/978-3-031-65491-6_2

(c) Some mammals, such as marsupials, ovulate once per year and produce a single litter. There is abrupt and total mortality of males after mating. The births in the population are synchronized within a day or two, which is related to the environment with a predictable 'bloom' of insects in a given season of the year.

(iii) A species is called iteroparous if it is characterized by multiple reproductive cycles over the course of its lifetime. Difference equations can model such populations if the breeding only occurs during short, regularly spaced breeding periods. It is typical for birds. For instance, females of the Greater Snow Geese lay eggs between 8th and 20th of June (peak occurs on 12th–17th of June), and practically all eggs hatch between 8th and 13th of July.

From the modelling point of view, it is important to distinguish between populations in which all adults die immediately after the breeding season and populations in which individuals can survive several generations. As we said earlier,

- monocarpic plants, cicadas, most insects and some fish species have non-overlapping generations,
- in birds' populations, such as the Snow Geese, we observe overlapping generations.

In the first case and the cases with overlapping generations but no significant differences between generations, we often can model the population by a single, first order in time equation. The occurrence of significant differences between generations leads to the so-called structured models, which will be discussed later.

2.1.2 *Exponential Growth*

Possibly, the simplest way to describe a population is to provide information on how its size (number of individuals) changes over time. If we take the census of adults in the population in fixed time intervals, say, immediately after the breeding season, then we obtain a sequence of numbers

$$P(0), P(1), \dots, P(k),$$

where $P(k)$ is the number of adults just after the k-th breeding season. This approach makes sense in the models described above, that is, if the population has well-defined breeding seasons.

The simplest assumption to make is that there is a functional dependence between subsequent generations

$$P(k+1) = R(P(k))P(k), \quad k = 0, 1, \dots, \tag{2.1.1}$$

where $R(P(k))$ is the (density-dependent) net growth rate. Then the simplest case of (2.1.1) is when $R(P) = R$, that is,

$$P(k+1) = RP(k), \quad k = 0, 1, \ldots. \tag{2.1.2}$$

The exponential (or Malthusian) equation (2.1.2) well describes completely homogeneous populations, with the characteristics of the environment and individual members' constant over time. Moreover, the ratio of females to males must remain the same in each breeding season. This allows for describing the population by only giving the number of females. Further, we assume that

- each member of the population produces, on average, the same number of offspring;
- each member has an equal chance of dying, and death can only occur after giving birth;
- age differences between members of the population can be ignored;
- the population is isolated—there is no immigration or emigration.

Under these assumptions, the net growth rate R is given as

$$R = 1 + \beta - \mu.$$

Here, β is the average number of offspring per member of the population in each season; it is called fecundity or fertility. The constant μ is the probability that an individual will die after giving birth and before the following census; it is called the per capita death rate. Thus, (2.1.2) can be written as

$$P(k+1) = P(k) + \beta P(k) - \mu P(k), \tag{2.1.3}$$

which expresses the basic principle of population modelling in discrete-time:

> The number of individuals at the $(k+1)$th census equals the size of the population at the kth census + the total number of births in the population between the censuses (all individuals present at the kth census survived to the breeding season) − the total number of deaths in the population between the censuses (only individuals present at the kth census can die).

Remark 2.1 In other words, model (2.1.3) describes the situation in which no neonates die before their first breeding season. It is an unrealistic assumption, but the model allows for some flexibility. We can accommodate the situation in which any individual can die between breeding seasons by modifying (2.1.3) as follows,

$$P(k+1) = P(k) + \beta(1-\mu_0)P(k) - \mu_0(1-\mu_1)P(k) - \mu_1 P(k), \tag{2.1.4}$$

where μ_0 is the probability that an individual will die from natural causes before the breeding season while μ_1 is the per capita death rate due to giving birth. The constant $\beta(1-\mu_0)$ is the (effective) per capita birth rate. It is a very important parameter for the survival of a population as it is not only important what the natural fertility of a female is, that is, how many offspring she can have each season, but also how many of them survive on average till the next breeding season. We note that (2.1.4) will describe a population with no overlapping generations if $\mu_1 = 1$, that is, if all adults die just after giving birth.

It is important to recognize that mathematically, (2.1.4) is the same as (2.1.2); the interpretation of R, however, changes. This becomes important when we try to validate the model using experimental data.

Changes due to migrations can be incorporated, but the structure of the equation may not persist—e.g., they are not necessarily proportional to the total population.

Clearly, the solution to (2.1.2) is

$$P(k) = R^k P(0), \qquad k = 0, 1, 2 \ldots, \tag{2.1.5}$$

and if $R < 1$, then the population decreases towards extinction, but with $R > 1$ it grows indefinitely. Such behaviour over long periods of time is not observed in any population, so it is clear that the model is oversimplified and requires corrections. However, it is often valid over short time intervals and can bring demographic insights.

Exercise 2.1.1 A population of birds on an island has constant per capita (effective) birth (β) and death (μ) rates. Also, a constant number I of birds migrate to the island each year. Assume the census is made after breeding when all migrating birds from the previous year have already arrived. Find the difference equation describing the growth of the birds' population on this island. Consider two cases:

(a) the immigrants breed in the year of their arrival as the resident birds, and
(b) the immigrants start breeding only a year after their arrival.

Exercise 2.1.2 In a certain type of female insect population, all the adults die before the eggs hatch. Each adult contributes a constant number of eggs b, of which a fraction f survive and develop into adult females. Set up a difference equation for the population after the kth hatching. What is the minimum number of eggs that should be laid by each female so that the population survives if $f = 20\%$?

Exercise 2.1.3 Suppose a population doubles its original size in 100 years and triples in 200 years. Can this population grow exponentially? Give a reason for your answer.

Exercise 2.1.4 Let a population grow at an exponential rate. Assume that we can only estimate the initial population $P(0)$ as $P_1 \leq P(0) \leq P_2$. Knowing the exact population $P(N)$ at some later time N, find the upper and lower bound for the growth rate R.

2.1.3 The Death Rate μ and the Average Lifespan of an Individual

Using the concept of the death rate, we can derive an important characteristic of a population—the expected lifespan of its individuals. Let the constant μ be the probability that an individual dies, for whatever reason, between the censuses (in model (2.1.4), $\mu = \mu_0(1-\mu_1)+\mu_1$) and denote by $p(k)$ the probability that an individual, born at $k=0$, is alive at time k (if you are not familiar with probabilities, think about $p(k)$ as the fraction of the population that survived till k). For an individual to be alive at time k, it had to be alive at time $k-1$, with probability $p(k-1)$, and could not die between $k-1$ and k, which happens with probability $1-\mu$. Hence,

$$p(k) = (1-\mu)p(k-1), \qquad p(0) = 1.$$

Again, it can be explained by saying that the surviving fraction at time k is the fraction alive at time $k-1$ decreased by the fraction that died in that period. Thus,

$$p(k) = (1-\mu)^k. \tag{2.1.6}$$

The average lifespan L is the expected duration of life. To find it we observe that to die exactly at time k, an individual must be alive at time $k-1$ and die in the interval $(k-1,k]$ that occurs with the probability $\mu p(k-1) = \mu(1-\mu)^{k-1}$. It also can be explained as follows: in the first year, the proportion μ of the population dies and $1-\mu$ survives, then after the second year, a proportion μ of them die, that is, $\mu(1-\mu)$ part of the initial population dies before the 2nd census while $(1-\mu)^2$ survives beyond the second birthday, etc. Since our assumptions do not specify the maximum duration of life, the average lifespan is given by

$$L = \mu\sum_{k=1}^{\infty} k(1-\mu)^{k-1} = -\mu\frac{d}{d\mu}\frac{1-\mu}{1-(1-\mu)} = \frac{1}{\mu}, \tag{2.1.7}$$

where we used

$$\sum_{k=1}^{\infty} kz^{k-1} = \frac{d}{dz}\sum_{k=1}^{\infty} z^k = \frac{d}{dz}\frac{z}{1-z}$$

for $z = 1-\mu$.

2.1.4 Basic Nonlinear Models

As we explained in Remark 2.1, in real populations, some of the β offspring produced by a female will not survive to be counted as adults in the next census. Let us denote by $S(P)$ the *survival rate*, that is, the fraction of newborns that survive till the next breeding season. If we consider, for simplicity, populations with no overlapping generations, then the Malthusian equation will be replaced by

$$P(k+1) = \beta S(P(k))P(k), \quad k = 0, 1, \ldots, \tag{2.1.8}$$

which may be alternatively written as

$$P(k+1) = R(P(k))P(k) = f(P(k)), \quad k = 0, 1, \ldots, \tag{2.1.9}$$

where $R(P)$ is the (effective) per capita production of a population of size P. If $S(P)$ is independent of P, then (2.1.8) is the same as (2.1.4) with $S = 1 - \mu_0$, that is, with the adjusted birth rate, and with $\mu_1 = 1$. However, in general, we have a density-dependent S that leads to nonlinear equations.

Let us discuss typical behaviours in such populations before embarking on basic examples. The survival rate S in (2.1.8) reflects the intraspecific (within-species) competition for some resource (typically, food or space) which is in short supply. The three main (idealized) forms of intraspecific competition are, see [27],

- *No competition*. Here, $S(P) = 1$ for all P.
- *Contest competition*. Here, there is a finite number of units of resource. Each individual who obtains one of these units survives to breed and produces β offspring in the subsequent generations, while all others die without producing offspring. Thus $S(P) = 1$ for $P \leq P_c$ and $S(P) = P_c/P$ for $P > P_c$ for some critical value P_c (equal to the availability of resources).
- *Scramble competition*. Here, each individual is assumed to get an equal share of a limited resource. If this amount is sufficient for the survival to breed, all individuals survive and produce β offspring each; if not, all die. Thus, $S(P) = 1$ for $P \leq P_C$ and $S(P) = 0$ if $P > P_C$ for a critical value P_C (P_C is different from P_c).

Such ideal situations do not occur in real populations: real data are not easily classified in terms of the contest or scramble competition. Threshold density is not usually seen; zero survival is unrealistic, at least for large populations. Thus, a similar classification is done based on the asymptotic behaviour of $S(P)$ (or $f(P)$) as $P \to \infty$.

1. Contest competition corresponds to *exact compensation*:

$$\lim_{P\to\infty} f(P) = c \tag{2.1.10}$$

for some constant c (or $S(P) \sim cP^{-1}$ for large P). This describes the situation when an increased mortality compensates exactly for any increase in population size. Thus, only the predetermined number of individuals in the population can survive.

2. The other case is when

$$S(P) \sim \frac{c}{P^b}, \quad P \to \infty. \tag{2.1.11}$$

Here we have

Under-compensation if $0 < b < 1$, that is, when the increased mortality less than compensates for the increase in the size of the population;

Over-compensation if $b > 1$.

In general, if $b \approx 1$, we say there is a contest; if b is large, we call it a scramble. Indeed, in the first case, $f(P)$ eventually levels out at a nonzero value for large populations, which indicates that the population stabilizes by rejecting too many newborns. On the other hand, for $b > 1$, $f(P)$ tends to zero for large populations, which indicates that the resources are over-utilized, leading to eventual extinction.

We note that the earlier discussed linear case, where $S(P) = 1 - \mu_0 > 0$, does not fit into this description, which was designed to model the environment's response to an increasing population. However, we see that the case $1 - \mu_0 < 1/\beta$ is over-compensatory, as the population eventually dies out. On the other hand, the case $1 - \mu_0 > 1/\beta$ is under-compensatory as the population grows indefinitely.

We introduce the most common nonlinear models.

Beverton–Holt Type Models

Let us look at the model (2.1.9)

$$P(k+1) = R(P(k))P(k), \quad k = 0, 1, \ldots,$$

where $R(P(k)) = \beta S(P(k))$. We should have $PS(P) \approx const$ for large P to exhibit compensatory behaviour. Also, for small P, we expect very small intra-species competition, hence $S(P)$ should be approximately one, so the growth should be exponential with the effective birth rate given by the fertility β. A simple function of this form is

$$S(P) = \frac{1}{1 + aP}$$

leading to

$$P(k+1) = \frac{\beta P(k)}{1 + aP(k)}. \tag{2.1.12}$$

If we introduce the concept of carrying capacity of the environment K and assume that the population, having reached K, will stay there, that is, if $P(k) = K$ for some k, then $P(k+m) = K$ for all $m \geq 0$; this gives

$$K(1 + aK) = \beta K, \tag{2.1.13}$$

leading to $a = (\beta - 1)/K$ and the resulting model, called the *Beverton–Holt model*, takes the form

$$P(k+1) = \frac{\beta P(k)}{1 + \frac{\beta - 1}{K} P(k)}. \tag{2.1.14}$$

As we said earlier, this model is compensatory.

A generalization of this model is called the *Hassell* or again *Beverton–Holt* model, and reads

$$P(k+1) = \frac{\beta P(k)}{(1 + aP(k))^b}. \tag{2.1.15}$$

It can exhibit any type of compensatory behaviour, depending on b. For $b > 1$, the model describes a scramble competition, while for $b = 1$ we have a contest.

Substitution $x(k) = aP(k)$ reduces the number of parameters giving

$$x(k+1) = \frac{\beta x(k)}{(1 + x(k))^b}, \tag{2.1.16}$$

which will be analysed later.

If the justification of the Beverton–Holt equation given in this section seems not very convincing, alternative derivations will be presented in Sects. 2.1.5 and 3.4.2.

Exercise 2.1.5 A model for insect population in which all adults are assumed to die before the next breeding season leads to a difference equation

$$P(k+1) = \frac{\lambda P(k)}{1 + aP(k)},$$

where λ and a are positive constants. Find this population's growth rate $R(P(k))$.

Do the same, assuming that only 50% of the adult population dies before the breeding season. Hint. Re-write the equation, taking into account that 50% of the population continues into the following cycle.

The Logistic Equation

The Beverton–Holt models are best applied to semelparous insect populations, but it also has been popular in the context of fisheries. For populations surviving to the next cycle, it is more informative to write the difference equation in the form

$$P(k+1) = P(k) + R(P(k))P(k), \tag{2.1.17}$$

so that the increase in the population is given by $R(P) = \beta S(P)$. Here, we assume for simplicity that no adults die (such deaths can be incorporated by subtracting the term $\mu P(k)$ from the right-hand side of the equation, but this will not change the structure of the equation). Below, we will show another way of incorporating death into the equation.

As with the Beverton–Holt model, we will use the Ockham razor principle to write the equation. Indeed, the function R can have different forms but must satisfy the requirements:

(a) Due to overcrowding, $R(P)$ must decrease as P increases until P reaches the carrying capacity K; then $R(K) = 0$ so that, as above, $P = K$ stops changing.
(b) Since for P much smaller than K there is small intra-species competition, we should observe an exponential growth of the population so that $R(P) \to \beta$ as $P \to 0$; here $\beta = R(0)$ is called the unrestricted birth rate of the population.

Constants β and K are usually determined experimentally.

The simplest function satisfying these requirements is a linear function which, to satisfy (a) and (b), must be chosen as

$$R(P) = -\frac{\beta}{K}P + \beta.$$

Substituting this formula into (2.1.17) yields the so-called discrete logistic equation

$$P(k+1) = P(k) + \beta P(k)\left(1 - \frac{P(k)}{K}\right), \tag{2.1.18}$$

which is still one of the most often used discrete equations of population dynamics.

In the context of an insect population, where there are no survivors from the previous generation, the above equation reduces to

$$P(k+1) = \beta P(k)\left(1 - \frac{P(k)}{K}\right). \tag{2.1.19}$$

Both equations are mathematically equivalent. Indeed,

$$P(k) + \beta P(k)\left(1 - \frac{P(k)}{K}\right) = (1+\beta)P(k)\left(1 - \frac{P(k)}{\frac{1+\beta}{\beta}K}\right),$$

which is the right hand side of (2.1.19) with just different constants. Conversely, the right hand side of (2.1.19) can be transformed to (2.1.18) provided $\beta > 1$.

While the above argument may seem to be of *bunny-out-of-the-hat* type, it could be justified by generalizing (2.1.3). Indeed, assume that the mortality μ is not constant but equals

$$\mu = \mu_0 + \mu_1 P,$$

where μ_0 corresponds to the death due to natural causes and μ_1 could be attributed to cannibalism, where one adult eats/kills on average μ_1 portion of the population. Then (2.1.3) can be written as

$$P(k+1) = (1+\beta-\mu_0)P(k)\left(1-\frac{P(k)}{\frac{1+\beta-\mu_0}{\mu_1}}\right), \tag{2.1.20}$$

which is (2.1.19) with β modified to $\beta - \mu_0$ and the modified carrying capacity $K = \frac{1+\beta-\mu_0}{\mu_1}$. A generalization of this equation, obtained by using a more flexible formula for the death rate,

$$\mu = \mu_0 + \mu_1 P^\theta, \quad \theta > 0,$$

is the discrete *Bernoulli equation*

$$P(k+1) = P(k) + \beta P(k)\left(1-\left(\frac{P(k)}{K}\right)^\theta\right). \tag{2.1.21}$$

Exercise 2.1.6 Suppose that a population grows according to the logistic equation

$$P(k+1) = P(k) + rP(k)\left(1-\frac{P(k)}{K}\right),$$

with r and K unknown. Show that $P(k+1) \neq P(k)$ for any k, unless $P(k) = K$. Using this result, show that three measurements of the population size, say, P_0, P_1, P_2 suffice to determine uniquely r and K. Write down the expressions for r and K.

Exercise 2.1.7 Show that the substitution

$$x(k) = \frac{\beta}{1+\beta}\frac{P(k)}{K}, \qquad \gamma = 1+\beta,$$

reduces (2.1.18) to a simpler form

$$x(k+1) = \gamma x(k)(1-x(k)). \tag{2.1.22}$$

The problem with the discrete logistic equation is that large (close to K) populations can become negative in the next step. For instance, for

$$P(k+1) = P(k) + 4P(k)(1 - P(k)),$$

beginning with $P(0) = 0.9$, we obtain $P(1) = 1.26$ and $P(2) = -0.0504$.

Although we could interpret a negative population as extinct, this may not be the behaviour that would actually happen. Indeed, the model was constructed to have $P = K$ as a stationary population. Thus, if we hit exactly K, the population survives, but if we even marginally overshoot, the population becomes extinct.

One way to avoid such problems with a negative population is to replace the density-dependent survival rate by considering only the positive part of $\left(1 - \frac{P(k)}{K}\right)$, that is,

$$S(P(k)) = \left(1 - \frac{P(k)}{K}\right)_+ \tag{2.1.23}$$

to take into account that S cannot be negative. However, this model also leads to population extinction if it exceeds K, which is not always realistic. In most cases, we expect some adjustments rather than the total and instantaneous extinction. We shall discuss other ways to circumvent this difficulty in the following subsection.

Here, to conclude, we only observe that the logistic equation, especially with S given by (2.1.23), is an extreme example of the scramble competition.

Ricker Equation

Here, we will try to find a model in which large values of $P(k)$ produce very small but still positive values of $P(k+1)$. Thus, a population well over the carrying capacity crashes to a very low level but survives. Let us find a way in which this can be modelled. Consider the per capita population change

$$\frac{\Delta P}{P} = R(P).$$

First we note that it is impossible for R to be less than -1—this would mean that the population of the next generation would be negative. We also need a decreasing R which is non-zero ($= \beta$) at 0. One such a function can be recovered from the Beverton–Holt model; another simple choice is an exponential, shifted down by 1, that is,

$$\frac{\Delta P}{P} = ae^{-bP} - 1,$$

which leads to

$$P(k+1) = aP(k)e^{-bP(k)}. \tag{2.1.24}$$

If, as before, we introduce the carrying capacity K and require it to give a constant population at each iteration, then we obtain

$$b = \frac{\ln a}{K}$$

and, letting for simplicity $\xi = \ln a$, we obtain the so-called *Ricker equation*

$$P(k+1) = P(k)e^{\xi(1-\frac{P(k)}{K})}, \tag{2.1.25}$$

see Fig. 2.1. We note that if $P(k) > K$, then $P(k+1) < P(k)$ and if $P(k) < K$, then $P(k+1) > P(k)$, see Fig. 2.2. The intrinsic growth rate β is given by $\beta = e^{\xi} - 1$ but, using the Maclaurin formula, for small ξ, we have $\beta \approx \xi$.

Exercise 2.1.8 Show that the Ricker model describes a scramble competition.

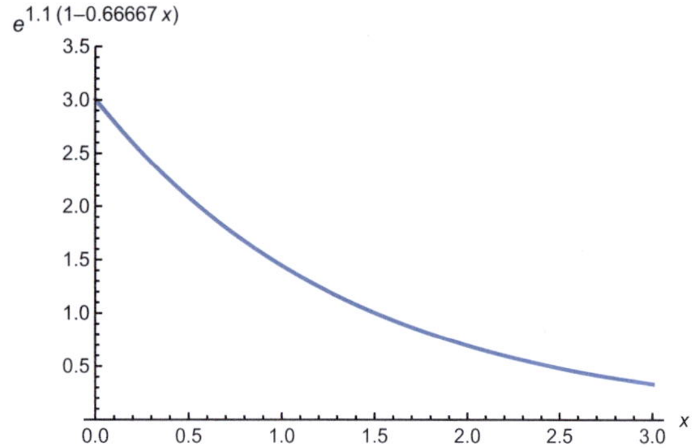

Fig. 2.1 The function $f(x) = e^{r(1-x/K)}$ for $r = 1.1$ and $K = 1.5$

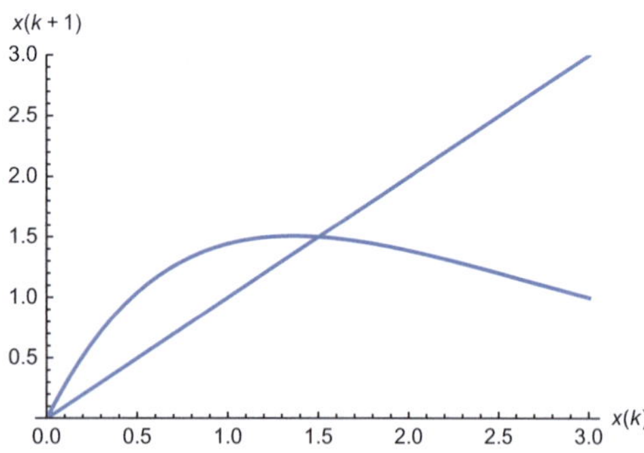

Fig. 2.2 The relation $x(k+1) = x(k)e^{\xi(1-x(k)/K)}$. The intersection of the function graph and the diagonal gives the carrying capacity K

2.1.5 *Discrete Nonlinear Population Models from First Principles*

The Ricker Equation in a Continuous Space Framework, [22, 81]

The way we have introduced the Ricker equation may seem to be completely *ad hoc*. It follows, however, that it can be derived from a set of assumptions describing a uniformly distributed population in which the reproductive success of an individual is adversely affected by other individuals staying in its close proximity. To do this, first, we must describe how to model the situation in which one may have various numbers of neighbours in a fixed neighbourhood. The simplest way to do this is to use the so-called *Poisson distribution.*

Poisson Distribution

Let us first assume that there are n individuals in a population and let p be the probability that one of them is in your neighbourhood. Under the usual assumption of independence of individuals occurring in the neighbourhood, the probability of having exactly r neighbours is given by the binomial formula

$$b(n, p; r) = \binom{n}{r} p^r (1-p)^{n-r}.$$

Indeed, the probability that exactly r given members of the population appear in our neighbourhood means that $n-r$ members must be outside it, which happens with the probability $p^r(1-p)^{n-r}$. However, since we are interested in any r members, we have to multiply this probability by the number of ways we can choose r members from n members of the population, that is, by $\binom{n}{r}$. Then, the average number of neighbours is $\omega = np$. This is intuitively clear and also can be checked by direct calculation,

$$\omega = \sum_{r=0}^{n} r \binom{n}{r} p^r (1-p)^{n-r} = np \sum_{r=0}^{n-1} \binom{n-1}{r-1} p^{r-1}(1-p)^{n-r} = np.$$

If we consider a potentially infinite population so that n grows to infinity in such a way that the average number of neighbours ω stays constant (so p goes to zero), then the probability of having exactly r neighbours is given by

$$\begin{aligned} p(r) &= \lim_{n\to\infty} b(n, \omega/n; r) = \lim_{n\to\infty} \frac{n!}{r!(n-r)!} \frac{\omega^r}{n^r} \left(1 - \frac{\omega}{n}\right)^{n-r} \\ &= \frac{\omega^r}{r!} \lim_{n\to\infty} \frac{n(n-1)\cdot \ldots \cdot (n-r+1)}{n^r} \left(1-\frac{\omega}{n}\right)^{-r} \left(\left(1-\frac{\omega}{n}\right)^{\frac{n}{\omega}}\right)^{\omega} = \frac{e^{-\omega}\omega^r}{r!}, \end{aligned}$$

which is called the *Poisson distribution*, see Fig. 2.3.

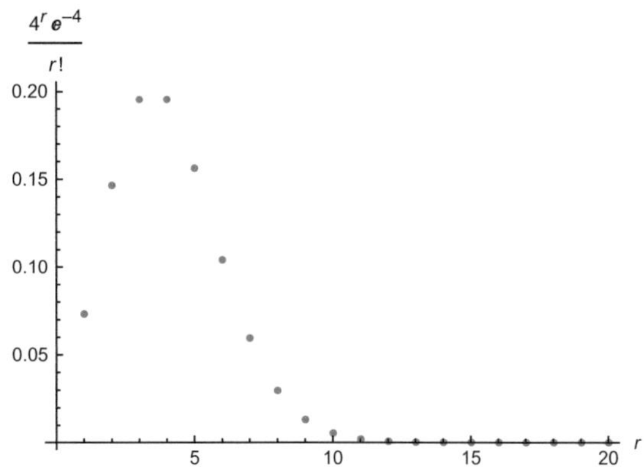

Fig. 2.3 Poisson distribution with $\omega = 4$

Returning to our problem, we consider a population the size of which at time k is given by $P(k)$. The standard growth equation is (2.1.1)

$$P(k+1) = R(P(k))P(k), \quad k = 0, 1, \ldots, \tag{2.1.26}$$

where R gives the average number of offspring in one cycle. As mentioned before, we assume that the number of offspring per individual decreases with the number of individuals living in its neighbourhood, say, a disc D of area s. A simple possible model is that the number of offspring per capita is given by bc^r where $b \geq 0, 0 < c < 1$, where r is the number of neighbours in D. If we assume that the population is uniformly distributed in an environment with area A, then the average number of individuals in D at time k is given by $sP(k)/A$. Then, using the Poisson distribution, the probability of having r neighbours in D is

$$\frac{(sP(k))^r e^{-\frac{sP(k)}{A}}}{A^r r!}$$

and the average number of offspring per individual is

$$R(P(k)) = be^{-\frac{sP(k)}{A}} \sum_{r=0}^{\infty} \frac{(csP(k))^r}{A^r r!} = be^{-\frac{s(1-c)P(k)}{A}}.$$

Hence, we obtain the Ricker model

$$P(k+1) = bP(k)e^{-\frac{s(1-c)P(k)}{A}}.$$

Comparing this expression with (2.1.25), we see that the carrying capacity K can be expressed as

$$K = \frac{A \ln b}{s(1-c)}.$$

Ricker and Beverton–Holt Models Via a Site-Based Approach, [22]

There is a general framework which allows the derivation of most of the single species unstructured models by assigning the probability distribution of the resource sites' occupation. Let us consider a habitat consisting of N discrete resource sites. At time n, a population of $P(n)$ individuals is distributed and reproduces. Let h_k be the proportion of sites accommodating k individuals. Generally, it will be a function of $P(n)$ and N. Once on site, the individuals reproduce, and the success of reproduction, that is, the number of offspring, denoted by $\phi(k)$, only depends on the number of individuals at the site. Then, the difference equation governing the growth of the population is

$$P(n+1) = N\sum_{k=0}^{\infty} h_k\phi(k). \tag{2.1.27}$$

To be able to make use of this equation, we must specify the site occupation function h_k and the maternity function $\phi(k)$.

We consider two types of site occupation functions: uniform and preferential. In what follows, we assume a large population with a large number of sites, so we can assume that the expected occupation of a site is equal to $P(n)/N$.

Uniform Distribution If individuals are uniformly distributed, then, as in the above section, the probability of finding k individuals at any given site is the Poisson distributed,

$$h_k = \frac{\left(\frac{P(n)}{N}\right)^k e^{-\frac{P(n)}{N}}}{k!},$$

and (2.1.27) can be written as

$$P(n+1) = Ne^{\frac{P(n)}{N}}\sum_{k=0}^{\infty}\frac{(P(n))^k}{N^k k!}\phi(k). \tag{2.1.28}$$

Preferential Distribution To shorten notation, denote $\omega = P(n)/N$. Assume that the sites are allocated at random a value, say $t \in \mathbb{R}_+$, representing, e.g., accessibility. We assume the random variable defined as the site's value has a continuous probability density f. Then, we assume that the average occupation of a site with value t is $t\omega$. If t was known, then, as before, the number h_k of site occupants would be given by Poisson distribution $(\omega t)^k e^{-\omega t}/k!$. Since, however, the value of t is not known, h_k will be given by

$$h_k = \int_0^{\infty} f(t)\frac{(\omega t)^k e^{-\omega t}}{k!}dt. \tag{2.1.29}$$

Fig. 2.4 Gamma distribution (2.1.30) with $\lambda = 2$

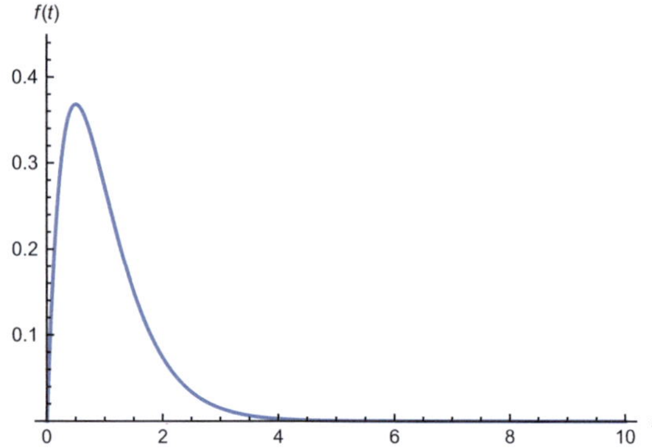

The function $f(t)$ is the probability density of the number of sites with value t and must be selected for each case. Quite often, it is assumed to be Gamma-distributed, that is,

$$f(t) = \frac{\lambda^\lambda}{\Gamma(\lambda)} t^{\lambda-1} e^{-\lambda t}, \tag{2.1.30}$$

where λ is a positive parameter. We see that it is a continuous density function having a similar shape to the Poisson distribution shown in Fig. 2.3, see Fig. 2.4.

Then

$$\begin{aligned} h_k &= \int_0^\infty f(t) \frac{(\omega t)^k e^{-\omega t}}{k!} dt = \frac{\lambda^\lambda \omega^k}{k! \Gamma(\lambda)} \int_0^\infty t^{\lambda+k-1} e^{-t(\lambda+\omega)} dt \\ &= \frac{\lambda^\lambda \omega^k}{k! \Gamma(\lambda)(\lambda+\omega)^{\lambda+k}} \int_0^\infty s^{\lambda+k-1} e^{-s} ds = \frac{\lambda^\lambda \omega^k \Gamma(k+\lambda)}{k! \Gamma(\lambda)(\lambda+\omega)^{\lambda+k}}, \end{aligned} \tag{2.1.31}$$

which is the so-called negative binomial distribution. Then

$$P(n+1) = N \frac{\lambda^\lambda}{\Gamma(\lambda)(\lambda + P(n)/N)^\lambda} \sum_{k=0}^\infty \frac{(P(n)/N)^k \Gamma(k+\lambda)}{k!(\lambda + P(n)/N)^k} \phi(k) \tag{2.1.32}$$

gives the formula for the growth of the population.

The next step is to specify the offspring outcome at each site.

Scramble Competition Let us assume that each site contains resources to support one individual. Then

$$\phi_k = \begin{cases} b & \text{if} \quad k = 1, \\ 0 & \text{otherwise}, \end{cases}$$

where b is the number of offspring a site containing only one individual produces. Substituting this into (2.1.28), we obtain

$$P(n+1) = bP(n)e^{-\frac{P(n)}{N}}, \tag{2.1.33}$$

which is the Ricker model. On the other hand, the negative binomial distribution gives

$$P(n+1) = b\frac{\lambda^{\lambda+1}P(n)}{(\lambda + P(n)/N)^{\lambda+1}}, \tag{2.1.34}$$

where we used $\Gamma(\lambda + 1) = \lambda\Gamma(\lambda)$. Since $\lambda > 0$, this is the Hassel model (2.1.15). Note that since we have scramble competition, we cannot get here the basic Beverton–Holt model, which is compensatory and thus describes a contest competition.

Contest Competition Again, we assume that each site can support one individual, but in contrast to the scramble competition, if more individuals are at the site, only one emerges victorious, and the others perish. Thus, the function ϕ_k is given by

$$\phi_k = \begin{cases} b \text{ if } k \geq 1, \\ 0 \text{ if } k = 0, \end{cases}$$

where, as before, b is the number of offspring produced by one individual. Then, the uniform distribution gives

$$P(n+1) = bNe^{-\frac{P(n)}{N}}\sum_{k=1}^{\infty}\frac{(P(n))^k}{N^k k!} = bN\left(1 - e^{-\frac{P(n)}{N}}\right), \tag{2.1.35}$$

which is the so-called Skellam model. It appears in the Nicholson–Bailey model in Sect. 7.1.1 but is not discussed in this book in detail.

Let us consider the negative binomial distribution. We have

$$P(n+1) = bN\frac{\lambda^{\lambda}}{\Gamma(\lambda)(\lambda + P(n)/N)^{\lambda}}\sum_{k=1}^{\infty}\frac{(P(n)/N)^k\Gamma(k+\lambda)}{k!(\lambda + P(n)/N)^k}. \tag{2.1.36}$$

Now, using the fact that $\Gamma(k+\lambda) = \lambda(\lambda+1)\cdot\ldots\cdot(\lambda+k-1)\Gamma(\lambda)$ and denoting $z = (P(n)/N)/(\lambda + P(n)/N)$, we get

$$\sum_{k=0}^{\infty}\frac{\Gamma(k+\lambda)}{k!}z^k = \Gamma(\lambda)\sum_{k=0}^{\infty}\frac{\lambda(\lambda+1)\cdot\ldots\cdot(\lambda+k-1)}{k!}z^k = \Gamma(\lambda)(1-z)^{-\lambda}.$$

Now,

$$1 - z = 1 - \frac{P(n)/N}{\lambda + P(n)/N} = \frac{\lambda}{\lambda + P(n)/N}$$

and thus (2.1.36) can be written as

$$\begin{aligned} P(n+1) &= bN \frac{\lambda^\lambda}{\Gamma(\lambda)(\lambda + P(n)/N)^\lambda} \left(\Gamma(\lambda)\lambda^{-\lambda}(\lambda + P(n)/N)^\lambda - \Gamma(\lambda)\right) \\ &= bN \left(1 - \frac{\lambda^\lambda}{(\lambda + P(n)/N)^\lambda}\right). \end{aligned}$$

If $\lambda = 1$, the above equation corresponds to the Beverton–Holt model (2.1.14). Indeed, in this case

$$P(n+1) = bN \left(1 - \frac{1}{(1 + P(n)/N)}\right) = \frac{bP(n)}{1 + P(n)/N}.$$

We see that the carrying capacity is given by $K = N(b-1)$, that is, it is proportional to the number of sites and the per capita birth rate above the simple reproduction.

Allee Type Equations

In all previous models with density-dependent growth rates, the bigger the population (or the higher the density), the slower the growth. However, in 1931, Warder Clyde Allee noticed that in small or dispersed populations, the individual's chances of survival decrease, which can lead to the extinction of the populations. This could be due to difficulties finding a mating partner or more difficult cooperation, e.g., organizing defence against predators. Models having this property can also be built within the considered framework by introducing two thresholds: the carrying capacity K and a parameter $0 < L < K$ at which the behaviour of the population changes so that $\Delta P/P < 0$ for $0 < P < L$ and $P > K$ and $\Delta P/P > 0$ for $L < P < K$. If, as before,

$$\frac{\Delta P}{P} = R(P),$$

and

$$P(k+1) = P(k) + P(k)R(P(k)),$$

then the required properties can be obtained by taking $R(P) \leq 0$ for $0 < P < L$ and $P > K$ and $R(P) \geq 0$ for $L < P < K$. To avoid negative populations we have to assume that $PR(P) + P \geq 0$, that is, $R(P) \geq -1$ for $0 \leq P \leq L$, see Fig. 2.5.

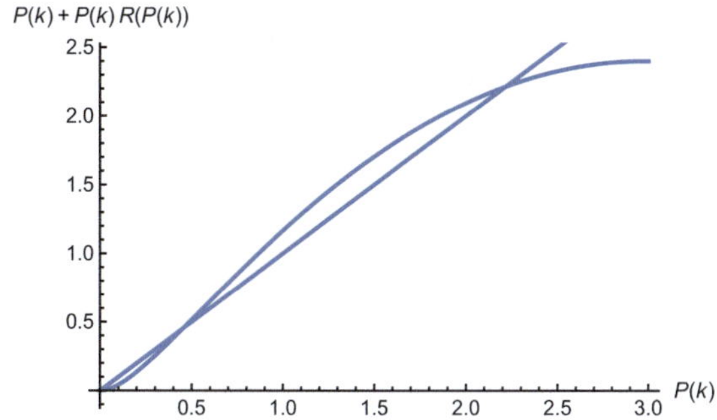

Fig. 2.5 The relation $P(k+1) = P(k) + P(k)R(P(k))$ yielding the Allee effect

Exercise 2.1.9 Show that if $R(P) = r(L - P)(P - K)$, where r is a parameter satisfying $0 < r < 1/LK$, then

$$P(k+1) = P(k)(1 + r(L - P(k))(P(k) - K)) \tag{2.1.37}$$

is an Allee-type model.

A more detailed analysis of this model can be found in Sect. 4.2.3.

An Allee-Type Model by Multiple Time Scales

Other models of this type can be derived by considering populations in which females search for mating partners or by introducing a generalist (that is, preying also on other species) predator. Here, we focus on the latter, while the former model, in continuous time, is discussed in a subsection of Sect. 4.1.2.

Assume that $N(k)$ describes a population of prey and $H(k)$ describes a population of active predators at time k. The total population of predators P is supposed to be constant in time and is subdivided into $H(k)$ active (hungry) predators and $S(k)$ satiated predators,

$$P = H(k) + S(k).$$

We assume that the prey population without the predator follows the logistic law with carrying capacity C and unrestricted growth rate $\lambda = \beta - \mu$. The predation process is modelled by the *law of mass action*. It is the simplest and thus most often used way of describing nonlinear interactions, and thus warrants a closer look. In the current context, it amounts to saying that one hungry predator in a unit of time can eat, on average, a fraction ξ of the prey population. Thus, H predators will consume ξNH prey. In real modelling, determining the coefficient ξ is paramount, including the probability of a predator meeting a prey, efficacy of killing, etc. Clearly, the model has many shortcomings, such as not considering the interactions between predators pursuing the same prey or the finite capacity of the predator's stomach.

Still, it has been successfully used in many situations. With this understanding, we write down the equation for the prey population as

$$N(k+1) = N(k) + \lambda N(k)\left(1 - \frac{N(k)}{C}\right) - \xi N(k)H(k).$$

We assume that the life cycle of predators is much longer than that of prey, and we neglect the vital processes in the predator population. Thus, the evolution of active predators is governed by the equation

$$\begin{aligned} H(k+1) &= H(k) + \sigma S(k) - \theta N(k)H(k) - \zeta H(k) \\ &= H(k) + \sigma(P - H(k)) - \theta N(k)H(k) - \zeta H(k). \end{aligned}$$

Here, σ is the proportion of the satiated predators that become hungry again (so that, as in (2.1.7), $1/\sigma$ is the average duration of satiation). The third term gives the number of active predators becoming satiated due to catching the prey from the prey population (note that we used a coefficient different than ξ—while ξNH is the total amount of prey killed in one cycle, this amount can feed more (or less) than ξNH predators). The last term gives the number of active predators becoming inactive due to predation on other species. If we introduce new variables as fractions of C and P, respectively, by $N = xC$ and $H = yP$, then we obtain the system

$$\begin{aligned} x(k+1) &= x(k) + \lambda x(k)(1 - x(k)) - \xi P x(k)y(k), \\ y(k+1) &= y(k) + \sigma(1 - y(k)) - \theta C x(k)y(k) - \zeta y(k). \end{aligned} \tag{2.1.38}$$

Note that the second equation only describes the changes in the number of predators due to hunting. If we assume that the prey's reproductive cycle is one year (e.g., antelopes), it is clear that the predator will hunt many times during this period. In other words, the state of y will change many times when k changes to $k+1$. It is then plausible to introduce another time scale for predators, say, n, and write

$$y(n+1) = y(n) + \sigma(1 - y(n)) - \theta C x(k)y(n) - \zeta y(n). \tag{2.1.39}$$

Note that here, $x(k)$ is treated as a constant, that is, we assume that the changes in the prey population due to hunting are negligible. We assume that y quickly settles at the equilibrium determined by solving

$$\bar{y} = \bar{y} + \sigma(1 - \bar{y}) - \theta C x(k)\bar{y} - \zeta\bar{y},$$

that is,

$$\bar{y} = \frac{\sigma}{\sigma + \theta C x(k) + \zeta}. \tag{2.1.40}$$

Substitution of (2.1.40) into the first equation of (2.1.38) gives

$$x(k+1) = x(k) + \lambda x(k)(1 - x(k)) - \frac{\sigma \xi P x(k)}{\sigma + \zeta + \theta C x(k)}, \tag{2.1.41}$$

which, returning to the old variable $x(k) = \dfrac{N(k)}{C}$, yields

$$N(k+1) = N(k)\left(1 + \lambda\left(1 - \frac{N(k)}{C} - \frac{A}{1 + BN(k)}\right)\right), \tag{2.1.42}$$

where $A = \dfrac{\sigma \xi P}{\lambda(\sigma + \zeta)}$ and $B = \dfrac{\theta}{\sigma + \zeta}$.

Example 2.2 Using the results of Sect. 4.2.3 to guide our choice, we consider the following coefficients: $\lambda = 1.2$, $\sigma = \zeta = 0.01$, $P = C = 1$, $\theta = 0.1$ and $\xi = 3$ so that (2.1.38) takes the form

$$\begin{aligned} x(k+1) &= x(k) + 1.2x(k)(1 - x(k)) - 3x(k)y(k), \\ y(k+1) &= y(k) + 0.01(1 - y(k)) - 0.1x(k)y(k) - 0.01y(k). \end{aligned} \tag{2.1.43}$$

Then the approximation (2.1.41) is given by

$$x(k+1) = x(k) + 1.2x(k)(1 - x(k)) - \frac{0.03x(k)}{0.02 + 0.1x(k)}. \tag{2.1.44}$$

Plotting the graph of

$$f(x) = 1.2x(1 - x) - \frac{0.03x}{0.02 + 0.1x}, \tag{2.1.45}$$

see Fig. 2.6, we get a graph of the required shape. Finally, in Fig. 2.7, we provide a numerical illustration that, in this case, the solution to (2.1.44) indeed is an approximation to the x-component of the solution to (2.1.43).

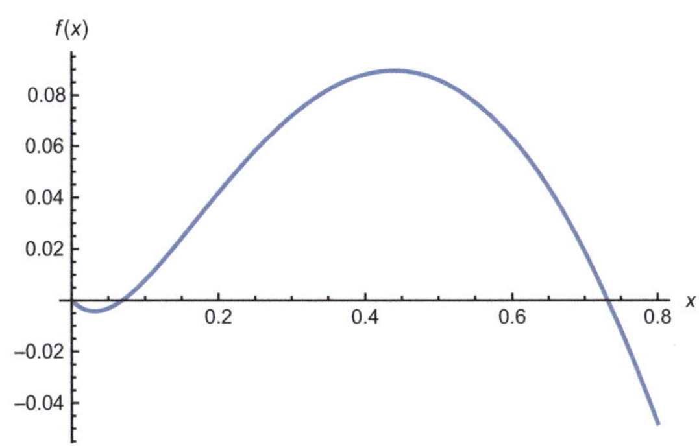

Fig. 2.6 Function f in (2.1.45); we see three equilibria, as required by the Allee model

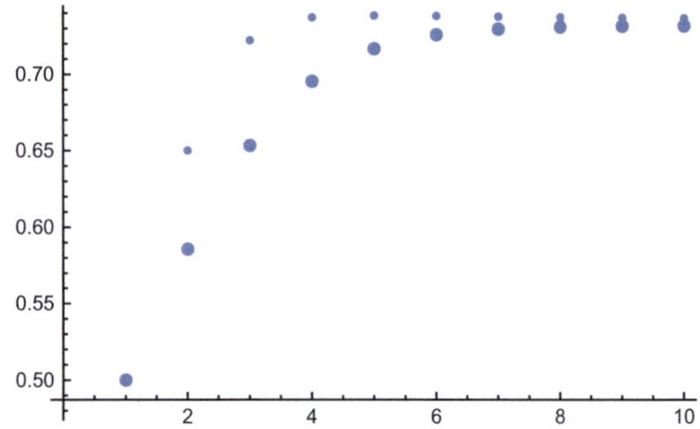

Fig. 2.7 Comparison of the solution to (2.1.44) (large dots) with the x-component of the solution to (2.1.43) (small dots). The initial conditions in both cases were $x(1) = 0.5$ and $y(1) = 0.1$

Exercise 2.1.10 Find another set of parameters which ensures that (2.1.42) gives the Allee-type dynamics. Discuss what other types of dynamics can (2.1.42) describe.

2.1.6 Some Explicitly Solvable Nonlinear Population Models

We complete this chapter by presenting some nonlinear models which can be explicitly solved by appropriate substitutions. The formula for solutions of linear equations plays an important role here.

Solvability of Linear Difference Equations

The simplest difference equations are those defining geometric and arithmetic progressions:

$$x(k+1) = ax(k),$$

and

$$y(k+1) = y(k) + a,$$

respectively, where a is a constant. The solutions of these equations are known to be

$$x(k) = a^k x(0),$$

and

$$y(k) = y(0) + ka.$$

We shall consider the generalization of both these equations, the general first-order difference equation,

$$x(k+1) = a(k)x(k) + g(k) \tag{2.1.46}$$

with an initial condition $x(0) = \mathring{x}$. Calculating the first few iterates, we obtain

$$\begin{aligned}
x(1) &= a(0)x(0) + g(0),\\
x(2) &= a(1)x(1) + g(1) = a(1)a(0)x(0) + a(1)g(0) + g(1),\\
x(3) &= a(2)x(2) + g(2) = a(2)a(1)a(0)x(0) + a(2)a(1)g(0) + a(2)g(1) + g(2),\\
x(4) &= a(3)x(3) + g(3)\\
&= a(3)a(2)a(1)a(0)x(0) + a(3)a(2)a(1)g(0) + a(3)a(2)g(1)\\
&\quad +a(3)g(2) + g(3).
\end{aligned}$$

At this moment, we have enough evidence to conjecture that the general form of the solution could be

$$x(k) = x(0)\prod_{j=0}^{k-1} a(j) + \sum_{j=0}^{k-1} g(j) \prod_{i=j+1}^{k-1} a(i), \tag{2.1.47}$$

where we adopted the convention that $\prod\limits_{i=k}^{k-1} = 1$. Similarly, we agree to put $\sum\limits_{i=j+1}^{j} = 0$.

Then

$$\begin{aligned}
x(k+1) &= a(k)x(k) + g(k) = a(k)\left(x(0)\prod_{r=0}^{k-1} a(r) + \sum_{r=0}^{k-1} g(r)\prod_{i=r+1}^{k-1} a(i)\right) + g(k)\\
&= x(0)\prod_{r=0}^{k} a(r) + a(k)\sum_{r=0}^{k-1} g(r)\prod_{i=r+1}^{k-1} a(i) + g(k)\\
&= x(0)\prod_{r=0}^{k} a(r) + \sum_{r=0}^{k-1} g(r)\prod_{i=r+1}^{k} a(i) + g(k)\prod_{i=k+1}^{k} a(i)\\
&= x(0)\prod_{r=0}^{k} a(r) + \sum_{r=0}^{k} g(r)\prod_{i=r+1}^{k} a(i),
\end{aligned}$$

which proves that (2.1.47) is valid for all $k \in \mathbb{N}$.

Two Special Cases

Two special cases of (2.1.46) appear in many applications. In the first, the equation is

$$x(k+1) = ax(k) + g(k), \tag{2.1.48}$$

with the value $x(0)$ given. In this case $\prod_{r=k_1}^{k_2} a(r) = a^{k_2-k_1+1}$ and (2.1.47) takes the form

$$x(k) = a^k x(0) + \sum_{r=0}^{k-1} a^{k-r-1} g(r). \tag{2.1.49}$$

The second case is a simpler form of (2.1.48), given by

$$x(k+1) = ax(k) + g, \tag{2.1.50}$$

with g independent of k. In this case, the sum in (2.1.49) can be evaluated in an explicit form, giving

$$x(k) = \begin{cases} a^k x(0) + g\frac{a^k-1}{a-1}, & \text{if } a \neq 1, \\ x(0) + gk, & \text{if } a = 1. \end{cases} \tag{2.1.51}$$

Exercise 2.1.11 Prove by induction that

$$\sum_{k=1}^{n} k = \frac{n(n+1)}{2},$$

$$\sum_{k=1}^{n} k^2 = \frac{n(n+1)(2n+1)}{6},$$

$$\sum_{k=1}^{n} k^3 = \left(\frac{n(n+1)}{2}\right)^2,$$

$$\sum_{k=1}^{n} k^4 = \frac{n(6n^4+15n^3+10n^2-1)}{30}.$$

Exercise 2.1.12 Use the above identities, whenever necessary, to solve the following difference equations,

$$y(k+1) = (k+1)y(k) + 2^k(k+1)!, \qquad y(0) = 1,$$
$$x(k+1) = 2x(k) + 3^k, \qquad x(1) = 0.5,$$
$$y(k+1) = \frac{k}{k+1}y(k) + 4, \qquad y(1) = c.$$

The Hassel–Beverton–Holt Model

We recall that the Beverton–Holt equation, Eq. (2.1.15), can be simplified to

$$x(k+1) = \frac{\beta x(k)}{(1+x(k))^b}. \tag{2.1.52}$$

While for general b this equation can display very rich dynamics, see Sect. 4.2.6, for $b = 1$ it can be solved explicitly. So, let us consider

$$x(k+1) = \frac{\beta x(k)}{1+x(k)}. \tag{2.1.53}$$

The substitution $y(k) = \frac{1}{x(k)}$ converts (2.1.53) into

$$y(k+1) = \frac{1}{\beta} + \frac{1}{\beta}y(k).$$

Using (2.1.51), we find

$$y(k) = \frac{1}{\beta}\frac{\beta^{-k}-1}{\beta^{-1}-1} + \beta^{-k}\mathring{y} = \frac{1-\beta^k}{\beta^k(1-\beta)} + \beta^{-k}\mathring{y}$$

if $\beta \neq 1$, and

$$y(k) = k + \mathring{y}$$

for $\beta = 1$. From these equations, we see that $x(k) \to \beta - 1$ if $\beta > 1$ and $x(k) \to 0$ if $\beta \leq 1$ as $k \to \infty$. It may be surprising that a population faces extinction if $\beta = 1$ (which corresponds to every individual giving birth to one offspring on average). However, the density-dependent factor causes some individuals to die between reproductive seasons, which means that the population with $\beta = 1$, in fact, decreases with every cycle.

The Logistic Equation

The discrete logistic equation generally does not admit closed-form solutions and displays rich dynamics. However, some special cases can be solved by an appropriate substitution. We will look at two such cases. First, consider

$$x(k+1) = 2x(k)(1-x(k)) \tag{2.1.54}$$

and use substitution $x(k) = 1/2 - y(k)$. Then

$$\frac{1}{2} - y(k+1) = 2\left(\frac{1}{2} - y(k)\right)\left(\frac{1}{2} + y(k)\right) = \frac{1}{2} - 2(y(k))^2,$$

so that $y(k+1) = 2(y(k))^2$. We see that if $y(0) = \mathring{y} = 0$, then $y(k) = 0$ for all $k \geq 1$. Furthermore, the solution $y(k)$ for $k \geq 1$ does not change if we change the $\mathring{y}$ sign. Thus, we can take $|\mathring{y}| > 0$ as the initial condition. Then $y(k) > 0$ for $k \geq 1$, and we can take the logarithm of both sides getting, for $k \geq 1$,

$$\ln y(k+1) = 2\ln y(k) + \ln 2,$$

which, upon the substitution $z(k) = \ln y(k)$, becomes the inhomogeneous linear equation

$$z(k+1) = 2z(k) + \ln 2.$$

Using (2.1.51), we find the solution to be

$$z(k) = 2^k \mathring{z} + \ln 2(2^k - 1).$$

Hence,

$$y(k) = e^{z(k)} = e^{2^k \ln|\mathring{y}|} e^{\ln 2(2^k-1)} = \mathring{y}^{2^k} 2^{2^k-1},$$

where we dropped the absolute value bars as we raise $\mathring{y}$ to even powers. Thus

$$x(k) = \frac{1}{2} - \left(\frac{1}{2} - \mathring{x}\right)^{2^k} 2^{2^k-1}.$$

We note that for $\mathring{x} = 1/2$ we have $x(k) = 1/2$ for all k, so that we obtain a constant solution. In other words, $x = 1/2$ is an equilibrium point of (2.1.54).

Another logistic equation which can be solved by substitution is

$$x(k+1) = 4x(k)(1-x(k)). \tag{2.1.55}$$

First we note that since $f(x) = 4x(1-x) \le 1$ for $0 \le x \le 1$, we have $0 \le x(k+1) \le 1$ if $x(k)$ has this property. Thus, assuming $0 \le \mathring{x} \le 1$, we can use the substitution

$$x(k) = \sin^2 y(k), \tag{2.1.56}$$

which yields

$$\begin{aligned} x(k+1) = \sin^2 y(k+1) &= 4\sin^2 y(k)(1-\sin^2 y(k)) \\ &= 4\sin^2 y(k)\cos^2 y(k) = \sin^2 2y(k). \end{aligned}$$

This gives the family of equations

$$y(k+1) = \pm 2y(k) + j\pi, \qquad j \in \mathbb{Z}.$$

However, bearing in mind that our aim is to find $x(k)$ given by (2.1.56) and using the periodicity and symmetry of the function $\sin^2$, we can discard $j\pi$ as well as the minus sign and focus on $y(k+1) = 2y(k)$. This is a geometric progression, and we get $y(k) = C2^k$, where $C \in \mathbb{R}$ is arbitrary, as the general solution. Hence

$$x(k) = \sin^2 C2^k,$$

where C is to be determined from $\mathring{x} = \sin^2 C$. What is remarkable in this example is that even though there is an explicit formula for the solution, the dynamics generated by (2.1.55) is very irregular (chaotic); see Sect. 4.2.5.

Exercise 2.1.13 Find the solutions to

$$P(k+1) = P(k) + P(k)\left(1 - \frac{P(k)}{30}\right), \qquad P(0) = 1,$$

and

$$P(k+1) = P(k) + 3P(k)\left(1 - \frac{P(k)}{20}\right), \qquad P(0) = 1.$$

Hint Use Exercise 2.1.7.

Chapter 3
Models with Continuous Time

As noted in Sect. 1.2.1, it appears impossible to model species' growth by differential equations since the population of any species always changes by integer amounts. Hence, the population of any species can never be a differentiable function of time. However, if the population is large and increases by one, then the change is very small compared to a given population. Thus, we make the approximation that large populations change continuously (and even in a differentiable way) in time, and we shall round it to the nearest one if the final answer is not an integer. A similar justification applies to our use of t as a real variable: in the absence of specific breeding seasons, reproduction can occur at any time, and for a sufficiently large population, it is natural to think of reproduction as occurring continuously.

In this chapter, we shall introduce continuous models, the derivation of which parallels the derivation of discrete models discussed above. We commence with the exponential growth.

3.1 Population Balance Equations

Let $P(t)$ denote the size of a population of a given isolated species at time t and let Δt be a small time interval. As in the discrete case, the population at time $t + \Delta t$ can be expressed as

$$P(t + \Delta t) - P(t) = \text{number of births in } \Delta t - \text{number of deaths in } \Delta t.$$

It is reasonable to assume that the number of births and deaths in a short time interval is proportional to the population at the beginning of this interval and proportional

J. Banasiak, *Introduction to Mathematical Methods in Population Theory*, Springer Undergraduate Mathematics Series,
https://doi.org/10.1007/978-3-031-65491-6_3

to the length of this interval, so that introducing birth and death rates β and μ, respectively, we obtain

$$P(t+\Delta t)-P(t)=\beta(t,P(t))P(t)\Delta t-\mu(t,P(t))P(t)\Delta t. \qquad (3.1.1)$$

Defining $r(t,P)$ to be the difference between the birth and death rates at time t and the population of size P, we obtain

$$P(t+\Delta t)-P(t)=r(t,P(t))\Delta t P(t).$$

If we fix Δt, taking it as a unit time interval, and drop the dependence on t in the above equation, then it becomes exactly (2.1.9) with $R(P)=1+r(P)$. Here, however, we assume that the change happens continuously, so dividing by Δt and passing with Δt to 0 we arrive at the continuous in time counterpart of (2.1.9),

$$\frac{dP}{dt}=r(t,P)P. \qquad (3.1.2)$$

This is a new type of equation—a differential equation which, in contrast to difference equations that, in principle, can be solved by recursion, requires a more sophisticated approach. While the general theory of solvability of differential equations is presented in Appendix B, we briefly discuss how to solve some basic equations we shall encounter in the next sections.

3.2 A Survey of Ordinary Differential Equations That Can Be Solved in a Closed Form

3.2.1 Separable Equations

Separable equations are those that can be written in the form

$$\frac{dy}{dt}=g(t)h(y), \qquad (3.2.3)$$

where g and h are known continuous functions. Firstly, we note that any constant function $y=\mathring{y}$, such that $h(\mathring{y})=0$, is a solution to (3.2.3), called a *stationary* or *equilibrium solution.*

We assume that $h(y)\neq 0$ in some interval to find a general solution. Dividing both sides of (3.2.3) by $h(y)$, we get

$$\frac{1}{h(y)}\frac{dy}{dt}=g(t). \qquad (3.2.4)$$

Observe that denoting by $H(y) = \int \frac{1}{h(y)} dy$ the antiderivative of $1/h$, we can write (3.2.3) in the form

$$\frac{d}{dt}(H(y(t))) = g(t).$$

Thus, upon integration we obtain

$$H(y(t)) = \int g(t)dt + c, \tag{3.2.5}$$

where c is an arbitrary constant of integration. The next step depends on the properties of H. If, for instance, $H : \mathbb{R} \to \mathbb{R}$ is monotonic and surjective, then we can find y explicitly for all values of t and c as

$$y(t) = H^{-1}\left(\int g(t)dt + c\right).$$

Otherwise, we have to do it locally, around the initial values. This is possible as we defined H only for that y for which $dH/dt \neq 0$. To explain this, we solve the initial value problem

$$\begin{aligned} \frac{dy}{dt} &= g(t)h(y), \\ y(t_0) &= \mathring{y}, \end{aligned} \tag{3.2.6}$$

where $h(\mathring{y}) \neq 0$. Then, by the continuity of h, we can assume that $h \neq 0$ on some interval containing $\mathring{y}$. Using the solution (3.2.5), we obtain

$$H(y(t)) = \int_{t_0}^{t} g(s)ds + c,$$

as long as $y(t)$ is such that $h(y(t)) \neq 0$. Hence,

$$H(\mathring{y}) = H(y(t_0)) = \int_{t_0}^{t_0} g(s)ds + c,$$

which, due to $\int_{t_0}^{t_0} g(s)ds = 0$, gives

$$c = H(\mathring{y}),$$

so that

$$H(y(t)) = \int_{t_0}^{t} g(s)ds + H(\mathring{y}).$$

If $(t_0 - \alpha, t_0 + \alpha)$ be the interval on which $h(y(t)) \neq 0$. Then, by definition,

$$\frac{dH}{dy} = \frac{1}{h} \neq 0$$

for t in this interval and hence H is invertible, giving

$$y(t) = H^{-1}\left(\int_{t_0}^{t} g(s)ds + H(\mathring{y})\right).$$

3.2.2 *Linear Ordinary Differential Equations of First Order*

The general first-order linear differential equation is of the form

$$\frac{dy}{dt} + a(t)y = b(t). \tag{3.2.7}$$

Functions a and b are known continuous functions of t. One of the methods of solving (3.2.7) is by multiplying both sides by a function, called an integrating factor, to convert the left-hand side of the equation into the derivative of a single function. Let us assume that ϕ is an integrating factor, that is, we must have

$$\frac{d}{dt}(\phi y) = \phi\frac{dy}{dt} + \frac{d\phi}{dt}y = \phi\frac{dy}{dt} + \phi a y.$$

Thus, we see that we can take as ϕ any solution of the equation

$$\frac{d\phi}{dt} = a(t)\phi.$$

It is a separable equation, the general solution of which is given by (3.2.5). Since we need only one such function, we may take

$$\phi(t) = \exp\left(\int a(t)dt\right).$$

Accordingly, with such a function, (3.2.7) can be written as

$$\frac{d}{dt}(\phi(t)y(t)) = \phi(t)b(t).$$

Finally,

$$\begin{aligned} y(t) &= \frac{1}{\phi(t)}\left(\int \phi(t)b(t)dt + c\right) \\ &= \exp\left(-\int a(t)dt\right)\left(\int b(t)\exp\left(\int a(t)dt\right)dt + c\right), \end{aligned} \quad (3.2.8)$$

where c is an arbitrary constant. It is worthwhile to note that the solution consists of two parts: the general solution to the reduced equation associated with (3.2.7)

$$c\exp\left(-\int a(t)dt\right),$$

and, what can be checked by direct differentiation, a particular solution to the full equation.

If we want to find a particular solution satisfying $y(t_0) = \mathring{y}$, then we write (3.2.8) using definite integrals

$$y(t) = \exp\left(-\int_{t_0}^{t} a(s)ds\right)\left(\int_{t_0}^{t} b(s)\exp\left(\int_{t_0}^{s} a(r)dr\right)ds + c\right)$$

and use the fact that $\int_{t_0}^{t_0} f(s)ds = 0$ for any function f. This shows that the part of the solution satisfying the nonhomogeneous equation,

$$y_b(t) = \exp\left(-\int_{t_0}^{t} a(s)ds\right)\int_{t_0}^{t} b(s)\exp\left(\int_{t_0}^{s} a(r)dr\right)ds,$$

equals zero at $t = t_0$. Thus

$$\mathring{y} = y(t_0) = c$$

and the solution to the initial value problem is given by

$$y(t) = \mathring{y}\exp\left(-\int_{t_0}^{t} a(s)ds\right) + \exp\left(-\int_{t_0}^{t} a(s)ds\right)\int_{t_0}^{t} b(s)\exp\left(\int_{t_0}^{s} a(r)dr\right)ds. \quad (3.2.9)$$

Exercise 3.2.1

(i) Show that every solution of the equation $y' + ay = be^{-ct}$, where a and c are positive constants and b is any real number, approaches zero as t approaches infinity.
(ii) Given the differential equation $y' + a(t)y = f(t)$ with $a(t)$ and $f(t)$ continuous for $-\infty < t < +\infty$, $a(t) \geq c > 0$, and $\lim_{t\to\infty} f(t) = 0$, show that its every solution $y(t)$ tends to zero as t approaches infinity.

3.2.3 *The Bernoulli Equation*

Several equations can be transformed into the linear equation. One that we often see in population models is the Bernoulli equation,

$$\frac{dy}{dt} + a(t)y + b(t)y^\theta = 0, \quad y(t_0) = \mathring{y}, \tag{3.2.10}$$

where a and b are continuous functions of time and θ is an arbitrary real number not equal to 0 or 1 (as then the equation would be again the linear equation). For $\theta > 0$, $y = 0$ is a stationary solution. Assuming thus $y \neq 0$, we multiply both sides of (3.2.10) by $y^{-\theta}$ to get

$$y^{-\theta}\frac{dy}{dt} + a(t)y^{1-\theta} + b(t) = 0. \tag{3.2.11}$$

If we let $z = y^{1-\theta}$, then

$$\frac{dz}{dt} = (1-\theta)y^{-\theta}\frac{dy}{dt}$$

and (3.2.11) takes the form

$$\frac{dz}{dt} + (1-\theta)a(t)z + (1-\theta)b(t) = 0, \quad z(t_0) = \mathring{y}^{1-\theta}, \tag{3.2.12}$$

which is a linear equation whose solution is given by a suitably modified (3.2.8) or (3.2.9).

3.2.4 *Equations of Homogeneous Type*

In differential equations, as in integration, a smart substitution can often convert a complicated equation into a manageable one. For some classes of differential equations, there are standard substitutions that transform them into separable equations. We shall discuss one such class in detail.

A differential equation that can be written in the form

$$\frac{dy}{dt} = f\left(\frac{y}{t}\right), \tag{3.2.13}$$

where f is a function of the single variable $z = y/t$, is said to be of *homogeneous type*. Note that in some textbooks, such equations are called *homogeneous equations* but this often creates confusion as the name homogeneous equation is generally used in another context.

To solve (3.2.13) we make the substitution

$$y = tz \tag{3.2.14}$$

where z is the new unknown function. Then, by the product rule for derivatives,

$$\frac{dy}{dt} = z + t\frac{dz}{dt}$$

and (3.2.13) becomes

$$z + t\frac{dz}{dt} = f(z),$$

or

$$t\frac{dz}{dt} = f(z) - z. \tag{3.2.15}$$

In (3.2.15), the variables are separable, so it can be solved as in Sect. 3.2.

3.2.5 Equations That Can Be Reduced to First Order Equations

Some higher-order equations can be reduced to equations of the first order. We shall discuss two such second-order equations.

Equations That Do Not Contain the Unknown Function

If we have the equation of the form

$$F(y'', y', t) = 0, \tag{3.2.16}$$

then the substitution $z = y'$ reduces this equation to an equation of the first order

$$F(z', z, t) = 0. \tag{3.2.17}$$

If we can solve this equation,

$$z = \phi(t, C),$$

where C is an arbitrary constant, then returning to the original unknown function y, we obtain another first-order equation

$$y' = \phi(t, C),$$

which is immediately solvable as

$$y(t) = \int \phi(t, C)dt + C_1.$$

Equations That Do Not Contain the Independent Variable
Let us consider the equation

$$F(y'', y', y) = 0, \tag{3.2.18}$$

that does not involve the independent variable t. Such an equation can also be reduced to a first-order equation; the idea, however, is a little more complicated. Firstly, we note that the derivative y' is uniquely defined by the function y. This means that we can write $y' = g(y)$ for some function g. Using the chain rule, we obtain

$$y'' = \frac{d}{dt}y' = \frac{dg}{dy}(y)\frac{dy}{dt} = y'\frac{dg}{dy}(y) = g(y)\frac{dg}{dy}(y). \tag{3.2.19}$$

Substituting (3.2.19) into (3.2.18) gives the first order equation with y as an independent variable

$$F\left(g\frac{dg}{dy}, g, y\right) = 0. \tag{3.2.20}$$

If we solve this equation in the form $g(y) = \phi(y, C)$, then to find y, we have to solve one more first-order equation with t as the independent variable

$$\frac{dy}{dt} = \phi(y, C).$$

Exercise 3.2.2 Recognize the type, find the general solutions, and determine the interval of their existence for the following differential equations.

$$
\begin{aligned}
y' + y\cos t &= 0,\\
(1+t^2)y' &= 1+y^2,\\
&\qquad Hint: Use\ \tan(x+y) = \frac{\tan x + \tan y}{1-\tan x \tan y}\ in\ the\ answer,\\
y' &= \frac{t^2+y^2}{yt},\\
y' &= 1-t+y^2-ty^2,\\
\cos y \sin t\, y' &= \sin y \cos t,\\
y' + t^2 y &= 1,\\
t^2 y' &= ty - y^2,\\
(1+t^2)y' + ty &= (1+t^2)^{5/2},\\
y'' + \frac{2t}{1+t^2}y' &= \frac{1}{1+t^2},\\
yy'' &= y'.
\end{aligned}
$$

3.3 Examples in Population Dynamics

3.3.1 Exponential Growth

To proceed with modelling, we must specify the form of r. As before, the simplest possible $r(t, P)$ is constant, and, in fact, such models are used in short-term population forecasting. So let us assume that $r(t, P(t)) = r$ so that

$$\frac{dP}{dt} = rP. \tag{3.3.21}$$

It is easy to check, see also Sect. 3.2 for a derivation, that the solution to (3.3.21) is given by

$$P(t) = P(t_0)e^{r(t-t_0)}, \tag{3.3.22}$$

where $P(t_0)$ is the size of the population at some fixed initial time t_0.

To be able to give some numerical illustration to this equation, we need the coefficient r and the population at some time t_0. Following [20], we use the data of the U.S. Department of Commerce: it was estimated that the Earth's population in 1965 was 3.34 billion and that the population was increasing at an average rate of 2% per year during the decade 1960–1970. Thus $P(t_0) = P(1965) = 3.34 \times 10^9$ with $r = 0.02$, and (3.3.22) takes the form

$$P(t) = (3.34 \times 10^9)e^{0.02(t-1965)}. \tag{3.3.23}$$

To test the accuracy of this formula, let us calculate when the population of the Earth is expected to double. To do this, we solve the equation

$$P(T + t_0) = 2P(t_0) = P(t_0)e^{0.02T},$$

thus

$$2 = e^{0.02T}$$

and

$$T = 50 \ln 2 \approx 34.6 \text{ years}.$$

This agrees with the estimated value of the Earth's population in 2000, which was 6.07 billion people. We see that it also agrees relatively well with the observed data if we don't go too far into the past, Fig. 3.1. On the other hand, if we try to extrapolate this model into the distant future, then we see that, say, in the year 2515, the population will reach almost 200,000 billion. To realize what this means, let us recall that the Earth's total surface area is 510,072,000 square kilometres, 70.8% of which is covered by water; thus, we have only 148,940,000 square kilometres to our disposal, and there will be only 0.7447 m^2 (86.3 cm $\times$ 86.3 cm) per person.

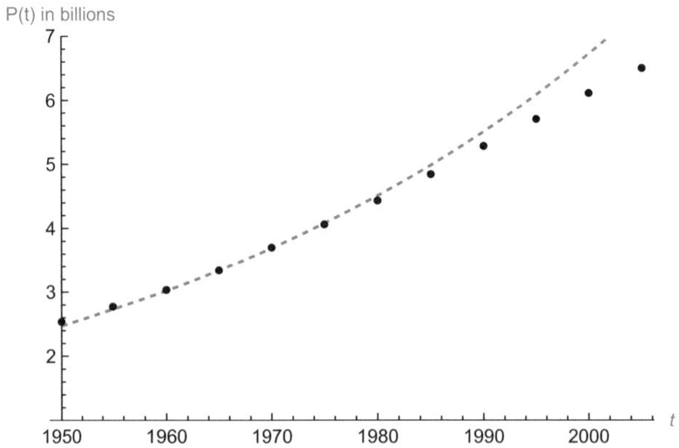

Fig. 3.1 Comparison of actual population figures (points) with those obtained from Eq. (3.3.23)

Therefore, we can only hope that this model is not valid for all times. Indeed, as for discrete models, it is observed that the linear model for population growth is often in good agreement with observations as long as the population is not too large. When the population gets very large (with respect to its habitat), such models cannot be very accurate since they don't reflect that the individual members have to compete with each other for the limited living space and other resources. It is reasonable that a given habitat can sustain only a finite number K of individuals, and the closer the population is to this number, the slower its growth.

Exercise 3.3.1 In 1979 and 1981, some yearling fish were seined in a breeding site, taken across the country and released in a lake. A total of only 435 fish survived these two trips. Yet, in 1999, the commercial net catch was 480 tons. Since the growth of this population was so fast, it is reasonable to assume that it obeyed the Malthusian law $\frac{dP}{dt} = rP$. Assuming that the average weight of a fish is 1.2 kg and that in 1999, approximately 10% of the population was caught, find the lower and upper bounds for r.

Exercise 3.3.2 Suppose a population is growing exponentially (with continuous time). Show that if the population doubles during the initial period T, it doubles during every time interval of length T. Write down the relation between the growth rate and the doubling time T.

Exercise 3.3.3 Radioactive substances spontaneously decay due to the emission of α particles. The mass of α particles is small compared to the sample of the radioactive material, so it is reasonable to assume that the mass decrease happens continuously. Experiments indicate that

$$rate\ of\ decrease\ is\ proportional\ to\ the\ mass\ of\ the\ sample\ still\ present$$

(a) Find the equation obeyed by the mass of the sample and write down the solution to this equation.
(b) Show that if the mass of the sample halves during the initial period T, then it halves during every time interval of length T. Write down the relationship between the decay rate and the time T, called the half-life time.
(c) Radioactive decay is used for dating samples of old materials, like fossils. The so-called Carbon-14 dating is based on the observation that the element carbon appears in nature as a mixture of stable Carbon-12 (C^{12}) and radioactive Carbon-14 (C^{14}), and the ratio between them remains constant throughout history. Thus, while an animal or a plant lives, the ratio C^{12}/C^{14} in its tissue is a known constant. When it dies, however, there is no new carbon absorbed by tissues and since the radioactive C^{14} decays, the ratio C^{12}/C^{14} increases as the amount of C^{14} decreases. The half-life of C^{14} is 5730 years.
   • In a sample of charcoal found in 2003 in a prehistoric cave, the ratio C^{12}/C^{14} was determined to be 14.5% of its original value. How many years ago did the wood grow?

Exercise 3.3.4 Another important process leading to an exponential decay model is the absorption of drugs from the bloodstream into the body tissues. The significant quantity to monitor is the drug concentration in the bloodstream, defined as the amount of drug per unit volume of blood. Observations show that the rate of absorption of the drug, which is equal to the rate of decrease of the concentration of the drug in the bloodstream, is proportional to the concentration of the drug in the bloodstream. Thus

rate of decrease of concentration is proportional to concentration.

(a) Write down the equation for the concentration of the drug.
(b) It is known that for a drug administered for asthma, the concentration in the blood below 5 mg/l has little effect, whereas undesirable side effects appear if the concentration exceeds 20 mg/l. Suppose a dose corresponding to the concentration of 14 mg/l is administered initially. The concentration satisfies the equation

$$\frac{dc}{dt} = -\frac{c}{6},$$

where time is measured in hours.

(i) Write down the concentration at time t.
(ii) Show that the second injection will need to be given after about 6 hours to prevent the concentration from becoming ineffective.
(iii) Given that the second injection also increases the concentration by 14 mg/l, how long is it before another injection is necessary?
(iv) What is the shortest safe time that the second injection (of 14 mg/l) may be given so that side effects do not occur?

Exercise 3.3.5 In many cases, the drug instead of being given in a series of injections, as described in the previous exercise, is fed into the blood continuously by a process called intravenous infusion. Experiments show that for a body of mass $W\,kg$, the concentration of the drug in blood when $M\,mg$ of the drug is present in the system is given by

$$c = \frac{2M}{W}\text{mg/l}.$$

(a) Derive the differential equation governing this process if the infusion rate is $D(t)$mg/hour.
(b) Find the concentration of the drug, if $D(t) = D$ is a constant.
(c) Find the limiting concentration of the drug as $t \to \infty$.
(d) Using the data from Exercise 3.3.4, find the permissible range of D for a 60 kg patient.

Death Rate and the Average Lifespan of Individuals—Continuous Case
As in the discrete case, see Sect. 2.1.3, the death rate is related to the average lifespan of individuals. The argument is similar, but some technicalities differ.

Consider a population in which individuals die at a constant rate μ,

$$P' = -\mu P.$$

Then, the probability that an individual dies in a time interval Δt is approximately equal to $\mu \Delta t$. Let $p(t)$ be the probability that the individual is alive at time t. Then the individual is alive at time $t + \Delta t$ provided it was alive at time t and did not die in the time interval $(t, t+\Delta t)$, that is, $p(t+\Delta t) = (1-\mu\Delta t)p(t)$ which, as above, yields

$$p' = -\mu p$$

with $p(0) = 1$ (expressing the fact that the individual was born, and thus alive, at $t = 0$) yielding $p(t) = e^{-\mu t}$. The average lifespan is given by

$$L = \int_0^\infty s m(s) ds,$$

where $m(s)$ is the probability (density) of dying exactly at age s. Since the probability of death between t and $t + \Delta t$, that is, being alive at t and not being alive at $t + \Delta t$, is given by

$$p(t) - p(t + \Delta t) = -\int_t^{t+\Delta t} \frac{d}{ds} p(s) ds = \int_t^{t+\Delta t} m(s) ds$$

(as the set of individuals alive at t is bigger that that of being alive at $t + \Delta t$), we have $m(s) = -p'(s)$ and

$$L = -\int_0^\infty s \frac{d}{ds} e^{-\mu s} ds = \mu \int_0^\infty s e^{-\mu s} ds = \frac{1}{\mu}. \tag{3.3.24}$$

Exercise 3.3.6 The assumption that the rate of death is constant is unrealistic—the death rate $\mu(a)$ of children should be higher than that of juveniles and adults, and, for elderly people, it should diverge to infinity for ages close to the maximum duration of life. Give examples of such functions if the maximum observed age is 100 years.

Find the average lifespan in such a population if

$$\mu(a) = \frac{1}{100 - a}.$$

3.3.2 Logistic Equation

Again, the simplest way to address drawbacks of a constant r is to take $r(t, P) = r(1 - \frac{P}{K})$, where $K > 0$. Then we obtain the so-called *continuous logistic model*

$$P' = rP\left(1 - \frac{P}{K}\right), \tag{3.3.25}$$

which proved to be one of the most successful models for describing a single species population. Alternatively, as in the discrete case, we can obtain (3.3.25) by taking in (3.1.1) a constant birth rate β, while introducing a density dependent mortality rate

$$\mu(P) = \mu_0 + \mu_1 P.$$

The increase in the population over a time interval Δt is given by

$$P(t + \Delta t) - P(t) = \beta P(t)\Delta t - \mu_0 P(t)\Delta t - \mu_1 P^2(t)\Delta t$$

which, upon dividing by Δt and passing with it to zero, gives

$$P' = (\beta - \mu_0)P - \mu_1 P^2$$

which is another form of (3.3.25).

A more general form of this equation is obtained by taking $\mu(P) = \mu_0 + \mu_1 P^\theta$ for some positive constant θ which leads to a continuous Bernoulli equation, see Sect. 3.2,

$$P' = (\beta - \mu_0)P - \mu_1 P^{\theta+1}. \tag{3.3.26}$$

Let us focus on the logistic equation (3.3.25). Since the right-hand side does not contain t, it is a separable equation which, unlike its discrete counterpart, can be solved explicitly, as explained in Sect. 3.2.

Let us start with some qualitative features. The constant in time solutions, corresponding to (2.1.13), are obtained by solving (3.3.26) with the right-hand side equal to zero. This gives $P = 0$ and $P = K$. In other words, the constant functions $P(t) = 0$ and $P(t) = K$ are equilibria. We shall discuss them in more detail in Sect. 4.1. We shall focus on solutions with an initial condition $P(t_0) > 0$. Then, from general theory (see Lemma 4.2), if $P(t_0) < K$, then $P(t)$ stays between

0 and K. Otherwise, we would have two different solutions passing through the intersection point. Similarly, if $P(t_0) > K$, then the solution always stays above K. With this information, we can proceed with solving the related Cauchy problem

$$P' = rP\left(1 - \frac{P}{K}\right), \tag{3.3.27a}$$

$$P(t_0) = \mathring{P}, \tag{3.3.27b}$$

with $\mathring{P} \neq K$ and $\mathring{P} \neq 0$. Due to biological interpretation, we assume $\mathring{P} > 0$. By the argument above, the right-hand side is never zero, and (3.3.27a) is equivalent to

$$\frac{1}{rP\left(1 - \frac{P}{K}\right)} P' = 1.$$

Integrating both sides from t_0 to t, and changing variables in the integral on the left-hand side and using (3.3.27b), we obtain

$$\frac{K}{r} \int\limits_{\mathring{P}}^{P} \frac{ds}{(K-s)s} = t - t_0.$$

To integrate the left-hand side, we use partial fractions

$$\frac{1}{(K-s)s} = \frac{1}{K}\left(\frac{1}{s} + \frac{1}{K-s}\right),$$

which gives

$$\frac{K}{r} \int\limits_{\mathring{P}}^{P} \frac{ds}{(K-s)s} = \frac{1}{r} \int\limits_{\mathring{P}}^{P} \left(\frac{1}{s} + \frac{1}{K-s}\right) ds = \frac{1}{r} \ln \frac{P}{\mathring{P}} \left| \frac{K - \mathring{P}}{K - P} \right|.$$

From the considerations preceding (3.3.27), if $0 < \mathring{P} < K$, then $P(t) < K$, while if $\mathring{P} > K$, then $P(t) > K$ for all $t > 0$. Therefore $(K - \mathring{P})/(K - P(t))$ is always positive and

$$r(t - t_0) = \ln \frac{P}{\mathring{P}} \frac{K - \mathring{P}}{K - P}.$$

Exponentiating, we get

$$e^{r(t-t_0)} = \frac{P(t)}{\mathring{P}} \frac{K - \breve{P}}{K - P(t)},$$

that is,

$$\mathring{P}(K - P(t))e^{r(t-t_0)} = P(t)(K - \mathring{P}).$$

Bringing all the terms involving P to the left-hand side and multiplying by -1, we get

$$P(t)\left(\mathring{P}e^{r(t-t_0)} + K - \mathring{P}\right) = \mathring{P}Ke^{r(t-t_0)},$$

thus finally

$$P(t) = \frac{\mathring{P}K}{\mathring{P} + (K - \mathring{P})e^{-r(t-t_0)}}. \tag{3.3.28}$$

Let us examine (3.3.28) to see whether it describes the population's behaviour we had in mind building the model (which helps to ensure that we have not made any mistake in solving the equation). First, observe that

$$\lim_{t\to\infty} P(t) = K,$$

hence, our model correctly reflects the initial assumption that K is the maximal capacity of the habitat. Next, we obtain

$$P' = \frac{r\mathring{P}K(K - \mathring{P})e^{-r(t-t_0)}}{(\mathring{P} + (K - \mathring{P})e^{-r(t-t_0)})^2},$$

thus, if $\mathring{P} < K$, the population monotonically increases, whereas if we start with a larger population, then the capacity of the habitat, then such a population will decrease until it reaches K. Also, using repeatedly (3.3.27a),

$$KP'' = r\frac{d}{dt}(P(K - P)) = P'(K - 2P) = P(K - P)(K - 2P)$$

from which it follows that, if we start from $\mathring{P} < K$, then the population curve is convex down for $P < K/2$ and convex up for $P > K/2$. Thus, as long as the population is small (less than half of the capacity K), the rate of growth increases, whereas for a larger population, the growth rate decreases. This results in the famous *logistic* or *S-shaped* curve which, for particular values of parameters $r = 0.2$, $K = 10$ and $t_0 = 0$, as the graph of

$$P(t) = \frac{10\mathring{P}}{\mathring{P} + (10 - \mathring{P})e^{-0.2t}},$$

is presented in Fig. 3.2.

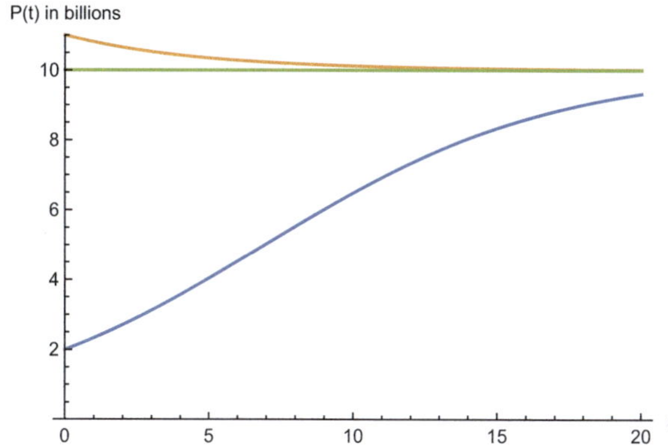

Fig. 3.2 Logistic curves with $\mathring{P} < K$ (blue line) and $\mathring{P} > K$ (orange line) for $K = 10$ and $r = 0.2$

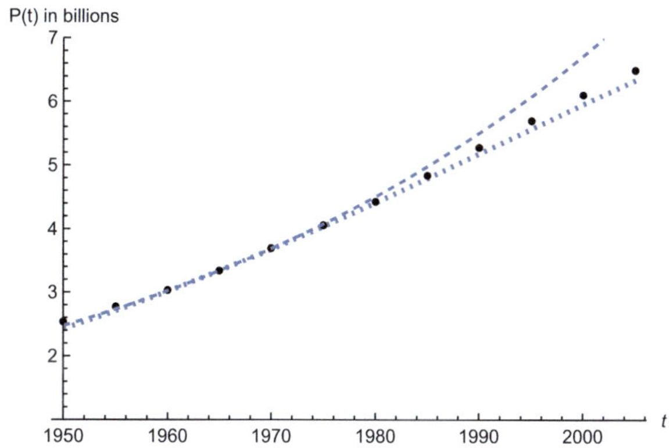

Fig. 3.3 Human population on Earth. Comparison of observational data (points), exponential growth (dashed line) and logistic growth (dotted line)

To show how the solution of the logistic equation compares with the real data and with the exponential growth, we take the experimental coefficients $K = 10.76$ billion and $r = 0.029$. Then, the logistic equation for the growth of the Earth's population will read

$$P(t) = \frac{\mathring{P}(10.76 \times 10^9)}{\mathring{P} + ((10.76 \times 10^9) - \mathring{P})e^{-0.029(t-t_0)}}.$$

We use this function with $\mathring{P} = 3.34 \times 10^9$ at $t_0 = 1965$. The comparison is shown in Fig. 3.3.

Despite its simplicity, the logistic model gives a fairly accurate long-term prediction, as illustrated in Table 3.1, see [20]. Nevertheless, despite the good fit, its reason should be further investigated to ensure it is not a coincidence. In [20], the author gives an example of Belgium, where the logistic model predictions proved well off the mark in the early twentieth century. A possible explanation of this fact may come from the realization that, due to its size, the carrying capacity of the US remained largely unchanged during the discussed period, while that of Belgium

Table 3.1 Comparison of the actual and predicted by the logistic model population of the US, see [25]

	Real	Predicted	Error	%
1790	3,929,000	3,929,000	0	0.0
1800	5,308,000	5,336,000	28,000	0.5
1810	7,240,000	7,228,000	−12,000	−0.2
1820	9,638,000	9,757,000	119,000	1.2
1830	12,866,000	13,109,000	243,000	1.9
1840	17,069,000	17,506,000	437,000	2.6
1850	23,192,000	23,192,000	0	0.0
1860	31,443,000	30,412,000	−1,031,000	−3.3
1870	38,558,000	39,372,000	814,000	2.1
1880	50,156,000	50,177,000	21,000	0.0
1890	62,948,000	62,769,000	−179,000	−0.3
1900	75,995,000	76,870,000	875,000	1.2
1910	91,972,000	91,972,000	0	0.0
1920	105,711,000	107,559,000	1,848,000	1.7
1930	122,775,000	123,124,000	3,498,000	0.3
1940	131,669,000	136,653,000	4,984,000	3.8
1950	150,697,000	149,053,000	−1,644,000	−1.1

significantly increased due to the acquisition of the African colonies (at the expense of the native inhabitants).

Exercise 3.3.7 Let a population grow according to the (continuous) logistic law with intrinsic growth rate r and let T be the time at which half of the limiting population K is achieved. Show that $P(t)$ is given by

$$P(t) = \frac{K}{1 + e^{-r(t-T)}}.$$

Exercise 3.3.8 Consider a fish population in a pond that evolves according to the logistic law (3.3.28). Assume that we know that the carrying capacity K of the pond is between 9000 and 11,000 fish and that we initially let 5000 fish into the pond. After one year, we estimate that there are between 6000 and 7000 fish in the pond. Estimate the intrinsic growth rate r.

Exercise 3.3.9 A population grows according to the logistic law, with a limiting population of $5 \cdot 10^8$ individuals. When the population is low, it doubles every 40 minutes. What will the population be after 2 hours if initially it is (a) 10^8, (b) 10^9?

Exercise 3.3.10 If we neglect high migration and homicide rates, the population of a certain city would satisfy the logistic law

$$P' = \frac{1}{25}P - \frac{1}{25 \cdot 10^6}P^2,$$

where t is measured in years,

(a) Modify this equation to take into account the fact that 12,000 people per year move from the city, 20,000 migrate to the city and 1000 people per year are murdered.
(b) Assume that the city's population was 4,200,000 in 2010. Find the population for all future times. What happens when $t \to \infty$?

Logistic Model with Time-Dependent Coefficients

In many cases, the parameters in (3.3.27a) depend on time, reflecting the environmental changes. Then we obtain

$$P' = r(t)P\left(1 - \frac{P}{K(t)}\right), \tag{3.3.29a}$$

$$P(t_0) = \mathring{P}. \tag{3.3.29b}$$

No longer is Eq. (3.3.29a) separable, but it fits into the theory of the Bernoulli equation, (3.2.10). Introducing $z(t) = \frac{1}{P(t)}$, we transform (3.3.29a) into

$$z'(t) = -r(t)z(t) + \alpha(t), \quad z(t_0) = \mathring{z} = \frac{1}{\mathring{P}}, \tag{3.3.30}$$

where $\alpha(t) = \frac{r(t)}{K(t)}$. Using (3.2.9), we have

$$z(t) = e^{-\int\limits_{t_0}^{t} r(\tau)d\tau}\left(\mathring{z} + \int_{t_0}^{t} \alpha(s)e^{\int_{t_0}^{s} r(\tau)d\tau} ds\right).$$

Returning to the original variable, we obtain

$$P(t) = \frac{\mathring{P}e^{\int_{t_0}^{t} r(\tau)d\tau}}{1 + \mathring{P}\int_{t_0}^{t} \frac{r(s)}{K(s)}e^{\int_{t_0}^{s} r(\tau)d\tau} ds}. \tag{3.3.31}$$

3.3.3 A Simplified Logistic Model

We have considered two basic demographic models, the Malthusian and logistic models. The drawback of the Malthusian model is that it can only describe very simple dynamics: the population decays to zero, exponentially grows to infinity, or else remains constant. On the other hand, the drawback of the logistic model is that it is nonlinear and thus may create additional difficulties in analysing more complicated systems it is a part of. For these reasons, some researchers use an intermediate model that takes the form

$$\frac{dP}{dt} = \Lambda - \mu P, \qquad P(0) = \mathring{P}, \tag{3.3.32}$$

where Λ is the total birth/recruitment rate and μ is the per capita death rate. This is a linear nonhomogeneous equation in P, which can be solved by introducing the integration factor, as in Sect. 3.2. Multiplying (3.3.32) by $e^{\mu t}$ and using the product rule, we get

$$\frac{d}{dt}\left(e^{\mu t}P\right) = e^{\mu t}\frac{dP}{dt} + \mu e^{\mu t}P = e^{\mu t}\Lambda.$$

Hence, by integrating both sides, we arrive at the solution

$$P(t) = \mathring{P}e^{-\mu t} + \frac{\Lambda}{\mu}(1 - e^{-\mu t}). \tag{3.3.33}$$

It is easy to see that all solutions to (3.3.32) converge to

$$P^* = \frac{\Lambda}{\mu}, \tag{3.3.34}$$

which means that it is the only (!) equilibrium point of (3.3.32) and, with the terminology developed in Chap. 4, Definition 4.1, it is globally asymptotically stable.

While mathematically convenient, model (3.3.32) does not make much sense in population theory. First, we observe that the population's growth rate, Λ, is positive even if the population is zero, that is, when there are no individuals to reproduce. Thus, it cannot be a model of a closed population as it would breach the population conservation law. It can describe, however, a population with constant immigration. Also, as shown in Fig. 3.4, even if we can fit this model to the observational data over a reasonably long time period, the fitted demographic parameters are unrealistic.

Exercise 3.3.11 In some classes of organisms, the birth rate is not proportional to the population size. Suppose that each member of the population requires a partner for reproduction, and each member relies on chance encounters to meet the mate. Let us denote the population at time t by $P(t)$. Assume that males and females are equally distributed in the population and each member meets on average aP other members (female or male) in a unit time. Also, assume that only a fraction β of encounters results in breeding one offspring.

(a) Write down the number of offspring produced over a short time Δt in this population.

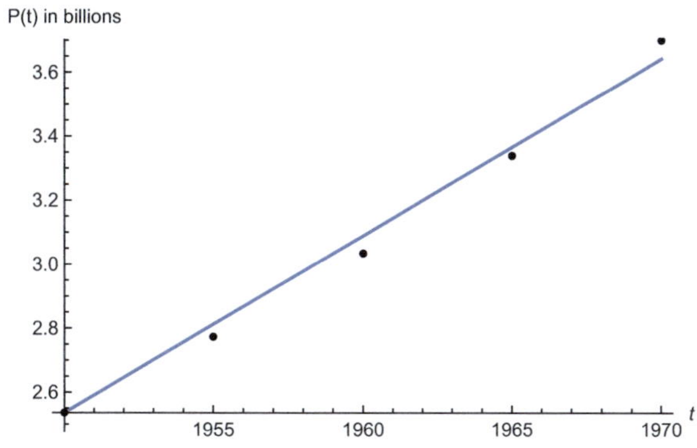

Fig. 3.4 World population alongside the simplified logistic model prediction. The value $P(1950) = 2.5565$ billion people. The least-square error is 0.0086 billion, so the fit is acceptable. However, the numerical values of the parameters are $\Lambda = 0.0553307$ billion and $\mu = 3.55438 \times 10^{-6}$. This gives the average lifespan of 0.28×10^6 years—completely unrealistic

(b) If the death rate in the population is still proportional to the size of the population, write down the equation describing the growth of this population.
(c) Modify the above equation if the population's ratio of males and females is a constant p.

Exercise 3.3.12 In the previous exercise, you should derive a differential equation of the form

$$P' = bP^2 - aP, \qquad P(0) = \mathring{P},$$

where $a, b > 0$ are some constants (identify a and b in the context of the previous exercise). Solve this initial value problem and show that if the initial population satisfies $0 < \mathring{P} < a/b$, then the population faces extinction, that is, $\lim_{t\to\infty} P(t) = 0$. What will happen if $\mathring{P} > a/b$?

3.3.4 *The Allee Effect Via the Holling Type Argument*

In Sect. 2.1.5, we introduced Allee-type models in discrete time by using a multiple time scale argument. The essence of the Allee effect is the presence of an additional death risk, high in very small and very large populations. Clearly, it can also occur in continuous time models. One can derive models like that in the framework of multiple scale models, as in the discrete-time, see [12, 14]. Here, however, we have decided to present another derivation based on the Holling-type argument, which, though developed for predator-prey systems, can be equally well adapted to single equations.

We consider a population evolving according to the logistic law, but additional deaths can occur when individuals must leave a refuge to find a mate. While outside, they are exposed to additional dangers. Hence, the total number of excessive deaths induced by this activity in some period of time T is

$$d = \mu_1 P T_{exp}, \tag{3.3.35}$$

where μ_1 is the additional rate of death due to this activity, P is the size of the population and T_{exp} is the time in the period of length T that an average individual is exposed to the dangers by staying outside. Now,

$$T = T_s + T_{exp},$$

where T_s is the time spent in a shelter. We assume that an individual's lifetime is divided between searching for a mate and caring for the offspring in the shelter. Hence, T_s can be obtained as the product of the number of successful matings in T times the average time spent looking after the offspring. Now, the number of successful matings is the product of the area searched in T_{exp}, the males' density and the mating efficiency. We assume that the ratio of females and males is constant so that the density of males is proportional to P. Further, the searched area is proportional to the search time T_{exp} so, combining all proportionality factors, we can write

$$T_s = aPT_{exp}$$

for some constant a. Hence,

$$T = aPT_{exp} + T_{exp},$$

or $T_{exp} = T/(1 + aP)$. Hence, using (3.3.35), we see that the additional death rate is given by

$$\frac{d}{T} = \frac{\mu_1 P}{1 + aP}.$$

An analogous argument can be used in a population exposed to a generalist predator, that is, a predator that can eat other prey, so that its numbers are not affected by the presence of the prey from this particular population. In this case, (3.3.35) is replaced by

$$d = \mu P T_{hunt},$$

which is the additional number of deaths inflicted by one predator. Here T_{hunt} is the hunting time of the predator. Then, as above,

$$T = T_{hunt} + T_{handl}d,$$

where T_{handl} is the handling time of a single prey. Hence,

$$T_{hunt} = \frac{T}{1 + \mu T_{handl} P}$$

and the additional rate of the decrease in the prey population is given by

$$\frac{dP}{dt} = -\frac{\mu N P}{1 + \mu T_{handl} P},$$

where N is the (constant) population of predators.

Let us include additional deaths into a logistic population modelled by the abovementioned mechanism. We get

$$\frac{dP}{dt} = \lambda P\left(1 - \frac{P}{C} - \frac{A}{1 + BP}\right), \tag{3.3.36}$$

with positive constants λ, C, A, B. Not always does the above equation describe the Allee effect. Let us recall that for this, the equation must have three equilibria, 0 and, say, $0 < L < K$, such that if the size of the population P satisfies $0 < P < L$, then P decreases to 0, if $L < P < K$, then P increases to K and if $P > K$, then the population decreases to K. We shall find the ranges of the parameters for which (3.3.36) indeed represents an Allee model in Sect. 4.1.

3.3.5 The Gompertz Model

One way to construct a population model is to find a rule on how the changes in the population impact the growth rate in (3.1.2). From this point of view, the logistic equation was constructed by assuming

$$\frac{dr}{dP} = -\frac{1}{K}, \quad r(K) = 0,$$

so that $r(P) = 1 - P/K$.

We can consider similar models by assuming that the population's growth slows down its growth by, for instance, blocking the access of individuals to resources. The simplest model of this type is

$$\frac{dr}{dP} = -\frac{\alpha}{P}, \quad r(K) = 0, \tag{3.3.37}$$

called the Gompertz model.

The above equation can be easily solved, giving

$$r(P) = \alpha \ln\left(\frac{K}{P}\right)$$

and thus, the population equation takes the form

$$\frac{dP}{dt} = \alpha \ln\left(\frac{K}{P}\right) P, \tag{3.3.38}$$

see Fig. 3.5.

One can derive another, possibly more instructive, form of this equation. Using (3.3.37) and the Chain Rule, we get

$$\frac{dr}{dt} = \frac{dr}{dP}\frac{dP}{dt} = -\frac{\alpha}{P} r P = -\alpha r.$$

Hence the growth rate decays exponentially as $r(t) = r_0 e^{-\alpha t}$ and we can write (3.3.38) as

$$\frac{dP}{dt} = (r_0 e^{-\alpha t})P = r_0 \cdot (e^{-\alpha t} P). \tag{3.3.39}$$

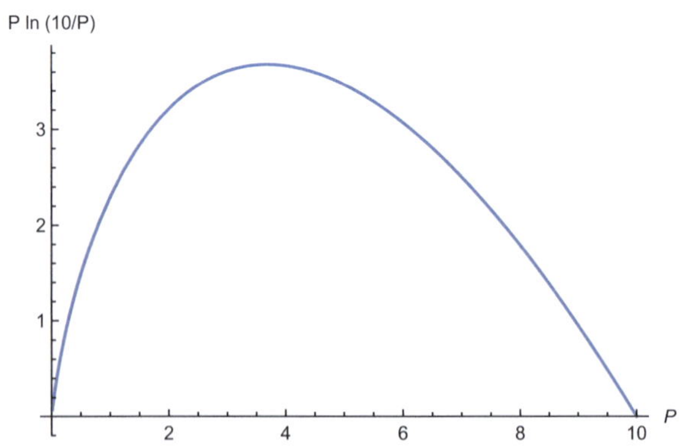

Fig. 3.5 The graph of the RHS of (3.3.38) with $\alpha = 1$ and $K = 1$; the dynamics is similar to the logistic dynamics but skewed towards smaller population values

Different places of brackets indicate different interpretations—the left one suggests that the growth rate is decreasing, while the right one suggests that the pool of fertile individuals is decreasing. That is why the model has been quite successful in modelling cancer growth in the early stages when, before angiogenesis, only the surface layers of the tumour cells receive nutrients; see, e.g., [62].

The equation can be solved by separation of variables, giving

$$P(t) = P(0)e^{\frac{r_0}{\alpha}}e^{-\frac{r_0}{\alpha}e^{-\alpha t}}. \tag{3.3.40}$$

By the construction of K, we have

$$\lim_{t\to\infty} P(t) = P(0)e^{\frac{r_0}{\alpha}} = K,$$

and we can rewrite the solution in terms of the carrying capacity as

$$P(t) = Ke^{e^{-\alpha t}\ln\frac{P(0)}{K}}. \tag{3.3.41}$$

3.4 From Discrete to Continuous Models and Back

3.4.1 Discretization of Continuous Models

As we have seen, continuous models are obtained by using the same principles as corresponding discrete models. In fact, a discrete model (represented by a difference equation) is often an intermediate step in deriving a corresponding differential equation. A question arises whether, under reasonable circumstances, discrete and continuous models are equivalent in the sense that they give the same solutions (or at least the same qualitative features of the solution) if both can be applied to a particular model.

There are several ways of discretizing differential equations. We shall use the two most common. The first one is similar to the standard numerical analysis practice of replacing the derivative with the difference quotient

$$\frac{df}{dt} \approx \frac{f(t+\Delta t) - f(t)}{\Delta t}.$$

Another one is based on the observation that solutions of autonomous equations display the so-called semigroup, or flow, property. To explain it this, let us denote by $x(t, x_0)$ the solution to the equation

$$x' = g(x), \qquad x(0) = \mathring{x};$$

then

$$x(t_1 + t_2, \mathring{x}) = x(t_1, x(t_2, \mathring{x})).$$

This follows from the uniqueness of solutions to differential equations (valid if g is at least Lipschitz continuous—see Theorem B.10 and further discussion in Sects. 4.1 and 6.2). Indeed, both sides of the equation give solutions that at time $t = t_2$ were at $x(t_2, \mathring{x})$. Thus, in particular,

$$x((n+1)\Delta t, \mathring{x}) = x(\Delta t, x(n\Delta t, \mathring{x})). \tag{3.4.42}$$

This indicates that the solution after $n+1$ time steps can be obtained as the solution after one time step with the initial condition given as the solution after n time steps. In other words, denoting $x(n) = x(n\Delta t, \mathring{x})$ we have

$$x(n+1) = f_{\Delta t}(x(n))$$

where f is an operation of getting the problem's solution at Δt with the initial condition as its argument.

In further applications, we shall take $\Delta t = 1$ and call f_1 the time-one map; e.g., [36, 52].

Exponential Growth

We begin with the exponential growth equation

$$P' = rP, \qquad P(0) = \mathring{P}, \tag{3.4.43}$$

having the solution

$$P(t) = \mathring{P}e^{rt}.$$

Euler Discretization

Let us denote by $(p(k))_{k\geq 0}$ the solution of the discrete equation obtained by the Euler discretization of (3.4.43) with $\Delta t = 1$,

$$p(k+1) - p(k) = rp(k),$$

having the solution

$$p(k) = (1+r)^k \mathring{P}.$$

Clearly, this solution does not coincide with the solution $\mathring{P}e^{rt}$ for any value $t = 1, 2, \ldots$. However, qualitatively, these solutions are similar since both grow exponentially, and one can be transformed into the other by rescaling the growth rate. In other words, the discrete Malthusian process with the growth rate given by R, see (2.1.5), coincides with the continuous Malthusian growth with the growth rate r at any time $t \in \mathbb{N}$ if, instead of $R = 1 + r$ as above, we take $R = e^r$. We note that both are approximately equal for small r since $e^r = 1 + r + \ldots$.

Time-One Map Discretization

On the other hand, consider the second discretization, which amounts to assuming that we take the census of the population in evenly spaced time moments $t_0 = 0, t_1 = 1, \ldots, t_k = k, \ldots$, so that

$$p(k) = P(k) = e^{rk}\mathring{P} = \left(e^r\right)^k \mathring{P}. \tag{3.4.44}$$

Comparing this equation with (2.1.5), we see that it corresponds to the discrete model with the intrinsic growth rate

$$R = e^r,$$

as before. With this discretization, we do not need any rescaling.

Thus, we can state that if we observe a continuously growing population in discrete unit time intervals and the observed (discrete) intrinsic growth rate is R, then the real (continuous) growth rate is given by $r = \ln R$.

Logistic Growth

Consider now the logistic equation

$$P' = rP\left(1 - \frac{P}{K}\right).$$

Euler Discretization for the Logistic Equation

The first discretization immediately produces the discrete logistic equation (2.1.18),

$$p(k+1) = p(k) + rp(k)\left(1 - \frac{p(k)}{K}\right),$$

the solutions of which, as we shall see, behave in a dramatically different way than those of the continuous equation, unlike in the exponential growth model.

We shall work with the simplified logistic differential equation

$$Y' = aY(1-Y), \qquad Y(0) = \mathring{y}. \tag{3.4.45}$$

From Sect. 2.1.6, see also Sect. 4.2.5, we know that for, say, $a = 4$, the dynamics of the corresponding discretization

$$y(n+1) = y(n) + 4y(n)(1-y(n)) \tag{3.4.46}$$

is very complex (chaotic), and thus, the latter cannot be used for numerical calculations of (3.4.45) as the solutions to (3.4.45) are monotonic, so the Euler discretization does not preserve the basic qualitative features. This is shown in Fig. 3.6. Let us, however, write down the complete Euler scheme,

$$y(n+1) = y(n) + a\Delta t y(n)(1-y(n)), \tag{3.4.47}$$

where $y(n) := y(n\Delta t)$ and $y(0) = \mathring{y}$. Then

$$y(n+1) = (1+a\Delta t)y(n)\left(1 - \frac{a\Delta t}{1+a\Delta t}y(n)\right).$$

The substitution

$$x(n) = \frac{a\Delta t}{1+a\Delta t}y(n) \tag{3.4.48}$$

reduces (3.4.47) to

$$x(n+1) = \gamma x(n)(1-x(n)), \tag{3.4.49}$$

where $\gamma = 1 + a\Delta t$. Thus, the parameter γ, which, as we shall see in Sect. 4.2.5, controls the long time behaviour of solutions to the discrete equation (3.4.49),

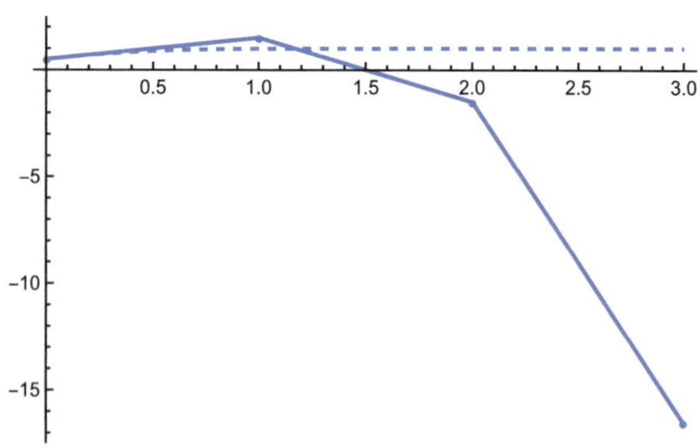

Fig. 3.6 Comparison of solutions to (3.4.45) with $a = 4$ (dashed) and (3.4.46)

depends on Δt and, by choosing a suitably small Δt, we can make the solutions of (3.4.49) mimic the behaviour of solutions to (3.4.45). Indeed, by taking $1+a\Delta t \leq 3$, we obtain the convergence of solutions $x(n)$ to the equilibrium

$$x = \frac{a\Delta t}{1 + a\Delta t}.$$

Here, we restrict ourselves to a numerical illustration of this claim and refer the proofs to Sect. 4.2.5. Reverting (3.4.48), we get the discrete approximation $y(n)$, which converges to 1, as the solution to (3.4.45). However, as seen in Fig. 3.7, this convergence is not monotonic, and this shows that the approximation is rather poor. This can be remedied by taking $1+a\Delta t \leq 2$, in which case the qualitative features of $y(t)$ and $y(n)$ are the same, see Fig. 3.8. This fact is discussed more in Remark 4.20.

These features of the discrete logistic model can, to a certain extent, be explained by interpreting it as a game between the population and the environment, in which the response of the environment to the population size $y(n)$ comes only after the full-time step, resulting in the population of $y(n + 1)$. It is then natural to expect that the system is more likely to lose stability if the response times are long. On

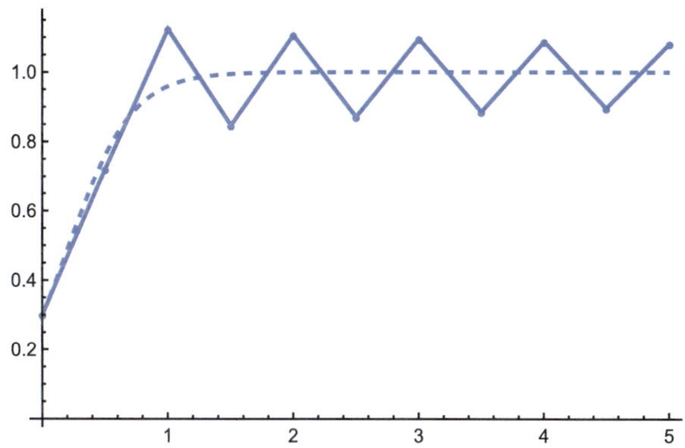

Fig. 3.7 Comparison of solutions to (3.4.45) with $a = 4$ (dashed) and (3.4.47) with $\Delta t = 0.5$

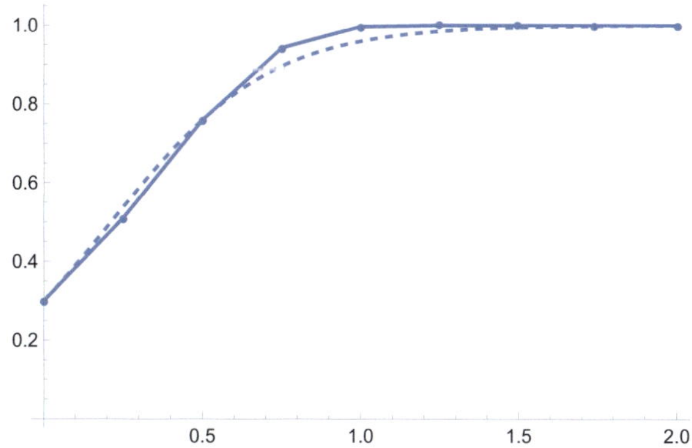

Fig. 3.8 Comparison of solutions to (3.4.45) with $a = 4$ (dashed) and (3.4.47) with $\Delta t = 0.25$

the other hand, in the continuous logistic model, the responses are instantaneous, resulting in its monotonic and smooth behaviour.

We note that the above problems can also be solved by introducing the so-called non-standard difference scheme; see, e.g., [72], which consists in replacing the derivatives and/or nonlinear terms with more sophisticated expressions which, though equivalent when the time step goes to 0, produce, nevertheless, qualitatively different discrete picture. In the case of the logistic equation, such a non-standard scheme can be constructed by replacing y^2 not by $y^2(n)$ but by $y(n)y(n+1)$, resulting

$$y(n+1) - y(n) = a\Delta t(y(n) - y(n)y(n+1)).$$

In general, such a substitution yields an implicit scheme, but in our case, the resulting recurrence can be solved for $y(n+1)$, producing

$$y(n+1) = \frac{(1+a\Delta t)y(n)}{1+a\Delta t y(n)},$$

and we recognize the Beverton–Holt equation with $\beta = 1 + a\Delta t$ (and $K = 1$). We have seen in Sect. 2.1.6 that $(y(n))_{n\in\mathbb{N}}$ monotonically converges to equilibrium, and, as we shall see below, it exactly follows the solution of the continuous logistic equation. In the spirit of the game interpretation of the model discussed above, this stability can be attributed to the fact that the environment response is based on the input $y(n)y(n+1)$ combining the previous and the current time instants, in contrast to $y^2(n)$ in the case of the Euler discretization above.

To use the time-one map discretization, we re-write (3.3.28) as

$$P(t) = \frac{\mathring{P}e^{rt}}{1 + \frac{e^{rt}-1}{K}\mathring{P}},$$

which, upon denoting $e^r = \beta$, gives the time-one map

$$P(1, \mathring{P}) = \frac{\mathring{P}\beta}{1 + \frac{\beta-1}{K}\mathring{P}},$$

which, as above, yields the Beverton–Holt model

$$p(k+1) = \frac{\beta p(k)}{1 + \frac{\beta-1}{K}p(k)},$$

with the discrete intrinsic growth rate related to the continuous one in the same way as in the exponential growth equation.

Exercise 3.4.1 A fish population is observed once per year. It is found that it satisfies the Beverton–Holt equation

$$p(k+1) = \frac{ap(k)}{b + cp(k)}$$

for some positive constants a, b, c. Find the continuous logistic equation governing the growth of this population and, hence, find the population's carrying capacity and intrinsic growth rate.

3.4.2 Discrete Equations in Continuous Time Models

In the previous section, we have seen that it is often difficult to describe processes occurring in continuous time using difference equations, which, on the other hand, are usually easier to handle. Here, we shall describe two situations in which it is possible. First, we discuss models with periodic coefficients. They yield to a discrete-time description, provided one is satisfied with only knowing the system's state in time instants, corresponding to the period of the coefficients. The second case concerns hybrid (sometimes called impulsive) models, when we have a continuous repetitive process interspersed with instantaneous events at evenly spaced time intervals.

Discrete Models of Seasonally Changing Populations

So far, we have considered models in which the laws governing the process are independent of time. In most real processes, we must consider phenomena that depend on time, such as the year's seasons. Here, we use the same modelling principles as in this chapter, but with time-dependent birth and death coefficients $\beta(t)$ and $\mu(t)$. We also consider emigration, which is also supposed to be proportional to the total population, and immigration, which is just a given flux of individuals into the system. Then, instead of (3.3.21), we have

$$P'(t) = (\beta(t) - \mu(t))P(t) - e(t)P(t) + c(t), \tag{3.4.50}$$

where e is a (time-dependent) per capita emigration rate and c is the global immigration rate. Below, we shall consider the simple linear model (3.4.50), referring to a more general theory of equations with periodic coefficients in Sect. 4.1.3.

Closed Systems

Here, we are interested in populations in which all coefficients change periodically with the same period, e.g., with the seasons of the year. As we shall see, contrary to naive expectations, in general, this assumption does not yield periodic solutions. We start with a closed population, that is, we do not consider emigration or immigration processes. As in Sect. 3, we define $r(t) = \beta(t) - \mu(t)$ to be the net growth rate of the population at time t, and assume that it is a periodic function with a period T. Under this assumption, we introduce the average growth rate of the population by

$$\bar{r} = \frac{1}{T}\int_0^T r(t)dt. \tag{3.4.51}$$

Hence, let us consider the initial value problem

$$P'(t) = r(t)P(t), \qquad P(t_0) = \mathring{P}, \tag{3.4.52}$$

the solution of which is given by

$$P(t) = \mathring{P}e^{\int_{t_0}^{t} r(s)ds}. \tag{3.4.53}$$

Since, by the periodicity of r,

$$\int_{t_0}^{t+T} r(s)ds = \int_{t_0}^{t} r(s)ds + \int_{t}^{t+T} r(s)ds = \int_{t_0}^{t} r(s)ds + \bar{r}T,$$

we have

$$P(t+T) = P(t)e^{\bar{r}T};$$

hence, the solution is not periodic. However, we may better describe the evolution by identifying its periodic component. In other words, let us try to find what is 'missing' in the function $R(t) := \int_{t_0}^{t} r(s)ds$ so that it is not periodic. We observe that

$$R(t+T) = \int_{t_0}^{t+T} r(s)ds = \int_{t_0}^{t} r(s)ds + \int_{t}^{t+T} r(s)ds = R(t) + \bar{r}T,$$

so

$$R(t+T) - \bar{r}(t+T-t_0) = R(t) - \bar{r}(t-t_0),$$

and therefore the function $R(t)$, complemented by $-\bar{r}(t-t_0)$, becomes periodic. Using this result, we can write

$$P(t) = \mathring{P}e^{\int\limits_{t_0}^{t} r(s)ds} = \mathring{P}e^{\bar{r}(t-t_0)}Q(t), \tag{3.4.54}$$

where

$$Q(t) = e^{\int\limits_{t_0}^{t} r(s)ds - \bar{r}(t-t_0)} \tag{3.4.55}$$

is a periodic function satisfying $Q(t_0) = 1$.

In particular, if we observe the population in discrete time intervals of length T, we get

$$p(k) := P(t_0 + kT) = \mathring{P}e^{\bar{r}kT}Q(t_0) = \mathring{P}[e^{\bar{r}T}]^k,$$

which is the discrete exponential model with the growth rate given by $e^{\bar{r}T}$.

Exercise 3.4.2 Assume that a population evolves according to (3.4.52), where r is a periodic function with period T. The population is observed in discrete time intervals of length T, and the observed sizes form a geometric progression with ratio R. Estimate the maximum and minimum of $r(t)$.

Exercise 3.4.3 Consider a population evolving according to the Malthusian equation in a periodically changing environment according to (3.4.52).

(a) For the population given by (3.4.52) with $r(t) = 2 + 3\sin 2\pi t$, where time is measured in years, find

  (i) the average growth rate per year;
  (ii) the solution to this equation;
  (iii) the periodic and aperiodic components of the solution.

(b) Write down the difference equation giving the growth of the population at times $t = 1, 2, \ldots, n, \ldots$. Is the population, on average, growing or declining on a year-to-year basis? What is the corresponding discrete growth rate?

Remark 3.1 We presented a simple case of the Floquet theory, see [33, 48], which deals with systems of linear equations with periodic coefficients. The number $e^{\bar{r}T}$ is called the Floquet multiplier, while the exponent $\bar{r}$ is called the Floquet exponent.

Since the function Q in (3.4.54) is periodic and continuous, it is bounded. Hence, the solutions $P(t)$ are bounded if the Floquet exponent is non-positive and tend to the stationary point $P = 0$ if the Floquet exponent is negative.

Open Systems

Consider next an open population described by

$$P'(t) = r(t)P(t) + c(t), \tag{3.4.56}$$

where $r(t) = \beta(t) - \mu(t) - e(t)$ and $c(t)$ are continuous and periodic functions with period T. Let the constant $\bar{r}$ and the periodic function $Q(t)$ be defined as in (3.4.51) and (3.4.55). Using the integrating factor, we find the general solution to (3.4.56) to be

$$P(t) = e^{\int_{t_0}^{t} r(s)ds} P(t_0) + e^{\int_{t_0}^{t} r(s)ds} \int_{t_0}^{t} e^{-\int_{t_0}^{u} r(s)ds} c(u)du. \tag{3.4.57}$$

A necessary condition for the existence of a periodic solution of period T is the existence of a solution $\bar{P}$ and of an initial condition P_o satisfying $P_o = \bar{P}(t_0) = \bar{P}(t_0 + T)$. We prove that, in this case, $\bar{P}$ is periodic. Using (3.4.57), we obtain

$$P_o = \bar{P}(t_0) = e^{\int_{t_0}^{t_0+T} r(s)ds} \bar{P}(t_0) + e^{\int_{t_0}^{t_0+T} r(s)ds} \int_{t_0}^{t_0+T} e^{-\int_{t_0}^{u} r(s)ds} c(u)du.$$

For simplicity, we assume $\bar{r} \neq 0$ (see [33] for a discussion of this case in full generality). By (3.4.55),

$$\bar{P}(t_0) = \frac{e^{\bar{r}T}}{1 - e^{\bar{r}T}} \int_{t_0}^{t_0+T} \frac{e^{-\bar{r}(u-t_0)}c(u)}{Q(u)} du.$$

Let us define $\hat{P}(t) = \bar{P}(t + T)$, where $\bar{P}(t_0) = P_o$. Then

$$\hat{P}'(t) = \bar{P}'(t + T) = r(t + T)\bar{P}(t + T) + c(t + T) = r(t)\hat{P}(t) + c(t)$$

and, since $\hat{P}(t_0) = \bar{P}(t_0 + T) = \bar{P}(t_0) = P_o$, the uniqueness of solutions of linear differential equations yields

$$\bar{P}(t + T) = \hat{P}(t) = \bar{P}(t)$$

for any $t \in \mathbb{R}$. Hence $\bar{P}$ is periodic.

We know that the general solution to an inhomogeneous linear equation can be expressed as a sum of an arbitrary solution of the inhomogeneous equation and the general solution of the homogeneous equation. Hence, since $\bar{P}$ is a solution of the inhomogeneous equation (3.4.56) and the general solution of its homogeneous version is given (3.4.54), we have

$$P(t) = Ke^{\bar{r}(t-t_0)}Q(t) + \bar{P}(t),$$

where $K = P(t_0) - P_o$. Finally,

$$P(t) = (P(t_0) - P_o)e^{\bar{r}(t-t_0)}Q(t) + \bar{P}(t). \tag{3.4.58}$$

This formula yields, in particular, that if $\bar{r} < 0$, then

$$\lim_{t\to\infty}(P(t) - \bar{P}(t)) = 0,$$

that is, with a negative average growth rate, an arbitrary solution is asymptotically periodic.

Exercise 3.4.4 Find the difference equation satisfied by the population described by (3.4.58) if it is observed at times kT, $k = 0, 1, \ldots$.

Exercise 3.4.5 Find a condition on c which ensures the existence of periodic solutions to (3.4.56) if $\bar{r} = 0$. Show that in such a case, any solution is periodic. Confirm your result by considering

$$P'(t) = P(t)\sin t + c(t), \quad P(0) = P_0,$$

with various functions $c(t)$.

Hybrid Models

In many cases, the observed process appears as a combination of two different ones: one occurs continually, while the other happens at discrete moments of time. For instance, breeding happens in evenly spaced intervals, but an unrelated death may happen at any time. Similarly to the periodic cases, such combined processes can be described by discrete equations. We shall consider two such models, one leading to the Beverton–Holt equation while the other provides yet another derivation of the Ricker equation from more basic principles.

The Beverton–Holt Equation Model

Consider a population which reproduces once a year and the length of the reproductive season is negligible. Let $P(n)$ be the population size in the nth year immediately after the reproductive season. During the year, outside the reproductive season, the individuals can only die so the size $p(t)$ of the population obeys the equation

$$p' = -\mu(p)p = -(\mu_0 + \mu_1 p)p, \quad p(0) = P(n), \quad \mu_0, \mu_1 > 0,$$

where t denotes the time elapsed since the end of the previous reproductive season (with one year as the time unit). Here, the death process is split into two parts: death from natural causes (ageing), represented by the death rate μ_0 and death due to overcrowding, the per capita rate of which is given by $\mu_1 p$. Thus, $p(1)$ gives the population after one year, immediately before the reproductive season. Then, after the reproductive season, the population enters the next year with $P(n+1) = \beta p(1)$ individuals, where β is the birth rate. To solve the above equation, we separate variables and, integrating, we obtain

$$-t = \int\limits_{P(n)}^{p(t)} \frac{ds}{s(\mu_0 + \mu_1 s)} = \frac{1}{\mu_0} \ln \frac{p(t)(\mu_0 + \mu_1 P(n))}{(\mu_0 + \mu_1 p(t))P(n)}.$$

Exponentiating and solving for $p(t)$ gives

$$p(t) = \frac{\mu_0 P(n)}{\mu_0 e^{\mu_0 t} + \mu_1 P(n)(e^{\mu_0 t} - 1)}$$

and, using the reproduction law,

$$P(n+1) = \beta p(1) = \frac{\beta \mu_0 P(n)}{\mu_0 e^{\mu_0} + \mu_1 P(n)(e^{\mu_0} - 1)}.$$

We can rewrite it as

$$P(n+1) = \frac{\beta e^{-\mu_0} P(n)}{1 + \frac{\mu_1(1 - e^{-\mu_0})}{\mu_0} P(n)}. \tag{3.4.59}$$

We find the equilibrium (carrying capacity) K as

$$K = \frac{\mu_0(\beta e^{-\mu_0} - 1)}{\mu_1(1 - e^{-\mu_0})},$$

hence (3.4.59) can be written in the form

$$P(n+1) = \frac{\beta e^{-\mu_0} P(n)}{1 + \frac{\beta e^{-\mu_0} - 1}{K} P(n)}. \tag{3.4.60}$$

This, again, is the Beverton–Holt model (2.1.14). We note, however, a difference: while in the classical Beverton–Holt model (2.1.14), it was assumed that there is no death from natural causes but only due to intra-species competition and that all adults die immediately after reproduction, here the number of offspring is adjusted by the factor $e^{-\mu_0}$ which is a yearly natural death rate resulting from the continuous time death process with the death rate μ_0, see (3.4.44).

The Ricker Equation From a Hybrid Model

We consider a model describing the size of a salmon population at the end of each spawning cycle, [70, Section 1.8]. We adopt the notation from the previous paragraph. However, the construction of the model is slightly different. For simplicity, we assume that the breeding occurs at yearly intervals, numbered $n = 0, 1, 2, \ldots$. The time between breeding seasons will be denoted by $t \in (0, 1)$. We assume that the adults live for a short time $t \in [0, \tau]$, with $\tau \ll 1$, after each breeding. This is the case for Pacific salmon (but not Atlantic salmon). So, let $P(n)$ denote the population of adult salmon immediately after the end of the spawning season n and thus at the beginning of the season $n + 1$. Let us look at this season. During spawning, the adult salmon produce larvae, which become adults by the end of the cycle, that is, at $t = 1$. The size of the population of larvae is denoted by $p_n(t)$ with $0 < t < 1$. Let us assume that the number of produced larvae is proportional to the adult population at the end of the previous season, that is,

$$p_n(0) = \beta P(n).$$

Not all of these larvae become adults. The prevalent cause during $0 < t < \tau$ is cannibalism by the adult salmon. Again, a simple assumption is that the rate at which the larvae are eaten is proportional to the adult population and the larvae population (law of mass action), but here we use $P(n)$ as the number of adults (we assume that during this short time, death from natural causes is negligible). Thus, for $t \in (0, \tau)$,

$$\frac{dp_n(t)}{dt} = -\lambda P(n) p_n(t), \tag{3.4.61}$$

where $\lambda > 0$ is a constant. Using the initial condition above, we find

$$p_n(t) = \beta P(n) e^{-\lambda P(n) t}. \tag{3.4.62}$$

Even when all adults are gone, not all larvae survive till the next breeding season. Again, the simple assumption is that a fraction γ of them will become adults. Thus

$$P(n+1) = \gamma p_n(\tau) = \gamma \beta P(n) e^{-\lambda \tau P(n)}, \tag{3.4.63}$$

which we recognize as the Ricker equation (2.1.24).

Let us modify the equation to cater for the Atlantic salmon. In this case, the adults do not die shortly after spawning. Assume instead that they die at a constant rate μ (per capita per year). In such a case, Eq. (3.4.61) is no longer correct as we cannot assume that the number $P(n)$ is constant. Let $P_n(t)$ be the amount of adults alive at time $t \in (0, 1)$ in the n-th cycle. Using the assumption, we find that

$$\frac{dP_n(t)}{dt} = -\mu P_n(t), \qquad P_n(0) = P(n),$$

so that

$$P_n(t) = e^{-\mu t} P(n). \tag{3.4.64}$$

No longer can we separate the natural death of larvae from the death through cannibalism. Hence, using the law of mass action and the exponential death rate γ as before, we rewrite (3.4.61) as

$$\frac{dp_n(t)}{dt} = -\lambda P(n) p_n(t) e^{-\mu t} - \gamma p_n(t), \qquad p_n(0) = \beta P(n). \tag{3.4.65}$$

This is a separable equation whose solution is

$$p_n(t) = \beta P(n) e^{-\frac{\lambda}{\mu} P(n)(1-e^{-\mu t})} e^{-\gamma t}. \tag{3.4.66}$$

Thus, using the fact that at the beginning of the $(n+1)$th cycle, the number of adults is given by the number of surviving larvae (changing into adults) and the surviving adults (note that the surviving adults give birth to larvae which are not adults!), we have

$$P(n+1) = p_n(1) + P_n(1).$$

Hence, we find

$$\begin{aligned} P(n+1) &= \beta P(n) e^{-\frac{\lambda}{\mu} P(n)(1-e^{-\mu})} e^{-\gamma} + e^{-\mu} P(n) \\ &= e^{-\mu} P(n)(\beta e^{\mu-\gamma} e^{-\frac{\lambda}{\mu} P(n)(1-e^{-\mu})} + 1) = e^{-\mu} P(n)(A e^{-BP(n)} + 1), \end{aligned}$$

for appropriate constants A and B.

Exercise 3.4.6 Provide the details of solving (3.4.65).

Exercise 3.4.7 Consider the following combination of continuous and discrete processes. In some populations, individuals breed once a year during a short reproductive season. Find the equation(s) governing the growth of the population and its solution if the breeding occurs according to the discrete exponential rule and the death rate is constant.

Exercise 3.4.8 This is a continuation of Exercise 3.3.4. Assuming that the injection is administered at times $0, T, 2T, ..., nT, ...,$ and each injection increases the concentration of the drug in blood by c_0, find the difference equation describing the concentration of the drug immediately after and immediately before the nth injection. Solve the resulting difference equations and determine the maximum and minimum concentrations as $n \to \infty$.

Chapter 4
Qualitative Theory for a Single Equation

In most cases, finding an explicit solution to a given differential, or difference, equation is impossible. However, the power of mathematics lies in the fact that one often can deduce properties of solutions and answer some relevant questions just by analyzing the right-hand side of the equation. If, as in Chap. 3, it describes the evolution of a population, it is often not necessary to know the population size at each time t. Rather, it is of interest to know some qualitative properties of the solution. Specifically, one of the most important questions biologists ask is whether the population goes extinct or whether there exists a positive value at which the population eventually stabilizes. Hence, particular solutions to the equations, called stationary or equilibrium solutions, informally mentioned in previous sections, play a unique role. Here, we shall make these concepts precise.

A typical problem in the theory of differential and difference equations is to determine whether the system is stable, that is, whether it will eventually settle at some state if we allow it to run for a sufficiently long time (which, in the case of difference equations, means many iterations).

In both difference and differential equations, by *equilibria* or *stationary solutions*, we understand solutions which are constant with respect to the independent variable (in the case of difference equations we also use the term *fixed point*). Since, however, in the differential equation

$$x' = f(x),$$

the right-hand side describes the rate of change of a given quantity, whereas in the difference equation

$$x(n+1) = f(x(n)),$$

J. Banasiak, *Introduction to Mathematical Methods in Population Theory*, Springer Undergraduate Mathematics Series,
https://doi.org/10.1007/978-3-031-65491-6_4

the right-hand side gives the quantity at time $n+1$ relative to the amount present at the previous observation, the theories are different and will be treated in separate subsections.

As we will see below, finding equilibria of equations is considerably easier than solving them. Thus, knowing that the system will converge to a particular equilibrium allows us to regard this equilibrium as an approximation of solutions originating in its neighbourhood.

4.1 Stability of Equilibria of Autonomous Differential Equations

Since in this section we shall also discuss equations with periodic right-hand-sides, let us begin with a general Cauchy problem,

$$x' = f(t, x), \qquad x(t_0) = \mathring{x}. \tag{4.1.1}$$

To have anything to discuss, we must ensure that (4.1.1) has solutions. This is settled by the Picard–Lindelöf theorem, Theorem B.10, which asserts that if f is sufficiently regular, then the initial value problem (4.1.1) has exactly one solution defined for t in some interval $(t_0 - \delta, t_0 + \delta)$, $\delta > 0$. By the discussion following Theorem B.10, the solution can be extended to the maximal interval of existence (t_-, t_+). The solution may be defined only on a finite interval. Still, if we can prove that it is bounded on each finite interval of its existence, then, by Theorem B.14, it is defined on $\mathbb{R}$.

To avoid dealing with technicalities, hereafter, we assume that f is an everywhere-defined function satisfying all the assumptions of the Picard–Lindelöf theorem on $\mathbb{R}$.

In many problems, it is important to emphasize the dependence of the solution on the initial conditions. Thus, we use the notion of the flow $\phi(t, t_0, \mathring{x})$ of (4.1.1), see also Sect. 6.2, which is the solution of the Cauchy problem (4.1.1).

Next, we define the stability of a solution. In defining stability, we tacitly assume that we deal only with solutions defined on $\overline{\mathbb{R}}_+$ (see the notation in Appendix A.3.1).

Definition 4.1 Let $t_0 \in \overline{\mathbb{R}}_+$ be an arbitrary initial time.

1. A solution $\phi(t, t_0, \mathring{x})$ is stable if for given $\epsilon > 0$ there is $\delta > 0$ (depending on ϵ but not on t_0) such that for any $\mathring{y}$, $|\mathring{x} - \mathring{y}| < \delta$ implies $|\phi(t, t_0, \mathring{x}) - \phi(t, t_0, \mathring{y})| < \epsilon$ for all $t > t_0$. If $\phi(t, t_0, \mathring{x})$ is not stable, then it is called unstable.
2. A solution $\phi(t, t_0, \mathring{x})$ is called asymptotically stable if it is stable and there is $\eta > 0$, independent of t_0, such that $|\mathring{x} - \mathring{y}| < \eta$ implies

$$\lim_{t\to\infty} (\phi(t, t_0, \mathring{x}) - \phi(t, t_0, \mathring{y})) = 0. \tag{4.1.2}$$

If $\eta = \infty$, then $\phi(t, t_0, \mathring{x})$ is called a global attractor.

We shall deal mostly with autonomous Cauchy problems

$$x' = f(x), \qquad x(t_0) = \mathring{x}. \tag{4.1.3}$$

In such a case, the initial time is irrelevant, and we only use $t_0 = 0$; then we write $\phi(t, 0, \mathring{x}) =: \phi(t, \mathring{x})$.

If (4.1.3) has a stationary solution $x(t) \equiv x^*$ that, by definition, is constant in time, then such a solution satisfies $x'(t) \equiv 0$ and consequently

$$f(x^*) = 0. \tag{4.1.4}$$

Conversely, if the equation $f(x) = 0$ has a solution, which we call an *equilibrium point* then, since f is independent of time, such a solution is a number, say x^*. If we now consider the constant function $x(t) \equiv x^*$, then $x'(t) \equiv 0$. Consequently,

$$x'(t) \equiv 0 = f(x^*),$$

and such a function is a stationary solution. Summarizing, equilibrium points are solutions to the algebraic equation (4.1.4) and, treated as constant functions, they are (the only) constant solutions to (4.1.3), called stationary or equilibrium solutions. Usually, we will not differentiate between these terms.

For autonomous equations, a particular role is played by the stability of the stationary solution, in which case $\phi(t, x^*) = x^*$ for any t. Then the asymptotic stability of x^*, formulated in Definition 4.1, can be expressed by saying that it must be stable and attracting,

$$\lim_{t\to\infty} \phi(t, \mathring{y}) = x^* \tag{4.1.5}$$

for all $\mathring{y}$ sufficiently close to x^*. If (4.1.5) occurs for any $\mathring{y}$ (in a given domain), then we say that x^* is globally asymptotically stable (GAS) (in that domain).

Equilibrium points play another important role in the theory of differential equations—they are the only limit points of bounded solutions as $t \to \pm\infty$. To make this precise, we begin with the following lemma.

Lemma 4.2 *If $\mathring{x}$ is not an equilibrium point of (4.1.3), then $\phi(t, \mathring{x})$ is never equal to an equilibrium point. In other words, $f(x(t, \mathring{x})) \neq 0$ for any t for which the solution exists.*

Proof An equilibrium point x^* generates a stationary solution, given by $x(t) \equiv x^*$. Thus, if $\phi(t_1, \mathring{x}) = x^*$ for some t_1, then $(t_1, \mathring{x})$ belongs to two different solutions, which contradicts the Picard–Lindelöf theorem. □

From the above lemma, it follows that if f has several equilibrium points, then the stationary solutions corresponding to these points divide the (x, t) plane into horizontal strips, having the property that any solution remains confined to one of them.

Exercise 4.1.1 Sketch all possible solutions to the problem

$$y' = y(y+1)(y-2).$$

Next, we shall formulate and prove a theorem that strengthens this observation.

Theorem 4.3 *Let $\phi(t, \mathring{x})$, $\mathring{x} \in \mathbb{R}$, be a non-stationary solution of (4.1.3) and let $I_{max} = (t_-, t_+)$ be its maximal interval of existence. Then $\phi(t, \mathring{x})$ is either a strictly decreasing or a strictly increasing function of t. Moreover, $\phi(t, \mathring{x})$ either diverges to $+\infty$ or to $-\infty$, or converges to an equilibrium point, as $t \to t_\pm$. In the latter case $t_\pm = \pm\infty$.*

Proof Assume that for some $t_* \in I_{max}$ the solution $x(t) := \phi(t, \mathring{x})$ has a local maximum or minimum $x_* = x(t_*)$. Since $x(t)$ is differentiable, we must have $x'(t_*) = 0$ but then $f(x_*) = 0$, which makes x_* an equilibrium point of f. This means that a non-stationary solution $\phi(t, \mathring{x})$ reaches an equilibrium in finite time, which contradicts Lemma 4.2. Thus, if $\phi(t, \mathring{x})$ is not a stationary solution, then it cannot attain local maxima or minima, and thus it must be either strictly increasing or strictly decreasing.

Since the solution is monotonic, it either diverges to $\pm\infty$ (depending on whether it increases or decreases) or converges to finite limits as $t \to t_\pm$. Let us focus on the right endpoint t_+ of I_{max}. If $\phi(t, \mathring{x})$ converges as $t \to t_+$, then $t_+ = \infty$, by Theorem B.14. Thus

$$\lim_{t \to +\infty} \phi(t, \mathring{x}) = \bar{x}.$$

Without compromising generality, we further assume that $\phi(t, \mathring{x})$ is an increasing function. Assume that $\bar{x}$ is not an equilibrium point. By continuity of ϕ with respect to t, the intermediate value theorem asserts that the values of $\phi(t, \mathring{x})$ must fill the interval $[\mathring{x}, \bar{x})$ as t runs from 0 to ∞. This interval cannot contain any equilibrium point as the existence of such points would violate Lemma 4.2. Thus, for any $x < \bar{x}$, $f(x)$ is strictly positive, and hence, by separating variables and integrating both sides of the equation, we obtain

$$t(x) - t(\mathring{x}) = \int_{\mathring{x}}^{x} \frac{ds}{f(s)}. \tag{4.1.6}$$

Passing with t to infinity (since $t(\bar{x}) = \infty$), we see that the left hand side becomes infinite, and so

$$\int_{\mathring{x}}^{\bar{x}} \frac{ds}{f(s)} = \infty.$$

By assumption, the interval of integration is finite so that for the integral to be infinite, we must have $f(s) = 0$ for some $s \in [\mathring{x}, \bar{x}]$. The only such point is $s = \bar{x}$; thus $\bar{x}$ is an equilibrium point. □

Remark 4.4 We note that Eq. (4.1.6) is of independent interest as it gives a formula for the blow-up time of the solution $\phi(t, \mathring{x})$. To wit, let the interval $[\mathring{x}, \infty)$ be free of equilibria and let $\phi(t, \mathring{x})$ be increasing for $t > 0$. Then $\lim_{t\to t_+} \phi(t, \mathring{x}) = \infty$ so that, by (4.1.6),

$$t_+ - t(\mathring{x}) = \int_{\mathring{x}}^{\infty} \frac{ds}{f(s)}$$

and, in particular, we see that if $1/f$ is integrable at $+\infty$ (precisely, if the improper integral above exists), then the maximal interval of existence is finite, and we have the blow-up of the solution in finite time. On the other hand, if $1/f$ is not integrable, then $t_+ = +\infty$. We note that the latter occurs if $f(s)$ does not grow faster than s as $s \to \infty$, for instance if f' is bounded on $\mathbb{R}$. On the other hand, if $f(s)$ behaves, say, as s^2 for large s, then the integral on the right-hand side is finite and thus $t_+ < \infty$.

Exercise 4.1.2 Let in the non-autonomous problem

$$y' = f(t, y), \quad y(0) = \mathring{y},$$

$f(t, y)$ satisfie

$$\underline{f}(y) \le f(t, y) \le \bar{f}(y), \quad t \ge 0,$$

with $\underline{f}(y)$ and $\bar{f}(y)$ satisfying the assumptions of the Picard–Lindelöf theorem and $\underline{f}(y) > 0$ for $y \ge y_0$. Provide upper and lower bounds for the blow-up time for y (if it exists).

Remark 4.5 Theorem 4.3 shows that for scalar differential equations with regular right-hand sides, the distinction between different properties of an equilibrium made in Definition 4.1 is superfluous. Indeed, if an equilibrium x^* is stable, then solutions originating close to it stay close to it. However, by Theorem 4.3, these solutions are monotonic. Hence, the solutions are closer to x^* than their initial conditions are. In particular, they must be bounded and, being monotonic, they must converge as $t \to \infty$. The proof of Theorem 4.3 shows that the limit point must be the equilibrium x^*. This implies that x^* is attracting and hence asymptotically stable. Also, by the monotonicity of solutions, any attracting equilibrium must be stable and thus asymptotically stable.

However, in higher-dimensional systems, there are equilibria with only some properties of Definition 4.1; see Example 6.23.

Remark 4.6 Theorem 4.3 is usually used in the following weaker form. Let f be a continuously differentiable function. Then the equilibrium x^* is stable provided $f'(x^*) < 0$ and unstable if $f'(x^*) > 0$. The proof is obvious—if $f'(x^*) < 0$, then $f'(x) < 0$ in some neighbourhood of x^*, by continuity of f'. Thus, $f > 0$ to the left and $f < 0$ to the right of x^* and any solution originating in such a left neighbourhood of x^* is increasing and must converge to x^*. Similarly, any solution originating in such a right neighbourhood of x^* is decreasing and must converge to x^*. Thus, x^* is asymptotically stable. An analogous argument shows that $f'(x^*) > 0$ means that x^* is unstable. However, Theorem 4.3 is much more general and allows us to ascertain stability or instability in the so-called nonhyperbolic cases, that is, when $f'(x^*) = 0$, by considering the sign of f to the left and the right of x^*.

4.1.1 Application to the Logistic Equation

Consider the Cauchy problem for the logistic equation

$$y' = y(1-y), \qquad y(0) = \mathring{y}, \tag{4.1.7}$$

where, for simplicity, we set $r = K = 1$. We have solved this problem in Sect. 3.3.2. Let us now get as much information as possible about the solutions to this problem without actually solving it. First, we observe that the right-hand side is given by $f(y) = y(1-y)$, which is a polynomial, and therefore at each point of $\mathbb{R}^2$ the assumptions of the Picard–Lindelöf theorem are satisfied, that is, only one solution of (4.1.7) passes through each point $(t_0, \mathring{y})$. However, f is not a globally Lipschitz function, so that this solution may be defined only on a finite time interval.

The equilibrium points are found by solving $y(1-y) = 0$, hence $y \equiv 0$ and $y \equiv 1$ are the only stationary solutions. Moreover, $f(y) < 0$ for $y < 0$ and $y > 1$ and $f(y) > 0$ for $0 < y < 1$. Hence, it follows from Lemma 4.2 that the solutions starting from $\mathring{y} < 0$ will stay strictly negative, those starting from $0 < \mathring{y} < 1$ will stay in this interval, and those with $\mathring{y} > 1$ will be larger than 1, for all times of their respective existence, as they cannot cross the equilibrium solutions. Then, from Theorem 4.3, we see that the solutions with negative initial conditions are decreasing and therefore tend to $-\infty$ when time increases. In fact, they blow up in finite time since, by integrating the equation, we obtain

$$t(y) = \int\limits_{\mathring{y}}^{y} \frac{d\eta}{\eta(1-\eta)}.$$

For $\mathring{y} < 0$ and $y < \mathring{y}$, passing with y to $-\infty$ on the right hand side, we obtain a finite positive time of the blow-up, as the integrand is negative.

Next, solutions with $0 < \mathring{y} < 1$ are bounded and, thus, are defined for all times. They are increasing, and thus, they must converge to the larger equilibrium, that is, $\lim_{t\to\infty} y(t, \mathring{y}) = 1$. Finally, if we start with $\mathring{y} > 1$, then $y(t, \mathring{y})$ decreases and thus is bounded from below, satisfying again $\lim_{t\to\infty} y(t, \mathring{y}) = 1$. The shape of the solution curves can be determined in the same way as in Sect. 3.3.2. By differentiating Eq. (4.1.7) with respect to time, we obtain

$$y'' = y'(1 - y) - yy' = y'(1 - 2y).$$

Since for each solution (apart from the stationary ones), y' has a a fixed sign, we see that an inflection point can exist only for solutions starting at $\mathring{y} \in (0, 1)$ and it occurs at $y = 1/2$, where the solution changes from being convex downward to being convex upward. In the two other cases, the second derivative is of constant sign, giving the solution convex upward for negative solutions and convex downward for solutions larger than one. The character of the equilibria and the type of solutions are illustrated in Figs. 4.1 and 4.2.

We see that we got the same picture as when solving the equation but with much less work.

Exercise 4.1.3 Consider a logistic equation with a constant rate of immigration

$$P' = rP\left(1 - \frac{P}{K}\right) - \alpha,$$

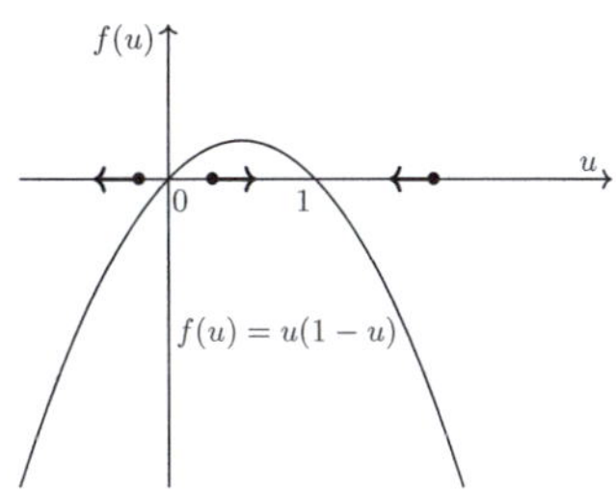

Fig. 4.1 Repelling and attractive character of the equilibria $x_1^* = 0$ and $x_2^* = 1$, respectively

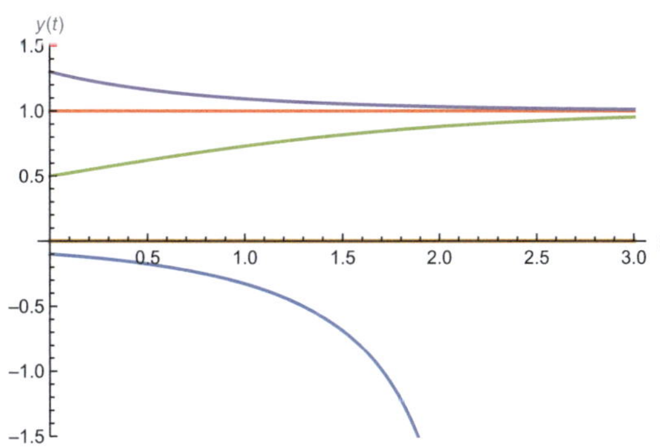

Fig. 4.2 Monotonic behaviour of solutions to (4.1.7) depends on the right hand side f of the equation

where r and K denote, as before, the intrinsic growth rate and the carrying capacity, respectively, and α is the constant emigration rate; we assume that all constants are positive. Find the equilibria and determine their stability.

Exercise 4.1.4 Consider a logistic model of population P with predation. We assume that the density of predation is constant, and the predation rate depends on the prey density according to $\rho(P) = \frac{aP^k}{1+bP^k}$. The predation rate for small prey populations depends on k (recall that we derived such a model with $k = 1$ in Sect. 3.3.4). For each of the equations

$$(i)\ P' = P(1-N) - \frac{2P}{1+16P}, \qquad (ii)\ P' = P(1-N) - \frac{\frac{1}{2}P}{1+P},$$

$$(iii)\ P' = P(1-N) - \frac{3P}{1+P},$$

find the number of equilibria and their stability. Explain the differences by properly interpreting the coefficients in the predation term.

4.1.2 Application to the Allee Model

Here we provide another application of the theory introduced above and determine constants A, B and C for which Eq. (3.3.36),

$$P' = \lambda P\left(1 - \frac{P}{C} - \frac{A}{1+BP}\right), \tag{4.1.8}$$

indeed describes the Allee effect. The equation is difficult to solve explicitly (though it is possible, as it is separable), and hence, we employ Theorem 4.3. The equilibria are solutions to

$$f(P) := P\left(1 - \frac{P}{C} - \frac{A}{1+BP}\right) = 0. \tag{4.1.9}$$

Clearly, $P \equiv 0$ is an equilibrium, so, in particular, any solution originating from $P(0) = \mathring{P} > 0$ satisfies $P(t) > 0$. We see that

$$f'(P) = 1 - \frac{2P}{C} - \frac{A}{(1+BP)^2} \tag{4.1.10}$$

and since $f'(0) = 1 - A$ we obtain that if $A > 1$, then $P = 0$ is an asymptotically stable equilibrium. By analysing the second derivative, we can also state that if $A = 1$ and $BC < 1$, $P = 0$ is semi-stable, attracting trajectories originating from positive initial conditions. Now, we can focus on the other equilibria. For (4.1.8) to describe an Allee model first we must show that

$$1 - \frac{P}{C} - \frac{A}{1 + BP} = 0 \tag{4.1.11}$$

has two positive solutions. It could be done directly, but then the calculations become messy so that we follow a more elegant approach of [93] and use the above equation to define a function $A(P)$ by

$$A(P) = \frac{1}{C}(C - P)(1 + BP)$$

and analyse it. It is an inverted parabola satisfying $A(0) = 1$, which takes its maximum at the point P^*, where

$$A'(P^*) = -\frac{1}{C} + B - \frac{2B}{C}P^* = 0.$$

This gives

$$P^* = \frac{BC - 1}{2B}$$

with the maximum

$$A^* := A(P^*) = \frac{(BC + 1)^2}{4BC}.$$

Now, the nonzero equilibria of (4.1.8) are the points at which the horizontal line $A = const$ cuts the graph of $A(P)$, see Fig. 4.3. First, if $BC < 1$, the stationary point P^* is negative. Thus, there are a positive and a negative solution for $0 < A < 1$, a negative and 0 solutions for $A = 1$, two negative solutions if $1 < A < A^*$, one (double) negative solution if $A = A^*$ and no solutions if $A > A^*$. If $BC = 1$, we have one positive, one negative solution for $0 < A < 1$, double 0 solution for $A = A^* = 1$ and no solutions for $A > 1$. Thus, in no case with $BC \leq 1$ can we expect the Allee-type behaviour. Let us focus then on the case $BC > 1$. Since $A > 0$, we have the following cases, see Fig. 4.3,

(a) if $0 < A < 1$, then there are two solutions to (4.1.11), but only one is positive while the other is negative;
(b) if $A = 1$, then there is one 0 and one positive solution to (4.1.11);
(c) if $1 < A < A^*$, then there are two distinct positive solutions to (4.1.11);
(d) if $A = A^*$, then there is a double positive solution to (4.1.11);
(e) if $A > A^*$, then there are no solutions to (4.1.11).

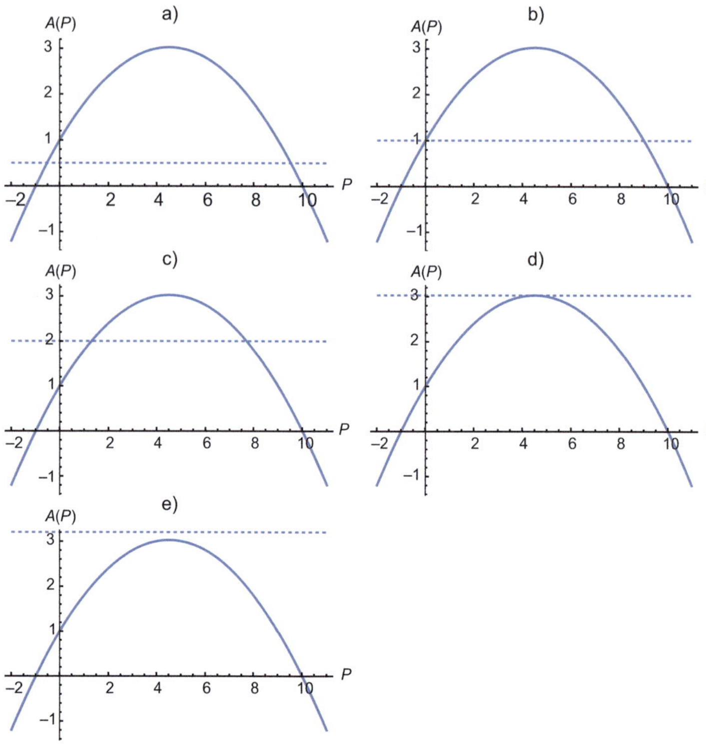

Fig. 4.3 The solutions to (4.1.9) as a function of A (the ordinate of the horizontal line), with $B = 1, C = 10$ and hence $BC > 1$; L is the smaller solution and K is the bigger one

To determine the stability of the equilibria, we re-write (4.1.8) in the following form

$$P' = \lambda P\left(1 - \frac{P}{C} - \frac{A}{1+BP}\right) = \frac{\lambda BP}{C(1+BP)}\left(-P^2 + P\frac{BC-1}{B} + \frac{C(1-A)}{B}\right)$$

$$= \frac{\lambda BP}{C(1+BP)}(P-L)(K-P). \tag{4.1.12}$$

Using the results of the first part of this section, we can describe the dynamics of (4.1.8) as follows. Let $BC > 1$. Then, see Fig. 4.4,

(a) For $0 < A < 1$, there is one negative, L, and two nonnegative equilibria of (4.1.8), 0 and K. Zero is unstable and, since the parabola crosses the line $A = const$ at $P = K$ from above, $P - L > 0$ and $K - P$ changes its sign from positive to negative so that $f(P)$ is decreasing at $P = K$, and hence K is asymptotically stable.
(b) At $A = 1$, the negative equilibrium L merges with 0. Zero becomes semi-stable (unstable for positive trajectories), and, as above, K is asymptotically stable.

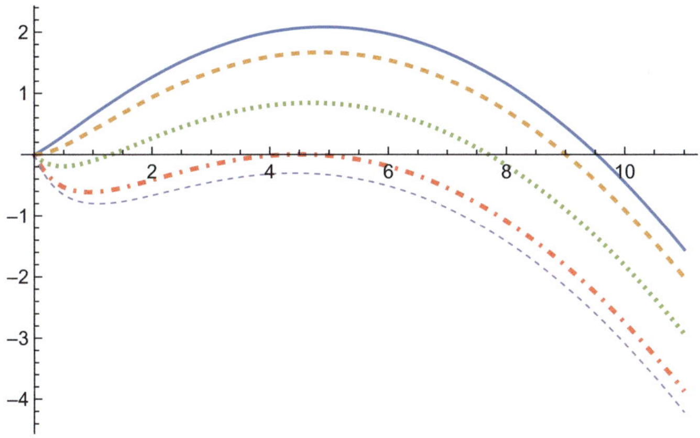

Fig. 4.4 The change of the equilibria to (4.1.8), when A runs from $A = 0.2$ (solid) through $A = 1$ (dashed), $A = 2$ (dotted), $A = 3.025$ (dotted-dashed) to $A = 3$ (thin dashed), where, as above, $B = 1$ and $C = 10$

(c) For $1 < A < A^*$, there are three nonnegative equilibria, 0 and $0 < L < K$. 0 becomes a stable equilibrium, $P - L$ changes sign from negative to positive as P passes through L with $K - P > 0$, so L is unstable. Finally, as P passes through K, $P - L$ stays positive while $K - P$ changes its sign from positive to negative, as above, and hence K is asymptotically stable.

(d) At $A = A^*$, there are two nonnegative equilibria, 0 and double $L = K$. Zero is stable, and $L = K$ becomes semistable, attracting trajectories from the right and repelling those from the left.

(e) For $A > A^*$, there is only one equilibrium at 0, which is globally attracting.

As mentioned above, if $BC \leq 1$, we cannot have two positive equilibria, so the Allee effect cannot occur. However, to complete the analysis, we note that if $0 < BC \leq 1$, then the only case in which there is a positive equilibrium K is if $0 < A < 1$, and in this case, K is asymptotically stable, while 0 is unstable. For all other cases, the only biologically relevant equilibrium is 0, and it is stable if $1 < A$, semistable (attracting positive trajectories) if $A = 1$ and $BC < 1$ and stable if $A = 1 = BC$. Summarizing, (4.1.8) describes the Allee effect if and only if

$$BC > 1 \quad \text{and} \quad 1 < A < \frac{(BC+1)^2}{4BC}. \tag{4.1.13}$$

In any other case with a positive equilibrium, the dynamics described by (4.1.8) is similar to those described by the logistic equation, see Fig. 4.5.

Another way of looking at the problem is to consider the number and stability of the equilibria as a function of a parameter. This approach is known as *bifurcation theory*. Here we focus on the case $BC > 1$, and select the parameter A, which can represent additional mortality over the mortality due to the overcrowding

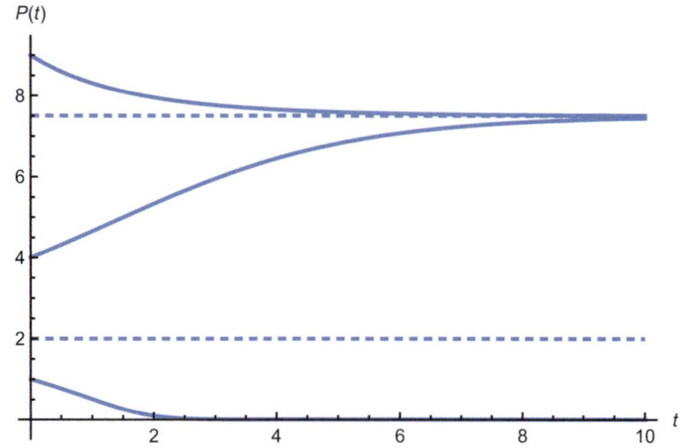

Fig. 4.5 Trajectories $P(t)$ of (4.1.8) for various initial conditions showing the Allee effect, where $A = 4$, $C = 10$, $B = 2$, $L = 2$ (lower dashed line), $K = 7.5$ (upper dashed line)

characteristic for the logistic model. Then, for small $A \in (0, 1)$, 0 is an unstable equilibrium, and K is stable, as in the logistic model. When A moves through 1, a new positive equilibrium L 'bifurcates' from 0 and the latter changes from repelling to attractive, while K stays attractive, and hence we are in the 'Allee region'. Finally, when A moves across A^*, K vanishes, and 0 becomes globally attractive—large mortality drives the population to extinction. The Allee phenomenon is of concern in many practical applications. For instance, if we try to eradicate a pest whose population can be modelled by an Allee-type equation, then it is enough to create conditions in which the size of the population will be below L; then the population will die out without any external intervention; this effect, for discrete models, will be discussed in Sect. 4.2.3. Similarly, if by overhunting or overfishing, we drive a population below L, then it will become extinct even if we stop its exploitation.

4.1.3 *Periodic Equations*

In Sect. 3.4.2, we discussed examples of periodic equations, which are a special case of non-autonomous equations

$$x' = f(t, x), \quad x(t_0) = \mathring{x}, \tag{4.1.14}$$

where f is 1-periodic, that is, it satisfies

$$f(t + 1, x) = f(t, x). \tag{4.1.15}$$

We assume that f is continuous on $\mathbb{R} \times \mathbb{R}$, locally Lipschitz continuous in x uniformly in t, see (B.4.28). We observe that by appropriately rescaling time, (4.1.14) with any T-periodic function f can be reduced to a 1-periodic case.

In non-autonomous equations, the initial point t_0 plays an important role, and thus we shall use the full notation $\phi(t, t_0, \mathring{x})$ to denote the solution passing through the point $(t_0, \mathring{x}) \in \mathbb{R} \times \mathbb{R}$, see Sect. 4.1.

Let us fix $(t_0, \mathring{x}) \in \mathbb{R} \times \mathbb{R}$ and let $\phi(t, t_0, \mathring{x})$ be a solution to (4.1.14). First, we note that the function $\phi_k(t, t_0, \mathring{x}) := \phi(t + k, t_0, \mathring{x})$ is a solution to the equation in (4.1.14) for any $k \in \mathbb{Z}$. Indeed, denoting $\tau = t + k$, by the Chain Rule and (4.1.15),

$$\begin{aligned}\frac{d\phi_k(t, t_0, \mathring{x})}{dt} &= \frac{d\phi(\tau, t_0, \mathring{x})}{d\tau} = f(\tau, \phi(\tau, t_0, \mathring{x}))\\ &= f(t + k, \phi(t + k, t_0, \mathring{x})) = f(t, \phi_k(t, t_0, \mathring{x})).\end{aligned}$$

This formula implies the invariance property of the flow, namely, since

$$\phi(t_0 + 1, t_0, \mathring{x}) = \phi(t_0, t_0, \phi(t_0 + 1, t_0, \mathring{x}))$$

and both $\phi(t + 1, t_0, \mathring{x})$ and $\phi(t, t_0, \phi(t_0 + 1, t_0, \mathring{x}))$ are solutions to (4.1.14), the uniqueness ensures

$$\phi(t + 1, t_0, \mathring{x}) = \phi(t, t_0, \phi(t_0 + 1, t_0, \mathring{x})) \tag{4.1.16}$$

for $t \in \mathbb{R}$. Then, we have the following lemma, whose proof depends on an argument already used in Sect. 3.4.2.

Lemma 4.7 *A solution $\phi(t, t_0, \mathring{x})$ to (4.1.14) is 1-periodic if and only if*

$$\phi(t_0 + 1, t_0, \mathring{x}) = \mathring{x}. \tag{4.1.17}$$

Proof If $\phi(t + 1, t_0, \mathring{x}) = \phi(t, t_0, \mathring{x})$, then, on evaluating both sides of the equation at $t = t_0$ we obtain $\phi(t_0 + 1, t_0, \mathring{x}) = \phi(t_0, t_0, \mathring{x}) = \mathring{x}$, by the properties of the flow. On the other hand, again by the properties of the flow, if $\phi(t_0 + 1, t_0, \mathring{x}) = \mathring{x}$, then, by (4.1.16),

$$\phi(t, t_0, \mathring{x}) = \phi(t, t_0, \phi(t_0 + 1, t_0, \mathring{x})) = \phi(t + 1, t_0, \phi(t_0, t_0, \mathring{x})) = \phi(t + 1, t_0, \mathring{x}).$$

□

Next, we consider the asymptotic behaviour of solutions to (4.1.14) as $t \to \pm\infty$. To simplify notation, we let $t_0 - 0$. We have the following result, [52, Theorem 4.11].

Theorem 4.8 *Let $\phi(t, 0, \mathring{x})$ be a solution to 1-periodic equation (4.1.14) that is bounded for $t \geq 0$. Then there is a periodic solution, say, $\Phi(t)$, defined by*

$$\lim_{k\to\infty} \phi(t + k, 0, \mathring{x}) = \Phi(t), \quad t \in [0, 1], \tag{4.1.18}$$

where the limit is monotonic and uniform, and

$$\lim_{t\to\infty} |\phi(t, 0, \mathring{x}) - \Phi(t)| = 0. \tag{4.1.19}$$

Similarly, if $\phi(t, 0, \mathring{x})$ is bounded for $t \leq 0$, then there is a periodic solution, say, $\Psi(t)$, defined by

$$\lim_{k\to\infty} \phi(t-k, 0, \mathring{x}) = \Psi(t), \quad t \in [0, 1], \tag{4.1.20}$$

where the limit is monotonic and uniform, and

$$\lim_{t\to-\infty} |\phi(t, 0, \mathring{x}) - \Psi(t)| = 0. \tag{4.1.21}$$

Proof As above, consider the sequence $\phi_k(t) = \phi(t+k, 0, \mathring{x})$ for $k = 0, 1, \ldots$. Since, as noted above, each ϕ_k is a solution to (4.1.14), the uniqueness of solutions ensures that the ϕ_ks do not intersect. This means that the sequence $(\phi_k(t))_{k\in\mathbb{N}_0}$ is monotone. More precisely, if $\phi_1(0) \geq \phi_0(0) = \mathring{x}$, then $(\phi_k)_{k\in\mathbb{N}_0}$ is increasing and if $\phi_1(0) \leq \phi_0(0) = \mathring{x}$, then $(\phi_k)_{k\in\mathbb{N}_0}$ is decreasing. Hence, if the sequence is bounded, it must monotonically converge pointwise to some function $\Phi(t)$ as $k \to \infty$. In particular, $\phi_k(0) \to \Phi(0)$ in a monotonic way. Now, using (4.1.16), we have

$$\phi_k(t) = \phi(t, 0, \phi(k, 0, \mathring{x})) = \phi(t, 0, \phi_k(0))$$

and from Theorem B.16 we see that

$$\Phi(t) = \phi(t, 0, \Phi(0)) = \lim_{k\to\infty} \phi(t, 0, \phi_k(0))$$

and the limit is uniform in $t \in [0, 1]$, see (B.4.50). In other words, $\Phi(t)$ is a solution to (4.1.14) on $[0, 1]$. Furthermore,

$$\Phi(1) = \lim_{k\to\infty} \phi(1, 0, \phi_k(0)) = \lim_{k\to\infty} \phi_{k+1}(0) = \Phi(0)$$

which shows, by Theorem B.14 and Lemma 4.7, that $\Phi(t)$ extends 1-periodically as a solution defined on $\mathbb{R}$. We note that this also means that $\Phi(t) = \phi(t, 0, \Phi(0))$ is an asymptotically stable periodic solution in the sense of Definition 4.1. Indeed, let $t \to \infty$. We can write $t = t_{k_n} + k_n$, where $t_{k_n} \in [0, 1)$ and $k_n \in \mathbb{N}$, $k_n \to \infty$ as $n \to \infty$. Then

$$\begin{aligned}
\lim_{t\to\infty} |\phi(t, 0, \mathring{x}) - \Phi(t)| &= \lim_{n\to\infty} \left|\phi(t_{k_n} + k_n, 0, \mathring{x}) - \Phi(t_{k_n})\right| \\
&= \lim_{n\to\infty} \left|\phi(t_{k_n}, 0, \phi_{k_n}(0)) - \Phi(t_{k_n})\right| \\
&\leq \lim_{n\to\infty} \max_{t\in[0,1]} \left|\phi(t, 0, \phi_{k_n}(0)) - \Phi(t)\right| = 0,
\end{aligned}$$

where the claim follows from the uniform convergence of $\phi(\cdot, 0, \phi_k(0))$ to $\Phi(\cdot)$ on $[0, 1]$. □

We explore further the argument used in the proof of Theorem 4.8, namely that the convergence of the sequence $(\phi_k(0))_{k\in\mathbb{N}_0}$ implies the convergence of the sequence of functions $(\phi_k(t))_{k\in\mathbb{N}_0}$ and, consequently, the asymptotic stability of the periodic solution $\Phi(t)$. If we recall the concept of a time-one map, introduced in Sect. 3.4.1, then the map $\Pi : \mathbb{R} \to \mathbb{R}$ defined by

$$\Pi(\mathring{x}) = \phi(1, 0, \mathring{x}) \tag{4.1.22}$$

offers an example of it and is related to the so-called Poincaré map, [4, 54]. Then we observe that

$$\phi_k(0) = \phi(k, 0, \mathring{x}) = \underbrace{\phi(1, 0, \phi(1, 0, \phi(1, \ldots, \phi(1, 0, \mathring{x}))))}_{k-times} = \Pi^k(\mathring{x}).$$

With this, we can restate Lemma 4.7 by saying that $\mathring{x}$ is an initial condition for a 1-periodic solution to (4.1.14) if and only if it is a fixed point of the Poincaré map (4.1.22). Furthermore, the proof of Theorem 4.8 shows that the stability of the periodic solution $\Phi(t)$ to (4.1.14) is determined by whether the iterates $\Pi^k(\mathring{x})$ converge to a fixed point of Π or not, or, using the terminology of the next section, by its stability. We shall illustrate this concept in the following example.

Example 4.9 Consider the 1-periodic logistic equation

$$P' = r(t)P\left(1 - \frac{P}{K(t)}\right), \quad P(0) = \mathring{P}, \tag{4.1.23}$$

with continuous and positive growth rate $r(t)$ and carrying capacity $K(t)$, whose solution is

$$P(t) = \frac{\mathring{P} e^{\int_0^t r(\tau)d\tau}}{1 + \mathring{P}\int_0^t \alpha(z) e^{\int_0^z r(\tau)d\tau} dz}, \tag{4.1.24}$$

where $\alpha(t) = \frac{r(t)}{K(t)}$, see (3.3.31). We see that (4.1.17) takes the form

$$\mathring{P} = \frac{\mathring{P} e^{\int_0^1 r(\tau)d\tau}}{1 + \mathring{P}\int_0^1 \alpha(z) e^{\int_0^z r(\tau)d\tau} dz},$$

with a unique different from zero solution given by

$$P^* = \frac{e^{\int_0^1 r(\tau)d\tau} - 1}{\int_0^1 \alpha(z) e^{\int_0^z r(\tau)d\tau} dz} > 0. \tag{4.1.25}$$

Thus, using Lemma 4.7, there is a unique nontrivial periodic solution to (4.1.14), namely (4.1.24), originating from the initial condition (4.1.25). This solution is globally asymptotically stable. This follows from the general theory as briefly described in [52, pp. 126–128]. Here, we provide a direct proof based on the properties of the Poincaré map associated with (4.1.23). For arbitrary $\mathring{P} > 0$ we have

$$\Pi^k(\mathring{P}) = \frac{\mathring{P} e^{\int_0^k r(\tau)d\tau}}{1 + \mathring{P} \int_0^k \alpha(z) e^{\int_0^z r(\tau)d\tau} dz}. \tag{4.1.26}$$

By the periodicity, we have

$$e^{\int_0^k r(\tau)d\tau} = \left(e^{\int_0^1 r(\tau)d\tau}\right)^k =: \rho^k, \qquad \rho > 1.$$

Further,

$$\int_0^k \alpha(z) e^{\int_0^z r(\tau)d\tau} dz = \sum_{j=1}^k \int_{(j-1)}^j \alpha(z) e^{\int_0^z r(\tau)d\tau} dz$$

and, using the periodicity of α and r,

$$\begin{aligned}
\int_{j-1}^j \alpha(z) e^{\int_0^z r(\tau)d\tau} dz &= \int_{j-1}^j \alpha(z) e^{\int_0^{j-1} r(\tau)d\tau} e^{\int_{j-1}^z r(\tau)d\tau} dz \\
&= \rho^{j-1} \int_{j-1}^j \alpha(z) e^{\int_{j-1}^z r(\tau)d\tau} dz \\
&= \rho^{j-1} \int_0^1 \alpha(y + j - 1) e^{\int_{j-1}^{y+j-1} r(\tau)d\tau} dy \\
&= \rho^{j-1} \int_0^1 \alpha(y) e^{\int_0^y r(\sigma+j-1)d\sigma} dy \\
&= \rho^{j-1} \int_0^1 \alpha(y) e^{\int_0^y r(\sigma)d\sigma} dy =: \rho^{j-1} A.
\end{aligned}$$

Hence,

$$\int_0^k \alpha(z) e^{\int_0^z r(\tau)d\tau} dz = \sum_{j=1}^k \rho^{j-1} A = \frac{\rho^k - 1}{\rho - 1} A.$$

Thus

$$\lim_{k\to\infty} \Pi^k(\mathring{P}) = \lim_{k\to\infty} \frac{\mathring{P}\rho^k}{1 + \mathring{P}\frac{\rho^k - 1}{\rho - 1}A} = \frac{\rho - 1}{A} = P^*.$$

Therefore, P^*, and the corresponding periodic solution, are globally attractive for $\mathring{P} > 0$.

Next, we find this periodic solution explicitly. The solution (4.1.24) with P^* given by (4.1.25) can be written as

$$P(t) = \frac{(\rho - 1)e^{\int_0^t r(\tau)d\tau}}{A + (\rho - 1)\int_0^t \alpha(z)e^{\int_0^z r(\tau)d\tau}dz}.$$

Let $t = k + t'$ for $t' \in [0, 1)$. Then, similarly to the above,

$$e^{\int_0^t r(\tau)d\tau} = \rho^k e^{\int_0^{t'} r(\tau)d\tau}$$

and

$$\begin{aligned}\int_0^t \alpha(z)e^{\int_0^z r(\tau)d\tau}dz &= \int_0^k \alpha(z)e^{\int_0^z r(\tau)d\tau}dz + \int_k^{k+t'} \alpha(z)e^{\int_0^z r(\tau)d\tau}dz \\ &= A\frac{\rho^k - 1}{\rho - 1} + \rho^k \int_0^{t'} \alpha(z)e^{\int_0^z r(\tau)d\tau}dz.\end{aligned}$$

Hence, for $t \in [k, k + 1)$, we have

$$\begin{aligned}P(t) &= \frac{(\rho - 1)\rho^k e^{\int_0^{t'} r(\tau)d\tau}}{A(\rho^k - 1) + (\rho - 1)\rho^k \int_0^{t'} \alpha(z)e^{\int_0^z r(\tau)d\tau}dz + A} \\ &= \frac{(\rho - 1)e^{\int_0^{t'} r(\tau)d\tau}}{A + (\rho - 1)\int_0^{t'} \alpha(z)e^{\int_0^z r(\tau)d\tau}dz} \\ &= \frac{(e^{\int_0^1 r(\tau)d\tau} - 1)e^{\int_0^{t'} r(\tau)d\tau}}{\int_0^1 \alpha(z)e^{\int_0^z r(\sigma)d\sigma}dz + (e^{\int_0^1 r(\tau)d\tau} - 1)\int_0^{t'} \alpha(z)e^{\int_0^z r(\tau)d\tau}dz}.\end{aligned}$$

4.2 Equilibrium Points of Difference Equations

Consider the autonomous first-order difference equation

$$x(k + 1) = f(x(k)), \qquad k \in \mathbb{N}_0, \tag{4.2.27}$$

with the initial condition $\mathring{x}$. In what follows, we shall assume that f is at least continuous. It is clear that the solution to (4.2.27) is given by iterations

$$x(k) = f(f(\dots f(\mathring{x}))) = f^k(\mathring{x}). \tag{4.2.28}$$

Henceforth, we will be using both notations.

A point x^* in the domain of f is said to be an *equilibrium point* of (4.2.27) if it is a fixed point of f, that is, if $f(x^*) = x^*$. In other words, the constant sequence $(x^*, x^*, \dots)$ is a stationary solution of (4.2.27). As in the case of differential equations, we shall not differentiate between these concepts here.

Example 4.10 Consider the logistic equation

$$x(k+1) = 3x(k)(1 - x(k)). \tag{4.2.29}$$

The equation for the equilibrium points is $x = 3x(1-x)$, which gives $x_0 = 0$ and $x_1 = 2/3$. Clearly, if $x(0) = 0$, then $x(k) = 0$ for any $k \in \mathbb{N}$. Similarly, if $x(0) = 2/3$, then $x(1) = 3 \cdot (2/3) \cdot (1 - 2/3) = 2/3$ and, iterating, $x(k) = 2/3$ for any $k \in \mathbb{N}$. The behaviour of the iterates for a nonequilibrium initial condition is illustrated in Fig. 4.6.

Graphically, an equilibrium is the x-coordinate of any point, where the graph of f intersects the diagonal $y = x$. This is the basis of the cobweb method of finding and analysing equilibria, described in the following subsection.

The following definition is the discrete analogue of Definition 4.1.

Definition 4.11

1. The equilibrium x^* is stable if for given $\epsilon > 0$ there is $\delta > 0$ such that for any x and for any $k > 0$, $|x - x^*| < \delta$ implies $|f^k(x) - x^*| < \epsilon$. If x^* is not stable, then it is called unstable (that is, x^* is unstable if there is $\epsilon > 0$ such that for any $\delta > 0$ there are x and k such that $|x - x^*| < \delta$ and $|f^k(x) - x^*| \geq \epsilon$.)
2. A point x^* is called attracting if there is $\eta > 0$ such that $|x - x^*| < \eta$ implies $\lim\limits_{k\to\infty} f^k(x) = x^*$. If $\eta = \infty$, then x^* is called a global attractor or globally attracting.
3. The point x^* is called an asymptotically stable equilibrium if it is stable and attracting. If $\eta = \infty$, then x^* is said to be a globally asymptotically stable equilibrium.

Remark 4.12 We note an important difference between discrete and continuous time systems. In Lemma 4.2, we showed that no non-equilibrium solution to a differential equation with a Lipschitz right-hand side can reach the equilibrium in finite time. On the other hand, such a behaviour is possible for solutions to difference equations; see Example 4.19. The points on a trajectory which reaches an equilibrium in finite time are called *eventual equilibria*.

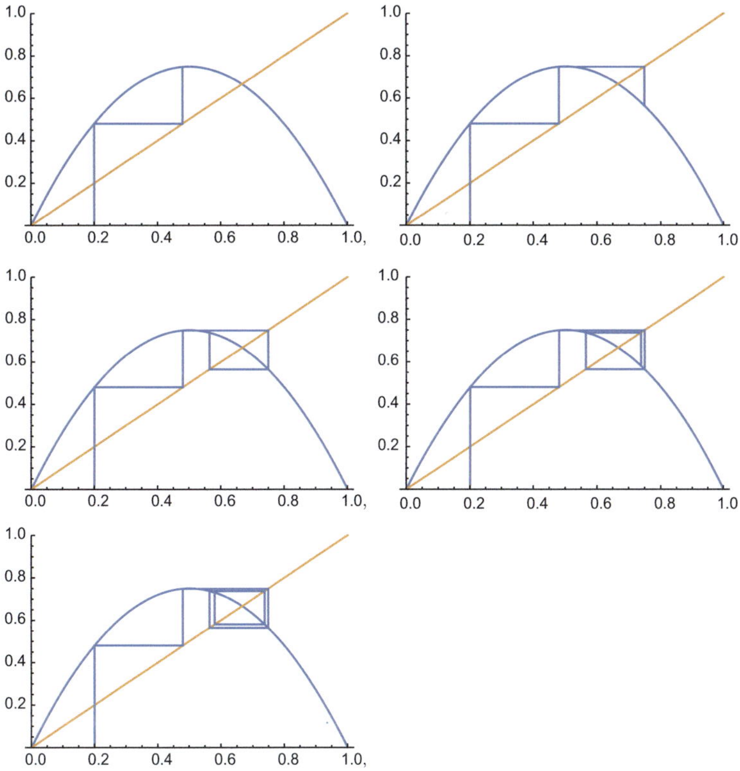

Fig. 4.6 Cobweb diagrams of the solution to the logistic difference equation $x(k+1) = 3x(k)(1 - x(k)$ converging to the fixed point $x^* = \frac{2}{3}$

4.2.1 The Cobweb Diagrams

We describe an important graphical method for visualising the behaviour of solutions of (4.2.27). Since $x(k+1) = f(x(k))$, we may draw a graph of f in the $(x(k), x(k+1))$ system of coordinates. Then, given $x(0) = \mathring{x}$, we pinpoint the value $x(1)$ by drawing a vertical line through $x(0)$ so that it also intersects the graph of f at $(x(0), x(1))$. Next, we draw a horizontal line from $(x(0), x(1))$ to meet the diagonal line $y = x$ at the point $(x(1), x(1))$. A vertical line drawn from the point $(x(1), x(1))$ will meet the graph of f at the point $(x(1), x(2))$. In this way, we may find any $x(k)$. This is illustrated in Fig. 4.6, where we present several steps of drawing the cobweb diagram for the logistic equation (4.2.29) with $\mathring{x} = 0.2$. Based on the diagram, we can conjecture that $x(1) = 2/3$ is an asymptotically stable equilibrium as the solution converges to it as k becomes large. However, to be sure, we need to develop analytical tools for analysing stability.

4.2.2 Analytic Criterion for Stability

Theorem 4.13 *Let x^* be an isolated equilibrium point of the difference equation*

$$x(k+1) = f(x(k)), \tag{4.2.30}$$

where f is continuously differentiable in some neighbourhood of x^*. Then,

(i) if $|f'(x^*)| < 1$, then x^* is asymptotically stable;
(ii) if $|f'(x^*)| > 1$, then x^* is unstable.

Proof Suppose $|f'(x^*)| < M < 1$. Then $|f'(x)| \le M < 1$ over some interval $J = (x^* - \gamma, x^* + \gamma)$ by the property of local preservation of sign for continuous functions, [31]. Let $\mathring{x} \in J$. We have

$$|x(1) - x^*| = |f(\mathring{x}) - f(x^*)|,$$

and, by the Mean Value Theorem, for some $\xi \in [\mathring{x}, x^*]$,

$$|f(\mathring{x}) - f(x^*)| = |f'(\xi)||\mathring{x} - x^*|.$$

Hence, $|x(1) - x^*| = |f(\mathring{x}) - f(x^*)| \le M|\mathring{x} - x^*|$. Since $M < 1$, the inequality shows that $x(1)$ is closer to x^* than $\mathring{x}$ and consequently $x(1) \in J$. By induction,

$$|x(k) - x^*| \le M^k|\mathring{x} - x^*|.$$

For given ϵ, define $\delta = \epsilon$. Then $|x(k) - x^*| < \epsilon$ for $k > 0$ provided $|\mathring{x} - x^*| < \delta$ (since $M < 1$). Furthermore, $x(k) \to x^*$ and $k \to \infty$ so that x^* is asymptotically stable.

To prove the second part of the theorem, we observe that, as in the first part, there is $\epsilon > 0$ such that on $J = (x^* - \epsilon, x^* + \epsilon)$ we have $|f'(x)| \ge M > 1$. Take an arbitrary $\delta > 0$ smaller than ϵ and choose x satisfying $|x - x^*| < \delta$. Again, using the Mean Value Theorem, we get $|f(x) - x^*| = |f'(\xi)||x - x^*|$ for some ξ between x^* and x so that $|f(x) - x^*| \ge M|x - x^*|$. We are done if $f(x)$ is outside J. If not, we can repeat the argument getting $|f^2(x) - x^*| \ge M^2|x - x^*|$, that is, $f^2(x)$ is further away from x^* than $f(x)$. If $f^2(x)$ is still in J, we continue the procedure till $|f^k(x) - x^*| \ge M^k|x - x^*| > \epsilon$ for some k, which must occur and M^k diverges to infinity as $n \to \infty$. □

The equilibrium x^* with $|f'(x^*)| \ne 1$ is called *hyperbolic*.

Remark 4.14 The geometric meaning of the conditions of Theorem 4.13 is that the equilibrium x^* is stable if the graph of f at x^* crosses the diagonal from above if f is increasing, and if f is decreasing, its graph must cross the diagonal from the region to the left of the perpendicular to the diagonal at x^* to the region to the right of it. As we shall see below, this principle also guides the behaviour of iterates in non-hyperbolic cases.

What happens if the equilibrium point x^* is not hyperbolic? To simplify the considerations, we assume that f is at least three times continuously differentiable in a neighbourhood of x^*. First, let us reflect on the geometry of the situation. We assume $f'(x^*) > 0$ in this discussion. The equilibrium x^* is stable if the graph of $y = f(x)$ is less steep than the graph of $y = x$, that is, if the graph of f crosses the line $y = x$ from above to below as x increases. This ensures that the cobweb iterations from the left are increasing and from the right are decreasing while converging to x^*. On the contrary, x^* is unstable if the graph of f crosses $y = x$ from below—then the cobweb iterations will move away from x^*. If $f'(x^*) = 1$, then the graph of f is tangent to the line $y = x$ at $x = x^*$, but the stability properties follow from the geometry. If $f''(x^*) \neq 0$, then f is convex (or concave) close to x^*, and the graph of f will be (locally) either entirely above or entirely below the line $y = x$. Therefore, the picture is the same as in the unstable case either to the left or right of x^*. Hence, x^* is unstable (remember that for instability, it is sufficient to display, for any neighbourhood of x^*, only one diverging sequence of iterations emanating from this neighbourhood).

Remark 4.15 To be more precise, in the case $f'(x^*) = 1$ and $f''(x^*) \neq 0$, there is always a one-sided neighbourhood of x^* such that the iterates starting in this neighbourhood will converge to x^* (right neighbourhood of $x^* = 1$ on the upper left graph of Fig. 4.7). Such an equilibrium is called *semi-stable*. In applications, an equilibrium often appears at the boundary of the domain of interest, e.g., $x^* = 0$ in population dynamics. If $x^* = 0$ is semi-stable from the right, it can be considered a

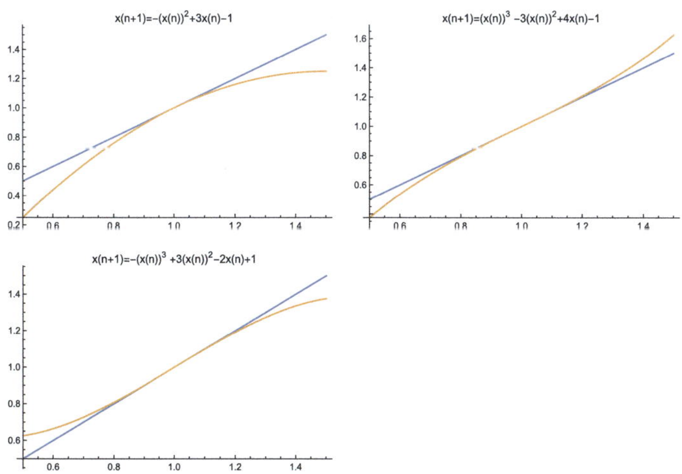

Fig. 4.7 Illustration of the three cases of Theorem 4.16: the stationary point at $x^* = 1$ and $f''(1) \neq 0$ (upper left), $f''(1) = 0$ and $f'''(1) > 0$ (upper right), $f''(1) = 0$ and $f'''(1) < 0$ (lower left)

stable equilibrium since the instability for negative initial values is irrelevant in that particular application.

On the other hand, if $f''(x^*) = 0$, then x^* is an inflection point and the graph of f crosses the line $y = x$. This case is essentially the same as when $|f'(x^*)| \neq 1$: the equilibrium is stable if the graph of f crosses $y = x$ from above and unstable if it does it from below. A quick reflection ascertains that the former occurs when $f'''(x^*) < 0$, while the latter if $f'''(x^*) > 0$, see Fig. 4.7. Summarizing, we have:

Theorem 4.16 *Let x^* be an isolated equilibrium with $f'(x^*) = 1$ and let f be at least three times continuously differentiable in a neighbourhood of x^*. Then,*

(i) if $f''(x^*) \neq 0$, then x^* is unstable,
(ii) if $f''(x^*) = 0$ and $f'''(x^*) > 0$, then x^* is unstable,
(iii) if $f''(x^*) = 0$ and $f'''(x^*) < 0$, then x^* is asymptotically stable.

The case of $f'(x^*) = -1$ is more difficult. First, we note that if $g(x) = -x+2x^*$, that is, if g is a linear function giving an equilibrium at $x = x^*$ with $g'(x^*) = -1$, then the iterations starting from $\mathring{x} \neq x^*$ produce a solution taking on only two values oscillating around x^*. Thus, if $-1 < f'(x^*) < 0$, then f passes from below the line $y = -x+2x^*$ to above as x increases. Hence, the stability follows from the fact that subsequent iterations oscillate around x^*, getting closer to x^* with each iteration. If, on the contrary, $f'(x^*) < -1$, then the oscillating iterations move away from x^*. If $f'(x^*) = -1$, then the graph of f crosses the line $y = x$ at a right angle. Hence, the stability depends on fine details of the shape of f close to x^*. Unfortunately, using an argument similar to the case with $f'(x^*) = 1$ and considering the relation of the graph of f with the graph of $y = -x + 2x^*$, only produces a partial result: x^* will be stable if $f''(x^*) = 0$ and $f'''(x^*) > 0$ (as then the graph of f will have the same shape as in the stable case, crossing the line $y = -x + 2x^*$ from below). However, the stability of x^* can be achieved in a more general situation. First, we note that x^* is also an equilibrium of $g(x) := f(f(x))$, and it is a stable equilibrium of f if and only if it is stable for g. This statement follows from the continuity of f: if x^* is stable for g, then $|g^k(\mathring{x}) - x^*| = |f^{2k}(\mathring{x}) - x^*|$ is small for $\mathring{x}$ sufficiently close to x^*. But then $|f^{2k+1}(\mathring{x}) - x^*| = |f(f^{2k}(\mathring{x})) - f(x^*)|$ is also small by continuity of f. The reverse is obvious. Since $g'(x) = f'(f(x))f'(x)$ with $g'(x^*) = 1$, we can apply Theorem 4.16 to the function g. The second derivative of g is given by

$$g''(x) = f''(f(x))[f'(x)]^2 + f'(f(x))f''(x)$$

and, since $f(x^*) = x^*$ and $f'(x^*) = -1$, we have $g''(x^*) = 0$. Using the chain rule again, we find

$$g'''(x^*) = -2f'''(x^*) - 3[f''(x^*)]^2.$$

Hence, we can state

Theorem 4.17 *Suppose that at an equilibrium point* x^* *we have* $f'(x^*) = -1$. *Define* $S(x^*) = -f'''(x^*) - 3(f''(x^*))^2/2$. *Then* x^* *is asymptotically stable if* $S(x^*) < 0$ *and unstable if* $S(x^*) > 0$.

Example 4.18 Consider the equation

$$x(k+1) = x^2(k) + 3x(k).$$

Solving $f(x) = x^2 + 3x = x$, we find that $x = 0$ and $x = -2$ are the equilibrium points. Since $f'(0) = 3 > 1$, we conclude that the equilibrium at $x = 0$ is unstable. Next, $f'(-2) = -1$. We calculate $f''(-2) = 2$ and $f'''(-2) = 0$ so that $S(-2) = -12 < 0$. Hence, $x = -2$ is an asymptotically stable equilibrium; see Fig. 4.8.

Example 4.19 Consider the equation $x(k+1) = Tx(k)$, where

$$T(x) = \begin{cases} 2x & \text{for } 0 \le x \le 1/2, \\ 2(1-x) & \text{for } 1/2 < x \le 1 \end{cases}$$

is the so-called tent map. There are two equilibrium points, 0 and 2/3. Looking for eventual equilibria is not as simple. Taking $x(0) = 1/8$, we find $x(1) = 1/4$,

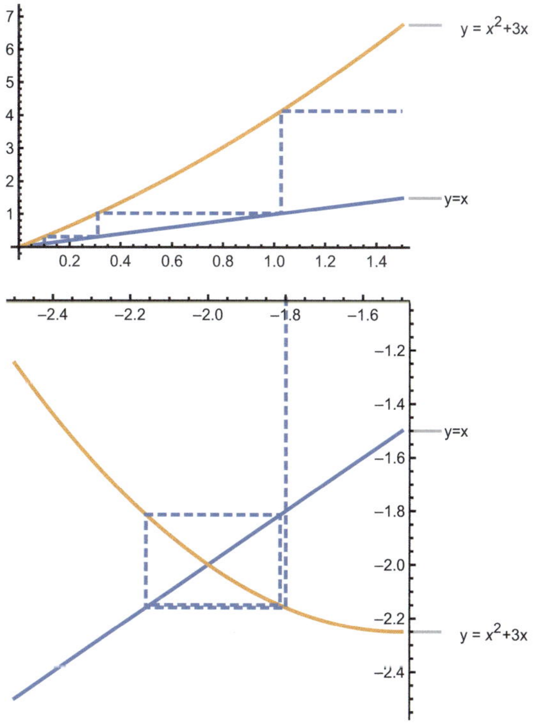

Fig. 4.8 An illustration for Example 4.18 showing the unstable character of the equilibrium $x^* = 0$ with the initial point $\mathring{x} = 0.1$ (top), and the stable character of the equilibrium $x^* = -2$ with $\mathring{x} = -1.8$ (bottom)

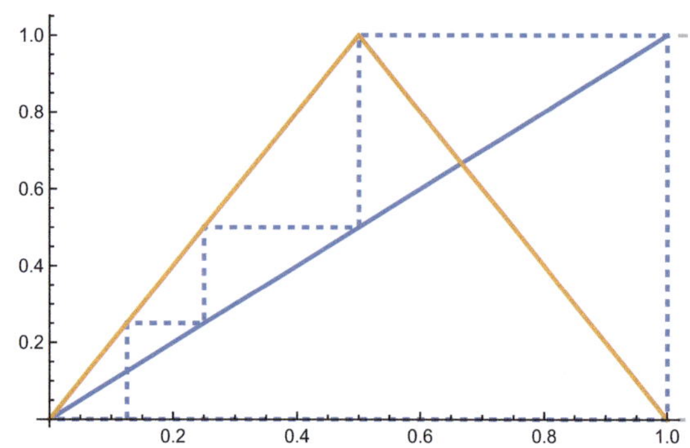

Fig. 4.9 Eventual equilibrium $x = 1/8$ for the tent map

$x(2) = 1/2$, $x(3) = 1$ and $x(4) = 0$, and hence $1/8$ (as well as $1/4$, $1/2$ and 1) are eventual equilibria, see Fig. 4.9. It can be checked that all points of the form $x = k/2^k$, $k \in \mathbb{N}$ satisfies $0 < k/2^k < 1$, are eventual equilibria.

Remark 4.20 We can provide a fine-tuning of the notion of stability by noting that if $f'(x^*) < 0$, then the solution moves in an oscillatory way around x^* and if $f'(x^*) > 0$, then it is monotonic. Indeed, consider (in a neighbourhood of x^* where $f'(x) < 0$) $f(x) - f(x^*) = f(x) - x^* = f'(\xi)(x - x^*)$, $\xi \in (x^*, x)$. Since $f' < 0$, $f(x) > x^*$ if $x < x^*$ and $f(x) < x^*$ if $x > x^*$, hence each iteration moves the point to the other side of x^*. If $|f'| < 1$ over this interval, then $f^k(x)$ converges to x^* in an oscillatory way, while if $|f'| > 1$, the iterations will move away from x^*, also in an oscillatory way.

Based on this observation, we may say that the equilibrium is oscillatory unstable or stable if $f'(x^*) < -1$ or $-1 < f'(x^*) < 0$, respectively, and monotonically stable or unstable depending on whether $0 < f'(x^*) < 1$ or $f'(x^*) > 1$, respectively.

What happens if $f'(x^*) = 0$? Clearly, $|f'(x)| < 1$ in some neighbourhood of x^*, and we shall restrict our attention hereafter to this neighbourhood. If this is not a local extremum, then in some neighbourhood of x^*, f is monotonic; hence, its derivative will not change sign. Thus, the behaviour of the iterates will be as above. Let, on the other hand, $f'(x) < 0$ in some one-sided neighbourhood of x^* and $f'(x) > 0$ on the other side. Then, if the iterates start in the latter, they will stay there, converging to x^*, while if the iterates start in the former, they will begin converging to x^* in an oscillatory way until they reach the neighbourhood in which the derivative is positive and then they will converge monotonically.

Exercise 4.2.1 The plant biomass $P(k)$ at time k is governed by the difference equation

$$P(k+1) = 2P(k) - \frac{3P(k)}{1 + 2P(k)}.$$

Find the equilibria of this equation and their stability. Draw cobweb diagrams.

Exercise 4.2.2 Consider a population described by the difference equation

$$P(k+1) = \frac{aP^2(k)}{b^2 + P^2(k)},$$

where $a > 0$. Assume $a^2 > 4b^2$.

Show that the population may be driven to extinction if its size becomes smaller than a critical value.

4.2.3 Some Applications

In this subsection, we shall present applications of the above considerations to the problems of sustainable fishing, biological pest control and the discrete Allee model.

Sustainable Fishing

Let us consider a population of fish living in a pond, which grows according to the logistic equation (2.1.18),

$$P(k+1) = P(k) + rP(k)\left(1 - \frac{P(k)}{K}\right).$$

This equation can only be solved in some cases; see Sect. 2.1.6. However, even without solving it, we can draw a conclusion of some importance for fisheries.

The basic idea of a sustainable economy is to find an optimal level of fishing: too much harvesting would deplete the fish population beyond recovery, and too little would provide insufficient return for the community. To maintain the population at a constant level, only the increase in the population should be harvested during any one season. In other words, the harvest should be $H(k) = P(k+1) - P(k)$. Using the logistic equation, we find

$$H(k) = rP(k)\left(1 - \frac{P(k)}{K}\right).$$

Hence, to maximize the harvest at each k, the population should be kept at the size $P(k) = P$ for which the right-hand side attains absolute maximum. We note that the right-hand side is a quadratic function, $f(P) = rP(1 - P/K)$. It is easy to find that the maximum is attained at $P = K/2$, that is, the population should be kept at around half of the environment's carrying capacity. Thus, the maximum sustainable harvest is

$$H = \frac{rK}{4}. \qquad (4.2.31)$$

The considered model does not include the actual fishing—it simply says that to sustain a population, we cannot harvest more fish than there are born and indicates the size of the population giving the maximum yield, but it does not specify how to organize the fishing. We can generalize the model to

$$P(k+1) = P(k) + rP(k)\left(1 - \frac{P(k)}{K}\right) - qEP(k), \tag{4.2.32}$$

where E is the so-called fishing effort, for instance, the number of fishing boats at sea, and q is the fishing efficiency, that is, the fraction of the population caught by one boat in the unit time. The concept of sustainable fishing is the same as above. We should find the amount of fish that can be caught during a year (more precisely, (4.2.32) models fishing done instantaneously at the end of the season) to maintain the population at a constant level and hence find the population for which the yield is optimal.

To find the possible constant population level for a given level of fishing, we solve

$$P = P + rP\left(1 - \frac{P}{K}\right) - qEP. \tag{4.2.33}$$

This gives $P = 0$ or

$$P^* = K\left(1 - \frac{qE}{r}\right). \tag{4.2.34}$$

The first solution is trivial and not interesting. The second solution is positive if

$$qE < r.$$

This is consistent with the first part, showing that the fishing rate cannot exceed the unrestricted net growth of the population. However, this model can furnish further information. Let us suppose that for the given fishing rate qE the population is kept at P^* given by (4.2.34). Then, the yield Y is given by

$$Y(qE) = qEN^* = qEK\left(1 - \frac{qE}{r}\right). \tag{4.2.35}$$

Thus, the yield is a quadratic function of the fishing rate qE; hence, its maximum can be found similarly to the first part of the section. The maximum of $Y(qE)$ is attained at $E = \frac{r}{2q}$, which gives the maximum sustainable yield as

$$Y_{max}(qE) = \frac{rK}{4},$$

as in (4.2.31). Clearly, if we increase the fishing effort, $E > r/2q$, then the yield will decrease (greed does not pay!). To understand why, let us summarize the mechanism described by the model. For a given fishing rate qE, we find that the equilibrium P^*, given by (4.2.34), is asymptotically stable provided

$$\left|\frac{d}{dP}\left(P + rP\left(1 - \frac{P}{K}\right) - qEP\right)\Big|_{P=P^*}\right| = \left|1 + r - \frac{2rP^*}{K} - qE\right| < 1.$$

Using (4.2.34), this gives

$$-1 < 1 - r + qE < 1$$

or

$$r - 2 < qE < r.$$

Thus, if the fishing rate satisfies the above condition, even if we begin with a different initial fish population, it will eventually stabilize at P^*. If the fishing rate qE is higher than $r/2$, then the population will stabilize at a lower value P^*, which the higher qE will not compensate. On the other hand, a smaller fishing rate leads to a larger P^*, but not large enough to balance qE in (4.2.35) to optimize Y.

Remark 4.21 One may wonder about the significance of the condition $r-2 < qE$, which puts an upper bound on r. Common sense would suggest that the higher the net growth rate, the better the yield. To explain this, we observe that then (4.2.33) can be written as

$$P(k+1) = (1 + r - qE)\, P(k)\left(1 - \frac{P}{\frac{K(1+r-qE)}{r}}\right).$$

The results of Sect. 4.2.5 show that the positive equilibrium of a discrete logistic equation is unstable if the unrestricted growth rate is larger than 3, which in our case corresponds to

$$1 + r - qE > 3 \quad \text{or} \quad r - 2 > qE.$$

Thus, the case $r - 2 > qE$ does not give a stable fish population, ensuring a sustainable constant yield.

Biological Pest Control

Assume we must deal with an insect population invading our plantation. The insects reproduce according to the Beverton–Holt model (2.1.12),

$$P(k+1) = \frac{\beta P(k)}{1 + aP(k)}, \tag{4.2.36}$$

where β is the natural fertility of insects while $a = (\beta - 1)/K$, where K is the capacity of the environment. As we know, see Sect. 2.1.6, if $\beta > 1$, then the population $P(k)$ tends to a nonzero equilibrium $K = (\beta - 1)/a$, that is, there is a nonzero stable population of insects. An ecological way of eradicating the pest is to decrease the birth rate, and one of the methods is to introduce several sterile insects S into the population. We assume that S is under our control, and we can keep the number of sterile insects constant. Though suppressed in the model, insects reproduce sexually; thus, the effective birth rate depends on the probability of finding a mate. If $P(k)$ is the number of fertile insects, then the probability of picking a fertile insect is $P(k)/(P(k) + S)$. Thus, the Beverton–Holt model can be modified as

$$P(k+1) = \beta P(k) \frac{P(k)}{S + P(k)} \frac{1}{1 + aP(k)} = \beta P(k) f(P(k)). \tag{4.2.37}$$

To find the equilibria of (4.2.37), we solve

$$P = \beta P \frac{P}{S + P} \frac{1}{1 + aP},$$

which leads to $P_1 = 0$ or the simplified equation

$$1 = \beta \frac{P}{S + P} \frac{1}{1 + aP}.$$

While the above equation can be solved for P^*, the approach used in Sect. 4.1.2 to analyse the continuous Allee model is a faster way to find the answer. Thus, we solve the above equation for S as a function of P, getting

$$S(P) = \frac{(\beta - 1 - aP)P}{1 + aP}. \tag{4.2.38}$$

We find that $S(0) = S((\beta - 1)/a) = 0$ and the derivative is given by $S'(P) = -1 + \beta/(1 + aP)^2$. Hence, the maximum in the interval $(0, (\beta - 1)/a)$ is attained at $P = (\sqrt{\beta} - 1)/a$; the maximum is $S_{\max} = (\sqrt{\beta} - 1)^2/a$. Thus, we conclude that if $S > S_{max}$, the only equilibrium of (4.2.37) is $P_1 = 0$.

Next, we find

$$\beta (Pf(P))' = \beta \frac{P^2(1 + aS) + 2PS}{((P + S)(1 + aP))^2}.$$

Hence, for any $S > 0$, $\beta (Pf(P))'|_{P=0} = 0$ and the equilibrium $P_1 = 0$ is asymptotically stable, while for $S = 0$, $\beta (Pf(P))'|_{P=0} = \beta$ and $P_1 = 0$ is unstable whenever $\beta > 1$.

For the other equilibria, let us first focus on $S > 0$. It is laborious to provide explicit estimates of the derivatives of $\beta P f(P)$ at the other two equilibria that exist for $0 < S < S_{\max}$, but still, we can determine their stability by using the discussion preceding Theorem 4.16. First we notice that $\beta(Pf(P))' > 0$ for all $P > 0$ (provided $S > 0$). This means that the derivative never equals -1. Let $0 < S < S_{\max}$. The curve $\beta Pf(P)$ starts at zero below the diagonal, and thus, at the smaller equilibrium, it must cross the diagonal from below. Hence, this equilibrium is unstable. At the second equilibrium, the curve crosses the diagonal from above and hence, since the curve is ascending, the derivative is between 0 and 1. Thus, this equilibrium is asymptotically stable. If $S = S_{\max}$, we have a tangent point, which means that the (unique) positive equilibrium is semi-stable—unstable from the left and stable from the right. If $S > S_{\max}$, the only equilibrium of the system is $P_1 = 0$, and it is globally attracting.

If $S = 0$, then we have one positive equilibrium that is attracting.

The possible situations are presented in Figs. 4.10 and 4.11.

In summary, without sterile insects, the pest population will settle at a non-zero level. We could introduce a number $S > S_{\max}$ of sterile insects to eradicate it.

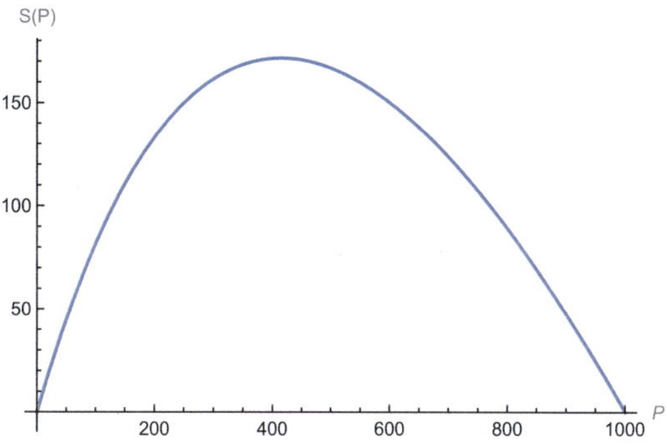

Fig. 4.10 The graph of $S(P)$, given by (4.2.38), for $\beta = 2$ and $a = 0.001$

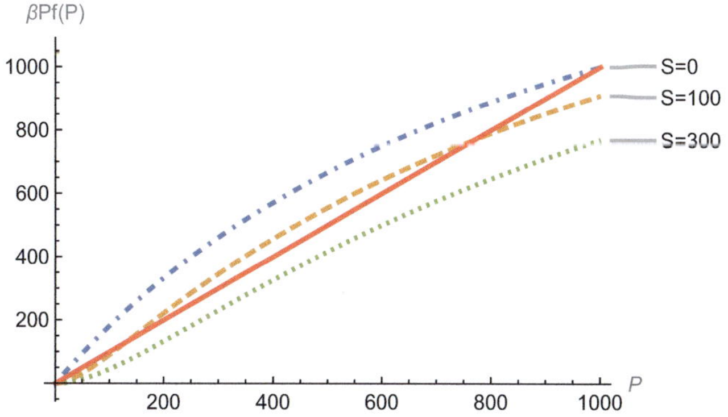

Fig. 4.11 The graphs of $\beta Pf(P)$, as in (4.2.37), for $\beta = 2$, $a = 0.001$ with $S = 300$ (dotted line), $S = 100$ (dashed line) and $S = 0$ (dot-dashed line)

Then, the pest population will converge to the extinction equilibrium, provided the number S of sterile insects is kept above $S_{\max}$ at each cycle (this may require our intervention as the insects die of natural causes). However, the Allee character of the models offers a smarter, more economical way to eradicate the pest. We can introduce a smaller number of sterile insects and then, by an additional intervention applied just once, drive the pest population below the smaller equilibrium—then it will also converge to the extinction equilibrium without any further external effort.

Exercise 4.2.3 Find the minimum number of sterile insects needed to drive the pest population to extinction if the population is described by (4.2.36) with $\beta = 3$ and the carrying capacity $K = 20000$.

The Discrete Allee Model (2.1.42)

Let us recall the model (2.1.42)

$$P(k+1) = P(k)\left(1 + \lambda\left(1 - \frac{P(k)}{C} - \frac{A}{1 + BP(k)}\right)\right).$$

The equilibria are determined by solving

$$P = P\left(1 + \lambda\left(1 - \frac{P}{C} - \frac{A}{1 + BP}\right)\right),$$

and we see that the equilibria are $0, L, K$ for appropriate parameters C, A, B, exactly as in the continuous case, see Sect. 4.1.2 and Eq. (4.1.9). To fix attention, we focus on the case with two positive equilibria $0 < L < K$, in which case we must have $1 < A < (BC + 1)^2/4BC$ and $BC > 1$ (see Sect. 4.1.2). In contrast to (3.3.36), it is not obvious that (2.1.42) describes a population, that is, whether the solution is nonnegative for a nonnegative initial condition. Here we shall find conditions ensuring that the interval $[0, K]$ is invariant under

$$g(P) := P\left(1 + \lambda\left(1 - \frac{P}{C} - \frac{A}{1 + BP}\right)\right),$$

that is, the solutions are nonnegative, and the population will never get larger than K. This rules out the possibility of K being an oscillatory stable equilibrium but simplifies the calculations. First, we look at nonnegativity. We can write

$$P(k+1) - P(k) = \lambda f(P(k)),$$

where f is defined in (4.1.9). Thus $P(k+1) > P(k)$, that is, $P(k)$ increases as long as $L < P(k) < K$. Hence, $P(k+1)$ is nonnegative as long as $P(k) \in (L, K)$. However, $P(k+1) < P(k)$ if $0 < P(k) < L$ and it could become negative. To avoid this, it suffices to ensure that

$$P(k) + \lambda f(P(k)) > 0$$

or

$$\lambda h(P) := \lambda \left(1 - \frac{P}{C} - \frac{A}{1 + BP}\right) > -1, \quad 0 < P < L.$$

Since

$$h'(P) = -\frac{1}{C} + \frac{AB}{(1 + BP)^2}$$

and, in our case, $A > 1$, $BC > 1$, we see that h has only one stationary point $\bar{P}$. Since $h(0) = 1 - A < 0$ and $h(L) = h(K) = 0$ for $0 < L < K$, we see that $L < \bar{P} < K$ and thus h is increasing on $[0, \bar{P})$. Hence, $\lambda h > -1$ on $(0, L)$ if and only if $\lambda h(0) \geq -1$, which translates into

$$A \leq \frac{1 + \lambda}{\lambda}. \tag{4.2.39}$$

The stability of the equilibria 0 and L can be determined using the continuous case. Indeed, writing (2.1.42) as

$$P(k + 1) = g(P(k)),$$

we see that

$$g(P) = P + \lambda f(P) = P + \lambda P h(P),$$

and thus

$$g'(P) = 1 + \lambda f'(P) = 1 + \lambda h(P) + \lambda P h'(P). \tag{4.2.40}$$

Hence

$$g'(0) = 1 + \lambda h(0) = 1 + \lambda(1 - A),$$

which gives stability provided

$$A < 1 + \frac{2}{\lambda}.$$

However, the monotonic stability, see Remark 4.20, coincides with the assumption (4.2.39)—otherwise, we would have oscillatory convergence and consequently negative solutions. Further, since $h'(L) > 0$ by the considerations above, (4.2.40) immediately yields $g'(L) > 1$ and hence the instability of L.

To show that $P(k+1) < K$ provided $0 < P(k) < K$, and also to determine the character of stability of K, requires some more work. First, observe that

$$h(P) < 1 - \frac{P}{C} =: \psi(P).$$

Since ψ is a monotonically decreasing function, having $P = C$ as the only zero, we obtain

$$L < K < C. \tag{4.2.41}$$

Indeed, $0 = h(K) < \psi(K)$ and (4.2.41) follows from monotonicity of ψ.

Next, using (4.1.12) we can write

$$\begin{aligned} g'(P) = 1 + \lambda f'(P) &= 1 + \frac{d}{dP}\left(\frac{\lambda BP}{C(1+BP)}(P-L)(K-P)\right) \\ &= 1 + \phi'(P)(P-L)(K-P) + \phi(P)(-2P+L+K), \end{aligned} \tag{4.2.42}$$

where $\phi(P) = \frac{\lambda BP}{C(1+BP)}$. It is easy to see that ϕ is a strictly increasing function on $[0, \infty)$, satisfying

$$0 \leq \phi(P) < \frac{\lambda}{C}. \tag{4.2.43}$$

Thus

$$g'(K) = 1 + \frac{\lambda BK}{C(1+BK)}(L-K),$$

and we obtain that $g'(K) \geq 0$ provided $\dfrac{\lambda(L-K)}{C} \geq -1$, that is,

$$\lambda \leq \frac{C}{K-L}. \tag{4.2.44}$$

We note that $\dfrac{C}{K-L} > 1$ by (4.2.41); thus, the Allee effect can occur in populations with positive intrinsic growth rates. If we look closer at (4.2.42), we note that the second term in the last equality is positive for $L < P < K$, while for the third one we have

$$\phi(P)(-2P+L+K) = \phi(P)(-2P+2K+L-K) > \phi(P)(L-K),$$

as $P < K$ and $\phi(P) > 0$.

Hence, by (4.2.43), and provided (4.2.44) holds,

$$g'(P) \geq 1 + \phi(P)(L - K) \geq 0, \quad L \leq P \leq K,$$

and thus

$$g(P) \leq g(K) = K$$

for any $P \in [L, K]$. Since we proved earlier that g transforms $[0, L]$ into $[0, L]$, we see that if (4.1.13), (4.2.39) and (4.2.44) are satisfied, then $g(P) \in [0, K]$ for $P \in [0, K]$ as required.

An example of parameters satisfying these conditions is provided in Example 2.2.

Exercise 4.2.4 For the equation

$$x(k+1) = x(k)(1 + r(x(k) - 1)(3 - x(k))),$$

find sufficient conditions on r for which the above equation represents an Allee model. Illustrate your results by the cobweb analysis.

4.2.4 Periodic Points and Cycles

Theorem 4.3 tells us that a solution to a scalar autonomous differential equation must be monotonic. We have already seen above that solutions to scalar autonomous difference equations can be oscillatory. In fact, such equations may admit periodic solutions, which cannot occur in the continuous case.

Definition 4.22 Let b be a point in the domain of f. Then,

(i) b is called a periodic point of f if $f^k(b) = b$ for some $k \in \mathbb{N}$, and the periodic orbit of b, $O(b) = \{b, f(b), f^2(b), \dots, f^{k-1}(b)\}$ is called a k-cycle,
(ii) b is called eventually k-periodic if, for some integer m, $f^m(b)$ is a k-periodic point.

Example 4.23 Consider $x(k+1) = T^2x(k)$, where T is the tent map introduced in Example 4.19. Then

$$T^2(x) = \begin{cases} 4x & \text{for } 0 \leq x \leq 1/4, \\ 2(1-2x) & \text{for } 1/4 < x \leq 1/2, \\ 4x - 2 & \text{for } 1/2 < x \leq 3/4, \\ 4(1-x) & \text{for } 3/4 < x \leq 1. \end{cases}$$

There are four equilibrium points, 0, 2/5, 2/3 and 4/5, two of which are equilibria of T. Hence $\{2/5, 4/5\}$ is the only 2-cycle of T, as seen in Fig. 4.12.

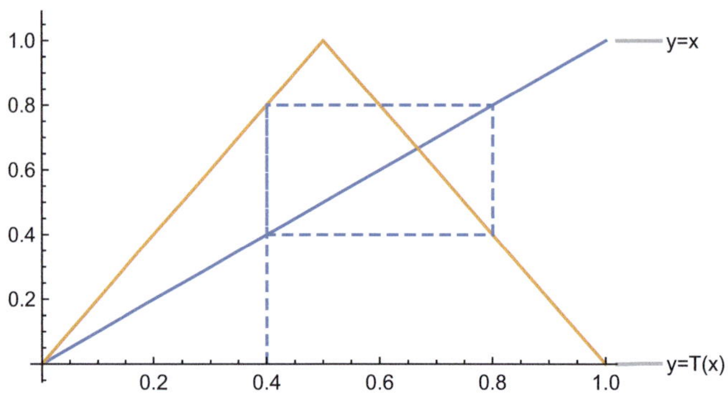

Fig. 4.12 2-Cycle for the tent map

Exercise 4.2.5 Find 3-cycles of the tent map.

Definition 4.24 Let b be a k-periodic point of f. Then b is said to be:

(i) stable, if it is a stable fixed point of f^k,
(ii) asymptotically stable, if it is an asymptotically stable fixed point of f^k,
(iii) unstable, if it is an unstable fixed point of f^k.

This definition, together with Theorem 4.13, yield the following classification of the stability of k-cycles.

Theorem 4.25 *Let $O(b) = \{x_0 = b, x(1) = f(b), \ldots, x(k-1) = f^{k-1}(b)\}$ be a k-cycle of a continuously differentiable function f. Then*

(i) The k-cycle $O(b)$ is asymptotically stable if

$$|f'(x_0)f'(x(1))\ldots f'(x(k-1))| < 1.$$

(ii) The k-cycle $O(b)$ is unstable if

$$|f'(x_0)f'(x(1))\ldots f'(x(k-1))| > 1.$$

Proof Follows from Theorem 4.13 by the Chain Rule applied to f^k. □

It follows that if b is k-periodic, then every point of its k-cycle $\{x(0) = b, x(1) = f(b), \ldots, x(k-1) = f^{k-1}(b)\}$ is also k-periodic. This follows from $f^k(f^r(b)) = f^r(f^k(b)) = f^r(b), r = 0, 1, \ldots, k-1$.

Note that the stability of b means that $|f^{nk}(x) - b| < \epsilon$ for all k, provided x is close enough to b. In other words, we are ensured only that the iterates of $f^k(x)$ will stay close to b. However, from the continuity of f it also follows that for $r = 1, \ldots, k-1$, $f^{nk+r}(x)$ will stay close to $f^{nk+r}(b) = f^r(b)$ if x is close enough to b.

Actually, a stronger result is valid.

Proposition 4.26 *Each point of the k-cycle $O(b) = \{x(0) = b, x(1) = f(b), \dots, x(k-1) = f^{k-1}(b)\}$ possesses the same stability property as b.*

Proof The proof would be easy if f was continuously invertible. Indeed, then the preimage of any small neighbourhood of $f^r(b)$ would be a small neighbourhood of b and starting the iterates from a small neighbourhood of $f^r(b)$ would be equivalent to starting from a small neighbourhood of b. Unfortunately, it is not possible for non-invertible functions (think about $f(x) = \sin x$ and preimage $\{x \in \mathbb{R};\ |\sin x| < \epsilon\}$—it is unbounded).

Hence, assume that b is stable and fix $0 < r < k$. The stability of b, the continuity of f^r and $f^r(b) = f^{k+r}(b)$ imply that for any ϵ we can find δ so that

$$\begin{aligned} |f^r f^{(n-1)k}(y) - f^r f^{(n-1)k}(b)| &= |f^{(n-1)k} f^r(y) - f^{(n-1)k} f^r(b)| \\ &= |f^{(n-1)k} f^r(y) - f^{nk} f^r(b)| < \epsilon, \end{aligned}$$

provided $|y - b| < \delta$. Next, by the continuity of f^{k-r}, we can find $\delta_1 > 0$ such that from $|x - f^r(b)| < \delta_1$ it follows that

$$\delta > |f^{k-r}(x) - f^{k-r} f^r(b)| = |f^{k-r}(x) - f^k(b)| = |f^{k-r}(x) - b|.$$

Taking now $y = f^{k-r}(x)$, we have $|y - b| < \delta$, hence

$$|f^{(n-1)k} f^k(x) - f^{nk} f^r(b)| = |f^{nk}(x) - f^{nk} f^r(b)| < \epsilon,$$

provided $|x - f^r(b)| < \delta_1$, which yields the stability of $f^r(b)$.

Asymptotic stability can be proved in a similar way. We know that we can find δ_1 such that $|x - f^r(b)| < \delta_1$ implies $f^{kn} f^{k-r}(x) \to b$ for $k \to \infty$. The continuity of f yields

$$f^r f^{kn} f^{k-r}(x) = f^{k(n+1)}(x) \to f^r(b)$$

but this is the same as $f^{kn}(x) \to f^r(b)$. Hence $f^r(b)$ is asymptotically stable.

Finally, let b be unstable. If $f^r(b)$ were stable, then $b = f^{k-r} f^r(b)$ would be stable, contradicting the first part of the proof. □

Exercise 4.2.6 Determine the stability of the 2-cycle of the tent map.

Exercise 4.2.7 Find 2-cycles of the map

$$x(k+1) = 1 - x^2(n)$$

and determine their stability.

4.2.5 Dynamics of the Logistic Equation

Consider the logistic equation

$$x(k+1) = F_\gamma(x(k)) := \gamma x(k)(1-x(k)), \quad x \in [0,1],\ \gamma > 0. \tag{4.2.45}$$

Our aim is to investigate how the properties of the equilibria of (4.2.45) depend on the parameter γ. The values of γ, for which there is a qualitative change of the properties of the equilibria, are called the *bifurcation points*.

To find the equilibrium points, we solve $F_\gamma(x^*) = x^*$, which gives

$$x^* = 0 \quad \text{and} \quad x^* = \frac{\gamma - 1}{\gamma}.$$

We investigate the stability of each point separately.

(a) For $x^* = 0$, we have $F'_\gamma(0) = \gamma$ and thus $x^* = 0$ is asymptotically stable for $0 < \gamma < 1$ and unstable for $\gamma > 1$. To investigate the stability for $\gamma = 1$, we find $F''_\gamma(0) = -2\gamma \neq 0$ and thus in this case $x^* = 0$ is unstable (semi-stable) by Theorem 4.16. However, the instability comes from the negative values of x, which we discarded from the domain. If we restrict our attention to the domain $[0, 1]$, then $x^* = 0$ is stable, as discussed in Remark 4.15.

(b) The equilibrium point $x^* = \dfrac{\gamma - 1}{\gamma}$ belongs to the domain $[0, 1]$ if and only if $\gamma > 1$. Here,

$$F'_\gamma\left(\frac{\gamma-1}{\gamma}\right) = 2 - \gamma \quad \text{and} \quad F''_\gamma\left(\frac{\gamma-1}{\gamma}\right) = -2\gamma.$$

Thus, using Theorems 4.13 and 4.16, we obtain that x^* is asymptotically stable if $1 < \gamma \leq 3$ and it is unstable if $\gamma > 3$.

Further, by Remark 4.20, we observe that for $1 < \gamma < 2$, the population approaches the carrying capacity monotonically from below. However, for $2 < \gamma \leq 3$ the population can go over the carrying capacity, but it will eventually stabilize around it, as shown in Fig. 4.6.

What happens for $\gamma > 3$? Consider 2-cycles. We have $F^2_\gamma(x) = \gamma^2 x(1-x)(1-\gamma x(1-x))$ so that we are looking for solutions to $\gamma^2 x(1-x)(1-\gamma x(1-x)) = x$, which can be re-written as

$$x(\gamma^3 x^3 - 2\gamma^3 x^2 + \gamma^2(1+\gamma)x + (1-\gamma^2)) = 0.$$

To simplify, we observe that any equilibrium is also a 2-cycle (and any k-cycle, for that matter). Thus, we can divide this equation by x and $x - (\gamma - 1)/\gamma$, getting

$$\gamma^2 x^2 - \gamma(\gamma + 1)x + \gamma + 1 = 0.$$

Solving this quadratic equation, we obtain a 2-cycle

$$x_{\pm} = \frac{(1+\gamma) \pm \sqrt{(\gamma-3)(\gamma+1)}}{2\gamma}. \tag{4.2.46}$$

Clearly, these points determine a 2-cycle provided $\gamma > 3$ (in fact, for $\gamma = 3$, these two points collapse into the equilibrium point $x^* = 2/3$. Thus, we see that when the parameter γ passes through $\gamma = 3$, the stable equilibrium becomes unstable and bifurcates into two 2-cycles.

Theorem 4.25 allows for determining the stability of 2-cycles. We have $F'_\gamma(x) = \gamma(1-2x)$ so the 2-cycle is stable, provided

$$-1 < \gamma^2(1-2x_+)(1-2x_-)) < 1. \tag{4.2.47}$$

Re-writing the product and using Viète's formulae, we have

$$1 - 2(x_+ + x_-) + 4x_+x_- = 1 - 2\frac{\gamma+1}{\gamma} + 4\frac{\gamma+1}{\gamma^2},$$

so that (4.2.47) is satisfied, provided $-1 < -\gamma^2 + 2\gamma + 4 < 1$ and, upon solving this inequality, we get $\gamma < -1$ or $\gamma > 3$ and $1 - \sqrt{6} < \gamma < 1 + \sqrt{6}$ which yields $3 < \gamma < 1 + \sqrt{6}$.

Similarly, we find that for $\gamma_1 = 1+\sqrt{6}$, the 2-cycle is still attracting but becomes unstable for $\gamma > \gamma_1$. These considerations are illustrated in Fig. 4.13 for $\gamma = 4$.

To find 4-cycles, we solve $F_\gamma^4(x) = 0$. However, the algebra becomes intractable, and one should resort to a computer. It turns out that there is a 4-cycle, when $\gamma > 1 + \sqrt{6}$, which is attracting for $1 + \sqrt{6} < \gamma < 3.544090\ldots =: \gamma_2$. When $\gamma = \gamma_2$, the 2^2-cycle bifurcates into a 2^3-cycle which is stable for $\gamma_2 \leq \gamma \leq \gamma_3 := 3.564407\ldots$. Continuing, we obtain a sequence of numbers (γ_n), such that the 2^k-cycle bifurcates into a 2^{k+1}-cycle passing through γ_k. In this particular case, $\lim_{k\to\infty} \gamma_n = \gamma_\infty = 3.57\ldots$. A remarkable observation, made by M.J. Feigenbaum, is that for any sufficiently smooth family F_γ of mappings of an interval into itself, the number

$$\delta = \lim_{k\to\infty} \frac{\gamma_k - \gamma_{k-1}}{\gamma_{k+1} - \gamma_k} = 4.6692016\ldots,$$

in general, does not depend on the family of maps, provided they have a single maximum. The interested reader could consult, e.g., [42] for further information on the dynamics of the logistic map.

Feigenbaum's result expresses the fact that the picture obtained for the logistic equation is, to a large extent, universal. What happens for γ_∞? Here, we find a densely interwoven region with periodic and irregular orbits. In particular, a 3-cycle appears and, by the celebrated theorem by Šarkovsky, see, e.g., [48, 86], this implies the existence of orbits of any period. In fact, we observe the emergence of the so-

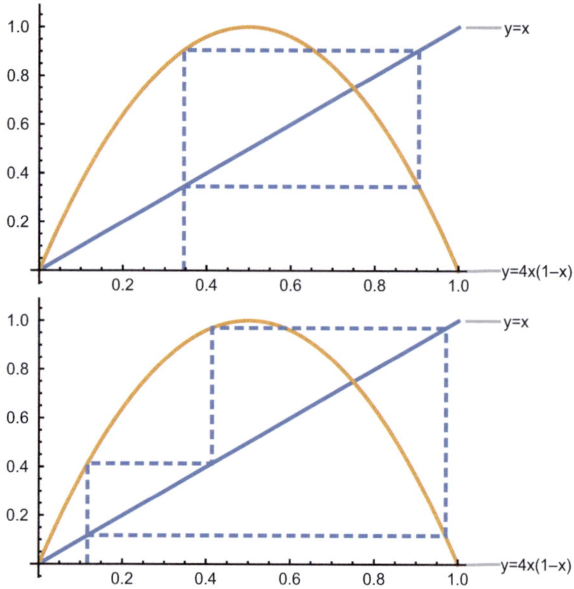

Fig. 4.13 2-Cycle $\left\{\frac{5-\sqrt{5}}{8}, \frac{5+\sqrt{5}}{8}\right\}$ (top) and 3-cycle {0.116978, 0.413176, 0.969846} (bottom) for the logistic map $F_4(x) = 4x(1-x)$

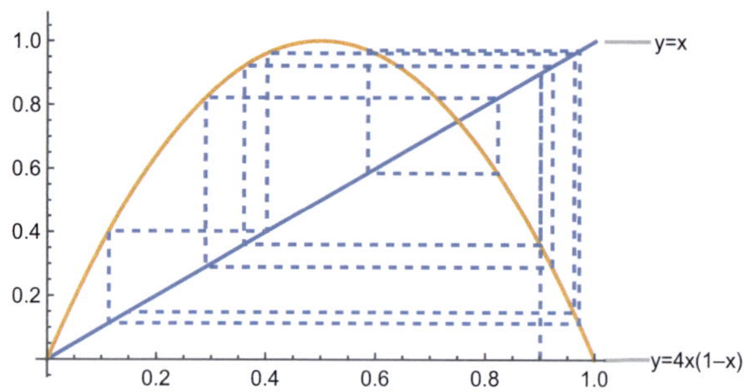

Fig. 4.14 Chaotic orbit for the logistic map $F_4(x) = 4x(1-x)$ with the initial condition $\mathring{x} = 0.9$

called chaotic dynamics, see Fig. 4.14. The discussion of this topic is beyond the scope of this book, and the reader is referred to, e.g., [42] (discrete dynamics) and [48, 90] (continuous dynamics) for a more detailed account of it.

Exercise 4.2.8 Show that 3-cycles appear if $\mu = 4$.

Exercise 4.2.9 Show that by a linear transformation, one can transform the quadratic equation

$$x(n+1) = x^2(n) + c, \quad c \in \mathbb{R}, \tag{4.2.48}$$

to a logistic equation

$$y(n+1) = \mu y(n)(1 - y(n)).$$

(i) Find the relation between c and μ.
(ii) Find the equilibria x_1^* and x_2^* of (4.2.48).
(iii) Find the values of c for which x_1^* is asymptotically stable or unstable.
(iv) Find the values of c for which x_2^* is asymptotically stable or unstable.

Exercise 4.2.10 Consider the Pacific salmon equation, (3.4.63),

$$P(n+1) = \gamma p_n(\tau) = \gamma \beta P(n) e^{-\lambda \tau P(n)}. \tag{4.2.49}$$

Find the equilibria of (4.2.49) and determine conditions under which they are stable/unstable. Then, find the conditions under which a 2-periodic solution exists and determine its stability. Hint. The existence of equilibria and two periodic solutions can be proved analytically. You can find the concrete thresholds of the parameters using computer simulations.

4.2.6 Stability in the Beverton–Holt Equation

We conclude with a brief description of the stability of equilibrium points for the Beverton–Holt equation which was discussed in Sects. 2.1.4 and 2.1.6. Let us recall this equation,

$$x(n+1) = f(x(n), \beta, b) = \frac{\beta x(n)}{(1+x(n))^b}.$$

Writing

$$x^*(1+x^*)^b = \beta x^*,$$

we find a steady state $x^* = 0$, and we observe that if $\beta \leq 1$, then this is the only steady state (at least for positive values of x). If $\beta > 1$, then there is another steady state given by $x^* = \beta^{1/b} - 1$. Evaluating the derivative at x^*, we have

$$f'(x^*, \beta, b) = \frac{\beta}{(1+x^*)^b} - \frac{\beta b x^*}{(1+x^*)^{b+1}} = 1 - b + \frac{b}{\beta^{1/b}}.$$

Clearly, with $\beta > 1$, we always have $f' < 1$. Hence, for monotone stability, we must have $1 - b + b\beta^{-1/b} > 0$, whereas oscillatory stability requires $-1 < 1 - b + b\beta^{-1/b} < 0$, see Figs. 4.15 and 4.16. Solving these inequalities, we obtain that the borderlines between different types of behaviour are given by

$$\beta = \left(\frac{b}{b-1}\right)^b \text{ and } \beta = \left(\frac{b}{b-2}\right)^b, \tag{4.2.50}$$

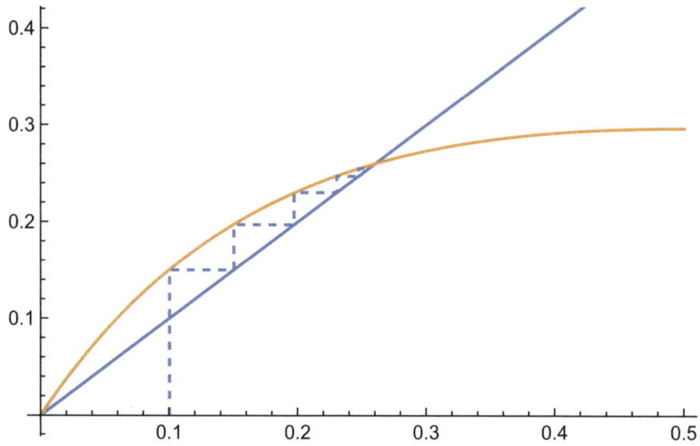

Fig. 4.15 Monotonic stability of the equilibrium for the Beverton–Holt model with $b = 3$ and $\beta = 2$; see Eq. (4.2.50)

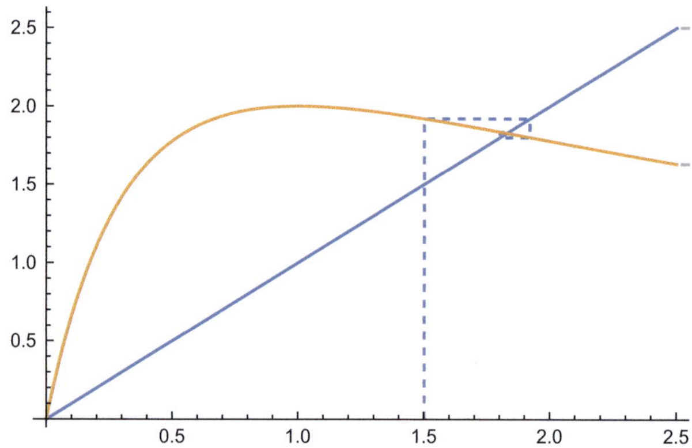

Fig. 4.16 Oscillatory stability of the equilibrium for the Beverton–Holt model with $b = 2$ and $\beta = 8$; see Eq. (4.2.50)

see Fig. 4.17. Let us consider the existence of 2-cycles; see Fig. 4.18. The second iteration of the map $H(x) = \beta x/(1+x)^b$ is given by

$$H(H(x)) = \frac{\beta^2 x(1+x)^{b^2-b}}{((1+x)^b + \beta x)^b},$$

so that 2-cycles can be obtained by solving $H(H(x)) = x$, which can be rewritten as

$$x\beta^2(1+x)^{b^2-b} = x((1+x)^b + \beta x)^b,$$

or, discarding $x = 0$ and taking the bth root,

$$(1+x)^{b-1}\beta^{\frac{2}{b}} = (1+x)^b + \beta x.$$

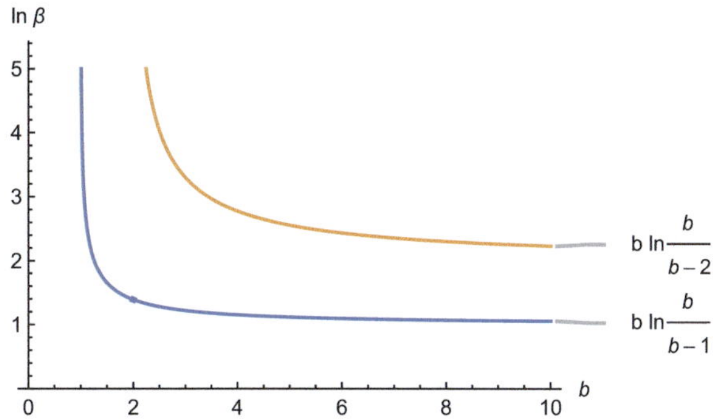

Fig. 4.17 Regions of stability of the Beverton–Holt model described by (4.2.50): monotonically stable below the bottom curve, oscillatory stable between the curves and unstable above the top curve

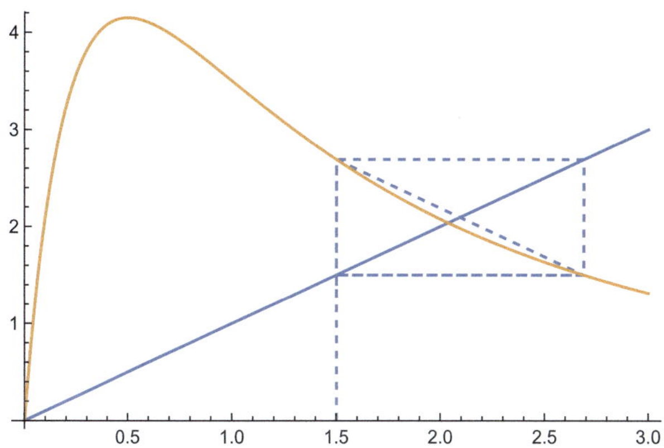

Fig. 4.18 2-Cycles for the Beverton–Holt model with $b = 3$ and $\beta = 28$; see Eq. (4.2.50)

Introducing the change of variables $z = 1 + x$, we see that we have to investigate the existence of positive roots of

$$\phi(z, b, \beta) = z^b - z^{b-1}\beta^{\frac{2}{b}} + \beta z - \beta.$$

In what follows, we drop the dependence of ϕ on b and β to shorten the notation. Clearly, we have $\phi(\beta^{\frac{1}{b}}) = 0$, since any equilibrium of H is also an equilibrium of H^2. First, let us consider $1 < b \leq 2$ (the case $b = 1$ yields an explicit solution obtained in Sect. 2.1.6).

We have

$$\phi'(z) = bz^{b-1} - (b-1)z^{b-2}\beta^{2/b} + \beta,$$
$$\phi''(z) = (b-1)z^{b-3}(bz + (2-b)\beta^{2/b}),$$

hence we see that $\phi'' > 0$ for all $z > 0$. Furthermore, $\phi(0) = -\beta < 0$. Hence, the region Ω, bounded from the left by the axis $z = 0$ and lying above the graph of ϕ

for $z > 0$, is convex. Thus, the z axis, being perpendicular to the axis $z = 0$, cuts the boundary of Ω in exactly two points, one being $(0, 0)$ and the other $(\beta^{\frac{1}{b}}, 0)$. Hence, there are no additional equilibria of H^2 for $1 < b < 2$; therefore, H does not have 2-cycles for $b \leq 2$.

Exercise 4.2.11 Prove the above statement directly if $b = 2$.

Consider next $b > 2$. Then ϕ'' has exactly one positive root

$$z_i = \frac{(b-2)\beta^{\frac{2}{b}}}{b}.$$

The fact that the equilibrium $x^* = \beta^{\frac{1}{b}} - 1$ loses stability at $\beta_{crit} = b^b(b-2)^{-b}$ suggests that a 2-cycle can appear, when β increases passing through this point.

The analysis of this case rests on the following observation.

Lemma 4.27 *Let g be a differentiable function on an interval $[\alpha, \gamma]$ such that $g(\alpha) > 0$, $g(\gamma) \geq 0$ and the only zero of g' is at $x = \alpha$ or at $x = \gamma$. Then $g(x) > 0$ for all $x \in [\alpha, \gamma)$.*

Proof Assume there is $\alpha < x_0 < \gamma$ where $g(x_0) \leq 0$. Contrary to the assumption, it cannot be the local minimum of g as we would have $g'(x_0) = 0$. Thus, for some $x_1 > x_0$ we have $g(x_1) < 0$ and therefore there is $x_2 \in (x_1, \gamma]$ with $g(x_2) = 0$ and, by Rolle's theorem, there is $x_3 < \gamma$ for which $g'(x_3) = 0$, again contradicting the assumption. □

Let us discuss the stable region $\beta \leq b^b(b-2)^{-b} =: R_{0,crit}$. Then $z_i \leq \beta^{\frac{1}{b}}$, that is, z_i is below the original equilibrium $\beta^{\frac{1}{b}}$; it is easy to see that, when β passes through the value $R_{0,crit}$, z_i passes through $\beta^{\frac{1}{b}}$ and moves on to take on values larger than the original equilibrium. Let us evaluate the first derivative at z_i,

$$\phi'\left(\frac{b-2}{b}\beta^{\frac{2}{b}}\right) = \beta\left(1 - \left(\frac{b-2}{b}\right)^{b-2}\beta^{\frac{(b-2)}{b}}\right).$$

Thus we see that $\phi'\left(\frac{b-2}{b}\beta^{\frac{2}{b}}\right) > 0$ provided $z_i < \beta^{\frac{1}{b}}$ and becomes negative as z_i moves through $\beta^{\frac{1}{b}}$. Further, we see that

$$\phi'(\beta^{\frac{1}{b}}) = \beta(b\beta^{-\frac{1}{b}} - (b-2))$$

and $\phi'(\beta^{\frac{1}{b}}) > 0$ provided $\beta < b^b(b-2)^{-b}$, that is, for $z_i < \beta^{\frac{1}{b}}$.

Thus, consider the case $\beta < b^b(b-2)^{-b}$. Since then $\phi'(0) = \beta > 0$, $\phi'\left(\frac{b-2}{b}\beta^{\frac{2}{b}}\right) > 0$, $\phi'(\beta^{\frac{1}{b}}) > 0$, $0 < \frac{b-2}{b}\beta^{\frac{2}{b}} < \beta^{\frac{1}{b}}$ and $\frac{b-2}{b}\beta^{\frac{2}{b}}$ is the only zero of ϕ'' on $[0, \beta^{\frac{1}{b}}]$, we can apply the lemma above (to $g = \phi'$ on the intervals $[0, z_i]$ and

$[z_i, \beta^{\frac{1}{b}}]$). Hence, ϕ' is positive on the interval $[0, \beta^{\frac{1}{b}}]$ and $\beta^{\frac{1}{b}}$ is the only zero of ϕ in this interval. Consider now the interval $[\beta^{\frac{1}{b}}, \infty)$. Since $\phi'(\beta^{\frac{1}{b}}) > 0$ and $\phi(z)$ tends to $+\infty$ for $z \to \infty$, for ϕ to have zeroes in this interval, it would have to have a local maximum, then take on 0, and then to have a local minimum before crossing the axis again to move to infinity. This would give another zero of ϕ'' between the local extrema, but then this zero would be greater than $\beta^{\frac{1}{b}}$, which is impossible.

For $\beta = b^b(b-2)^{-b}$, the points z_i and $\beta^{\frac{1}{b}}$ coalesce and, analogously, we see that there is only one zero of ϕ.

Let us consider the case with $\beta > b^b(b-2)^{-b}$. Then $\phi'(\beta^{\frac{1}{b}}) < 0$ and hence ϕ takes on negative values for $z > \beta^{\frac{1}{b}}$. Since, however, $\phi(z)$ tends to $+\infty$ for $z \to \infty$, there must be $z^* > \beta^{\frac{1}{b}}$, for which $\phi(z^*) = 0$. Also, by $\phi'(\beta^{\frac{1}{b}}) < 0$, $\phi(z) > 0$ in a left neighbourhood of $\beta^{\frac{1}{b}}$ and hence there must be another zero $1 < z^\# < \beta^{\frac{1}{b}}$, by $\phi(1) = 1 - \beta^{\frac{1}{b}} < 0$. Since $\beta^{\frac{1}{b}} - 1$ and 0 were the only equilibria of H, $x^* = z^* - 1$ and $x^\# = z^\# - 1 > 0$ must give a 2-cycle.

Figure 4.19 shows, for $b = 3$, how the point z_i moves with β through the equilibrium point $z = 3$ to produce new zeros of ϕ, giving rise to 2-cycles. With much more, mainly computer-aided, work, we can establish that, as with the logistic equation, we obtain period doubling and transition to chaos.

Experimental results are in quite good agreement with the model; see [27]. Most models fell into the stable region. On the other hand, it is obvious that a high reproductive ratio R_0 and highly over-compensating density dependence (large b) are capable of provoking periodic or chaotic fluctuations in the population density. This can be demonstrated mathematically (before the advent of the mathematical theory of chaos, it was assumed that these irregularities were of a stochastic nature) and it is observed in the fluctuations of the populations of the Colorado beetle.

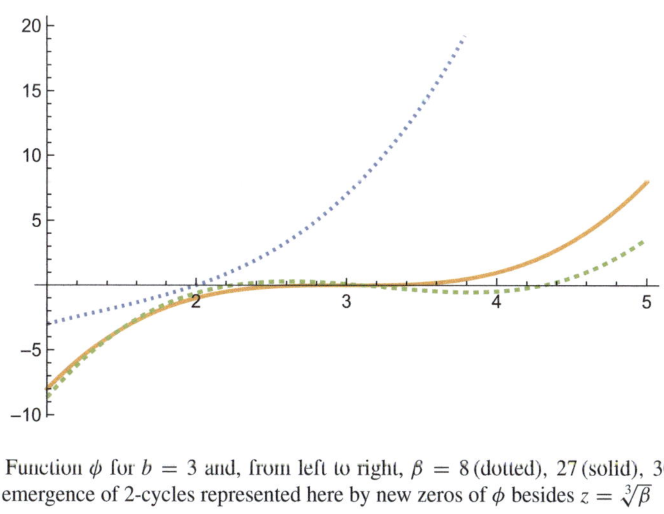

Fig. 4.19 Function ϕ for $b = 3$ and, from left to right, $\beta = 8$ (dotted), 27 (solid), 30 (dashed); notice the emergence of 2-cycles represented here by new zeros of ϕ besides $z = \sqrt[3]{\beta}$

The question of whether chaotic behaviours exist in ecology is still an area of active debate. Observational time series are always finite and inherently noisy; thus, it can be argued that a regular model can always fit these data. However, in several laboratory host-parasitoid systems, good fits were obtained between the data and chaotic mathematical models, and therefore, it is reasonable to treat these systems as chaotic.

Part II
Models with Discrete Structure

Chapter 5
Linear Models with Discrete Structure

5.1 Introducing Structure

The Malthusian model clearly is hugely oversimplified, and its improvements may go in many directions. Earlier, we discussed models still with an aggregated description but with variable and nonlinear coefficients. Another option is to consider a relevant internal structure of the population. This could be the age and a related differentiation in birth and death rates. Other possibilities include size or geographical structure, which also may impact death and birth rates. Let us start by revisiting the classical Fibonacci's problem of rabbits.

Fibonacci's Rabbits

In his famous book, *Liber Abaci*, published in 1202, Leonardo of Pisa, called Fibonacci, formulated the following problem:

> A certain man put a pair of rabbits in a place surrounded on all sides by a wall. How many rabbits can be produced from that pair in a year if it is supposed that each pair begets a new pair every month, which becomes productive from the second month on?

To fix attention, we assume that we take the monthly census just after the births for this month took place, and the rabbits were newly born at the beginning of the experiment. Usually, the problem is modelled as the initial value problem for a second-order difference equation

$$v(k+2) = v(k+1) + v(k), \quad v(0) = 1, v(1) = 1, \tag{5.1.1}$$

where $v(k)$ is the number of pairs of rabbits after k months.

Example 5.1 It is clear that (5.1.1) as a model describing a population of rabbits is oversimplified: rabbits do not die, they are always fertile as soon as they are mature, etc. However, there are biological phenomena for which (5.1.1) provides an exact

J. Banasiak, *Introduction to Mathematical Methods in Population Theory*, Springer Undergraduate Mathematics Series,
https://doi.org/10.1007/978-3-031-65491-6_5

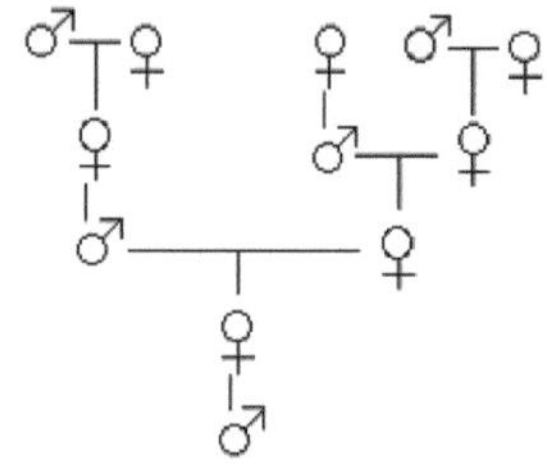

Fig. 5.1 The family tree of a drone

fit. One of them is the family tree of honeybees. Honeybees live in colonies, and one of their unusual features is that not every bee has two parents. To be more precise, let us describe a colony in more detail. First, in any colony, there is a special female called the queen. Further, there are female worker bees, but they produce no eggs. Finally, there are male drones who do no work but fertilize the queen's eggs. Drones are borne from the queen's unfertilised eggs; thus, they have a mother but no father. On the other hand, the females are born when the queen has mated with a male, and so have two parents. In Fig. 5.1, we present a family tree of a drone. It is clear that the number of ancestors kth generations earlier exactly satisfies (5.1.1).

Exercise 5.1.1

(i) Suppose that in the population of rabbits described in the Fibonacci model, each pair of rabbits gives birth to two pairs. Derive the equation governing the growth of the population, assuming that all other assumptions remain unchanged.
(ii) Suppose that in the population of rabbits described in the Fibonacci model, the rabbits mature only after two months. Derive the equation governing the growth of the population, assuming that all other assumptions remain unchanged.
(iii) Still with Fibonacci. Assume that from the first litter of each pair, only one pair survives, and from each subsequent litter, we have two pairs. Derive the equation governing the growth of the population.

Our aim here is not to reheat the classical Fibonacci equation but rather to use it to explain a more general structure.

5.1.1 *Structured Models in Discrete Time*

Fibonacci model is an example of an *age-structured population model*: in this particular case, each month, the population is represented by two classes of rabbits, adults $v_1(k)$ and juveniles $v_0(k)$. Thus the vector

$$\boldsymbol{v}(k) = \begin{pmatrix} v_0(k) \\ v_1(k) \end{pmatrix}$$

describes the state of the population at time k. Since the number of juvenile (one-month-old) pairs in month $k+1$ is equal to the number of adults in the month k and the number of adults is the number of adults from the previous month and the number of juveniles from the previous month (who became adults). In other words

$$\begin{aligned} v_0(k+1) &= v_1(k), \\ v_1(k+1) &= v_1(k) + v_0(k), \end{aligned} \tag{5.1.2}$$

or, in a more compact form,

$$\boldsymbol{v}(k+1) = \mathcal{L}\boldsymbol{v}(k) := \begin{pmatrix} 0 & 1 \\ 1 & 1 \end{pmatrix} \boldsymbol{v}(k). \tag{5.1.3}$$

The vector $\boldsymbol{v}(k)$ can be found by iterations

$$\boldsymbol{v}(k+1) = \mathcal{L}(\mathcal{L}\boldsymbol{v}(k-1)),$$

that is,

$$\boldsymbol{v}(k) = \mathcal{L}^k \boldsymbol{v}(0), \quad k = 1, 2, \ldots, . \tag{5.1.4}$$

Leslie Matrices

How do we generalize this? Assume that we are tracking only females and not pairs, and that census is taken immediately after the reproductive period (the length of which is negligible). Further, assume that there is the oldest age class n and no individual can stay in an age class for more than one time period (which is **not** the case for Fibonacci rabbits, where we allowed adults to stay adults forever). We introduce the year-to-year survival probability s_i and the age-dependent maternity function m_i, that is, s_i is the probability of survival from age i to age $i+1$ and each individual of age i produces m_i offspring on average. Thus, say, in the kth breeding season, we have $v_i(k)$ individuals of age i, s_i of them survive to the $(k+1)$th breeding season, that is, to age $i+1$, producing on average $f_i v_i(k) := m_{i+1} s_i v_i(k)$ offspring. Thus, the surviving individuals of all ages will produce

$$v_0(k+1) = \sum_{i=0}^{n-1} f_i v_i(k) = \sum_{i=0}^{n-1} m_{i+1} s_i v_i(k)$$

offspring. In this case, the evolution of the population can be described by the system of difference equations

$$\boldsymbol{v}(k+1) = \mathcal{L}\boldsymbol{v}(k),$$

where $\mathcal{L}$ is the $n \times n$ matrix

$$\mathcal{L} := \begin{pmatrix} f_0 & f_1 & \cdots & f_{n-2} & f_{n-1} \\ s_0 & 0 & \cdots & 0 & 0 \\ 0 & s_1 & \cdots & 0 & 0 \\ \vdots & \vdots & \cdots & \vdots & \vdots \\ 0 & 0 & \cdots & s_{n-2} & 0 \end{pmatrix}. \tag{5.1.5}$$

The matrix of the form (5.1.5) is referred to as a *Leslie matrix*.

A generalization of the Leslie matrix can be obtained by assuming that a fraction τ_i of the ith population stays in the same population. This gives the matrix

$$\mathcal{L} := \begin{pmatrix} f_0 + \tau_0 & f_1 & \cdots & f_{n-2} & f_{n-1} \\ s_0 & \tau_1 & \cdots & 0 & 0 \\ 0 & s_1 & \cdots & 0 & 0 \\ \vdots & \vdots & \cdots & \vdots & \vdots \\ 0 & 0 & \cdots & s_{n-2} & \tau_{n-1} \end{pmatrix}. \tag{5.1.6}$$

Such matrices are called *Usher matrices*, see, e.g., [3]. We note that the matrix of the Fibonacci process is an Usher matrix.

In most cases $f_i \neq 0$ only if $\alpha \leq i \leq \beta$ where $[\alpha, \beta]$ is the fertile period. For example, a typical mammal population has three stages: immature (pre-breeding), breeding and post-breeding. If we perform a census every year, then naturally, a fraction of each class remains in the same class. The transition matrix, in this case, is given by

$$\mathcal{L} := \begin{pmatrix} \tau_0 & f_1 & 0 \\ s_0 & \tau_1 & 0 \\ 0 & s_1 & \tau_2 \end{pmatrix}. \tag{5.1.7}$$

On the other hand, in many insect populations, reproduction occurs only in the final stage of life, and in such a case, $f_i = 0$ unless $i = n$.

Projection Matrices

Leslie matrices fit into a more general mathematical structure describing the evolution of populations divided into states or subpopulations, not necessarily age-related. Matrices resulting from such modelling, that is, describing changes in a structured population from one generation to another due to migrations between structural states and (generalized) birth processes, are called *projection matrices*. For example, we can consider clusters of cells divided into classes with respect to their size; cancer cells, divided into classes based on the number of copies of a particular gene responsible for its drug resistance, or a population divided into subpopulations depending on the geographical patch they occupy at a particular moment. Let us suppose we have n states. Each individual in a given state j contributes on average, say, a_{ij} individuals in state j per unit of time. Typically, this is due to state i individuals:

- migrating to the ith subpopulation with probability p_{ij};
- contributing to the birth of an individual in the ith subpopulation with probability b_{ij};
- dying with probability d_j, that is, surviving with probability $1 - d_j$.

It must be born in mind that the exact interpretation of the coefficients can vary depending on the way of taking the census and when the death and migrations occur in the process, see Remark 2.1.

Other choices and interpretations are, however, also possible. For instance, if we consider a size-structured population of clusters of cells divided into subpopulations according to their size i, a j-cluster can split into several smaller clusters, contributing thus to 'births' of clusters in subpopulations indexed by $i < j$. Hence, a_{ij} are non-negative numbers such that $a_{ij} = 0$ for $i \leq j$. Denoting, as before, by $v_i(k)$ the number of individuals at time k in state i, with $\boldsymbol{v}(k) = (v_1(k), \ldots, v_n(k))$, we have

$$\boldsymbol{v}(k+1) = \mathcal{A}\boldsymbol{v}(k), \tag{5.1.8}$$

where

$$\mathcal{A} := \begin{pmatrix} a_{11} & a_{12} & \cdots & a_{1\,n-1} & a_{1n} \\ 0 & a_{22} & \cdots & a_{2\,n-1} & a_{2n} \\ \vdots & \vdots & \cdots & \vdots & \vdots \\ 0 & 0 & \cdots & 0 & a_{nn} \end{pmatrix} \tag{5.1.9}$$

Thus

$$\boldsymbol{v}(k) = \mathcal{A}^k \boldsymbol{v}_0,$$

where $\boldsymbol{v}_0$ is the initial distribution of the population between the subpopulations.

Example 5.2 ([79, 92]) Any chromosome ends with a *telomere* protecting it against damage during DNA replication. Recurring cell divisions can shorten telomeres' length, and this process is considered responsible for cell ageing. If the telomere is too short, the cell cannot divide, which explains why many cell types can undergo only a finite number of divisions. Let us consider a simplified model of telomere shortening. The length of a telomere is a natural number from 0 to n, so cells with telomere of length i are in subpopulation i. A cell from subpopulation i can die with probability μ_i and divide (into two daughters). Any daughter can have a telomere of length i with probability a_i and of length $i-1$ with probability $1-a_i$. Cells of zero-length telomere cannot divide and thus will die sometime later. To find coefficients of the transition matrix, we see that the average production of an offspring with telomere of length i by a parent of the same class is

$$2a_i^2 + 2a_i(1 - a_i) = 2a_i,$$

(2 daughters with telomere of length i produced with probability a_i^2 and one daughter with telomere of length $i - 1$ produced with probability $2a_i(1 - a_i)$). Similarly, the average production of daughters with length $i-1$ telomere is $2(1-a_i)$. However, to have offspring, the cell must have survived from one census to another, which happens with probability $1 - \mu_i$. Hence, defining $r_i = 2a_i(1 - \mu_i)$ and $d_i = 2(1 - a_i)(1 - \mu_i)$, we have

$$\mathcal{A} := \begin{pmatrix} 0 & d_1 & 0 & \cdots & 0 \\ 0 & r_1 & d_2 & \cdots & 0 \\ \vdots & \vdots & \vdots & \ldots & \vdots \\ 0 & 0 & 0 & \cdots & r_n \end{pmatrix}. \tag{5.1.10}$$

The model can be modified to make it closer to reality by allowing, for instance, the shortening of telomeres by different lengths or consider models with more telomeres in a cell and with probabilities depending on the length of all of them.

Markov Matrices

A particular version of (5.1.9) is obtained when we assume that the total population has a constant size so that no individual dies and no new individual can appear so that the only changes occur due to migration between states. In other words, $b_{ij} = d_j = 0$ for any $1 \leq i, j \leq n$ and thus $a_{ij} = p_{ij}$ is the fraction of j-th subpopulation, which, on average, moves to the i-th subpopulation or, using a probabilistic language, probabilities of such a migration. Then, in addition to the constraint $p_{ij} \geq 0$, we must have $p_{ij} \leq 1$. Moreover, since the total number of individuals contributed by the state j to all other states must equal the number of individuals in this state, we must have

$$v_j = \sum_{1 \leq i \leq n} p_{ij} v_j,$$

so

$$\sum_{1 \leq i \leq n} p_{ij} = 1. \tag{5.1.11}$$

In other words, the sum of entries in each column must be equal to 1. This expresses the fact that each individual must be in one of the n states at any time.

Matrices of this form are called *Markov matrices*.

We can check that, indeed, this condition ensures that the size of the population is constant. Indeed, the size of the population at time k is $N(k) = v_1(k)+\ldots+v_n(k)$ so that

$$\begin{aligned} N(k+1) &= \sum_{1\le i\le n} v_i(k+1) = \sum_{1\le i\le n}\left(\sum_{1\le j\le n} p_{ij}v_j(k)\right) \\ &= \sum_{1\le j\le n} v_j(k)\left(\sum_{1\le i\le n} p_{ij}\right) = \sum_{1\le j\le n} v_j(k) = N(k). \end{aligned}$$

Exercise 5.1.2 Suppose a forest is composed of two species of trees, with A_k and B_k denoting the number of each species in the forest in year k. When a tree dies, a new one grows in its place but may be of a different species. Suppose that A are long living, with only 1% dying on average per year; on the other hand, an average 5% of B trees die each year. However, it is more likely that a B tree will take over a free spot; say, 75% of free spots goes to the species B and only 25% to the species A. Find the projection matrix for this process and show that the total number of trees does not change, that is, the process is a Markov process.

5.1.2 Transition Matrices for Continuous Time Processes

Let us consider a model with a population divided into n subpopulations but with continuous transitions between them. Note that this, in a natural way, excludes age-structured populations discussed earlier, as those models were constructed assuming discrete time. Continuous time age structure population models require a slightly different approach and will be considered later.

Let $v_i(t)$ denote the number of individuals in the subpopulation i at time t and consider the change in the size of this population in a small time interval Δt. Over this interval, an individual from a jth subpopulation can undergo the same processes as in the discrete case, that is,

- move to ith subpopulation with (approximate) probability $p_{ij}\Delta t$;
- contribute to the birth of an individual in ith subpopulation with probability $b_{ij}\Delta t$;
- die with probability $d_j\Delta t$.

Thus, the number of individuals in class i at time $t+\Delta t$ equals

> the number of individuals in class i at time t − the number of deaths in class i + the number of births in class i due to interactions with individuals in all other classes + the number of

individuals who migrated to class i from all other classes − the number of individuals who migrated from class i to all other classes,

or, mathematically,

$$v_i(t+\Delta t) = v_i(t) - d_i \Delta t v_i(t) + \sum_{j=1}^{n} b_{ij} \Delta t v_j(t)$$

$$+ \sum_{\substack{j=1 \\ j\neq i}}^{n} \left(p_{ij}\Delta t v_j(t) - p_{ji}\Delta t v_i(t)\right), \quad i = 1, \ldots, n. \tag{5.1.12}$$

To make the notation more compact, we denote $q_{ij} = b_{ij} + p_{ij}$ for $i \neq j$ and

$$q_{ii} = b_{ii} - d_i - \sum_{\substack{j=1 \\ j\neq i}}^{n} p_{ji}.$$

Using this notation in (5.1.12), dividing by Δt and passing to the limit with $\Delta t \to 0$, we obtain

$$v_i'(t) = \sum_{j=1}^{n} q_{ij} v_j(t), \quad i = 1, \ldots, n, \tag{5.1.13}$$

or

$$\boldsymbol{v}' = \mathcal{Q}\boldsymbol{v}, \tag{5.1.14}$$

where $\mathcal{Q} = (q_{ij})_{1\leq i,j\leq n}$.

Let us reflect on the similarities and differences between continuous and discrete time models. To simplify the discussion, we shall focus on processes with no births or deaths events: $b_{ij} = d_j = 0$ for $1 \leq i, j \leq n$. As in the discrete-time model, the total size of the population at any given time t is given by $N(t) = v_1(t) + \ldots + v_n(t)$. Then, the rate of change of N is given by

$$N'(t) = \sum_{1\leq i\leq n} v_i'(t) = \sum_{i=1}^{n}\left(\sum_{j=1}^{n} q_{ij}v_j(t)\right) = \sum_{i=1}^{n} q_{ii}v_i(t) + \sum_{i=1}^{n}\left(\sum_{j=1, j\neq i}^{n} q_{ij}v_j(t)\right)$$

$$= -\sum_{i=1}^{n} v_i(t)\left(\sum_{j=1, j\neq i}^{n} p_{ji}\right) + \sum_{i=1}^{n}\left(\sum_{j=1, j\neq i}^{n} p_{ij}v_j(t)\right)$$

$$
= -\sum_{i=1}^{n} v_i(t) \left(\sum_{j=1, j\neq i}^{n} p_{ji} \right) + \sum_{j=1}^{n} v_j(t) \left(\sum_{i=1, i\neq j}^{n} p_{ij} \right)
$$

$$
= -\sum_{i=1}^{n} v_i(t) \left(\sum_{j=1 j\neq i}^{n} p_{ji} \right) + \sum_{i=1}^{n} v_i(t) \left(\sum_{j=1, j\neq i}^{n} p_{ji} \right) = 0,
$$

where we used the fact that i, j are dummy variables and the change of order of summation can be justified as follows

$$
\sum_{i=1}^{n} \left(\sum_{\substack{j=1 \\ j\neq i}}^{n} p_{ij} v_j \right) = \sum_{i=1}^{n} \left(\sum_{j=1}^{n} p_{ij} v_j \right) - \sum_{i=1}^{n} p_{ii} v_i = \sum_{j=1}^{n} \left(\sum_{i=1}^{n} p_{ij} v_j \right) - \sum_{j=1}^{n} p_{jj} v_j
$$

$$
= \sum_{j=1}^{n} v_j \left(\sum_{i=1}^{n} p_{ij} - p_{jj} \right) = \sum_{j=1}^{n} v_j \left(\sum_{\substack{i=1 \\ i\neq j}}^{n} p_{ij} \right).
$$

Hence, $N(t) = N(0)$ for all time, and the process is conservative.

To a certain extent, a continuous process can be compared to the increments in the discrete time process, defined as

$$
\boldsymbol{v}(k+1) - \boldsymbol{v}(k) = (-I + \mathcal{P})\boldsymbol{v}(k) =: \begin{pmatrix} -1 + p_{11} & p_{12} & \cdots & p_{1n} \\ p_{21} & -1 + p_{22} & \cdots & p_{2n} \\ \vdots & \vdots & \cdots & \vdots \\ p_{n1} & p_{n2} & \cdots & -1 + p_{nn} \end{pmatrix} \boldsymbol{v}(k), \tag{5.1.15}
$$

so that the 'increment' matrix has the property that each column adds up to zero due to (5.1.11). However, it is important to remember that the coefficients p_{ij} in the continuous case are not probabilities, and thus, they do not add up to 1. In fact, they can be arbitrary numbers and represent probability rates with $p_{ij}\Delta t$ being approximate interstate transition probabilities.

5.2 Long-Term Behaviour of Structured Population Models

The main interest in population theory is to determine the long-time structure of the population.

5.2.1 *Numerical Example of the Asynchronous Exponential Growth (AEG) Property*

Before we embark on mathematical analysis, let us consider two numerical examples which indicate what we should expect from the models.

Example 5.3 Let us consider a population divided into three classes, the evolution of which is modelled by the Leslie matrix

$$\mathcal{L} = \begin{pmatrix} 2 & 1 & 1 \\ 0.5 & 0 & 0 \\ 0 & 0.4 & 0 \end{pmatrix},$$

so that the population $\boldsymbol{v} = (v_1, v_2, v_3)$ evolves according to

$$\boldsymbol{v}(k+1) = \mathcal{L}\boldsymbol{v}(k), \quad k = 0, 1, 2 \ldots,$$

or

$$\boldsymbol{v}(k) = \mathcal{L}^k \mathring{\boldsymbol{v}},$$

where $\mathring{\boldsymbol{v}}$ is an initial distribution of the population. In Fig. 5.2, we observe that each component grows very fast with k. However, if we compare growth of $v_1(k)$ with $v_2(k)$ and of $v_2(k)$ with $v_3(k)$ (see Fig. 5.3) we see that the ratios stabilize quickly around 4.5 in the first case and around 5.62 in the second case. This suggests that there is a scalar function $f(k)$ and a vector $\boldsymbol{e} = (e_1, e_2, e_3) = (25.29, 5.62, 1)$ such that for large k

$$\boldsymbol{v}(k) \approx f(k)\boldsymbol{e}. \tag{5.2.16}$$

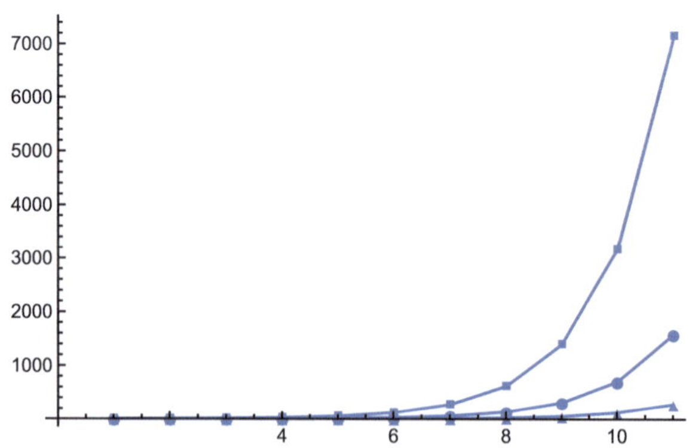

Fig. 5.2 Evolution of $v_1(k)$ (circles), $v_2(k)$ (squares) and $v_3(k)$ (rhombi) for the initial distribution $\mathring{\boldsymbol{v}} = (1, 0, 3)$ and $k = 1, \ldots, 12$, in Example 5.3

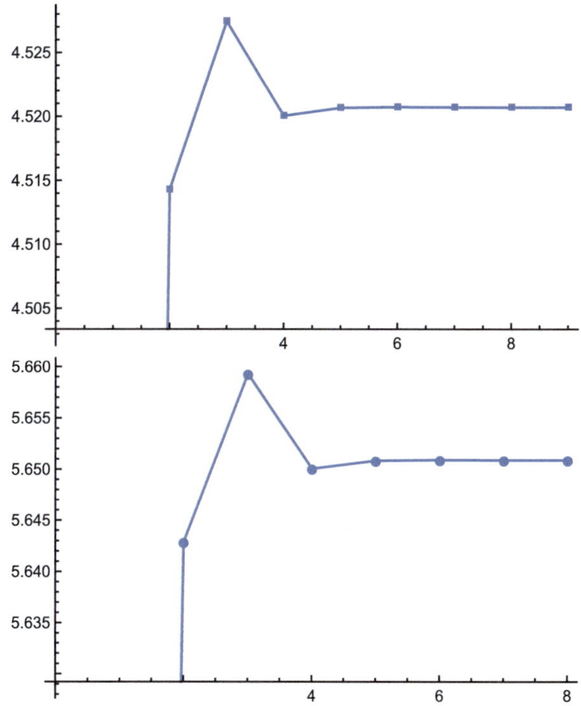

Fig. 5.3 Evolution of $v_1(k)/v_2(k)$ (top) and $v_2(k)/v_3(k)$ (bottom) for the initial distribution $\mathring{\boldsymbol{v}} = (1, 0, 3)$ and $k = 1, \ldots, 10$, in Example 5.3

Let us consider another initial condition, say, $\mathring{v} = (2, 1, 4)$ and make the same comparison; see Fig. 5.4. It turns out that the ratios stabilize at the same level, which further suggest that $\boldsymbol{e}$ does not depend on the initial condition so that (5.2.16) can be refined to

$$\boldsymbol{v}(k) \approx f_1(k) g(\mathring{\boldsymbol{v}}) \boldsymbol{e}, \quad k \to \infty \tag{5.2.17}$$

where g is a linear function. Anticipating the development of the theory, it can be proved that $f_1(k) = \lambda^k$, where λ is the largest eigenvalue of $\mathcal{L}$, $\boldsymbol{e}$ is the eigenvector corresponding to λ and $g(\boldsymbol{x}) = \boldsymbol{g} \cdot \boldsymbol{x}$ with $\boldsymbol{g}$ being the eigenvector of the transpose matrix corresponding to λ. In our case, $\lambda \approx 2.26035$ and the ratios $v_i(k)/\lambda^k$ stabilize as seen in Fig. 5.5.

The situation in which the structure of the population after long time does not depend on the initial condition but only on the intrinsic properties of the model (here, the leading eigenvalue) is called the asynchronous exponential growth (AEG) property.

Not all Leslie matrices have this property.

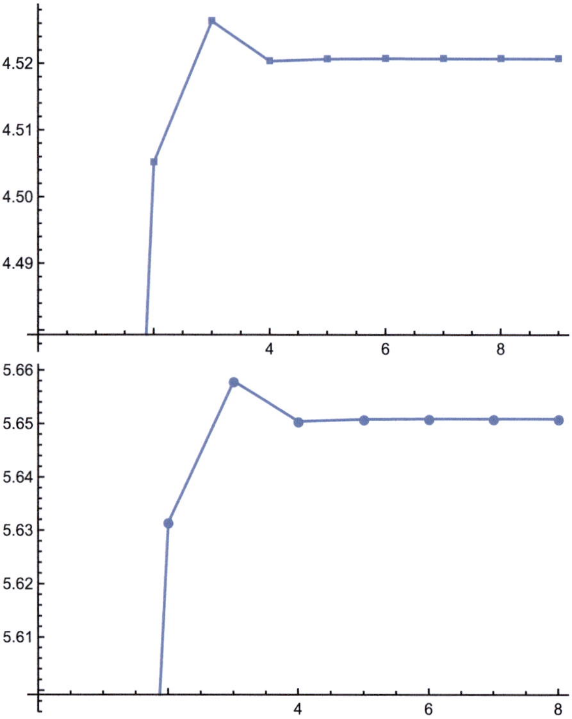

Fig. 5.4 Evolution of $v_1(k)/v_2(k)$ (top) and $v_2(k)/v_3(k)$ (bottom) for the initial distribution $\mathring{\boldsymbol{v}} = (2, 1, 4)$ and $k = 1, \ldots, 10$, in Example 5.3

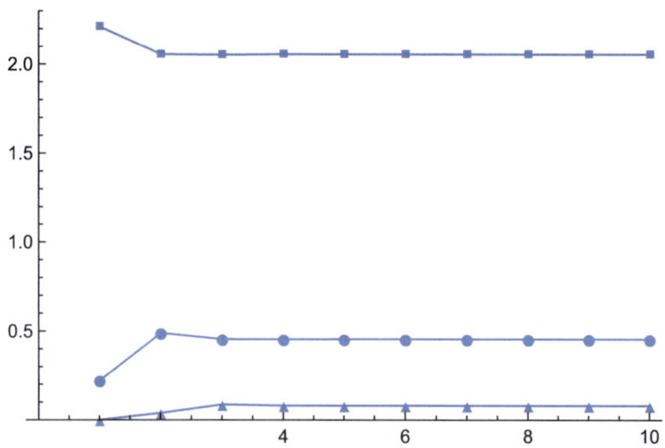

Fig. 5.5 Evolution of $v_1(k)/\lambda^k)$ (circles), $v_2(k)/\lambda^k$ (squares) and $v_3(k)/\lambda^k$ (rhombi) for the initial distribution $\mathring{\boldsymbol{v}} = (1, 0, 3)$ and $k = 1, \ldots, 10$, in Example 5.3

Example 5.4 Consider a Leslie matrix given by

$$\mathcal{L} = \begin{pmatrix} 0 & 0 & 3 \\ 0.5 & 0 & 0 \\ 0 & 0.4 & 0 \end{pmatrix},$$

and hence, a population evolving according to

$$\boldsymbol{v}(k) = \mathcal{L}^k \mathring{\boldsymbol{v}},$$

with $\mathring{\boldsymbol{v}} = (2, 3, 4)$. The solution is given in Fig. 5.6.

Fig. 5.6 Evolution of $v_1(k)$ (top) and $v_2(k)$ (middle) and $v_3(k)$ (bottom) for the initial distribution $\mathring{v} = (2, 3, 4)$ and $k = 1, \ldots, 10$, in Example 5.4

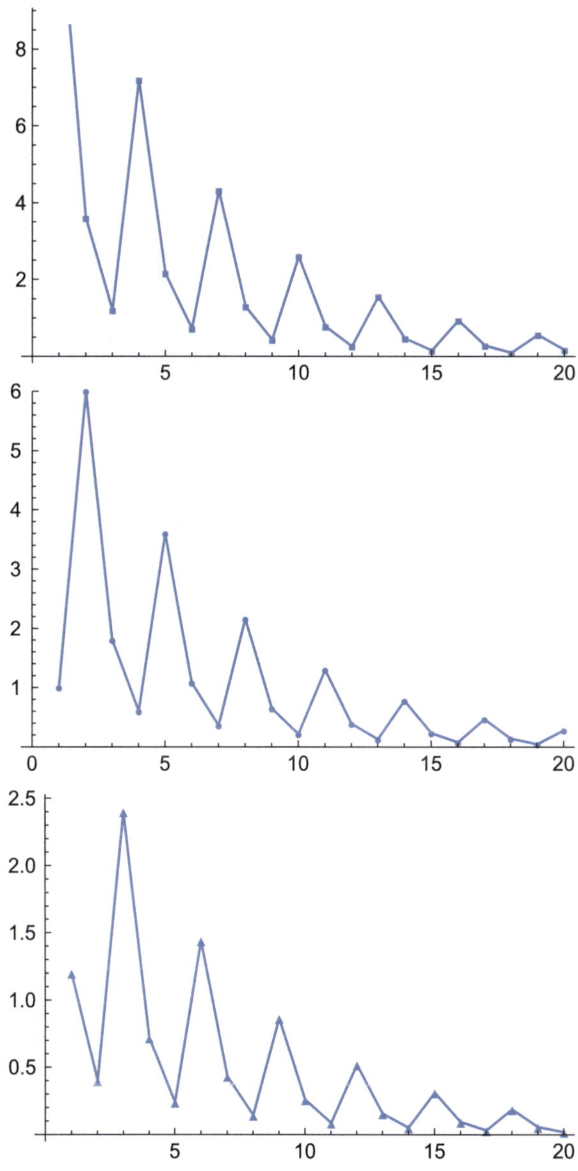

The picture differs completely from that obtained in Example 5.3. We observe some patterns, but the ratios do not tend to a fixed limit but oscillate, as shown in Fig. 5.7. As above, this can be explained using eigenvalues: indeed, the eigenvalues are given by $\lambda_1 = 0.843433$, $\lambda_2 = -0.421716 + 0.730434i$, $\lambda_2 = -0.421716 - 0.730434i$ and we can check that $|\lambda_1| = |\lambda_2| = |\lambda_3| = 0.843433$, and thus, we do not have the dominant eigenvalue. The question we will try to answer in the next chapter is what population features are responsible for such a behaviour.

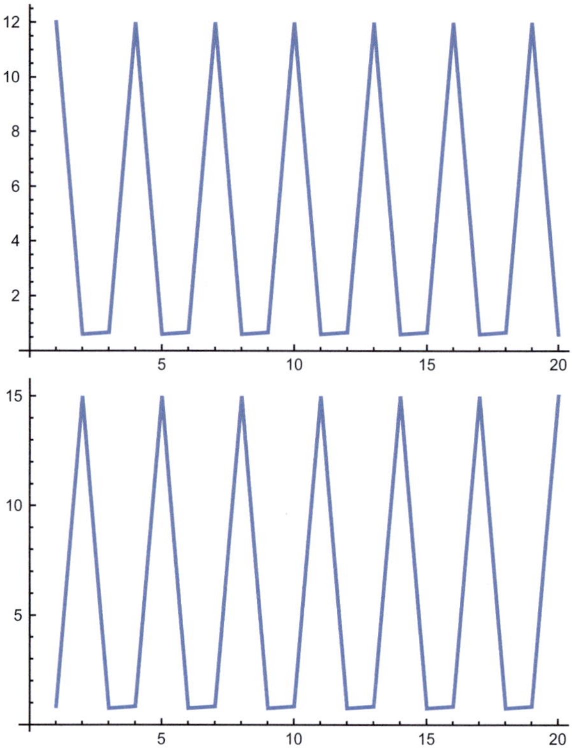

Fig. 5.7 Evolution of $v_1(k)/v_2(k)$ (top) and $v_2(k)/v_3(k)$ (bottom) for the initial distribution $\mathring{\boldsymbol{v}} = (2, 3, 4)$ and $k = 1, \ldots, 10$, in Example 5.4

The next example shows that structured population models in continuous time have similar properties.

Example 5.5 Consider the following problem

$$\boldsymbol{v}' = \mathcal{A}\boldsymbol{v}, \tag{5.2.18}$$

where

$$\mathcal{A} = \begin{pmatrix} -1 & 1 & 1 \\ 0.5 & -0.5 & 0 \\ 0 & 0.4 & -1 \end{pmatrix}.$$

We consider this equation with the initial conditions $\mathring{\boldsymbol{v}} = (1, 0, 3)$ and $\mathring{\boldsymbol{v}} = (2, 1, 4)$.

As before, we see that the components grow fast, see Fig. 5.8, but $v_1(t)/v_2(t)$ and $v_2(t)/v_3(t)$ stabilize quickly around 1.57631 and 0.970382, respectively, see Fig. 5.9, and these ratios are independent of the initial conditions. Thus,

$$\boldsymbol{v}(t) \approx f(t)g(\mathring{\boldsymbol{v}})\boldsymbol{e} \tag{5.2.19}$$

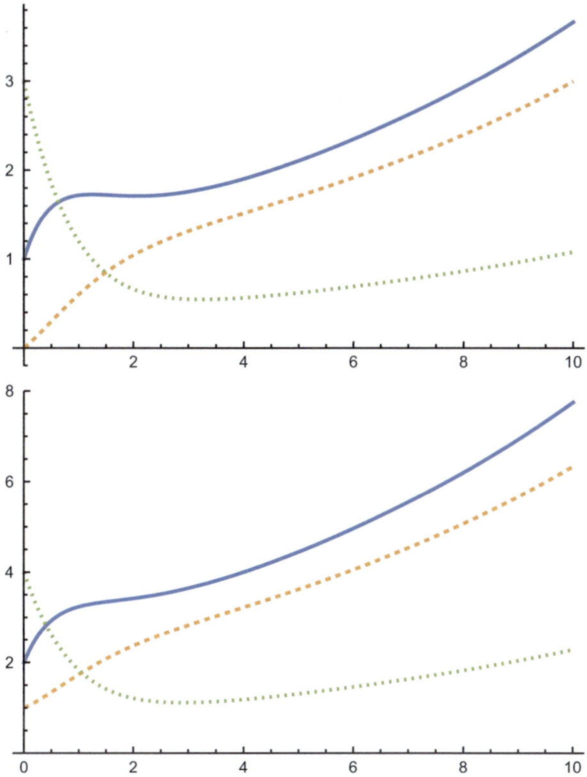

Fig. 5.8 Solutions $v_1(t)$ (continuous), $v_2(t)$ (dashed) and $v_3(t)$ (dotted) for the initial condition $\mathring{\boldsymbol{v}} = (1, 0, 3)$ (left) and $\mathring{\boldsymbol{v}} = (2, 1, 4)$ (right), in Example 5.5

for large t, where $\boldsymbol{e} = (0.755, 0.617, 0.222)$ and g is a scalar linear function of $\mathring{\boldsymbol{v}}$. As illustrated in Fig. 5.10, $f(t) = e^{0.1116t}$ and the number 0.1116 is the eigenvalue with the largest real part of $\mathcal{A}$.

Interestingly, the behaviour observed in Example 5.4 cannot occur in continuous time.

The question of whether any matrix with nonnegative entries in discrete time and with positive off-diagonal entries gives rise to such behaviour and, if not, what models lead to AEG is quite delicate and requires invoking the Frobenius–Perron theorem, which will be discussed in the next chapter. First, however, we have to review general properties of solutions to systems of difference and differential equations.

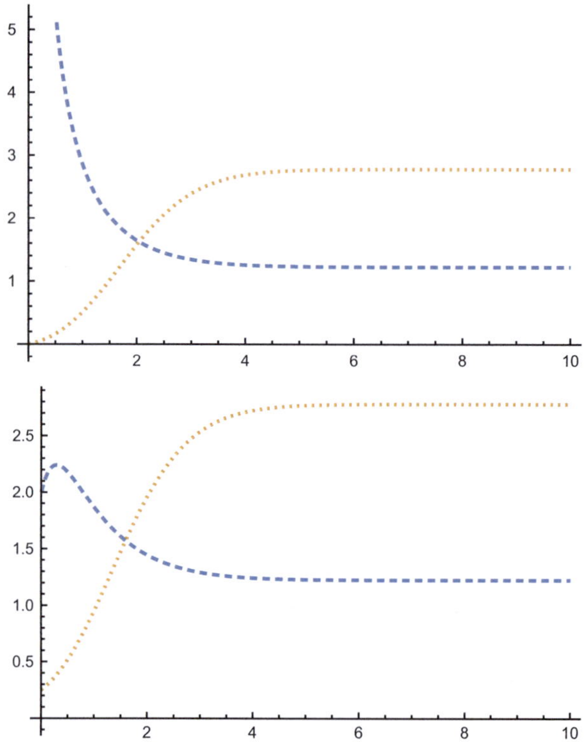

Fig. 5.9 Evolution of $v_1(t)/v_2(t)$ (dashed) and $v_2(t)/v_3(t)$ (dotted) for the initial distributions $\mathring{\boldsymbol{v}} = (1, 0, 3)$ (top) and $\mathring{\boldsymbol{v}} = (2, 1, 4)$ (bottom), in Example 5.5

5.3 Systems of Linear Differential and Difference Equations

To explain and be able to predict similar behaviour in a population models, first, we discuss basic facts about the solvability of initial value problems for systems of difference

$$\boldsymbol{y}(k+1) = \mathcal{A}\,\boldsymbol{y}(k), \quad \boldsymbol{y}(0) = \mathring{\boldsymbol{y}}, \tag{5.3.20}$$

and differential,

$$\boldsymbol{y}' = \mathcal{A}\,\boldsymbol{y}, \quad \boldsymbol{y}(0) = \mathring{\boldsymbol{y}}, \tag{5.3.21}$$

equations. Here $\boldsymbol{y} = (y_1, \ldots, y_d) \in \mathbb{R}^d$ and $\mathcal{A} = (a_{ij})_{1\leq i,j\leq d}$ is an $d \times d$ matrix. To make further progress, we have to formalize several statements made in the previous sections and, in particular, the meaning of the approximate equality (5.2.19). For this, we refer the reader to the results presented in Appendix A.

In what follows, we shall work in the d dimensional vector space $X = \mathbb{R}^d$ or $X = \mathbb{C}^d$ (even if the solutions must be real to make sense in the population context, it is often more convenient to use complex numbers to write them in a compact

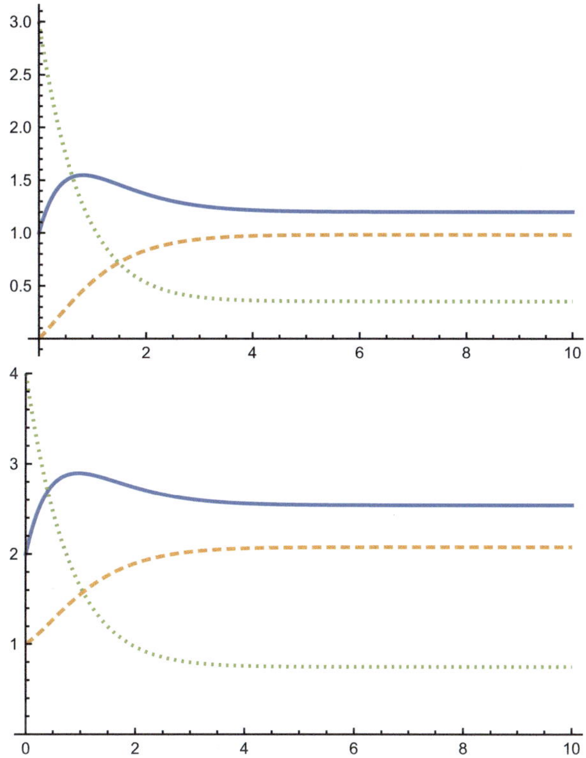

Fig. 5.10 Evolution $v_1(t)/e^{0.1116t}$ (continuous), $v_2(t)/e^{0.1116t}$ (dashed) and $v_3(t)/e^{0.1116t}$ (dotted) for the initial distributions $\mathring{\boldsymbol{v}} = (1, 0, 3)$ (top) and $\mathring{\boldsymbol{v}} = (2, 1, 4)$ (bottom), in Example 5.5

form). We use any norm $\|\cdot\|$ in X, which will be specified should the need arise. For matrices, we use the operator norm (A.2.8), which also can be specified if the norm in the underlying space X is given.

5.3.1 Systems of Difference Equations I

We are interested in solving

$$\mathbf{y}(k+1) = \mathcal{A}\mathbf{y}(k), \tag{5.3.22}$$

where $\mathcal{A}$ is an $d \times d$ matrix $\mathcal{A} = (a_{ij})_{1 \le i,j \le d}$, that is,

$$\mathcal{A} = \begin{pmatrix} a_{11} & \dots & a_{1d} \\ \vdots & & \vdots \\ a_{d1} & \dots & a_{dd} \end{pmatrix},$$

and $\mathbf{y}(k) = (y_1(k), \dots, y_d(k))$.

Equation (5.3.22) is usually supplemented by the initial condition $\mathbf{y}(0) = \mathring{\mathbf{y}}$. As noted earlier, by induction, the solution to (5.3.20) is given by

$$\mathbf{y}(k) = \mathcal{A}^k \mathring{\mathbf{y}}, \quad k = 1, 2, \ldots. \tag{5.3.23}$$

The problem with (5.3.23) is that it is rather difficult to give an explicit form of $\mathcal{A}^k$.

Since $\mathbb{R}^d$ is d-dimensional, it is enough to find d linearly independent vectors $\mathbf{v}^i$, $i = 1, \ldots, d$, for which $\mathcal{A}^k \mathbf{v}^i$ can be easily evaluated. Assume for a moment that such vectors have been found. Then, for arbitrary $\mathring{\mathbf{x}} \in \mathbb{R}^d$, we can find constants $c_1, \ldots, c_d$, such that

$$\mathring{\mathbf{x}} = c_1 \mathbf{v}^1 + \ldots + c_d \mathbf{v}^d.$$

Precisely, let $\mathcal{V}$ be the matrix having vectors $\mathbf{v}^i$ as its columns

$$\mathcal{V} = \begin{pmatrix} | & \ldots & | \\ \mathbf{v}^1 & \ldots & \mathbf{v}^d \\ | & \ldots & | \end{pmatrix}. \tag{5.3.24}$$

Note, that $\mathcal{V}$ is invertible as the vectors $\mathbf{v}^i$ are linearly independent. Denoting $\mathbf{c} = (c_1, \ldots, c_d)$, we obtain

$$\mathbf{c} = \mathcal{V}^{-1} \mathring{\mathbf{x}}. \tag{5.3.25}$$

Thus, for an arbitrary $\mathring{\mathbf{x}}$ we have

$$\mathcal{A}^k \mathring{\mathbf{x}} = \mathcal{A}^k (c_1 \mathbf{v}^1 + \ldots + c_d \mathbf{v}^d) = c_1 \mathcal{A}^k \mathbf{v}^1 + \ldots + c_d \mathcal{A}^k \mathbf{v}^d. \tag{5.3.26}$$

Now, if we denote by $\mathcal{A}_k$ the matrix whose columns are vectors $\mathcal{A}^k \mathbf{v}^1, \ldots, \mathcal{A}^k \mathbf{v}^d$, then we can write

$$\mathcal{A}^k \mathring{\mathbf{x}} = \mathcal{A}_k \mathbf{c} = \mathcal{A}_k \mathcal{V}^{-1} \mathring{\mathbf{x}}. \tag{5.3.27}$$

Hence, the problem is to find linearly independent vectors $\mathbf{v}^i$, $i = 1, \ldots, d$, on which powers of $\mathcal{A}$ can be easily evaluated. We shall use eigenvalues and eigenvectors for this purpose. Firstly, note that if $\mathbf{v}^1$ is an eigenvector of $\mathcal{A}$ corresponding to an eigenvalue λ_1, that is, $\mathcal{A}\mathbf{v}^1 = \lambda_1 \mathbf{v}^1$, then, by induction

$$\mathcal{A}^k \mathbf{v}^1 = \lambda_1^k \mathbf{v}^1.$$

Therefore, if we have d linearly independent eigenvectors $\mathbf{v}^1, \ldots, \mathbf{v}^d$, corresponding to eigenvalues $\lambda_1, \ldots, \lambda_d$ (not necessarily distinct), then from (5.3.26) we obtain

$$\mathcal{A}^k \mathring{\mathbf{x}} = c_1 \lambda_1^k \mathbf{v}^1 + \ldots + c_d \lambda_d^k \mathbf{v}^d.$$

with $c_1, \ldots, c_n$ given by (5.3.25), or

$$\mathcal{A}^k \mathring{x} = \begin{pmatrix} | & \cdots & | \\ \lambda_1^k \boldsymbol{v}^1 & \ldots & \lambda_d^k \boldsymbol{v}^d \\ | & \cdots & | \end{pmatrix} \mathcal{V}^{-1} \boldsymbol{x}. \tag{5.3.28}$$

5.3.2 *Systems of Differential Equations I*

Non-autonomous Systems

Considerations of the previous paragraph, to some extent, can be repeated for systems of differential equations. Though our main interest lies in autonomous systems

$$\boldsymbol{y}'(t) = \mathcal{A}\boldsymbol{y}(t), \tag{5.3.29}$$

where $\boldsymbol{y}(t) = (y_1(t), \ldots, y_d(t))$ and, as before, $\mathcal{A} = (a_{ij})_{1\leq i,j\leq d}$ is an $d \times d$ matrix with constant entries; occasionally, we will need some results on non-autonomous and non-homogeneous systems

$$\begin{aligned} y_1'(t) &= a_{11}(t)y_1(t) + a_{12}(t)y_2(t) + \ldots + a_{1d}(t)y_d(t) + g_1(t), \\ &\vdots \\ y_d'(t) &= a_{d1}(t)y_1(t) + a_{d2}(t)y_2(t) + \ldots + a_{dd}(t)y_d(t) + g_d(t), \end{aligned} \tag{5.3.30}$$

where $\boldsymbol{g} = (g_1, \ldots, g_d)$ is a continuous function on $\mathbb{R}$ and

$$\mathcal{A}(t) = (a_{ij}(t))_{1\leq i,j\leq d} = \begin{pmatrix} a_{11}(t) & \ldots & a_{1d}(t) \\ \vdots & \ddots & \vdots \\ a_{d1}(t) & \ldots & a_{dd}(t) \end{pmatrix}$$

is a continuous matrix function on $\mathbb{R}$. System (5.3.30) is considered together with the following initial conditions

$$\boldsymbol{y}(t_0) = \mathring{\boldsymbol{y}}. \tag{5.3.31}$$

The question of solvability and uniqueness of (5.3.30), (5.3.31) is not as easy as in the case of difference equations, but it follows from the Picard–Lindelöf theorem, Theorem B.10.

The main role is played by solutions of (5.3.30) with $\boldsymbol{g} = 0$, that is, homogeneous systems,

$$\boldsymbol{y}'(t) = \mathcal{A}(t)\boldsymbol{y}(t). \tag{5.3.32}$$

We summarize the relevant properties of solutions to (5.3.32) in the following theorem.

Theorem 5.6 ([20, 30, 33])

(i) *There exists one and only one solution to the initial value problem (5.3.32), (5.3.31), which is defined for all* $t \in \mathbb{R}$.
(ii) *The set* $\boldsymbol{X}$ *of all solutions to (5.3.32) is a linear space of dimension* d.
(iii) *If* $\mathbf{y}^1(t), \ldots, \mathbf{y}^r(t)$ *are linearly independent solutions of (5.3.32) and let* $t_0 \in \mathbb{R}$. *Then,* $\{\mathbf{y}^1(t), \ldots, \mathbf{y}^r(t)\}$ *is a linearly independent set of functions if and only if* $\{\mathbf{y}^1(t_0), \ldots, \mathbf{y}^r(t_0)\}$ *is a linearly independent set of vectors in* $\mathbb{R}^d$.

An important consequence of *(iii).* is that solutions starting from linearly independent initial conditions remain linearly independent. Note that this is not necessarily the case in systems of difference equations. This property requires $\mathcal{A}$ to be nonsingular.

Exercise 5.3.1 Let $\boldsymbol{X}$ be the vector space of all continuous functions defined on $(-\infty, \infty)$ with values in $\mathbb{R}^d$. Let $\boldsymbol{x}^1(t), \ldots, \boldsymbol{x}^k(t)$ be arbitrary functions in $\boldsymbol{X}$.

(i) Show that if $\boldsymbol{x}^1(t_0), \ldots, \boldsymbol{x}^k(t_0)$ are linearly dependent vectors in $\mathbb{R}^d$, then $\boldsymbol{x}^1(t), \ldots, \boldsymbol{x}^k(t)$ are linearly dependent functions in $\boldsymbol{X}$.
(ii) Is it true that if $\boldsymbol{x}^1(t_0), \ldots, \boldsymbol{x}^k(t_0)$ are linearly independent vectors in $\mathbb{R}^d$, then $\boldsymbol{x}^1(t), \ldots, \boldsymbol{x}^k(t)$ are linearly independent functions in $\boldsymbol{X}$? Give reasons for your answer. How does this result relate to Theorem 5.6 *(iii).*?

This theorem implies that if we consider solutions originating from d linearly independent initial vectors, these solutions form a basis for all solutions to (5.3.32). Thus, the following definition makes sense.

Definition 5.7

(i) Any basis of the space of solutions is called a fundamental set of solutions (5.3.32).
(ii) Any matrix

$$\mathcal{E}(t) = \begin{pmatrix} | & \ldots & | \\ \mathbf{y}^1(t) & \ldots & \mathbf{y}^d(t) \\ | & \ldots & | \end{pmatrix}, \tag{5.3.33}$$

whose columns are vectors from a fundamental set of solutions is called a fundamental matrix of (5.3.32).

We see that $\mathcal{E}(t)$ satisfies the matrix version of (5.3.32),

$$\mathcal{E}'(t) = \mathcal{A}(t)\mathcal{E}(t). \tag{5.3.34}$$

Proposition 5.8 *Matrix* $\mathcal{E}(t)$ *is invertible for any* $t \in \mathbb{R}$.

Proof The result follows since $\mathcal{E}(t)$ is an $d \times d$ matrix and its columns are linearly independent. □

Let $\boldsymbol{C}$ be an arbitrary vector so that $\boldsymbol{y}(t) = \mathcal{E}(t)\boldsymbol{C}$ is a solution to (5.3.32) satisfying

$$\boldsymbol{y}(t_0) = \mathcal{E}(t_0)\boldsymbol{C}.$$

If we require $\boldsymbol{y}(t_0) = \mathring{\boldsymbol{y}}$, then

$$\boldsymbol{C} = \mathcal{E}^{-1}(t_0)\mathring{\boldsymbol{y}},$$

and hence, if $\mathcal{E}(t)$ is a fundamental matrix for (5.3.32), then the solution to the Cauchy problem (5.3.32), (5.3.31) is given by

$$\boldsymbol{y}(t) = \mathcal{E}(t)\mathcal{E}^{-1}(t_0)\mathring{\boldsymbol{y}}. \tag{5.3.35}$$

Exercise 5.3.2 For each of the systems below, determine whether the given solutions are a basis for the set of solutions

(a)

$$\boldsymbol{y}'(t) = \begin{pmatrix} 0 & -1 \\ -1 & 0 \end{pmatrix} \boldsymbol{y}, \qquad \boldsymbol{y}_1(t) = \begin{pmatrix} e^t \\ e^{-t} \end{pmatrix}, \quad \boldsymbol{y}_2(t) = \begin{pmatrix} e^t \\ -e^{-t} \end{pmatrix}.$$

(b)

$$\boldsymbol{y}'(t) = \begin{pmatrix} -3 & -2 & 3 \\ 1 & 0 & -3 \\ 1 & -2 & -1 \end{pmatrix} \boldsymbol{y}, \qquad \boldsymbol{y}_1(t) = \begin{pmatrix} e^{-2t} \\ \epsilon^{-2t} \\ e^{-2t} \end{pmatrix},$$

$$\boldsymbol{y}_2(t) = \begin{pmatrix} e^{2t} \\ -e^{2t} \\ e^{2t} \end{pmatrix}, \quad \boldsymbol{y}_3(t) = \begin{pmatrix} -e^{-4t} \\ \epsilon^{-4t} \\ e^{-4t} \end{pmatrix}.$$

Let us turn our attention to the Cauchy problem for the non-homogeneous system of equations

$$\boldsymbol{y}' = \mathcal{A}(t)\boldsymbol{y} + \boldsymbol{g}(t), \tag{5.3.36}$$

$$\boldsymbol{y}(0) = \mathring{\boldsymbol{y}}. \tag{5.3.37}$$

As we noted above, the general solution to the homogeneous equation ($\boldsymbol{g}(t) \equiv 0$) is given by

$$\boldsymbol{y}_h(t) = \mathcal{E}(t)\boldsymbol{u},$$

where $\mathcal{E}(t)$ is a fundamental matrix and $\boldsymbol{u}$ is an arbitrary vector. Using the technique of variation of parameters, we will be looking for the solution in the form

$$\boldsymbol{y}(t) = \mathcal{E}(t)\boldsymbol{u}(t) = u_1(t)\boldsymbol{y}^{\boldsymbol{1}}(t) + \ldots + u_d(t)\boldsymbol{y}^{\boldsymbol{d}}(t), \tag{5.3.38}$$

where $\boldsymbol{u}(t) = (u_1(t), \ldots, u_d(t))$ is a vector-function to be determined so that (5.3.38) satisfies (5.3.36). Thus, substituting (5.3.38) into (5.3.36), we obtain

$$\mathcal{E}'(t)\boldsymbol{u}(t) + \mathcal{E}(t)\boldsymbol{u}'(t) = \mathcal{A}(t)\mathcal{E}(t)\boldsymbol{u}(t) + \boldsymbol{g}(t),$$

where we used the fact that since $\mathcal{E}(t)$ is a fundamental matrix, and hence $\mathcal{E}'(t) = \mathcal{A}(t)\mathcal{E}(t)$. Therefore,

$$\mathcal{E}(t)\boldsymbol{u}'(t) = \boldsymbol{g}(t).$$

By Proposition 5.8, $\mathcal{E}(t)$ is invertible, hence

$$\boldsymbol{u}'(t) = \mathcal{E}^{-1}(t)\boldsymbol{g}(t)$$

and thus

$$\boldsymbol{u}(t) = \int \mathcal{E}^{-1}(s)\boldsymbol{g}(s)ds + \boldsymbol{C},$$

for some constant vector $\boldsymbol{C}$. Finally, we obtain the general solution to (5.3.36) as

$$\boldsymbol{y}(t) = \mathcal{E}(t)\boldsymbol{C} + \mathcal{E}(t)\int \mathcal{E}^{-1}(s)\boldsymbol{g}(s)ds, \tag{5.3.39}$$

while the solution to the Cauchy problem (5.3.36), (5.3.37) is given by

$$\boldsymbol{y}(t) = \mathcal{E}(t)\mathcal{E}^{-1}(0)\mathring{\boldsymbol{y}} + \mathcal{E}(t)\int_0^t \mathcal{E}^{-1}(s)\boldsymbol{g}(s)ds. \tag{5.3.40}$$

If we take the solutions $\boldsymbol{y}^{\boldsymbol{i}}(t)$ satisfying $\boldsymbol{y}^{\boldsymbol{i}}(0) = \boldsymbol{e}_i = (0, \ldots, 1, \ldots, 0)$, with 1 at the ith place, then the solution $\boldsymbol{y}(t)$ of (5.3.32) satisfying $\boldsymbol{y}(0) = \mathring{\boldsymbol{y}}$ can be represented as

$$\boldsymbol{y}(t) = \mathcal{E}(t)\mathring{\boldsymbol{y}}, \tag{5.3.41}$$

which satisfies $\mathcal{E}(0) = \mathcal{I}$ (the identity matrix). The matrix $\mathcal{E}(t)$ can be found explicitly if $\mathcal{A}$ is a constant matrix, to which case we return now.

Autonomous Case

We proceed as in the discrete case by assuming that we can find d linearly independent vectors $\boldsymbol{v}^{i}$, $i = 1, \ldots, d$, for which $\mathcal{E}(t)\boldsymbol{v}^{i}$ can be easily evaluated. Then, for arbitrary $\mathring{\boldsymbol{y}} \in \mathbb{R}^{d}$ we can find constants $c_1, \ldots, c_d$, such that

$$\mathring{\boldsymbol{y}} = c_1\boldsymbol{v}^{\mathbf{1}} + \ldots + c_d\boldsymbol{v}^{\boldsymbol{d}},$$

that is, denoting $\boldsymbol{c} = (c_1, \ldots, c_d)$,

$$\boldsymbol{c} = \mathcal{V}^{-1}\mathring{\boldsymbol{x}}, \tag{5.3.42}$$

where $\mathcal{V}$ was defined in (5.3.24). Thus, for an arbitrary $\mathring{\boldsymbol{y}}$, we have

$$\mathcal{E}(t)\mathring{\boldsymbol{y}} = \mathcal{E}(t)(c_1\boldsymbol{v}^{\mathbf{1}} + \ldots + c_d\boldsymbol{v}^{\boldsymbol{d}}) = c_1\mathcal{E}(t)\boldsymbol{v}^{\mathbf{1}} + \ldots + c_d\mathcal{E}(t)\boldsymbol{v}^{\boldsymbol{d}}. \tag{5.3.43}$$

Now, if we denote by $\mathcal{E}_{\boldsymbol{v}}(t)$ the matrix whose columns are vectors $\mathcal{E}(t)\boldsymbol{v}^{\mathbf{1}}, \ldots,$ $\mathcal{E}(t)\boldsymbol{v}^{\boldsymbol{d}}$, then we can write

$$\mathcal{E}(t)\mathring{\boldsymbol{y}} = \mathcal{E}_{\boldsymbol{v}}(t)\boldsymbol{c} = \mathcal{E}_{\boldsymbol{v}}(t)\mathcal{V}^{-1}\mathring{\boldsymbol{y}}. \tag{5.3.44}$$

Hence, again, the problem lies in finding linearly independent vectors $\boldsymbol{v}^{i}$, $i = 1, \ldots, d$, on which $\mathcal{E}$ can be easily evaluated. Mimicking the scalar case, see Sect. 3.2.2, let us consider $\boldsymbol{y}(t) = e^{\lambda t}\boldsymbol{v}$ for some vector $\boldsymbol{v} \in \mathbb{R}^{d}$. Since

$$(e^{\lambda t}\boldsymbol{v})' = \lambda e^{\lambda t}\boldsymbol{v}$$

and

$$\mathcal{A}(e^{\lambda t}\boldsymbol{v}) = e^{\lambda t}\mathcal{A}\boldsymbol{v},$$

as $e^{\lambda t}$ is a scalar, $\boldsymbol{y}(t) = e^{\lambda t}\boldsymbol{v}$ is a solution to (5.3.21) if and only if

$$\mathcal{A}\boldsymbol{v} = \lambda\boldsymbol{v}, \tag{5.3.45}$$

or in other words, $\boldsymbol{y}(t) = e^{\lambda t}\boldsymbol{v}$ is a solution if and only if $\boldsymbol{v}$ is an eigenvector of $\mathcal{A}$ corresponding to the eigenvalue λ.

Thus, for each eigenvector $\boldsymbol{v}^{j}$ of $\mathcal{A}$ with eigenvalue λ_j we have a solution $\boldsymbol{y}^{j}(t) = e^{\lambda_j t}\boldsymbol{v}^{j}$. By Theorem 5.6, these solutions are linearly independent if and only if the vectors $\boldsymbol{v}^{j}$ are linearly independent in $\mathbb{R}^{d}$. Thus, if we can find d linearly independent eigenvectors of $\mathcal{A}$ with eigenvalues $\lambda_1, \ldots, \lambda_d$ (not necessarily distinct), then the general solution of (5.3.56) is of the form

$$\boldsymbol{y}(t) = c_1 e^{\lambda_1 t}\boldsymbol{v}^{\mathbf{1}} + \ldots + c_d e^{\lambda_d t}\boldsymbol{v}^{\boldsymbol{d}}. \tag{5.3.46}$$

with $c_1, \ldots, c_d$ given by (5.3.25), or

$$\mathcal{E}(t)\mathring{\mathbf{y}} = \begin{pmatrix} | & \cdots & | \\ e^{\lambda_1 t}\boldsymbol{v}^1 & \ldots & e^{\lambda_d t}\boldsymbol{v}^d \\ | & \cdots & | \end{pmatrix} \mathcal{V}^{-1}\mathring{\mathbf{y}}. \tag{5.3.47}$$

Unfortunately, in many cases, there are insufficiently many eigenvectors to generate all solutions.

5.3.3 Eigenvalues, Eigenvectors and Associated Eigenvectors

Let $\mathcal{A}$ be an $d \times d$ matrix. We say that a number λ (real or complex) is an *eigenvalue* of $\mathcal{A}$ is there exists a non-zero solution to the equation

$$\mathcal{A}\boldsymbol{v} = \lambda\boldsymbol{v}. \tag{5.3.48}$$

Such a solution is called an *eigenvector* of $\mathcal{A}$. The set of eigenvectors corresponding to a given eigenvalue is a vector subspace. Equation (5.3.48) is equivalent to the homogeneous system $(\mathcal{A} - \lambda\mathcal{I})\boldsymbol{v} = 0$, where $\mathcal{I}$ is the identity matrix, therefore λ is an eigenvalue of $\mathcal{A}$ if and only if the determinant of $\mathcal{A}$ satisfies

$$det(\mathcal{A} - \lambda\mathcal{I}) = \begin{vmatrix} a_{11} - \lambda \ldots & a_{1d} \\ \vdots & \vdots \\ a_{d1} & \ldots a_{dd} - \lambda \end{vmatrix} = 0. \tag{5.3.49}$$

Evaluating the determinant, we obtain a polynomial in λ of degree d. This polynomial is also called the *characteristic polynomial* of the system (5.3.29) (or (5.3.22)). We shall denote this polynomial by $p(\lambda)$. The Fundamental Theorem of Algebra, [50, Section V.13], states that there are exactly d, possibly complex, roots of $p(\lambda)$.

Exercise 5.3.3 Show that:

(a) of $\mathcal{A}$ is invertible, then the eigenvalues of $\mathcal{A}^{-1}$ are the reciprocals of the eigenvalues of $\mathcal{A}$,
(b) if λ is an eigenvalue of $\mathcal{A}$, then λ^n is an eigenvalue of $\mathcal{A}^n$,
(c) if μ is an eigenvalue of $\mathcal{A}^n$, then at least one $\sqrt[n]{\mu}$ is an eigenvalue of $\mathcal{A}$, and give an example that not all roots $\sqrt[n]{\mu}$ must be eigenvalues of $\mathcal{A}$.

The set of eigenvalues is called the *spectrum* of $\mathcal{A}$ and denoted by $\sigma(\mathcal{A})$. Important roles in applications are played by the spectral radius of $\mathcal{A}$, defined as

$$r(A) = \max\{|\lambda| : \lambda \in \sigma(\mathcal{A})\}, \tag{5.3.50}$$

and by the spectral bound of $\mathcal{A}$,

$$s(A) = \max\{\Re\lambda : \lambda \in \sigma(\mathcal{A})\}. \quad (5.3.51)$$

Some of the roots of the characteristic polynomial can be multiple, so that, in general, $p(\lambda)$ factorizes into

$$p(\lambda) = (\lambda_1 - \lambda)^{n_1} \cdot \ldots \cdot (\lambda_k - \lambda)^{n_k}, \quad (5.3.52)$$

with $n_1 + \ldots + n_k = d$. It is also worthwhile to note that since the coefficients of the polynomial are real, complex roots always appear in conjugate pairs, that is, if $\lambda_j = \xi_j + i\omega_j$ is a characteristic root, then so is $\bar{\lambda}_j = \xi_j - i\omega_j$. Thus, eigenvalues are the roots of the characteristic polynomial of $\mathcal{A}$. The exponent n_i appearing in the factorization (5.3.52) is called the *algebraic multiplicity* of λ_i. For each eigenvalue λ_i there corresponds an eigenvector $\boldsymbol{v}^i$ and eigenvectors corresponding to distinct eigenvalues are linearly independent. The set of all eigenvectors corresponding to λ_i spans a subspace, called the *eigenspace* corresponding to λ_i which we will denote by $\tilde{E}_{\lambda_i}$,

$$\tilde{E}_{\lambda_i} := \mathcal{L}in\{\boldsymbol{v} : \boldsymbol{v} \text{ is an eigenvector belonging to } \lambda_i\}.$$

The dimension of $\tilde{E}_{\lambda_i}$ is called the *geometric multiplicity* of λ_i. In general, algebraic and geometric multiplicities are different with geometric multiplicity being at most equal to the algebraic one. Thus, in particular, if λ_i is a single root of the characteristic polynomial, then the eigenspace corresponding to λ_i is one-dimensional. In this case, we say that λ_i is a *simple eigenvalue.* If the algebraic and geometric multiplicities of λ_i are equal, we say that it is a *semi-simple eigenvalue.* We have the following result.

Lemma 5.9 *Eigenvectors of $\mathcal{A}$ belonging to different eigenvalues are linearly independent.*

Proof We prove the lemma by induction. The statement is obvious for two eigenvectors as

$$c_1\boldsymbol{v}^1 + c_2\boldsymbol{v}^2 = 0,$$

with $|c_1| + |c_2| \neq 0$, implies that one of the eigenvectors is a scalar multiple of the other, and thus it is an eigenvector belonging to the same eigenvalue. Assume now that

$$\boldsymbol{v}^{i_k} = c_1\boldsymbol{v}^{i_1} + \cdots c_{k-1}\boldsymbol{v}^{i_{k-1}} \quad (5.3.53)$$

for some selection of eigenvectors and not simultaneously vanishing constants. Multiplying both sides by $\mathcal{A}$, we get

$$\lambda_{i_k} \boldsymbol{v}^{i_k} = \lambda_{i_1} c_1 \boldsymbol{v}^{i_1} + \cdots \lambda_{i_{k-1}} c_{k-1} \boldsymbol{v}^{i_{k-1}}. \tag{5.3.54}$$

If $\lambda_{i_k} = 0$, then the eigenvalues on the right-hand side are different from zero, and hence the set $\{\boldsymbol{v}^{i_1}, \ldots, \boldsymbol{v}^{i_{k-1}}\}$ is linearly dependent. If $\lambda_{i_k} \neq 0$, then dividing both sides of (5.3.54) by λ_{i_k} and subtracting from (5.3.53), we get

$$0 = (1 - \lambda_{i_1}\lambda_{i_k}^{-1}) c_1 \boldsymbol{v}^{i_1} + \cdots (1 - \lambda_{i_{k-1}}\lambda_{i_k}^{-1}) c_{k-1} \boldsymbol{v}^{i_{k-1}}.$$

Since the eigenvalues are different, we again get linear dependence of $\{\boldsymbol{v}^{i_1}, \ldots, \boldsymbol{v}^{i_{k-1}}\}$. Thus, using induction, we reach a contradiction. □

Using this lemma, we see that if the geometric multiplicities of eigenvalues add up to d, then we have d linearly independent eigenvectors, and these eigenvectors form a basis for $\mathbb{R}^d$. Hence, (5.3.47) gives the general solution to (5.3.29). In particular, this happens if all eigenvalues are single roots of the characteristic polynomial.

If this is not the case, then we do not have sufficiently many eigenvectors to span $\mathbb{R}^d$ and if we need a basis for $\mathbb{R}^d$, then we have to find additional linearly independent vectors. A procedure that can be employed here, and that will be very useful in our treatment of systems of difference and differential equations, is to find solutions to equations of the form $(\mathcal{A} - \lambda_i \mathcal{I})^k \boldsymbol{v} = 0$ for $1 < k \leq n_i$, where n_i is the algebraic multiplicity of λ_i. Precisely speaking, if λ_i has algebraic multiplicity n_i and if

$$(\mathcal{A} - \lambda_i \mathcal{I})\boldsymbol{v} = 0$$

has only $\nu_i < n_i$ linearly independent solutions, then we consider the equations

$$(\mathcal{A} - \lambda_i \mathcal{I})^j \boldsymbol{v} = 0$$

for $j \geq 1$. It is clear that the solutions of these equations form an increasing set, and, by Theorem C.4, the process terminates at worst at

$$(\mathcal{A} - \lambda_i \mathcal{I})^{n_i} \boldsymbol{v} = 0,$$

with n_i linearly independent solutions. In practice, we first find linearly independent eigenvectors, and if there are fewer than n_i of them, we look for solutions to $(\mathcal{A} - \lambda_i \mathcal{I})^2 \boldsymbol{v} = 0$, selecting the ones that are not eigenvectors (then they must be linearly independent of eigenvectors) and continue with higher powers until we find n_i linearly independent solutions. In each step, say j, we select solutions that are independent of the solutions obtained in step $j - 1$; for this it is enough to find solutions to $(\mathcal{A} - \lambda_i \mathcal{I})^j \boldsymbol{v} = 0$ that satisfy $(\mathcal{A} - \lambda_i \mathcal{I})^{j-1} \boldsymbol{v} \neq 0$.

Vectors $\boldsymbol{v}$ obtained in this way for a given λ_i are called *generalized* or *associated eigenvectors* corresponding to λ_i and they span an n_i dimensional subspace called a *generalized* or *associated eigenspace* corresponding to λ_i, denoted hereafter by E_{λ_i}. Using Theorem C.4 again, we see that Lemma 5.9 is valid also for associated eigenvectors, that is, associated eigenvectors belonging to different eigenvalues are linearly independent (see Theorem 5.21 for a refinement of this result).

Now, we show how to apply the concepts discussed above to solve systems of difference and differential equations.

5.3.4 Systems of Difference Equations II

Let us return to the system

$$\boldsymbol{y}(k+1) = \mathcal{A}\boldsymbol{y}(k), \quad \boldsymbol{y}(0) = \mathring{\boldsymbol{y}}.$$

As discussed, we need to find formulae for $\mathcal{A}^k\boldsymbol{v}$ for a selected d linearly independent vectors $\boldsymbol{v}$. Let us take as these independent vectors the collection of all eigenvectors and associated eigenvectors of $\mathcal{A}$. We know that if $\boldsymbol{v}^{\boldsymbol{i}}$ is an eigenvector associated to an eigenvalue λ^i, then $\mathcal{A}^k\boldsymbol{v}^{\boldsymbol{i}} = \lambda_i^k\boldsymbol{v}^{\boldsymbol{i}}$. Thus, the question is whether $\mathcal{A}^k$ can be effectively evaluated on associated eigenvectors.

Let $\boldsymbol{v}^{\boldsymbol{j}}$ be an associated eigenvector found as a solution to $(\mathcal{A} - \lambda_i\mathcal{I})^j\boldsymbol{v}^{\boldsymbol{j}} = 0$ with $j \leq n_i$. Then, using the binomial expansion, we find

$$\begin{aligned}\mathcal{A}^k\boldsymbol{v}^{\boldsymbol{j}} &= (\lambda_i\mathcal{I} + \mathcal{A} - \lambda_i\mathcal{I})^k\boldsymbol{v}^{\boldsymbol{j}} = \sum_{r=0}^{k}\lambda_i^{k-r}\binom{k}{r}(\mathcal{A} - \lambda_i\mathcal{I})^r\boldsymbol{v}^{\boldsymbol{j}} \\ &= \Big(\lambda_i^k\mathcal{I} + k\lambda_i^{k-1}(\mathcal{A} - \lambda_i\mathcal{I}) + \ldots \\ &\quad + \frac{k!}{(j-1)!(k-j+1)!}\lambda_i^{k-j+1}(\mathcal{A} - \lambda_i\mathcal{I})^{j-1}\Big)\boldsymbol{v}^{\boldsymbol{j}},\end{aligned} \tag{5.3.55}$$

where

$$\binom{k}{r} = \frac{k!}{r!(k-r)!}$$

is the Newton symbol. It is important to note that (5.3.55) is a finite sum for any k; it terminates at most at the term $(\mathcal{A} - \lambda_1\mathcal{I})^{n_i-1}$, where n_i is the algebraic multiplicity of λ_i.

We shall illustrate these considerations with several examples.

Example 5.10 Find $\mathcal{A}^k$ for

$$\mathcal{A} = \begin{pmatrix} 4 & 1 & 2 \\ 0 & 2 & -4 \\ 0 & 1 & 6 \end{pmatrix}.$$

We start with finding eigenvalues of $\mathcal{A}$. The characteristic equation

$$p(\lambda) = \begin{vmatrix} 4-\lambda & 1 & 2 \\ 0 & 2-\lambda & -4 \\ 0 & 1 & 6-\lambda \end{vmatrix} = (4-\lambda)(16-8\lambda+\lambda^2) = (4-\lambda)^3 = 0$$

gives the eigenvalue $\lambda = 4$ of algebraic multiplicity 3. To find eigenvectors corresponding to $\lambda = 4$, we solve

$$(\mathcal{A} - 4\mathcal{I})\boldsymbol{v} = \begin{pmatrix} 0 & 1 & 2 \\ 0 & -2 & -4 \\ 0 & 1 & 2 \end{pmatrix} \begin{pmatrix} v_1 \\ v_2 \\ v_3 \end{pmatrix} = \begin{pmatrix} 0 \\ 0 \\ 0 \end{pmatrix}.$$

Thus, v_1 is arbitrary and $v_2 = -2v_3$ so that the eigenspace is two-dimensional, spanned by

$$\boldsymbol{v}^1 = \begin{pmatrix} 1 \\ 0 \\ 0 \end{pmatrix}, \qquad \boldsymbol{v}^2 = \begin{pmatrix} 0 \\ -2 \\ 1 \end{pmatrix}.$$

Therefore

$$\mathcal{A}^k \boldsymbol{v}^1 = 4^k \begin{pmatrix} 1 \\ 0 \\ 0 \end{pmatrix}, \qquad \mathcal{A}^k \boldsymbol{v}^2 = 4^k \begin{pmatrix} 0 \\ -2 \\ 1 \end{pmatrix}.$$

To find the associated eigenvector, we consider

$$(\mathcal{A} - 4\mathcal{I})^2\boldsymbol{v} = \begin{pmatrix} 0 & 1 & 2 \\ 0 & -2 & -4 \\ 0 & 1 & 2 \end{pmatrix} \begin{pmatrix} 0 & 1 & 2 \\ 0 & -2 & -4 \\ 0 & 1 & 2 \end{pmatrix} \begin{pmatrix} v_1 \\ v_2 \\ v_3 \end{pmatrix} = \begin{pmatrix} 0 & 0 & 0 \\ 0 & 0 & 0 \\ 0 & 0 & 0 \end{pmatrix} \begin{pmatrix} v_1 \\ v_2 \\ v_3 \end{pmatrix} = \begin{pmatrix} 0 \\ 0 \\ 0 \end{pmatrix}.$$

Any vector solves this equation, so we have to take a vector that is not an eigenvector. Possibly the simplest choice is

$$\boldsymbol{v}^3 = \begin{pmatrix} 0 \\ 0 \\ 1 \end{pmatrix}.$$

Thus, by (5.3.55)

$$
\begin{aligned}
\mathcal{A}^k \boldsymbol{v}^3 &= \left(4^k \mathcal{I} + k4^{k-1}(\mathcal{A} - 4\mathcal{I})\right) \boldsymbol{v}^3 \\
&= \left(\begin{pmatrix} 4^k & 0 & 0 \\ 0 & 4^k & 0 \\ 0 & 0 & 4^k \end{pmatrix} + k4^{k-1} \begin{pmatrix} 0 & 1 & 2 \\ 0 & -2 & -4 \\ 0 & 1 & 2 \end{pmatrix} \right) \begin{pmatrix} 0 \\ 0 \\ 1 \end{pmatrix} = \begin{pmatrix} 2k4^{k-1} \\ -k4^k \\ 4^k + 2k4^{-1} \end{pmatrix}.
\end{aligned}
$$

To find an explicit expression for $\mathcal{A}^k$, we use (5.3.27). In our case

$$
\mathcal{A}_k = \begin{pmatrix} 4^k & 0 & 2k4^{k-1} \\ 0 & -2 \cdot 4^k & -k4^k \\ 0 & 4^k & 4^k + 2k4^{k-1} \end{pmatrix}.
$$

Further,

$$
\mathcal{V} = \begin{pmatrix} 1 & 0 & 0 \\ 0 & -2 & 0 \\ 0 & 1 & 1 \end{pmatrix},
$$

so that

$$
\mathcal{V}^{-1} = \begin{pmatrix} 1 & 0 & 0 \\ 0 & -\frac{1}{2} & 0 \\ 0 & \frac{1}{2} & 1 \end{pmatrix}.
$$

Therefore

$$
\mathcal{A}^k = \mathcal{A}_k \mathcal{V}^{-1} = \begin{pmatrix} 4^k & k4^{k-1} & 2k4^{k-1} \\ 0 & 4^k - 2k4^{k-1} & -k4^k \\ 0 & k4^{k-1} & 4^k + 2k4^{k-1} \end{pmatrix}.
$$

The next example shows how to deal with complex eigenvalues. We recall that if $\lambda = \xi + i\omega$ is a complex eigenvalue, then also its complex conjugate $\bar{\lambda} = \xi - i\omega$ is an eigenvalue, as the characteristic polynomial $p(\lambda)$ has real coefficients. Eigenvectors $\boldsymbol{v}$ corresponding to a complex eigenvalue λ will be complex vectors, that is, vectors with complex entries. Thus, we can write

$$
\boldsymbol{v} = \begin{pmatrix} v_1^1 + i v_1^2 \\ \vdots \\ v_n^1 + i v_n^2 \end{pmatrix} = \begin{pmatrix} v_1^1 \\ \vdots \\ v_n^1 \end{pmatrix} + i \begin{pmatrix} v_1^2 \\ \vdots \\ v_n^2 \end{pmatrix} = \Re \boldsymbol{v} + i \Im \boldsymbol{v}.
$$

Since $(\mathcal{A} - \lambda\mathcal{I})\boldsymbol{v} = \mathbf{0}$, taking complex conjugate of both sides and using the fact that matrices $\mathcal{A}$ and $\mathcal{I}$ have only real entries, we see that

$$\overline{(\mathcal{A} - \lambda\mathcal{I})\boldsymbol{v}} = (\mathcal{A} - \bar{\lambda}\mathcal{I})\bar{\boldsymbol{v}} = \mathbf{0}$$

so that the complex conjugate $\bar{\boldsymbol{v}}$ of the eigenvector $\boldsymbol{v}$ is an eigenvector corresponding to the eigenvalue $\bar{\lambda}$. Since $\lambda \neq \bar{\lambda}$, as we assumed that λ is complex, the eigenvectors $\boldsymbol{v}$ and $\bar{\boldsymbol{v}}$ are linearly independent, and thus we obtain two linearly independent complex-valued solutions $\lambda^k\boldsymbol{v}$ and $\bar{\lambda}^k\bar{\boldsymbol{v}}$. Since taking real and imaginary parts are linear operations,

$$\Re(\lambda^k\boldsymbol{v}) = \frac{\lambda^k\boldsymbol{v} + \bar{\lambda}^k\bar{\boldsymbol{v}}}{2}, \qquad \Im(\lambda^k\boldsymbol{v}) = \frac{\lambda^k\boldsymbol{v} - \bar{\lambda}^k\bar{\boldsymbol{v}}}{2i},$$

both $\Re(\lambda^k\boldsymbol{v})$ and $\Im(\lambda^k\boldsymbol{v})$ are real valued solutions. To find explicit expressions for them, we write $\lambda = re^{i\phi}$ where $r = |\lambda|$ and $\phi = \arg\lambda$. Then

$$\lambda^k = r^k e^{ik\phi} = r^k(\cos k\phi + i\sin k\phi)$$

and

$$\begin{aligned}\Re(\lambda^k\boldsymbol{v}) &= r^k(\cos k\phi\Re\boldsymbol{v} - \sin k\phi\Im\boldsymbol{v}),\\ \Im(\lambda^k\boldsymbol{v}) &= r^k(\sin k\phi\Re\boldsymbol{v} + \cos k\phi\Im\boldsymbol{v}).\end{aligned}$$

Example 5.11 Find $\mathcal{A}^k$ if

$$\mathcal{A} = \begin{pmatrix} 1 & -5 \\ 1 & -1 \end{pmatrix}.$$

We have

$$\begin{vmatrix} 1-\lambda & -5 \\ 1 & -1-\lambda \end{vmatrix} = \lambda^2 + 4$$

so that $\lambda_{1,2} = \pm 2i$. Taking $\lambda_1 = 2i$, we find the corresponding eigenvector by solving

$$\begin{pmatrix} 1-2i & -5 \\ 1 & -1-2i \end{pmatrix}\begin{pmatrix} v_1 \\ v_2 \end{pmatrix} = \begin{pmatrix} 0 \\ 0 \end{pmatrix}.$$

Thus

$$\boldsymbol{v}^1 = \begin{pmatrix} 1+2i \\ 1 \end{pmatrix}$$

and

$$x(k) = \mathcal{A}^k v^1 = (2i)^k \begin{pmatrix} 1+2i \\ 1 \end{pmatrix}.$$

To find real-valued solutions, we take real and imaginary parts of $x(k)$. Since $i = \cos\frac{\pi}{2} + i\sin\frac{\pi}{2}$ we have, by de Moivre's formula,

$$(2i)^k = 2^k \left(\cos\frac{\pi}{2} + i\sin\frac{\pi}{2}\right)^k = 2^k \left(\cos\frac{k\pi}{2} + i\sin\frac{k\pi}{2}\right).$$

Therefore

$$\Re x(k) = 2^k \left(\cos\frac{k\pi}{2}\begin{pmatrix}1\\1\end{pmatrix} - \sin\frac{k\pi}{2}\begin{pmatrix}2\\0\end{pmatrix}\right),$$

$$\Im x(k) = 2^k \left(\cos\frac{k\pi}{2}\begin{pmatrix}2\\0\end{pmatrix} + \sin\frac{k\pi}{2}\begin{pmatrix}1\\1\end{pmatrix}\right).$$

The initial values for $\Re x(k)$ and $\Im x(k)$ are, respectively, $\begin{pmatrix}1\\1\end{pmatrix}$ and $\begin{pmatrix}2\\0\end{pmatrix}$. Since $\mathcal{A}^k$ is a real matrix, we have $\Re\mathcal{A}^k v^1 = \mathcal{A}^k \Re v^1$ and $\Im\mathcal{A}^k v^1 = \mathcal{A}^k \Im v^1$, thus

$$\mathcal{A}^k\begin{pmatrix}1\\1\end{pmatrix} = 2^k\left(\cos\frac{k\pi}{2}\begin{pmatrix}1\\1\end{pmatrix} - \sin\frac{k\pi}{2}\begin{pmatrix}2\\0\end{pmatrix}\right) = 2^k\begin{pmatrix}\cos\frac{k\pi}{2} - 2\sin\frac{k\pi}{2}\\ \cos\frac{k\pi}{2}\end{pmatrix}$$

and

$$\mathcal{A}^k\begin{pmatrix}2\\0\end{pmatrix} = 2^k\left(\cos\frac{k\pi}{2}\begin{pmatrix}2\\0\end{pmatrix} + \sin\frac{k\pi}{2}\begin{pmatrix}1\\1\end{pmatrix}\right) = 2^k\begin{pmatrix}2\cos\frac{k\pi}{2} + \sin\frac{k\pi}{2}\\ \sin\frac{k\pi}{2}\end{pmatrix}.$$

To find $\mathcal{A}^k$, we use again (5.3.27). In our case

$$\mathcal{A}_k = 2^k\begin{pmatrix}\cos\frac{k\pi}{2} - 2\sin\frac{k\pi}{2} & 2\cos\frac{k\pi}{2} + \sin\frac{k\pi}{2}\\ \cos\frac{k\pi}{2} & \sin\frac{k\pi}{2}\end{pmatrix}.$$

Further

$$\mathcal{V} = \begin{pmatrix}1 & 2\\ 1 & 0\end{pmatrix},$$

so that

$$\mathcal{V}^{-1} = -\frac{1}{2}\begin{pmatrix}0 & -2\\ -1 & 1\end{pmatrix}.$$

Therefore

$$\mathcal{A}^k = \mathcal{A}_k \mathcal{V}^{-1} = -2^{k-1} \begin{pmatrix} -2\cos\frac{k\pi}{2} - \sin\frac{k\pi}{2} & 5\sin\frac{k\pi}{2} \\ -\sin\frac{k\pi}{2} & -2\cos\frac{k\pi}{2} + \sin\frac{k\pi}{2} \end{pmatrix}.$$

Exercise 5.3.4 Find general solutions to the following systems of difference equations

$$\begin{pmatrix} x_1(k+1) \\ x_2(k+1) \\ x_3(k+1) \end{pmatrix} = \begin{pmatrix} 1 & -2 & -2 \\ 0 & 0 & -1 \\ 0 & 2 & 3 \end{pmatrix} \begin{pmatrix} x_1(k) \\ x_2(k) \\ x_3(k) \end{pmatrix},$$

$$\begin{pmatrix} x_1(k+1) \\ x_2(k+1) \\ x_3(k+1) \end{pmatrix} = \begin{pmatrix} \frac{3}{2} & \frac{1}{2} & \frac{1}{2} \\ \frac{1}{2} & \frac{5}{2} & -\frac{1}{2} \\ 0 & 1 & 2 \end{pmatrix} \begin{pmatrix} x_1(k) \\ x_2(k) \\ x_3(k) \end{pmatrix},$$

$$\begin{pmatrix} x_1(k+1) \\ x_2(k+1) \\ x_3(k+1) \end{pmatrix} = \begin{pmatrix} 1 & 1 & 0 \\ -1 & 1 & 0 \\ 1 & 0 & 1 \end{pmatrix} \begin{pmatrix} x_1(k) \\ x_2(k) \\ x_3(k) \end{pmatrix}.$$

5.3.5 *Systems of Differential Equations II*

Let us return to the system

$$\boldsymbol{y}'(t) = \mathcal{A}\boldsymbol{y}(t), \quad \boldsymbol{y}(0) = \mathring{\boldsymbol{y}}. \tag{5.3.56}$$

As before, we aim to find d linearly independent solutions of (5.3.56). We do not have a natural expression for the solution matrix $\mathcal{E}(t)$ as for the difference system. If all eigenvalues are simple, then we have a sufficient number of eigenvectors to define $\mathcal{E}(t)$ by (5.3.47). The same formula is valid if multiple eigenvalues exist, but each eigenvalue's algebraic and geometric multiplicities are the same. However, it still remains to find a formula for $\mathcal{E}(t)$, when $\mathcal{A}$ has less than d linearly independent eigenvectors.

Recall that for a single equation $y' = ay$, where a is a constant, the general solution is given by $y(t) = e^{at}C$, where C is a constant. Similarly, we would like to say that the general solution to (5.3.56) is $\boldsymbol{y} = e^{\mathcal{A}t}\boldsymbol{v}$, where $\boldsymbol{v}$ is any constant vector in $\mathbb{R}^d$. The problem is that we do not know what it means to evaluate the exponential of a matrix. However, let us recall that the exponential of a number can be evaluated as the power (Maclaurin) series

$$e^x = 1 + x + \frac{x^2}{2} + \frac{x^3}{3!} + \cdots + \frac{x^k}{k!} + \cdots,$$

where the only involved operations on the argument x are additions, scalar multiplications and taking integer powers. Then, as explained in detail in Sect. A.3, we can define

$$e^{\mathcal{A}} = \mathcal{I} + \mathcal{A} + \frac{1}{2}\mathcal{A}^2 + \frac{1}{3!}\mathcal{A}^3 + \cdots + \frac{1}{k!}\mathcal{A}^k + \cdots, \tag{5.3.57}$$

and the series above converges absolutely and thus has most of the properties of a classical power series. We note that we can only use the infinite sum to evaluate $e^{\mathcal{A}}$ in very simple cases. For example, if we take

$$\mathcal{A} = \begin{pmatrix} \lambda & 0 & 0 \\ 0 & \lambda & 0 \\ 0 & 0 & \lambda \end{pmatrix} = \lambda\mathcal{I},$$

then

$$\mathcal{A}^k = \lambda^k \mathcal{I}^k = \lambda^k \mathcal{I},$$

and

$$\begin{aligned} e^{\lambda\mathcal{I}} &= \mathcal{I} + \lambda\mathcal{I} + \frac{\lambda^2}{2}\mathcal{I} + \frac{\lambda^3}{3!}\mathcal{I} + \cdots + \frac{\lambda^k}{k!} + \cdots \\ &= \left(1 + \lambda + \frac{\lambda^2}{2} + \frac{\lambda^3}{3!} + \cdots + \frac{\lambda^k}{k!} + \cdots\right)\mathcal{I} = e^{\lambda}\mathcal{I}. \end{aligned}$$

Unfortunately, in most cases, finding the explicit form for $e^{\mathcal{A}}$ directly is very difficult. Nevertheless, we will be able to avoid dealing with the infinite series by showing that, in fact, the infinite sum in (5.3.57) can always be replaced by a finite one.

To justify the algebraic manipulations below, we note the following algebraic properties of matrix exponentials. First,

$$e^{\mathcal{A}+\mathcal{B}} = e^{\mathcal{A}}e^{\mathcal{B}} \tag{5.3.58}$$

provided the matrices $\mathcal{A}$ and $\mathcal{B}$ commute, $\mathcal{A}\mathcal{B} = \mathcal{B}\mathcal{A}$. Then, taking $\mathcal{B} = -\mathcal{A}$,

$$\left(e^{\mathcal{A}}\right)^{-1} = e^{-\mathcal{A}}.$$

Furthermore, defining a function of t by

$$e^{t\mathcal{A}} = \mathcal{I} + t\mathcal{A} + \frac{t^2}{2}\mathcal{A}^2 + \frac{t^3}{3!}\mathcal{A}^3 + \cdots + \frac{t^k}{k!}\mathcal{A}^k + \cdots, \tag{5.3.59}$$

and, differentiating it with respect to t we find, see (A.3.12),

$$\frac{d}{dt}e^{t\mathcal{A}} = e^{t\mathcal{A}}\mathcal{A},$$

proving thus that $\boldsymbol{y}(t) = e^{t\mathcal{A}}\boldsymbol{v}$ is a solution to our system of equations for any constant vector $\boldsymbol{v}$.

As mentioned, it is generally difficult to directly find the explicit form of $e^{t\mathcal{A}}$. However, we can always find d linearly independent vectors $\boldsymbol{v}$ for which the series $e^{t\mathcal{A}}\boldsymbol{v}$ is finite. This is based on the following two observations. First, since $\lambda\mathcal{I}$ and $\mathcal{A} - \lambda\mathcal{I}$ commute, by (5.3.58) we have

$$e^{t\mathcal{A}}\boldsymbol{v} = e^{t(\mathcal{A}-\lambda\mathcal{I})}e^{t\lambda\mathcal{I}}\boldsymbol{v} = e^{\lambda t}e^{t(\mathcal{A}-\lambda\mathcal{I})}\boldsymbol{v},$$

for any $\boldsymbol{v} \in \mathbb{R}^d$. Second, if $(\mathcal{A} - \lambda\mathcal{I})^m\boldsymbol{v} = \mathbf{0}$ for some m, then

$$(\mathcal{A} - \lambda\mathcal{I})^r\boldsymbol{v} = 0, \tag{5.3.60}$$

for all $r \geq m$. This follows from

$$(\mathcal{A} - \lambda\mathcal{I})^r\boldsymbol{v} = (\mathcal{A} - \lambda\mathcal{I})^{r-m}[(\mathcal{A} - \lambda\mathcal{I})^m\boldsymbol{v}] = 0.$$

Consequently, for such a $\boldsymbol{v}$

$$e^{t(\mathcal{A}-\lambda\mathcal{I})}\boldsymbol{v} = \boldsymbol{v} + t(\mathcal{A} - \lambda\mathcal{I})\boldsymbol{v} + \cdots + \frac{t^{m-1}}{(m-1)!}(\mathcal{A} - \lambda\mathcal{I})^{m-1}\boldsymbol{v}$$

and

$$e^{t\mathcal{A}}\boldsymbol{v} = e^{\lambda t}e^{t(\mathcal{A}-\lambda\mathcal{I})}\boldsymbol{v} = e^{\lambda t}\left(\boldsymbol{v} + t(\mathcal{A} - \lambda\mathcal{I})\boldsymbol{v} + \cdots + \frac{t^{m-1}}{(m-1)!}(\mathcal{A} - \lambda\mathcal{I})^{m-1}\boldsymbol{v}\right). \tag{5.3.61}$$

Thus, to find all solutions to (5.3.56), it is sufficient to find d independent vectors $\boldsymbol{v}$ satisfying (5.3.60) for some scalars λ. To check the consistency of this method with our previous consideration, we observe that if $\lambda = \lambda_1$ is a single eigenvalue of $\mathcal{A}$ with a corresponding eigenvector $\boldsymbol{v}^1$, then $(\mathcal{A} - \lambda_1\mathcal{I})\boldsymbol{v}^1 = 0$, thus m of (5.3.60) is equal to 1. Consequently, the sum in (5.3.61) terminates after the first term, and we obtain

$$\boldsymbol{y}_1(t) = e^{\lambda_1 t}\boldsymbol{v}^1$$

in accordance with (5.3.46). From our discussion of eigenvalues and eigenvectors, see Sect. 5.3.56, it follows that if λ_i is a multiple eigenvalue of $\mathcal{A}$ of algebraic multiplicity n_i and the geometric multiplicity ν_i is less then n_i, that is, there is less than n_i linearly independent eigenvectors corresponding to λ_i, then the missing independent vectors can be found by solving successively equations $(\mathcal{A}-\lambda_i\mathcal{I})^k\boldsymbol{v} = \mathbf{0}$ with k running at most up to n_i.

Remark 5.12 Let us mention here that the exponential function $e^{t\mathcal{A}}$ has been introduced just as a guideline to explain how the formula (5.3.61) was arrived at. Once we have this formula, we can directly check that it gives a solution to (5.3.56). Indeed,

$$
\begin{aligned}
&\frac{d}{dt}e^{\lambda t}\left(\boldsymbol{v} + t(\mathcal{A}-\lambda\mathcal{I})\boldsymbol{v} + \cdots + \frac{t^{m-1}}{(m-1)!}(\mathcal{A}-\lambda\mathcal{I})^{m-1}\boldsymbol{v}\right)\\
&= \lambda e^{\lambda t}\left(\boldsymbol{v} + t(\mathcal{A}-\lambda\mathcal{I})\boldsymbol{v} + \cdots + \frac{t^{m-1}}{(m-1)!}(\mathcal{A}-\lambda\mathcal{I})^{m-1}\boldsymbol{v}\right)\\
&\quad + e^{\lambda t}\left((\mathcal{A}-\lambda\mathcal{I})\boldsymbol{v} + t(\mathcal{A}-\lambda\mathcal{I})^2\boldsymbol{v} + \cdots + \frac{t^{m-2}}{(m-2)!}(\mathcal{A}-\lambda\mathcal{I})^{m-1}\boldsymbol{v}\right)\\
&= \lambda e^{\lambda t}\left(\boldsymbol{v} + t(\mathcal{A}-\lambda\mathcal{I})\boldsymbol{v} + \cdots + \frac{t^{m-1}}{(m-1)!}(\mathcal{A}-\lambda\mathcal{I})^{m-1}\boldsymbol{v}\right)\\
&\quad + e^{\lambda t}(\mathcal{A}-\lambda\mathcal{I})\left(\boldsymbol{v} + t(\mathcal{A}-\lambda\mathcal{I})^2\boldsymbol{v} + \cdots + \frac{t^{m-2}}{(m-2)!}(\mathcal{A}-\lambda\mathcal{I})^{m-2}\boldsymbol{v}\right)\\
&= \lambda e^{\lambda t}\frac{t^{m-1}}{(m-1)!}(\mathcal{A}-\lambda\mathcal{I})^{m-1}\boldsymbol{v}\\
&\quad + e^{\lambda t}\mathcal{A}\left(\boldsymbol{v} + t(\mathcal{A}-\lambda\mathcal{I})^2\boldsymbol{v} + \cdots + \frac{t^{m-2}}{(m-2)!}(\mathcal{A}-\lambda\mathcal{I})^{m-2}\boldsymbol{v}\right)\\
&= \mathcal{A}e^{\lambda t}\frac{t^{m-1}}{(m-1)!}(\mathcal{A}-\lambda\mathcal{I})^{m-1}\boldsymbol{v} - (\mathcal{A}-\lambda\mathcal{I})e^{\lambda t}\frac{t^{m-1}}{(m-1)!}(\mathcal{A}-\lambda\mathcal{I})^{m-1}\boldsymbol{v}\\
&\quad + e^{\lambda t}\mathcal{A}\left(\boldsymbol{v} + t(\mathcal{A}-\lambda\mathcal{I})^2\boldsymbol{v} + \cdots + \frac{t^{m-2}}{(m-2)!}(\mathcal{A}-\lambda\mathcal{I})^{m-2}\boldsymbol{v}\right)\\
&\quad - e^{\lambda t}\frac{t^{m-1}}{(m-1)!}(\mathcal{A}-\lambda\mathcal{I})^{m}\boldsymbol{v}\\
&\quad + e^{\lambda t}\mathcal{A}\left(\boldsymbol{v} + t(\mathcal{A}-\lambda\mathcal{I})^2\boldsymbol{v} + \cdots + \frac{t^{m-1}}{(m-1)!}(\mathcal{A}-\lambda\mathcal{I})^{m-1}\boldsymbol{v}\right)\\
&= e^{\lambda t}\mathcal{A}\left(\boldsymbol{v} + t(\mathcal{A}-\lambda\mathcal{I})^2\boldsymbol{v} + \cdots + \frac{t^{m-1}}{(m-1)!}(\mathcal{A}-\lambda\mathcal{I})^{m-1}\boldsymbol{v}\right),
\end{aligned}
$$

where we used $(\mathcal{A}-\lambda\mathcal{I})^m\boldsymbol{v} = 0$.

We illustrate the theory with a few examples.

Example 5.13 Find the general solution to

$$\mathbf{y}'(t) = \begin{pmatrix} 1 & -1 & 4 \\ 3 & 2 & -1 \\ 2 & 1 & -1 \end{pmatrix} \mathbf{y}(t).$$

To obtain the eigenvalues, we calculate the characteristic polynomial

$$\begin{aligned} p(\lambda) = det(\mathcal{A} - \lambda\mathcal{I}) &= \begin{vmatrix} 1-\lambda & -1 & 4 \\ 3 & 2-\lambda & -1 \\ 2 & 1 & -1-\lambda \end{vmatrix} \\ &= -(1+\lambda)(1-\lambda)(2-\lambda) + 12 + 2 - 8(2-\lambda) + (1-\lambda) - 3(1+\lambda) \\ &= -(1+\lambda)(1-\lambda)(2-\lambda) + 4\lambda - 4 = (1-\lambda)(\lambda-3)(\lambda+2), \end{aligned}$$

so that the eigenvalues of $\mathcal{A}$ are $\lambda_1 = 1$, $\lambda_2 = 3$ and $\lambda_3 = -2$. All the eigenvalues have the algebraic multiplicity one, so they should give rise to 3 linearly independent eigenvectors.

(i) $\lambda_1 = 1$: we seek a nonzero vector $\boldsymbol{v}$ such that

$$(\mathcal{A} - 1\mathcal{I})\boldsymbol{v} = \begin{pmatrix} 0 & -1 & 4 \\ 3 & 1 & -1 \\ 2 & 1 & -2 \end{pmatrix} \begin{pmatrix} v_1 \\ v_2 \\ v_3 \end{pmatrix} = \begin{pmatrix} 0 \\ 0 \\ 0 \end{pmatrix}.$$

Thus

$$-v_2 + 4v_3 = 0, \qquad 3v_1 + v_2 - v_3 = 0, \qquad 2v_1 + v_2 - 2v_3 = 0,$$

and we get $v_2 = 4v_3$ and $v_1 = -v_3$ from the first two equations, and the third is automatically satisfied. Thus, we obtain the eigenspace corresponding to $\lambda_1 = 1$ containing all the vectors of the form

$$\boldsymbol{v}^{\mathbf{1}} = C_1 \begin{pmatrix} -1 \\ 4 \\ 1 \end{pmatrix},$$

where C_1 is any constant, and the corresponding solutions

$$\mathbf{y}^{\mathbf{1}}(t) = C_1 e^t \begin{pmatrix} -1 \\ 4 \\ 1 \end{pmatrix}.$$

(ii) $\lambda_2 = 3$: we seek a nonzero vector $\boldsymbol{v}$ such that

$$(\mathcal{A} - 3\mathcal{I})\boldsymbol{v} = \begin{pmatrix} -2 & -1 & 4 \\ 3 & -1 & -1 \\ 2 & 1 & -4 \end{pmatrix} \begin{pmatrix} v_1 \\ v_2 \\ v_3 \end{pmatrix} = \begin{pmatrix} 0 \\ 0 \\ 0 \end{pmatrix}.$$

Hence

$$-2v_1 - v_2 + 4v_3 = 0, \qquad 3v_1 - v_2 - v_3 = 0, \qquad 2v_1 + v_2 - 4v_3 = 0.$$

Solving for v_1 and v_2 in terms of v_3 from the first two equations gives $v_1 = v_3$ and $v_2 = 2v_3$. Consequently, vectors of the form

$$\boldsymbol{v}^2 = C_2 \begin{pmatrix} 1 \\ 2 \\ 1 \end{pmatrix}$$

are eigenvectors corresponding to the eigenvalue $\lambda_2 = 3$ and the function

$$\boldsymbol{y}^2(t) = e^{3t} \begin{pmatrix} 1 \\ 2 \\ 1 \end{pmatrix}$$

is the second solution of the system.

(iii) $\lambda_3 = -2$: we have to solve

$$(\mathcal{A} + 2\mathcal{I})\boldsymbol{v} = \begin{pmatrix} 3 & -1 & 4 \\ 3 & 4 & -1 \\ 2 & 1 & 1 \end{pmatrix} \begin{pmatrix} v_1 \\ v_2 \\ v_3 \end{pmatrix} = \begin{pmatrix} 0 \\ 0 \\ 0 \end{pmatrix}.$$

Thus

$$3v_1 - v_2 + 4v_3 = 0, \qquad 3v_1 + 4v_2 - v_3 = 0, \qquad 2v_1 + v_2 + v_3 = 0.$$

Again, solving for v_1 and v_2 in terms of v_3 from the first two equations gives $v_1 = -v_3$ and $v_2 = v_3$ so that each vector

$$\boldsymbol{v}^3 = C_3 \begin{pmatrix} -1 \\ 1 \\ 1 \end{pmatrix}$$

is an eigenvector corresponding to the eigenvalue $\lambda_3 = -2$. Consequently, the function

$$\boldsymbol{y}^{\mathbf{3}}(t) = e^{-2t}\begin{pmatrix} -1 \\ 1 \\ 1 \end{pmatrix}$$

is the third solution of the system. These solutions are linearly independent since the vectors $\boldsymbol{v}^{\mathbf{1}}, \boldsymbol{v}^{\mathbf{2}}, \boldsymbol{v}^{\mathbf{3}}$ are linearly independent as eigenvectors corresponding to distinct eigenvalues; see Lemma 5.9. Therefore, every solution is of the form

$$\boldsymbol{y}(t) = C_1 e^{t}\begin{pmatrix} -1 \\ 4 \\ 1 \end{pmatrix} + C_2 e^{3t}\begin{pmatrix} 1 \\ 2 \\ 1 \end{pmatrix} + C_3 e^{-2t}\begin{pmatrix} -1 \\ 1 \\ 1 \end{pmatrix}.$$

If we have a simple, complex eigenvalue λ with an eigenvector $\boldsymbol{v}$ then, as explained before Example 5.11, $\bar{\lambda}$ is also an eigenvalue with corresponding eigenvector $\bar{\boldsymbol{v}}$. Thus, we have two linearly independent (complex) solutions

$$\boldsymbol{z}^{\mathbf{1}}(t) = e^{\lambda t}\boldsymbol{v}, \qquad \boldsymbol{z}^{\mathbf{2}}(t) = e^{\bar{\lambda} t}\bar{\boldsymbol{v}} = \overline{\boldsymbol{z}^{\mathbf{1}}}(t).$$

Since the sum and the difference of any two solutions are, again solutions by taking

$$\boldsymbol{y}^{\mathbf{1}}(t) = \frac{\boldsymbol{z}^{\mathbf{1}}(t) + \boldsymbol{z}^{\mathbf{2}}(t)}{2} = \frac{\boldsymbol{z}^{\mathbf{1}}(t) + \overline{\boldsymbol{z}^{\mathbf{1}}}(t)}{2} = \Re \boldsymbol{z}^{\mathbf{1}}(t)$$

and

$$\boldsymbol{y}^{\mathbf{2}}(t) = \frac{\boldsymbol{z}^{\mathbf{1}}(t) - \boldsymbol{z}^{\mathbf{2}}(t)}{2i} = \frac{\boldsymbol{z}^{\mathbf{1}}(t) - \overline{\boldsymbol{z}^{\mathbf{1}}}(t)}{2i} = \Im \boldsymbol{z}^{\mathbf{1}}(t),$$

we obtain two real-valued (and linearly independent) solutions. To find explicit formulae for $\boldsymbol{y}^{\mathbf{1}}(t)$ and $\boldsymbol{y}^{\mathbf{2}}(t)$, we write

$$\begin{aligned} \boldsymbol{z}^{\mathbf{1}}(t) &= e^{\lambda t}\boldsymbol{v} = e^{\xi t}(\cos\omega t + i\sin\omega t)(\Re\boldsymbol{v} + i\Im\boldsymbol{v}) \\ &= e^{\xi t}(\cos\omega t\,\Re\boldsymbol{v} - \sin\omega t\,\Im\boldsymbol{v}) + ie^{\xi t}(\cos\omega t\,\Im\boldsymbol{v} + \sin\omega t\,\Re\boldsymbol{v}) \\ &= \boldsymbol{y}^{\mathbf{1}}(t) + i\,\boldsymbol{y}^{\mathbf{2}}(t). \end{aligned}$$

Summarizing, if λ and $\bar{\lambda}$ are simple complex roots of the characteristic equation with complex eigenvectors $\boldsymbol{v}$ and $\bar{\boldsymbol{v}}$, respectively, then the we can use two real linearly independent solutions

$$\begin{aligned} \mathbf{y}^{\mathbf{1}}(t) &= e^{\xi t}(\cos\omega t\, \Re\boldsymbol{v} - \sin\omega t\, \Im\boldsymbol{v}), \\ \mathbf{y}^{\mathbf{2}}(t) &= e^{\xi t}(\cos\omega t\, \Im\boldsymbol{v} + \sin\omega t\, \Re\boldsymbol{v}). \end{aligned} \tag{5.3.62}$$

Example 5.14 Solve the initial value problem

$$\mathbf{y}'(t) = \begin{pmatrix} 1 & 0 & 0 \\ 0 & 1 & -1 \\ 0 & 1 & 1 \end{pmatrix} \mathbf{y}(t), \qquad \mathbf{y}(0) = \begin{pmatrix} 1 \\ 1 \\ 1 \end{pmatrix}.$$

The characteristic polynomial of the coefficient matrix is given by

$$\begin{aligned} p(\lambda) = det(\mathcal{A} - \lambda\mathcal{I}) &= \begin{vmatrix} 1-\lambda & 0 & 0 \\ 0 & 1-\lambda & -1 \\ 0 & 1 & 1-\lambda \end{vmatrix} \\ &= (1-\lambda)^3 + (1-\lambda) = (1-\lambda)(\lambda^2 - 2\lambda + 2), \end{aligned}$$

so that we have eigenvalues $\lambda_1 = 1$ and $\lambda_{2,3} = 1 \pm i$.

It is immediate that

$$\boldsymbol{v}^{\mathbf{1}} = \begin{pmatrix} 1 \\ 0 \\ 0 \end{pmatrix}$$

is an eigenvector corresponding to $\lambda_1 = 1$, and thus we obtain a solution to the system in the form

$$\mathbf{y}^{\mathbf{1}}(t) = e^t \begin{pmatrix} 1 \\ 0 \\ 0 \end{pmatrix}.$$

Let us take now the complex eigenvalue $\lambda_2 = 1 + i$. We have to solve

$$(\mathcal{A} - (1+i)\mathcal{I})\boldsymbol{v} = \begin{pmatrix} -i & 0 & 0 \\ 0 & -i & -1 \\ 0 & 1 & -i \end{pmatrix} \begin{pmatrix} v_1 \\ v_2 \\ v_3 \end{pmatrix} = \begin{pmatrix} 0 \\ 0 \\ 0 \end{pmatrix}.$$

Thus

$$-iv_1 = 0, \qquad -iv_2 - v_3 = 0, \qquad v_2 - iv_3 = 0.$$

The first equation gives $v_1 = 0$ and the other two yield $v_2 = i v_3$ so that each vector

$$\boldsymbol{v}^2 = C_2 \begin{pmatrix} 0 \\ i \\ 1 \end{pmatrix}$$

is an eigenvector corresponding to the eigenvalue $\lambda_2 = 1 + i$. Consequently, we obtain a complex-valued solution

$$z(t) = e^{(1+i)t} \begin{pmatrix} 0 \\ i \\ 1 \end{pmatrix}.$$

To obtain real-valued solutions, we separate z into real and imaginary parts:

$$e^{(1+i)t} \begin{pmatrix} 0 \\ i \\ 1 \end{pmatrix} = e^t (\cos t + i \sin t) \left(\begin{pmatrix} 0 \\ 0 \\ 1 \end{pmatrix} + i \begin{pmatrix} 0 \\ 1 \\ 0 \end{pmatrix} \right)$$

$$= e^t \left(\cos t \begin{pmatrix} 0 \\ 0 \\ 1 \end{pmatrix} - \sin t \begin{pmatrix} 0 \\ 1 \\ 0 \end{pmatrix} + i \sin t \begin{pmatrix} 0 \\ 0 \\ 1 \end{pmatrix} + i \cos t \begin{pmatrix} 0 \\ 1 \\ 0 \end{pmatrix} \right)$$

$$= e^t \begin{pmatrix} 0 \\ -\sin t \\ \cos t \end{pmatrix} + i e^t \begin{pmatrix} 0 \\ \cos t \\ \sin t \end{pmatrix}.$$

Thus, we obtain two real solutions

$$\boldsymbol{y}^2(t) = e^t \begin{pmatrix} 0 \\ -\sin t \\ \cos t \end{pmatrix}, \quad \boldsymbol{y}^3(t) = e^t \begin{pmatrix} 0 \\ \cos t \\ \sin t \end{pmatrix},$$

and the general solution to our original system is given by

$$\boldsymbol{y}(t) = C_1 e^t \begin{pmatrix} 1 \\ 0 \\ 0 \end{pmatrix} + C_2 e^t \begin{pmatrix} 0 \\ -\sin t \\ \cos t \end{pmatrix} + C_3 e^t \begin{pmatrix} 0 \\ \cos t \\ \sin t \end{pmatrix}.$$

These solutions are linearly independent, as their initial values,

$$\begin{pmatrix} 1 \\ 0 \\ 0 \end{pmatrix}, \quad \begin{pmatrix} 0 \\ 0 \\ 1 \end{pmatrix}, \quad \begin{pmatrix} 0 \\ 1 \\ 0 \end{pmatrix},$$

are independent. To find the solution to our initial value problem, we set $t = 0$ and solve for C_1, C_2 and C_3 the system

$$\begin{pmatrix}1\\1\\1\end{pmatrix} = C_1\begin{pmatrix}1\\0\\0\end{pmatrix} + \begin{pmatrix}0\\0\\1\end{pmatrix} + C_3\begin{pmatrix}0\\1\\0\end{pmatrix} = \begin{pmatrix}C_1\\C_2\\C_3\end{pmatrix}.$$

Thus $C_1 = C_2 = C_3 = 1$ and finally

$$\mathbf{y}(t) = e^t\begin{pmatrix}1\\0\\0\end{pmatrix} + e^t\begin{pmatrix}0\\-\sin t\\\cos t\end{pmatrix} + e^t\begin{pmatrix}0\\\cos t\\\sin t\end{pmatrix} = e^t\begin{pmatrix}1\\\cos t - \sin t\\\cos t + \sin t\end{pmatrix}.$$

The last of our examples here deals with multiple eigenvalues.

Example 5.15 Find three linearly independent solutions of the differential equation

$$\mathbf{y}'(t) = \begin{pmatrix}1&1&0\\0&1&0\\0&0&2\end{pmatrix}\mathbf{y}(t).$$

The characteristic polynomial is given by

$$p(\lambda) = det(\mathcal{A} - \lambda\mathcal{I}) = \begin{vmatrix}1-\lambda & 1 & 0\\0 & 1-\lambda & 0\\0 & 0 & 2-\lambda\end{vmatrix} = (1-\lambda)^2(2-\lambda),$$

so $\lambda_1 = 1$ is an eigenvalue of the algebraic multiplicity 2 and $\lambda_2 = 2$ is an eigenvalue of multiplicity 1.

(i) $\lambda = 1$: We seek all non-zero vectors such that

$$(\mathcal{A} - 1\mathcal{I})\boldsymbol{v} = \begin{pmatrix}0&1&0\\0&0&0\\0&0&1\end{pmatrix}\begin{pmatrix}v_1\\v_2\\v_3\end{pmatrix} = \begin{pmatrix}0\\0\\0\end{pmatrix}.$$

This implies that $v_2 = v_3 = 0$ and v_1 is arbitrary so that the corresponding solution is given by

$$\mathbf{y}^{\mathbf{1}}(t) = C_1 e^t\begin{pmatrix}1\\0\\0\end{pmatrix}.$$

However, this is only one solution, and $\lambda_1 = 1$ has algebraic multiplicity 2, so we must look for one more solution. To this end, we consider

$$(\mathcal{A} - 1\mathcal{I})^2 \boldsymbol{v} = \begin{pmatrix} 0 & 1 & 0 \\ 0 & 0 & 0 \\ 0 & 0 & 1 \end{pmatrix} \begin{pmatrix} 0 & 1 & 0 \\ 0 & 0 & 0 \\ 0 & 0 & 1 \end{pmatrix} \begin{pmatrix} v_1 \\ v_2 \\ v_3 \end{pmatrix} = \begin{pmatrix} 0 & 0 & 0 \\ 0 & 0 & 0 \\ 0 & 0 & 1 \end{pmatrix} \begin{pmatrix} v_1 \\ v_2 \\ v_3 \end{pmatrix} = \begin{pmatrix} 0 \\ 0 \\ 0 \end{pmatrix},$$

so $v_3 = 0$ and both v_1 and v_2 arbitrary. The set of all solutions here is a two-dimensional space given by

$$\begin{pmatrix} v_1 \\ v_2 \\ 0 \end{pmatrix} = v_1 \begin{pmatrix} 1 \\ 0 \\ 0 \end{pmatrix} + v_2 \begin{pmatrix} 0 \\ 1 \\ 0 \end{pmatrix}, \qquad v_1, v_2 \in \mathbb{R}.$$

We have to select from this subspace a vector that is not a solution to $(\mathcal{A} - \lambda\mathcal{I})\boldsymbol{v} = \mathbf{0}$. Since for the later the solutions are scalar multiples of the vector $(1, 0, 0)$, we see that the vector $(0, 1, 0)$ is not of this form and consequently can be taken as the second independent vector corresponding to the eigenvalue $\lambda_1 = 1$. Hence

$$\begin{aligned} \boldsymbol{y}^2(t) &= e^t \, (\mathcal{I} + t(\mathcal{A} - \mathcal{I})) \begin{pmatrix} 0 \\ 1 \\ 0 \end{pmatrix} = e^t \left(\begin{pmatrix} 0 \\ 1 \\ 0 \end{pmatrix} + t \begin{pmatrix} 0 & 1 & 0 \\ 0 & 0 & 0 \\ 0 & 0 & 1 \end{pmatrix} \begin{pmatrix} 0 \\ 1 \\ 0 \end{pmatrix} \right) \\ &= e^t \begin{pmatrix} 0 \\ 1 \\ 0 \end{pmatrix} + te^t \begin{pmatrix} 1 \\ 0 \\ 0 \end{pmatrix} = e^t \begin{pmatrix} t \\ 1 \\ 0 \end{pmatrix}. \end{aligned}$$

(ii) $\lambda = 2$: We seek solutions to

$$(\mathcal{A} - 2\mathcal{I})\boldsymbol{v} = \begin{pmatrix} -1 & 1 & 0 \\ 0 & -1 & 0 \\ 0 & 0 & 0 \end{pmatrix} \begin{pmatrix} v_1 \\ v_2 \\ v_3 \end{pmatrix} = \begin{pmatrix} 0 \\ 0 \\ 0 \end{pmatrix}.$$

We obtain $v_1 = v_2 = 0$, while v_3 is arbitrary so that the corresponding solutions are of the form

$$\boldsymbol{y}^3(t) = C_3 e^{2t} \begin{pmatrix} 0 \\ 0 \\ 1 \end{pmatrix}.$$

Thus, we have found three linearly independent solutions.

Exercise 5.3.5 Determine all vectors $\boldsymbol{y}^0$ for which the solution of the initial value problem

$$\boldsymbol{y}'(t) = \begin{pmatrix} 1 & 0 & -2 \\ 0 & 1 & 0 \\ 1 & -1 & -1 \end{pmatrix} \boldsymbol{y}, \qquad \boldsymbol{y}(0) = \boldsymbol{y}^0;$$

is a periodic function of time.

Exercise 5.3.6 Suppose that $\mathcal{A}^2 = \alpha\mathcal{A}$ for some constant α. Find $e^{t\mathcal{A}}$.

Exercise 5.3.7 Let

$$\mathcal{A} = \begin{pmatrix} 0 & 1 \\ -1 & 0 \end{pmatrix}.$$

Show that $\mathcal{A}^2 = -\mathcal{I}$ and hence that

$$e^{\mathcal{A}t} = \begin{pmatrix} \cos t & \sin t \\ -\sin t & \cos t \end{pmatrix}.$$

Exercise 5.3.8 Find $\mathcal{A}$ if

$$e^{\mathcal{A}t} = \begin{pmatrix} 2e^{2t} - e^t & e^{2t} - e^t & e^t - e^{2t} \\ e^{2t} - e^t & 2e^{2t} - e^t & e^t - e^{2t} \\ 3e^{2t} - 3e^t & 3e^{2t} - 3e^t & 3e^t - 2e^{2t} \end{pmatrix}.$$

5.3.6 *Fundamental Matrix and Nonhomogeneous Linear Systems*

Following Theorem 5.6, let us suppose that we have d linearly independent solutions $\boldsymbol{y}^1(t), \ldots, \boldsymbol{y}^d(t)$ of (5.3.29) and, as in (5.3.41), consider the fundamental matrix

$$\mathcal{E}(t) = \begin{pmatrix} y_1^1(t) & \ldots & y_1^d(t) \\ \vdots & \ddots & \vdots \\ y_d^1(t) & \ldots & y_d^d(t) \end{pmatrix}.$$

By the results of Sect. 5.3.5, we know that for a given initial vector $\mathring{\boldsymbol{y}}$ the solution to the autonomous system (5.3.29) with initial condition (5.3.31) is given by

$$\boldsymbol{y}(t) = e^{t\mathcal{A}} \mathring{\boldsymbol{y}}$$

on the one hand, and, on the other hand, using Theorem 5.6, by

$$\boldsymbol{y}(t) = C_1\boldsymbol{y}^{\mathbf{1}}(t) + \cdots + C_d\boldsymbol{y}^{\boldsymbol{d}}(t) = \mathcal{E}(t)\boldsymbol{C},$$

where

$$\mathring{\boldsymbol{y}} = \mathcal{E}(0)\boldsymbol{C},$$

compare (5.3.42). Hence, using (5.3.35) and the uniqueness of solutions,

$$e^{t\mathcal{A}} = \mathcal{E}(t)\mathcal{E}^{-1}(0). \tag{5.3.63}$$

Consider now the non-homogeneous system

$$\begin{aligned} \boldsymbol{y}' &= \mathcal{A}\boldsymbol{y} + \boldsymbol{g}(t), \\ \boldsymbol{y}(0) &= \mathring{\boldsymbol{y}}, \end{aligned}$$

in the autonomous case. The solution formula (5.3.40) becomes much simpler if we take $e^{t\mathcal{A}}$ as the fundamental matrix because in such a case $\mathcal{E}^{-1}(t) = \left(e^{t\mathcal{A}}\right)^{-1} = e^{-t\mathcal{A}}$, that is, to calculate the inverse of $e^{t\mathcal{A}}$ it is enough to replace t by $-t$. Then the solution (5.3.40) takes the form

$$\boldsymbol{y}(t) = e^{t\mathcal{A}}\mathring{\boldsymbol{x}} + \int_0^t e^{(t-s)\mathcal{A}}\boldsymbol{g}(s)ds. \tag{5.3.64}$$

This formula is usually referred to as the variation of constants formula.

Exercise 5.3.9 Find general solutions to the following systems of equations.

(a)

$$\boldsymbol{y}' = \begin{pmatrix} 3 & 2 & 4 \\ 2 & 0 & 2 \\ 4 & 2 & 3 \end{pmatrix}\boldsymbol{y};$$

(b)

$$\boldsymbol{y}' = \begin{pmatrix} 1 & 0 & 0 \\ 3 & 1 & -2 \\ 2 & 2 & 1 \end{pmatrix}\boldsymbol{y};$$

(c)

$$y' = \begin{pmatrix} -1 & -1 & 0 \\ 0 & -1 & 0 \\ 0 & 0 & -2 \end{pmatrix} y.$$

Exercise 5.3.10 Find the solutions to the following initial value problems

(a)

$$y'(t) = \begin{pmatrix} 3 & 1 & -1 \\ 1 & 3 & -1 \\ 3 & 3 & -1 \end{pmatrix} y(t), \qquad y(0) = \begin{pmatrix} 1 \\ -2 \\ -1 \end{pmatrix};$$

(b)

$$y'(t) = \begin{pmatrix} -3 & 0 & 2 \\ 1 & -1 & 0 \\ -2 & -1 & 0 \end{pmatrix} y(t), \qquad y(0) = \begin{pmatrix} 0 \\ -1 \\ -2 \end{pmatrix};$$

(c)

$$y'(t) = \begin{pmatrix} -1 & 1 & 2 \\ -1 & 1 & 1 \\ -2 & 1 & 3 \end{pmatrix} y(t), \qquad y(0) = \begin{pmatrix} 1 \\ 0 \\ 1 \end{pmatrix};$$

(d)

$$y'(t) = \begin{pmatrix} 2 & -5 \\ 1 & -2 \end{pmatrix} y(t) + \begin{pmatrix} \sin t \\ \tan t \end{pmatrix}, \qquad y(0) = \begin{pmatrix} 0 \\ 0 \end{pmatrix};$$

(e)

$$y'(t) = \begin{pmatrix} 2 & 0 & 1 \\ 0 & 2 & 0 \\ 0 & 1 & 3 \end{pmatrix} y(t) + e^{2t} \begin{pmatrix} 1 \\ 0 \\ 1 \end{pmatrix}, \qquad y(0) = \begin{pmatrix} 1 \\ 1 \\ 1 \end{pmatrix}.$$

5.3.7 *Higher Order Difference Equations*

Once we know how to solve systems of difference (or differential) equations, it is easy to adopt the theory to cater for higher-order scalar equations. We shall describe only the case of difference equations, while the analogous approach to differential equations of higher order is presented in Remark 6.2.

First, consider the linear difference equation of order d:

$$y(k+n) + a_1 y(k+n-1) + \ldots + a_n y(k) = 0, \qquad n \geq 0, \tag{5.3.65}$$

where $a_1, \ldots, a_n$ are known numbers. This equation determines the values of $y(N)$, $N > n$ by n preceding values of $y(k)$. Thus, it is clear that to be able to solve this equation, that is, to start the recurrence procedure, we need n initial values $y(0), y(1), \ldots, y(n-1)$. Equation (5.3.65) can be written as a system of first-order equations of dimension n. We let

$$\begin{aligned} z_1(k) &= y(k), \\ z_2(k) &= y(k+1) = z_1(k+1), \\ z_3(k) &= y(k+2) = z_2(k+1), \\ &\vdots \; \vdots \; \vdots, \\ z_n(k) &= y(k+n-1) = z_{n-1}(k-1), \end{aligned} \tag{5.3.66}$$

hence, we obtain the system

$$\begin{aligned} z_1(k+1) &= z_2(k), \\ z_2(k+1) &= z_3(k), \\ &\vdots \; \vdots \; \vdots, \\ z_{n-1}(k+1) &= z_n(k), \\ z_n(k+1) &= -a_n z_1(k) - a_2 z_2(k) \ldots - a_1 z_n(k), \end{aligned}$$

or, in matrix notation,

$$z(k+1) = \mathcal{A} z(k),$$

where $z = (z_1, \ldots, z_n)$, and

$$\mathcal{A} = \begin{pmatrix} 0 & 1 & 0 & \ldots & 0 \\ 0 & 0 & 1 & \ldots & 0 \\ \vdots & \vdots & \vdots & \vdots & \vdots \\ -a_n & -a_{n-1} & -a_{n-2} & \ldots & -a_1 \end{pmatrix}.$$

The matrix $\mathcal{A}$ often is called the companion matrix of Eq. (5.3.65). It is clear that the initial values $y(0), \ldots, y(n-1)$ give the initial vector $\mathring{z} = (y(0), \ldots, y(n-1))$.

Next, we observe that the eigenvalues of $\mathcal{A}$ can be obtained by solving the equation

$$p_n(\lambda) = \begin{vmatrix} -\lambda & 1 & 0 & \dots & 0 \\ 0 & -\lambda & 1 & \dots & 0 \\ \vdots & \vdots & \vdots & \vdots & \vdots \\ -a_n & -a_{n-1} & -a_{n-2} & \dots & -a_1 - \lambda \end{vmatrix} = 0.$$

It is not obvious how to directly calculate the determinant. Here, we present one method that gives the determinant as a solution to a difference equation. Later, in Remark 5.37, we will present another way, which is more related to interpreting the problem. Expanding the determinant along the first row, we find

$$\begin{aligned} p_n(\lambda) &= -\lambda \begin{vmatrix} -\lambda & 1 & 0 & \dots & 0 \\ 0 & -\lambda & 1 & \dots & 0 \\ \vdots & \vdots & \vdots & \vdots & \vdots \\ -a_{n-1} & -a_{n-2} & -a_{n-3} & \dots & -a_1 - \lambda \end{vmatrix} \\ &\quad - \begin{vmatrix} 0 & 1 & 0 & \dots & 0 \\ 0 & 0 & 1 & \dots & 0 \\ \vdots & \vdots & \vdots & \vdots & \vdots \\ -a_n & -a_{n-2} & -a_{n-3} & \dots & -a_1 - \lambda \end{vmatrix} \\ &= -\lambda p_{n-1}(\lambda) + (-1)^n a_n. \end{aligned}$$

This is a linear difference equation, and we can solve it using (2.1.49). Taking into account that $p_1(\lambda) = -a - 1 - \lambda$ and adjusting (2.1.49) to cater for the fact that we start with $n = 1$, we obtain

$$\begin{aligned} p_n(\lambda) &= (-\lambda)^{n-1}(a_1 + \lambda) + \sum_{k=1}^{n-1} (-\lambda)^{n-k-1}(-1)^{k+1} a_{k+1} \\ &= (-1)^n(\lambda^n + \lambda^{n-1}a_1 + \dots + a_n). \end{aligned} \tag{5.3.67}$$

We note that the characteristic polynomial of the companion matrix can be obtained by just replacing $y(k + n - i)$ in (5.3.65) by λ^{n-i}, $i = 0, \dots, n$. Consequently, solutions of higher order equations can be obtained by solving the associated first order systems, but there is no need to repeat the whole procedure. In fact, to solve an $n \times n$ system we have to construct n linearly independent vectors $\boldsymbol{v}^1, \dots, \boldsymbol{v}^n$ so that the solution is given by $\boldsymbol{z}^1(k) = \mathcal{A}^k \boldsymbol{v}^1, \dots, \boldsymbol{z}^n(k) = \mathcal{A}^k \boldsymbol{v}^n$ and the coordinates of each $\boldsymbol{z}^i$ are products of λ_i and polynomials in k of degree strictly smaller than the algebraic multiplicity of λ_i. Thus, to obtain n_i solutions of the higher order equation corresponding to the eigenvalue λ_i, by (5.3.66), we take only the first coordinates of all $\boldsymbol{z}^i(k)$ that correspond to λ_i. On the other hand, we must have here n_i linearly independent scalar solutions of this form, and therefore, we can use the set $\{\lambda_i^k, k\lambda_i^k, \ldots, k^{n_i-1}\lambda_i^k\}$ as a basis for the set of solutions corresponding to λ_i, and the union of such sets over all eigenvalues to obtain a basis for the set of all solutions.

Example 5.16 Consider the Fibonacci equation (5.1.1) written as

$$y(k+2) = y(k+1) + y(k), \tag{5.3.68}$$

to be consistent with the notation of the present chapter. Introducing new variables $z_1(k) = y(k)$ and $z_2(k) = y(k+1) = z_1(k+1)$, so that $y(k+2) = z_2(k+1)$, we re-write the equation as the system

$$\begin{aligned} z_1(k+1) &= z_2(k), \\ z_2(k+1) &= z_1(k) + z_2(k); \end{aligned}$$

note that it is not the same form as (5.1.3). The eigenvalues of the matrix

$$\mathcal{A} = \begin{pmatrix} 0 & 1 \\ 1 & 1 \end{pmatrix}$$

are obtained by solving the equation

$$\begin{vmatrix} -\lambda & 1 \\ 1 & 1-\lambda \end{vmatrix} = \lambda^2 - \lambda - 1 = 0;$$

they are $\lambda_{1,2} = \frac{1\pm\sqrt{5}}{2}$. Since the eigenvalues are distinct, we immediately obtain that the general solution of (5.3.68) is given by

$$y(n) = C_1 \left(\frac{1+\sqrt{5}}{2}\right)^n + C_2 \left(\frac{1-\sqrt{5}}{2}\right)^n. \tag{5.3.69}$$

Let us find the particular solution that satisfies the initial conditions $y(0) = 1$, $y(1) = 2$ (corresponding to one pair of adult rabbits initially). We substitute these values and get the system of equations for C_1 and C_2

$$\begin{aligned} 1 &= C_1 + C_2, \\ 2 &= C_1 \frac{1+\sqrt{5}}{2} + C_2 \frac{1-\sqrt{5}}{2}, \end{aligned}$$

the solution of which are $C_1 = 1 + 3\sqrt{5}/5$ and $C_2 = -3\sqrt{5}/5$.

Exercise 5.3.11 Find general solutions to the following higher-order equations

$$\begin{aligned}
&(a) \quad x(k+2) + 8x(k+1) + 12x(k) = 0,\\
&(b) \quad x(k+2) - 5x(k+1) + 4x(k) = 4^k,\\
&(c) \quad x(k+2) + 4x(k) = 0.
\end{aligned}$$

Exercise 5.3.12 Solve the initial value problem

$$x(k+3) - 7x(k+2) + 16x(k+1) - 12x(k) = 0, \quad x(0) = 0, x(1) = 1, x(2) = 1.$$

5.3.8 *Spectral Decomposition*

One of the main questions in modelling natural phenomena is the long-term behaviour of the described system and, in particular, the stability of its equilibria. While the concept of stability is the same as in scalar systems, see Definitions 4.1 and 4.11 with an obvious replacement of the modulus by a norm in $\mathbb{R}^d$, linear systems allow for a much more precise description of the behaviour of their solutions for large times. This is achieved by separating the dominant terms of the solution using spectral decomposition, which we shall describe below.

If $\boldsymbol{v}$ is an eigenvector of a matrix $\mathcal{A}$ belonging to an eigenvalue λ, then the one dimensional eigenspace space $\tilde{E}_\lambda$ has an important property of being *invariant* under $\mathcal{A}$ as well as under $\mathcal{A}^k$ and $e^{t\mathcal{A}}$, that is, if $\boldsymbol{y} \in \tilde{E}_\lambda$, then $\mathcal{A}\boldsymbol{y} \in \tilde{E}_\lambda$ (and $\mathcal{A}^k\boldsymbol{y}, e^{t\mathcal{A}}\boldsymbol{y} \in \tilde{E}_\lambda$ for all $k = 1, 2, \ldots$, and $t > 0$, respectively). In fact, in this case, $\boldsymbol{y} = \alpha\boldsymbol{v}$ for some $\alpha \in \mathbb{R}$ and

$$\mathcal{A}\boldsymbol{y} = \alpha\mathcal{A}\boldsymbol{v} = \alpha\lambda\boldsymbol{v} \in \tilde{E}_\lambda.$$

Similarly, $\mathcal{A}^k\boldsymbol{y} = \lambda^k\alpha\boldsymbol{v} \in \tilde{E}_\lambda$ and $e^{t\mathcal{A}}\boldsymbol{y} = e^{\lambda t}\alpha\boldsymbol{v} \in \tilde{E}_\lambda$. Thus, if $\mathcal{A}$ is diagonalizable, then the evolution governed by $\mathcal{A}$ can be decomposed into n independent scalar evolutions occurring in eigenspaces of $\mathcal{A}$. Multiple eigenvalues are more complicated since the one-dimensional spaces spanned by generalized eigenvectors are not invariant under $\mathcal{A}$. However, we can show that each generalized eigenspace spanned by all eigenvectors and generalized eigenvectors corresponding to the same eigenvalue is invariant under $\mathcal{A}$.

We start with the following property of E_{λ_i}, which is important in this context.

Lemma 5.17 *Let* $\{\boldsymbol{v}^1, \ldots, \boldsymbol{v}^{n_i}\}$ *be eigenvectors and associated eigenvectors constructed in Sect. 5.3.3 and let* $E_{\lambda_i} = \mathcal{L}in\{\boldsymbol{v}^1, \ldots, \boldsymbol{v}^{n_i}\}$ *be the corresponding generalized eigenspace corresponding to an eigenvalue* λ_i *and let* $\boldsymbol{v} \in E_{\lambda_i}$ *satisfy*

$$(\mathcal{A} - \lambda_i \mathcal{I})^k \boldsymbol{v} = 0,$$

for some $1 < k < n_i$, *and* $(\mathcal{A} - \lambda_i \mathcal{I})^{k-1} \boldsymbol{v} \neq 0$. *Then* $\boldsymbol{v}$ *satisfies*

$$(\mathcal{A} - \lambda_i \mathcal{I}) \boldsymbol{v} = \boldsymbol{w},$$

where $(\mathcal{A} - \lambda_i \mathcal{I})^{k-1} \boldsymbol{w} = 0$ *and hence* $\boldsymbol{w} \in Ker(\mathcal{A} - \lambda_i \mathcal{I})^{k-1} \subset E_{\lambda_i}$. *Moreover,*

$$(\mathcal{A} - \lambda_i \mathcal{I})^{k-1} \boldsymbol{v} = \boldsymbol{z},$$

where $\boldsymbol{z}$ *is an eigenvector.*

Proof Let $\{\boldsymbol{v}^1, \ldots, \boldsymbol{v}^{n_i}\}$ be grouped so that the first ν_i elements: $\{\boldsymbol{v}^1, \ldots, \boldsymbol{v}^{\nu_i}\}$ are the eigenvectors, $\{\boldsymbol{v}^\rho\}_{\nu_i+1 \leq \rho \leq r'}$, are not eigenvectors which satisfy $(\mathcal{A} - \lambda \mathcal{I})^2 \boldsymbol{v}^\rho = 0$, etc. Then, by the construction of Sect. 5.3.3 and by Theorem C.4, $E_{\lambda_i} = Ker(\mathcal{A} - \lambda_i \mathcal{I}) \cup \ldots \cup Ker(\mathcal{A} - \lambda_i \mathcal{I})^{n_i}$ with $Ker(\mathcal{A} - \lambda_i \mathcal{I})^j \subset Ker(\mathcal{A} - \lambda_i \mathcal{I})^{j+1}$, $1 \leq j < n_i$.

Let $\boldsymbol{v}$ satisfy

$$(\mathcal{A} - \lambda_i \mathcal{I})^k \boldsymbol{v} = 0, \tag{5.3.70}$$

for some $1 < k < n_i$ with $(\mathcal{A} - \lambda_i \mathcal{I})^{k-1} \boldsymbol{v} = \boldsymbol{z} \neq 0$. Thus,

$$(\mathcal{A} - \lambda_i \mathcal{I}) \boldsymbol{z} = 0,$$

and $\boldsymbol{z}$ is an eigenvector. Writing (5.3.70) as

$$(\mathcal{A} - \lambda_i \mathcal{I})^{k-1} (\mathcal{A} - \lambda_i \mathcal{I}) \boldsymbol{v} = 0$$

and recognizing that $\boldsymbol{w} = (\mathcal{A} - \lambda_i \mathcal{I}) \boldsymbol{v} \neq 0$ (otherwise $(\mathcal{A} - \lambda_i \mathcal{I})^{k-1} \boldsymbol{v} = 0$ for any $k > 1$), we see that

$$(\mathcal{A} - \lambda_i \mathcal{I})^{k-1} \boldsymbol{w} = 0$$

and hence $\boldsymbol{w} \in Ker(\mathcal{A} - \lambda_i \mathcal{I})^{k-1} \subset E_{\lambda_i}$. □

An important corollary of this lemma is

Corollary 5.18 *Each generalized eigenspace E_{λ_i} of $\mathcal{A}$ is invariant under $\mathcal{A}$, that is, for any $\boldsymbol{v} \in E_{\lambda_i}$ we have $\mathcal{A}\boldsymbol{v} \in E_{\lambda_i}$. It is also invariant under $\mathcal{A}^k$, $k = 1, 2, \ldots$, and $e^{t\mathcal{A}}$, $t > 0$.*

Proof We use the representation of E_{λ_i} obtained in the previous lemma. Indeed, let $\boldsymbol{x} = \sum_{j=1}^{n_i} a_j \boldsymbol{v}^{\boldsymbol{j}}$ be an arbitrary element of E_{λ_i}, where $\{\boldsymbol{v}^{\boldsymbol{1}}, \ldots, \boldsymbol{v}^{\boldsymbol{n_i}}\}$ was ordered as in the proof of the previous lemma. Then

$$(\mathcal{A} - \lambda_i \mathcal{I})\boldsymbol{x} = \sum_{j=1}^{n_i} a_j (\mathcal{A} - \lambda_i \mathcal{I})\boldsymbol{v}^{\boldsymbol{j}},$$

where, by (5.3.42),

$$\boldsymbol{x} = \begin{pmatrix} x_1 \\ \vdots \\ x_{n_i} \end{pmatrix} = \begin{pmatrix} | & \ldots & | \\ \boldsymbol{v}^{\boldsymbol{1}} & \ldots & \boldsymbol{v}^{\boldsymbol{n_i}} \\ | & \ldots & | \end{pmatrix} \begin{pmatrix} a_1 \\ \vdots \\ a_{n_i} \end{pmatrix} = \begin{pmatrix} | & \ldots & | \\ \boldsymbol{v}^{\boldsymbol{1}} & \ldots & \boldsymbol{v}^{\boldsymbol{n_i}} \\ | & \ldots & | \end{pmatrix} \boldsymbol{a}, \tag{5.3.71}$$

and $\boldsymbol{a}$ is uniquely determined by $\boldsymbol{x}$, $\boldsymbol{a} = \boldsymbol{a}(\boldsymbol{x})$, since $\boldsymbol{x} \in \mathcal{L}in\{\boldsymbol{v}^{\boldsymbol{1}}, \ldots, \boldsymbol{v}^{\boldsymbol{n_i}}\}$.

By construction, $(\mathcal{A} - \lambda_i \mathcal{I})\boldsymbol{v}^{\boldsymbol{j}} = \boldsymbol{w}^{\boldsymbol{j}}$ where $\boldsymbol{w}^j \in Ker(\mathcal{A} - \lambda_i \mathcal{I})^{j-1} \subset E_{\lambda_i}$. In particular, $(\mathcal{A} - \lambda_i \mathcal{I})\boldsymbol{v}^{\boldsymbol{j}} = 0$ for $1 \leq j \leq \nu_i$, (eigenvectors). Thus

$$\mathcal{A}\boldsymbol{x} = \lambda_i \boldsymbol{x} - \sum_{j > \nu_i} a_j \boldsymbol{w}^{\boldsymbol{j}} \in E_{\lambda_i}, \tag{5.3.72}$$

which ends the proof of the first part.

Now, if any subspace X is invariant under $\mathcal{A}$, then

$$\mathcal{A}^2 X = \mathcal{A}(\mathcal{A}X) \subset \mathcal{A}X \subset X, \tag{5.3.73}$$

and, by induction, it is invariant under $\mathcal{A}^k$ for any $k \in \mathbb{N}$. The same argument works for $e^{t\mathcal{A}}$ by (5.3.59) and the closedness of any finite-dimensional linear normed space, see Corollary A.6. □

This result suggests that the evolution governed by $\mathcal{A}$ in both discrete and continuous cases can be broken into several simpler and independent pieces occurring in each generalized eigenspace. To write this in proper mathematical terms, we need to introduce some notation.

Let us recall that for any set of d linearly independent vectors $\{\boldsymbol{v}^{\boldsymbol{1}}, \ldots, \boldsymbol{v}^{\boldsymbol{d}}\}$, we have the representations

$$\mathcal{A}^k \mathring{\boldsymbol{x}} = \begin{pmatrix} | & \ldots & | \\ \mathcal{A}^k \boldsymbol{v}^{\boldsymbol{1}} & \ldots & \mathcal{A}^k \boldsymbol{v}^{\boldsymbol{d}} \\ | & \ldots & | \end{pmatrix} \mathcal{V}^{-1} \mathring{\boldsymbol{x}}, \tag{5.3.74}$$

and

$$e^{t\mathcal{A}}\mathring{\boldsymbol{x}} = \begin{pmatrix} | & \dots & | \\ e^{t\mathcal{A}}\boldsymbol{v}^{\boldsymbol{1}} & \dots & e^{t\mathcal{A}}\boldsymbol{v}^{\boldsymbol{d}} \\ | & \dots & | \end{pmatrix} \mathcal{V}^{-1}\mathring{\boldsymbol{x}}, \tag{5.3.75}$$

where

$$\mathcal{V} = \begin{pmatrix} | & \dots & | \\ \boldsymbol{v}^{\boldsymbol{1}} & \dots & \boldsymbol{v}^{\boldsymbol{d}} \\ | & \dots & | \end{pmatrix}. \tag{5.3.76}$$

Following our considerations, we select the vectors $\boldsymbol{v}^{\boldsymbol{1}}, \dots, \boldsymbol{v}^{\boldsymbol{d}}$ to be eigenvectors and generalized eigenvectors of $\mathcal{A}$ as then the entries of the solution matrices can be evaluated explicitly with relative ease. We want to split these expressions into generalized eigenspaces.

Let us introduce the matrix

$$\mathcal{P}_i = \begin{pmatrix} 0 \dots & | & \dots 0 \\ 0 \dots & \boldsymbol{v}^{\boldsymbol{i}} & \dots 0 \\ 0 \dots & | & \dots 0 \end{pmatrix} \begin{pmatrix} | & \dots & | \\ \boldsymbol{v}^{\boldsymbol{1}} & \dots & \boldsymbol{v}^{\boldsymbol{d}} \\ | & \dots & | \end{pmatrix}^{-1}, \tag{5.3.77}$$

and note that, for $\boldsymbol{x} = c_1\boldsymbol{v}^{\boldsymbol{1}} + \dots + c_n\boldsymbol{v}^{\boldsymbol{d}}$, $\mathcal{P}_i\boldsymbol{x} = c_i\boldsymbol{v}^{\boldsymbol{i}}$, that is, $\mathcal{P}_i$ selects the part of $\boldsymbol{x}$ along $\boldsymbol{v}^{\boldsymbol{i}}$. It is easy to see that

$$\mathcal{P}_i^2 = \mathcal{P}_i, \qquad \mathcal{P}_i\mathcal{P}_j = 0. \tag{5.3.78}$$

Matrices with such properties are called *projections*. In particular, $\mathcal{P}_i$ is a projection onto $\boldsymbol{v}^{\boldsymbol{i}}$. Clearly,

$$\mathcal{I} = \sum_{i=1}^{d} \mathcal{P}_i,$$

however, $\mathcal{A}\mathcal{P}_i\boldsymbol{x} = c_i\mathcal{A}\boldsymbol{v}^{\boldsymbol{i}}$ is in the span of $\boldsymbol{v}^{\boldsymbol{i}}$ only if $\boldsymbol{v}^{\boldsymbol{i}}$ is an eigenvector. Thus, as we said earlier, this decomposition is not useful unless all $\boldsymbol{v}^{\boldsymbol{i}}$s are eigenvectors.

On the other hand, if we consider operators

$$\mathcal{P}_{\lambda_i} = \sum_{\{j:\, \boldsymbol{v}^{\boldsymbol{j}} \in E_{\lambda_i}\}} \mathcal{P}_{\lambda_i, j}, \tag{5.3.79}$$

where $\mathcal{P}_{\lambda_i,j}$ is the projection onto $\boldsymbol{v}^j \in E_{\lambda_i}$, then such operators again will be projections. This follows from (5.3.78) by termwise multiplication. These projections are called *spectral projections*. Recall that $\sigma(\mathcal{A})$ denotes the spectrum of $\mathcal{A}$, that is, the set of all its eigenvalues. The decomposition

$$\mathcal{I} = \sum_{\lambda\in\sigma(\mathcal{A})} \mathcal{P}_\lambda \tag{5.3.80}$$

is called the *spectral resolution of identity*. One can then see that the solution $\boldsymbol{a}$ to (5.3.71) is given by

$$\boldsymbol{a} = \begin{pmatrix} a_1 \\ \vdots \\ a_{n_i} \end{pmatrix} = \mathcal{P}_{\lambda_i}\boldsymbol{x} = \begin{pmatrix} \mathcal{P}_{\lambda_i,1}\boldsymbol{x} \\ \vdots \\ \mathcal{P}_{\lambda_i,n_i}\boldsymbol{x} \end{pmatrix}. \tag{5.3.81}$$

In particular, if all eigenvalues are simple (or semi-simple), we obtain the spectral decomposition of $\mathcal{A}$ in the form

$$\mathcal{A} = \sum_{\lambda\in\sigma(\mathcal{A})} \lambda\mathcal{P}_\lambda,$$

and, for $\mathcal{A}^k$ and $e^{t\mathcal{A}}$,

$$\mathcal{A}^k = \sum_{\lambda\in\sigma(\mathcal{A})} \lambda^k\mathcal{P}_\lambda, \tag{5.3.82}$$

and

$$e^{t\mathcal{A}} = \sum_{\lambda\in\sigma(\mathcal{A})} e^{\lambda t}\mathcal{P}_\lambda, \tag{5.3.83}$$

which is another way of writing (5.3.28) and (5.3.47), respectively.

In the general case, we use (5.3.80) to write

$$\mathcal{A}\boldsymbol{x} = \sum_{\lambda\in\sigma(A)} \mathcal{A}\mathcal{P}_\lambda\boldsymbol{x}, \tag{5.3.84}$$

where, by Corollary 5.18, we have $\mathcal{A}\mathcal{P}_\lambda\boldsymbol{x} \in E_\lambda$. Thus, using (5.3.78), we get $\mathcal{P}_{\lambda_i}\mathcal{A}\mathcal{P}_{\lambda_j} = 0$ for $i \neq j$. Using (5.3.79), we obtain the commutativity relation

$$\mathcal{P}_\mu\mathcal{A}\boldsymbol{x} = \mathcal{P}_\mu\mathcal{A}\mathcal{P}_\mu\boldsymbol{x} = \mathcal{A}\mathcal{P}_\mu\boldsymbol{x}, \quad \mu \in \sigma(\mathcal{A}).$$

Thus, (5.3.84) defines a decomposition of the action of $\mathcal{A}$ into non-overlapping invariant subspaces E_λ, $\lambda \in \sigma(\mathcal{A})$, which is called the *spectral decomposition* of $\mathcal{A}$.

To give spectral decomposition of $\mathcal{A}^k$ and $e^{t\mathcal{A}}$, generalizing (5.3.82) and (5.3.83), we observe that, by Corollary 5.18, also $\mathcal{A}^k\mathcal{P}_\lambda \boldsymbol{x} \in E_\lambda$ and $e^{t\mathcal{A}}\mathcal{P}_\lambda \boldsymbol{x} \in E_\lambda$. Therefore

$$\mathcal{A}^k \boldsymbol{x} = \sum_{\lambda\in\sigma(\mathcal{A})} \mathcal{A}^k\mathcal{P}_\lambda \boldsymbol{x} = \sum_{\lambda\in\sigma(\mathcal{A})} \lambda^k \boldsymbol{p}_\lambda(k)\boldsymbol{x}, \tag{5.3.85}$$

and

$$e^{t\mathcal{A}} \boldsymbol{x} = \sum_{\lambda\in\sigma(\mathcal{A})} e^{\lambda t}\mathcal{P}_\lambda \boldsymbol{x} = \sum_{\lambda\in\sigma(\mathcal{A})} e^{\lambda t} \boldsymbol{q}_\lambda(t)\boldsymbol{x}, \tag{5.3.86}$$

where $\boldsymbol{p}_\lambda$ and $\boldsymbol{q}_\lambda$ are polynomials in k and, respectively, in t, of degree strictly smaller than the algebraic multiplicity of λ, and with matrix coefficients, given, respectively, by (5.3.72) and (5.3.73) with the coefficients $\boldsymbol{a}$ determined by (5.3.81).

Exercise 5.3.13 Whenever there are multiple eigenvalues, identify $\boldsymbol{p}_\lambda(k)$ and $\boldsymbol{q}_\lambda(t)$ in solutions to Exercises 5.3.4, and 5.3.9 and 5.3.10, respectively.

Returning to our main problem, that is, the long time behaviour of iterates $\mathcal{A}^k$ and the exponential function $e^{t\mathcal{A}}$, then, from (5.3.85) and (5.3.86), we see that on each generalized eigenspace the long time behaviour of $\mathcal{A}^k$ (respectively, of $e^{t\mathcal{A}}$) is determined by λ^k (respectively, $e^{t\lambda}$), possibly multiplied by a polynomial of degree smaller than the algebraic multiplicity of λ. Thus, the overall long-time behaviour of the system is driven by the eigenvalues with the largest absolute values in the case of the discrete-time system (5.3.22) and, since $|e^{\lambda t}| = e^{\Re\lambda}$, by the eigenvalues with the largest real part.

Remark 5.19 We emphasize that often used rule of thumb that solutions of linear systems behave as the powers of eigenvalues with the largest absolute values in the case of the discrete-time system (5.3.22) or as $e^{\lambda t}$ with λ having the largest real part is valid only if the corresponding eigenvalues are simple (or semisimple). Otherwise, we need to consider the polynomials (5.3.72) or, respectively, (5.3.73). In particular, all solutions of discrete systems with $r(\mathcal{A}) = \max\{|\lambda|;\ \lambda \in \sigma(\mathcal{A})\} = 1$ or continuous systems with $s(\mathcal{A}) = \max\{\Re\lambda;\ \lambda \in \sigma(\mathcal{A})\} = 0$ are bounded only if the eigenvalues on which the above maxima are attained are semisimple.

Observe, however, that for any $0 < \delta < 1$, $\beta > 0$ and any $p \in \mathbb{N}$, we have

$$\begin{aligned}\lim_{k\to\infty} k^p \delta^{-k} &= 0,\\ \lim_{t\to\infty} t^p e^{-\beta t} &= 0.\end{aligned} \tag{5.3.87}$$

Hence, for any p and $\alpha > \lambda$, there is M_α such that for all $t \geq 0$

$$k^p \lambda^k \leq M_\alpha \alpha^k, \tag{5.3.88a}$$

$$t^p e^{\lambda t} \leq M_\alpha e^{\alpha t}, \tag{5.3.88b}$$

and thus for any $\boldsymbol{x} \in \mathbb{R}^d$,

$$\lim_{k\to\infty} \mathcal{A}^k \boldsymbol{x} = 0, \tag{5.3.89}$$

provided $r(\mathcal{A}) < 1$, and

$$\lim_{t\to\infty} e^{t\mathcal{A}} \boldsymbol{x} = 0, \tag{5.3.90}$$

if $s(\mathcal{A}) < 0$.

Exercise 5.3.14 Prove (5.3.87) and (5.3.88).

There are, however, situations when the 'largest' eigenvalues are simple, and they fully determine the long-term behaviour of dynamical systems, as we observed in Examples 5.3 and 5.5, where, in each case, there was a real positive simple eigenvalue, say, $\lambda_{\max}$, satisfying $\lambda_{\max} > |\lambda|$ in the discrete-time case, or $\lambda_{\max} > \Re\lambda$ in continuous time, for any other $\lambda \in \sigma(\mathcal{A})$. Such an eigenvalue is called the *principal* or *dominant* eigenvalue. In such a case, for any initial condition $\mathring{\boldsymbol{x}}$ for which $\mathcal{P}_{\lambda_{\max}} \mathring{\boldsymbol{x}} \neq 0$, we have

$$\mathcal{A}^k \mathring{\boldsymbol{x}} \approx c_{\max} \lambda_{\max}^k \boldsymbol{v}^{\mathbf{max}}$$

for large k in discrete time or, in continuous time,

$$e^{t\mathcal{A}} \mathring{\boldsymbol{x}} \approx c_{\max} e^{\lambda_{\max} t} \boldsymbol{v}^{\mathbf{max}},$$

for large t, where $\boldsymbol{v}^{\mathbf{max}}$ is an eigenvector belonging to $\lambda_{\max}$.

The important question is how to efficiently determine $c_{\max}$ (and possibly other relevant coefficients of the spectral decomposition). Clearly, in this case, $c_{\max}(\boldsymbol{x})\boldsymbol{v}^{\mathbf{max}} = \mathcal{P}_{\lambda_{\max}} \boldsymbol{x}$, but the definition of $\mathcal{P}_{\lambda_{\max}}$ involves knowing all eigenvectors and associated eigenvectors of $\mathcal{A}$ and thus, it is not particularly handy. Here, we shall describe an alternative method that is particularly useful if we only need to know the dominant behaviour of the system.

Let us recall that the transposed matrix $\mathcal{A}^T$ is obtained from $\mathcal{A}$ by interchanging columns and rows, and it satisfies

$$\langle \mathcal{A}^T \boldsymbol{x}^*, \boldsymbol{y} \rangle = \langle \boldsymbol{x}^*, \mathcal{A}\boldsymbol{y} \rangle$$

for any $\boldsymbol{x}^*, \boldsymbol{y} \in \mathbb{R}^d$, where $\langle \boldsymbol{x}^*, \boldsymbol{y} \rangle = \sum_{i=1}^d x_i^* y_i$ is the dot or inner product (in $\mathbb{R}^d$). Matrices $\mathcal{A}$ and $\mathcal{A}^T$ have the same eigenvalues and, though eigenvectors and associated eigenvectors are different (unless $\mathcal{A}$ is symmetric), the structure of the generalized eigenspaces corresponding to the same eigenvalue is identical (that is, the geometric multiplicities of λ are equal, and we have the same number of associated eigenvectors solving $(\mathcal{A} - \lambda\mathcal{I})^\nu \boldsymbol{v} = 0$ and $(\mathcal{A}^T - \lambda\mathcal{I})^\nu \boldsymbol{v}^* = 0$). This follows from the the fact that the determinant, nullity and rank of a matrix and its transpose are the same, see, e.g., [41, Section 2.4].

Remark 5.20 It is easy to see that $\mathcal{A}^T \boldsymbol{w}^* = \lambda \boldsymbol{w}^*$ if and only if $\boldsymbol{w}^* \mathcal{A} = \lambda \boldsymbol{w}^*$ thus the eigenvectors of $\mathcal{A}^T$ are often called left eigenvectors of $\mathcal{A}$ (and thus the eigenvectors of $\mathcal{A}$ are called right eigenvectors.)

Theorem 5.21 *Let E_λ and $E^*_{\lambda^*}$ be generalized eigenspaces of, respectively, $\mathcal{A}$ and $\mathcal{A}^T$, corresponding to different eigenvalues $\lambda \neq \lambda^*$. If $\boldsymbol{v}^* \in E^*_{\lambda^*}$ and $\boldsymbol{v} \in E_\lambda$, then*

$$\langle \boldsymbol{v}^*, \boldsymbol{v} \rangle = 0. \tag{5.3.91}$$

Moreover, if $0 \neq \boldsymbol{v} \in E_\lambda$, then there is $\boldsymbol{v}^* \in E^*_\lambda$ such that

$$\langle \boldsymbol{v}^*, \boldsymbol{v} \rangle = 1. \tag{5.3.92}$$

Proof We can assume that $\lambda^* \neq 0$, since if $\lambda^* = 0$, then $\lambda \neq 0$, and we can repeat the calculations below, starting with λ instead of λ^*. We begin with $\boldsymbol{v} \in E_\lambda$ and $\boldsymbol{v}^* \in E^*_{\lambda^*}$ being eigenvectors. Then

$$\langle \boldsymbol{v}^*, \boldsymbol{v} \rangle = \frac{1}{\lambda^*}\langle \mathcal{A}^T \boldsymbol{v}^*, \boldsymbol{v} \rangle = \frac{1}{\lambda^*}\langle \boldsymbol{v}^*, \mathcal{A}\boldsymbol{v} \rangle = \frac{\lambda}{\lambda^*}\langle \boldsymbol{v}^*, \boldsymbol{v} \rangle.$$

Thus, $\left(\frac{\lambda}{\lambda^*} - 1\right)\langle \boldsymbol{v}^*, \boldsymbol{v} \rangle = 0$ and, since $\lambda \neq \lambda^*$, we must have $\langle \boldsymbol{v}^*, \boldsymbol{v} \rangle = 0$. Next we assume, that $\boldsymbol{v}^*$ is an eigenvector and $\boldsymbol{v}$ is an associated eigenvector which solves $(\mathcal{A} - \lambda\mathcal{I})^k \boldsymbol{v} = 0$ for some $k > 1$. Then, by Lemma 5.17, $(\mathcal{A} - \lambda\mathcal{I})\boldsymbol{v} = \boldsymbol{v}'$, where $(\mathcal{A} - \lambda\mathcal{I})^{k-1}\boldsymbol{v}' = 0$. We adopt induction assumption that $\langle \boldsymbol{v}^*, \boldsymbol{v}' \rangle = 0$ for any $\boldsymbol{v}'$ which satisfies $(\mathcal{A} - \lambda\mathcal{I})^{k-1}\boldsymbol{v}' = 0$. Then, as above

$$\langle \boldsymbol{v}^*, \boldsymbol{v} \rangle = \frac{1}{\lambda^*}\langle \boldsymbol{v}^*, \mathcal{A}\boldsymbol{v} \rangle = \frac{1}{\lambda^*}\langle \boldsymbol{v}^*, \lambda\boldsymbol{v} + \boldsymbol{v}' \rangle = \frac{\lambda}{\lambda^*}\langle \boldsymbol{v}^*, \boldsymbol{v} \rangle,$$

and the proof follows as before. Finally, let $(\mathcal{A}^T - \lambda^*\mathcal{I})^k \boldsymbol{v}^* = 0$ for some $k > 1$. Then $\lambda^* \boldsymbol{v}^* = \mathcal{A}^T \boldsymbol{v}^* - \tilde{\boldsymbol{v}}^*$, where $(\mathcal{A}^T - \lambda^*\mathcal{I})^{k-1}\tilde{\boldsymbol{v}}^* = 0$. We can adopt the induction assumption that $\langle \tilde{v}^*, v \rangle = 0$ for any $\tilde{v}^*$ satisfying $(\mathcal{A}^* - \lambda^* \mathcal{I})^{k-1} \tilde{v}^* = 0$. Then

$$\langle v^*, v \rangle = \frac{1}{\lambda^*} \langle \lambda^* v^*, v \rangle = \frac{1}{\lambda^*} \langle \mathcal{A}^T v^* - \tilde{v}^*, v \rangle = \frac{1}{\lambda^*} \langle \mathcal{A}^T v^*, v \rangle = \frac{\lambda}{\lambda^*} \langle v^*, v \rangle,$$

which ends the proof as above.

Finally, consider $v \in E_\lambda$ and assume

$$\langle v^*, v \rangle = 0$$

for any $v^* \in E^*_\lambda$. By the previous part of the proof,

$$\langle w^*, v \rangle = 0$$

for any $w^* \in E^*_{\lambda^*}$, where $\lambda \neq \lambda^* \in \sigma(\mathcal{A})$. Since, by construction and Theorem C.4, the set of all eigenvectors and generalized eigenvectors spans $\mathbb{R}^d$,

$$\langle x, v \rangle = 0$$

for all $x \in \mathbb{R}^d$, we get $v = 0$, which contradicts the assumption. Hence, there is $\tilde{v}^* \in E^*_\lambda$ such that

$$\langle \tilde{v}^*, v \rangle \neq 0$$

and thus (5.3.92) follows by taking

$$v^* = \frac{\tilde{v}^*}{\langle \tilde{v}^*, v \rangle}.$$

□

The usefulness of the above theorem becomes clear if we additionally assume that the bases $\{v^1_\lambda, \dots, v^l_\lambda\}$ and $\{v^{*1}_\lambda, \dots, v^{*l}_\lambda\}$ in E_λ and E^*_λ, respectively, are also biorthonormal, that is, they satisfy

$$\langle v^i_\lambda, v^{*j}_\lambda \rangle = \delta_{ij}, \quad 1 \leq i, j \leq l, \tag{5.3.93}$$

where δ_{ij} is the Kronecker delta, that is, $\delta_{ij} = 1$ if $i = j$ and zero otherwise. This can be done by a procedure similar to the Gram–Schmidt orthogonalization, see [59].

This observation, together with Theorem 5.21, shows that there exist biorthonormal bases

$$B = \{v^1_{\lambda_1}, \dots, v^{n_1}_{\lambda_1}, \dots, v^1_{\lambda_k}, \dots, v^{n_k}_{\lambda_k}\}, \quad B^* = \{v^{*1}_{\lambda_1}, \dots, v^{*n_1}_{\lambda_1}, \dots, v^{*1}_{\lambda_k}, \dots, v^{*n_k}_{\lambda_k}\}$$

in $\mathbb{R}^d$ consisting of eigenvectors and generalized eigenvectors of $\mathcal{A}$ and $\mathcal{A}^T$ belonging to, respectively, eigenvalues $\lambda_1, \ldots, \lambda_k$ with algebraic multiplicities $n_1, \ldots, n_k, n_1 + \cdots + n_k = d$. This implies a more explicit version of the spectral decomposition (5.3.80): for any $\boldsymbol{x} \in \mathbb{R}^d$,

$$\boldsymbol{x} = \langle \boldsymbol{v}^{*1}_{\lambda_1}, \boldsymbol{x}\rangle \boldsymbol{v}^{1}_{\lambda_1} + \cdots + \langle \boldsymbol{v}^{*n_k}_{\lambda_k}, \boldsymbol{x}\rangle \boldsymbol{v}^{n_k}_{\lambda_k}. \tag{5.3.94}$$

Exercise 5.3.15 Find biorthonormal bases for the matrices $\mathcal{A}$ introduced in Examples 5.10 and 5.15.

Exercise 5.3.16 Let $E_\lambda = \mathcal{L}in\{\boldsymbol{v}^1, \boldsymbol{v}^2\}$ and $E^*_\lambda = \mathcal{L}in\{\boldsymbol{v}^{*1}, \boldsymbol{v}^{*2}\}$ be associated eigenspaces of some matrix $\mathcal{A}$ belonging to the same eigenvalue λ.

(i) Show that no $\boldsymbol{w}^* \in E^*_\lambda$ satisfies $\langle \boldsymbol{w}^*, \boldsymbol{w}\rangle = 0$ for all $\boldsymbol{w} \in E_\lambda$. Hint. Use the argument from the end of the proof of Theorem 5.21.
(ii) Construct biorthonormal bases in E_λ and E^*_λ.

Remark 5.22 Once we have the biorthonormal bases B and B^*, it is easy to find the inverse of the matrix

$$\mathcal{V} = \begin{pmatrix} | & \cdots & | \\ \boldsymbol{v}^1_{\lambda_1} & \cdots & \boldsymbol{v}^{n_k}_{\lambda_k} \\ | & \cdots & | \end{pmatrix}.$$

Indeed, by the definition of B^*,

$$\begin{pmatrix} \text{---} & \boldsymbol{v}^{*1}_{\lambda_1} & \text{---} \\ \vdots & \vdots & \vdots \\ \text{---} & \boldsymbol{v}^{*n_k}_{\lambda_k} & \text{---} \end{pmatrix} \begin{pmatrix} | & \cdots & | \\ \boldsymbol{v}^1_{\lambda_1} & \cdots & \boldsymbol{v}^{n_k}_{\lambda_k} \\ | & \cdots & | \end{pmatrix} = \mathcal{I},$$

hence, as both matrices are invertible,

$$\mathcal{V}^{-1} = \begin{pmatrix} \text{---} & \boldsymbol{v}^{*1}_{\lambda_1} & \text{---} \\ \vdots & \vdots & \vdots \\ \text{---} & \boldsymbol{v}^{*n_k}_{\lambda_k} & \text{---} \end{pmatrix}. \tag{5.3.95}$$

In summary, to determine the long-term time behaviour of a population described by either discrete, $\boldsymbol{y}(k+1) = \mathcal{A}\boldsymbol{y}$, or continuous system, $\boldsymbol{y}' = \mathcal{A}\boldsymbol{y}$, we have to

1. Find eigenvalues of $\mathcal{A}$ and determine whether there is the dominant eigenvalue, that is, a simple real eigenvalue, say, $\lambda_{\max}$ satisfying $\lambda_{\max} > |\lambda|$ in discrete time, or $\lambda_{\max} > \Re\lambda$ in continuous time, for any other $\lambda \in \sigma(\mathcal{A})$.
2. If this is the case, we find the eigenvector $\boldsymbol{v}^{\max}$ of $\mathcal{A}$ and $\boldsymbol{v}^{*\max}$ of $\mathcal{A}^*$ corresponding to $\lambda_{\max}$.

3. The long time behaviour of the population is then described by

$$\mathcal{A}^k \boldsymbol{x} \approx \lambda_{\max}^k \langle \boldsymbol{v}^{*\,\mathbf{max}}, \boldsymbol{x}\rangle \boldsymbol{v}^{\mathbf{max}} \tag{5.3.96}$$

for large k in the discrete case or, in the continuous case, by

$$e^{t\mathcal{A}} \boldsymbol{x} \approx e^{\lambda_{\max} t} \langle \boldsymbol{v}^{*\,\mathbf{max}}, \boldsymbol{x}\rangle \boldsymbol{v}^{\mathbf{max}} \tag{5.3.97}$$

for large time, for any initial distribution of the population satisfying $\langle \boldsymbol{v}^{*\,\mathbf{max}}, \boldsymbol{x}\rangle \neq 0$.

Of course, if 1. is not satisfied, then the system's long-term behaviour may be more complicated, as seen in Example 5.4.

We illustrate this result by finding the long-term behaviour of solutions to the system discussed in Example 5.13.

Example 5.23 Consider

$$\boldsymbol{y}' = \begin{pmatrix} 1 & -1 & 4 \\ 3 & 2 & -1 \\ 2 & 1 & -1 \end{pmatrix} \boldsymbol{y}.$$

The eigenvalues of $\mathcal{A}$ are $\lambda_1 = 1$, $\lambda_2 = 3$ and $\lambda_3 = -2$. We found eigenvectors corresponding to this eigenvalues to be

$$\boldsymbol{v}^{\mathbf{1}} = \begin{pmatrix} -1 \\ 4 \\ 1 \end{pmatrix}, \quad \boldsymbol{v}^{\mathbf{2}} = \begin{pmatrix} 1 \\ 2 \\ 1 \end{pmatrix}, \quad \boldsymbol{v}^{\mathbf{3}} = \begin{pmatrix} -1 \\ 1 \\ 1 \end{pmatrix},$$

and the general solution

$$\boldsymbol{y}(t) = C_1 e^{t} \begin{pmatrix} -1 \\ 4 \\ 1 \end{pmatrix} + C_2 e^{3t} \begin{pmatrix} 1 \\ 2 \\ 1 \end{pmatrix} + C_3 e^{-2t} \begin{pmatrix} -1 \\ 1 \\ 1 \end{pmatrix}.$$

Clearly, the dominant eigenvalue is $\lambda_2 = 3$, hence writing

$$\boldsymbol{y}(t) = e^{3t} \left(C_2 \begin{pmatrix} 1 \\ 2 \\ 1 \end{pmatrix} + C_1 e^{-2t} \begin{pmatrix} -1 \\ 4 \\ 1 \end{pmatrix} + C_3 e^{-5t} \begin{pmatrix} -1 \\ 1 \\ 1 \end{pmatrix} \right). \tag{5.3.98}$$

we see that for large times,

$$y(t) \approx e^{3t} C_2 \begin{pmatrix} 1 \\ 2 \\ 1 \end{pmatrix}, \quad (5.3.99)$$

where C_2 depends on the initial condition.

The transposed matrix is given by

$$\mathcal{A}^T = \begin{pmatrix} 1 & 3 & 2 \\ -1 & 2 & 1 \\ 4 & -1 & -1 \end{pmatrix},$$

and the eigenvector $\boldsymbol{v}^{*2}$ belonging to $\lambda_2 = 3$ can be calculated by solving

$$(\mathcal{A}^T - 3\mathcal{I})\boldsymbol{v} = \begin{pmatrix} -2 & 3 & 2 \\ -1 & -1 & 1 \\ 4 & -1 & -4 \end{pmatrix} \begin{pmatrix} v_1 \\ v_2 \\ v_3 \end{pmatrix} = \begin{pmatrix} 0 \\ 0 \\ 0 \end{pmatrix}.$$

Hence, we get $v_2 = 0$ and $v_1 = v_3$. Thus, $\boldsymbol{v}^{2*} = (1, 0, 1)$ and we can check that, indeed, $\langle \boldsymbol{v}^{*2}, \boldsymbol{v}^1 \rangle = \langle \boldsymbol{v}^{*2}, \boldsymbol{v}^2 \rangle = 0$. Then, multiplying (5.3.98) by $\boldsymbol{v}^{*2}$, we obtain

$$\langle \boldsymbol{v}^{*2}, \boldsymbol{y}(t) \rangle = C_2 e^{\lambda_2 t} \langle \boldsymbol{v}^{*2}, \boldsymbol{v}^2 \rangle$$

so, taking $t = 0$ we have

$$\langle \boldsymbol{v}^{*2}, \mathring{\boldsymbol{y}} \rangle = C_2 \langle \boldsymbol{v}^{*2}, \boldsymbol{v}^2 \rangle,$$

and $C_2 = \frac{1}{2}(\mathring{y}_1 + \mathring{y}_3)$. Clearly, long time picture of evolution given by (5.3.99) will not be realized if $\mathring{\boldsymbol{y}}$ is orthogonal to $\boldsymbol{v}^{*2}$.

Example 5.24 Returning to Fibonacci rabbits, we see that the eigenvalues of $\mathcal{L}$ are exactly

$$\lambda_{1,2} = r_{\pm} = \frac{1 \pm \sqrt{5}}{2}$$

and clearly, $\lambda_1 = \frac{1+\sqrt{5}}{2}$ is the dominant eigenvalue. The eigenvector associated with this eigenvalue is $\boldsymbol{v}^1 = (\lambda_1, 1) = \left(\frac{\sqrt{5}+1}{2}, 1\right)$ and this gives the stable age structure. Moreover, the matrix $\mathcal{L}$ is symmetric and thus, eigenvectors of $\mathcal{L}^T$ are the same as of $\mathcal{L}$. Thus, for large k,

$$\boldsymbol{v}(k) = \begin{pmatrix} v_1(k) \\ v_0(k) \end{pmatrix} \approx C_1 r_+^k \begin{pmatrix} \frac{\sqrt{5}+1}{2} \\ 1 \end{pmatrix},$$

where

$$C_1 = \frac{2\left(v_1(0)\frac{\sqrt{5}+1}{2} + v_0(0)\right)}{5+\sqrt{5}}$$

as $\langle \boldsymbol{v}^1, \boldsymbol{v}^1 \rangle = \frac{5+\sqrt{5}}{2}$.

Taking, for instance, the initial condition discussed in Sect. 5.1: $v_1(0) = 0, v_0(0) = 1$, we find $C_1 = \frac{2}{5+\sqrt{5}}$. If we like to estimate the growth of the whole population, we obtain

$$y(k) = v_1(k) + v_0(k) \approx \frac{2}{5+\sqrt{5}}\left(\frac{\sqrt{5}+1}{2} + 1\right) r_+^k = \frac{3+\sqrt{5}}{5+\sqrt{5}} r_+^k = \frac{1+\sqrt{5}}{2\sqrt{5}} r_+^k,$$

in accordance with (5.3.69).

Example 5.25 Consider a population with individuals living up to two years in which juveniles and adults can reproduce with effective maternity rates of juveniles and adults per capita, respectively, 1 and 4. Assume further that the survival rate of juveniles is $1/2$. Find the formula for the evolution of this population.

The question amounts to finding iterates $\mathcal{L}^k$ of the Leslie matrix

$$\mathcal{L} = \begin{pmatrix} 1 & 4 \\ \frac{1}{2} & 0 \end{pmatrix}.$$

We start with finding eigenvalues of $\mathcal{L}$ by solving the characteristic equation

$$p(\lambda) = \begin{vmatrix} 1-\lambda & 4 \\ \frac{1}{2} & -\lambda \end{vmatrix} = \lambda^2 - \lambda - 2 = 0.$$

Hence, the eigenvalues are $\lambda = 2$ and -1, both of algebraic multiplicity 1. Thus, the eigenvalues are simple. To find eigenvectors corresponding to $\lambda_1 = 2$, we solve

$$(\mathcal{L} - 2\mathcal{I})\boldsymbol{v} = \begin{pmatrix} -1 & 4 \\ \frac{1}{2} & -2 \end{pmatrix}\begin{pmatrix} v_1 \\ v_2 \end{pmatrix} = \begin{pmatrix} 0 \\ 0 \end{pmatrix}.$$

Thus, v_2 is arbitrary, say $v_2 = 1$, and therefore $v_1 = 4$, so that

$$\boldsymbol{v}^1 = \begin{pmatrix} 4 \\ 1 \end{pmatrix}.$$

Repeating the calculations for $\lambda_2 = -1$, we obtain

$$(\mathcal{L} + 1\mathcal{I})\boldsymbol{v} = \begin{pmatrix} 2 & 4 \\ \frac{1}{2} & 1 \end{pmatrix}\begin{pmatrix} v_1 \\ v_2 \end{pmatrix} = \begin{pmatrix} 0 \\ 0 \end{pmatrix}.$$

Again, we take $v_2 = 1$ and therefore $v_1 = -2$, so that

$$\boldsymbol{v}^2 = \begin{pmatrix} -2 \\ 1 \end{pmatrix}.$$

Therefore,

$$\mathcal{L}^k \boldsymbol{v}^1 = 2^k \begin{pmatrix} 4 \\ 1 \end{pmatrix}, \qquad \mathcal{L}^k \boldsymbol{v}^2 = (-1)^k \begin{pmatrix} -2 \\ 1 \end{pmatrix}.$$

To find explicit expression for $\mathcal{L}^k$, we use (5.3.27); the solution originating from $\mathring{\boldsymbol{x}} = (\mathring{x}_1, \mathring{x}_2)$ is given by

$$\mathcal{L}^k \mathring{\boldsymbol{x}} = c_1 2^k \begin{pmatrix} 4 \\ 1 \end{pmatrix} + c_2(-1)^k \begin{pmatrix} -2 \\ 1 \end{pmatrix} = \begin{pmatrix} 4 \cdot 2^k & -2 \cdot (-1)^k \\ 2^k & (-1)^k \end{pmatrix} \begin{pmatrix} c_1 \\ c_2 \end{pmatrix}.$$

The constants c_1 and c_2 are determined from $\mathring{\boldsymbol{x}} = c_1 \boldsymbol{v}^1 + c_2 \boldsymbol{v}^2$ so that

$$\begin{pmatrix} c_1 \\ c_2 \end{pmatrix} = \mathcal{V}^{-1} \mathring{\boldsymbol{x}} = \begin{pmatrix} \frac{1}{6} & \frac{1}{3} \\ -\frac{1}{6} & \frac{2}{3} \end{pmatrix} \mathring{\boldsymbol{x}},$$

where

$$\mathcal{V} = \begin{pmatrix} | & | \\ \boldsymbol{v}^1 & \boldsymbol{v}^2 \\ | & | \end{pmatrix} = \begin{pmatrix} 4 & -2 \\ 1 & 1 \end{pmatrix}.$$

Finally

$$\mathcal{L}^k \mathring{\boldsymbol{x}} = \begin{pmatrix} 4 \cdot 2^k & -2 \cdot (-1)^k \\ 2^k & (-1)^k \end{pmatrix} \begin{pmatrix} \frac{1}{6} & \frac{1}{3} \\ -\frac{1}{6} & \frac{2}{3} \end{pmatrix} \mathring{\boldsymbol{x}},$$

where the matrix on the left-hand side is the matrix $\mathcal{A}_k$, introduced in (5.3.27). Now, we see that for large k, we have

$$\mathcal{L}^k \mathring{\boldsymbol{x}} = \begin{pmatrix} x_1(k) \\ x_2(k) \end{pmatrix} \approx c_1 2^k \begin{pmatrix} 4 \\ 1 \end{pmatrix} = 2^k \begin{pmatrix} 4 \\ 1 \end{pmatrix} \left(\frac{1}{6} \mathring{x}_1 + \frac{1}{3} \mathring{x}_2 \right).$$

To provide a better interpretation of the above formula, we note that the total population approximately evolves as

$$P(k) = x_1(k) + x_2(k) \approx 2^k \left(\frac{1}{6} \mathring{x}_1 + \frac{1}{3} \mathring{x}_2 \right) (4 + 1)$$

and therefore, the fractions of juveniles and adults in the total population approximately are given by, respectively,

$$J = \frac{x_1(k)}{P(k)} \approx \frac{4}{5}, \qquad A = \frac{x_2(k)}{P(k)} = \frac{1}{5}.$$

Thus, the eigenvector of the dominant eigenvalue normalized with respect to the total population,

$$(J, A) = \frac{1}{5}(4, 1)$$

gives the so-called stable age profile of the population.

The next example shows how to deal with complex eigenvalues.

Example 5.26 Consider a population with individuals living up to 3 years in which only the eldest generation reproduces with an effective birth rate of 32 offspring per individual. Assume further that the survival rate of juveniles is $3/4$ and of yearlings to the 2-year-old is $1/3$. Find the formula for the evolution of this population.

Here we have

$$\mathcal{L} = \begin{pmatrix} 0 & 0 & 32 \\ \frac{3}{4} & 0 & 0 \\ 0 & \frac{1}{3} & 0 \end{pmatrix},$$

and the eigenvalues are determined by solving

$$\begin{vmatrix} -\lambda & 0 & 32 \\ \frac{3}{4} & -\lambda & 0 \\ 0 & \frac{1}{3} & -\lambda \end{vmatrix} = -\lambda^3 + 8 = 0.$$

Thus,

$$\lambda_1 = 2, \qquad \lambda_2 = 2e^{\frac{2\pi i}{3}} = 2\left(\frac{-1}{2} + i\frac{\sqrt{3}}{2}\right), \qquad \lambda_3 = \bar{\lambda}_2 = 2\left(\frac{-1}{2} - i\frac{\sqrt{3}}{2}\right).$$

We find general form of eigenvectors $\boldsymbol{v}^i$ corresponding to eigenvalues λ_i, $i = 1, 2, 3$, by solving

$$\begin{pmatrix} -\lambda_i & 0 & 32 \\ \frac{3}{4} & -\lambda_i & 0 \\ 0 & \frac{1}{3} & -\lambda_i \end{pmatrix} \begin{pmatrix} v_1^i \\ v_2^i \\ v_3^i \end{pmatrix} = \begin{pmatrix} 0 \\ 0 \\ 0 \end{pmatrix}.$$

Taking $v_3^i = 1$, we find $v_2^i = 3\lambda_i$ and $v_1^i = 4\lambda_i^2$ (note that the first equation of the system gives the eigenvalue equation). Thus, we can write

$$\boldsymbol{v}^{\mathbf{1}} = \begin{pmatrix} 16 \\ 6 \\ 1 \end{pmatrix}.$$

Then

$$\boldsymbol{v}^{\mathbf{2}} = \begin{pmatrix} 16e^{\frac{4\pi i}{3}} \\ 6e^{\frac{2\pi i}{3}} \\ 1 \end{pmatrix} = \begin{pmatrix} 16\left(\frac{-1}{2} - i\frac{\sqrt{3}}{2}\right) \\ 6\left(\frac{-1}{2} + i\frac{\sqrt{3}}{2}\right) \\ 1 \end{pmatrix}$$

and

$$\boldsymbol{v}^{\mathbf{3}} = \overline{\boldsymbol{v}^{\mathbf{2}}} = \begin{pmatrix} 16e^{-4\pi i/3} \\ 6e^{-2\pi i/3} \\ 1 \end{pmatrix} = \begin{pmatrix} 16\left(\frac{-1}{2} + i\frac{\sqrt{3}}{2}\right) \\ 6\left(\frac{-1}{2} - i\frac{\sqrt{3}}{2}\right) \\ 1 \end{pmatrix}.$$

One could write the general solution of the problem as

$$\mathcal{L}^k\mathring{\boldsymbol{x}} = \begin{pmatrix} x_1(k) \\ x_2(k) \\ x_3(k) \end{pmatrix} = c_1 2^k \boldsymbol{v}^{\mathbf{1}} + c_2 2^k e^{2k\pi i/3}\boldsymbol{v}^{\mathbf{2}} + c_3 2^k e^{-2k\pi i/3}\overline{\boldsymbol{v}^{\mathbf{2}}}, \tag{5.3.100}$$

where $\mathring{\boldsymbol{x}} = c_1\boldsymbol{v}^{\mathbf{1}} + c_2\boldsymbol{v}^{\mathbf{2}} + c_3\boldsymbol{v}^{\mathbf{3}}$, but the expressions above are complex, while we require real solutions. Following the general theory, see Example 5.14 and the preceding discussion, we take advantage of the fact that complex solutions appear as a complex conjugate pair. To find explicit real expressions for the eigenvalues, we proceed as in Sect. 5.3.4 and write $\lambda = re^{i\phi}$ where $r = |\lambda|$ and $\phi = \arg\lambda$. Then,

$$\lambda^k = r^n e^{in\phi} = r^k(\cos k\phi + i\sin k\phi),$$

and

$$\Re(\lambda^k\boldsymbol{v}) = r^k(\cos k\phi\Re\boldsymbol{v} - \sin k\phi\Im\boldsymbol{v}),$$
$$\Im(\lambda^k\boldsymbol{v}) = r^k(\sin k\phi\Re\boldsymbol{v} + \cos k\phi\Im\boldsymbol{v}).$$

In our case

$$\Re\lambda_2^k\boldsymbol{v}^{\mathbf{2}} = 2^k\left(\cos\frac{2k\pi}{3}\begin{pmatrix} -8 \\ -3 \\ 1 \end{pmatrix} - \sin\frac{2k\pi}{3}\begin{pmatrix} -8\sqrt{3} \\ 3\sqrt{3} \\ 0 \end{pmatrix}\right)$$

and

$$\Im\lambda_2^k \boldsymbol{v}^{\mathbf{2}} = 2^k \left(\sin \frac{2k\pi}{3} \begin{pmatrix} -8 \\ -3 \\ 1 \end{pmatrix} + \cos \frac{2k\pi}{3} \begin{pmatrix} -8\sqrt{3} \\ 3\sqrt{3} \\ 0 \end{pmatrix} \right).$$

The initial values for $\Re\lambda^k \boldsymbol{v}^{\mathbf{2}}$ and $\Im\lambda^k \boldsymbol{v}^{\mathbf{2}}$ are, respectively, $\begin{pmatrix} -8 \\ -3 \\ 1 \end{pmatrix}$ and $\begin{pmatrix} -8\sqrt{3} \\ 3\sqrt{3} \\ 0 \end{pmatrix}$. Since $\mathcal{L}^k$ is a real matrix, we have $\Re\mathcal{L}^k \boldsymbol{v}^{\mathbf{2}} = \mathcal{L}^k \Re \boldsymbol{v}^{\mathbf{2}}$ and $\Im\mathcal{L}^k \boldsymbol{v}^{\mathbf{2}} = \mathcal{L}^k \Im \boldsymbol{v}^{\mathbf{2}}$, thus

$$\begin{aligned} \mathcal{L}^k \begin{pmatrix} -8 \\ -3 \\ 1 \end{pmatrix} &= 2^k \left(\cos \frac{2k\pi}{3} \begin{pmatrix} -8 \\ -3 \\ 1 \end{pmatrix} - \sin \frac{2k\pi}{3} \begin{pmatrix} -8\sqrt{3} \\ 3\sqrt{3} \\ 0 \end{pmatrix} \right) \\ &= 2^k \begin{pmatrix} -8\left(\cos \frac{2k\pi}{3} - \sqrt{3} \sin \frac{2k\pi}{3}\right) \\ -3\left(\cos \frac{2k\pi}{3} + \sqrt{3} \sin \frac{2k\pi}{3}\right) \\ \cos \frac{2k\pi}{3} \end{pmatrix} \end{aligned}$$

and

$$\begin{aligned} \mathcal{L}^k \begin{pmatrix} -8\sqrt{3} \\ 3\sqrt{3} \\ 0 \end{pmatrix} &= 2^k \left(\sin \frac{2k\pi}{3} \begin{pmatrix} -8 \\ -3 \\ 1 \end{pmatrix} + \cos \frac{2k\pi}{3} \begin{pmatrix} -8\sqrt{3} \\ 3\sqrt{3} \\ 0 \end{pmatrix} \right) \\ &= 2^k \begin{pmatrix} -8\left(\sin \frac{2k\pi}{3} + \sqrt{3} \cos \frac{2k\pi}{3}\right) \\ -3\left(\sin \frac{2k\pi}{3} - \sqrt{3} \cos \frac{2k\pi}{3}\right) \\ \sin \frac{2k\pi}{3} \end{pmatrix}. \end{aligned}$$

Thus, we see that (5.3.100) can be written using only real expressions as

$$\begin{aligned} \mathcal{L}^k \mathring{\boldsymbol{x}} = c_1 2^k \begin{pmatrix} 16 \\ 6 \\ 1 \end{pmatrix} &+ c_2 2^k \begin{pmatrix} -8\left(\cos \frac{2k\pi}{3} - \sqrt{3} \sin \frac{2k\pi}{3}\right) \\ -3\left(\cos \frac{2k\pi}{3} + \sqrt{3} \sin \frac{2k\pi}{3}\right) \\ \cos \frac{2k\pi}{3} \end{pmatrix} \\ &+ c_3 2^k \begin{pmatrix} -8\left(\sin \frac{2k\pi}{3} + \sqrt{3} \cos \frac{2k\pi}{3}\right) \\ -3\left(\sin \frac{2k\pi}{3} - \sqrt{3} \cos \frac{2k\pi}{3}\right) \\ \sin \frac{2k\pi}{3} \end{pmatrix}, \end{aligned}$$

where $\boldsymbol{c} = (c_1, c_2, c_3)$ are solutions to

$$\mathring{\boldsymbol{x}} = \mathcal{V}\boldsymbol{c}$$

with

$$\mathcal{V} = \begin{pmatrix} 16 & -8 & -8\sqrt{3} \\ 6 & -3 & 3\sqrt{3} \\ 1 & 1 & 0 \end{pmatrix}.$$

The final example illustrates a case when we do not have enough eigenvectors.

Example 5.27 Consider a population divided into two classes: reproductive and post-reproductive. Assume that the reproductive season lasts exactly one year and the effective fertility is 0.5. Assume further that 20% of this class survives the first year and moves to the postreproductive class. The individuals can live indefinitely, but every year, 50% of individuals in the postreproductive class die. Write down the Usher matrix for this population and find its eigenvalues and eigenvectors.

Here we have

$$\mathcal{U} = \begin{pmatrix} 0.5 & 0 \\ 0.2 & 0.5 \end{pmatrix}.$$

The characteristic equation is

$$p(\lambda) = \begin{vmatrix} 0.5 - \lambda & 0 \\ 0.2 & 0.5 - \lambda \end{vmatrix} = \lambda^2 - \lambda + 0.25 = 0,$$

which gives double eigenvalue $\lambda = 0.5$ so that the algebraic multiplicity of λ is 2. Let us find eigenvectors and associated eigenvectors. To find eigenvectors corresponding to $\lambda_1 = 2$, we solve

$$(\mathcal{U} - 0.5\mathcal{I})\boldsymbol{v} = \begin{pmatrix} 0 & 0 \\ 0.2 & 0 \end{pmatrix} \begin{pmatrix} v_1 \\ v_2 \end{pmatrix} = \begin{pmatrix} 0 \\ 0 \end{pmatrix},$$

which gives $\boldsymbol{v}^2 = (0, 1)$; this is the only eigenvector so that the eigenvalue is not semisimple. Hence, we proceed to find the associated eigenvector by solving

$$(\mathcal{U} - 0.5\mathcal{I})^2\boldsymbol{v} = \begin{pmatrix} 0 & 0 \\ 0 & 0 \end{pmatrix} \begin{pmatrix} v_1 \\ v_2 \end{pmatrix} = \begin{pmatrix} 0 \\ 0 \end{pmatrix}.$$

Any vector satisfies the above equation. We select the simplest vector linearly independent of $\boldsymbol{v}^2 = (0, 1)$ which is $\boldsymbol{v}^1 = (1, 0)$. In this way, we obtain $\mathcal{V} = \mathcal{I}$.

Thus we can write, for $\mathring{\boldsymbol{x}} = (\mathring{x}_1, \mathring{x}_2)$,

$$\begin{aligned}
\mathcal{U}^k \mathring{\boldsymbol{x}} &= \mathring{x}_1 \mathcal{U}^k \boldsymbol{v}^1 + \mathring{x}_2 \mathcal{U}^k \boldsymbol{v}^2 = \mathring{x}_1 (0.5\mathcal{I} + (\mathcal{U} - 0.5\mathcal{I}))^k \boldsymbol{v}^1 + \mathring{x}_2 (0.5)^k \boldsymbol{v}^2 \\
&= \mathring{x}_1 \left((0.5)^k \boldsymbol{v}^1 + k(0.5)^{k-1} \begin{pmatrix} 0 & 0 \\ 0.2 & 0 \end{pmatrix} \boldsymbol{v}^1 \right) + \mathring{x}_2 (0.5)^k \boldsymbol{v}^2 \\
&= \mathring{x}_1 \left((0.5)^k \begin{pmatrix} 1 \\ 0 \end{pmatrix} + k(0.5)^{k-1} \begin{pmatrix} 0 \\ 0.2 \end{pmatrix} \right) + \mathring{x}_2 (0.5)^k \begin{pmatrix} 0 \\ 1 \end{pmatrix} = (0.5)^k \boldsymbol{p}_{0.5}(k) \mathring{\boldsymbol{x}},
\end{aligned}$$

where

$$\boldsymbol{p}_{0.5}(k) = k \begin{pmatrix} 0 & 0 \\ 0.4 & 0 \end{pmatrix} + \begin{pmatrix} 1 & 0 \\ 0 & 1 \end{pmatrix}$$

is a first order spectral polynomial in k, see (5.3.85).

Exercise 5.3.17 Consider a population in which no individual can live longer than three years. Assume that all newborns survive the first year, half of the yearlings survive the next year, and half of the 2-year-olds live till the 3rd year. Further, assume that the natural maternity per capita rate for juveniles is 1; for yearlings, it is four, and 2-year-olds do not reproduce. Find the Leslie matrix for this population, derive the formula for the stable age profile and determine the population's long-term behaviour.

Exercise 5.3.18 Consider a plant living up to two years. In the first year, young plants produce seeds, which develop into new plants in the following year. Seeds produced by two young plants produce one new plant. A quarter of new plants survive winter and produce seeds the following year. On the other hand, seeds produced from old plants give two new plants each. Suggest a model describing this population and discuss the population's behaviour as time goes to infinity, assuming that the initial population consists of six young and three old plants.

Exercise 5.3.19 Describe in words a population governed by the Usher matrix

$$\mathcal{L} = \begin{pmatrix} s_1 m & 0 \\ s_2 & s_3 \end{pmatrix}, \quad 0 < s_1, s_2, s_3 \le 1, m \ge 0.$$

Find the birth rates m that ensure the survival of the population.

Exercise 5.3.20 Determine long-term behaviour of solutions to

$$y' = \begin{pmatrix} 3 & 2 & 4 \\ 2 & 0 & 2 \\ 4 & 2 & 3 \end{pmatrix} y;$$

$$y' = \begin{pmatrix} -3 & 0 & 2 \\ 1 & -1 & 0 \\ -2 & -1 & 0 \end{pmatrix} y;$$

$$y' = \begin{pmatrix} 1 & 0 & -2 \\ 0 & 1 & 0 \\ 1 & -1 & -1 \end{pmatrix} y;$$

$$y' = \begin{pmatrix} 1 & -1 & 4 \\ 3 & 2 & -1 \\ 2 & 1 & -1 \end{pmatrix} y.$$

In each case, determine initial conditions which do not yield the determined typical growth rate.

Exercise 5.3.21 Determine the long-term behaviour of solutions to the following systems of difference equations

$$\begin{pmatrix} x_1(k+1) \\ x_2(k+1) \\ x_3(k+1) \end{pmatrix} = \begin{pmatrix} 1 & -2 & -2 \\ 0 & 0 & -1 \\ 0 & 2 & 3 \end{pmatrix} \begin{pmatrix} x_1(k) \\ x_2(k) \\ x_3(k) \end{pmatrix};$$

$$\begin{pmatrix} x_1(k+1) \\ x_2(k+1) \\ x_3(k+1) \end{pmatrix} = \begin{pmatrix} 1 & 1 & 0 \\ -1 & 1 & 0 \\ 1 & 0 & 1 \end{pmatrix} \begin{pmatrix} x_1(k) \\ x_2(k) \\ x_3(k) \end{pmatrix}.$$

In each case, find initial conditions which do not yield the determined typical growth rate.

Exercise 5.3.22 Determine the long-term behaviour of solutions to the following higher-order equations

$$x(k+2) + 8x(k+1) + 12x(k) = 0;$$

$$x(k+2) + 4x(k) = 0;$$

$$x(k+3) - 7x(k+2) + 16x(k+1) - 12x(k) = 0.$$

In each case, find initial conditions which do not yield the determined typical growth rate.

5.4 Positive Linear Dynamical Systems—Frobenius–Perron Theorem

5.4.1 Problems with General Linear Models—The Natchez Structure

Many societies are divided into classes, the membership of which is largely hereditary. Such a class structure is essential in maintaining power and controlling the distribution of resources by an appropriately defined elite. One of the ways to prevent the watering down of the elite is the practice of endogamy, that is, marrying within one's own class. Some societies, however, practice an open class system to prevent stagnation of the structure. Following [67], we describe and analyse one of the most famous societies of such a type—the civilization of Natchez, see [1, 66, 91].

Natchez were Native Americans who lived in the lower Mississippi in North America. The early accounts about them came from the Spanish explorer Hernando de Soto in 1542. Most of the information about Natchez is due to their contacts with French colonisers who first met them in 1682. The civilisation ceased to exist after the so-called Natchez massacre in 1731, when they were defeated and then dispersed or were enslaved.

Natchez created a complex system of open class structure based on exogamous marriages so that the power was passed between people born to different social classes. Society was divided into two main classes: the nobility and Commoners (in some papers, they were called Stinkers). The nobility was further divided into subclasses (casts): Suns, Nobles and Honoured People. A member of nobility could only marry a Commoner. According to [91], a person's social status and class were determined matrilineally; that is, the children of female Suns, Nobles, or Honoureds kept the status of their mothers. However, the children of male Suns and Nobles did not become Commoners, as noble exogamy and matrilineal descent would appear to dictate, but rather moved one class below the class of their fathers. In other words, children of male Suns became Nobles, while children of male Nobles became Honoured.

We summarize the permissible marriages and inherited statuses in Table 5.1 below; see also [67]. An empty place intersection of the row and a column in the table means that such a marriage is not permitted; if it is permitted, the entry at the intersection indicates the status of the offspring. To analyse the evolution of this

Table 5.1 Possible marriages in the Natchez population and the status of their offspring

Mother/Father	Sun	Noble	Honoured	Commoner
Sun				Sun
Noble				Noble
Honoured				Honoured
Commoner	Noble	Honoured	Commoner	Commoner

population, we shall create a mathematical model based on Table 5.1. To make it feasible, we adopt the following simplifying assumptions.

1. There is the same number of males and females in each class in each generation.
2. Each person marries only once, and the spouse is from the same generation.
3. Each pair has exactly one son and one daughter.

Since we have the same number of males and females in each generation, we will only track the males in our model. Hence, let $x_i(k)$ denote the number of males in the class i and in the generation k, where the classes are numbered as follows: 1—Sun, 2—Noble, 3—Honoured, 4—Commoner. Let us consider the class distribution in the generation $k+1$. Since a Sun son can only be born to a Sun mother and there is no other way to become a Sun, using the fact that the number of female Suns equals the number of male Suns, we can write

$$x_1(k+1) = x_1(k).$$

Next, a Noble son can only be born to a Sun father or a Noble mother; using the parity of males and females in the Noble class, we get

$$x_2(k+1) = x_1(k) + x_2(k).$$

An Honoured son is either a descendent of a Noble father or an Honoured mother; hence, as before,

$$x_3(k+1) = x_2(k) + x_3(k).$$

Finally, the number of male offspring in the Commoner class is equal to the number of Commoner males who are not married to females from the nobility plus the number of sons of Commoner mothers and Honoured fathers (remember that the son of a Commoner father and the Honoured mother is Honoured but then the son of a Commoner mother and Honoured father is a Commoner). Hence, using again the fact that the numbers of males and females in each class are equal, we arrive at

$$x_4(k+1) = -x_1(k) - x_2(k) + x_4(k).$$

Writing these equations in a matrix form, we obtain

$$\begin{pmatrix} x_1(k+1) \\ x_2(k+1) \\ x_3(k+1) \\ x_4(k+1) \end{pmatrix} = \begin{pmatrix} 1 & 0 & 0 & 0 \\ 1 & 1 & 0 & 0 \\ 0 & 1 & 1 & 0 \\ -1 & -1 & 0 & 0 \end{pmatrix} \begin{pmatrix} x_1(k) \\ x_2(k) \\ x_3(k) \\ x_4(k) \end{pmatrix}. \tag{5.4.101}$$

Let us denote the coefficient matrix by $\mathcal{A}$. The matrix form allows for expressing $\mathcal{A}^k$ in an explicit form. We write $\mathcal{A} = \mathcal{I} + \mathcal{B}$, where

$$\mathcal{B} = \begin{pmatrix} 0 & 0\,0\,0 \\ 1 & 0\,0\,0 \\ 0 & 1\,0\,0 \\ -1 & -1\,0\,0 \end{pmatrix}.$$

It is easy to see that $\mathcal{B}$ is nilpotent. Indeed,

$$\mathcal{B}^2 = \begin{pmatrix} 0\,0\,0\,0 \\ 0\,0\,0\,0 \\ 1\,0\,0\,0 \\ -1\,0\,0\,0 \end{pmatrix}, \qquad \mathcal{B}^3 = \begin{pmatrix} 0\,0\,0\,0 \\ 0\,0\,0\,0 \\ 0\,0\,0\,0 \\ 0\,0\,0\,0 \end{pmatrix}.$$

Hence, using the Newton formula, we get

$$\mathcal{A}^k = (\mathcal{I} + B)^k = \mathcal{I} + k\mathcal{B} + \frac{k(k-1)}{2}\mathcal{B}^2,$$

and, using the formulae for the powers of $\mathcal{B}$,

$$\mathcal{A}^k = \begin{pmatrix} 1 & 0\,0\,0 \\ k & 1\,0\,0 \\ \frac{k(k-1)}{2} & k\;1\;0 \\ \frac{-k(k+1)}{2} & -k\;0\;1 \end{pmatrix}.$$

Hence we obtain

$$\begin{aligned} \boldsymbol{x}(k) = (&x_1(0), kx_1(0) + x_2(0), \frac{1}{2}k(k-1)x_1(0) + kx_2(0) + x_3(0), \\ &-\frac{1}{2}k(k+1)x_1(0) - kx_2(0) + x_4(0)). \end{aligned}$$

In particular, if $x_1(0) = x_2(0) = 0$, then the class structure of the population does not change as we have

$$\boldsymbol{x}(k) = (0, 0, x_3(0), x_4(0)).$$

If, however, we originally had some members of either the Sun or Noble class, then the size of the Commoner class would shrink and become negative in finite time. Indeed, e.g., solving for k

$$-\frac{1}{2}k(k+1)x_1(0) - kx_2(0) + x_4(0) = 0,$$

we obtain a positive solution

$$k_0 = \sqrt{(1+2p)^2 + 8q} - (1+2p),$$

where $p = x_2(0)/x_1(0)$ and $q = x_4(0)/x_1(0)$. For instance, if the numbers of Suns and Nobles are equal ($p = 1$) and the number of Commoners is five times the number of Suns ($q = 5$), we obtain $k_0 = 4$, and then the number of Commoners will become 0 after four generations. Thus, nobody from the nobility will be able to marry and have offspring, and thus, the population will become extinct. Since, according to the historical records, the Natchez civilisation survived for several hundred years, the model in the presented form cannot be correct.

We recognize, however, that we have made several assumptions to simplify the model. We can modify some of them to get a more realistic population evolution model by avoiding its quick collapse. Let us try to adjust the birth coefficients. Indeed, in many societies, it is observed that the birth rate depends on the status of the parents. Thus, let us address the following question: Can we find a birth rate for each combination of parents in the Natchez community which will result in its stable class distribution? Here, we partly follow [67].

To simplify the analysis without compromising the general picture, we restrict ourselves to three classes, which are labelled **A**, **B** and **C** and, following Table 5.1, we construct the intermarriage scheme summarized in Table 5.2. The interpretation of the table is the same as before. The Roman letter at the intersection of a column and a row indicates the class of the offspring coming from the mother in the row and father in the column, and the Greek letter α_i gives the average number of male offspring from such a marriage. Further, as before, we assume that

1. the numbers of males and females are equal in each class and each generation;
2. there are no inter-generation marriages.

By $x_1(k)$, $x_2(k)$, $x_3(k)$, we denote the number of male members of the population in the class, **A**, **B** and **C** in the kth generation. Then

$$x_1(k+1) = \alpha_1 x_1(k). \tag{5.4.102}$$

Similarly, since a child in **B** may be born either to a father from class **A** (and mother from class **C**) or to a mother from class **B** (and father from **C**), we have

$$x_2(k+1) = \alpha_2 x_1(k) + \alpha_3 x_2(k).$$

Table 5.2 Intermarriage scheme in a simplified Natchez-like community

Mother/Father	**A**	**B**	**C**
A			**A** α_1
B			**B** α_3
C	**B** α_2	**C** α_4	**C** α_5

Finally, a child of class **C** may be born to a father of class **B** or father of class **C**, except for those who married females of class **A** or **B**. Hence,

$$x_3(k+1) = \alpha_4 x_2(k) + \alpha_5 \left(x_3(k) - x_1(k) - x_2(k)\right).$$

In matrix form

$$\begin{pmatrix} x_1(k+1) \\ x_2(k+1) \\ x_3(k+1) \end{pmatrix} = \begin{pmatrix} \alpha_1 & 0 & 0 \\ \alpha_2 & \alpha_3 & 0 \\ -\alpha_5 & (\alpha_4 - \alpha_5) & \alpha_5 \end{pmatrix} \begin{pmatrix} x_1(k) \\ x_2(k) \\ x_3(k) \end{pmatrix}, \tag{5.4.103}$$

or, in compact form,

$$\boldsymbol{x}(k+1) = \mathcal{A}\boldsymbol{x}(k).$$

Before we analyze the model, let us check that in the case $\alpha_1 = \alpha_2 = \alpha_3 = \alpha_4 = \alpha_5 = 1$, the model reproduces the behaviour of the full, four class, Natchez model. In such a case, we have

$$\mathcal{A} = \begin{pmatrix} 1\,0\,0 \\ 1\,1\,0 \\ -1\,0\,1 \end{pmatrix}.$$

As in the previous case, we write $\mathcal{A} = \mathcal{I} + \mathcal{B}$ and see that $\mathcal{B}^2 = 0$. Hence, $\mathcal{A}^k = \mathcal{I} + k\mathcal{B}$, or

$$\mathcal{A}^k = \begin{pmatrix} 1\,0\,0 \\ k\,1\,0 \\ -k\,0\,1 \end{pmatrix}.$$

Therefore, $x_3(k) = -kx_1(0) + x_3(0)$ and thus $x_1(0) > 0$ yields $x_3(k) < 0$ for $k > x_3(0)/x_1(0)$. The simplified system displays the same collapse as the original Natchez system.

Let us return to the model (5.4.103). Following the analysis of Sect. 5.3.8, we will try to find a positive eigenvector of $\mathcal{A}$ associated with the largest positive eigenvalue. Such a vector would correspond to a stable population structure, which, once attained, would not change in time. The total population would change as powers of the eigenvalue. In this way, if the structure is perturbed, the model will return to a stable distribution. Moreover, we must ensure that in our model, we always have enough class **C** members to provide spouses to the higher classes. Since the matrix $\mathcal{A}$ is triangular, its eigenvalues are given by the diagonal elements, that is, eigenvalues are α_1, α_3 and α_5. First, let $\alpha_5 > \alpha_1$ and $\alpha_5 > \alpha_3$; that is, α_5 is the dominant eigenvalue of $\mathcal{A}$. This assumption corresponds to maximum fertility in the lowest class. To find the corresponding eigenvector, we solve

$$\begin{pmatrix} (\alpha_1 - \alpha_5) & 0 & 0 \\ \alpha_2 & (\alpha_3 - \alpha_5) & 0 \\ \alpha_5 & (\alpha_4 - \alpha_5) & 0 \end{pmatrix} \begin{pmatrix} x_1 \\ x_2 \\ x_3 \end{pmatrix} = \begin{pmatrix} 0 \\ 0 \\ 0 \end{pmatrix}.$$

Since $x_1 = 0$, there is no positive eigenvector. Moreover, this leads to $x_2 = 0$ and arbitrary x_3, so the only possible long-term structure is given by $(0, 0, 1)$; the population will only consist of Commoners. While it is certainly possible, we are interested in the community's structure overall survival, which is impossible if α_5 is the dominant eigenvalue. A similar outcome is obtained if we assume that α_3 is the dominant eigenvalue. Here also $x_1 = 0$ and though under additional assumption $\alpha_4 > \alpha_5$ we obtain a nonnegative eigenvector $\left(0, 1, \frac{\alpha_4-\alpha_5}{\alpha_3-\alpha_5}\right)$, the structure is nevertheless destroyed.

The last option is to assume that α_1 is a dominant eigenvalue. Thus, let $\alpha_1 > \alpha_3$ and $\alpha_1 > \alpha_5$. The eigenvector corresponding to α_1 satisfies

$$\begin{pmatrix} 0 & 0 & 0 \\ \alpha_2 & (\alpha_3 - \alpha_1) & 0 \\ -\alpha_5 & (\alpha_4 - \alpha_5) & (\alpha_5 - \alpha_1) \end{pmatrix} \begin{pmatrix} x_1 \\ x_2 \\ x_3 \end{pmatrix} = \begin{pmatrix} 0 \\ 0 \\ 0 \end{pmatrix}.$$

The second equation yields

$$x_2 = \frac{\alpha_2}{(\alpha_1 - \alpha_3)} x_1.$$

Since $\alpha_1 > \alpha_3$, we see that $x_2 > 0$ provided $x_1 > 0$. Then, from the last equation, we get

$$x_3 = \frac{1}{\alpha_1 - \alpha_5}\left(-\alpha_5 + \alpha_2 \frac{\alpha_4 - \alpha_5}{\alpha_1 - \alpha_3}\right) x_1.$$

Since $\alpha_1 > \alpha_5$ and $\alpha_1 > \alpha_3$, we have $x_3 > 0$ if and only if

$$-\alpha_5(\alpha_1 - \alpha_3) + \alpha_2(\alpha_4 - \alpha_5) > 0.$$

In other words, the eigenvector corresponding to the dominant eigenvalue α_1 is positive, provided

$$\begin{cases} \alpha_1 > \alpha_3, \\ \alpha_1 > \alpha_5, \\ \alpha_2(\alpha_4 - \alpha_5) > \alpha_5(\alpha_1 - \alpha_3). \end{cases} \tag{5.4.104}$$

We observe that a necessary condition for the last inequality to be satisfied is $\alpha_4 > \alpha_5$. In particular, the inequality will be satisfied if α_5 is sufficiently small. Thus, we have a positive stable distribution if the birth rate in the **C**-class is small.

To complete our considerations, we must show that in each generation, we have sufficiently many members of class **C** for the members of classes **B** and **A** to marry. Thus, we require that the components of the stable distribution vector additionally satisfy

$$x_3 \geq x_1 + x_2. \tag{5.4.105}$$

Substituting here the formulae for x_2 and x_3, which were derived earlier, we obtain the inequality

$$\frac{(\alpha_4 - \alpha_5)}{(\alpha_1 - \alpha_5)} - \frac{\alpha_5(\alpha_1 - \alpha_3)}{\alpha_2(\alpha_1 - \alpha_5)} - \frac{(\alpha_1 - \alpha_3)}{\alpha_2} - 1 \geq 0. \tag{5.4.106}$$

We can satisfy this inequality by taking sufficiently large α_4. In particular, we must have

$$\alpha_4 > \alpha_5. \tag{5.4.107}$$

Thus, to ensure that in the stable population distribution, we have a sufficient number of members of class **C**, we should ensure that α_4; that is, the birth rate of in marriages of mothers of class **C** and fathers of class **B**, is sufficiently large. Of course, this is not a necessary condition. Any set of birth rates satisfying (5.4.105), in addition to (5.4.104), will ensure (5.4.105).

Example 5.28 The birth coefficients $\alpha_1 = 1.2$, $\alpha_2 = 0.9$, $\alpha_3 = 1.1$, $\alpha_4 = 1.4$ and $\alpha_5 = 0.8$ satisfy (5.4.106). Then the stable population structure is given by

$$\boldsymbol{x} = (1, 9, 11.5).$$

We observe that the provided analysis is not complete. It only shows that there is a structure of the Natchez society that can persist stably. However, we do not know whether any initial population will eventually stabilize at the structure determined by the eigenvector found above. Fortunately, the problem allows for a more detailed solution, confirming the results obtained by analysing the dominant eigenvalue and the corresponding eigenvector. First, we observe that the first equation in (5.4.103) is independent of the other two, and the second does not depend on the third one. Then we obtain

$$x_1(k) = \alpha_1^k \mathring{x}_1 \tag{5.4.108}$$

and, using (2.1.49),

$$x_2(k) = \alpha_3^k \mathring{x}_2 + \alpha_2 \mathring{x}_1 \frac{\alpha_1^k - \alpha_3^k}{\alpha_1 - \alpha_3}. \tag{5.4.109}$$

Finally, using (5.4.108) and (5.4.109) and, again, (2.1.49), we have

$$\begin{aligned} x_3(k) &= \alpha_5^k \mathring{x}_3 + \mathring{x}_2(\alpha_4 - \alpha_5) \sum_{i=0}^{k-1} \alpha_5^{k-i-1} \alpha_3^i \\ &\quad + \mathring{x}_1 \frac{\alpha_2(\alpha_4 - \alpha_5}{\alpha_1 - \alpha_3} \left(\sum_{i=0}^{k-1} \alpha_5^{k-i-1} \alpha_1^i - \sum_{i=0}^{k-1} \alpha_5^{k-i-1} \alpha_3^i \right) \\ &\quad - \alpha_5 \mathring{x}_1 \sum_{i=0}^{k-1} \alpha_5^{k-i-1} \alpha_1^i \\ &= \alpha_5^k \mathring{x}_3 + \frac{\alpha_3^k - \alpha_5^k}{\alpha_3 - \alpha_5} (\alpha_4 - \alpha_5) \left(\mathring{x}_2 - \mathring{x}_1 \frac{\alpha_2}{\alpha_1 - \alpha_3} \right) \\ &\quad + \frac{\alpha_1^k - \alpha_5^k}{\alpha_1 - \alpha_5} \left(\frac{\alpha_2(\alpha_4 - \alpha_5)}{\alpha_1 - \alpha_3} - \alpha_5 \right) \mathring{x}_1. \end{aligned} \tag{5.4.110}$$

Assuming that (5.4.107) is satisfied, we see that for $x_3(k)$ to be positive, we need (5.4.104) to be satisfied as well as

$$\frac{\mathring{x}_1}{\mathring{x}_2} < \frac{\alpha_1 - \alpha_3}{\alpha_2}. \tag{5.4.111}$$

Hence, though in general some non-negative initial conditions may lead to a negative solution, that is, to the collapse of the society, positive initial conditions satisfying (5.4.111) ensure that the population will stay positive for all times and the distribution of class sizes will eventually stabilize at the vector

$$\left(1, \frac{\alpha_2}{(\alpha_1 - \alpha_3)}, \frac{1}{\alpha_1 - \alpha_5} \left(-\alpha_5 + \alpha_2 \frac{\alpha_4 - \alpha_5}{\alpha_1 - \alpha_3} \right) \right).$$

Note the relation of the above formula with condition (5.4.111). The latter states that for the solution to remain positive, the ratio of the initial populations in classes **A** and **B** must be smaller than the ratio of these populations in the stable distribution vector. The analysis is, however, still not complete. The reason is that though the model (5.4.103) properly reflects the assumptions, we cannot claim that the model determines the assumptions uniquely. For instance, Eq. (5.4.102) only specifies how many offspring the class **A** produces in each cycle, but it does not tell that the offspring must result from the marriage of a class **A** mother and class **C** father.

In other words, we do not know whether

$$x_3(k) \geq x_1(k) + x_2(k) \tag{5.4.112}$$

is satisfied with any k; that is, we do not know that society can obey its marriage scheme in each cycle. For this, using (5.4.108)–(5.4.110), we have to prove that

$$\begin{aligned}
&\alpha_5^k \mathring{x}_3 + \frac{\alpha_3^k - \alpha_5^k}{\alpha_3 - \alpha_5}(\alpha_4 - \alpha_5)\left(\mathring{x}_2 - \mathring{x}_1 \frac{\alpha_2}{\alpha_1 - \alpha_3}\right) \\
&\quad + \frac{\alpha_1^k - \alpha_5^k}{\alpha_1 - \alpha_5}\left(\frac{\alpha_2(\alpha_4 - \alpha_5)}{\alpha_1 - \alpha_3} - \alpha_5\right)\mathring{x}_1 \\
&\qquad \geq \alpha_3^k \mathring{x}_2 + \alpha_2 \mathring{x}_1 \frac{\alpha_1^k - \alpha_3^k}{\alpha_1 - \alpha_3} + \alpha_1^k \mathring{x}_1.
\end{aligned} \tag{5.4.113}$$

It is natural to assume that the initial conditions satisfy $\mathring{x}_3 \geq \mathring{x}_1 + \mathring{x}_2$. We see that if we prove (5.4.113) for $\mathring{x}_3 = \mathring{x}_1 + \mathring{x}_2$, then it will be valid for any $\mathring{x}_3 \geq \mathring{x}_1 + \mathring{x}_2$. So, let $\mathring{x}_3 = \mathring{x}_1 + \mathring{x}_2$ and, denoting $q = \mathring{x}_2/\mathring{x}_1$, we must find conditions so that

$$\begin{aligned}
\Psi_k(q) &:= (\alpha_3^k - \alpha_5^k)\frac{\alpha_5 - \alpha_3 + 1}{\alpha_3 - \alpha_5} q + \alpha_5^k - \alpha_1^k - \frac{\alpha_2(\alpha_3^k - \alpha_5^k)}{(\alpha_3 - \alpha_5)(\alpha_1 - \alpha_3)} \\
&\quad + \frac{\alpha_1^k - \alpha_5^k}{\alpha_1 - \alpha_5}\left(\frac{\alpha_2(\alpha_4 - \alpha_5)}{\alpha_1 - \alpha_3} - \alpha_5\right) - \alpha_2 \frac{\alpha_1^k - \alpha_3^k}{\alpha_1 - \alpha_3} \geq 0
\end{aligned}$$

for all $q \geq \frac{\alpha_2}{\alpha_1 - \alpha_3}$ and $k \geq 1$.

We re-write the previous equation as

$$\begin{aligned}
\Psi_k(q) &= \frac{\alpha_3^k - \alpha_5^k}{\alpha_3 - \alpha_5}\left((\alpha_5 - \alpha_3 + 1)q - \frac{\alpha_2}{\alpha_1 - \alpha_3}\right) \\
&\quad + \frac{\alpha_2(\alpha_1^k - \alpha_5^k)}{\alpha_1 - \alpha_3}\left(\frac{\alpha_4 - \alpha_5}{\alpha_1 - \alpha_5} - \frac{\alpha_5(\alpha_1 - \alpha_3)}{\alpha_2(\alpha_1 - \alpha_5)} - \frac{\alpha_1 - \alpha_3}{\alpha_2}\right) - \alpha_2 \frac{\alpha_1^k - \alpha_3^k}{\alpha_1 - \alpha_3}
\end{aligned}$$

and, using (5.4.106),

$$\begin{aligned}
\Psi_k(q) &\geq \frac{\alpha_3^k - \alpha_5^k}{\alpha_3 - \alpha_5}\left((\alpha_5 - \alpha_3 + 1)q - \frac{\alpha_2}{\alpha_1 - \alpha_3}\right) + \frac{\alpha_2}{\alpha_1 - \alpha_3}(\alpha_3^k - \alpha_5^k) \\
&\geq \frac{\alpha_3^k - \alpha_5^k}{\alpha_3 - \alpha_5}\left((\alpha_5 - \alpha_3 + 1)\left(q - \frac{\alpha_2}{\alpha_1 - \alpha_3}\right)\right).
\end{aligned}$$

The last line is nonnegative whenever $\alpha_5 - \alpha_3 + 1 \geq 0$ (since $q \geq \frac{\alpha_2}{\alpha_1 - \alpha_3}$).

We note that a complete characterization of the cone of initial conditions which ensure the survival of the Natchez population and its social structure is given in [11].

Exercise 5.4.1 Carry out the same analysis if the population consists of two classes, Sun and Commoners, a member of the Sun class can marry only a commoner, and the child of such a couple becomes a Sun.

5.4.2 *Positive Dynamical Systems*

If a given difference, differential equation, or a system of equations, are to describe the evolution of a population; that is, if the solution is the population size or density, then clearly solutions emanating from non-negative data must stay non-negative. If we deal with systems of equations, then non-negativity must be understood in the sense defined in Sect. A.3.1, that is, coordinate-wise. We note that there are models admitting negative solutions, such as the discrete logistic equation (see discussion preceding (2.1.23) or the Natchez population in the previous section), but then the moment the solution becomes negative is interpreted as the extinction of the population, and the model ceases to be applicable for later times.

Let us first consider processes occurring in discrete time.

Proposition 5.29 *The solution* $\mathbf{y}(k)$ *of*

$$\mathbf{y}(k+1) = \mathcal{A}\mathbf{y}(k), \quad \mathbf{y}(0) = \mathring{\mathbf{y}} \geq 0,$$

satisfies $\mathbf{y}(k) \geq 0$ *for any* $k = 1, \ldots,$ *if and only if* $\mathcal{A} \geq 0$.

Proof The 'if' part is easy. We have $y_i(k) = \sum_{j=1}^{d} a_{ij} y_j(k-1)$ for $k \geq 1$ so if $a_{ij} \geq 0$ and $\mathring{y}_j \geq$ for $i, j = 1, \ldots, d$, then $y_i(1) \geq 0$ for all $i = 1, \ldots, d$, and the extension for $k > 1$ follows by induction.

On the other hand, assume that $a_{ij} < 0$ for some i and j, and consider $\mathring{\mathbf{y}} = \boldsymbol{e}_j = (0, \ldots, 0, 1, 0, \ldots, 0)$, where 1 is in the jth place. Then $\mathcal{A}\mathring{\mathbf{y}} = (a_{1j}, \ldots, a_{ij}, \ldots, a_{dj})$ so the output is not non-negative. Thus, the condition $\mathcal{A} \geq 0$ is also necessary. □

The proof of an analogous result in continuous time is slightly more involved.

Proposition 5.30 *The solution* $\mathbf{y}(t)$ *of*

$$\mathbf{y}'(t) = \mathcal{A}\mathbf{y}(t), \quad \mathbf{y}(0) = \mathring{\mathbf{y}} \geq 0,$$

satisfies $\mathbf{y}(t) \geq 0$ *for any* $t > 0$ *if and only if* $\mathcal{A}$ *has non-negative off-diagonal entries.*

Proof First let us consider $\mathcal{A} \geq 0$. Then, using representation (5.3.59),

$$e^{t\mathcal{A}} = \mathcal{I} + t\mathcal{A} + \frac{t^2}{2}\mathcal{A}^2 + \frac{t^3}{3!}\mathcal{A}^3 + \cdots + \frac{t^k}{k!}\mathcal{A}^k + \cdots,$$

and by the previous proposition, we see that $e^{t\mathcal{A}} \geq 0$ (note that the limit of any non-negative sequence is non-negative). Next, we observe that for any real a and $\mathring{\boldsymbol{y}} \in \mathbb{R}^d$, the function $\boldsymbol{y}(t) = e^{at}e^{t\mathcal{A}}\mathring{\boldsymbol{y}}$ is non-negative and satisfies the equation

$$\boldsymbol{y}' = a\boldsymbol{y} + \mathcal{A}\boldsymbol{y} = (a\mathcal{I} + \mathcal{A})\boldsymbol{y}.$$

Hence, if the diagonal entries of $\mathcal{A}$, a_{ii}, are negative, then, denoting $r = \max_{1\leq i\leq d}\{-a_{ii}\}$, we find that $\tilde{\mathcal{A}} = r\mathcal{I} + \mathcal{A} \geq 0$. Using the first part of the proof, we see that

$$e^{t\mathcal{A}} = e^{-rt}e^{t\tilde{\mathcal{A}}} \geq 0. \tag{5.4.114}$$

To prove the converse, let us write

$$e^{t\mathcal{A}} = \mathcal{E}(t) = \begin{pmatrix} \epsilon_{11}(t) & \dots & \epsilon_{1d}(t) \\ \vdots & \ddots & \vdots \\ \epsilon_{d1}(t) & \dots & \epsilon_{dd}(t) \end{pmatrix},$$

so $\epsilon_{ij}(t) \geq 0$ for all $i, j = 1, \dots, d$, and consider $\mathcal{E}(t)\boldsymbol{e}_i = (\epsilon_{1i}(t), \dots, \epsilon_{ii}(t), \dots, \epsilon_{di}(t))$. Then

$$\begin{aligned}(a_{1i}, \dots, a_{ii}, \dots, a_{di}) &= \mathcal{A}\mathcal{E}(t)\boldsymbol{e}_i|_{t=0} = \left.\frac{d}{dt}\mathcal{E}(t)\boldsymbol{e}_i\right|_{t=0} \\ &= \lim_{h\to 0^+}\left(\frac{\epsilon_{1i}(h)}{h}, \dots, \frac{\epsilon_{ii}(h) - 1}{h}, \dots, \frac{\epsilon_{di}(h)}{h}\right),\end{aligned}$$

so that $a_{ji} \geq 0$ for $j \neq i$. □

Remark 5.31 Matrices satisfying the assumptions of Proposition 5.30, also if the entries are time-dependent, are called Metzler matrices. They are discussed in detail in Sect. C.3.2, and they play an essential role in many applications described in Chap. 8. It is also important to note that also in the non-autonomous case, they generate non-negative dynamical systems by Theorem B.21.

Exercise 5.4.2 Consider matrices $\mathcal{A}$:

$$a)\ \begin{pmatrix} 1 & 2 & 2 \\ 0 & 0 & 1 \\ 0 & 2 & 3 \end{pmatrix}, \quad b)\ \begin{pmatrix} -1 & 2 & 2 \\ 0 & 0 & 1 \\ 0 & 2 & -3 \end{pmatrix}, \quad c)\ \begin{pmatrix} 1 & -2 & 2 \\ 0 & 0 & 1 \\ 0 & 2 & 3 \end{pmatrix}.$$

(i) Which of these matrices yield non-negative solutions to $\mathbf{y}(k+1) = \mathcal{A}\mathbf{y}(k)$ for non-negative initial condition.
(ii) Which of these matrices yield non-negative solutions to $\mathbf{y}' = \mathcal{A}\mathbf{y}$ for non-negative initial conditions.

Justify your claims by displaying solutions that are not non-negative or non-negative initial data whenever applicable.

5.4.3 Classification of Projection Matrices

The long-time behaviour of $(\mathcal{A}^k)_{k\geq 1}$ is fully determined by whether $\mathcal{A}$ is a primitive irreducible, imprimitive irreducible or a reducible matrix. These concepts are also easily interpreted in the context of population dynamics.

Definition 5.32 For a matrix $\mathcal{A} = (a_{ij})_{1\leq i,j\leq d}$, we say that there is an arc from i to j if $a_{ij} > 0$; a path from i to j is a sequence of arcs starting from i and ending in j in which the endpoint of each arc (apart from the last) is the beginning of the subsequent arc. A loop is a path from i to itself.

We say that a non-negative matrix is irreducible if, for each i and j, there is a path from i to j. Otherwise, we say that it is *reducible*.

This definition can easily be interpreted in terms of graphs. A *graph* is a nonempty finite set of vertices and (possibly empty) set of edges (edge can be interpreted as an unordered pair of vertices).

A directed graph or a *digraph* is a graph with directed edges (a directed edge is then an ordered pair of vertices); see Fig. 5.11. As with matrices, a *path* in a graph is a finite sequence of directed edges $((i_1, i_2), (i_2, i_3), \ldots, (i_{k-1}, i_k))$ in which no vertex is repeated apart from possibly $i_1 = i_k$; in the latter case the path is called a loop. We say that a digraph is strongly connected if, for any pair of vertices i and j, a path connects them.

By the *incidence matrix* associated with a nonnegative matrix $\mathcal{A}$ we understand the matrix $\mathcal{D} = (d_{ij})_{1\leq i,j\leq d}$ where $d_{ij} = 1$ if $a_{ij} > 0$ and $d_{ij} = 0$ otherwise. There is a one-to-one correspondence between diagraphs and incidence matrices (up to a permutation). For a given $\mathcal{D}$, we take $\{1, \ldots, d\}$ as the set of vertices, and we draw a directed edge from j to i whenever $d_{ij} > 0$. Conversely, given a digraph with d vertices, we number them $\{1, \ldots, d\}$ and set $d_{ij} = 1$ whenever there is an edge from j to i.

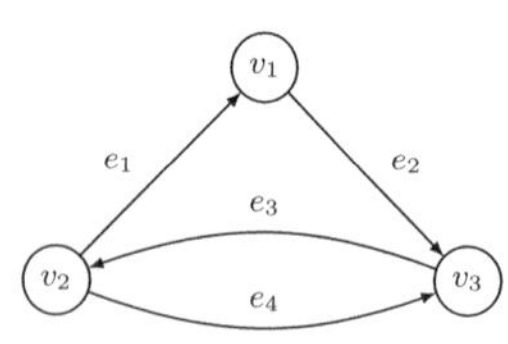

Fig. 5.11 A digraph; here $e_1 = (2, 1), e_2 = (1, 3), e_3 = (3, 2)$ and $e_4 = (2, 3)$

Then, it is relatively easy to prove that a matrix $\mathcal{A}$ is irreducible if and only if its incidence matrix $\mathcal{D}$ is irreducible. Then one can prove that $\mathcal{A}$ is irreducible if and only if the digraph corresponding to the incidence matrix of $\mathcal{A}$, is strongly connected.

Some notation must precede an equivalent but more algebraic condition. We write

$$\mathcal{A}^k = (a_{ij}^{(k)})_{1 \le i,j \le d}.$$

It is easy to see that

$$a_{ij}^{(k)} = \sum_{1 \le i_r \le d, r=1,\ldots,k-1} a_{i,i_1} \cdot a_{i_1 i_2} \cdot \ldots \cdot a_{i_{k-1} j} \tag{5.4.115}$$

If some $a_{ii_1} \cdot a_{i_1 i_2} \cdot \ldots \cdot a_{i_{k-1} j} \neq 0$, then, there is a path starting from j and passing through $i_{k-1}, \ldots, i_1$ to i. Since the matrix elements are nonnegative, for $a_{ij}^{(k)}$ to be non-zero, it is enough that there exists at least one such path. Thus, $\mathcal{A}$ is irreducible if for each pair (i, j) there is k such that $a_{i,j}^{(k)} > 0$.

Remark 5.33 It is easy to see that it is enough to consider $k \le d - 1$. Indeed, if we have a positive term,

$$a_{ii_1} \cdot a_{i_1 i_2} \cdot \ldots \cdot a_{i_{k-1} j} > 0,$$

where $k \ge d$, and this is the shortest path joining i and j, then the set of indices $\{i_1, \ldots, i_{k-1}, j\}$ would contain the set $\{1, \ldots, d\}$, that is, all available indices would appear in the set. But then we can assume $i \neq i_1$ as otherwise we would be able to remove a_{ii} from the product. Thus, $i = i_r$ for some $r \neq 1$. This means that there must be entry $a_{i_{r-1} i_r}$ and again, we could remove the product $a_{ii_1} \cdot \ldots \cdot a_{i_{r-1} r} = a_{i_r i_1} \cdot \ldots \cdot a_{i_{r-1} r}$ and hence our assumption that the shortest path has length larger than $d - 1$ is false.

If the matrix $\mathcal{A}$ is not irreducible, then we say that it is *reducible*. Thus, a matrix is reducible if the associated graph is not strongly connected, that is, if there are vertices i and j such that i is not accessible from j. An equivalent definition is that $\mathcal{A}$ is reducible if, by simultaneous permutation of rows and columns, it can be brought to the form

$$\begin{pmatrix} \mathcal{A} & \mathbf{0} \\ \mathcal{B} & \mathcal{C} \end{pmatrix},$$

where $\mathcal{A}$ and $\mathcal{C}$ are square matrices, see [46, Chapter III §4].

Returning to the age-structured population dynamics, a matrix is irreducible if each stage i can contribute to any other stage j. E.g., the Usher matrix

$$\begin{pmatrix} 0 & 0 & 1 & 0 \\ 1 & 0 & 0 & 0 \\ 0 & 1 & 0 & 0 \\ 0 & 0 & 1 & 1 \end{pmatrix}$$

is reducible as the last state cannot contribute to any other state and fertile individuals are concentrated in only one state.

Irreducible matrices are subdivided into two further classes. An irreducible matrix $\mathcal{A}$ is called *primitive* if

$$\mathcal{A}^k > 0,$$

otherwise it is called *imprimitive*.

Note the difference between irreducibility and primitivity. For irreducibility we require that for each (i, j) there is k such that $a_{ij}^{(k)} > 0$ but for primitivity there must be k such that $a_{ij}^{(k)} > 0$ for all (i, j).

In population dynamics, if the population has a single reproductive stage, then its projection matrix is imprimitive. E.g., the matrix

$$\begin{pmatrix} 0 & 0 & 1 \\ 1 & 0 & 0 \\ 0 & 1 & 0 \end{pmatrix},$$

describing a semelparous population is imprimitive.

Exercise 5.4.3 Consider matrices

$$(i) \begin{pmatrix} 1 & 0 & 1 \\ 1 & 0 & 0 \\ 0 & 1 & 0 \end{pmatrix}, \quad (ii) \begin{pmatrix} 0 & 0 & 1 & 0 & 0 \\ 1 & 0 & 0 & 0 & 0 \\ 0 & 1 & 0 & 0 & 0 \\ 0 & 0 & 0 & 0 & 0 \\ 0 & 1 & 0 & 1 & 0 \end{pmatrix}, \quad (iii) \begin{pmatrix} 0 & 0 & 1 \\ 1 & 0 & 0 \\ 0 & 1 & 0 \end{pmatrix}, \quad (iv) \begin{pmatrix} 1 & 0 & 1 \\ 1 & 0 & 0 \\ 0 & 1 & 0 \end{pmatrix}.$$

By any method, determine which matrix is reducible, irreducible but imprimitive, or primitive.

The Perron–Frobenius theorem can be summarized as follows.

Theorem 5.34 *Let $\mathcal{A}$ be a nonnegative matrix.*

1. *There exists a real nonnegative eigenvalue $\lambda_{\max} = r(\mathcal{A})$ such that $\lambda_{\max} \geq |\lambda|$ for any $\lambda \in \sigma(\mathcal{A})$. There is an eigenvector (called the Perron eigenvector) corresponding to $\lambda_{\max}$, which is real and nonnegative.*

2. *If, in addition, $\mathcal{A}$ is irreducible, then $\lambda_{\max}$ is simple and strictly positive, $\lambda_{\max} \geq |\lambda|$ for $\lambda \in \sigma(\mathcal{A})$. The eigenvector corresponding to $\lambda_{\max}$ may be chosen to be strictly positive.*
   - *(i) If $\mathcal{A}$ is additionally primitive, then $\lambda_{\max} > |\lambda|$.*
   - *(ii) If $\mathcal{A}$ is imprimitive, then there are $h - 1$ (h is the imprimitivity index, see Sect. C.3.) eigenvalues $\lambda_j = \lambda_{\max} e^{2\pi i \frac{j}{h}}$, $j = 1, \ldots, h-1$, with $\lambda_{\max} = |\lambda_j|$.*

The proof of this result is given in Sect. C.3. Here, we show its application in a population context. Suppose that our population is divided into n age (or other) classes and the state of the population is given by the vector $\boldsymbol{x} = (x_1, \ldots, x_n)$ giving the number of individuals (or density) in each class. Let $\mathring{\boldsymbol{x}} \geq 0$ denote the initial distribution of the population among the classes. Then

$$\boldsymbol{x}(k) = \mathcal{A}^k \mathring{\boldsymbol{x}}$$

is the distribution after k periods, and

$$P(k, \mathring{\boldsymbol{x}}) = \|\mathcal{A}^k \mathring{\boldsymbol{x}}\|_1 = \sum_{i=1}^{n} (\mathcal{A}^k \mathring{\boldsymbol{x}})_i = \sum_{i=1}^{n} x_i(k) = \|\boldsymbol{x}(k)\|_1$$

is the total population at time k evolving from the initial distribution $\mathring{\boldsymbol{x}}$.

If $\mathcal{A}$ is nonnegative, irreducible or primitive, then the transpose $\mathcal{A}^T$ has the same property. Let $r := \lambda_{\max}$ be the dominant eigenvalue of both matrices and $\boldsymbol{v}$ and $\boldsymbol{v}^*$ be the corresponding strictly positive eigenvectors of, respectively, $\mathcal{A}$ and $\mathcal{A}^T$, corresponding to $\lambda_{\max}$. We normalize $\boldsymbol{v}$ so that $\|\boldsymbol{v}\|_1 = 1$ and $\boldsymbol{v}^*$ so that $\langle \boldsymbol{v}^*, \boldsymbol{v} \rangle = 1$.

Combining the Perron–Frobenius theorem with the spectral decomposition (5.3.94), we arrive at the following result.

Theorem 5.35 (Fundamental Theorem of Demography) *Suppose that the projection matrix $\mathcal{A}$ is irreducible and primitive and let r be the strictly positive dominant eigenvalue of $\mathcal{A}$, $\boldsymbol{v}$ the strictly positive eigenvector of $\mathcal{A}$ and $\boldsymbol{v}^*$ strictly positive eigenvector of $\mathcal{A}^T$ belonging to r. Then, for any $\mathring{\boldsymbol{x}} \geq 0$,*

(a) $\mathcal{A}$ has the AEG property

$$\lim_{k \to \infty} r^{-k} \mathcal{A}^k \mathring{\boldsymbol{x}} = \langle \boldsymbol{v}^*, \mathring{\boldsymbol{x}} \rangle \boldsymbol{v}. \tag{5.4.116}$$

(b)

$$\lim_{k \to \infty} \frac{\boldsymbol{x}(k)}{P(k, \mathring{\boldsymbol{x}})} = \frac{\mathcal{A}^k \mathring{\boldsymbol{x}}}{P(k, \mathring{\boldsymbol{x}})} = \boldsymbol{v}. \tag{5.4.117}$$

(c) If $r < 1$

$$\lim_{k\to\infty} P(k, \mathring{\boldsymbol{x}}) = 0,$$

and

$$\lim_{k\to\infty} P(k, \mathring{\boldsymbol{x}}) = \infty$$

if $r > 1$.

Proof (a) We use (5.3.85), (5.3.96) and Theorem 5.34 1.(i) to get

$$\mathcal{A}^k \mathring{\boldsymbol{x}} = \sum_{\lambda\in\sigma(\mathcal{A})} \lambda^k \boldsymbol{p}_\lambda(k)\mathring{\boldsymbol{x}} = r^k \langle \boldsymbol{v}^*, \mathring{\boldsymbol{x}}\rangle \boldsymbol{v} + \sum_{\lambda\in\sigma(\mathcal{A})\setminus\{r\}} \lambda^k \boldsymbol{p}_\lambda(k)\mathring{\boldsymbol{x}}. \tag{5.4.118}$$

By the primitivity of $\mathcal{A}$, $r > |\lambda|$ for any $\lambda \in \sigma(\mathcal{A}) \setminus \{r\}$ and, since $p_\lambda(k)$ are polynomials in k, we have

$$\left\| \left(\frac{\lambda}{r}\right)^k \boldsymbol{p}_\lambda(k)\mathring{\boldsymbol{x}} \right\| \to 0, \qquad k \to \infty,$$

and (a) is proved.

For (b), we see that, by (a),

$$\lim_{k\to\infty} \frac{P(k, \mathring{\boldsymbol{x}})}{r^k} = \lim_{k\to\infty} \frac{\|\mathcal{A}^k \mathring{\boldsymbol{x}}\|}{r^k} = \lim_{k\to\infty} \left\| \frac{\mathcal{A}^k \mathring{\boldsymbol{x}}}{r^k} \right\| = |\langle \boldsymbol{v}^*, \mathring{\boldsymbol{x}}\rangle| > 0. \tag{5.4.119}$$

Hence, by (a) and (5.4.119),

$$\lim_{k\to\infty} \frac{\mathcal{A}^k \mathring{\boldsymbol{x}}}{P(k, \mathring{\boldsymbol{x}})} = \lim_{k\to\infty} \frac{r^{-k}\mathcal{A}^k \mathring{\boldsymbol{x}}}{r^{-k} P(k, \mathring{\boldsymbol{x}})} = \boldsymbol{v},$$

which gives (5.4.117).

To prove (c), we observe that

$$P(k, \mathring{\boldsymbol{x}}) = \left\| \mathcal{A}^k \left(\frac{\mathring{\boldsymbol{x}}}{P(k-1, \mathring{\boldsymbol{x}})} \right) \right\| P(k-1, \mathring{\boldsymbol{x}}) = \rho_{k-1} P(k-1, \mathring{\boldsymbol{x}}),$$

where

$$\begin{aligned}\rho_{k-1} &= \left\| \mathcal{A}^k \left(\frac{\mathring{\boldsymbol{x}}}{P(k-1, \mathring{\boldsymbol{x}})} \right) \right\| \\ &= r \left\| \frac{r^{-k}\mathcal{A}^k \mathring{\boldsymbol{x}}}{r^{-(k-1)} P(k-1, \mathring{\boldsymbol{x}})} \right\| \to r \frac{|\langle \boldsymbol{v}^*, \mathring{\boldsymbol{x}}\rangle| \|\boldsymbol{v}\|}{|\langle \boldsymbol{v}^*, \mathring{\boldsymbol{x}}\rangle|} = r, \quad \text{as} \quad k \to \infty,\end{aligned}$$

by (a), (b), and the normalization of $\boldsymbol{v}$, $\boldsymbol{v}^*$. Thus, if $r < 1$, then we can pick $\bar{r} < 1$ such that $\rho_{k-1} \leq \bar{r}$ for all k larger than some k_0 and

$$P(k_0 + i, \mathring{\boldsymbol{x}}) \leq \bar{r}^i P(k_0, \mathring{\boldsymbol{x}}) \to 0 \quad \text{as} \quad i \to \infty.$$

Similarly, if $r > 1$, then we can pick $\tilde{r} > 1$ such that $\rho_{k-1} \geq \tilde{r}$ for all k larger than some k_0 and

$$P(k_0 + i, \mathring{\boldsymbol{x}}) \geq \tilde{r}^i P(k_0, \mathring{\boldsymbol{x}}) \to \infty \quad \text{as} \quad i \to \infty,$$

as $P(k_0, \mathring{\boldsymbol{x}}) \neq 0$ for any finite k_0. Indeed, otherwise, from nonnegativity, we would have $\boldsymbol{x}(k_0) = 0$ and thus $\boldsymbol{x}(k) = 0$ for $k \geq k_0$, contradicting (a). □

If we note that for each $1 \leq i \leq n$,

$$\frac{(\mathcal{A}^k \boldsymbol{x})_i}{P(k, \boldsymbol{x})}$$

is the fraction of the population in the state i at time k, then the result above states that for large times, the fraction of the population in the state i approximately is given by the i coordinate of the Perron eigenvector and is independent of the initial distribution $\boldsymbol{x}$. Moreover, $\boldsymbol{v}$ is approached (or departed from) at an exponential rate, hence the name *asynchronous exponential growth.*

5.4.4 Example—Irreducible Case

Let us consider the Leslie matrix (5.1.5),

$$\mathcal{L} := \begin{pmatrix} f_0 & f_1 & \cdots & f_{n-2} & f_{n-1} \\ s_0 & 0 & \cdots & 0 & 0 \\ 0 & s_1 & \cdots & 0 & 0 \\ \vdots & \vdots & \ddots & \vdots & \vdots \\ 0 & 0 & \cdots & s_{n-2} & 0 \end{pmatrix}, \tag{5.4.120}$$

and find under what conditions the population described by $\mathcal{L}$ exhibits asynchronous exponential growth.

First, we observe that we need all $s_i \neq 0$, $0 \leq i \leq n-2$ for irreducibility. Indeed, if for some i the coefficient $s_i = 0$, there would be no path from $k \leq i$ to $k > i$. In other words, there would be no way of reaching the age $k > i$. Assuming this, $\mathcal{L}$ is irreducible if and only if $f_{n-1} > 0$. Clearly, if $f_{n-1} = 0$, then there is no communication from class $n - 1$ to any other class; thus $\mathcal{L}$ is reducible. Now, let $f_{n-1} > 0$ and pick an (i, j). If $j < i$, then there is a path $(j, j + 1)...(i - 1, i)$ ensured by the survival coefficients $s_j, s_{j+1}, \ldots s_{i-1}$. If $j \geq i$, then the survival coefficients ensure that we reach the last class $n - 1$, then, since $f_{n-1} > 0$, we reach the class 0 and then we arrive at i by ageing, that is $(j, j + 1) \ldots (n - 2, n - 1)(n - 1, 0)(0, 1), \ldots (i - 1, i)$.

The question of primitivity is more complicated. Let us first assume that $f_j > 0$ for $j = 0, \ldots, n - 1$, that is, any age group can reproduce. Let us consider arbitrary initial state j. Then there is an arc between j and 0 ($a_{0j} = f_j > 0$) and then from state 0, one can reach any state i in exactly i steps ($s_0 s_1 \cdot \ldots \cdot s_{i-1}$). Thus, there is a path joining j and i of length $i + 1$, which still depends on the target state. However, there is an arc from 0 to itself, so we can wait at 0 for any number of steps. In particular, we can wait for $n - (i + 1)$ steps so that j can be connected with i in n steps. In other words,

$$s_{i-1} \cdot \ldots \cdot s_1 s_0 f_0 \cdot \ldots \cdot f_0 f_j > 0$$

where f_0 occurs $n - i$ times. Hence $\mathcal{L}^n > 0$.

Remark 5.36 The above argument shows that any irreducible matrix in which at least one diagonal entry is not equal to zero is primitive.

This result requires too much—typically, young individuals cannot reproduce. Hence, we will strengthen this result with a more detailed analysis of the characteristic equation of the Leslie matrix, which can be calculated explicitly and has its own biological interpretation.

Let us start with $n = 2$. Then the characteristic polynomial $p_2(\lambda)$ equals

$$p_2(\lambda) = \lambda^2 - \lambda f_0$$

and for $n = 3$ we get

$$p_3(\lambda) = det \begin{pmatrix} f_0 - \lambda & f_1 & f_2 \\ s_0 & -\lambda & 0 \\ 0 & s_1 & -\lambda \end{pmatrix} = -\lambda^3 + \lambda^2 f_0 + \lambda f_1 s_0 + f_2 s_0 s_1. \tag{5.4.121}$$

This suggests the inductive assumption

$$p_{n-1}(\lambda) = (-1)^{n-1} \left(\lambda^{n-1} - \sum_{k=0}^{n-2} \lambda^{n-2-k} f_k \prod_{i=0}^{k-1} s_i \right),$$

with $\prod_{i=0}^{-1} = 1$. Then,

$$p_n(\lambda) = det \begin{pmatrix} f_0 - \lambda & f_1 & \cdots & f_{n-2} & f_{n-1} \\ s_0 & -\lambda & \cdots & 0 & 0 \\ 0 & s_1 & \cdots & 0 & 0 \\ \vdots & \vdots & \cdots & \vdots & \vdots \\ 0 & 0 & \cdots & s_{n-2} & -\lambda \end{pmatrix}$$

$$= -\lambda p_{n-1} + (-1)^{2n-1} s_{n-2} (-1)^n f_{n-1} \prod_{i=0}^{n-1} s_i$$

$$= (-1)^n \left(\lambda^n - \sum_{k=0}^{n-1} \lambda^{n-1-k} f_k \prod_{i=0}^{k-1} s_i \right).$$

This equation can be simplified if we remember that $s_i = l_{i+1}/l_i$ with $l_0 = 1$, where l_i is the probability of surviving from birth to age i and $f_i = m_{i+1} s_i$. Thus

$$f_k \prod_{i=0}^{k-1} s_i = m_{k+1} s_k s_{k-1} \cdot \ldots \cdot s_0 = m_{k+1} l_{k+1}$$

and the characteristic equation for a Leslie matrix can be written as

$$\sum_{k=1}^{n} \lambda^{n-k} m_k l_k = \lambda^n. \tag{5.4.122}$$

Remark 5.37 (Alternative Derivation of the Characteristic Equation of a Leslie Matrix and Its Eigenvectors) Consider the eigenvalue-eigenvector equation for a Leslie matrix

$$\mathcal{L} \boldsymbol{v} = \begin{pmatrix} f_0 & f_1 & \cdots & f_{n-2} & f_{n-1} \\ s_0 & 0 & \cdots & 0 & 0 \\ 0 & s_1 & \cdots & 0 & 0 \\ \vdots & \vdots & \cdots & \vdots & \vdots \\ 0 & 0 & \cdots & s_{n-2} & 0 \end{pmatrix} \begin{pmatrix} v_0 \\ v_1 \\ v_2 \\ \vdots \\ v_{n-1} \end{pmatrix} = \lambda \begin{pmatrix} v_0 \\ v_1 \\ v_2 \\ \vdots \\ v_{n-1} \end{pmatrix}.$$

The equations from the second row down read

$$s_0 v_0 = \lambda v_1, \quad s_1 v_1 = \lambda v_2, \quad \ldots \quad s_{n-2} v_{n-2} = \lambda v_{n-1}.$$

Taking $v_0 = 1$, we obtain

$$v_1 = \frac{s_0}{\lambda}, \quad v_2 = \frac{s_0 s_1}{\lambda^2}, \quad \dots \quad v_{n-1} = \frac{s_0 s_1 \dots s_{n-2}}{\lambda^{n-1}}.$$

Now, the first row gives the equation

$$\lambda = \left(f_0 + \frac{f_1 s_0}{\lambda} + \frac{f_2 s_0 s_1}{\lambda^2} + \dots + \frac{f_{n-1} s_0 s_1 \dots s_{n-2}}{\lambda^{n-1}} \right).$$

Again, we use $s_i = l_{i+1}/l_i$, where l_i is the probability of survival till the $i + 1$st reproductive cycle from birth (thus s_i is the conditional probability of survival to the next reproductive cycle if one survived till i from birth) and $f_i = m_{i+1} s_i$ to rewrite the above as

$$1 = \left(\frac{m_1 l_1}{\lambda} + \frac{m_2 l_2}{\lambda^2} + \frac{m_3 l_3}{\lambda^3} + \dots + \frac{m_n l_n}{\lambda^n} \right),$$

where we used $l_0 = 1$. This equation is called the discrete Euler–Lotka equation.

We can use a result, [71, p. 679], that allows finding the imprimitivity index without calculating the eigenvalues. Namely, let

$$0 = det(\lambda I - \mathcal{L}) = \lambda^n + a_{n_1} \lambda^{n_1} + \dots a_{n_i} \lambda^{n_i} \tag{5.4.123}$$

with $n > n_1 > \dots > n_i$, $a_{n_k} \neq 0, k = 1, \dots, i$, be the characteristic equation of $\mathcal{L}$. It follows that if $\mathcal{L}$ is imprimitive of index h, then h is the greatest common divisor of $n - n_1, n_1 - n_2, \dots, n_{i-1} - n_i$. Using the criterion mentioned above, a Leslie matrix is irreducible and primitive if, e.g., the fertility of the oldest generation m_n is not zero (irreducibility) and two subsequent generations have nonzero fertility (primitivity).

However, the relative simplicity of the characteristic equation allows us to strengthen this result even more without invoking the result in [71]. In fact, we prove that $\mathcal{L}$ is imprimitive if and only if the maternity function is periodic, that is, if the greatest common divisor of the ages of positive reproduction, called the period, is greater than 1. For instance, the sequence $m_2, m_4, m_6 \dots$, has period 2. In particular, the period is equal to the imprimitivity index.

The proof follows from the criterion above, but we shall provide a direct instructive proof. Indeed, suppose that

$$\lambda_j = re^{i\theta}, \quad \theta \neq 2\pi n,$$

is a negative or complex root to

$$\psi(\lambda) = \lambda^{-1} m_1 l_1 + \ldots + \lambda^{n-1} m_{n-1} l_{n-1} + \lambda^{-n} m_n l_n = \sum_{k=1}^{n} \lambda^{-k} m_k l_k = 1. \tag{5.4.124}$$

Then

$$\sum_{k=1}^{n} r^{-k} e^{-ik\theta} l_k m_k = 1 \tag{5.4.125}$$

or, taking real and imaginary parts,

$$\sum_{k=1}^{n} r^{-k} \cos(k\theta) l_k m_k = 1, \tag{5.4.126}$$

$$\sum_{k=1}^{n} r^{-k} \sin(k\theta) l_k m_k = 0. \tag{5.4.127}$$

If m_k is periodic, then the only nonzero terms correspond to multiples of d, $m_{k_1 d}, m_{k_2 d}, m_{k_3 d}, \ldots$. Taking $\theta_j = 2\pi j/d$, $j = 0, 1, \ldots, d-1$, we see $\cos k_l d\theta_j = 1$, $\sin k_l d\theta_j = 0$, and so, if the above equations are satisfied by r, they are also satisfied by any $\lambda_j = re^{i\theta_j}$.

If m_k is aperiodic, with $m_{k_i} \neq 0$, $i \in I \subset \{1, \ldots, n\}$, then there is no $\theta \neq 0$ for which $\cos k_i\theta = 1$ for all k_i. Indeed, otherwise there is $\theta \in (0, 2\pi)$ such that

$$\cos k_1\theta = 1.$$

This implies

$$\theta = 2\pi \frac{l}{k_1} = 2\pi \frac{j}{d},$$

where $l < k_1$ is an integer and j and d are relatively prime integers, so that $0 < j \leq d-1$ (note that if $j = 0$, then $\theta = 0$). But then

$$k_i\theta = k_i \frac{2\pi j}{d} = 2\pi l_i$$

for some integer l_i so that

$$k_l = l_l \frac{d}{j}.$$

However, k_i is an integer and j and d are relatively prime, so l_i must be divisible by j. Hence

$$k_i = r_i d$$

for some integer r_i, $i \in I$. Thus, whatever θ, for some k we must have $\cos k\theta < 1$. But then, if (5.4.126) was satisfied, we would have

$$\sum_{k=1}^{n} |\lambda_j|^{-k} l_k m_k > 1.$$

On the other hand, since

$$\sum_{k=1}^{n} r^{-k} l_k m_k = 1,$$

we obtain $|\lambda_j| < r$.

Remark 5.38 Equation (5.4.124) is the characteristic equation of the so-called *Lotka renewal equation* and can be derived directly. Let B_k be the number of births at time k. These births can be divided into two classes: one attributed to females born between time 0 and k and the other due to females alive at time $k = 0$. Females of age i at time k were born at time $k - i$. The number of females born at $k - i$ is given by B_{k-i}. The number of them that survive till the age k is $l_i B_{k-i}$, and thus the number of births at time k by females of age i is $l_i B_{k-i} m_i$. Thus, this contribution is

$$\sum_{i=1}^{k} B_{k-i} l_i m_i.$$

To find the contribution of the females present at time $k = 0$, we begin by taking the number of females of age i present at $k = 0$, $x_i(0)$. These females must live till the age $k + i$, that is, we must take survival rate till $k + i$, l_{k+i}, conditioned upon the female having survived till i. Hence we see that $x_i(0) l_{k+i}/li$ females survived till time k. These gave birth to $m_{k+i} x_i(0) l_{k+i}/l_i$ new females. To find the number of all births due to females older than k, we again sum over all ages. However, no female survives beyond n, so the summation terminates at $n - k$ (no female older than $n - k$ at time $k = 0$ will survive till k). Combining these two formulae, we get

$$B_k = \sum_{i=1}^{k} B_{k-i} l_i m_i + G_k, \tag{5.4.128}$$

where

$$G_k = \sum_{i=1}^{n-k} x_i(0)\frac{l_{k+i}}{l_i}m_{k+i}.$$

For large k, $k > n$, Eq. (5.4.128) reduces to

$$B_k = \sum_{i=1}^{k} B_{k-i}l_i m_i$$

which is a constant coefficient difference equation, the characteristic equation of which is exactly (5.4.124).

Exercise 5.4.4 Consider matrices

$$(i)\ \mathcal{L} = \begin{pmatrix} 0.2 & 3 \\ 0.3 & 0.3 \end{pmatrix}, \qquad (ii)\ \mathcal{L} = \begin{pmatrix} 0.11 & 0.15 & 0.15 \\ 0.3 & 0 & 0 \\ 0 & 0.6 & 0.6 \end{pmatrix}.$$

In each case,

(a) find the number of iterations needed to connect any two states of the described population and, hence, determine the type of each matrix, and
(b) determine the long-term behaviour of the described population and, in particular, the stable age profile whenever it exists.

Exercise 5.4.5 Consider a population divided into three age groups of equal length. The first group consists of immature individuals who cannot reproduce. Due to high mortality, only one in sixteen juveniles becomes an adult. Each adult in the second group produces 31 offspring on average, and one in two adults reaches the senior group. In the latter, the individuals can still reproduce, but each senior produces, on average, 15 offspring. Build a model describing such a population and find the population's age distribution after ten breeding seasons, assuming that we originally had ten juveniles, four adults and two seniors.

5.4.5 *Reducible Case*

Let us consider a more complicated case where the fertility is restricted to some interval $[n_1, n_2]$, that is, when $f_j > 0$ only for $j \in [n_1, n_2]$. As we noted earlier, if $n_2 < n$, the matrix cannot be irreducible as there is no communication between the postreproductive and reproductive stages. Consequently, if we start only with individuals in postreproductive age, the population will die out in a finite time.

Nevertheless, if $n_1 < n_2$ then the population still displays asynchronous exponential growth, albeit slightly modified, as explained below.

To analyse this model, we note that since we cannot move from stages with $j > n_2$ to earlier stages, the population with $j \leq n_2$ evolves independently from the postreproductive part (but feeds into it.) Assume that $n_1 < n_2$ and introduce the restricted matrix

$$\tilde{\mathcal{L}} = \begin{pmatrix} f_0 & f_1 & \cdots & f_{n_2-1} & f_{n_2} \\ s_0 & 0 & \cdots & 0 & 0 \\ 0 & s_1 & \cdots & 0 & 0 \\ \vdots & \vdots & \ddots & \vdots & \vdots \\ 0 & 0 & \cdots & s_{n_2-1} & 0 \end{pmatrix},$$

and the matrix, which provides a (one-way) link from the reproductive to postreproductive stages is given by

$$\mathcal{R} = \begin{pmatrix} 0 & \cdots & s_{n_2} & 0 & \cdots & 0 & 0 \\ \vdots & \cdots & \vdots & \vdots & \ddots & \vdots & \vdots \\ 0 & \cdots & 0 & 0 & \cdots & s_{n-2} & 0 \end{pmatrix}.$$

For the matrix $\tilde{\mathcal{L}}$, $f_{n_2} > 0$ and $f_{n_2-1} > 0$, and we can apply the considerations of the previous section and Theorem 5.35. Thus, there are $r > 0$ and vectors $\boldsymbol{v} = (v_0, \dots v_{n_2})$ and $\boldsymbol{v}^* = (v_0^*, \dots v_{n_2}^*)$ such that $\tilde{\mathcal{L}}\boldsymbol{v} = r\boldsymbol{v}$ and

$$\lim_{k\to\infty} r^{-k}\boldsymbol{x}(k+1) = \lim_{k\to\infty} r^{-k}\tilde{\mathcal{L}}^k\mathring{\boldsymbol{x}} = \boldsymbol{v}\langle\boldsymbol{v}^*, \mathring{\boldsymbol{x}}\rangle, \quad 0 \leq \mathring{\boldsymbol{x}} \in \mathbb{R}^{n_2}. \tag{5.4.129}$$

For $n_2 \leq j < n, k \geq 0$, we have $x_{j+1}(k+1) = s_j x_j(k)$. Hence, starting from $x_{n_2}(k)$ we get $x_{n_2+i}(k+i) = c_i x_{n_2}(k)$, where $c_i = s_{n_2+i-1} \cdot \ldots \cdot s_{n_2}$, as long as $i \leq n - n_2 - 1$. So

$$\lim_{k\to\infty} r^{-k}x_{n_2+i}(k+i) = c_i v_{n_2}\langle\boldsymbol{v}^*, \mathring{\boldsymbol{x}}\rangle, \quad 0 \leq \mathring{\boldsymbol{x}} \in \mathbb{R}^{n_2},$$

and hence, changing $k+i$ into k,

$$\lim_{k\to\infty} r^{-k}x_{n_2+i}(k) = c_i r^{-i} v_{n_2}\langle\boldsymbol{v}^*, \mathring{\boldsymbol{x}}\rangle, \quad 0 \leq \mathring{\boldsymbol{x}} \in \mathbb{R}^{n_2},$$

for any $i = 1, \dots, n - n_2 - 1$.

Hence, we see that the formula (5.3.96) is satisfied if we take

$$\begin{aligned} \boldsymbol{v} &= (v_0, \dots v_{n_2}, c_1 r^{-1} v_{n_2}, \dots, c_{n-n_2-1} r^{-(n-n_2-1)} v_{n_2}), \\ \boldsymbol{v}^* &= (v_0^*, \dots v_{n_2}^*, 0, \dots, 0). \end{aligned}$$

Finally, we observe that if only one f_j is positive (semelparous population), then we do not have asynchronous exponential growth. Indeed, in this case, starting from the initial population in one class, we will have a cohort of individuals in the same age group moving through the system. We have observed such a behaviour in Example 5.4.

Exercise 5.4.6 Let a population be described by the following Leslie matrix with a postreproductive class,

$$\mathcal{L} = \begin{pmatrix} 0 & 2 & 0 \\ 0.5 & 0 & 0 \\ 0 & 0.5 & 0.1 \end{pmatrix}.$$

Determine the long-term behaviour of this population.

5.5 Birth-and-Death Type Problems

Here, we shall show that models introduced in Sect. 5.1.2 appear also in a different context. We extend them to describe denumerable number of states. Thus, the model becomes infinite-dimensional, making its analysis more complex. However, we refrain from entering into detailed mathematical analysis, for which we refer the reader to, e.g., [13, 21], treating the results of this section as illustrative examples of applications of differential equations.

Consider a population of $N(t)$ individuals at time t. We allow stochasticity to intervene in the process so that $N(t)$ becomes a random variable. Accordingly, we denote by

$$p_n(t) = P\{N(t) = n\}, \quad n = 1, 2, \ldots, \tag{5.5.130}$$

the probability that the population has n individuals at time t.

5.5.1 *Birth Process*

At first, we consider only births and assume that individuals give birth to new ones independently of others. For a single individual, we assume that for some small Δt

$$P\{1 \text{ birth in } (t, t+\Delta t] | N(t) = 1\} = \beta \Delta t + o(\Delta t), \tag{5.5.131}$$

$$P\{\text{more than 1 birth in } (t, t+\Delta t] | N(t) = 1\} = o(\Delta t), \tag{5.5.132}$$

$$P\{0 \text{ births in } (t, t+\Delta t] | N(t) = 1\} = 1 - \beta \Delta t + o(\Delta t), \tag{5.5.133}$$

where $P\{a|b\}$ denotes the conditional probability of the event a if the event b occurred. If we have n individuals, then one birth will occur if exactly one of them gives birth to one offspring, and the remaining $n-1$ produce 0 newborns. This can happen in n ways. Thus

$$P\{1 \text{ birth in } (t, t+\Delta t]|N(t)=n\}$$
$$= n(\beta\Delta t + o(\Delta t))(1-\beta\Delta t + o(\Delta t))^{n-1} = n\beta\Delta t + o(\Delta t). \quad (5.5.134)$$

Similarly, more than one birth can occur if one individual gives births to more than one offspring or at least two individuals give birth to one new one. Considering all possible combinations, we end up with a finite sum, each term of which is multiplied by Δt or its higher powers. Thus

$$P\{\text{more than 1 birth in } (t, t+\Delta t]|N(t)=n\} = o(\Delta t). \quad (5.5.135)$$

Finally, no birth occurs if no individual produces an offspring, that is,

$$P\{0 \text{ births in } (t, t+\Delta t]|N(t)=n\} = (1-\beta\Delta t + o(\Delta t))^n = 1 - n\beta\Delta t + o(\Delta t). \quad (5.5.136)$$

Using the law of total probability, we can set up the equation describing the evolution of $p_n(t)$. There can be n individuals at time $t+\Delta t$ if there were $n-1$ individuals at time t and one birth occurred in $(t, t+\Delta t)$ or if there were n individuals and zero births occurred, or less than $n-1$ individuals and the corresponding number of births, more than one, occurred. However, the last event occurs with probability $o(\Delta t)$ and will be omitted. Hence,

$$p_n(t+\Delta t) = p_{n-1}(t)P\{1 \text{ birth in } (t, t+\Delta t]|N(t)=n-1\}$$
$$+ p_n(t)P\{0 \text{ births in } (t, t+\Delta t]|N(t)=n\}, \quad (5.5.137)$$

that is, using the above formulae

$$p_n(t+\Delta t) = (n-1)\beta\Delta t p_{n-1}(t) + (1-n\beta\Delta t)p_n(t) + o(\Delta t). \quad (5.5.138)$$

After some algebra, we get

$$\frac{p_n(t+\Delta t) - p_n(t)}{\Delta t} = -n\beta p_n(t) + (n-1)\beta p_{n-1}(t) + \frac{o(\Delta t)}{\Delta t},$$

and passing to the limit,

$$\frac{dp_n(t)}{dt} = -n\beta p_n(t) + (n-1)\beta p_{n-1}(t). \quad (5.5.139)$$

This is an infinite chain of differential equations which must be supplemented by an initial condition. The population at $t = 0$ had to have some number of individuals, say, n_0. Hence,

$$p_n(0) = \begin{cases} 1 \text{ for } n = n_0, \\ 0 \text{ for } n \neq n_0. \end{cases} \tag{5.5.140}$$

Since this is purely birth process, $p_n(t) = 0$ for $t > 0$ and $n < n_0$.

Since the rate of change of p_n depends only on itself and on the preceding $p_{n-1}(t)$, we have

$$\frac{dp_{n_0}(t)}{dt} = -n_0\beta p_{n_0}(t), \tag{5.5.141}$$

so that

$$p_{n_0}(t) = e^{-\beta n_0 t}.$$

For $p_{n_0+1}(t)$ we obtain the nonhomogeneous equation

$$\frac{dp_{n_0+1}(t)}{dt} = -(n_0 + 1)\beta p_{n_0+1}(t) + \beta n_0 e^{-\beta n_0 t}.$$

Using integrating factor $e^{\beta(n_0+1)t}$ we obtain

$$\frac{d}{dt}\left(p_{n_0+1}(t)e^{\beta(n_0+1)t}\right) = \beta n_0 e^{\beta t},$$

that is,

$$p_{n_0+1}(t) = (n_0 e^{\beta t} + C)e^{-\beta(n_0+1)t}$$

so, using the initial condition $p_{n_0+1}(0) = 0$, we obtain

$$p_{n_0+1}(t) = n_0(1 - e^{-\beta t})e^{-\beta n_0 t}$$

In general, it can be proved that

$$p_{n_0+m}(t) = \binom{n_0 + m - 1}{n_0 - 1} e^{-\beta n_0 t}(1 - e^{-\beta t})^m.$$

Indeed, we proved the validity of the formula for $m = 1$. Next

$$\frac{dp_{n_0+m+1}(t)}{dt} = -(n_0+m+1)\beta p_{n_0+m+1}(t)+\beta \binom{n_0 + m - 1}{n_0 - 1} e^{-\beta n_0 t}(1-e^{-\beta t})^m.$$

and, as before

$$\frac{d}{dt}\left(p_{n_0+m+1}(t)e^{\beta(n_0+m+1)t}\right) = \beta\binom{n_0+m-1}{n_0-1}e^{\beta t}(1-e^{-\beta t})^m.$$

Integrating

$$\begin{aligned}
p_{n_0+m+1}(t)e^{\beta(n_0+m+1)t} &= C + \beta(n_0+m)\binom{n_0+m-1}{n_0-1} \\
&\quad \times \int e^{\beta(m+1)t}(1-e^{-\beta t})^m dt \\
&= C + \beta(n_0+m)\binom{n_0+m-1}{n_0-1}\int e^{\beta t}(e^{\beta t}-1)^m dt \\
&= C + (n_0+m)\binom{n_0+m-1}{n_0-1}\int u^m du \\
&= C + \frac{n_0+m}{m+1}\binom{n_0+m-1}{n_0-1}(e^{\beta t}-1)^{m+1} \\
&= C + \binom{n_0+m}{n_0-1}(e^{\beta t}-1)^{m+1}.
\end{aligned}$$

Using the initial condition $p_{n_0+m+1}(0) = 0$, we find $C = 0$ and so

$$p_{n_0+m+1}(t) = \binom{n_0+m}{n_0-1}e^{-\beta n_0 t}(1-e^{-\beta t})^{m+1}.$$

Exercise 5.5.1 Assume that the process begins with the initial condition (5.5.140). Prove that the expected size of the population and its variance are given by

$$\begin{aligned}
E(N(t)) &= n_0 e^{\beta t}, \\
Var(N(t)) &= n_0(1-e^{-\beta t})e^{2\beta t}.
\end{aligned}$$

Hint. The variance is defined as

$$Var[N(t)] = E[N^2(t)] - (E[N(t)])^2.$$

Exercise 5.5.2 Derive the system of equations describing the death-only process modifying (5.5.131)–(5.5.133) so that instead of birth, only death can occur with the probability

$$P\{1 \text{ death in } (t, t+\Delta t]|N(t) = 1\} = \delta\Delta t + o(\Delta t)$$

in $(t, \Delta t]$. Assume that there are n_0 individuals present at the start of a simple death process. Let T be the extinction time for this process. What is the probability density function for T? What is the mean time to extinction?

5.5.2 *Birth-and-Death System*

The obvious drawback of the system discussed in the previous section is that individuals never die. We can easily remedy this by combining the results of Exercise 5.5.2 with the birth process. Thus,

$$P\{1 \text{ birth in } (t, t+\Delta t] | N(t) = 1\} = \beta \Delta t + o(\Delta t), \tag{5.5.142}$$

$$P\{1 \text{ death in } (t, t+\Delta t] | N(t) = 1\} = \delta \Delta t + o(\Delta t), \tag{5.5.143}$$

$$P\{\text{no change in } (t, t+\Delta t] | N(t) = 1\} = 1 - (\beta + \delta)\Delta t + o(\Delta t). \tag{5.5.144}$$

Possibility of more than one birth or death occurring in $(t, t + \Delta t]$ is assumed to be of order $o(\Delta t)$ and will be omitted in the discussion.

As before, we assume that in the population of n individuals births and deaths occur independently. The probability of one birth is given by

$$\begin{aligned} &P\{1 \text{ birth in } (t, t+\Delta t] | N(t) = n\} \\ &\quad = n(\beta \Delta t + o(\Delta t))(1 - (\beta + \delta)\Delta t + o(\Delta t))^{n-1} = n\beta \Delta t + o(\Delta t). \end{aligned} \tag{5.5.145}$$

Similarly, the probability of one (net) death in the population is

$$\begin{aligned} &P\{1 \text{ death in } (t, t+\Delta t] | N(t) = n\} \\ &\quad = n(\delta \Delta t + o(\Delta t))(1 - (\beta + \delta)\Delta t + o(\Delta t))^{n-1} = n\delta \Delta t + o(\Delta t), \end{aligned} \tag{5.5.146}$$

and, finally,

$$\begin{aligned} &P\{\text{no change in } (t, t+\Delta t] | N(t) = n\} \\ &\quad = (1 - (\beta + \delta)\Delta t + o(\Delta t))^n = 1 - n(\beta + \delta)\Delta t + o(\Delta t). \end{aligned} \tag{5.5.147}$$

Now, we can set up the equation describing the evolution of $p_n(t)$. Using the law of total probability, we get as before

$$p_n(t) = (n-1)\beta \Delta t p_{n-1} + (n+1)\delta \Delta t p_{n+1} + (1 - n(\beta + \delta)\Delta t) p_n(t) + o(\Delta t), \tag{5.5.148}$$

and, finally,

$$p_n'(t) = -n(\beta + \delta)p_n(t) + (n-1)\beta p_{n-1}(t) + (n+1)\delta p_{n+1}(t). \tag{5.5.149}$$

This system has to be supplemented by the initial condition

$$p_n(0) = \begin{cases} 1 \text{ for } n = n_0, \\ 0 \text{ for } n \neq n_0. \end{cases} \tag{5.5.150}$$

Remark 5.39 Equations similar to (5.5.149) can occur in many other ways, not necessarily describing stochastic processes. In general, we can consider a population consisting of individuals differentiated by a single feature, e.g., we can consider cells having n copies of a particular gene. Here, $u_n(t)$ will be the number of individuals having n copies of this gene. Due to mutations or other environmental influences, the number of genes can increase or decrease. We may assume that in a sufficiently small period, only one change may occur. Denoting by β_n and δ_n the rates of increase (resp. decrease) of the number of genes if there are n of them, by the argument used above, we have

$$u_n'(t) = -(\beta_n + \delta_n)u_n(t) + \delta_{n+1}u_{n+1}(t) + \beta_{n-1}u_{n-1}(t), \quad n \geq 0.$$

Contrary to (5.5.139), the system (5.5.149) is much more difficult to solve. Even proving that there is a solution to it is a highly nontrivial exercise; see [13, 21]. In what follows, we assume that $(p_0(t), p_1(t), \ldots)$ exists and describes a probability, that is, for all $t \geq 0$, $p_n(t) \geq 0$ and

$$\sum_{n=0}^{\infty} p_n(t) = 1. \tag{5.5.151}$$

Then, we will be able to find formulae for p_n by the generating function method. We define

$$F(t, x) = \sum_{n=0}^{\infty} p_n(t)x^n.$$

By (5.5.151), the generating function is defined in the closed circle $|x| \leq 1$ and analytic in $|x| < 1$. The generating function has the following interpretation in this context.

(1) The probability of extinction at time t, $p_0(t)$, is given by

$$p_0(t) = F(t, 0). \tag{5.5.152}$$

(2) The probabilities $p_n(t)$ are given by

$$p_n(t) = \frac{1}{n!} \frac{\partial^n F}{\partial x^n}\bigg|_{x=0}. \tag{5.5.153}$$

If $F(t, x)$ is analytic in a slightly larger circle containing $x = 1$, we can use F to find other useful quantities. The expected value of $N(t)$ at time t is defined by

$$E(N(t)) = \sum_{n=0}^{\infty} n p_n(t).$$

On the other hand,

$$\frac{\partial F}{\partial x}(t, x) = \sum_{n=0}^{\infty} n p_n(t) x^{n-1},$$

so that

$$E[N(t)] = \sum_{n=0}^{\infty} n p_n(t) = \frac{\partial F}{\partial x}\bigg|_{(t,x)=(t,1)}. \tag{5.5.154}$$

Similarly, the variance is defined by

$$Var[N(t)] = E[N^2(t)] - (E[N(t)])^2.$$

On the other hand,

$$\frac{\partial^2 F}{\partial x^2}(t, x)\bigg|_{x=1} - \sum_{n=0}^{\infty} n(n-1) p_n(t) - E[N^2(t)] - E[N(t)].$$

Combining these formulae, we get

$$Var[N(t)] = \left(\frac{\partial^2 F}{\partial x^2} + \frac{\partial F}{\partial x} - \left(\frac{\partial F}{\partial x} \right)^2 \right)\bigg|_{(t,x)=(t,1)}. \tag{5.5.155}$$

Let us find the equation satisfied by F. Using (5.5.149) and $p_{-1} = 0$, we have

$$\begin{aligned}\frac{\partial F}{\partial t}(t,x) &= \sum_{n=0}^{\infty} n\frac{dp_n}{dt}(t) = -(\beta+\delta)\sum_{n=0}^{\infty} np_n(t)x^n\\ &\quad +\beta\sum_{n=0}^{\infty}(n-1)p_{n-1}(t)x^n + \delta\sum_{n=0}^{\infty}(n+1)p_{n+1}(t)x^n\\ &= -(\beta+\delta)x\frac{\partial F}{\partial x}(t,x) + \beta x^2\frac{\partial F}{\partial x}(t,x) + \delta\frac{\partial F}{\partial x}(t,x).\end{aligned}$$

That is, to find F, we have to solve the equation

$$\frac{\partial F}{\partial t} = \left(\beta x^2 - (\beta+\delta)x + \delta\right)\frac{\partial F}{\partial x}, \tag{5.5.156}$$

supplemented by the initial condition

$$F(0,x) = x^{n_0}.$$

The equation can be solved by the method of characteristics; see Appendix D. It is a homogeneous equation so that F is constant along characteristics, which are given as solutions to

$$x' = -(\beta x - \delta)(x-1),$$

see (D.1.13), that is,

$$\begin{aligned}-t + C &= \int\frac{dt}{(\beta x-\delta)(x-1)} = \frac{1}{\beta-\delta}\left(-\int\frac{dx}{x-\frac{\delta}{\beta}} + \int\frac{dx}{x-1}\right)\\ &= \frac{1}{\beta-\delta}\ln\left|\frac{x-1}{x-\frac{\delta}{\beta}}\right|,\end{aligned}$$

provided $\beta \neq \delta$ and $x \neq 1, \frac{\delta}{\beta}$. This gives

$$\left|\frac{\beta x-\delta}{x-1}\right| = Ce^{rt},$$

where $r = \beta - \delta$. Thus, we have a general solution

$$F(t,x) = G\left(e^{-rt}\left|\frac{\beta x - \delta}{x-1}\right|\right),$$

where G is an arbitrary function. Using the initial condition, we get

$$x^{n_0} = G\left(\left|\frac{\beta x - \delta}{x-1}\right|\right).$$

Assume $x < \min\{1, \frac{\delta}{\beta}\}$ or $x > \max\{1, \frac{\delta}{\beta}\}$ so that we can drop absolute value bars. Solving

$$s = \frac{\beta x - \delta}{x-1},$$

we get

$$x = \frac{s-\delta}{s-\beta},$$

so that

$$G(s) = \left(\frac{s-\delta}{s-\beta}\right)^{n_0}.$$

Thus, the solution is given by

$$F(x,t) = \left(\frac{e^{-rt}\frac{\beta x-\delta}{x-1} - \delta}{e^{-rt}\frac{\beta x-\delta}{x-1} - \beta}\right)^{n_0} = \left(\frac{e^{rt}\delta(1-x) + (\beta x - \delta)}{e^{rt}\beta(1-x) + (\beta x - \delta)}\right)^{n_0}. \tag{5.5.157}$$

Consider the zeroes of the denominator,

$$x = \frac{e^{rt} - \frac{\delta}{\beta}}{e^{rt} - 1}.$$

If $\frac{\delta}{\beta} < 1$, then $r > 0$ and we see that $x > 0$ and, as $t \to \infty$, x moves from $+\infty$ to 1 and thus F is analytical in the circle stretching from the origin to the first singularity, which is bigger than 1 for any finite t. If $\frac{\delta}{\beta} > 1$, then $r < 0$ and x above is again positive and moves from infinity to $\frac{\delta}{\beta} > 1$ so again F is analytic in a circle with a radius bigger than 1. Since we know that the generating function (defined by the series) coincides with F defined above for $|x| < \min\{1, \frac{\delta}{\beta}\}$, by the principle of analytic continuation, the generation function coincides with F in the whole domain of its analyticity (note that this is not necessarily a solution of the equation (5.5.156) outside this region as we have removed the absolute value bars).

Consider now the case $\beta = \delta$. Then the characteristic equation is

$$x' = -\beta(x-1)^2,$$

solving which we obtain

$$\frac{1}{x-1} = \beta t + \xi,$$

or

$$\xi = \frac{1 - x\beta t + \beta t}{x-1}.$$

Hence, the general solution is given by

$$F(t,x) = G\left(\frac{1 - x\beta t + \beta t}{x-1}\right)$$

for some function G. Using the initial condition, we have

$$x^{n_0} = G\left(\frac{1}{x-1}\right).$$

Defining

$$s = \frac{1}{x-1},$$

that is,

$$x = 1 + \frac{1}{s}.$$

Hence

$$G(s) = \left(1 + \frac{1}{s}\right)^{n_0}.$$

Therefore

$$F(t,x) = \left(1 + \frac{x-1}{1 - x\beta t + \beta t}\right)^{n_0} = \left(\frac{\beta t + (1-\beta t)x}{1 - x\beta t + \beta t}\right)^{n_0}.$$

Summarizing,

$$F(t,x) = \begin{cases} \left(\frac{e^{rt}\delta(1-x)+(\beta x-\delta)}{e^{rt}\beta(1-x)+(\beta x-\delta)}\right)^{n_0} & \text{if } \beta \neq \delta, \\ \left(\frac{\beta t+(1-\beta t)x}{1-x\beta t+\beta t}\right)^{n_0} & \text{if } \beta = \delta. \end{cases} \tag{5.5.158}$$

Let us complete this section by evaluating some essential parameters. The probability of extinction at time t is given by

$$p_0(t) = F(t,0) = \begin{cases} \left(\frac{\delta(e^{rt}-1)}{e^{rt}\beta-\delta}\right)^{n_0} & \text{if } \beta \neq \delta, \\ \left(\frac{\beta t}{1+\beta t}\right)^{n_0} & \text{if } \beta = \delta. \end{cases} \tag{5.5.159}$$

Hence, the asymptotic probability of extinction is given by

$$\lim_{t\to\infty} p_0(t) = \begin{cases} \left(\frac{\delta}{\beta}\right)^{n_0} & \text{if } \beta > \delta, \\ 1 & \text{if } \beta \leq \delta. \end{cases} \tag{5.5.160}$$

We note that even for positive net growth rates $\beta > \delta$, the probability of extinction is non-zero. Populations with small initial numbers are especially susceptible to extinction.

To derive the expected size of the population, we use (5.5.154). We have

$$E[N(t)] = \left.\frac{\partial F}{\partial x}\right|_{(t,x)=(t,1)} = n_0 \frac{(-e^{rt}\delta+\beta)(\beta-\delta)+\beta(e^{rt}-1)(\beta-\delta)}{(\beta-\delta)^2} = n_0 e^{rt}.$$

We must find the second derivative of F to get the variance. It is given by

$$\begin{aligned} &\frac{\partial^2 F}{\partial x^2}(t,x) \\ &= n_0 \left(\frac{x\beta-\delta+e^{rt}(1-x)\delta}{e^{rt}(1-x)\beta+x\beta-\delta}\right)^{n_0-1} \\ &\left(-\frac{2(\beta-e^{rt}\beta)(\beta-e^{rt}\delta)}{(e^{rt}(1-x)\beta+x\beta-\delta)^2} + \frac{2(\beta-e^{rt}\beta)^2(x\beta-\delta+e^{rt}(1-x)\delta)}{(e^{rt}(1-x)\beta+x\beta-\delta)^3}\right) \\ &+ (n_0-1)n_0 \left(\frac{x\beta-\delta+e^{rt}(1-x)\delta}{e^{rt}(1-x)\beta+x\beta-\delta}\right)^{n_0-2} \\ &\left(\frac{\beta-e^{rt}\delta}{e^{rt}(1-x)\beta+x\beta-\delta} - \frac{(\beta-e^{rt}\beta)(x\beta-\delta+e^{rt}(1-x)\delta)}{(e^{rt}(1-x)\beta+x\beta-\delta)^2}\right)^2. \end{aligned}$$

Hence,

$$
\begin{aligned}
Var[N(t)] &= \left(\frac{\partial^2 F}{\partial x^2} + \frac{\partial F}{\partial x} - \left(\frac{\partial F}{\partial x}\right)^2 \right)\Bigg|_{(t,x)=(t,1)} \\
&= n_0 \left(\frac{2(\beta - e^{rt}\beta)^2}{(\beta-\delta)^2} - \frac{2(\beta - e^{rt}\beta)(\beta - e^{rt}\delta)}{(\beta-\delta)^2} \right) \\
&\quad + n_0 \left(-\frac{\beta - e^{rt}\beta}{\beta-\delta} + \frac{\beta - e^{rt}\delta}{\beta-\delta} \right) \\
&\quad + (-1+n_0)n_0 \left(-\frac{\beta - e^{rt}\beta}{\beta-\delta} + \frac{\beta - e^{rt}\delta}{\beta-\delta} \right)^2 \\
&\quad - {n_0}^2 \left(-\frac{\beta - e^{rt}\beta}{\beta-\delta} + \frac{\beta - e^{rt}\delta}{\beta-\delta} \right)^2 \\
&= \frac{e^{rt}(-1+e^{rt})n_0(\beta+\delta)}{\beta-\delta}
\end{aligned}
$$

for $\beta \neq \delta$, while for $\beta = \delta$ we obtain

$$V(t) = 2n_0\beta t.$$

Exercise 5.5.3 Consider a birth-and-death process with immigration, where (5.5.142)–(5.5.144) replaced by

$$P\{N(t) \text{ increases by 1 in } (t, t+\Delta t] | N(t) = 1\} = \beta\Delta t + I\Delta t + o(\Delta t), \tag{5.5.161}$$

$$P\{N(t) \text{ decreases by 1 in } (t, t+\Delta t] | N(t) = 1\} = \delta\Delta t + o(\Delta t), \tag{5.5.162}$$

$$P\{\text{no change in } (t, t+\Delta t] | N(t) = 1\} = 1 - (\beta+\delta)\Delta t + o(\Delta t). \tag{5.5.163}$$

Assume the mortality rate exceeds the birth rate, $\delta > \beta$.

1. Derive differential equations for the probabilities.
2. Derive a partial differential equation for the probability generating function.
3. Find the probability generating function for the equilibrium probability distribution by setting the time derivative in the partial differential equation equal to zero. Solve the corresponding ordinary differential equation. Choose the generating function so that the corresponding probabilities sum up to 1.
4. What is the expected population size at equilibrium?
5. What is the variance of the equilibrium population sizes?
6. What are the odds that the population is of size zero at equilibrium?

Chapter 6
Continuous Time Non-linear Models for Interacting Species and Age-Structured Populations

The conservation law framework for populations interacting and undergoing internal changes in continuous time was introduced in Sect. 1.2.3. In this chapter, we will discuss several constitutive relations that specify the systems of conservation laws that describe populations in epidemiological or ecological contexts. Thus, we begin by introducing several models representative of the field, and then we develop multidimensional versions of the tools discussed for scalar models in Sect. 4.1.

6.1 Models

Models introduced in Sect. 1.2.3 are derived from conservation laws such as the mass or population balance. As a result, these models lead to particular systems of ordinary differential equations, often called compartmental systems. Linear systems of this type were the subject of the previous chapter; in particular, see the modelling in Sect. 5.1.2. Models of this type describe the dynamics of many processes in pharmacokinetics, metabolism, epidemiology, ecology, and other areas. Now, we will present two models of this type whose analysis will be accomplished later after introducing the necessary tools.

J. Banasiak, *Introduction to Mathematical Methods in Population Theory*, Springer Undergraduate Mathematics Series,
https://doi.org/10.1007/978-3-031-65491-6_6

6.1.1 Basic Models from Ecology

Let us return to the general model of two interacting species introduced in Sect. 1.2.3 and recall the system describing the interactions depicted in Fig. 1.3,

$$\begin{aligned} P_1' &= \text{Intrinsic evolution of } P_1 \text{ without } P_2 + \text{Impact of } P_2 \text{ on } P_1, \\ P_2' &= \text{Intrinsic evolution of } P_2 \text{ without } P_1 + \text{Impact of } P_1 \text{ on } P_2, \end{aligned} \tag{6.1.1}$$

where P_1 and P_2 typically will be the densities of some populations over a certain area. For the first terms on the right-hand side of each equation, we can use a suitable model describing the evolution of a single species. In more complicated situations, parameters in these models, such as the carrying capacity, can depend on the other species.

The interaction terms require more attention. The simplest terms would be linear terms, e.g.,

$$\begin{aligned} P_1' &= r_1 P_1 \left(1 - \frac{P_1}{K_1}\right) + a_{12} P_2, \\ P_2' &= r_2 P_2 \left(1 - \frac{P_2}{K_2}\right) + a_{21} P_1. \end{aligned} \tag{6.1.2}$$

This system could describe two populations with logistic vital dynamics each, where the additional rate of change of one population due to the other would be directly proportional to the latter's density.

It is important to understand the interaction terms. We could distinguish:

- competition, when $a_{12}, a_{21} < 0$, as the presence of each species harms the other;
- predator-prey interaction, when $a_{12} > 0$, $a_{21} < 0$, as the presence of species 2 (prey) has a positive impact on the species 1 (predator) and the presence of species 1 harms the species 2;
- mutualism, when $a_{12}, a_{21} > 0$, as the presence of each species has a positive impact on the other.

In the predator-prey case, the linear model above implies that the predator eats the same amount of prey per unit time, irrespective of the prey density. This is unrealistic, so the coefficients a_{ij} should depend on the density of the ith species. The simplest assumption is that a single predator will consume a proportion of the available prey, leading to the

$$\alpha_{21} P_1 P_2, \quad \alpha_{21} < 0,$$

that is, the law of mass action. The term is borrowed from chemical kinetics, where it is assumed that the rate of reactions is proportional to the product of the concentrations of substrates. Hence, we have a family of mass-action models of interactions between two species:

$$\begin{aligned} P_1' &= P_1(\beta_1 + \mu_1 P_1 + \alpha_{12} P_2), \\ P_2' &= P_2(\beta_2 + \mu_2 P_2 + \alpha_{21} P_1), \end{aligned} \tag{6.1.3}$$

using which we can model various types of interactions and vital dynamics. For instance, assuming predator-prey interactions, ($\alpha_{12} > 0, \alpha_{21} < 0$), we have $\beta_2 > 0, \mu_2 < 0$ if the prey population follows the logistic vital dynamics in the absence of predator. On the other hand, if we are to model a specialist predator, $\beta_1 < 0, \mu_1 \leq 0$, but for a generalist, we can have $\beta_1 > 0, \mu_1 < 0$.

Lotka–Volterra Equation

One of the most famous predator-prey models is the Lotka–Volterra model, which was created to explain why, in a period of reduced fishing during World War I, the number of sharks substantially increased. We shall describe it on using the original example of small fish–shark interaction.

To introduce the model, we consider two populations, small fish and sharks, with the following influencing factors considered.

(i) Populations of fish and sharks display exponential growth when considered in isolation. However, the growth rate of sharks in the absence of fish is negative due to the lack of food.
(ii) Fish is preyed upon by sharks, resulting in the decline in fish population. Each shark is assumed to eat a constant fraction of the fish population.
(iii) The population of sharks increases if there is more fish. The additional number of sharks is proportional to the number of available fish.
(iv) Fish and sharks are being fished indiscriminately, that is, the number of sharks and fish caught by fishermen is directly proportional to the present populations of fish and sharks, respectively, with the same proportionality constant.

If we denote by x and y the sizes of fish and shark populations, then arguing as above we obtain

$$\begin{aligned} x' &= (r - f)x - \alpha xy, \\ y' &= -(s + f)y + \beta xy, \end{aligned} \tag{6.1.4}$$

where α, β, r, s, f are positive constants.

Exercise 6.1.1 An orchard is infested by a population of insects. The insects are preyed upon by a type of beetle. The orchard owner decides to use a pesticide that kills a fixed fraction of insects and beetles. Derive the system of equations describing the change in the population of beetles and insects using a similar argument to the one used for the Lotka–Volterra system.

Exercise 6.1.2 Modify the Lotka–Volterra system by assuming that the fish population grows according to the logistic equation.

Exercise 6.1.3 Modify the Lotka–Volterra system by assuming that there is a competition for food among sharks, each shark killing a constant fraction of the shark population.

Exercise 6.1.4 Build a model describing the evolution of four populations: A of herbivores, B of insects, C of predators preying on both insects and herbivores and D of plants being the food for both insects and herbivores. We assume that insects pollinate the plants, thus helping in their development, and herbivores compete with insects for food.

Exercise 6.1.5 Build a model describing the evolution of populations of three species A, B and C, where the species A is a predator preying on C, B is omnivorous eating also carrion and C is a herbivore. Thus, apart from intraspecies competition, individuals from B and C compete with each other for food. Moreover, B-individuals eat leftovers of victims of A.

Holling II Type Models

It is clear that the mass action law also is not realistic: for instance, it implies that the predator could eat an arbitrary amount of prey in a unit time (if the density of prey is large enough). To be more realistic, we must incorporate at least some saturation effect. We describe one such model, given by Holling II functional response (mass action is termed Holling type I response). A similar model for single species dynamics was introduced in Sect. 3.3.4.

We assume that the density P of prey consumed by a single predator in time T is proportional to the prey density x and the time spent on hunting T_h,

$$P = cxT_h.$$

The mass action law assumes that $T_h = T$, that is, the predator does not stop hunting while devouring the prey. While it is sometimes possible (e.g., adult salmon eating its offspring), in most cases, the time T_e spent on eating the prey is positive. If P is the number of prey caught in time T, then the time used to consume it is PT_s and thus $T_h = T - PT_s$. Hence,

$$P = cx(T - PT_s)$$

and the density of prey eaten per unit time per predator is

$$\frac{P}{T} = \frac{cx}{1 + cT_s x}.$$

Multiplying by the density of predators, we obtain the predation term as

$$-c_1\frac{xy}{1 + c_2 x},$$

with positive constants c_1, c_2. An example of a model with such a functional response is the McArthur–Rosenzweig model, see, e.g., [23, Chapter 5],

$$\begin{aligned} x' &= x\left(r\left(1-\frac{x}{K}\right)-\frac{ay}{b+x}\right), \qquad x(0)=\mathring{x}, \\ y' &= y\left(\frac{cx}{b+x}-d\right), \qquad y(0)=\mathring{y}. \end{aligned} \tag{6.1.5}$$

Here, the prey and predator densities are x and y, respectively. In the absence of predators, the prey grows logistically, with a carrying capacity K and natural growth rate r, and the predators in the absence of prey die out at an exponential rate, as in the Lotka–Volterra model. The presence of predators has a detrimental effect on the prey growth rate. Conversely, prey benefits the predators, subject to some saturation modelled by the Holling II functional response. The constants a and c are the biomass conversion rates, and they differ due to the effects of predator–prey interactions on the respective populations. Other types of functional response are also possible; see [23, Chapter 5].

Exercise 6.1.6 We consider another ecological application concerning water pollution in lakes.

(a) We start with one lake. Suppose a lake is situated in an industrial region, and the industry dumps waste products into this lake, pumping polluted water at a constant rate of r_1 (say, cubic meters per day). We assume that the concentration of waste in this water is a known constant c_1. Moreover, there is a river flowing fresh water into the lake at a rate r_2 and another river flowing from the lake carrying away the polluted water at a rate r_3.

   (i) Find the relation between the flow rates that ensures that the volume of water in the lake is constant (disregard rainfall and evaporation).
   (ii) Derive the equation governing the change in concentration of the pollution in the lake, assuming that the lake is well stirred so that the pollutant is uniformly mixed throughout the lake. Hint. Write the conservation of mass equation for the amount of the pollutant.

(b) Generalize the model to include one more lake downstream from the first. Assume that Lake 1 supplies Lake 2 with the polluted water at a rate r_4, there is no industry surrounding Lake 2, there is fresh water flowing to Lake 2 at a rate r_5 and a river flowing out from Lake 2 at a rate r_6.

6.1.2 *Basic Epidemiological Models*

A broad class of interaction models describe infectious diseases. Though they can be fitted into the framework described in the previous section, they have a specific structure that often allows for the development of specialized tools for their analysis. Due to this, and also due to their importance, they warrant a more detailed description.

Epidemiological Terminology

Following [69, Chapter 1], we provide basic notions of epidemiology. An infectious disease is an illness caused by a microbial agent. The microbial agent can be:

bacteria: tuberculosis, pneumonia;
virus: HIV, influenza;
fungus: dermatomycosis;
parasite: malaria, bilharzia;
toxic protein or prions: Creutzfeldt-Jakob disease (mad cow disease).

Communicable diseases are infectious diseases that can be transmitted from one infectious person to another, directly or indirectly. There are infectious diseases, such as tetanus, that are not communicable. Transmittable diseases are infectious diseases that can be transmitted from one person to another by unnatural routes. For instance, mad cow disease can be passed from one person to another only through a surgical intervention.

For modelling purposes, we distinguish the following types of transmission:

direct: when the pathogen is transmitted from one person to another by personal contact, such as sexually transmitted diseases, influenza, smallpox, measles, chickenpox, TB;
vector: when the pathogen is transmitted by a vector such as mosquito, tick or snail, that includes malaria, dengue, Zika, Lyme disease;
environmental: when a human is infected by a pathogen present in the environment, water or food, such as cholera, salmonella, or stomach flu;
vertical: mother-to-child transmission, such as HIV.

The following terminology is essential in epidemiological research:

Susceptible individuals: a member of a population who is at risk of becoming infected;
Exposed individuals: susceptible individuals that made a potentially disease-transmitting contact and may or may not develop the disease;
Infected and infectious individuals: if a pathogen establishes itself in an individual, the individual becomes infected. An infected individual who can transmit the disease is called infectious;
Latent individuals: individuals who are infected but not yet infectious;
Latent period: the time from infection to the moment the individual becomes infectious;
Incubation period: the period between exposure to the pathogen to the onset of symptoms of the disease;
Incidence: the number of individuals who become ill during a specified time;
Prevalence: the number of people with the disease at a specific time.

First Models

SIRS/SEIRS Model

We begin with a simple model of a nonlethal disease in a homogeneous population divided into three classes (compartments): susceptible S, infective I and recovered R. Sometimes, we include another class of latent or exposed individuals who were in contact with an infectious agent but are not yet infectious (note that we assume this is the only way of getting infectious—we do not assume any vertical transmission). This division can be made more detailed depending on the disease and the purpose of the model. Let

λ = the force of infection, that is, the rate at which susceptibles become infected;
μ = the death rate;
δ = the rate at which the disease develops after infection;
ν = the recovery rate;
γ = the rate of immunity loss;
B = the birth rate of the population.

The flow between the compartments is shown in Fig. 6.1 and, based on this figure, we build the following system of equations,

$$
\begin{aligned}
S' &= B(N) - \lambda S + \gamma R - \mu S,\\
E' &= \lambda S - \delta E - \mu E,\\
I' &= \delta E - \nu I - \mu I,\\
R' &= \nu I - \gamma R - \mu R.
\end{aligned}
\tag{6.1.6}
$$

In the simplified version, when we ignore any demography, assume that the individuals become infectious immediately after getting infected and that the recovered do not lose immunity, we arrive at the classical SIR model

$$
\begin{aligned}
S' &= -\lambda S,\\
I' &= \lambda S - \nu I,\\
R' &= \nu I.
\end{aligned}
\tag{6.1.7}
$$

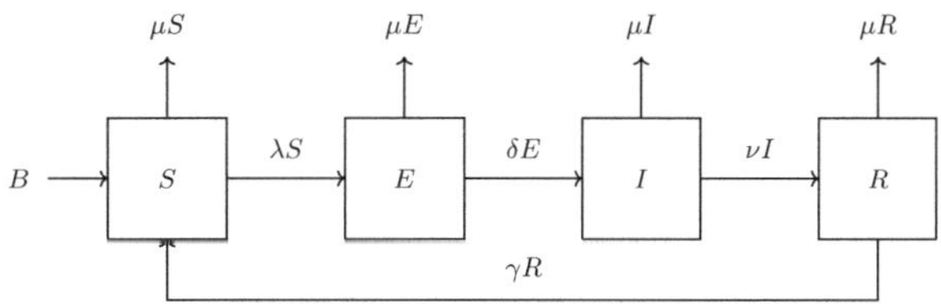

Fig. 6.1 Compartments in SEIRS model

Exercise 6.1.7 Formulate an epidemiological model when:

(i) individuals become infected immediately after infection and susceptible immediately after recovery,
(ii) individuals become infected immediately after infection and recover with immunity, which wanes after some time.

Parameter Interpretation

In Sect. 3.3.1, we have shown that $\frac{1}{\mu}$ gives an individual's expected lifespan. In the same way, $\frac{1}{\nu}$ is the average duration of the disease, $\frac{1}{\delta}$ is the average duration of the latent period, and $\frac{1}{\gamma}$ is the average duration of the acquired immunity.

For a directly transmitted pathogen, the force of infection λ is the product of

1. the contact rate;
2. the proportion of these contacts that are with infective;
3. the proportion of such contacts that result in infection.

How should we model λ using the above guidelines? A simple assumption would be the mass action law—a single infective meets a fraction $c_1 S$ of susceptibles in a unit time and infects a fraction c_2 of those met:

$$c_2 c_1 S I.$$

It may be a reasonable assumption if the population densities are low, but for large densities, we need to take into account that contacts take time so that there can be only a finite time of contacts in a unit time interval. Also, the growing number of infections causes saturation due to the decreasing probability of meeting a still noninfected person. Saturation may also be caused by satiation, which plays a role in sexual transmission, but also in blood meals taken by mosquitoes.

Hence, let us derive a reasonable form of λ, see [38]. In a population of size/density N, we define $C(N)$ as the fraction of the population engaged in a contact at any given time. Then $NC(N)$ (precisely $P = \frac{1}{2}NC(N)$) is the number of pairs in the population at any given time. Since the probability of choosing at random a pair consisting of a susceptible and an infective is

$$\frac{S}{N}\frac{I}{N},$$

the density of pairing that potentially can lead to infection is

$$C(N)\frac{SI}{N}.$$

The function $C(N)$ should be nonnegative, nondecreasing, linear in N for small N and having a limit (≤ 1) as $N \to \infty$. Let us try to derive such a function using a Holling-type argument. First, in a population of size N we introduce the number of singles X and pairs P so that

$$N = X + 2P, \tag{6.1.8}$$

so that

$$S' = -2\beta P \frac{SI}{N^2},$$

where β is the product of the probability that a contact between an infective and susceptible results in an infection and the contact rate. Assume that an individual can be single or form a pair and that the contact lasts some time T_h. Denote by Z the total number contacts over some time T and let $Y = Z/T$, the number of contacts per unit time. As in the Holling derivation, we have

$$Z = \rho X T_s = \rho X(T - ZT_h), \tag{6.1.9}$$

where ρ is a constant. However, contrary to the predator-prey model, where prey was unlimited, here the number of available singles is limited by time—singles are available only when they are not engaged in another contact. For a given single, in time T, it is available for $T - ZT_h$, and thus, for the fraction $1 - YT_h$ of time. In other words, any given single at any given time is available for contact with probability $p = 1 - YT_h$. Thus, the expected number of available singles at any given time is given by

$$\begin{aligned} X &= \sum_{k=0}^{N} k \binom{N}{k} p^k (1-p)^{N-k} = Np \sum_{l=0}^{N-1} \binom{N-1}{l} p^l (1-p)^{N-1-l} \\ &= Np = N(1 - YT_h). \end{aligned}$$

Hence, from (6.1.9),

$$\frac{Z}{T} = Y = \rho N(1 - YT_h)^2.$$

Using again (6.1.9), we find that the average number of pairs is

$$P = \frac{1}{2} NYT_h = \frac{\rho T_h}{2} N^2 (1 - YT_h)^2 =: \frac{\nu}{2} X^2.$$

By (6.1.8),

$$P = \frac{\nu}{2}(N - 2P)^2,$$

or

$$P = \frac{2\nu N + 1 \pm \sqrt{4\nu N + 1}}{4\nu},$$

and we have to select the negative sign,

$$P = \frac{2\nu N + 1 - \sqrt{4\nu N + 1}}{4\nu},$$

to keep $2P < N$. We re-write this as

$$P = \frac{\nu N^2}{2\nu N + 1 + \sqrt{4\nu N + 1}}.$$

Finally,

$$S' = -\beta C(N)\frac{SI}{N} = -\frac{2\nu\beta N}{2\nu N + 1 + \sqrt{4\nu N + 1}}\frac{SI}{N}.$$

However, most often, we use one of the following simplifications:

- low density approximation, $N \to 0$, gives $C(N) \sim N$, which leads to the mass action transmission rate $\beta\nu SI$;
- large density approximation, $N \to \infty$, gives a constant approximation of $C(N)$, $C(N) \sim 1$, which leads to the frequency-dependent, or standard, transmission rate $\beta IS/N$.

The force of infection is related to one of the most important parameters in epidemiology, the basic reproduction number.

Definition 6.1 The basic reproduction number R_0 is the number of infections that one infectious individual will introduce in a population consisting only of susceptible individuals.

Exercise 6.1.8 Find R_0 for models (6.1.6) and (6.1.7).

As we shall see later, it is closely related to the stability of the so-called disease-free equilibrium point, that is, of the scenario where the population has natural immunity, allowing it to eliminate introduced diseases.

Malaria Model

Next, we consider a more involved model of a vector-borne disease. Here, the infection does not occur through contact between an infectious and susceptible individual, but the pathogenic microorganisms are transmitted from an infected individual to a susceptible individual by an arthropod or other agent. Thus, the transmission of vector-borne diseases to humans depends on three factors: (1) the pathogenic agent, (2) the arthropod vector, and (3) the human host. The pathogenic agent in the malaria model discussed here is the Plasmodium parasite. Its life cycle is complex, both in the mosquito vector and the human host, but we will not discuss it here. The parasite is transmitted by female mosquitoes, who feed on blood necessary for reproduction. When a mosquito feeds on blood, the parasite enters the host's bloodstream and infects it. Thus, the infection requires a female mosquito first to bite an infective human, acquiring the parasite, and then biting a susceptible human to infect him/her. Mathematical models of vector-borne diseases typically consider the vectors' and humans' dynamics and only occasionally the dynamics of the parasite.

A typical diagram of the humans' and vectors' dynamics is presented in Fig. 6.2, leading to

$$\begin{aligned}
S_h' &= B_h(N_h) - \lambda_h S_h + \gamma_h R_h - \mu_h S_h,\\
E_h' &= \lambda_h S_h - \delta_h E_h - \mu_h E_h,\\
I_h' &= \delta_h E_h - \nu_h I - \mu_h I_h - \mu_d I_h,\\
R_h' &= \nu_h I_h - \gamma_h R_h - \mu_h R_h,\\
S_v' &= B_v(N_v) - \lambda_v S_v - \mu_v S_v,\\
E_v' &= \lambda_v S_v - \delta_v E_v - \mu_v E_v,\\
I_v' &= \delta_v E_v - \mu_v I_v.
\end{aligned} \tag{6.1.10}$$

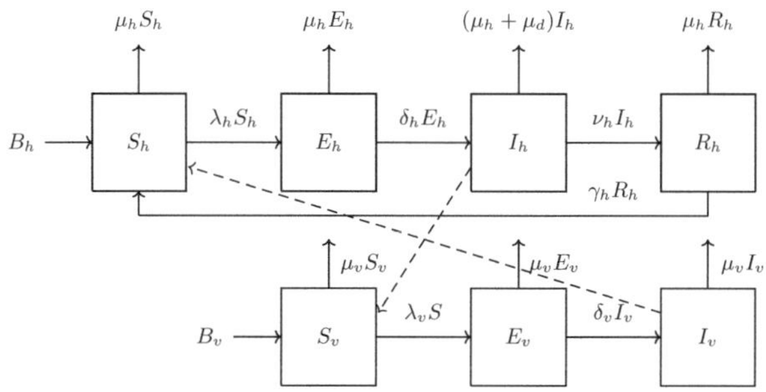

Fig. 6.2 Compartmental model of a lethal vector-borne disease

The state variables and parameter symbols are as in (6.1.6) with subscripts h and v referring to the human and vector populations; additionally, μ_d denotes the disease-induced death rate.

To model the infection rates, we introduce the following parameters:

σ_v = the number of bites on a human a mosquito could make per unit of time if humans were freely available.
σ_h = the maximum number of mosquito bites a human can have per unit time.
β_{hv} = probability of transmitting the infection from an infectious mosquito to a susceptible human, given that contact between the two occurs.
β_{vh} = probability of transmitting the infection from an infectious human to a susceptible mosquito, given that contact between the two occurs.

Then, [29], the infection rates are given by

$$\lambda_h = b_h(N_h, N_v)\beta_{hv}\frac{I_v}{N_v} \quad \text{and} \quad \lambda_v = b_v(N_v, N_h)\beta_{vh}\frac{I_h}{N_h}. \tag{6.1.11}$$

In other words, λ_h is the product of $b_h(N_h, N_v)$, the number of mosquito bites a human can have per unit time, the probability of the transmission of the infection, β_{hv} and the probability that the bite comes from an infected mosquito, $\frac{I_v}{N_v}$. Similarly, λ_v is the product of the number of bites on humans a mosquito has per unit time, $b_v(N_v, N_h)$ and the probability that the bite comes from an infectious human and that the transmission occurs. To model the number of bites, we first define the total number of bites that occur per unit time, $b(N_h, N_v)$, so that

$$b(N_h, N_v) = b_h(N_h, N_v)N_h = b_v(N_h, N_v)N_v.$$

In [29], the authors use the formula

$$b(N_h, N_v) = \frac{\sigma_v N_v \sigma_h N_h}{\sigma_v N_v + \sigma_h N_h} \tag{6.1.12}$$

without derivation. Here we derive (6.1.12) using the Holling-type argument, see [17]. In time T, the total number of bites received by humans can be written as

$$b(N_h, N_v)T = \sigma_h T_{av} N_h, \tag{6.1.13}$$

where T_{av} is the time available for the mosquitoes to bite. Thus

$$T = T_{av} + T_{nav}.$$

Table 6.1 Typical cases of the numbers of bites

	$b_h(N_h, N_v)$	$b_v(N_v, N_h)$
General model	$\frac{\sigma_v N_v \sigma_h}{\sigma_v N_v + \sigma_h N_h}$	$\frac{\sigma_v \sigma_h N_h}{\sigma_v N_v + \sigma_h N_h}$
Small $\frac{N_v}{N_h}$	$\frac{\sigma_v N_v}{N_h}$	σ_v
Small $\frac{N_h}{N_v}$	σ_h	$\frac{\sigma_h N_h}{N_v}$

Now, a mosquito cannot bite for some time after a meal. Precisely, in time T a mosquito has $\frac{\sigma_h N_h T_{av}}{N_v}$ meals, and the mosquito is not available for $\frac{1}{\sigma_v}$ after each meal. Thus,

$$T = T_{av} + \frac{\sigma_h N_h T_{av}}{\sigma_v N_v}.$$

Solving for T_{av} and substituting it into (6.1.13) gives (6.1.12).

Hence,

$$\lambda_h = \frac{\sigma_v \sigma_h}{\sigma_v N_v + \sigma_h N_h} \beta_{hv} I_v \quad \text{and} \quad \lambda_v = \frac{\sigma_v \sigma_h}{\sigma_v N_v + \sigma_h N_h} \beta_{vh} I_h. \tag{6.1.14}$$

Following [29], in Table 6.1 we list the typical cases for b_h and b_v. The arguments leading to the above formulae are based on the first-order Maclaurin expansion, e.g., for $b_h(N_h, N_v)$, we have

$$\frac{\sigma_v N_v \sigma_h}{\sigma_v N_v + \sigma_h N_h} = \frac{\sigma_v \sigma_h \frac{N_v}{N_h}}{\sigma_v \frac{N_v}{N_h} + \sigma_h} = \frac{\sigma_v N_v}{N_h} \frac{\sigma_h}{\sigma_v \frac{N_v}{N_h} + \sigma_h} = \frac{\sigma_v N_v}{N_h} \left(1 + \frac{\sigma_v N_v}{\sigma_h N_h} + \dots\right)$$

for small $\frac{N_v}{N_h}$, and

$$\frac{\sigma_v N_v \sigma_h}{\sigma_v N_v + \sigma_h N_h} = \frac{\sigma_h}{1 + \frac{\sigma_h N_h}{\sigma_v N_v}} = \sigma_h \left(1 + \frac{\sigma_h N_h}{\sigma_v N_v} + \dots\right)$$

for small $\frac{N_h}{N_v}$. Hence, the formulae in the second column are obtained by taking the first terms of the respective expansions.

Therefore, we obtain

$$\lambda_h = \sigma_v \beta_{hv} \frac{I_v}{N_h} \quad \text{and} \quad \lambda_v = \sigma_v \beta_{vh} \frac{I_h}{N_h} \tag{6.1.15}$$

in an environment with few mosquitoes, and

$$\lambda_h = \sigma_h \beta_{hv} \frac{I_v}{N_v} \quad \text{and} \quad \lambda_v = \sigma_h \beta_{vh} \frac{I_h}{N_v} \tag{6.1.16}$$

in mosquito-infested regions.

6.2 Flow of an Autonomous Equation

The examples introduced in the previous section fit into the framework of Cauchy problems for autonomous systems of equations in $\mathbb{R}^d$,

$$\boldsymbol{x}' = \boldsymbol{f}(\boldsymbol{x}), \tag{6.2.17}$$

$$\boldsymbol{x}(0) = \mathring{\boldsymbol{x}}. \tag{6.2.18}$$

To simplify the considerations here, we assume that $\boldsymbol{f}$ satisfies the assumptions of the Picard theorem, Theorem B.10, on a suitable open subset $\Omega \subseteq \mathbb{R}^d$ (typically, $\Omega = \mathbb{R}^d_+$) and the solutions exist in Ω for all $t \in I$, where $I = (-\infty, +\infty)$ (though often it is sufficient to have $I = [0, +\infty)$ as we are interested in the future). These facts, however, usually require proof.

Remark 6.2 As in the case of difference equations, described in Sect. 5.3.7, a higher-order differential equation

$$z^{(n)} = F(t, z, z', z'', \dots, z^{(n-1)}) = 0, \quad n \in \mathbb{N}, \tag{6.2.19}$$

where, for simplicity, we consider only equations solved with respect to the highest derivative, and F is a sufficiently regular nonlinear function of t and $z \in \mathbb{R}^k$ can always be written as a system of first-order equations (the converse, in general, is not true). Similarly to Sect. 5.3.7, we introduce new unknown variables $x_1(t) = z(t), x_2(t) = z'(t), x_n(t) = z^{(n-1)}(t)$ so that $x_1'(t) = z'(t) = x_2(t), x_2'(t) = z''(t) = x_3(t), \dots$, and (6.2.19) converts into

$$\begin{aligned} x_1' &= x_2, \\ x_2' &= x_3, \\ \vdots &\ \vdots\ \vdots \\ x_n' &= F(t, x_1, \dots, x_n). \end{aligned} \tag{6.2.20}$$

Clearly, by solving this system, we simultaneously obtain the solution of (6.2.19) by taking $z(t) = x_1(t)$. It should be noted, however, that if we have tools for solving directly (6.2.19), then they should be used first as by solving (6.2.20), we obtain simultaneously all derivatives of the solution to (6.2.19) which are not necessarily required.

Exercise 6.2.1 Convert the simultaneous system of second-order equations

$$\begin{aligned} y'' + 3z' + 2y &= 0, \\ z'' + 3y' + 2z &= 0, \end{aligned}$$

into a 4×4 system of first-order equations.

Exercise 6.2.2 Let $y(t)$ be a solution of the equation $y'' + y' + y = 0$. Show that $\boldsymbol{x}(t) = (y(t), y'(t))$ is a solution to the system

$$\boldsymbol{x}'(t) = \begin{pmatrix} 0 & 1 \\ -1 & -1 \end{pmatrix} \boldsymbol{x}.$$

Conversely, let $\boldsymbol{x}(t) = (x_1(t), x_2(t))$ be a solution to the above system. Show, that $y(t) = x_1(t)$ is a solution to $y'' + y' + y = 0$.

The concept of a flow was introduced in Sect. 4.1 in the scalar case. Here, it will play a much more critical role, and thus, we need to examine its properties in more detail. Let us recall that the flow generated by (6.2.17) is the map

$$\mathbb{R} \times \Omega \ni (t, \mathring{\boldsymbol{x}}) \to \phi(t, \mathring{\boldsymbol{x}}) \in \Omega$$

where $\boldsymbol{x}(t) = \phi(t, \mathring{\boldsymbol{x}})$ is the solution to (6.2.17) satisfying $\boldsymbol{x}(0) = \mathring{\boldsymbol{x}}$. We note the following important properties of the flow:

$$\phi(0, \boldsymbol{x}) = \boldsymbol{x}, \tag{6.2.21}$$

$$\phi(t_1, \phi(t_2, \boldsymbol{x})) = \phi(t_1 + t_2, \boldsymbol{x}) \tag{6.2.22}$$

for any $t_1, t_2 \in \mathbb{R}$ and $\boldsymbol{x} \in \Omega$.

Remark 6.3 If $I = [0, +\infty)$, the flow is sometimes referred to as semi-flow and (6.2.22) is to hold for $t_1, t_2 \geq 0$. We will not, however, emphasize this distinction as the meaning usually will be clear from the context.

Property (6.2.22) follows from the simple lemma, whose proof is a repetition of the argument used in Sect. 3.4.2 and which we recall here for further reference.

Lemma 6.4 *If $\boldsymbol{x}(t)$ is a solution to* (6.2.17), *then for any c the function $\hat{\boldsymbol{x}}(t) = \boldsymbol{x}(t + c)$ also satisfies this equation.*

Proof Define $\tau = t + c$ and use the chain rule for $\hat{x}$. We get

$$\hat{\boldsymbol{x}}'_t(t) = \boldsymbol{x}'_t(t + c) = \boldsymbol{x}'_\tau(\tau)\tau'_t = \boldsymbol{x}'_\tau(\tau) = \boldsymbol{f}(\boldsymbol{x}(\tau)) = \boldsymbol{f}(\boldsymbol{x}(t + c)) = \boldsymbol{f}(\hat{\boldsymbol{x}}(t)).$$

□

Then (6.2.22) follows if we recognize that in terms of the flow, we can rephrase the lemma by saying that $\boldsymbol{u}(t) := \phi(t + t_2, \boldsymbol{x})$ is the solution of (6.2.17), with $\boldsymbol{u}(0) = \phi(t_2, \boldsymbol{x})$, and thus, by the uniqueness part of Theorem B.10 and the definition of the flow, we must have $\boldsymbol{u}(t) = \phi(t, \phi(t_2, \boldsymbol{x}))$.

Remark 6.5 Let us recall that if we need solutions satisfying the initial condition at $t = t_0 \neq 0$, then we use the notation $\phi(t, t_0, \boldsymbol{x})(-\ \phi(t - t_0, \boldsymbol{x}))$ so that the first property above is $\phi(t_0, t_0, \boldsymbol{x}) = \boldsymbol{x}$.

Remark 6.6 Theorem B.16, expressed in terms of the flow, states that the flow is a continuous function with respect to the initial condition (even Lipschitz continuous—set $g = 0$ in (B.4.48) and apply the Gronwall inequality).

Example 6.7 Consider the scalar initial value problem

$$x' = ax, \qquad x(0) = \mathring{x}.$$

In this case, the solution can be found explicitly, and the flow is given by

$$\phi(t, 0) = \mathring{x}e^{at},$$

and clearly

$$\phi(t, \mathring{y}) - \phi(t, \mathring{x}) = (\mathring{y} - \mathring{x})e^{at}.$$

Hence, indeed, it is a Lipschitz continuous function with respect to the initial condition (uniformly in t on bounded time intervals).

6.2.1 *Further Stability Concepts*

Similarly to the scalar equations, Chap. 4, there are no general methods of solving (6.2.17), (6.2.18) but, in most cases, having exact solutions is irrelevant. Again, assuming that we consider a system of two equations describing the sizes of two populations, the pertinent questions are:

1. Do there exist values ξ_1 and ξ_2 at which the populations coexist in a steady state? That is to say, are there positive numbers ξ_1 and ξ_2 such that $\boldsymbol{x}(t) = (x_1(t), x_2(t)) \equiv (\xi_1, \xi_2)$ is a solution to (6.2.17)? Such values, if they exist, are *equilibrium points* of (6.2.17).
2. Suppose the two species can coexist in an equilibrium, and suddenly, a few members of one or both species are introduced to the environment. Will the new solution $\boldsymbol{y} = (\boldsymbol{y}_1(t), \boldsymbol{y}_2(t))$ remain close to the equilibrium for large times? Or maybe these extra few members will give one of the populations an advantage so that it will annihilate the other population?
3. Suppose that $\boldsymbol{x}_1$ and $\boldsymbol{x}_2$ have arbitrary values at $t = 0$. What happens for large times? Will one population emerge victorious, or will the struggle for existence continue forever, or will it end in a draw?

So, mathematically speaking, we are interested in the existence of equilibria of (6.2.17) and their stability or, more generally, in the long-time behaviour of the solutions if there are no equilibria, that is, whether the solutions exhibit some regular behaviour, like, e.g., periodicity.

The stability of solutions and equilibria is defined as in the scalar case, see Definition 4.1, which we recall here in the multidimensional setting of autonomous systems (6.2.17).

Definition 6.8 The solution $\boldsymbol{x}(t)$ of (6.2.17) is (Lyapunov) stable if every other solution $\boldsymbol{y}(t)$ that starts sufficiently close to $\boldsymbol{x}(t)$ will remain close to it at all times. Precisely, $\boldsymbol{x}(t)$ is stable if for any ϵ there is δ such that for any solution $\boldsymbol{y}$ of (6.2.17) from

$$\|\boldsymbol{x}(0) - \boldsymbol{y}(0)\| \leq \delta,$$

it follows that

$$\|\boldsymbol{x}(t) - \boldsymbol{y}(t)\| \leq \epsilon$$

for any $t > 0$. We say that $\boldsymbol{x}(t)$ is attracting if there is δ such that if $\|\boldsymbol{x}(0) - \boldsymbol{y}(0)\| \leq \delta$, then

$$\lim_{t\to\infty} \|\boldsymbol{y}(t) - \boldsymbol{x}(t)\| = 0.$$

Moreover, we say that $\boldsymbol{x}(t)$ is asymptotically stable, if it is Lyapunov stable and attracting.

The main interest in applications is to determine the stability of stationary solutions, that is, when $\boldsymbol{x}(t) = \boldsymbol{x}^*$ for all $t \geq 0$, where

$$\boldsymbol{f}(\boldsymbol{x}^*) = 0$$

or, in terms of the flow, when

$$\phi(t, \boldsymbol{x}^*) = \boldsymbol{x}^*.$$

Remark 6.9 The concept that is particularly useful in two-dimensional systems is that of nullclines. Nullclines on the plane are the curves along which the derivatives of the coordinates of the solution, and hence the coordinates of the vector field $\boldsymbol{f}$, vanish. In other words, nullclines consist of turning points of the coordinates of the solution. The equilibria are given as the intersections of nullclines.

Example 6.10 Find all equilibria of the system of differential equations

$$\begin{aligned} x_1' &= 1 - x_2, \\ x_2' &= x_1^3 + x_2. \end{aligned} \tag{6.2.23}$$

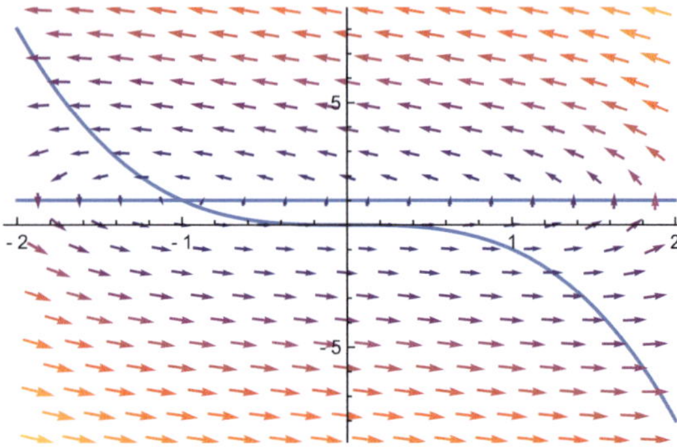

Fig. 6.3 Vector field, nullclines and the equilibrium of (6.2.23)

We have to solve the system of algebraic equations

$$0 = 1 - x_2,$$
$$0 = x_1^3 + x_2.$$

Hence, the nullclines are given by

$$x_2 = 1,$$
$$x_2 = -x_1^3,$$

see Fig. 6.3. For the equilibrium, we have $x_2 = 1$ and hence $x_1 = -1$ and the only equilibrium solution is

$$\boldsymbol{x}^* = \begin{pmatrix} -1 \\ 1 \end{pmatrix}.$$

Exercise 6.2.3 Find all equilibria of the given systems of differential equations.

(i)

$$x_1' = x_1 - x_1^2 - 2x_1x_2,$$
$$x_2' = 2x_2 - 2x_2^2 - 3x_1x_2.$$

(ii)

$$x_1' = x_1x_2^2 - x_1,$$
$$x_2' = x_1 \sin \pi x_2.$$

Example 6.11 (Linear Systems) Based on the results in Chap. 5, the stability of solutions can be fully resolved for linear systems. Consider

$$\boldsymbol{x}' = \mathcal{A}\boldsymbol{x}, \tag{6.2.24}$$

with an invertible matrix $\mathcal{A}$. Then, $\boldsymbol{x}^* = 0$ is the only equilibrium. Firstly, we observe that any solution $\boldsymbol{x}(t)$ of (6.2.24) is stable if and only if $\boldsymbol{x}^* = 0$ is stable. To show this, consider any two solutions, $\boldsymbol{x}(t)$ and $\boldsymbol{y}(t)$, of (6.2.24) and define $z(t) = \boldsymbol{x}(t) - \boldsymbol{y}(t)$ also satisfies (6.2.24). Therefore, if $\boldsymbol{x}^* = 0$ is stable, then $z(t)$ remains close to zero for all $t > 0$ if $z(0)$ is sufficiently small. This, however, implies that $\boldsymbol{x}(t)$ will remain close to $\boldsymbol{y}$ if $\boldsymbol{x}(0)$ is sufficiently close to $\boldsymbol{y}(0)$. A similar argument applies to the asymptotic stability. Conversely, let $\boldsymbol{x}^* = 0$ be unstable. Then, for arbitrarily small $\boldsymbol{h}(0)$ the corresponding solution $\boldsymbol{h}(t)$ moves a fixed distance from the origin for large t. Then, for any $\boldsymbol{y}(t)$, the function $z(t) = \boldsymbol{y}(t) + \boldsymbol{h}(t)$ is again a solution to (6.2.24) which sways away from $\boldsymbol{y}(t)$ for large t. Thus, any solution is unstable.

Hence, using the spectral decomposition (5.3.86), any solution to (6.2.24) is stable if and only if $s(\mathcal{A}) = \max\{\Re\lambda;\ \lambda \in \sigma(\mathcal{A})\} \leq 0$ and any $\lambda \in \sigma(\mathcal{A})$ with $\Re\lambda = 0$ is simple or semi-simple. Similarly, any solution to (6.2.24) is asymptotically stable if and only if $s(\mathcal{A}) = \max\{\Re\lambda;\ \lambda \in \sigma(\mathcal{A})\} < 0$.

We illustrate the possible types of the long-term behaviour of linear systems on an example of two-dimensional systems (6.2.24), where $\boldsymbol{x}(t) = (x_1(t), x_2(t))$ and

$$\mathcal{A} = \begin{pmatrix} a & b \\ c & d \end{pmatrix}.$$

We shall assume that $\mathcal{A}$ is invertible, that is, $ad - bc \neq 0$. In such a case $\boldsymbol{x}^* = (0, 0)$ is the only equilibrium point of (6.2.24). Otherwise, we can have lines or the plane consisting of equilibria.

The phase portrait is fully determined by the eigenvalues of the matrix $\mathcal{A}$. Let us briefly describe all possible cases, as determined by the theory of Sect. 5.3. The general solution can be obtained as a linear combination of two linearly independent solutions. To find them, we have to find first the eigenvalues of $\mathcal{A}$, that is, solutions to

$$(\lambda - \lambda_1)(\lambda - \lambda_2) = (a - \lambda)(d - \lambda) - bc = \lambda^2 - \lambda(d + a) + ad - bc = \lambda^2 - tr\,\mathcal{A} + det\,\mathcal{A},$$

where $tr\,\mathcal{A} = d + a$ is the trace of $\mathcal{A}$.

Remark 6.12 Using Viète's formulae, we see that $s(\mathcal{A}) < 0$, and hence so $(0, 0)$ is asymptotically stable if and only if $tr\,\mathcal{A} < 0$ and $det\,\mathcal{A} > 0$.

Note that by the assumption on invertibility, $\lambda = 0$ is not an eigenvalue of $\mathcal{A}$. We have the following possibilities:

(a) $\lambda_1 \neq \lambda_2$. In this case, each eigenvalue must be simple, and therefore, we have two linearly independent eigenvectors $\boldsymbol{v}^1$, $\boldsymbol{v}^2$. The expansion $e^{t\mathcal{A}}\boldsymbol{v}^i$ for $i = 1, 2$, terminates after the first term. We distinguish two cases.

- If λ_1, λ_2 are real numbers, then the general solution is given simply by

$$\boldsymbol{x}(t) = c_1 e^{\lambda_1 t}\boldsymbol{v}^1 + c_2 e^{\lambda_2 t}\boldsymbol{v}^2. \tag{6.2.25}$$

- If λ_1, λ_2 are complex numbers, then the above formula still gives a general solution, but the functions above are complex, and we would rather prefer the solution to be real. To achieve this, we note that λ_1, λ_2 must be necessarily complex conjugate $\lambda_1 = \xi + i\omega$, $\lambda_2 = \xi - i\omega$, where ξ and ω are real. It can also be proved that the associated eigenvectors $\boldsymbol{v}^1$ and $\boldsymbol{v}^2$ are also complex conjugate. Let $\boldsymbol{v}^1 = \boldsymbol{u} + i\boldsymbol{v}$; then the real-valued general solution is given by

$$\boldsymbol{x}(t) = c_1 e^{\xi t}(\boldsymbol{u}\cos\omega t - \boldsymbol{v}\sin\omega t) + c_2 e^{\xi t}(\boldsymbol{u}\sin\omega t + \boldsymbol{v}\cos\omega t). \tag{6.2.26}$$

This solution can be written in a more compact form

$$\boldsymbol{x}(t) = e^{\xi t}\left(A_1\cos(\omega t - \phi_1), A_2\cos(\omega t - \phi_2)\right), \tag{6.2.27}$$

for some choice of constants $A_1, A_2 > 0$ and ϕ_1, ϕ_2.

(b) $\lambda_1 = \lambda_2 = \lambda$. There are two cases to distinguish.

- There are two linearly independent eigenvectors $\boldsymbol{v}^1$ and $\boldsymbol{v}^2$ corresponding to λ. In this case, the general solution is given by

$$\boldsymbol{x}(t) = e^{\lambda t}(c_1\boldsymbol{v}^1 + c_2\boldsymbol{v}^2). \tag{6.2.28}$$

- If there is only one eigenvector, then following the discussion in Sect. 5.3.3, we must find a vector $\boldsymbol{v}^2$ satisfying $(\lambda I - \mathcal{A})\boldsymbol{v}^2 \neq 0$ and $(\lambda I - \mathcal{A})^2\boldsymbol{v}^2 = 0$. However, since we are in the two-dimensional space, the latter is satisfied by any vector $\boldsymbol{v}^2$, see Proposition C.4, and, since the eigenspace is one dimensional, from

$$(\lambda I - \mathcal{A})^2\boldsymbol{v}^2 = (\lambda I - \mathcal{A})(\lambda I - \mathcal{A})\boldsymbol{v}^2 = 0,$$

it follows that $(\lambda I - \mathcal{A})\boldsymbol{v}^2 = k\boldsymbol{v}^1$. Thus, the formula for $e^{\mathcal{A}t}e^{t\mathcal{A}}\boldsymbol{v}^2$ simplifies as

$$e^{t\mathcal{A}}\boldsymbol{v}^2 = e^{\lambda t}\left(\boldsymbol{v}^2 + t(\lambda I - \mathcal{A})\boldsymbol{v}^2\right) = e^{\lambda t}\left(\boldsymbol{v}^2 + kt\boldsymbol{v}^1\right).$$

Thus, the general solution in this case can be written as

$$\boldsymbol{x}(t) = e^{\lambda t}\left((c_1 + c_2 kt)\boldsymbol{v^1} + c_2\boldsymbol{v^2}\right). \tag{6.2.29}$$

Remark 6.13 Before we embark on describing phase portraits, let us observe that if we change the direction of time in (6.2.24): $\tau = -t$ and $z(\tau) = \boldsymbol{x}(-\tau) = \boldsymbol{x}(t)$, then we obtain

$$z'_\tau = -\mathcal{A}z$$

and the eigenvalues of $-\mathcal{A}$ are precisely the negatives of the eigenvalues of $\mathcal{A}$. Thus, the orbits of solutions corresponding to systems governed by $\mathcal{A}$ and $-\mathcal{A}$ or, equivalently, with eigenvalues that differ only by sign, are the same, the only difference being the direction in which they are traversed.

We are now in a position to describe all possible phase portraits of (6.2.24). Again, we have to go through several cases.

(i) $\lambda_2 < \lambda_1 < 0$. Let $\boldsymbol{v^1}$ and $\boldsymbol{v^2}$ be the eigenvectors of $\mathcal{A}$ with eigenvalues λ_1 and λ_2, respectively. In the $x_1 - x_2$ plane, we draw two lines l_1, l_2 in the direction of $\boldsymbol{v^1}$ and $\boldsymbol{v^2}$, respectively, emanating from the origin, as shown in Fig. 6.4. Observe first that $\boldsymbol{x}(t) = ce^{\lambda_i t}\boldsymbol{v^i}$, $i = 1, 2$, are the solutions to (6.2.24) for any choice of a non-zero constant c and, as they are parallel to $\boldsymbol{v^i}$, the orbits are the half-lines of l_1, l_2 (depending on the sign of the constant c) and all these orbits

Fig. 6.4 Stable node

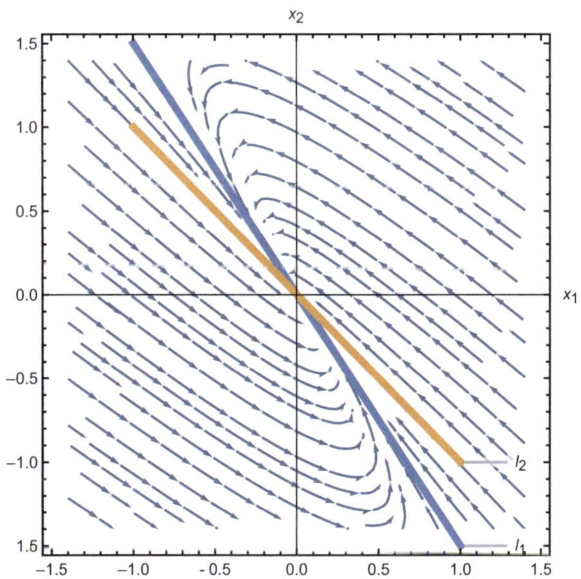

are traced towards the origin as $t \to \infty$. Since every solution $\boldsymbol{x}(t)$ of (6.2.24) can be written as

$$\boldsymbol{x}(t) = c_1 e^{\lambda_1 t} \boldsymbol{v}^{\mathbf{1}} + c_2 e^{\lambda_2 t} \boldsymbol{v}^{\mathbf{2}} \tag{6.2.30}$$

for some choice of constants c_1 and c_2. Since $\lambda_1, \lambda_2 < 0$, every solution tends to $(0, 0)$ as $t \to \infty$, and so every orbit approaches the origin for $t \to \infty$. We can prove an even stronger fact—as $\lambda_2 < \lambda_1$, the second term in (6.2.30) becomes negligible for large t and therefore the tangent of the orbit of $\boldsymbol{x}(t)$ approaches the direction of the positive half line of l_1 if $c_1 > 0$ and the negative one if $c_1 < 0$. Thus, every orbit except that with $c_1 = 0$ approaches the origin along the same line. Such an equilibrium point is called a *stable node*. If we have $0 < \lambda_1 < \lambda_2$, then by Remark 6.13, the orbits of (6.2.24) will have the same shape as in case (i) but the arrows will be reversed so that the origin will repel all the orbits and the orbits will be unbounded as $t \to \infty$. Such an equilibrium point is called an *unstable node*.

(ii) $\lambda_1 = \lambda_2 = \lambda < 0$. In this case, the phase portrait of (6.2.24) depends on whether $\mathcal{A}$ has one or two linearly independent eigenvectors. In the latter case, that is, when λ is semi-simple, the general solution is given (see (b) above) by

$$\boldsymbol{x}(t) = e^{\lambda t}(c_1 \boldsymbol{v}^{\mathbf{1}} + c_2 \boldsymbol{v}^{\mathbf{2}}),$$

so that orbits are half-lines parallel to $c_1 \boldsymbol{v}^{\mathbf{1}} + c_2 \boldsymbol{v}^{\mathbf{2}}$. These half-lines cover every direction of the $x_1 - x_2$ plane and, since $\lambda < 0$, each solution will converge to $(0, 0)$ along the respective line. Thus, the phase portrait looks as in Fig. 6.5. If

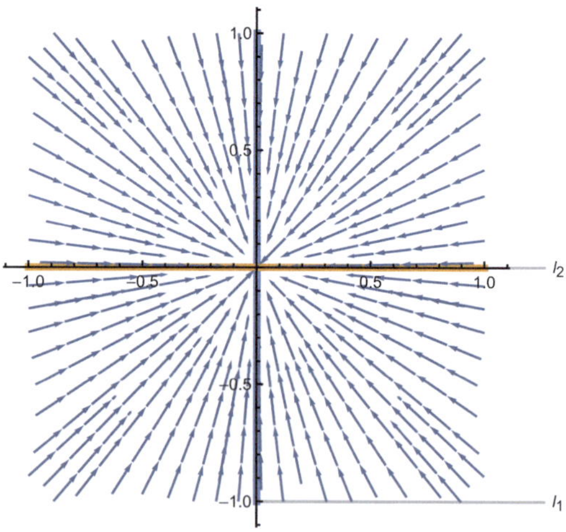

Fig. 6.5 Stable degenerate node when λ is semisimple

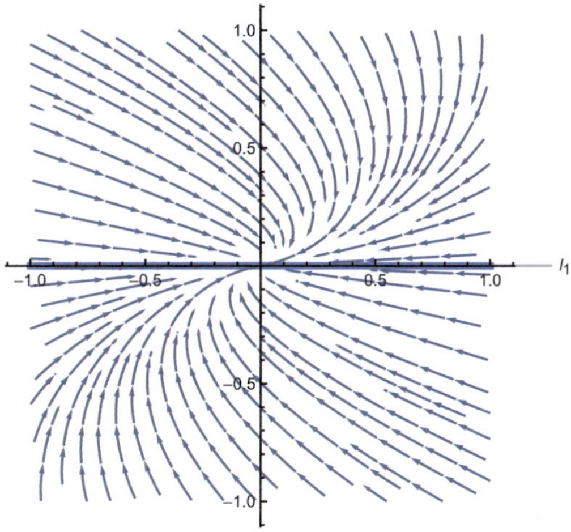

Fig. 6.6 Stable degenerate node when λ has algebraic multiplicity 2

there is only one independent eigenvector corresponding to λ, then, by (6.2.29),

$$\boldsymbol{x}(t) = e^{\lambda t}\left((c_1 + c_2kt)\boldsymbol{v}^{\mathbf{1}} + c_2\boldsymbol{v}^{\mathbf{2}}\right)$$

for some choice of constants c_1, c_2 and k. Every solution approaches $(0, 0)$ as $t \to \infty$. Putting $c_2 = 0$, we obtain two half-line orbits $c_1e^{\lambda t}\boldsymbol{v}^{\mathbf{1}}$ but, contrary to the case (i), there are no other half-line orbits. In addition, the term $c_1\boldsymbol{v}^{\mathbf{1}} + c_2\boldsymbol{v}^{\mathbf{2}}$ becomes small in comparison with $c_2kt\boldsymbol{v}^{\mathbf{1}}$ as $t \to \infty$ so that the orbits approach the origin in the direction of $\pm\boldsymbol{v}^{\mathbf{1}}$. The phase portrait is presented in Fig. 6.6. The equilibrium in both cases is called the *stable degenerate node*. If $\lambda_1 = \lambda_2 > 0$, then again by Remark 6.13, the picture in this case will be the same as in Figs. 6.5 and 6.6, but with the direction of the arrows reversed. Such an equilibrium is called an *unstable degenerate node*.

(iii) $\lambda_1 < 0 < \lambda_2$. As in case (i), in the $x_1 - x_2$ plane, we draw lines l_1 and l_2 that emanate from the origin and are parallel to the eigenvectors $\boldsymbol{v}^{\mathbf{1}}$ and $\boldsymbol{v}^{\mathbf{2}}$, respectively, as shown in Fig. 6.7. Any solution is given by

$$\boldsymbol{x}(t) = c_1e^{\lambda_1 t}\boldsymbol{v}^{\mathbf{1}} + c_2e^{\lambda_2 t}\boldsymbol{v}^{\mathbf{2}}$$

for some choice of c_1 and c_2. Again, the half-lines of l_1 are the orbits of the solutions $c_1e^{\lambda_1 t}\boldsymbol{v}^{\mathbf{1}}$ with $c_1 > 0$ and $c_1 < 0$, and the half-lines of l_2 of the solutions for $c_2e^{\lambda_2 t}\boldsymbol{v}^{\mathbf{2}}$ with $c_2 > 0$ and $c_2 < 0$, respectively. However, the direction of arrows is different on each pair of half-lines, while the solution $c_1e^{\lambda_1 t}\boldsymbol{v}^{\mathbf{1}}$ converges towards $(0, 0)$ along l_1 as $t \to \infty$, the solution $c_2e^{\lambda_2 t}\boldsymbol{v}^{\mathbf{2}}$ becomes unbounded moving along l_2, as $t \to \infty$. Next, we observe that if $c_1 \neq 0$, then for large t the second term $c_2e^{\lambda_2 t}\boldsymbol{v}^{\mathbf{2}}$ becomes negligible and so the solution becomes unbounded as $t \to \infty$, with the asymptotes given by the half-lines of l_2, respectively. Similarly, for $t \to -\infty$ the term $c_1e^{\lambda_1 t}\boldsymbol{v}^{\mathbf{1}}$

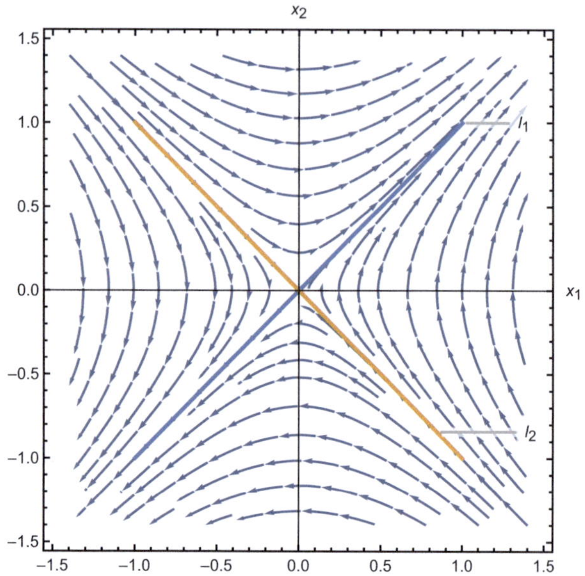

Fig. 6.7 A saddle point

becomes negligible, and the solution again escapes to infinity, but this time with asymptotes being the half-lines of l_1. Thus, the phase portrait, given in Fig. 6.7, resembles a saddle near $x_1 = x_2 = 0$ and, not surprisingly, such an equilibrium point is called a *saddle*. The case $\lambda_2 < 0 < \lambda_1$ is, of course, symmetric.

(iv) $\lambda_1 = \xi + i\omega, \lambda_2 = \xi - i\omega$. In (6.2.27), we derived the solution in the form

$$\boldsymbol{x}(t) = e^{\xi t}\left(A_1\cos(\omega t - \phi_1), A_2\cos(\omega t - \phi_2)\right).$$

We have to distinguish three cases:

(α) If $\xi = 0$, then

$$x_1(t) = A_1\cos(\omega t - \phi_1), \qquad x_2(t) = A_2\cos(\omega t - \phi_2),$$

both are periodic functions with period $2\pi/\omega$ and x_1 varies between $-A_1$ and A_1, while x_2 varies between $-A_2$ and A_2. Consequently, the orbit of any solution $\boldsymbol{x}(t)$ is a closed curve containing the origin inside and the phase the portrait has the form presented in Fig. 6.8. For this reason, we say that the equilibrium of (6.2.24) is a *centre* when the eigenvalues of $\mathcal{A}$ are purely imaginary. The direction of the arrows must be determined from the equation. The simplest way of doing this is to check the sign of x_2' when $x_2 = 0$. If at $x_2 = 0$ and $x_1 > 0$, we have $x_2' > 0$, then all orbits are traversed in the anticlockwise direction.

Fig. 6.8 Centre

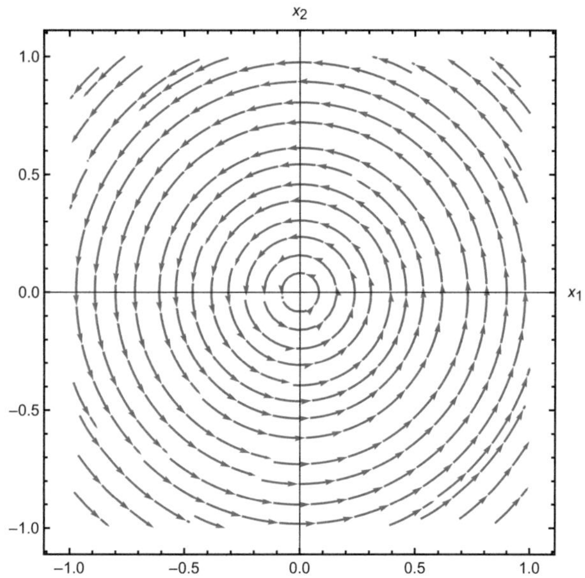

Fig. 6.9 Stable focus

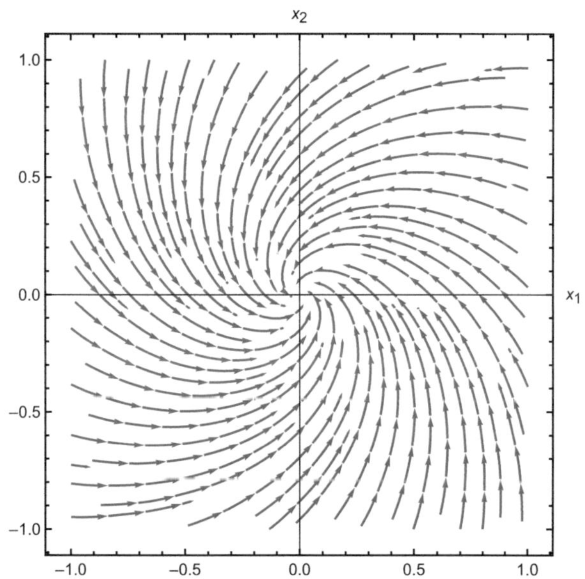

(β) If $\xi < 0$, then the factor $e^{\xi t}$ forces the solution to come closer to zero at every turn so that the solution spirals into the origin, giving the picture presented in Fig. 6.9. The orientation of the spiral must be again determined directly from the equation. Such an equilibrium point is called a *stable focus*.

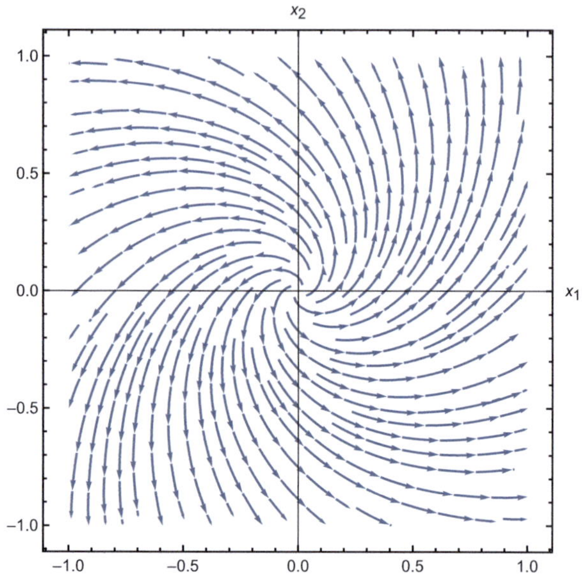

Fig. 6.10 Unstable focus

(γ) If $\xi > 0$, then the factor $e^{\xi t}$ forces the solution to spiral outwards, creating the picture shown in Fig. 6.10. Such an equilibrium point is called an *unstable focus*.

Exercise 6.2.4 Show that the real-valued solution of the 2×2 system, given by

$$\boldsymbol{y}(t) = c_1 e^{\xi t}(\boldsymbol{u}\cos\omega t - \boldsymbol{v}\sin\omega t) + c_2 e^{\xi t}(\boldsymbol{u}\sin\omega t + \boldsymbol{v}\cos\omega t), \qquad (6.2.31)$$

where $\xi \pm i\omega$ are the eigenvalues and $\boldsymbol{u} \pm i\boldsymbol{v}$ ($\boldsymbol{u}, \boldsymbol{v}$—real) are the corresponding eigenvectors, can be written as

$$\boldsymbol{y}(t) = e^{\xi t}\left(A_1\cos(\omega t - \phi_1), A_2\cos(\omega t - \phi_2)\right),$$

for some choice of constants $A_1, A_2 > 0$ and ϕ_1, ϕ_2.

An asymptotically stable stationary point is called a *sink* or an *attracting equilibrium*. On the contrary, if there is a neighbourhood of a stationary point such that any trajectory starting from this neighbourhood diverge from it, then such a stationary point is called a *repeller* or a *source*.

Definition 6.14 Suppose that $\boldsymbol{x}^*$ is an asymptotically stable stationary point of Eq. (6.2.17). The set

$$D_{\boldsymbol{x}^*} := \{\boldsymbol{x} : \lim_{t\to\infty}\phi(t, \boldsymbol{x}) = \boldsymbol{x}^*\} \qquad (6.2.32)$$

is called the domain of asymptotic stability (or the basin of attraction) of $\boldsymbol{x}^*$. We say that $\boldsymbol{x}^*$ is globally asymptotically stable in $D_{\boldsymbol{x}^*}$.

6.2.2 *The Phase Space and Orbits*

In this section, we shall give rudiments of the geometric theory of differential equations. This theory aims to obtain as complete a description as possible of all solutions to the autonomous system of differential equations (6.2.17) without solving it explicitly but by analysing geometric properties of its orbits. To explain the latter, we note that every solution $\boldsymbol{x}(t) = (x_1(t), \dots, x_d(t))$, $t \in I$, defines a curve in the $d+1$-dimensional space $(x_1, \dots, x_d, t)$.

Example 6.15 The solution $\boldsymbol{x}(t) = (x_1(t), x_2(t))$, where $x_1(t) = \cos 5t$ and $x_2(t) = \sin 5t$ of the system

$$\begin{aligned} x_1' &= -5x_2, \\ x_2' &= 5x_1, \end{aligned} \tag{6.2.33}$$

with $x_1(0) = 1, x_2(0) = 0$, describes a helix in the (x_1, x_2, t) space, see Fig. 6.11.

The foundation of the geometric theory of differential equations is the observation that every solution $\boldsymbol{x}(t)$, $t_0 \leq t \leq t_1$, of (6.2.17), (6.2.18) also describes a curve in the d-dimensional space, that is, as t runs from t_0 to t_1, the points $(x_1(t), \dots, x_d(t))$ trace out a curve in this space. We call this curve the integral curve of (6.2.17). Typically, we are interested more in the integral curve as the a geometric object consisting of points visited by the solution. In such a case, we talk about the orbit, or the trajectory, of the solution $\boldsymbol{x}(t)$. The formal definition is given below.

Definition 6.16 The set

$$\Gamma_{\boldsymbol{x}_0} - \{\boldsymbol{x} \in \mathbb{R}^d : \boldsymbol{x} = \phi(t, \boldsymbol{x}_0), t \subset \mathbb{R}\}$$

Fig. 6.11 The graph of $\boldsymbol{x}(t)$ in $\mathbb{R}^3$

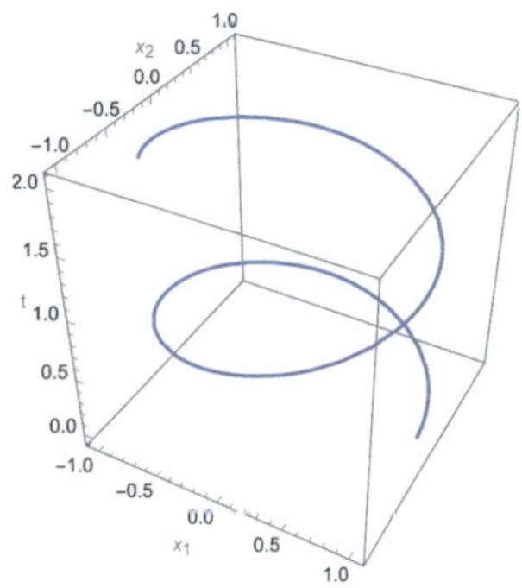

is called the trajectory, or orbit, of the flow through $\boldsymbol{x}_0$. We shall drop it from the notation if $\boldsymbol{x}_0$ plays no role in the considerations. We will often use the positive (or negative) half-orbits defined as

$$\Gamma^{\pm}_{\boldsymbol{x}_0} = \{\boldsymbol{x} \in \mathbb{R}^d : \boldsymbol{x} = \phi(t, \boldsymbol{x}_0), t \gtreqless 0\}.$$

The d-dimensional $\boldsymbol{x}$-space, in which the orbits are situated, is called the phase space of the solutions of (6.2.17).

Remark 6.17 As with semi-flows, we shall often make no distinction between trajectories and half-trajectories if the context is clear.

Remark 6.18 Note that the orbit of an equilibrium solution reduces to a point.

Example 6.19 The solution of the previous example, $x_1(t) = \cos 5t, x_2(t) = \sin 5t$, traces out the unit circle $x_1^2(t) + x_2^2(t) = 1$ when t runs from 0 to 2π, hence the unit circle is the orbit of this solution, see Fig. 6.12. If t runs from 0 to ∞, then the pair $(\cos 5t, \sin 5t)$ trace out this circle infinitely often.

Example 6.20 The function $\boldsymbol{x}(t) = (x_1(t), x_2(t))$, where $x_1(t) = e^{-t}\cos 5t$ and $x_2(t) = e^{-t}\sin 5t$, $-\infty < t < \infty$, solves the system

$$\begin{aligned} x_1' &= -x_1 - 5x_2, \\ x_2' &= 5x_1 - x_2. \end{aligned} \tag{6.2.34}$$

Since the distance from a point on a trajectory to the origin is given by $r^2(t) = x_1^2(t) + x_2^2(t) = e^{-2t}$, we see that the orbit of this solution is a spiral traced towards the origin as t runs towards ∞; see Figs. 6.13 and 6.14.

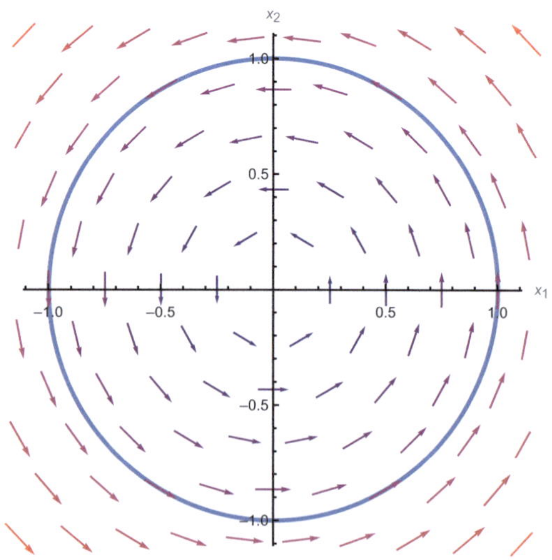

Fig. 6.12 The trajectory of the solution $\boldsymbol{x}(t)$ (solid line) and the vector field of (6.2.33)

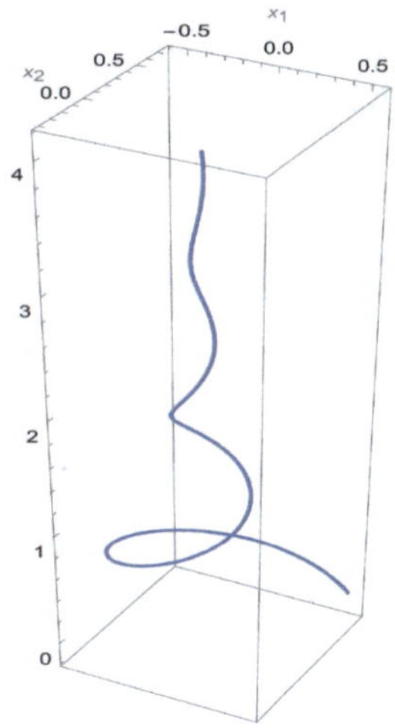

Fig. 6.13 The graph of the solution (6.2.34)

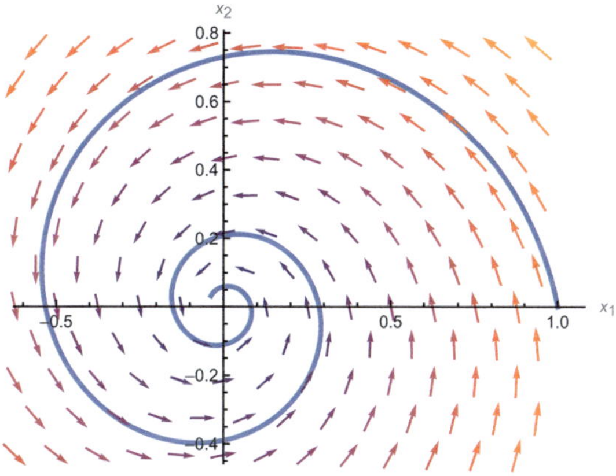

Fig. 6.14 The trajectory of the solution $\boldsymbol{x}(t)$ (solid line) and the vector field of (6.2.34)

Exercise 6.2.5 Verify that $x_1(t), x_2(t)$ given below are solutions of the respective systems of equations and find their orbits.

(i) $x_1(t) = 1 + t, x_2(t) = \cos t^2$,

$$x_1' = 1,$$
$$x_2' = 2(1 - x_1)\sin(1 - x_1)^2.$$

(ii) $x_1(t) = \tan t, x_2(t) = \tan(\tan t)$,

$$x_1' = 1 + x_1^2,$$
$$x_2' = (1 + x_1^2)\sec^2 x_1.$$

Another important concept is that of an invariant set of the flow.

Definition 6.21 A set $M \subset \mathbb{R}^d$ is called invariant with respect to the flow $\phi(t, \boldsymbol{x})$ generated by the system (6.2.17) if and only if $\phi(t, \boldsymbol{x}) \in M$ for any $\boldsymbol{x} \in M$ and $t \in \mathbb{R}$. In other words, M is invariant if $\Gamma_{\boldsymbol{x}} \subset M$ for any $\boldsymbol{x} \in M$.

A set M is forward (resp. backward) invariant if, for any $\boldsymbol{x} \in M$ we have $\Gamma_{\boldsymbol{x}}^+ \subset M$ (resp. $\Gamma_{\boldsymbol{x}}^- \subset M$).

Remark 6.22 As before, we usually do not distinguish between invariant and positively invariant sets if it does not lead to a misunderstanding.

Example 6.23 Consider the system

$$\begin{aligned} x_1' &= x_1 - x_2 - x_1(x_1^2 + x_2^2) + \frac{x_1 x_2}{\sqrt{x_1^2 + x_2^2}}, \\ x_2' &= x_1 + x_2 - x_2(x_1^2 + x_2^2) - \frac{x_1^2}{\sqrt{x_1^2 + x_2^2}}. \end{aligned} \tag{6.2.35}$$

Multiplying the first equation by x_1 and the second by x_2 and adding, we obtain

$$rr' = \frac{1}{2}(x_1^2 + x_2^2)' = x_1^2 + x_2^2 - (x_1^2 + x_2^2)^2 = r^2(1 - r^2).$$

Using $x_1 = r \cos\theta$, $x_2 = r \sin\theta$, we get $\tan\theta = \frac{x_2}{x_1}$ (at least for $x_1 \neq 0$) and hence

$$\frac{1}{\cos^2\theta}\theta' = \frac{x_2' x_1 - x_1' x_2}{x_1^2} = \frac{1 - \cos\theta}{\cos^2\theta} = \frac{2\sin^2\frac{\theta}{2}}{\cos^2\theta}.$$

So, disregarding $r = 0$, in the polar form we obtain

$$\begin{aligned} r' &= r(1 - r^2), \\ \theta' &= 2\sin^2\frac{\theta}{2}. \end{aligned} \tag{6.2.36}$$

We observe that $r' = 0$ when $r = 0$ or $r = 1$, and $\theta' = 0$, when $\theta = 0$. Also, $\theta' > 0$ for all $\theta \neq 0$. Thus, the half-line $\theta = 0$ is an invariant line, and all trajectories move around to approach this half-line from below. Conversely, in the r direction, the trajectories tend to $r = 1$ (unless initially $r = 0$). Hence, the system has two stationary points: the origin $r = 0$ and the point $(1, 0)$. It also has two invariant curves: half-line $\theta = 0$ and the circle $r = 1$. All other trajectories tend to $(1, 0)$. Consider any small neighbourhood of $(1, 0)$. Points with $\theta \leq 0$ will immediately tend to $(1, 0)$ without leaving this neighbourhood. On the other hand, points with $\theta > 0$ will make a circuit along the invariant curve $r = 1$ before tending to $(1, 0)$ from below. Hence, $(1, 0)$ is attracting but not Lyapunov stable; see Fig. 6.15.

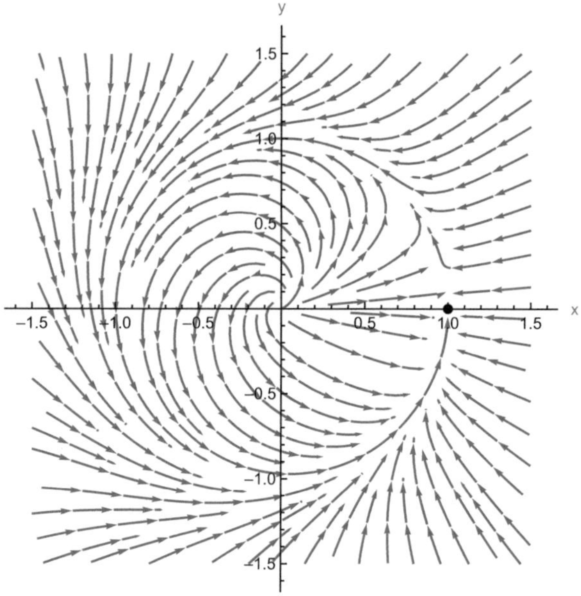

Fig. 6.15 Streamlines of (6.2.35) and the equilibrium $(1, 0)$; one can see that trajectories starting in the upper half-plane make the full revolution along the unit circle before reaching $(1, 0)$

One of the advantages of considering the orbits of the solutions rather than the solutions themselves is that it is often possible to find the orbit explicitly without prior knowledge of the solution. We shall describe this idea for a two-dimensional system

$$\begin{aligned} x_1' &= f_1(x_1, x_2), \\ x_2' &= f_2(x_1, x_2). \end{aligned} \tag{6.2.37}$$

Let $(x_1(t), x_2(t))$ be a solution of (6.2.17) defined in a neighbourhood of a point $\bar{t}$. If, e.g., $x_1'(\bar{t}) \neq 0$, then we can solve $x_1 = x_1(t)$ getting $t = t(y_1)$ in some neighbourhood of $\bar{x}_1 = x_1(\bar{t})$. Thus, for t near $\bar{t}$, the orbit of the solution $(x_1(t), x_2(t))$ is given as the graph of $x_2 = x_2(t(x_1))$. Next, using the chain rule and the inverse function theorem

$$\frac{dx_2}{dx_1} = \frac{dx_2}{dt}\frac{dt}{dx_1} = \frac{x_2'}{x_1'} = \frac{f_2(x_1, x_2)}{f_1(x_1, x_2)}.$$

Thus, the orbits of the solution $x_1 = x_1(t), x_2(t) = x_2(t)$ of (6.2.37) are the solution curves of the first-order scalar equation

$$\frac{dx_2}{dx_1} = \frac{f_2(x_1, x_2)}{f_1(x_1, x_2)}, \tag{6.2.38}$$

and therefore, to find the orbit of a solution, there is no need to solve (6.2.37). We only have to solve a single first-order scalar equation (6.2.38).

Remark 6.24 Certainly, (6.2.38) can be solved only for $f_1(x_1, x_2) \neq 0$. However, if this condition is not satisfied at isolated points, where $f_2(x_1, x_2) \neq 0$, then it simply means that the trajectory is parallel to the x_2-axis and the solution to (6.2.38) can be extended across such points.

The same procedure reduces an d-dimensional system to an $d - 1$-dimensional one in a neighbourhood of a point where, say, $f_i(x_1, \ldots, x_d) \neq 0$:

$$\frac{dx_j}{dx_i} = \frac{f_j(\mathbf{x})}{f_i(\mathbf{x})}, \qquad i \neq j = 1, \ldots, d.$$

Example 6.25 The orbits of the system of differential equations

$$\begin{aligned} x_1' &= x_2^2, \\ x_2' &= x_1^2, \end{aligned} \tag{6.2.39}$$

are the solution curves of the scalar equation

$$\frac{dx_2}{dx_1} = \frac{x_1^2}{x_2^2}.$$

This is a separable equation, and it is easy to see that every solution is of the form $x_2 = (x_1^3 + c)^{1/3}$, c constant. Thus, the orbits are the curves $y_2 = (y_1^3 + c)^{1/3}$.

Remark 6.26 A solution curve of (6.2.38) is an orbit of (6.2.37) if and only if $x_1' \neq 0$ and $x_2' \neq 0$ simultaneously along the solution. If a solution curve of (6.2.38) passes through an equilibrium point of (6.2.37), where $x_1'(\bar{t}) = 0$ and $x_2'(\bar{t}) = 0$ for some $\bar{t}$, then the entire solution curve is not a solution but rather, it is a union of several distinct orbits. For example, consider the system of differential equations

$$\begin{aligned} x_1' &= x_2(1 - x_1^2 - x_2^2), \\ x_2' &= -x_1(1 - x_1^2 - x_2^2). \end{aligned} \tag{6.2.40}$$

The solution curves of the scalar equation

$$\frac{dx_2}{dx_1} = -\frac{x_1}{x_2}$$

are the family of concentric circles $x_1^2 + x_2^2 = c^2$. Observe, however, that to get the latter equation, we have assumed $x_1^2 + x_2^2 = 1$, as each point of this circle is an

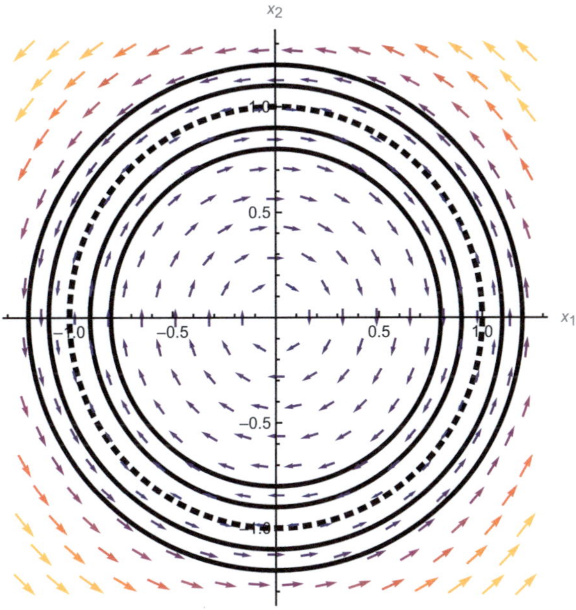

Fig. 6.16 The trajectories of the solution $\boldsymbol{x}(t)$ (solid), the circle $x_1^2 + x_2^2 = 1$ consisting of equilibria (squares) and the vector field of (6.2.40); note the change of the direction of the field while crossing the unit circle

equilibrium point of (6.2.40). Thus, the orbits of (6.2.40) are the circles $x_1^2+x_2^2 = c^2$ for $c \neq 1$, and each unit circle point is the orbit of a stationary solution; see Fig. 6.16.

Similarly, the full answer for the system (6.2.39) of the previous example is that $x_2 = (x_1^3 + c)^{1/3}$ are orbits for $c \neq 0$ as then neither solution curve passes through the only equilibrium point $(0, 0)$. For $c = 0$ the solution curve $x_2 = x_1$ consists of the equilibrium point $(0, 0)$ and two orbits $x_2 = x_1$ for $x_1 > 0$ and $x_1 < 0$.

Note that, in general, it is impossible to solve (6.2.38) explicitly. Hence, usually, we cannot find the equation of orbits in a closed form. Nevertheless, it is still possible to obtain an accurate description of all orbits of (6.2.17). In fact, the system (6.2.17) provides us with explicit information about how fast and in which direction the solution is moving at each point of the trajectory. In fact, as the orbit of the solution $(x_1(t), x_2(t))$ is in fact a curve of which $(x_1(t), x_2(t))$ is a parametric description, $(x_1'(t), x_2'(t)) = (f_1(x_1, x_2), f_2(x_1, x_2))$ is the tangent vector to the orbit at the point (x_1, x_2) showing, moreover, the direction at which the orbit is traversed. In particular, the orbit is vertical at each point (x_1, x_2) where $f_1(x_1, x_2) = 0$ and $f_2(x_1, x_2) \neq 0$ and it is horizontal at each point (x_1, x_2) where $f_1(x_1, x_2) \neq 0$ and $f_2(y_1, y_2) = 0$. As we noted earlier, each point (x_1, x_2), where $f_1(x_1, x_2) = 0$ and $f_2(x_1, x_2) = 0$, gives an equilibrium solution, and the orbit reduces to this point.

Exercise 6.2.6 Find the orbits of each of the following systems. Sketch them, indicating the direction in which they are traversed.

(i)

$$x_1' = x_2,$$
$$x_2' = -x_1 :$$

(ii)

$$x_1' = x_2(1 + x_1 + x_2),$$
$$x_2' = -x_1(1 + x_1 + x_2) :$$

(iii)

$$x_1' = 2x_1x_2,$$
$$x_2' = x_1^2 - x_2^2.$$

6.2.3 *Qualitative Properties of Orbits*

We shall now prove two orbit properties crucial to analyzing the system (6.2.17).

Theorem 6.27 *Assume that the assumptions of Theorem B.10 are satisfied. Then*

(i) there exists one and only one orbit through every point $\mathbf{x^0} \in \mathbb{R}^d$. In particular, if the orbits of two solutions $\mathbf{x}(t)$ and $\mathbf{y}(t)$ have one point in common, then they must be identical.

(ii) Let $\mathbf{x}(t)$ be a solution to (6.2.17). If for some $T > 0$ and some $t_0 \in \mathbb{R}$, we have $\mathbf{x}(t_0 + T) = \mathbf{x}(t_0)$, then $\mathbf{x}(t + T) = \mathbf{x}(t)$ for all t. In other words, if a solution $\mathbf{x}(t)$ returns to its starting value after a time $T > 0$, then it must be periodic (that is, it must repeat itself over every time interval of length T).

Proof

(i) Let $\boldsymbol{x}^{\mathbf{0}}$ be any point in $\mathbb{R}^d$. Then, from Theorem B.10, we know that there is a solution of the problem $\boldsymbol{x}' = \boldsymbol{f}(\boldsymbol{x}), \boldsymbol{x}(0) = \boldsymbol{x}^{\mathbf{0}}$ and the orbit of this solution passes through $\boldsymbol{x}^{\mathbf{0}}$ from the definition of the orbit. Assume now that there is another orbit passing through $\boldsymbol{x}^{\mathbf{0}}$, that is, there is a solution $\boldsymbol{y}(t)$ satisfying $\boldsymbol{y}(t_0) = \boldsymbol{x}^{\mathbf{0}}$ for some t_0. From Lemma 6.4 we know that $\hat{\boldsymbol{y}}(t) = \boldsymbol{y}(t + t_0)$ is also a solution. However, this solution satisfies $\hat{\boldsymbol{y}}(0) = \boldsymbol{y}(t_0) = \boldsymbol{x}^{\mathbf{0}}$, that is, the same initial condition as $\boldsymbol{x}(t)$. Using the uniqueness part of Theorem B.10, we must then have $\boldsymbol{x}(t) = \hat{\boldsymbol{y}}(t) = \boldsymbol{y}(t + t_0)$ for all t for which the solutions are defined.

This implies that the orbits are identical. In fact, if $\boldsymbol{\xi}$ is an element of the orbit of $\boldsymbol{x}$, then for some t' we have $\boldsymbol{x}(t') = \boldsymbol{\xi}$. However, we have also $\boldsymbol{\xi} = \boldsymbol{y}(t' + t_0)$ so that $\boldsymbol{\xi}$ belongs to the orbit of $\boldsymbol{y}(t)$. Conversely, if $\boldsymbol{\xi}$ belongs to the orbit of $\boldsymbol{y}$ so that $\boldsymbol{\xi} = \boldsymbol{y}(t'')$ for some t'', then, by $\boldsymbol{\xi} = \boldsymbol{y}(t'') = \boldsymbol{x}(t'' - t_0)$, we see that $\boldsymbol{\xi}$ belongs to the orbit of $\boldsymbol{x}$.

(ii) Assume that for some numbers t_0 and $T > 0$ we have $\boldsymbol{x}(t_0) = \boldsymbol{x}(t_0 + T)$. The function $\boldsymbol{y}(t) = \boldsymbol{x}(t+T)$ is again a solution satisfying $\boldsymbol{y}(t_0) = \boldsymbol{x}(t_0+T) = \boldsymbol{x}(t_0)$, thus, from Theorem B.10, $\boldsymbol{x}(t) = \boldsymbol{y}(t)$ for all t for which they are defined and therefore $\boldsymbol{x}(t) = \boldsymbol{x}(t + T)$ for all such t.

□

Example 6.28 A curve in the shape of 8 cannot be an orbit. In fact, suppose that the solution passes through the intersection point at some time t_0, then completing the first loop, it returns after time T, that is, we have $\boldsymbol{x}(t_0) = \boldsymbol{x}(t_0+T)$. From (ii), it follows then that this solution is periodic, that is, it must follow the same loop again and cannot switch to the other loop.

Corollary 6.29 *A solution $\boldsymbol{x}(t)$ of (6.2.17) is periodic if and only if its orbit is a closed curve in $\mathbb{R}^d$.*

Proof Assume that $\boldsymbol{x}(t)$ is a periodic solution of (6.2.17) of period T, that is $\boldsymbol{x}(t) = \boldsymbol{x}(t + T)$. If we fix t_0, then, as t runs from t_0 to $t_0 + T$, the point $\boldsymbol{x}(t)$ traces a curve, say C, from $\boldsymbol{\xi} = \boldsymbol{x}(t_0)$ back to the same point $\boldsymbol{\xi}$ without intersections and, if t runs from $-\infty$ to ∞, the curve C is traced infinitely many times.

Conversely, suppose that the orbit C is a closed curve (containing no equilibrium points). Assume the contrary. Then the orbit is parametrically described by $\boldsymbol{x}(t)$, $-\infty < t < \infty$ in a one-to-one way (as otherwise we would have $\boldsymbol{x}(t') = \boldsymbol{x}(t'')$ for some $t' \neq t''$ and, by the previous theorem, the solution would be periodic). Consider a sequence $(t_n)_{n\in\mathbb{N}}$ with $t_n \to \infty$. Since the sequence $\boldsymbol{x}(t)$ is bounded, we find a subsequence $t'_n \to \infty$ such that $\boldsymbol{x}(t'_n) \to \boldsymbol{x} \in C$. Then $\boldsymbol{x} = \boldsymbol{x}(t_0)$ for some finite t_0 since C does not contain equilibria. Consider a neighbourhood of $\boldsymbol{x}(t_0)$ which is the image of some interval $(t_0 - \epsilon, t_0 + \epsilon)$. Since $\boldsymbol{x}(t'_n) \to \boldsymbol{x}(t_0)$, $t'_n \in (t_0 - \epsilon, t_0 + \epsilon)$ for sufficiently large n which contradicts $t'_n \to \infty$. □

Example 6.30 We show that every solution $z(t)$ of the second order differential equation

$$z'' + z + z^3 = 0$$

is periodic. Following Remark 6.2.20, we convert this equation into a system: let $z = x_1$ and

$$x'_1 = x_2,$$
$$x'_2 = -x_1 - x_1^3.$$

The only equilibrium is (0, 0). The orbits are the solution curves of the equation

$$\frac{dx_2}{dx_1} = -\frac{x_1 + x_1^3}{x_2},$$

as long as $x_2 \neq 0$, so that

$$\frac{x_2^2}{2} + \frac{x_1^2}{2} + \frac{x_1^4}{4} = c^2$$

is the equation of orbits, with the equilibrium point (0, 0) if $c = 0$. By writing the above equation in the form

$$\frac{x_2^2}{2} + \left(\frac{x_1^2}{2} + \frac{1}{2}\right)^2 = c^2 + \frac{1}{4},$$

we see that for each $c \neq 0$ it describes a closed curve consisting of two branches $x_2 = \pm\frac{1}{\sqrt{2}}\sqrt{4c^2 + 1 - (x^2 + 1)^2}$ that stretch between $x_1 = \pm\sqrt{1 + \sqrt{4c^2 + 1}}$. Consequently, every solution is a periodic function.

Exercise 6.2.7 Show that every orbit of

$$x' = 4y,$$
$$y' = -9x,$$

is an ellipse.

Exercise 6.2.8 Prove that every solution of the system

$$x' = ye^{1+x^2+y^2},$$
$$y' = -xe^{1+x^2+y^2},$$

is periodic.

Exercise 6.2.9 Prove that all solutions $z(t)$ of

$$z'' + z^3 = 0$$

are periodic.

Exercise 6.2.10 Show that $x = 0, y > 0$ and $y = 0, x > 0$ are trajectories of the system

$$x' = x^2 + y \sin x,$$
$$y' = -1 + xy + \cos y.$$

Hence, all solutions that start in the first quadrant must remain there for all time (both backwards and forwards).

Exercise 6.2.11 Show that all solutions of

$$x' = -1 - y + x^2,$$
$$y' = x + xy,$$

that start inside the unit circle $x^2 + y^2 = 1$ must remain there for all time. Hint. Compute $\frac{d(x^2+y^2)}{dt}$.

We conclude this section by presenting another result on what an orbit can/cannot look like. We begin with the following lemma.

Lemma 6.31 *Assume that* $\boldsymbol{x} : [0, \infty) \to \mathbb{R}^d$ *is differentiable with* $\boldsymbol{x}'$ *uniformly continuous on* $[0, \infty)$. *If* $\boldsymbol{x}$ *satisfies* $\lim_{t\to\infty} \boldsymbol{x}(t) = \boldsymbol{x}^*$ *for some* $\boldsymbol{x}^*$, *then* $\lim_{t\to\infty} \boldsymbol{x}'(t) = 0$.

Proof Let us take an arbitrary $\epsilon > 0$. From uniform continuity of $\boldsymbol{x}$ we know that there is $h > 0$ such that

$$\|\boldsymbol{x}'(t_1) - \boldsymbol{x}'(t_2)\| < \epsilon$$

if $|t_1 - t_2| < h$. Further, from the Mean Value Theorem, for each $i = 1, \ldots, d$, there is $0 \leq \tau_i(t) < h$ such that

$$x_i(t+h) - x_i(t) = x_i'(t+\tau(t))h = x_i'(t)h + (x_i'(t+\tau(t)) - x_i'(t))h.$$

Also, for the ϵ and h selected above and any $i = 1, \ldots, d$, we can find t_0 such that for all $t \geq t_0$ we have

$$|x_i(t+h) - x_i(t)| \leq |x_i(t+h) - x_{i,0}| + |x_i(t) - x_{i,0}| < 2\epsilon h.$$

Hence

$$|x_i'(t)| \leq h^{-1}|x_i(t+h) - x_i(t)| + |(x_i'(t+\tau(t)) - x_i'(t))| < 3\epsilon.$$

Thus $\lim_{t\to\infty} \boldsymbol{x}'(t) = 0$. □

Then, we have

Corollary 6.32 *Suppose that a solution* $\boldsymbol{x}(t)$ *of (6.2.17) with Lipschitz continuous* f *approaches a vector* $\boldsymbol{x}^*$ *as* $t \to \infty$. *Then* $\boldsymbol{x}^*$ *is an equilibrium point of (6.2.17).*

Proof If $\lim_{t\to\infty} \boldsymbol{x}(t) = \boldsymbol{x}^*$. Then, its trajectory is bounded and closed. Hence, there is a global Lipschitz constant for $\boldsymbol{f}$ on the trajectory and, by $\boldsymbol{x}'(t) = \boldsymbol{f}(\boldsymbol{x}(t))$, $\boldsymbol{x}'$ is uniformly continuous on $[0, \infty)$. Therefore, by Lemma 6.31 and the continuity of $\boldsymbol{f}$

$$0 = \lim_{t\to\infty} \boldsymbol{x}'(t) = \lim_{t\to\infty} \boldsymbol{f}(\boldsymbol{x}(t)) = \boldsymbol{f}\left(\lim_{t\to\infty} \boldsymbol{x}(t)\right) = \boldsymbol{f}(\boldsymbol{x}^*).$$

□

The geometrical meaning of Corollary 6.32 is that no trajectory of a dynamical system (with Lipschitz right-hand-side) can end in 'mid-air'. It can be considered a multidimensional extension of Theorem 4.3.

6.2.4 Applications of Basic Phase-Plane Analysis

In this section, we shall show how even such simple results as presented above can be used to determine the long-term behaviour of solutions. We shall focus on models from epidemiology or ecology. Hence, the non-negativity of solutions originating from the non-negative initial conditions is one of the basic requirements for the well-posedness of (6.2.17), (6.2.18). We shall describe here standard steps which should be followed to prove it.

1. If $\boldsymbol{f}$ is (locally) Lipschitz continuous, then Theorem B.10 ensures that a unique solution to (6.2.17), (6.2.18) exists at least on $(-\delta, \delta)$ for some $\delta > 0$.
2. If $\boldsymbol{f}$ satisfies the assumptions of Theorem B.21, the solution is non-negative as long as it exists in forward time, that is, at least on $(0, \delta)$.
3. Since $\boldsymbol{x}(t) = (x_1(t), \ldots, x_d(t))$ describes a structured population, we often try to construct a scalar differential equation or inequality governing the total size of the population $N(t) = x_1(t) + \ldots + x_d(t)$.
4. If $N(t)$ exists on $[0, T)$, $\delta < T \geq +\infty$, from the non-negativity we obtain

$$0 \leq x_i(t) \leq N(t), \quad i = 1, \ldots, d,$$

   for as long as $x_i(t)$ exists (at least on $(0, \delta)$). Hence, by Theorem B.14, $\boldsymbol{x}(t)$ exists and is positive on $[0, T)$ (and exists globally if $T = +\infty$.)

We will show how to implement these steps in the following examples.

A Simple SIR Model

It is customary to discard demographical processes for short-lasting diseases, such as flu or common cold. If the disease induces immunity, at least in the time covered by the model, one of the simplest models is the SIR Kermack–McKendrick model

$$\begin{aligned} S' &= -\beta S I, \\ I' &= \beta S I - \nu I, \\ R' &= \nu I. \end{aligned} \tag{6.2.41}$$

As we see, we use the mass action transmission rate. The total population at time t is given by $N(t) = S(t) + I(t) + R(t)$ and, by adding the equations in (6.2.41), we obtain

$$N' = 0,$$

hence $N(t) = N(0)$, reflecting the assumption that there are no demographic processes included in the model.

The dynamics of the model can be fully analysed without any sophisticated tools.

Step 1. The model is well-posed, that is, for every $(S(0), I(0), R(0)) = (S_0, I_0, R_0)$, there is exactly one solution defined at least on some interval $(-\tau, \tau)$, $\tau > 0$. This follows from the Picard–Lindelöf theorem, Theorem B.10. We shall be interested in $t \geq 0$. Once we know that there is a solution $(S(t), I(t), R(t))$, $t \in (0, \tau)$, we can prove that it is positive provided $S(0), I(0)$ and $R(0)$ are positive. It follows directly from Theorem B.21, but it is instructive to show this from first principles. Indeed, for instance, for S, we see that it satisfies the linear equation

$$S'(t) = -\beta I(t) S(t),$$

and hence $S(t) = S(0)e^{-\beta \int_0^t I(s)ds}$, where I is a known function. Hence, $S(t) > 0$ as long as $I(t)$ is defined. The same argument works for I, and then R is nondecreasing as I is nonnegative; thus $R(t) \geq R(0) > 0$. Thus, we have $0 < S(t), I(t), R(t) < N(0)$ for t in any interval on which the solution is defined and hence, by Theorem B.14, the solution is defined globally for $t \geq 0$.

Step 2. We see that $S' \leq 0$, that is, S is nonincreasing and bounded from below. Since it is defined for all $t \geq 0$, we have

$$\lim_{t\to\infty} S(t) =: S_\infty.$$

Similarly, R is bounded from above and nondecreasing, hence

$$\lim_{t\to\infty} R(t) =: R_\infty.$$

Further, since $S(t) + I(t) + R(t) = N(0)$, we must have

$$\lim_{t\to\infty} I(t) =: I_\infty.$$

However, the number of infected individuals can increase or decrease depending on the sign of $\beta S(t) - \nu$. In particular, if $\beta S(0) - \nu > 0$, or

$$\frac{\beta S(0)}{\nu} > 1,$$

then the number of infectives initially will increase. Then we say that we have an outbreak or epidemic.

The number $\beta S(0)/\nu$ has an important interpretation. The coefficient β gives the probability of infecting a susceptible in a unit time induced by one infective, whereas $1/\nu$ is the average time an infective remains infectious. The number of susceptibles at the beginning is $S(0) = N(0) = N$. Thus, according to Definition 6.1, we have

$$R_0 = \frac{\beta N}{\nu}.$$

Next, we estimate the limits. First, observe that $I(t) \neq 0$ for any finite t_0 if $I(0) \neq 0$. Otherwise, $(S(t_0), 0, R(t_0))$ would be a constant solution to the problem taking the same value as $(S(t), I(t), R(t))$ at $t = t_0$, contradicting the uniqueness of solutions. Hence $t \to R(t)$ is strictly increasing and we can consider $t = t(R)$ on $[R(0), R_\infty)$. Thus, we can define $S(R) = S(t(R))$ and, using the Chain Rule and the derivative of the inverse function formula, we get

$$\frac{dS}{dR} = \frac{dS}{dt}\frac{dt}{dR} = -\frac{\beta SI}{\nu I} = -\frac{\beta}{\nu}S.$$

Hence

$$S(R) = S(R(0))e^{-\frac{\beta}{\nu}(R-R(0))} \geq S(R(0))e^{\frac{\beta}{\nu}R(0)}e^{-\frac{\beta}{\nu}N(0)} > 0$$

for any R. Therefore, we must have $S_\infty > 0$, which shows in particular that no epidemic can infect all susceptibles.

Let us consider I_∞. Integrating the first equation in (6.1.7) from 0 to ∞, we obtain

$$S_\infty - S_0 = \int_0^\infty S'(t)dt = -\beta\int_0^\infty S(t)I(t)dt \leq -\beta S_\infty \int_0^\infty I(t)dt.$$

In other words

$$\int_0^\infty I(t)dt \le \frac{S_0 - S_\infty}{\beta S_\infty} < \infty.$$

Since we know that I_∞ exists and is nonnegative, we must have $I_\infty = 0$.

Exercise 6.2.12 Prove the above statement.

Step 3. We note that (6.1.7) is, in fact, a two-dimensional system, and we can find orbits in the (S, I) plane. The two first equations are independent of R and can be solved separately so that R can be found from $R = N - S - I$. So, let us focus on

$$\begin{aligned} S' &= -\beta SI, \\ I' &= \beta SI - \nu I. \end{aligned} \tag{6.2.42}$$

From the above discussion, we know that S is a monotonic function of t for $S, I > 0$ and hence it can be inverted, yielding $t = t(S)$, so that, as in (6.2.38), we can write the equation for $I = I(S)$ as

$$\frac{dI}{dS} = \frac{dI}{dt}\frac{dt}{dS} = \frac{\beta SI - \nu I}{-\beta SI} = -1 + \frac{\nu}{\beta S}.$$

Separation of variables and integration yields

$$I - I(0) = S(0) - S + \frac{\nu}{\beta} \ln \frac{S}{S(0)}.$$

In particular, using that fact that $\lim_{t\to\infty} I(t) = 0$ and $\lim_{t\to\infty} S(t) = S_\infty$, we obtain

$$-I(0) = S(0) - S_\infty + \frac{\nu}{\beta} \ln \frac{S_\infty}{S(0)}$$

or

$$\frac{\beta}{\nu} = \frac{\ln \frac{S(0)}{S_\infty}}{S(0) + I(0) - S_\infty}. \tag{6.2.43}$$

Let us draw a few conclusions. First, since S is a nonincreasing function, $S(t) \ge S_\infty$ for any $t \ge 0$. Thus we obtain

$$S_\infty \le S(0) + I(0).$$

An important piece of information is the maximum number of infectives. This occurs for $I' = 0$ or at $I(0)$. We see that $I' = 0$ if $S = \nu/\beta$ (and hence, this can only occur if $S(0) > \nu/\beta$ since S is decreasing). Thus

$$I_{\max} = I(0) + S(0) - \frac{\nu}{\beta} + \frac{\nu}{\beta}\ln\frac{\nu}{\beta S(0)}. \tag{6.2.44}$$

An SIR Model with Demography

In many situations, when we model

- diseases of duration comparable with the life-span of individuals,
- lethal diseases,
- epidemics,
- long-term impact of the disease,

the demography of the affected population plays a significant role in the evolution of the disease and has to be taken into account. To illustrate this role, we combine the SIR model described above with the model introduced (3.3.32). While this affine model has many drawbacks, indicated in Sect. 3.3.3, it can be accepted for relatively short periods, and it is appealing and thus commonly used due to its mathematical simplicity. We assume that there is no vertical transmission, that is, all individuals are born susceptible, and we model the force by using the law of mass action. Then we have

$$\begin{aligned} S' &= \Lambda - \beta SI - \mu S, \\ I' &= \beta SI - \nu I - \mu I, \\ R' &= \nu I - \mu R. \end{aligned} \tag{6.2.45}$$

The evolution of the total population $N = S + I + R$ is given by (3.3.32), that is,

$$N' = \Lambda - \mu N,$$

and hence $N(t)$ is given by

$$N(t) = \mathring{N}e^{-\mu t} + \frac{\Lambda}{\mu}(1 - e^{-\mu t}). \tag{6.2.46}$$

Contrary to the previous cases, $N(t)$ is not constant.

Non-dimensionalization

As before, the first two equations are independent of R; we shall work with

$$\begin{aligned} S' &= \Lambda - \beta SI - \mu S, \\ I' &= \beta SI - \nu I - \mu I, \end{aligned} \tag{6.2.47}$$

and the number of recovered is given by $R(t) = N(t) - S(t) - I(t)$. To find the basic reproduction number, we must realize that we deal with a changing population here, so the definition requires some modification. We refer R_0 to the stationary value of the disease-free population, which is given by $N^* = \Lambda/\mu$, see (3.3.34). Hence

$$R_0 = \frac{\Lambda\beta}{\mu(\nu+\mu)}, \tag{6.2.48}$$

where the factor $\frac{1}{\nu+\mu}$ gives the average time an individual stays in the infective class.

Next, we nondimensionalize the system to reduce the number of parameters. Also, since we expect R_0 to play a significant role in the model, we try to express as many parameters as possible as functions of R_0. First, we observe that both sides must have dimension [number of people]/time. Usually, to nondimensionalize, we choose new variables as ratios of the old variables to some typical quantities of the same dimension. Here, we have an individual's average length of stay in the infective compartment $1/(\nu+\mu)$ for time and Λ/μ for the population size. Thus, we define new time τ by

$$\tau := (\nu+\mu)t$$

and $\hat{S}(\tau) = S(t)$, $\hat{I}(\tau) = I(t)$. This gives

$$\hat{S}'_\tau = \frac{1}{\nu+\mu}S'_t, \quad \hat{I}'_\tau = \frac{1}{\nu+\mu}I'_t.$$

Then, we introduce

$$x(\tau) := \frac{\mu\hat{S}(\tau)}{\Lambda}, \quad y(\tau) := \frac{\mu\hat{I}(\tau)}{\Lambda},$$

which are the sizes of susceptible and infections relative to the stationary population. This allows for writing (6.2.47) in the form

$$\begin{aligned} x' &= \rho(1-x) - R_0 xy, \\ y' &= (R_0 x - 1)y, \end{aligned} \tag{6.2.49}$$

where

$$\rho := \frac{\mu}{\nu + \mu}.$$

We consider (6.2.49), writing t instead of τ as rescaling time does not change the long-term behaviour.

Basic Results

Since the right-hand side of the system is polynomial, it is locally (but not globally) Lipschitz continuous. Thus, we have local solvability. We follow the steps described at the beginning of Sect. 6.2.4 to show the global existence of non-negative solutions.

As mentioned above, it is instructive to give a direct proof of the non-negativity of solutions even though it is easy to see that the assumptions of Theorem B.21 are satisfied.

We see that the equation for y is in the form

$$y' = \phi(t)y,$$

where ϕ is a given function (as we know that the solution to (6.2.49) exists), $y(t) = \mathring{y}e^{\int_0^t \phi(s)ds} > 0$ (provided $\mathring{y} > 0$).

The equation for x is not of this form, but we can argue as follows. We have

$$x' = -x\psi(t) + \rho,$$

where $\psi(t) = \rho + R_0 y(t)$ is a known function. So, the variation of constants formula, see (5.3.64), gives

$$x(t) = \mathring{x}e^{-\int_0^t \psi(s)ds} + \rho e^{-\int_0^t \psi(s)ds} \int_0^t e^{\int_0^r \psi(s)ds} dr \geq 0.$$

Hence, $x(t), y(t) \geq 0$ as long as they exist and are positive for positive initial conditions. But

$$(x + y)' = x' + y' = \rho(1 - x) - y,$$

or, setting $z = x + y$,

$$z' = -\rho z - (1 - \rho)y + \rho.$$

Since $y(t) \geq 0$ and $\rho \leq 1$, we can write

$$z' \leq -\rho z + \rho,$$

that is

$$(ze^{\rho t})' \leq \rho e^{\rho t},$$

that, upon integration, yields

$$z(t) \leq e^{-\rho t}\mathring{z} + (1 - e^{-\rho t}) = 1 + (\mathring{z} - 1)e^{-\rho t}.$$

From this we see that if $\mathring{z} \leq 1$ (equivalently, $S(0) + I(0) \leq \Lambda/\mu$), then $z(t) \leq 1$ (equivalently, $N(t) \leq \Lambda/\mu$). On the other hand, if $\mathring{z} > 1$, then

$$z(t) \leq 1 + (\mathring{z} - 1)e^{-\rho t} \leq \mathring{z},$$

as the second term is decreasing with the maximum attained at $t = 0$. Moreover, as the right-hand side of the inequality converges to 1, we obtain

$$\limsup_{t\to\infty} z(t) \leq 1. \tag{6.2.50}$$

Hence, by Definition 6.21, the region

$$V = \{(x, y) : x \geq 0, y \geq 0, x + y \leq 1\} \tag{6.2.51}$$

is positively invariant under the flow generated by (6.2.49). Also, due to (6.2.50), $\overline{\mathbb{R}}^2_+$ is a basin of attraction for V.

From these considerations, we obtain that the solution $(x(t), y(t))$ exists globally. Next, we consider the equilibria of (6.2.49):

$$\begin{aligned} 0 &= \rho(1 - x) - R_0 xy, \\ 0 &= (R_0 x - 1)y. \end{aligned} \tag{6.2.52}$$

Solving, gives $(x_0^*, y_0^*) = (1, 0)$, or

$$(S_0^*, I_0^*) = \left(\frac{\Lambda}{\mu}, 0\right),$$

which is the disease-free equilibrium. We say this is a boundary equilibrium as it is placed on the boundary of the feasible region $\overline{\mathbb{R}}^2_+$. Further, if $y \neq 0$, then we obtain $x = \frac{1}{R_0}$ and $y = \rho\left(1 - \frac{1}{R_0}\right)$. Thus, we have the endemic equilibrium

$$(x^*, y^*) = \left(\frac{1}{R_0}, \rho\left(1 - \frac{1}{R_0}\right)\right). \tag{6.2.53}$$

We start with a basic phase-plane analysis of (6.2.49) to understand its dynamics better and begin by sketching the nullclines of (6.2.49).

1. x-nullcline: $x' = 0$ if and only if

$$\rho(1-x) - R_0 xy = 0,$$

that is,

$$y = \frac{\rho}{R_0}\frac{1-x}{x}. \tag{6.2.54}$$

2. y-nullcline: $y' = 0$ if and only if

$$y = 0 \text{ or } x = \frac{1}{R_0}.$$

The intersections of the nullclines give the equilibria. The x-nullcline intersects the y-nullcline $y = 0$ at $x = 1$, as expected. Also, the x-nullcline intersects the y-nullcline $x = \frac{1}{R_0}$ at $y = \rho\left(1 - \frac{1}{R_0}\right)$ and this equilibrium is positive only if $R_0 > 1$.

Here, we consider $R_0 \leq 1$ since the case $R_0 > 1$ requires more sophisticated tools and thus will be considered later, in Example 6.100.

First, let $R_0 < 1$. There are two cases to consider. We see that the tangent to the x-nullcline at DFE is given by

$$\left.\frac{dy}{dx}\right|_{x=1} = -\frac{\rho}{R_0} = -\frac{\mu}{\beta}\frac{1}{\frac{\Lambda}{\mu}}.$$

Thus, if $R_0 > \rho$, then the x-nullcline passes through the interior of V, crosses the line $x + y = 1$ and divides V into two subsets

$$V_1 = \left\{(x, y) : x \geq 0, y \geq 0, y \leq \min\left\{1 - x, \frac{\rho}{R_0}\frac{1-x}{x}\right\}\right\}$$

and

$$V_2 = \left\{(x, y) : \frac{\rho}{R_0}\frac{1-x}{x} \leq y \leq 1 - x\right\}.$$

We observe that $x' > 0$ and $y' < 0$ in the interior of V_1 and $x' < 0$ and $y' < 0$ in the interior of V_2.

Let us assume that we have a trajectory in the interior of V_1. We use the argument of Lemma B.20. By the invariance of V, the trajectory cannot escape through $y = 0$, $x = 0$, or the piece of the boundary given by $y = 1 - x$, so it can only escape through the nullcline. The direction of the vector field along the x-nullcline, that is, the tangent to the trajectory, is given by $\boldsymbol{t} = (x', y') = (0, (R_0 x - 1)y(x))$, where $y(x)$ is given by (6.2.54). At the same time, the normal vector at this part of the boundary, pointing inward V_1, is given by $\boldsymbol{n} = \left(-\frac{R_0 x^2}{\rho}, -1\right)$. We see that

$$\boldsymbol{t} \cdot \boldsymbol{n} = (1 - R_0 x) y(x).$$

However, $y > 0$ along the nullcline (apart from the value at DFE and $x < 1/R_0$ since $R_0 < 1$. Thus, $\boldsymbol{t} \cdot \boldsymbol{n} > 0$ along the nullcline, and thus the trajectory cannot escape V_1.

Since $x' > 0$ and $y' < 0$ in V_1, $x(t)$ is increasing and bounded, while $y(t)$ is decreasing and bounded. Hence, there is $(x_0, y_0) \in V_1$ such that

$$\lim_{t\to\infty} x(t) = x_0, \qquad \lim_{t\to\infty} y(t) = y_0.$$

We observe that the assumptions of Corollary 6.32 are satisfied, and hence (x_0, y_0) is an equilibrium. Since in this case $(1, 0)$ is the only equilibrium, we have $(x_0, y_0) = (0, 0)$.

Next, if we have a solution originating in V_2, it cannot escape through $x + y = 1$, so it will either stay in V_2 or enter V_1. If it stays in V_2, then, as before, the derivatives x' and y' are of constant sign, and, as the solutions are bounded, they must converge within V_2. Based on the same argument, the solution must converge to the DFE. On the other hand, if the solution enters V_1, we can apply the argument of the first part.

Finally, assume $z(0) > 1$ so the trajectory starts outside V. By (6.2.50), $\limsup_{t\to\infty} z(t) \leq 1$. There are two cases to consider. If $\liminf_{t\to\infty} z(t) < 1$, then for some t_0 we must have $z(t_0) < 1$ and thus the trajectory is inside V, and the previous argument applies (by the flow property). On the other hand, if $\liminf_{t\to\infty} z(t) = 1$, then $\lim_{t\to\infty} z(t) = 1$. The argument below considers that we may have $R_0 = 1$; thus, the y nullcline could pass through the DFE. First we observe that along the line $x = 1$ the normal pointing to the right is given by $\boldsymbol{n} = (1, 0)$ while the field is given by $\boldsymbol{t} = \left(\rho\left(1 - \frac{1}{R_0}\right) - y, 0\right)$, so

$$\boldsymbol{t} \cdot \boldsymbol{n} = \rho\left(1 - \frac{1}{R_0}\right) - y < 0,$$

provided $y > 0$, as $R_0 \leq 1$. Since, however, $y = 0$ is a trajectory, no trajectory can move from the region $\{x \leq 1/R_0\}$ to $\{x > 1/R_0\}$. Hence, if $z(0) > 1$ with $x(0) \leq 1/R_0$, we have $y'(t) < 0$ for $t > 0$ and thus $y(t)$ converges. Since $z(t)$ converges, so must $x(t)$. So, the last case to consider is when $x(0) > 1/R_0$. Then, at least initially, $y' > 0$. If $x(t)$ converges, then $y(t)$ converges, and the limit must be the DFE. Hence, $y(t)$ must converge to 0. Assume that $y(t)$ does not converge to 0. Then $y(t_n) \geq c > 0$ for some $t_n \to \infty$. But since $x(t) + y(t) = z(t) \to 1$, then $x(t_n) < 1$ for sufficiently large t_n and hence $(x(t), y(t))$ enters the region $\{x \leq 1/R_0\}$, to which the previous argument applies.

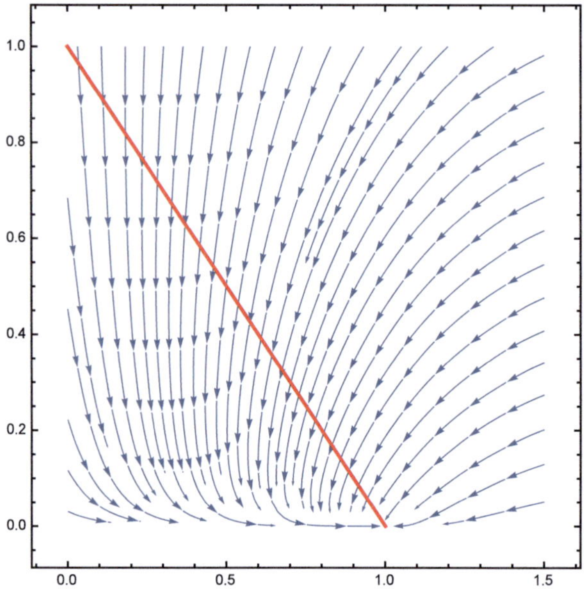

Fig. 6.17 Streamlines for $R_0 = 0.5$ and the line $x + y = 1$ (red)

The second subcase is when $R_0 \leq \rho$. Then the x-nullcline $y(x)$ stays outside V (on account of $\frac{dy}{dx}(x) < \left.\frac{dy}{dx}\right|_{x=1}$ on $(0, 1)$) and thus the set V_2 of the previous part becomes empty and the argument of the first part applies.

These considerations are illustrated in Figs. 6.17, 6.18 and 6.19.

As mentioned above, for the case $R_0 > 1$, the reader is referred to Example 6.100.

Basic Ecological Models

In this section, we consider two-dimensional models describing predator-prey type interactions, which fit into the framework of the so-called Rosenzweig–McArthur models, an example of which is (6.1.5). Their general form is

$$
\begin{aligned}
x' &= xf(x) - xy\phi(x), \\
y' &= -g(y)y + xy\psi(x),
\end{aligned}
\tag{6.2.55}
$$

where x denotes the size/density of the prey population, and y is the size/density of the predator population. In the absence of predators ($y \equiv 0$), this model reduces to $x' = xf(x)$; hence $xf(x)$ represents the growth rate of the prey species. The term $xy\phi(x)$ is called the predator functional response and $x\phi(x)$ is the number of prey consumed by one predator in a unit time. In the predator equation, $g(y)$ is the

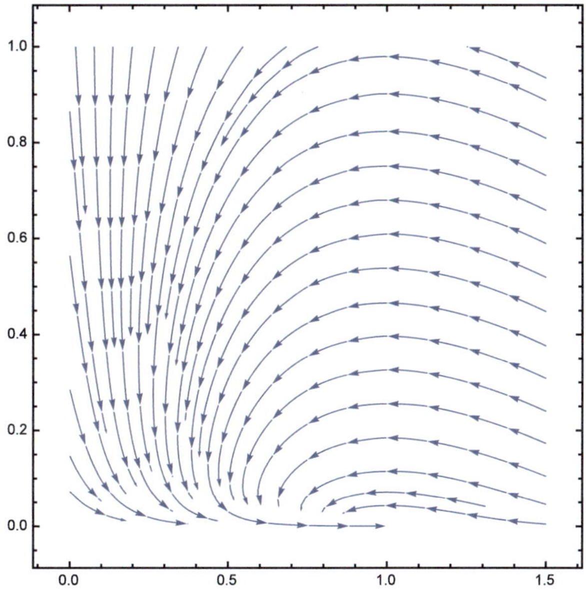

Fig. 6.18 Streamlines for $R_0 = 1$

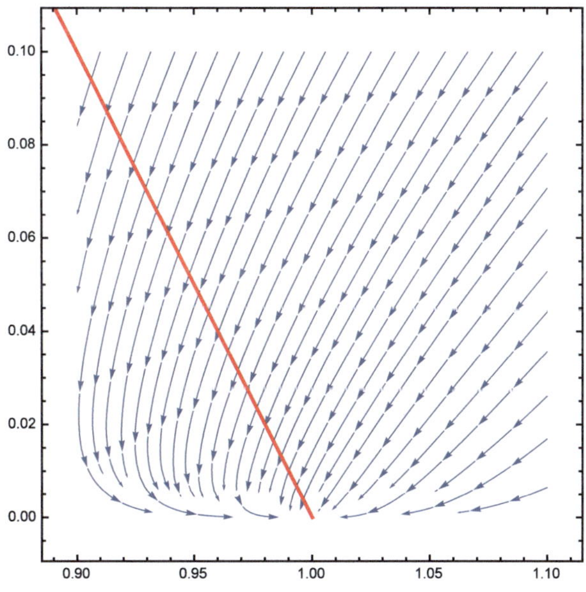

Fig. 6.19 Close up of the streamlines in the neighbourhood of DFE for $R_0 = 1$; note that there are trajectories with $x(t) > 1$ for all t

density-dependent predator mortality rate and $xy\psi(x)$ gives the rate of conversion of prey into the predator biomass (in terms of the predator population size).

A Model with Density-Dependent Mortality Rates

Considering (6.2.55) with $\phi(x) \equiv b/x$, $f(x) = ax - ex^2$, $g(y) = -cy - fy^2$ and $\psi(x) = d$, we obtain

$$\begin{aligned} x' &= ax - bxy - ex^2, \\ y' &= -cy + dxy - fy^2, \end{aligned} \tag{6.2.56}$$

where a, b, e, c, d, f are positive constants. This system can describe the evolution of two species x and y in an environment of limited capacity (indicated by the logistic growth of prey in the absence of predator), where the species y depends on the species x for its survival. Assume that

$$\frac{c}{d} > \frac{a}{e}. \tag{6.2.57}$$

We prove that every solution $(x(t), y(t))$ of (6.2.56), with $x(0), y(0) > 0$ approaches the equilibrium solution

$$x = \frac{a}{e}, \quad y = 0,$$

as t approaches infinity. The ecological interpretation of this result is that the predator's population will always go extinct, which can be attributed to the density-dependent predator's mortality term $-fy^2$.

As a first step, we show that the solutions with positive initial data must stay positive, that is, the orbit of any solution originating in the first quadrant must stay in this quadrant. Again, this follows directly from Theorem B.21, but we present a direct proof that sheds light on some additional aspects of the dynamics of the species. First, let us observe that by putting $y(t) \equiv 0$, we obtain the logistic equation for x that can be solved, giving

$$x(t) = \frac{a\mathring{x}}{e\mathring{x} + (a - e\mathring{x})e^{-at}},$$

where $\mathring{x} \geq 0$. The orbits of these solutions are the equilibrium point $(0, 0)$, the segment $0 \leq x < \frac{a}{e}$, $y = 0$ for $\mathring{x} < \frac{a}{e}$, the equilibrium point $\left(\frac{a}{e}, 0\right)$ for $\mathring{x} = \frac{a}{e}$ and the segments $\frac{a}{e} < x < \infty$, $y = 0$ for $\mathring{x} > \frac{a}{e}$. Thus, the non-negative x-semiaxis is the union of these four orbits. Similarly, putting $x(t) \equiv 0$ we obtain the equation

$$y' = -cy - fy^2.$$

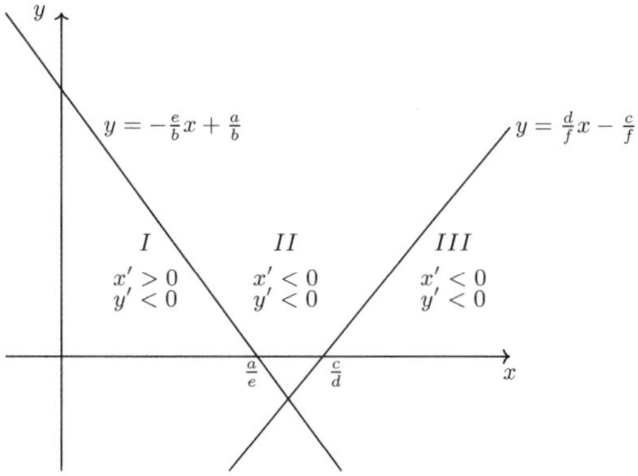

Fig. 6.20 Nontrivial nullclines of (6.2.56)

To use Theorem 4.3, we observe that the equilibrium points of this equation are $y = 0$ and $y = -\frac{c}{f}$ so that there are no equilibria on the positive y-semiaxis and $-cy - fy^2 < 0$ for $y > 0$. Therefore, any solution with an initial value $\mathring{y} > 0$ will decrease converging to 0, and the semiaxis $y > 0$ is a single orbit of (6.2.56). Thus, if a solution of (6.2.56) left the first quadrant, its orbit would cross one of the orbits the positive semiaxes consist of, which is precluded by Theorem 6.27. We note that this argument gives a stronger result than Theorem B.21, namely that any solution starting in the interior of the first quadrant will stay strictly positive as the orbits cannot even touch the semiaxes.

In the next step, we use the nullclines to divide the first quadrant into regions where the derivatives x' and y' are of a fixed sign. Clearly, $x \equiv 0$ and $y \equiv 0$ are the nullclines, but we are interested in the nontrivial nullclines, denoted by l_1 and l_2 in Fig. 6.20. The line l_1 is determined by $y = -\frac{e}{b}x + \frac{a}{b}$ so that $x' > 0$ in region I and $x' <$ in regions II and III. The line l_2 is given by $y = \frac{d}{f}x - \frac{c}{f}$ and $y' < 0$ in regions I and II, and $y' > 0$ in region III. In the following sequence of observations, we describe the behaviour of solutions in each region. The approach is taken from [25, Section 4.11] and provides an alternative for the argument used in Lemma B.20.

Observation 1 *Any solution to (6.2.56) which starts in the region I at $t = t_0$ will remain in this region for all $t > t_0$ and ultimately approach the equilibrium $x = \frac{a}{e}$, $y = 0$.*

Proof If the solution $(x(t), y(t))$ leaves region I at some time $t = t^*$, then $x'(t^*) = 0$, since the only way to leave this region is to cross the line l_1. Differentiation of the first equation in (6.2.56) gives

$$x'' = ax' - bx'y - bxy' - 2exx',$$

so that at $t = t^*$ we obtain

$$x''(t^*) = -bx(t^*)y'(t^*).$$

Since $y'(t^*) < 0, x''(t^*) > 0$, which means that $x(t^*)$ is a local minimum. However, $x(t)$ reaches this point from region I, where it is increasing, which is a contradiction. Thus, the solution $(x(t), y(t))$ stays in the region I for all times $t \geq t_0$. However, any solution staying in I must be bounded and $x' > 0$ and $y' < 0$ so that $x(t)$ is increasing, and $y(t)$ is decreasing and therefore, they must tend to finite limits. By Proposition 6.32, this limit must be an equilibrium point. The only equilibria are $(0, 0)$ and $\left(\frac{a}{e}, 0\right)$ and the solution cannot tend to the former as $x(t)$ is positive and increasing. Thus, any solution starting in region I for some time $t = t_0$ tends to the equilibrium $\left(\frac{a}{e}, 0\right)$ as $t \to \infty$. □

Observation 2 *Any solution of (6.2.56) that starts in region III at time t_0 must leave this region at some later time.*

Proof Suppose that a solution $(x(t), y(t))$ stays in region III for all time $t \geq 0$. Since the sign of both derivatives x' and y' is fixed, $x(t)$ decreases and $y(t)$ increases. Thus, $x(t)$ must tend to a finite limit, while $y(t)$ cannot escape to infinity as the only way this could be achieved would be if also $x(t)$ tended to infinity, which is impossible. Thus, $(x(t), y(t))$ tend to a finite limit that, by Proposition 6.32, has to be an equilibrium. However, there are no equilibria in region III, and thus, the solution must leave this region at some time. □

Observation 3 *Any solution of (6.2.56) that starts in region II at time $t = t_0$ and remains in this region for all $t \geq t_0$ must approach the equilibrium solution $x = \frac{a}{e}, y = 0$.*

Proof Suppose that a solution $(x(t), y(t))$ stays in region II for all $t \geq t_0$. Then both $x(t)$ and $y(t)$ are decreasing and since the region is bounded from below, we see that this the solution must converge to an equilibrium point; in this case necessarily $\left(\frac{a}{e}, 0\right)$. □

Observation 4 A solution cannot enter region III from region II.

Proof This case is similar to Observation 1. Indeed, if a solution crosses l_2 from II to III at $t = t^*$, then $y'(t^*) = 0$ but then, from the second equation of (6.2.56),

$$y''(t^*) = dy(t^*)x'(t^*) < 0,$$

so that $y(t^*)$ is a local maximum. This is, however, impossible, as $y(t)$ is decreasing in region II.

To summarise, see Fig. 6.21, if the initial values are in regions I or II, the solution tends to the equilibrium $\left(\frac{a}{e}, 0\right)$ as $t \to \infty$, by Observations 1,3 and 4. If the solution starts from region III, then, at some moment, it must enter region II, and we can apply the previous argument to claim again that the solution will eventually approach the equilibrium $\left(\frac{a}{e}, 0\right)$. Finally, if a solution starts on l_1, it must immediately enter region I as $y' < 0$ and $x' < 0$ in region II (if the solution ventured into II from l_1, then either x' or y' would have to be positive somewhere in II). Similarly, any solution starting from l_2 must immediately enter region II. Thus,

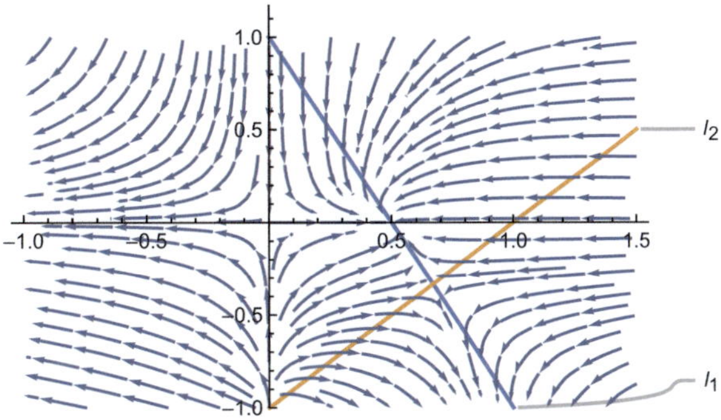

Fig. 6.21 The vector field and the nontrivial nullclines of (6.2.56) with $a = b = c = d = f = 1$ and $e = 2$. We see that the trajectories converge to the equilibrium $\left(\frac{1}{2}, 0\right)$, which is at the intersection of l_1 and the trivial nullcline $y = 0$

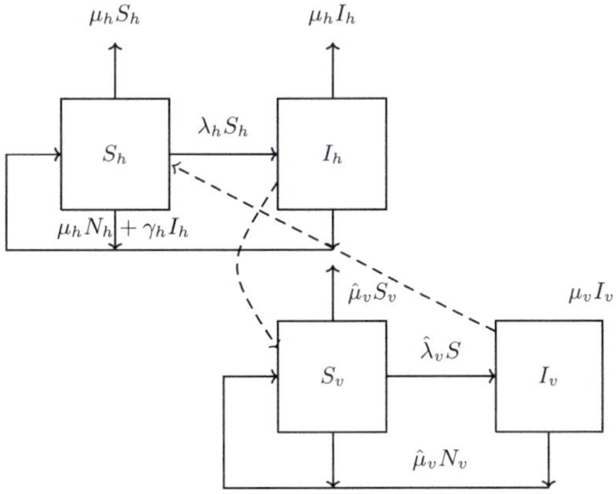

Fig. 6.22 The $S_h I_h S_h S_v I_v$ model

all solutions starting in the first quadrant (with strictly positive initial data) will converge to $\left(\frac{a}{e}, 0\right)$ as $t \to \infty$. □

Exercise 6.2.13 Prove the analogous result in the case

$$\frac{c}{d} = \frac{a}{e}.$$

Exercise 6.2.14 (Ross-Macdonald Model) We consider a simplification of (6.1.10), see Fig. 6.2, where it is assumed that the infectives recover without immunity, but the vital dynamics of the host population are neglected, as presented in Fig. 6.22. The vector population evolves according to Malthusian law, with equal death and birth rates, so both populations are constant. Further, use the explicit form of the forces of infection corresponding, according to Table 6.1, to a small, relatively to humans, vector population, in which case, see (6.1.15),

$$\lambda_h = b_h(N_h, N_v)\beta_{hv}\frac{I_v}{N_v} = \sigma_v\beta_{hv}\frac{N_v}{N_h}\frac{I_v}{N_v} =: \beta_h\frac{I_v}{N_h},$$

$$\lambda_v = b_v(N_h, N_v)\beta_{vh}\frac{I_h}{N_h} = \sigma_v\beta_{vh}\frac{I_h}{N_h} =: \beta_v\frac{I_h}{N_h},$$

where σ_v, σ_v are the mosquitoes' biting rates. These lead to

$$\begin{aligned} S_h' &= \gamma_h I_h - \frac{\beta_h}{N_h} S_h I_v, \\ I_h' &= \frac{\beta_h}{N_h} S_h I_v - \gamma_h I_h, \\ S_v' &= \mu_v (N_v - S_v) - \frac{\beta_v}{N_h} S_v I_h, \\ I_v' &= \frac{\beta_v}{N_h} S_v I_h - \mu_v I_v, \end{aligned} \tag{6.2.58}$$

with $(S_h(0), I_h(0), S_v(0), I_v(0)) = (\mathring{I}_h, \mathring{S}_h, \mathring{S}_v, \mathring{I}_v)$. Since the total populations are constant, it is reasonable to transform (6.2.58) to

$$\begin{aligned} I_h' &= \frac{\beta_h}{N_h} I_v (N_h - I_h) - \gamma_h I_h, \qquad & I_h(0) = \mathring{I}_h, \\ I_v' &= \frac{\beta_v}{N_h} I_h (N_v - I_v) - \mu_v I_v, \qquad & I_v(0) = \mathring{I}_v, \end{aligned} \tag{6.2.59}$$

We normalize (6.2.59) by introducing

$$x = \frac{I_h}{N_h}, \quad y = \frac{I_v}{N_v},$$

and $m = \frac{N_v}{N_h}$, which yields

$$\begin{aligned} x' &= m\beta_h y(1 - x) - \gamma_h x, \qquad & x(0) = \mathring{x}, \\ y' &= \beta_v x(1 - y) - \mu_v y, \qquad & y(0) = \mathring{y}. \end{aligned} \tag{6.2.60}$$

- Show that the basic reproduction number R_0 is given by

$$R_0 = \frac{m\beta_h \beta_v}{\gamma_h \mu_v}. \tag{6.2.61}$$

- Show that the set $[0, 1] \times [0, 1]$ is invariant under the flow of (6.2.60).
- Draw the isoclines of (6.2.60) for various values of the parameters and determine the ranges of the parameters for which an endemic equilibrium exists.
- By analyzing the vector field, show that if the endemic equilibrium exists, it is globally asymptotically stable in $[0, 1] \times [0, 1]$.

The Lotka–Volterra Model

Next, we discuss the Lotka–Volterra model, briefly discussed in Sect. 6.1.1. Let us recall that it is given by

$$\begin{aligned} x' &= (r-f)x - \alpha xy, \\ y' &= -(s+f)y + \beta xy, \end{aligned} \tag{6.2.62}$$

where α, β, r, s, f are positive constants. If the rate of harvesting f is zero, it is again a predator–prey system with the interactions between the populations modelled by the mass action law but with a Malthusian growth (resp. decay) in the prey x (resp. predator y) populations. All parameters are positive, and r is the growth rate of the prey in the absence of predators, s is the death rate of predators in the absence of prey (the population of specialist predators dies out without supply of the sole food source—prey), whereas αx is the number of prey caught by one predator in unit time and β is the biomass conversion rate measuring the benefit to the growth rate of encountering one prey.

Let us first consider the model without fishing,

$$\begin{aligned} x' &= rx - \alpha xy, \\ y' &= -sy + \beta xy. \end{aligned} \tag{6.2.63}$$

Observe that there are two equilibria

$$(x_1^*, y_1^*) = (0,0) \quad \text{and} \quad (x_2^*, y_2^*) = \left(\frac{s}{\beta}, \frac{r}{\alpha}\right). \tag{6.2.64}$$

The first solution is not interesting as it corresponds to the total extinction. We also observe that we have two other solutions,

$$(x(t), y(t)) = (c_1 e^{rt}, 0) \quad \text{and} \quad (x(t), y(t)) = (0, c_2 e^{-st}),$$

which correspond to the situation when one of the species is extinct. Thus, both positive x and y semi-axes are orbits and, by Theorem 6.27 (i), any orbit starting in the first quadrant will stay there or, in other words, any solution with positive initial data will remain strictly positive at all times.

As in (6.2.38), the orbits of (6.2.63) are the solution curves of the first order separable equation

$$\frac{dy}{dx} = \frac{y(-s+\beta x)}{x(r-\alpha y)}, \quad x \neq 0,\ y \neq \frac{r}{\alpha}, \tag{6.2.65}$$

as long as the numerator and denominator are not zero simultaneously, that is, the trajectory is not an equilibrium point (6.2.64). Separating variables and integrating we obtain

$$r \ln y - \alpha y + s \ln x - \beta x = k,$$

which can be written as

$$\frac{y^r}{e^{\alpha y}} \frac{x^s}{e^{\beta x}} = K, \tag{6.2.66}$$

for some constants k and K. Next, prove that the curves defined by (6.2.66) are closed. For this, we shall show that for each x from a certain open interval (x_m, x_M) we have exactly two solutions $y_m(x)$ and $y_M(x)$ and that these two solutions tend to the same limits as x approaches x_m and x_M.

First, let us define $f(y) = y^r e^{-\alpha y}$ and $g(x) = x^s e^{-\beta x}$. Due to the positivity of all coefficients, we see that $f(0) = 0$, $\lim_{y\to\infty} f(y) = 0$ and $f(y) > 0$ for $y > 0$. Further,

$$f'(y) = y^{r-1} e^{-\alpha y}(r - \alpha y),$$

so that f is increasing from 0 to $y = \frac{r}{\alpha}$, where it attains global maximum, say M_2, and then starts to decrease monotonically to 0. Similarly, $g(0) = \lim_{x\to\infty} g(x) = 0$, $g(x) > 0$ for $x > 0$ and it attains global maximum, say M_1, at $x = \frac{s}{b}$. We have to analyze the solvability of

$$f(y)g(x) = K.$$

Firstly, there are no solutions if $K > M_1 M_2$, and for $K = M_1 M_2$ we have the equilibrium solution $x_2^* = \frac{s}{\beta}$, $y_2^* = \frac{r}{\alpha}$. Thus, we consider $K = \lambda M_2$ with $\lambda < M_1$. Let us write this equation as

$$f(y) = \frac{\lambda}{g(x)} M_2. \tag{6.2.67}$$

From the shape of the graph g, we find that the equation $g(x) = \lambda$ has no solution if $\lambda > M_1$ but then $\frac{\lambda}{g(x)} \geq \frac{\lambda}{M_1} > 1$ so that (6.2.67) is not solvable. If $\lambda = M_1$, then we have again the equilibrium solution. Finally, for $\lambda < M_1$ there are two solutions x_m and x_M satisfying $x_m < \frac{s}{\beta} < x_M$. Now, for x satisfying $x_m < x < x_M$ we have $\frac{\lambda}{g(x)} < 1$ and therefore for such x, Eq. (6.2.67) has two solutions $y_m(x)$ and $y_M(x)$ satisfying $y_m < \frac{r}{\alpha} < y_M$, again based on the shape of the graph of f. Moreover, if x moves towards either x_m or x_M, then both solutions y_m and y_M move towards $\frac{r}{\alpha}$, that is the set of points satisfying (6.2.67) is a closed curve. In summary, the orbits are closed curves encircling the equilibrium solution $\left(\frac{s}{\beta}, \frac{r}{\alpha}\right)$ and are traversed in the anticlockwise direction. Thus, the solutions are periodic in time. The evolution can be described as follows. Suppose we start with initial values $x > \frac{s}{\beta}$, $y < \frac{r}{\alpha}$, that is, in the lower right quarter of the orbit. Then, the solution will move right and

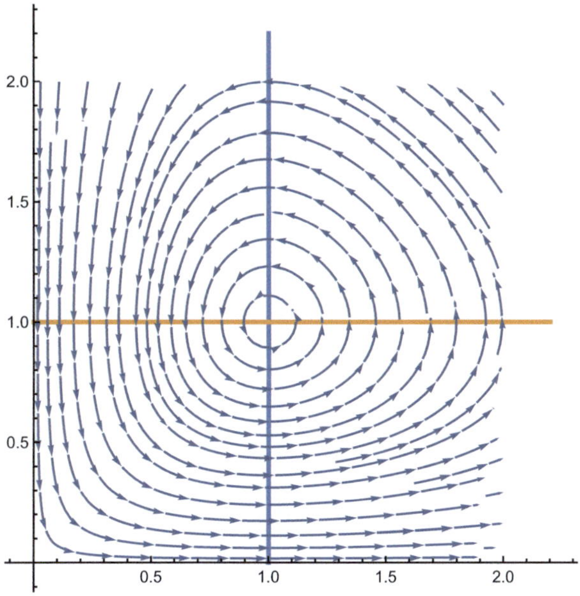

Fig. 6.23 The vector field and the nontrivial nullclines of (6.2.63) with $r = \alpha = s = \beta = 1$; the trajectories are closed curves around the equilibrium point $(1, 1)$, so the solutions are periodic

up until the prey population reaches maximum x_M. Because there is a lot of prey, the number of predators will still be growing, but then the number of prey will start decreasing, slowing down the predator population growth. The decrease in the prey population will eventually bring the growth of the predator's population to a stop at the maximum y_M. From now on, the number of predators will decrease. Still, the depletion of the prey population from the previous period will continue to prevail till the population reaches the minimum x_m, when it will start to take advantage of the decreasing number of predators and will start to grow; this growth will, however, slow down when the population of predators will reach its minimum. However, the number of prey will increase beyond the point when the number of predators is the smallest till the growing number of predators will, eventually, cause the prey population to start decreasing after having reached its peak at x_M and the cycle will repeat itself, as illustrated in Fig. 6.23.

Now, we are ready to explain the observational data that showed, recall, that in a period of reduced indiscriminate fishing during World War I, the number of sharks substantially increased.

Considering the full model (6.2.62), we see that the vital parameters r and s change to $r - f$ and $s + f$ but the structure of the system remains the same so that the nontrivial equilibrium becomes

$$\left(\frac{s+f}{\beta}, \frac{r-f}{\alpha}\right). \tag{6.2.68}$$

Thus, with a moderate amount of fishing ($f < r$), at the equilibrium, there are more fish x and fewer sharks y in comparison with the no-fishing situation. Thus, if we reduce fishing, the equilibrium moves towards a larger number of sharks and a lower number of fish. Of course, this is true for the equilibrium situation, which does not necessarily correspond to reality. Still, as the orbits are closed curves around the equilibrium, we can expect that the amounts of fish and sharks in a non-equilibrium situation will change similarly. We can confirm this hypothesis by comparing the average numbers of sharks and fish over the full cycle. For any function f, its average over an interval (a, b) is defined as

$$\bar{f} = \frac{1}{b-a}\int_a^b f(t)dt,$$

so that the average numbers of fish and sharks over one cycle are given by

$$\overline{x} = \frac{1}{T}\int_0^T x(t)dt, \qquad \overline{y} = \frac{1}{T}\int_0^T y(t)dt.$$

It turns out that these averages can be calculated explicitly. Dividing the first equation of (6.2.63) by x gives $\frac{x'}{x} = r - \alpha y$. Integrating both sides, we get

$$\frac{1}{T}\int_0^T \frac{x'(t)}{x(t)}dt = \frac{1}{T}\int_0^T (r - \alpha y(t))dt.$$

The left-hand side can be evaluated as

$$\int_0^T \frac{x'(t)}{x(t)}dt = \ln x(T) - \ln x(0) = 0,$$

on account of the periodicity of x. Hence,

$$\frac{1}{T}\int_0^T (r - \alpha y(t))dt = 0,$$

and

$$\overline{y} = \frac{r}{\alpha}. \tag{6.2.69}$$

In the same way,

$$\overline{x} = \frac{s}{\beta}, \tag{6.2.70}$$

so that the average values of x and y are exactly the equilibrium values. Thus, we can state that introducing fishing is more beneficial to prey than predators, as the average numbers of prey increases while and the average number of predators will decrease in accordance with (6.2.68), while reducing fishing will have the opposite effect of increasing the number of predators and decreasing the number of prey.

Exercise 6.2.15 Consider the orchard introduced in Exercise 6.1.1, which is infested by a population of aphids that are preyed upon by a type of beetle. Is it wise to try to get rid of aphids by using pesticide that kills indiscriminately a certain percentage of both aphids and beetles?

Exercise 6.2.16 Suppose there are two different species of prey, x_1 and x_2, which are the food of a single predator x_3.

(i) Derive the system for the changes in the populations of the species by using modelling assumptions analogous to those used to derive the Lotka–Volterra equation.
(ii) Find all the equilibrium solutions of this system and thus show that in any equilibrium state, at least one species of prey must be extinct unless the coefficients satisfy certain condition. What is this condition?

Exercise 6.2.17 Consider the Lotka–Volterra system modified so that the fish population in the absence of sharks grows logistically (see Exercise 6.1.2). How would fishing affect the populations in this case?

6.3 Advanced Stability Analysis

According to Theorem 4.3, dynamical systems generated by one-dimensional versions of (6.2.17) have a very simple long-term behaviour. Let us recall the possible scenarios. Assume that x^* is a single equilibrium point of f with $f(x) < 0$ for $x < x^*$ and $f(x) > 0$ for $x > x^*$. If the initial condition satisfies $\mathring{x} < x^*$, then $\phi(t, \mathring{x})$ decreases so it diverges either to $-\infty$ or to an equilibrium point. Since no equilibrium point is smaller than $\mathring{x}$, the solution must diverge to $-\infty$. Similarly, for $\mathring{x} > x^*$ we see that $\phi(t, \mathring{x})$ must diverge to infinity. Hence, x^* is a source (repeller).

Conversely, assuming that x^* is a single stationary point of f with $f(x) > 0$ for $x < x^*$ and $f(x) < 0$ for $x > x^*$, we see that if $\mathring{x} < x^*$, then $\phi(t, \mathring{x})$ increases so, being bounded, it must converge to x^*. Similarly, for $\mathring{x} > x^*$, $\phi(t, \mathring{x})$ must decrease, converging again to x^*. Hence, x^* is attracting. If there are more equilibrium points, then the behaviour of the solution is a combination of the above scenarios. Assume, for example, that f has two zeros $x_1 < x_2$, and it is positive for $x < x_1$, negative for $x_1 < x < x_2$ and again positive for $x > x_2$. Thus, for $\mathring{x} < x_1$, $\phi(t, \mathring{x})$ increases converging to x_1, for $x_1 < \mathring{x} < x_2$ we have $\phi(t, \mathring{x})$ decreasing and converging to x_1 and, finally, for $\mathring{x} > x_2$, $\phi(t, \mathring{x})$ increases to infinity.

6.3.1 *Stability by Linearization*

The question of stability of linear systems (6.2.24) is completely settled by the spectral decomposition; see Example 6.11. To recall, by considering the formulae (5.3.86) for solutions, we ascertain that the equilibrium point at the origin is (asymptotically stable) if all the eigenvalues have negative real parts and it is unstable if at least one eigenvalue has positive real part. The case of eigenvalues with zero real parts is more complicated as in higher dimensions, we can have multiple complex eigenvalues. Here, again, from (5.3.86), we can see that if for each eigenvalue with zero real part which has algebraic multiplicity k, there are k linearly independent eigenvectors, then the solution is stable. However, if the geometric and algebraic multiplicities of at least one such eigenvalue are different, then, in the solution corresponding to this eigenvalue, a polynomial in t will appear, which will cause the solution to be unstable.

Next, we shall use these results for the nonlinear autonomous system in $\mathbb{R}^d$

$$\boldsymbol{x}' = \boldsymbol{f}(\boldsymbol{x}), \tag{6.3.71}$$

with a differentiable right-hand side. Let us first note the following result.

Lemma 6.33 *If $\boldsymbol{f}$ is differentiable in some neighbourhood of $\boldsymbol{y}^0$, then*

$$\boldsymbol{f}(\boldsymbol{x} + \boldsymbol{y}^0) = \boldsymbol{f}(\boldsymbol{y}^0) + \mathcal{A}\boldsymbol{x} + \boldsymbol{g}(\boldsymbol{x}), \tag{6.3.72}$$

where

$$\mathcal{A} = \begin{pmatrix} \frac{\partial f_1}{\partial x_1}(\boldsymbol{y}^0) & \cdots & \frac{\partial f_1}{\partial x_d}(\boldsymbol{y}^0) \\ \vdots & \ddots & \vdots \\ \frac{\partial f_d}{\partial x_1}(\boldsymbol{y}^0) & \cdots & \frac{\partial f_d}{\partial x_d}(\boldsymbol{y}^0) \end{pmatrix}, \tag{6.3.73}$$

and $\boldsymbol{g}(\boldsymbol{x})/\|\boldsymbol{x}\|$ is continuous in some neighbourhood of $\boldsymbol{y}^0$ and vanishes at $\boldsymbol{x} = \boldsymbol{y}^0$.

Proof The matrix $\mathcal{A}$ has constant entries so that $\boldsymbol{g}$ defined by

$$\boldsymbol{g}(\boldsymbol{x}) = \boldsymbol{f}(\boldsymbol{x} + \boldsymbol{y}^0) - \boldsymbol{f}(\boldsymbol{y}^0) - \mathcal{A}\boldsymbol{x}$$

is a continuous function of $\boldsymbol{x}$. Hence, $\boldsymbol{g}(\boldsymbol{x})/\|\boldsymbol{x}\|$ is also continuous for $\boldsymbol{x} \neq \boldsymbol{0}$. Using now Taylor's formula for each component of $\boldsymbol{f}$ we obtain

$$f_i(\boldsymbol{x} + \boldsymbol{y}^0) = f_i(\boldsymbol{y}^0) + \frac{\partial f_i}{\partial x_1}(\boldsymbol{y}^0)x_1 + \ldots + \frac{\partial f_i}{\partial x_d}(\boldsymbol{y}^0)x_d + R_i(\boldsymbol{x}), \quad i = 1, \ldots, n,$$

where, for each i, the remainder R_i satisfies, [6, Remark 5.9],

$$|R_i(x)| \leq M(\|\boldsymbol{x}\|)\|\boldsymbol{x}\|$$

and M tends to zero is $\|\boldsymbol{x}\| \to 0$. Thus,

$$\boldsymbol{g}(\boldsymbol{x}) = (R_1(\boldsymbol{x}), \ldots, R_d(\boldsymbol{x}))$$

and

$$\frac{\|\boldsymbol{g}(\boldsymbol{x})\|}{\|\boldsymbol{x}\|} \leq M(\|\boldsymbol{x}\|) \to 0$$

as $\|\boldsymbol{x}\| \to 0$ and, $\boldsymbol{f}(\boldsymbol{y}^0) = 0$, the lemma is proved. □

If $\boldsymbol{y}^0 = \boldsymbol{x}^*$ is an equilibrium of (6.3.71), then $\boldsymbol{f}(\boldsymbol{y}^0) = \boldsymbol{f}(\boldsymbol{x}^*) = 0$ and the linear system

$$\boldsymbol{x}' = \mathcal{A}\boldsymbol{x}$$

is called the linearization of (6.4.153) around the equilibrium $\boldsymbol{x}^*$. The matrix $\mathcal{A}$, defined in (6.3.73), is called the Jacobi matrix and is usually denoted by $\mathcal{J}$; see Sect. B.1.

Exercise 6.3.1 Write the system

$$x_1' = x_1 + 6x_2 + x_1x_2,$$
$$x_2' = 4x_1 + 3x_2 - x_1^2,$$

in the form

$$\boldsymbol{x}' = \mathcal{B}\boldsymbol{x} + \boldsymbol{g}(\boldsymbol{x})$$

where

$$\mathcal{B}\boldsymbol{x} = \begin{pmatrix} -\mu & 0 \\ 0 & \lambda \end{pmatrix}$$

$\mu, \lambda > 0$ and $\boldsymbol{g}$ is a quadratic function of two variables. As in the linear case, the equilibrium (0, 0) is also called a saddle.

Theorem 6.34 *Suppose that $\boldsymbol{f}$ is a differentiable function in some neighbourhood of an isolated equilibrium point $\boldsymbol{x}^*$. Then,*

1. *the equilibrium point $\boldsymbol{x}^*$ is asymptotically stable if all the eigenvalues of the matrix $\mathcal{A}$ have negative real parts, that is, if the equilibrium solution $\boldsymbol{x}(t) = 0$ of the linearized system is locally asymptotically stable;*
2. *the equilibrium point $\boldsymbol{x}^*$ is unstable if at least one eigenvalue has a positive real part;*
3. *if all eigenvalues of $\mathcal{A}$ have non-negative real parts, but at least one of them has real part equal to 0, then the stability of the equilibrium point $\boldsymbol{x}^*$ of the nonlinear system (6.4.153) cannot be determined from the stability of its linearization.*

Proof To prove 1., we use the variation of constants formula (5.3.64) applied to (6.4.153) written in the form of Lemma 6.33 for $\boldsymbol{y}(t) = \boldsymbol{x}(t) + \boldsymbol{x}^*$:

$$\boldsymbol{x}' = \boldsymbol{y}' = \boldsymbol{f}(\boldsymbol{y}) = \boldsymbol{f}(\boldsymbol{x} + \boldsymbol{x}^*) = \mathcal{A}\boldsymbol{x} + \boldsymbol{g}(\boldsymbol{x}). \tag{6.3.74}$$

Thus,

$$\boldsymbol{x}(t) = e^{t\mathcal{A}}\mathring{\boldsymbol{x}} + \int\limits_0^t e^{(t-s)\mathcal{A}}\boldsymbol{g}(\boldsymbol{x}(s))ds.$$

Since, by assumption, $s(\mathcal{A}) < 0$, Remark 5.19 implies that for any $\alpha \in (0, -s(\mathcal{A}))$ there is $K \geq 1$ such that

$$\|e^{t\mathcal{A}}\mathring{\boldsymbol{x}}\| \leq Ke^{-\alpha t}\|\mathring{\boldsymbol{x}}\|, \quad t \geq 0.$$

From the assumption on $\boldsymbol{g}$, for any $\epsilon > 0$ we find $\delta > 0$ such that if $\|\boldsymbol{x}\| \leq \delta$, then

$$\|\boldsymbol{g}(\boldsymbol{x})\| \leq \epsilon\|\boldsymbol{x}\|. \tag{6.3.75}$$

Assuming for a moment that for $0 \leq s \leq t$ we can keep $\|\boldsymbol{x}(s)\| \leq \delta$, we can write

$$\|\boldsymbol{x}(t)\| \leq \|e^{\mathcal{A}t}\mathring{\boldsymbol{x}}\| + \int\limits_0^t \|e^{\mathcal{A}(t-s)}\boldsymbol{g}(\boldsymbol{x}(s))\|ds \leq Ke^{-\alpha t}\mathring{\boldsymbol{x}} + K\epsilon\int\limits_0^t e^{-\alpha(t-s)}\|\boldsymbol{x}(s)\|ds$$

or, multiplying both sides by $e^{\alpha t}$ and setting $z(t) = e^{\alpha t}\|\boldsymbol{x}(t)\|$,

$$z(t) \leq K\|\mathring{\boldsymbol{x}}\| + K\epsilon\int\limits_0^t z(s)ds. \tag{6.3.76}$$

Using Gronwall's lemma, we obtain thus

$$\|\boldsymbol{x}(t)\| = e^{-\alpha t} z(t) \leq K\|\mathring{\boldsymbol{x}}\| e^{(K\epsilon-\alpha)t},$$

providing $\|\boldsymbol{x}(s)\| \leq \delta$ for all $0 \leq s \leq t$. Let us take $\epsilon \leq \alpha/2K$, then the above can be written as

$$\|\boldsymbol{x}(t)\| \leq K\|\mathring{\boldsymbol{x}}\| e^{-\frac{\alpha t}{2}}. \tag{6.3.77}$$

Assume now that $\|\boldsymbol{x}(0)\| < \delta/K \leq \delta$, where δ was fixed for $\epsilon \leq \alpha/2K$. Then $\|\mathring{\boldsymbol{x}}\| < \delta$ and, by continuity, $\|\boldsymbol{x}(t)\| \leq \delta$ for some time. Let $\boldsymbol{x}(t)$ be defined on some interval I and $t_1 \in I$ be the first time for which $\|\boldsymbol{x}(t)\| = \delta$. Then, for $t \leq t_1$ we have $\|\boldsymbol{x}(t)\| \leq \delta$ so that for all $t \leq t_1$ we can use (6.3.77) getting, in particular,

$$\|\boldsymbol{x}(t_1)\| \leq \delta e^{-\frac{\alpha t_1}{2}} < \delta,$$

which is a contradiction. Thus $\|\boldsymbol{x}(t)\| < \delta$ if $\|\mathring{\boldsymbol{x}}\| < \delta_1$ in the whole interval of existence but then, if the interval was finite, we could extend the solution to a larger interval as the solution is bounded at the endpoint and the same procedure would ensure that the solution remains bounded by δ on the larger interval. Thus, the extension can be carried out for all the values of $t \geq 0$ and the solution exists for all t and satisfies $\|\boldsymbol{x}(t)\| \leq \delta$ for all $t \geq 0$. Consequently, (6.3.77) holds for all t and the solution $\boldsymbol{x}(t)$ converges exponentially to 0 as $t \to \infty$, proving the asymptotic stability of the stationary solution $\boldsymbol{x}^*$.

We shall prove 2. only in the two-dimensional case to avoid some algebraic technicalities; see [54, Chapter 9 §2]. Consider the solution $\boldsymbol{x}(t) = \phi(t, \mathring{\boldsymbol{x}})$ of (6.3.71) originating at $\boldsymbol{x}(0) = \mathring{\boldsymbol{x}}$. Let us recall that to prove that $\boldsymbol{x}^* = 0$ is unstable, it is enough to show that there is $\epsilon >$ such that for any $\delta > 0$, there is $\mathring{\boldsymbol{x}}$ and $t_0 < \infty$ such that $\|\mathring{\boldsymbol{x}}\| \leq \delta$ and $\|\phi(t_0, \mathring{\boldsymbol{x}})\| \geq \epsilon$.

Thus, let λ_1 and λ_2 be the eigenvalues of $\mathcal{A}$. There are two possible cases.

(i) Both $\Re\lambda_1, \Re\lambda_2 > 0$. If we replace t by $\tau = -t$, then (6.3.74) takes the form

$$\frac{dz}{d\tau} = -\mathcal{A}z - \boldsymbol{g}(z). \tag{6.3.78}$$

Since the eigenvalues of $-\mathcal{A}$ have negative real parts, the equilibrium $z = 0$ is locally asymptotically stable. This means, in particular, that there is $\epsilon > 0$ such that if $\|\mathring{z}\| = \epsilon$, then the solution $z(\tau) = \phi(\tau, \mathring{z})$ approaches $z = 0$ as $\tau \to \infty$.

So, let us fix arbitrary $\delta > 0$ and $\mathring{z}$ with $\|\mathring{z}\| = \epsilon$. Then there is $\hat{\tau}$ such that $\|\phi(\hat{\tau}, \mathring{z})\| = \delta$. Consider now (6.3.74) with $\mathring{\boldsymbol{x}} = \phi(\hat{\tau}, \mathring{z})$. Then, taking $\hat{t} = -\hat{\tau}$,

$$\boldsymbol{x}(\hat{t}) = \phi(\hat{t}, \phi(\hat{\tau}, \mathring{z})) = \phi(-\hat{\tau}, \phi(\hat{\tau}, \mathring{z})) = \mathring{z}.$$

Thus, for any $\delta > 0$, we can find an initial condition $\mathring{\boldsymbol{x}}$ such that the solution $\boldsymbol{x}(t)$ emanating from this initial condition will be at a fixed distance ϵ from the origin. Hence, the equilibrium $\boldsymbol{x}^* = 0$ is unstable.

(ii) Next, assume that only one eigenvalue has a positive real part. This implies that the eigenvalues are real; thus, the second must be non-positive. In other words, we have $\lambda_1 \leq 0 < \lambda_2$. Then, by a linear change of coordinates, (6.3.74) can be written in the form

$$\begin{aligned} y_1' &= \lambda_1 y_1 + g_1(y_1, y_2), \\ y_2' &= \lambda_2 y_2 + g_2(y_1, y_2), \end{aligned} \tag{6.3.79}$$

where the function $\boldsymbol{g} = (g_1, g_2)$ has the required property (6.3.75). We observe that, in principle, the solution can contract along the y_1 axis, so we have to somehow enhance the expansion along the y_2-axis. We use an argument similar to the classical Lyapunov approach discussed later. Consider the function

$$V(y_1, y_2) = \frac{1}{2}(y_2^2 - y_1^2). \tag{6.3.80}$$

This is a hyperboloid with the level curves

$$\frac{1}{2}(y_2^2 - y_1^2) = c, \quad c \geq 0, \tag{6.3.81}$$

being hyperbolas in the sector $|y_2| \geq |y_1|$ (with two intersecting lines $y_2 = \pm y_1$ for $c = 0$). Notice that the larger the value of c in (6.3.81), the higher the level curve. Let us consider how V behaves along the solutions to (6.3.74). From (6.3.75) we see that for any $m > 0$ there is ϵ such that

$$\|\boldsymbol{g}(\boldsymbol{y})\| \leq m\|\boldsymbol{y}\|, \tag{6.3.82}$$

provided $\|\boldsymbol{y}(t)\| < \epsilon$. Let us assume that $\|\boldsymbol{y}\| < \epsilon$. Then, using the Cauchy-Schwarz inequality,

$$\begin{aligned} -\|\boldsymbol{y}\|\|\boldsymbol{g}\| &\leq -\sqrt{y_1^2 + y_2^2}\sqrt{g_1^2 + g_2^2} \leq \langle (y_1, y_2), (-g_1, g_2)\rangle \\ &= y_2 g_2 - y_1 g_1 \leq \sqrt{y_1^2 + y_2^2}\sqrt{g_1^2 + g_2^2} = \|\boldsymbol{y}\|\|\boldsymbol{g}\|, \end{aligned}$$

we have

$$\begin{aligned} V'(\boldsymbol{y}(t)) &= y_2 y_2' - y_1 y_1' = \lambda_2 y_2^2 + y_2 g_2(y_1, y_2) - \lambda_1 y_1^2 - y_1 g_1(y_1, y_2) \\ &\geq \lambda_2 y_2^2 - \|\boldsymbol{g}\|\|\boldsymbol{y}\| - \lambda_1 y_1^2 \\ &\geq \lambda_2 y_2^2 - m\|\boldsymbol{y}\|^2 - \lambda_1 y_1^2 = (\lambda_2 - m)y_2^2 - (\lambda_1 + m)y_1^2 \geq (\lambda_2 - 2m)y_2^2, \end{aligned}$$

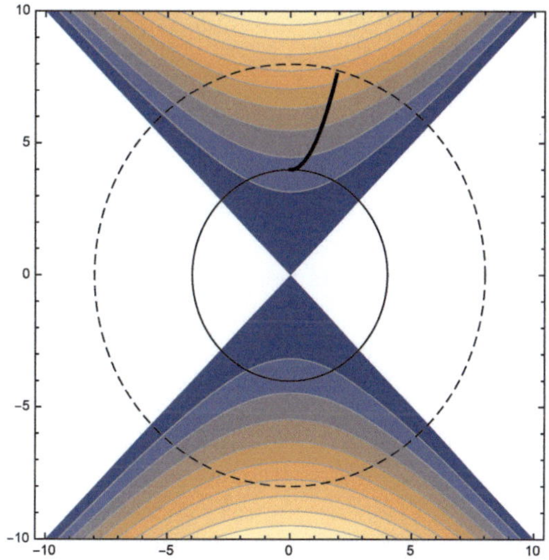

Fig. 6.24 Level sets of V, the circles $\|\boldsymbol{y}\| = \delta$ and $\|\boldsymbol{y}\| = \epsilon$, and the solution $\boldsymbol{y}(t)$

as $-\lambda_1 \geq 0$. Consider now the set

$$\Omega = \{(y_1, y_2);\;\; y_2 > |y_1|\},$$

that is, the wedge above lines $y_2 = \pm y_1$; see Fig. 6.24. Now, chose $m < \lambda_2/2$ and corresponding ϵ so that (6.3.82) is satisfied. Further, consider the ball $B_\epsilon = \{\boldsymbol{y} :\; \|y\| < \epsilon\}$. Then in $\Omega \cap B_\epsilon$ we have $V(\boldsymbol{y}) > 0$ and

$$V'(\boldsymbol{y}(t)) \geq (\lambda_2 - 2m)y_2^2 > 0.$$

Let us consider arbitrary $0 < \delta < \epsilon$ and $\boldsymbol{y}_0 \in \Omega$ with $\|\mathring{\boldsymbol{y}}\| = \delta$. Then $y_{0,2} > \delta/\sqrt{2}$. Further, $\mathring{\boldsymbol{y}} \in V^{-1}(c_0)$ for some $c_0 > 0$. Since $V(\boldsymbol{y}(t))$ is strictly increasing as long as $\boldsymbol{y}(t) \in \Omega \cap B_\epsilon$, we see that with increasing t the solution $\boldsymbol{y}(t)$ will move upward, through the level curves $V^{-1}(c)$ with an increasing c. This shows, in particular, that $\boldsymbol{y}(t)$ cannot reach $y_2 = \pm y_1$ from within $\Omega \cap B_\epsilon$. Hence $y_2(t) > \delta/\sqrt{2}$, implying $dV(\boldsymbol{y}(t))/dt \geq (\lambda_1 - m)\delta/\sqrt{2} > 0$. In other words, $c(t) = V(\boldsymbol{y}(t))$ is a strictly increasing function such that $\boldsymbol{y}(t) \in V^{-1}(c(t))$ for $t \in \boldsymbol{y}^{-1}(B_\epsilon)$. If we assume that $\|\boldsymbol{y}(t)\| < \epsilon$ for all t, then $c(t)$ diverges to infinity as $t \to \infty$ since $c'(t)$ is bounded away from zero by a constant. But then $\|\boldsymbol{y}(t)\| \to \infty$ since $\|\boldsymbol{y}(t)\| \geq \inf \|V^{-1}(c(t))\| = \sqrt{2c(t)}$. We arrived at a contradiction, and thus $\|\boldsymbol{y}(t)\| = \epsilon$ in finite time.

To prove 3., it is enough to exhibit two systems with the same linear part and different behaviour of solutions. Let us consider

$$\begin{aligned} x_1' &= x_2 \quad x_1(x_1^2 + x_2^2), \\ x_2' &= -x_1 - x_2(x_1^2 + x_2^2), \end{aligned}$$

with the linearized system given by

$$x_1' = x_2,$$
$$x_2' = -x_1.$$

The eigenvalues of the linearized system are $\pm i$. To analyze the behaviour of the solutions to the non-linear system, let us multiply the first equation by x_1 and the second by x_2 and add them together to get

$$\frac{1}{2}(x_1^2 + x_2^2)' = -(x_1^2 + x_2^2)^2.$$

Solving this equation, we obtain

$$x_1^2 + x_2^2 = \frac{c}{1 + 2ct},$$

where $c = x_1^2(0) + x_2^2(0)$. Thus $x_1^2(t) + x_2^2(t)$ approaches 0 as $t \to \infty$ and $x_1^2(t) + x_2^2(t) < x_1^2(0) + x_2^2(0)$ for any $t > 0$, and we can conclude that the equilibrium point 0 is asymptotically stable.

Consider now the system

$$x_1' = x_2 + x_1(x_1^2 + x_2^2),$$
$$x_2' = -x_1 + x_2(x_1^2 + x_2^2),$$

with the same linear part and thus with the same eigenvalues. As above, we obtain that

$$x_1^2 + x_2^2 = \frac{c}{1 - 2ct},$$

with the same meaning for c. Thus, any solution with non-zero initial condition blows up at the time $t = 1/2c$, and therefore the equilibrium solution 0 is unstable. □

Exercise 6.3.2 Find all equilibria of the systems of differential equations

(a)

$$x_1' = 1 - x_1 x_2,$$
$$x_2' = x_1 - x_2^3;$$

(b)

$$x_1' = \sin(x_1 + x_2),$$
$$x_2' = e^{x_1} - 1;$$

(c)

$$x_1' = x_1^2 + x_2^2 - 1,$$
$$x_2' = x_1^2 - x_2^2;$$

(d)

$$x_1' = x_1^2 + x_2^2 - 1,$$
$$x_2' = 2x_1x_2;$$

(e)

$$x_1' = \tan(x_1 + x_2),$$
$$x_2' = x_1 + x_1^3,$$

and determine whether they are stable or unstable, if possible.

Exercise 6.3.3 Consider the system

$$\begin{aligned} x_1' &= x_2, \\ x_2' &= x_1 + 2x_1^3. \end{aligned} \tag{6.3.83}$$

(i) Show that the equilibrium solution $(0, 0)$ of the linearized system is a saddle and draw the phase portrait of the linearized system.
(ii) Find orbits of (6.3.83) and then draw its phase portrait.
(iii) Show that there are exactly two orbits of (6.3.83) on which $(x(t), y(t)) \to (0, 0)$ as $t \to \infty$. Similarly, show that there are exactly two orbits of (6.3.83) on which $(x(t), y(t)) \to (0, 0)$ as $t \to -\infty$.

Example 6.35 (Local Stability of SEIR Model) Let us consider the SEIR model

$$\begin{aligned} S' &= \Lambda - \beta SI - \mu S, \\ E' &= \beta SI - (\mu + \gamma)E, \\ I' &= \gamma E - (\mu + \nu)I, \\ R' &- \nu I - \mu R, \end{aligned} \tag{6.3.84}$$

which is a version of (6.1.6) with the simplified logistic demography model (3.3.32) and permanent immunity ($\gamma = 0$); it is an extension of (6.2.45) that includes the exposed class. As before, Λ is a constant recruitment rate, β is the transmission coefficient, μ is the constant death rate and ν is the recovery rate (that is, $1/\nu$ is the average infectious time of an individual); the new parameter γ is the rate at which the exposed individuals become infective.

Exercise 6.3.4 Show that (6.3.84) has global positive and bounded solutions in $\overline{\mathbb{R}}^4_+$.

Our main task is to determine the long-term behaviour of solutions. As usual, we begin with finding the equilibria. For this, we solve

$$\begin{aligned}0 &= \Lambda - \beta SI - \mu S,\\ 0 &= \beta SI - (\mu+\gamma)E,\\ 0 &= \gamma E - (\mu+\nu)I,\\ 0 &= \nu I - \mu R.\end{aligned} \tag{6.3.85}$$

Clearly, we have the disease-free equilibrium (DFE)

$$\boldsymbol{E}_0 = \left(\frac{\Lambda}{\mu}, 0, 0, 0\right). \tag{6.3.86}$$

Now, we can define the basic reproduction number R_0. The full susceptible population at the equilibrium is given by $\frac{\Lambda}{\mu}$, and one infective will infect $\frac{\beta\Lambda}{\mu}$ susceptibles in the unit time. Now, the probability that an infected person will become infective is $\frac{\gamma}{\mu+\gamma}$ (recall that the average length of stay in the exposed class is $\frac{1}{\mu+\gamma}$ and the fraction γ of the exposed population becomes infective in a unit time), and it stays infective for $\frac{1}{\mu+\nu}$ units of time. Thus

$$R_0 = \frac{\Lambda\beta\gamma}{(\mu+\gamma)(\mu+\nu)\mu}. \tag{6.3.87}$$

To find the endemic equilibria, we see that

$$E = \frac{\mu+\nu}{\gamma} I \tag{6.3.88}$$

and then, taking into account that in an endemic equilibrium $I \neq 0$,

$$S = \frac{(\mu+\gamma)(\mu+\nu)}{\gamma\beta}. \tag{6.3.89}$$

Then, from the first equation,

$$I = \frac{\Lambda}{\beta S} - \frac{\mu}{\beta} = \frac{\mu}{\beta}(R_0 - 1). \tag{6.3.90}$$

Re-writing the equilibrium in terms of R_0, we see that if $R_0 > 1$, then (6.3.84) has a unique endemic equilibrium

$$\begin{aligned} \boldsymbol{E}_1 &= (S^*, E^*, I^*, R^*) \\ &= \left(\frac{(\mu+\gamma)(\mu+\nu)}{\gamma\beta}, \frac{\mu+\nu}{\gamma}\frac{\mu}{\beta}(R_0 - 1), \frac{\mu}{\beta}(R_0 - 1), \frac{\nu}{\beta}(R_0 - 1) \right). \end{aligned} \tag{6.3.91}$$

To investigate local stability, we find that the Jacobi matrix of (6.3.84) is given by

$$\mathcal{J}(S, I) = \begin{pmatrix} -\beta I - \mu & 0 & -\beta S & 0 \\ \beta I & -(\mu+\gamma) & \beta S & 0 \\ 0 & \gamma & -(\mu+\nu) & 0 \\ 0 & 0 & \nu & -\mu \end{pmatrix}. \tag{6.3.92}$$

At the DFE, we have

$$\mathcal{J}(\boldsymbol{E}_0) = \begin{pmatrix} -\mu & 0 & -\beta\frac{\Lambda}{\mu} & 0 \\ 0 & -(\mu+\gamma) & \beta\frac{\Lambda}{\mu} & 0 \\ 0 & \gamma & -(\mu+\nu) & 0 \\ 0 & 0 & \nu & -\mu \end{pmatrix}. \tag{6.3.93}$$

Expanding the characteristic equation $det(\mathcal{J}(\boldsymbol{E}_0) - \lambda\mathcal{I}) = 0$, we have

$$\begin{aligned} &(\lambda+\mu)^2 \begin{vmatrix} -(\mu+\gamma+\lambda) & \beta\frac{\Lambda}{\mu} \\ \gamma - (\mu+\nu+\lambda) & \end{vmatrix} \\ &= (\lambda+\mu)^2 \left((\mu+\gamma+\lambda)(\mu+\nu+\lambda) - \frac{\gamma\beta\Lambda}{\mu} \right) = 0. \end{aligned}$$

The quadratic equation in brackets can be written as

$$\lambda^2 + (2\mu+\nu+\gamma)\lambda + (\mu+\gamma)(\mu+\nu)(1 - R_0) = 0.$$

Hence, if $R_0 < 1$, the equation has either two negative real roots or two complex conjugate roots with negative real parts. If $R_0 > 1$, then we have real roots of the opposite sign. Hence, if $R_0 < 1$, the disease-free equilibrium is locally asymptotically stable, and if $R_0 > 1$, then the disease-free equilibrium is unstable. The stability of the DFE when $R_0 = 1$ cannot be determined from the linearization.

Let us turn our attention to the endemic equilibrium $\boldsymbol{E}_1$. The characteristic equation is given by

$$det(\mathcal{J}(\boldsymbol{E}_1) - \lambda \mathcal{I})$$

$$= \begin{vmatrix} -\beta I^* - \mu - \lambda & 0 & -\beta S^* & 0 \\ \beta I^* & -(\mu + \gamma + \lambda) & \beta S^* & 0 \\ 0 & \gamma & -(\mu + \nu + \lambda) & 0 \\ 0 & 0 & \nu & -(\mu + \lambda) \end{vmatrix} = 0. \quad (6.3.94)$$

Adding the first row to the second and expanding by the last column yields

$$(\mu + \lambda) \begin{vmatrix} -\beta I^* - \mu - \lambda & 0 & -\beta S^* \\ -(\mu + \lambda) & -(\mu + \gamma + \lambda) & 0 \\ 0 & \gamma & -(\mu + \nu + \lambda) \end{vmatrix} = 0.$$

Hence, we obtain an eigenvalue $\lambda = -\mu$ and, from the determinant, using $\beta\gamma S^*(\mu + \lambda) = (\mu + \gamma)(\mu + \nu)(\mu + \lambda)$, the cubic equation

$$(\beta I^* + \mu + \lambda)(\mu + \gamma + \lambda)(\mu + \nu + \lambda) = (\mu + \gamma)(\mu + \nu)(\mu + \lambda).$$

The signs of the real parts of the eigenvalues can be determined from the Routh–Hurwitz criterion, but this can often be done more simply by some smart observation. By inspection, we see that $\lambda = -\mu$ is not a root of the above; hence, dividing by the right-hand side, we get

$$\left(1 + \frac{\beta I^*}{\mu + \lambda}\right)\left(1 + \frac{\lambda}{\mu + \gamma}\right)\left(1 + \frac{\lambda}{\mu + \nu}\right) = 1. \quad (6.3.95)$$

Now, if $z \in \mathbb{C}$, then $|1 + z| = \sqrt{(1 + \Re z)^2 + \Im z^2} \geq 1$ provided $\Re z \geq 0$ with strict equality if $\Re z > 0$. Now,

$$\Re \frac{\beta I^*}{\mu + \lambda} = \frac{\beta I^*(\mu + \Re\lambda)}{|\mu + \lambda|^2} > 0$$

if $\Re\lambda \geq 0$. Hence

$$\left|\left(1 + \frac{\beta I^*}{\mu + \lambda}\right)\left(1 + \frac{\lambda}{\mu + \gamma}\right)\left(1 + \frac{\lambda}{\mu + \nu}\right)\right| > 1$$

as long as $\Re\lambda \geq 0$. Thus (6.3.95) cannot have solutions with $\Re\lambda \geq 0$, and the endemic equilibrium is locally asymptotically stable if $R_0 > 1$.

Example 6.36 (Principle of Competitive Exclusion) So far, we have discussed predator-prey models, where one species has a detrimental effect on the other while the latter's presence benefits the former. Quite often, however, several species function in the same environmental niche, competing for the same resources, so the existence of each species has a negative impact on any other species. Based on experimental results, G. F. Gause, [47], formulated the principle of competitive exclusion, which states that species competing for the same resource cannot coexist, that is, only one species, best fit for the environment, survives, while the others will die out. While the universality of this principle is arguable, we shall provide a mathematical justification for it.

Let us consider the following model

$$\begin{aligned} x' &= x(a - b\Phi(x, y)), \\ y' &= y(c - d\Phi(x, y)), \end{aligned} \tag{6.3.96}$$

where we take

$$\Phi(x, y) = \alpha x + \beta y \tag{6.3.97}$$

to model the crowding effects. We assume all parameters, $a, b, c, d, \alpha, \beta$, are positive. The parameters a and c are the natural growth rates in the absence of the crowding effects, and b and d represent the sensitivities of the species to them.

Exercise 6.3.5 Show the existence of unique positive global (forward) in time solutions to (6.3.96).

The equilibria of (6.3.96) are given by solutions to

$$\begin{aligned} 0 &= x(a - b(\alpha x + \beta y)), \\ 0 &= y(c - d(\alpha x + \beta y)), \end{aligned} \tag{6.3.98}$$

giving

$$\boldsymbol{e}_0^* = (0, 0), \quad \boldsymbol{e}_1^* = \left(\frac{a}{b\alpha}, 0\right), \quad e_2^* = \left(0, \frac{c}{d\beta}\right),$$

or the line

$$y = \frac{a}{b\beta} - \frac{\alpha}{\beta}x,$$

if $ad - bc = 0$. The Jacobi matrix of (6.3.96) is given by

$$\mathcal{J} = \begin{pmatrix} a - 2b\alpha x - b\beta y & -b\beta x \\ -d\alpha y & c - d\alpha x - 2d\beta y \end{pmatrix},$$

which gives

$$\mathcal{J}(\boldsymbol{e}_0^*) = \begin{pmatrix} a & 0 \\ 0 & c \end{pmatrix}, \quad \mathcal{J}(\boldsymbol{e}_1^*) = \begin{pmatrix} -a & -\frac{\beta a}{\alpha} \\ 0 & c - \frac{ad}{b} \end{pmatrix}, \quad \mathcal{J}(\boldsymbol{e}_2^*) = \begin{pmatrix} a - \frac{bc}{d} & 0 \\ -\frac{\alpha c}{\beta} & -c \end{pmatrix}.$$

Since all three matrices are triangular, their eigenvalues are the diagonal elements, and immediately we obtain that $\boldsymbol{e}_0^*$ is unstable, $\boldsymbol{e}_1^*$ is locally asymptotically stable if $ad - cb > 0$, see Fig. 6.25 (right), and $\boldsymbol{e}_2^*$ is locally asymptotically stable if $ad - cb < 0$. The case $ad - cb = 0$ cannot be determined by Theorem 6.34 as one of the eigenvalues becomes 0. To examine the behaviour of solutions in this case, we follow [79, Rozdział, II.1]. From (6.3.96) we obtain

$$\begin{aligned} \frac{1}{b}\left(\frac{d}{dt}\ln x - a\right) - \frac{1}{d}\left(\frac{d}{dt}\ln y - c\right) &= \frac{1}{b}\left(\frac{x'}{x} - a\right) - \frac{1}{b}\left(\frac{y'}{y} - c\right) \\ &= \Phi(x, y) - \Phi(x, y) = 0, \end{aligned}$$

hence

$$\frac{d}{dt}\ln\frac{x^d}{y^b} = ad - bc,$$

that is,

$$\frac{x^d}{y^b} = Ce^{(ad-bc)t} \tag{6.3.99}$$

for some constant $C > 0$. This shows that if $ad - cb = 0$, then

$$x(t) = Cy^{\frac{b}{d}}(t) \tag{6.3.100}$$

with a positive C, which, for a given trajectory, satisfies

$$\frac{\mathring{x}^d}{\mathring{y}^b} = C.$$

Thus

$$y' = y(c - Cd\alpha y^{\frac{b}{d}} - d\beta y). \tag{6.3.101}$$

Since $1 + \frac{b}{d} > 1$, the right-hand of the equation is continuously differentiable and Theorem 4.3 is available. The equilibria of (6.3.101) are $y = 0$ and the solutions of

$$h(y) = c - Cd\alpha y^{\frac{b}{d}} - d\beta y = 0.$$

We see that $h(0) = c$, $\lim_{y\to\infty} h(y) = -\infty$ and, since $h'(y) < 0$ for $y > 0$, h is strictly decreasing on $\mathbb{R}_+$. Hence there is a unique y^*, dependent on $(\mathring{x}, \mathring{y})$, such that $h(y^*) = 0$. We see that

$$\frac{d}{dy}(yh(y))\bigg|_{y=0} = c > 0.$$

Thus, $y = 0$ is repelling, which is consistent with the result for the full system. On the other hand,

$$\frac{d}{dy}(yh(y))\bigg|_{y=y^*} = h(y^*) + y^*h'(y^*) = y^*h'(y^*) < 0$$

and hence y^* is globally attracting. Using (6.3.100), we also have

$$\lim_{t\to\infty} x(t) = C \lim_{t\to\infty} y^{\frac{b}{d}}(t) = C(y^*)^{\frac{b}{d}}.$$

Thus, we obtain a manifold of equilibria of (6.3.96),

$$(x^*, y^*) = \left(\frac{\mathring{x}^d}{\mathring{y}^b}(y^*)^{\frac{b}{d}}, y^*\right),$$

where y^* is the unique solution to $h(y) = 0$, see Fig. 6.25 (left). If we consider the solution $\phi(t, (\mathring{x}, \mathring{y}))$ of (6.3.96), we have

$$\lim_{t\to\infty} \phi(t, (\mathring{x}, \mathring{y})) = \left(\frac{\mathring{x}^d}{\mathring{y}^b}(y^*)^{\frac{b}{d}}, y^*\right).$$

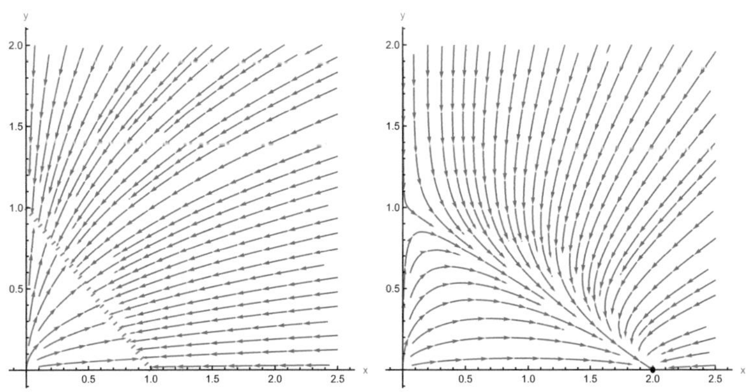

Fig. 6.25 Streamlines of (6.3.96) with $a = b = 2, c = d = 1, \alpha = \beta = 1$ and the line of equilibria $y = -x + 1$ (left), and $a = 2$, $b = c = d = 1$, $\alpha = \beta = 1$ and the equilibrium $(2, 0)$ (right)

Exercise 6.3.6 Consider the generalization of (6.3.96) in the form

$$\begin{aligned} x' &= x(a - \Phi_1(x, y)), \\ y' &= y(c - \Phi_2(x, y)), \end{aligned} \tag{6.3.102}$$

where

$$\Phi_i(x, y) = \alpha_i x + \beta_i y, \quad i = 1, 2.$$

Find the possible equilibria and determine their stability. Is the principle of competitive exclusion still valid in this case?

Exercise 6.3.7 Prove that the equilibrium $\boldsymbol{e}_1^*$ is globally asymptotically stable if $ad - cb > 0$ by the approach used for model (6.2.56). Alternatively, you can use (6.3.99) to show that $y(t)$ converges to zero and then deduce that $x(t)$ must converge to $\frac{a}{b\alpha}$.

6.4 Stability Through the Lyapunov Function

At the end of the eighteenth century, J. L. Lagrange pioneered the study of mechanical systems by proving that an isolated equilibrium point of a conservative system is stable if its potential energy attains local minimum at this point and it is unstable if it attains local maximum; see, e.g., [98, Section 5.3]. There have been several extensions of this result throughout the nineteenth century, but all of them were confined to mechanical systems. It was a great contribution of A.M. Lyapunov to realize that these ideas can be applied to arbitrary dynamical systems and that the energy can be replaced by an arbitrary function having energy-like properties but not necessarily related to any predefined physical characteristics of the system. His work, initially published in Russian and, though translated to French, received little attention for many years, possibly until the publication of [64].

6.4.1 *First Stability Theorem*

Consider again the system (6.2.17) in $\mathbb{R}^d$. Suppose it has an isolated equilibrium at $\boldsymbol{x}^*$. Then, by using the substitution $\boldsymbol{y} = \boldsymbol{x} - \boldsymbol{x}^*$, we can write it as

$$\boldsymbol{y}' = (\boldsymbol{x} - \boldsymbol{x}^*)' = \boldsymbol{x}' = \boldsymbol{f}(\boldsymbol{y} + \boldsymbol{x}^*) =: \tilde{\boldsymbol{f}}(\boldsymbol{y}), \tag{6.4.103}$$

we obtain an equivalent system for which $\boldsymbol{x} = 0$ becomes an isolated equilibrium. Thus, there is no loss of generality to consider (6.2.17) with $\boldsymbol{x}^* = 0$ as its equilibrium.

Let Ω be an open neighbourhood of 0 and let $V : \Omega \to \mathbb{R}$ be a continuously differentiable function and consider the composition $V((t))$, where $\boldsymbol{x}(t)$ is the global (forward) solution of (6.2.17). By the Chain Rule, we find

$$\frac{dV(\boldsymbol{x}(t))}{dt} = \nabla V(\boldsymbol{x}(t)) \cdot \boldsymbol{x}'(t) = \nabla V(\boldsymbol{x}(t)) \cdot \boldsymbol{f}(\boldsymbol{x}(t)) = \sum_{i=1}^{d} \frac{\partial V(\boldsymbol{x}(t))}{\partial x_i} f_i(\boldsymbol{x}(t)). \tag{6.4.104}$$

This shows that we can define the derivative of V along trajectories without explicit knowledge of them by

$$V'(\boldsymbol{x}) = \nabla V(\boldsymbol{x}) \cdot \boldsymbol{f}(\boldsymbol{x}) = \sum_{i=1}^{d} \frac{\partial V(\boldsymbol{x})}{\partial x_i} f_i(\boldsymbol{x}). \tag{6.4.105}$$

Example 6.37 Let us consider a system

$$\boldsymbol{x}' = \boldsymbol{f}(\boldsymbol{x})$$

with $\boldsymbol{f}$ being a potential field, that is, there is a scalar function V satisfying

$$\boldsymbol{f}(\boldsymbol{x}) = -\nabla V(\boldsymbol{x}).$$

In general, not every vector field has a potential. An exception is offered by one-dimensional fields, where we have

$$V(x) = -\int_a^x f(z)dz$$

for some fixed a (the potential of a field is determined up to a constant). We note that since $V' = -f$, the stationary points of V correspond to equilibria of f. Furthermore, if $t \to x(t)$ is any solution of the equation $x' = f(x)$, then we have

$$V_t'(x(t)) = V_x'(x(t))x_t'(t) = -f(x(t))f(x(t)) = -f^2(x(t)) < 0,$$

so that $V(x(t))$ strictly decreases along trajectories. In other words, the point $x(t)$ moves always in the direction of decreasing V and thus equilibria corresponding to the minima of V are asymptotically stable and corresponding to the maxima are unstable.

In this section, we shall discuss a generalization of the above concept.

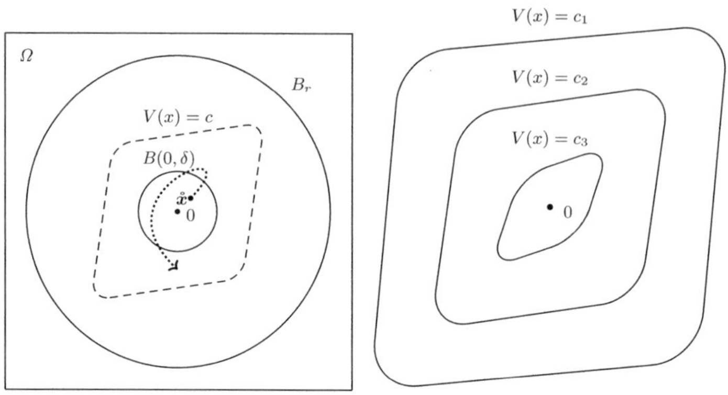

Fig. 6.26 Illustration of Theorem 6.39—the trajectory emanating from the ball B_δ cannot leave the set $\Omega_c = \{x : V(x) \leq c\}$ (indicated here by the dashed line) containing this ball, and hence from the ball B_r, containing Ω_c (left) and level curves $\{x : V(x) = c\}$ of V for $c_3 < c_2 < c_1$ (right); in general, the level curves of V need not be closed, see Example 6.55

Definition 6.38 A continuously differentiable function V on $\Omega \ni 0$ is called a Lyapunov function for (6.2.17) if

1. $V(0) = 0$ and $V(\boldsymbol{x}) > 0$ on Ω;
2. $V' \leq 0$ on Ω.

Theorem 6.39 *Assume that there exists a Lyapunov function defined on a neighbourhood Ω of an equilibrium $\boldsymbol{x} = 0$ of system (6.2.17). Then $\boldsymbol{x} = 0$ is stable.*

Proof Consider the situation illustrated in Fig. 6.26. There is a ball $B_r \subset \Omega$ (centred at 0 with radius r) such that $0 < V(\boldsymbol{x})$ on $B_r \setminus \{0\}$ and $V' \geq 0$ on B_r. Let us take $0 \neq \mathring{\boldsymbol{x}} \in B_r$ and consider the flow $\phi(t, \mathring{\boldsymbol{x}})$. Let $[0, t_{max})$ be the maximal interval of existence of the trajectory. At this moment, we do not know whether t_{max} is finite or not. Since V is decreasing along trajectories, we have

$$0 < V(\phi(t, \mathring{\boldsymbol{x}})) \leq V(\mathring{\boldsymbol{x}}), \quad t \in [0, t_{max}),$$

where the left-hand side inequality follows from the fact that $\phi(t, \mathring{\boldsymbol{x}}) \neq 0$ (by Theorem 6.27(i)) and strict positivity of V away from 0). Let $\mu = \min_{\|\boldsymbol{x}\|=r} V(\boldsymbol{x})$. Since $V(\boldsymbol{x}) \to 0$ as $\|\boldsymbol{x}\| \to 0$, we can find ball $B(0, \delta)$ with $\delta < r$ such that $V(\boldsymbol{x}) < \mu$ for $\boldsymbol{x} \in B(0, \delta)$. Then, for $\|\mathring{\boldsymbol{x}}\| < \delta$ we have

$$0 < V(\phi(t, \mathring{\boldsymbol{x}})) \leq V(\mathring{\boldsymbol{x}}) < \mu, \quad t \in [0, t_{max}),$$

(t_{max} here is not necessarily the same as above). By the definition of μ and continuity of the flow, $\|\phi(t, \mathring{\boldsymbol{x}})\| \leq r$ for $[0, t_{max})$. Indeed, otherwise there would be $t' > 0$ with $\|\phi(t', \mathring{\boldsymbol{x}})\| > r$ and, by continuity, for some t'' we would have $\|\phi(t'', \mathring{\boldsymbol{x}})\| = r$ so that $V(\phi(t'', \mathring{\boldsymbol{x}})) \geq \mu$.

This means, by Theorem B.14, that $t_{max} = \infty$ and, at the same time, yields stability as r was arbitrary. □

Remark 6.40 An important consequence of the above proof is the invariance of the sublevel sets of Lyapunov functions,

$$\Omega_c = \{\boldsymbol{x} : V_c(\boldsymbol{x}) \leq c\}, \tag{6.4.106}$$

as long as $\Omega_c \subset \Omega$. More precisely, no trajectory originating at $\mathring{\boldsymbol{x}}$ with $V(\mathring{\boldsymbol{x}}) < c$ cannot cross the surface $V(\mathring{\boldsymbol{x}}) = c$ as at such a point we would have $V(\phi(t', \mathring{\boldsymbol{x}})) = c$ for some $t' > 0$, contradicting the monotonicity of V along trajectories.

Remark 6.41 From the proof of Theorem 6.39, we see that the crucial role is played by the monotonicity of V and the assumption of its differentiability was used solely as a handy tool to establish it. Thus, the theorem is valid even if V is just a continuous function whose monotonicity along trajectories can be proved by other means. This observation is also the basis for a version of Theorem 6.39 for discrete systems.

Remark 6.42 The standard definition of the Lyapunov function and the formulation of Theorem 6.39 require the equilibrium $\boldsymbol{x}^*$ to be in the interior of an open set Ω. We have seen, however, that in many cases, the equilibria are located at the boundary of the region of interest; see Examples 6.35 and 6.36. We observe that the proof of Theorem 6.39 can still be carried out provided $\boldsymbol{x}^*$ is at the boundary of a closed invariant region Ω of (6.2.17). Indeed, the neighbourhood of $\boldsymbol{x}^*$, in this case, is bounded by the part of the sphere $\boldsymbol{x} = r$ inside Ω and the boundary of Ω, hence the trajectory can leave this neighbourhood neither by crossing the sphere as in the proof of Theorem 6.39 (see Remark 6.40), nor by crossing the boundary of Ω by the invariance; see [51, Theorem 2.1] or [19, Podrozdział 5.4], as illustrated in Fig. 6.27.

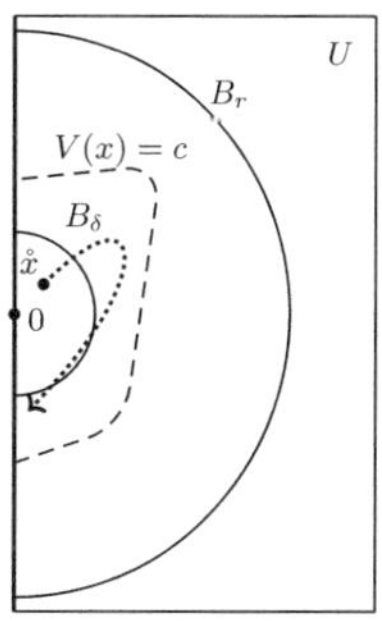

Fig. 6.27 The case of a boundary equilibrium

The main drawback of Lyapunov's method is that there are rules for constructing an appropriate Lyapunov function for a given dynamical system. Here, we provide a few examples of basic constructions.

Example 6.43 In his 1937 paper, [45], R. Fisher proposed equation

$$f_t = f_{xx} + rf(1-f), \quad x \in \mathbb{R}, t \in \mathbb{R}_+, \tag{6.4.107}$$

where $r \in \mathbb{R}_+$ is a constant, to describe the spatial spread of an advantageous allele. Since then, as a combination of random movement with logistic growth, this equation has become an extremely popular model of a spatial spread of a species in ecology; see, e.g., [73, Sections 11.2 & 13.2]. Of special interest are solutions of the form

$$f(x,t) = u(x-ct), \tag{6.4.108}$$

called travelling wave solutions. Here, u is a function of one variable having finite limits at $\pm\infty$, and $c > 0$ is a constant called the wave speed, see [73, Chapter 13]. Substituting (6.4.108) into (6.4.107), we get

$$u'' + cu' + ru(1-u) = 0. \tag{6.4.109}$$

This motivates us to consider a more general equation

$$u'' + cu' + g(u) = 0, \tag{6.4.110}$$

where g is a continuous function for $|u| < k$, with some constant $k > 0$, and changes sign at $u = 0$; in particular, by continuity, $g(0) = 0$. This allows for a significant extension of the range of applications. For example, if the independent variable is time, $c = 0$ and $g(u) = \omega^2 u$, (6.4.110) models a harmonic oscillator of frequency ω, and if $g(u) = \sin u$, it describes a simple pendulum with speed-dependent damping.

Writing the equation as a system, we get

$$\begin{aligned} x_1' &= x_2, \\ x_2' &= -cx_2 - g(x_1). \end{aligned} \tag{6.4.111}$$

It is clear that $(0,0)$ is an isolated equilibrium point. To construct a Lyapunov function, we employ mechanical interpretation of the model to find the system's energy. If we think of g as the restoring force of a spring, the potential energy of the particle at a displacement $u = x_1$ from equilibrium is given by

$$\int_0^{x_1} g(\sigma)d\sigma.$$

On the other hand, the kinetic energy is

$$\frac{1}{2}x_2^2,$$

as $x_2 = u'$, which is the velocity of the particle. This suggests to take the total energy of the system as a Lyapunov function

$$V(x_1, x_2) = \frac{1}{2}x_2^2 + \int_0^{x_1} g(\sigma)d\sigma.$$

This function is defined on the region

$$\Omega = \{(x_1, x_2) : |x_1| < k, x_2 \in \mathbb{R}\}.$$

Clearly, V is positive definite on Ω. Let us calculate the derivative of V along trajectories. We have

$$V'(x_1, x_2) = x_2x_2' + g(x_1)x_1' = (-cx_2 - g(x_1)x_2 + g(x_1)x_2 = -cx_2^2.$$

Thus, V is a Lyapunov function for (6.4.111) and the equilibrium at (0, 0) is stable for (6.4.110).

If $c = 0$, we can prove more. In this case, $V'(\boldsymbol{x}) = 0$, that is, V is constant on trajectories. In other words, for any $\boldsymbol{x}_0 = (x_{1,0}, x_{2,0}) \in \Omega$, we obtain

$$V(\phi(t, \boldsymbol{x}_0)) = V(\boldsymbol{x}_0)$$

for any t. Thus, the orbits are given by the implicit equation

$$\frac{1}{2}x_2^2 + \int_0^{x_1} g(\sigma)d\sigma = V(\boldsymbol{x}_0).$$

Because of the hypotheses on g, the integral is positive for both $x_1 > 0$ and $x_1 < 0$; moreover it is an increasing function of $|x_1|$, and is zero at $x_1 = 0$. On the other hand, $V(\boldsymbol{x}_0) \to 0$ as $\|\boldsymbol{x}_0\| \to 0$. This means that, for sufficiently small $\|\boldsymbol{x}_0\|$ (such that $V(\boldsymbol{x}_0) < \sup_{|x_1|<k} \int_0^{x_1} g(\sigma)d\sigma$), the orbits are closed and symmetric with respect to the x_2 axis; thus, solutions are periodic. Hence, in this case, (0, 0) is stable but not asymptotically stable. The question of asymptotic stability of (0, 0) if $c > 0$ requires a more sophisticated theory, which will be developed next; see Example 6.60.

It is rare to find a Lyapunov function in one go. If the left-hand side of (6.2.17) is polynomial, we often use the method of undetermined coefficients.

Example 6.44 Consider the system

$$\begin{aligned} x_1' &= x_2, \\ x_2' &= -cx_2 - ax_1 - bx_1^3, \end{aligned} \tag{6.4.112}$$

where a, b, c are positive constants. We are looking for a Lyapunov function as a polynomial in two variables. Let us try

$$V(x_1, x_2) = \alpha x_1^2 + \beta x_1^4 + \gamma x_2^2,$$

with $\alpha, \beta, \gamma > 0$. Clearly, $V(\boldsymbol{x}) > 0$ for $\boldsymbol{x} \neq 0$. Differentiating V along trajectories, we have

$$\begin{aligned} V'(\boldsymbol{x}) &= (2\alpha x_1 + 4\beta x_1^3)x_1' + 2\gamma x_2 x_2' \\ &= (2\alpha x_1 + 4\beta x_1^3)x_2 + 2\gamma x_2(-cx_2 - ax_1 - bx_1^3) \\ &= (2\alpha - 2\gamma a)x_1 x_2 + (4\beta - 2\gamma b)x_1^3 x_2 - 2\gamma c x_2^2. \end{aligned}$$

Since $c, \gamma > 0$, the last term is non-positive. The first two terms are more difficult, but we can choose free parameters α and β. Fixing $\gamma > 0$ and setting

$$\alpha = a\gamma, \quad \beta = \frac{\gamma b}{2} = \frac{\alpha b}{2a},$$

we obtain

$$V'(\boldsymbol{x}) = -2\gamma c x_2^2 \leq 0.$$

Hence, $V(\boldsymbol{x})$ is a Lyapunov function on any open bounded set of $\mathbb{R}^2$ which contains $(0, 0)$, and hence $(0, 0)$ is a stable equilibrium point.

A polynomial Lyapunov function does not always work for systems with polynomial right-hand sides.

Example 6.45 Consider again the Lotka–Volterra system (6.2.63),

$$\begin{aligned} x' &= rx - \alpha xy, \\ y' &= -sy + \beta xy, \end{aligned} \tag{6.4.113}$$

with the nontrivial equilibrium given by

$$(x_2^*, y_2^*) = \left(\frac{s}{\beta}, \frac{r}{\alpha}\right). \tag{6.4.114}$$

We have already shown that the first quadrant is forward invariant with respect to the flow generated by (6.4.113). We found that

$$r \ln y - \alpha y + s \ln x - \beta x = k$$

is the equation of orbits surrounding (x_2^*, y_2^*). In other words, the function

$$\tilde{V}(x, y) = r \ln y - \alpha y + s \ln x - \beta x, \quad x > 0, y > 0,$$

is constant along trajectories. Indeed, we have

$$\begin{aligned} \frac{d\tilde{V}(x(t), y(t))}{dt} &= r\frac{y'(y)}{y(t)} - \alpha y'(t) + s\frac{x'(y)}{x(t)} - \beta x'(t) \\ &= r(-s + \beta x(t)) - \alpha(-sy(t) + \beta x(t)y(t)) \\ &\quad + s(r - \alpha y(t)) - \beta(rx(t) - \alpha x(t)y(t)) = 0. \end{aligned}$$

Hence, $\tilde{V}$ satisfies condition 2. of Definition 6.38, but does not satisfy condition 1. of this definition, so it is not a Lyapunov function as $\tilde{V}(x_2^*, y_2^*) \neq 0$. Consider, however,

$$\begin{aligned} V(x, y) = \tilde{V}(x_2^*, y_2^*) - \tilde{V}(x, y) - &= -r \ln y + \alpha y - s \ln x \\ &+ \beta x + r \ln y_2^* - \alpha y_2^* - s \ln x_2^* + \beta x_2^* \\ &= -r \ln \frac{\alpha y}{r} + \alpha y - r - s \ln \frac{\beta x}{s} + \beta x - s. \end{aligned} \tag{6.4.115}$$

With $z = \frac{\alpha y}{r}$, we have

$$-r \ln \frac{\alpha y}{r} + \alpha y - r = -r(\ln z - z + 1) \geq 0$$

with equality attained only at $z = 1$. Since the same argument applies to $s \ln \frac{\beta x}{s} - \beta x + s$, V is a Lyapunov function for (6.4.113) at the equilibrium (x_2^*, y_2^*), confirming that it is stable.

Exercise 6.4.1 Show that $f(z) = \ln z - z + 1$ satisfies $f(1) = 0$ and $f(z) < 0$ for any $z > 0$ and $z \neq 1$.

The first Lyapunov theorem, Theorem 6.39, ensures stability but not asymptotic stability. Examples 6.43 and 6.45 show that the problem is created by trajectories along which V is constant, as these give rise to periodic orbits that prevent asymptotic stability. The following theorem shows that, indeed, preventing the possibility of V being constant in a neighbourhood of zero solves the problem, at least partially. The full theory requires, however, more sophisticated tools. First, we introduce further terminology related to the behaviour of the flow over a long period of time.

6.4.2 Limit Sets

Let us recall that the definition and notation for trajectories were introduced in Definition 6.16.

Definition 6.46 The ω-limit set of the trajectory $\Gamma_{\boldsymbol{x}_0}$ is the set of all points $\boldsymbol{p} \in \mathbb{R}^d$ such that there is a sequence $(t_k)_{k\in\mathbb{N}}$ such that $t_k \to \infty$ as $k \to \infty$ for which

$$\lim_{k\to\infty} \phi(t_k, \boldsymbol{x}_0) = \boldsymbol{p}.$$

Similarly, the α-limit set of the trajectory $\Gamma_{\boldsymbol{x}_0}$ is the set of all points $\boldsymbol{q} \in \mathbb{R}^d$ such that there is a sequence $(t_k)_{k\in\mathbb{N}}$ such that $t_k \to -\infty$ as $k \to \infty$ for which

$$\lim_{k\to\infty} \phi(t_k, \boldsymbol{x}_0) = \boldsymbol{q}.$$

Since for a given equation (6.2.17) any trajectory is uniquely determined by any point on it (and conversely), sometimes we shall use the notation $\omega(\boldsymbol{x}_0)$ instead of $\omega(\Gamma_{\boldsymbol{x}_0})$ (and the same for α-limit sets).

Example 6.47 If $\boldsymbol{v}_0$ is an equilibrium, then $\Gamma_{\boldsymbol{v}_0} = \{\boldsymbol{v}_0\} = \omega(\Gamma_{\boldsymbol{v}_0}) = \alpha(\Gamma_{\boldsymbol{v}_0})$. The equilibria are the only ω and α limit sets of scalar equations.

Similarly, if Γ is a periodic trajectory with period T, then $\omega(\Gamma) = \alpha(\Gamma) = \Gamma$ since for any $\boldsymbol{x} = \boldsymbol{x}(\tau) \in \Gamma$, we have $\boldsymbol{x}(\tau) = \lim_{k\to\infty} \boldsymbol{x}(\tau + kT)$ with $t_k = \tau + kT \to \infty$.

Example 6.48 Consider

$$\begin{aligned} x_1' &= x_1 - x_2 - x_1(x_1^2 + x_2^2), \\ x_2' &= x_1 + x_2 - x_2(x_1^2 + x_2^2). \end{aligned} \tag{6.4.116}$$

As in Example 6.23, we convert (6.4.116) into the polar form by multiplying the first equation by x_1, the second by x_2 and, adding them together, we obtain

$$rr' = r^2(1 - r^2).$$

Further, using $\tan\theta = \frac{x_2}{x_1}$ (at least for $x_1 \neq 0$), we get

$$\frac{1}{\cos^2\theta}\theta' = \frac{x_2'x_1 - x_1'x_2}{x_1^2} = \frac{x_1^2 + x_2^2}{r^2\cos^2\theta} = \frac{1}{\cos^2\theta}.$$

Hence, disregarding the trivial solution $r = 0$, we obtain the system

$$\begin{aligned} r' &= r(1 - r^2), \\ \theta' &= 1. \end{aligned}$$

Since $r' > 0$ if $r \in (0, 1)$ and $r' < 0$ if $r > 1$, trajectories which start with $0 < r < 1$ and $r > 1$ tend to $r = 1$ which, since $\theta' \neq 0$, is a periodic orbit. The origin $r = 0$ is a stationary point and so $\omega(\{r = 0\}) = \alpha(\{r = 0\}) = (0, 0)$. If $r \neq 0$, then

$$\omega(\Gamma_{(r,\theta)}) = \{(r, \theta) : r = 1\},$$

and

$$\alpha(\Gamma_{(r,\theta)}) = \begin{cases} \{(r, \theta) : r = 0\} & \text{if } r < 1, \\ \text{does not exist} & \text{if } r > 1, \end{cases}$$

see Fig. 6.28.

Lemma 6.49 *ω-Limit sets have the following properties.*

1. *If Γ is bounded, then $\omega(\Gamma)$ is non-empty.*
2. *An ω-limit set is closed.*
3. *If $\mathbf{y} \in \omega(\Gamma)$, then $\Gamma_{\mathbf{y}} \subset \omega(\Gamma)$, that is, $\omega(\Gamma)$ is invariant.*
4. *If Γ is bounded, then $\omega(\Gamma)$ is connected.*
5. *If $z \in \omega(\Gamma)$ and $\Gamma \cap \omega(\Gamma') \neq \emptyset$, then $z \in \omega(\Gamma')$.*
6. *If $\Gamma_{\mathbf{y}}$ is bounded, then $\phi(t, \mathbf{y}) \to \omega(\Gamma_{\mathbf{y}})$ as $t \to \infty$ in the sense that for each $\epsilon > 0$ there is $t > 0$ such that for every $t' > t$ there is $\mathbf{p} \in \omega(\Gamma_{\mathbf{y}})$ (possibly depending on t) which satisfies $\|\phi(t', \mathbf{y}) - \mathbf{p}\| < \epsilon$. In particular, if for some set Ω, $\phi(t, \mathbf{y}) \to \Omega$ as $t \to \infty$, then $\omega(\Gamma_{\mathbf{y}}) \subset \Omega$.*

The same properties are valid for α-limit sets.

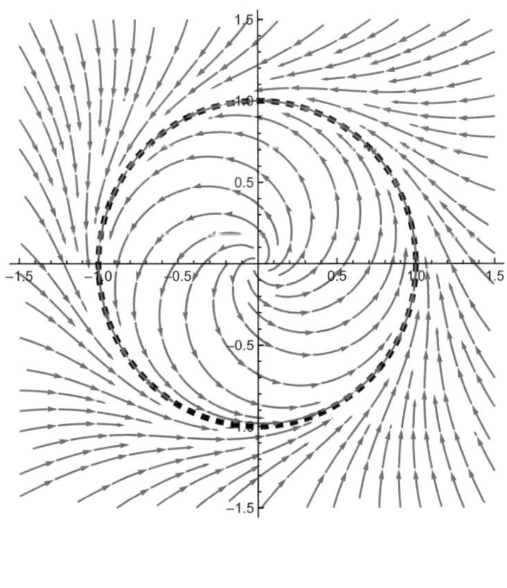

Fig. 6.28 Streamlines of (6.4.116)—the trajectories different from (0, 0) tend to the unit circle as $t \to \infty$; also, by reverting the direction of the arrows, we see that the trajectories tend to 0 as $t \to -\infty$ if $r < 1$ and do not have a limit if $r > 1$

Proof

1. Let $\boldsymbol{x} \in \Gamma$ and take $\boldsymbol{x}_k = \phi(k, \boldsymbol{x}), k \in \mathbb{N}$. We can assume that all terms in the sequence $(\boldsymbol{x}_k)_{k\in\mathbb{N}}$ are distinct. Indeed, otherwise, we would have $\boldsymbol{x}(p) = \boldsymbol{x}(s)$ for some $p \neq s$ and the trajectory would be periodic (or constant) and hence $\Omega(\Gamma) = \Gamma \neq \emptyset$; see Remark 6.28. Then, the sequence $(\boldsymbol{x}_k)_{k\in\mathbb{N}}$ is bounded and, by the multidimensional version of the Bolzano–Weierstrass theorem, must have a converging subsequence $(\boldsymbol{x}_{k_l})_{l\in\mathbb{N}}$, also consisting of distinct terms, with $\lim_{l\to\infty} \boldsymbol{x}_{k_l} = \boldsymbol{y}$. Let $\boldsymbol{x}_{k_l} = \phi(t_{k_l}, \boldsymbol{x})$, $t_{k_l} \in \mathbb{N}$. Then $(t_{k_l})_{l\in\mathbb{N}}$ is an infinite sequence of different natural numbers, and hence it must diverge to infinity. Indeed, otherwise, this sequence would contain a bounded subsequence, that is, there would be a bounded set of distinct natural numbers, which is impossible. Hence, $\boldsymbol{y} \in \omega(\Gamma)$.
2. Let $(\boldsymbol{p}_k)_{k\in\mathbb{N}}$ be a sequence of points of $\omega(\Gamma_{\boldsymbol{x}})$ for some $\boldsymbol{x}$, converging to $\boldsymbol{p} \in \mathbb{R}^n$. We must prove that there is a sequence $t_k \to \infty$ such that $\phi(t_k, \boldsymbol{x}) \to \boldsymbol{p}$. We proceed as follows. For $\epsilon = 1/2$ we can find $\boldsymbol{p}_{k_1}$ and t_{k_1} such that

$$\|\boldsymbol{p} - \phi(t_{k_1}, \boldsymbol{x})\| \le \|\boldsymbol{p} - \boldsymbol{p}_{k_1}\| + \|\boldsymbol{p}_{k_1} - \phi(t_{k_1}, \boldsymbol{x})\| \le \frac{1}{2} + \frac{1}{2} = 1.$$

   Similarly, for $\epsilon = 1/4$ we can find $\boldsymbol{p}_{k_2}$ and $t_{k_2} > t_{k_1} + 1$ such that

$$\|\boldsymbol{p} - \phi(t_{k_2}, \boldsymbol{x})\| \le \|\boldsymbol{p} - \boldsymbol{p}_{k_2}\| + \|\boldsymbol{p}_{k_2} - \phi(t_{k_2}, \boldsymbol{x})\| \le 1/4 + 1/4 = 1/2,$$

   and, by induction, for any given l we can find $\boldsymbol{p}_{k_l}$ and $t_{k_l} > t_{k_{l-1}} + 1$ such that

$$\|\boldsymbol{p} - \phi(t_{k_l}, \boldsymbol{x})\| \le \|\boldsymbol{p} - \boldsymbol{p}_{k_l}\| + \|\boldsymbol{p}_{k_l} - \phi(t_{k_l}, \boldsymbol{x})\| \le 1/2k + 1/2k = 1/l.$$

   Since the sequence $(t_{k_l})_{l\in\mathbb{N}}$ is infinite and increasing by one at each step, it must diverge to infinity. Thus, $\boldsymbol{p} \in \omega(\Gamma_{\boldsymbol{x}})$.
3. Let $\boldsymbol{y} \in \omega(\Gamma_{\boldsymbol{x}})$ and consider arbitrary $\boldsymbol{q} \in \Gamma_{\boldsymbol{y}}$. Thus $\boldsymbol{q} = \phi(t', \boldsymbol{y})$. Since $\boldsymbol{y}$ is in the ω-limit set,

$$\boldsymbol{y} = \lim_{t_k\to\infty} \phi(t_k, \boldsymbol{x}).$$

   But

$$\phi(t_k + t', \boldsymbol{x}) = \phi(t', \phi(t_k, \boldsymbol{x})) \to \phi(t', \boldsymbol{y}) = \boldsymbol{q},$$

   by the continuity of the flow with respect to the initial value, Theorem B.16. Since $t' + t_k \to \infty$, we obtain that $\boldsymbol{q} \in \omega(\Gamma_{\boldsymbol{y}})$ and since $\boldsymbol{q} \in \omega(\Gamma_{\boldsymbol{x}})$ was arbitrary, the thesis follows.
4. By 1. and 2. and the definition, $\omega(\Gamma)$ is bounded and closed. Thus, if it was not connected, then $\omega(\Gamma) = A \cup B$ with A, B closed bounded and a distance $d > 0$ apart. Fix $\boldsymbol{x} \in \Gamma$ and chose $\boldsymbol{v} \in A$. There is $t_k \to \infty$ such that $\boldsymbol{x}_k := \phi(t_k, \boldsymbol{x}) \to$

$\boldsymbol{v}$. Thus, for sufficiently large k, the distance between $\boldsymbol{x}_k$ and B is not less than $3d/4$. Let $\boldsymbol{w} \in B$. For each $\boldsymbol{x}_k$ we can find $t'_k > 0$ such that the distance between $\boldsymbol{y}_k = \phi(t'_k, \boldsymbol{x}_k) = \phi(t_k + t'_k, \boldsymbol{x})$ and $\boldsymbol{w}$ is not greater than $d/4$, thus the distance between A and $\boldsymbol{y}_k$ is at least $3d/4$. On the other hand, from the continuity of the flow, for each k, there is a point $z_k = \phi(t''_k, \boldsymbol{x})$ whose distance from A is $d/2$. The set of such points is bounded and thus has a convergent subsequence whose limit $z \in \omega(\Gamma)$ is at a distance $d/2$ from A, which contradicts the assumption that the components of $\omega(\Gamma)$ are d apart.

5. Let $z \in \omega(\Gamma)$, $\boldsymbol{y} \in \Gamma$ and $\boldsymbol{y} = \lim_{t'_k \to \infty} \phi(t'_k, \boldsymbol{x})$ with $\boldsymbol{x} \in \Gamma'$, so that $\boldsymbol{y} \in \Gamma \cap \omega(\Gamma')$. Let us fix $\epsilon > 0$. From the continuity of the flow with respect to the initial condition (Theorem B.16), we know that for a given ϵ and T, there is $\delta_{T,\epsilon}$ such that $\|\phi(t, \boldsymbol{x}_1) - \phi(t, \boldsymbol{x}_2)\| < \epsilon$ provided $\|\boldsymbol{x}_1 - \boldsymbol{x}_2\| < \delta_{T,\epsilon}$ for all $0 \leq t \leq T$.

   For this given $\epsilon > 0$ we find t_k such that $\|z - \phi(t_k, \boldsymbol{y})\| < \epsilon$ and also t'_k such that $\|\boldsymbol{y} - \phi(t'_k, \boldsymbol{x})\| < \delta_{t_k,\epsilon}$ (since $\boldsymbol{y} \in \omega(\Gamma')$). Hence

$$\|z - \phi(t_k, \phi(t'_k, \boldsymbol{x}))\| \leq \|z - \phi(t_k, \boldsymbol{y})\| + \|\phi(t_k, \boldsymbol{y}) - \phi(t_k, \phi(t'_k, \boldsymbol{x}))\| < 2\epsilon$$

   but, since $\phi(t_k, \phi(t'_k, \boldsymbol{x}_0)) = \phi(t_k + t'_k, \boldsymbol{x})$, we see that $z \in \omega(\Gamma')$ (we note that the sequence $\tau_k := t_k + t'_k$ can be made increasing with k and thus convergent to ∞ as t_k and t'_k are).

6. Suppose that the statement is false, that is, there is $\epsilon > 0$ and a sequence $t_k \to \infty$ with $\|\phi(t_k, \boldsymbol{y}) - \boldsymbol{p}\| > \epsilon$ for all $\boldsymbol{p} \in \omega(\Gamma_{\boldsymbol{y}})$. This means that the sequence $\phi(t_k, \boldsymbol{y})$ stays at least ϵ away from $\omega(\Gamma_{\boldsymbol{y}})$. But the orbit is bounded, so, by 1., there is a converging subsequence $\phi(t_{k_l}, \boldsymbol{y})$ of this sequence which, by the definition of $\omega(\Gamma_{\boldsymbol{y}})$, must converge to it. The last statement of point 6. follows from the fact that if $\phi(t, \boldsymbol{y}) \to \Omega$ for $t \to \infty$, then Ω must contain all limits of converging sequences of the form $\phi(t_n, \boldsymbol{y})$, $t_n \to \infty$, and hence, by the definition of ω-limit set, $\omega(\Gamma_{\boldsymbol{y}}) \subset \Omega$.

□

6.4.3 Asymptotic Stability Through Lyapunov Functions

Definition 6.50 A continuously differentiable function V on $\Omega \ni 0$ is called a strict Lyapunov function for (6.2.17) if

1. $V(0) = 0$ and $V(\boldsymbol{x}) > 0$ on Ω;
2.

$$V' < 0 \quad \text{on} \quad \Omega \setminus \{0\}. \tag{6.4.117}$$

Theorem 6.51 *If there exists a strict Lyapunov function defined on a neighbourhood Ω of an isolated equilibrium $\boldsymbol{x} = 0$ of system (6.2.17), then $\boldsymbol{x} = 0$ is asymptotically stable.*

Proof Using the notation of the proof of Theorem 6.39, we consider $\phi(t, \mathring{\boldsymbol{x}})$ with $\|\mathring{\boldsymbol{x}}\| < \delta$ so that the solution stays in the closed ball $\overline{B_r}$ which does not contain other equilibria. Since this set is closed, by Lemma 6.49, point 1., we have sequence $(t_k)_{k\in\mathbb{N}}$, $t_n \to \infty$ as $n \to \infty$ such that $\phi(t_k, \mathring{\boldsymbol{x}}) \to z \in \overline{B_r}$. We have to prove that $z = 0$. To this end, first observe that $V(\phi(t, \mathring{\boldsymbol{x}})) > V(z)$ for all t as V decreases along trajectories and $V(\phi(t_k, \mathring{\boldsymbol{x}})) \to V(z)$ by continuity of V. If $z \neq 0$, we consider $\phi(t, z)$ which must satisfy $V(\phi(t, z)) < V(z)$. By continuity of the flow with respect to the initial condition and the continuity of V, if $\boldsymbol{x}$ is close enough to z, then for some $t > 0$ (not necessarily all) we will have also $V(\phi(t, \boldsymbol{x})) < V(z)$. We take $\boldsymbol{x} = \phi(t_k, \mathring{\boldsymbol{x}})$ for t_k large enough. But then we obtain

$$V(z) > V(\phi(t, \phi(t_k, \mathring{\boldsymbol{x}}))) = V(\phi(t + t_k, \mathring{\boldsymbol{x}})) > V(z),$$

which is a contradiction. Thus $z = 0$. This shows asymptotic stability. Indeed, if there was a sequence $(t_k)_{k\in\mathbb{N}}$ converging to infinity, for which $\phi(t, \mathring{\boldsymbol{x}})$ was not converging to zero (that is, staying outside some ball B_{r_0}), $r_0 < r$, then, as above, we could pick a subsequence of $\phi(t_k, \mathring{\boldsymbol{x}})$ converging to some z which, by the above, must be 0. □

Remark 6.52 As in Remark 6.42, Theorem 6.51 works, with obvious modifications, if $\boldsymbol{x}^*$ is at the boundary of a closed invariant region Ω of (6.2.17); see again [51, Theorem 2.1] or [19, Podrozdział 5.4].

Example 6.53 Consider a scalar equation

$$x' = f(x) := ax^p + g(x),$$

$p \in \mathbb{N}$ and $g(x) = O(x^{p+1})$ in some neighbourhood of $x = 0$. We will show that the equilibrium $x = 0$ is asymptotically stable if p is odd and $a < 0$, and it is unstable, if p odd and $a > 0$ or p is even and $a \neq 0$.

From the definition of the O-symbol, we get $|g(x)| \leq M|x|^{p+1}$ for some M and sufficiently small $|x|$. By assumption, in a neighbourhood of zero, we have sign $f(x)$ = sign ax^p. Consider first the case when $a < 0$ and odd p and take the Lyapunov function $V(x) = \frac{1}{2}x^2$. Then

$$V'(x) = x(ax^p + g(x)) \leq ax^{p+1} + k|x|^{p+2}$$

and the term ax^{p+1} is dominant for small $|x|$. Since $p + 1$ is even, V' is negative close to $x = 0$ and hence $x = 0$ is asymptotically stable.

Next, consider the case when p is still odd but $a > 0$. Then sign $f(x)$ = sign x^p and solutions originating from a neighbourhood of the point $x = 0$ will always move away from it.

Finally, if p is even, such behaviour will occur to the left from $x = 0$ for $a < 0$ and to the right if $a > 0$. Hence also in this case $x = 0$ is an unstable equilibrium.

Example 6.54 Consider the system

$$\begin{aligned} x' &= -x + x^2 - 2xy, \\ y' &= -2y - 5xy + y^2. \end{aligned} \tag{6.4.118}$$

The point $(0, 0)$ clearly is a stationary point. Let us investigate its stability. Let us try the simplest Lyapunov function

$$V(x, y) = \frac{1}{2}(x^2 + y^2).$$

We obtain

$$\begin{aligned} V'(x, y) &= xx' + yy' = x(-x + x^2 - 2xy) + y(-2y - 5xy + y^2) \\ &= -x^2(1 - x + 2y) - y^2(2 + 5x - y). \end{aligned}$$

Hence V is a strict Lyapunov function in $\Omega := \{(x, y) : 2+5x-y > 0, 1-x+2y > 0\}$. It is an open set containing $(0, 0)$, so the origin is asymptotically stable.

Exercise 6.4.2 Consider the Lorenz system

$$\begin{aligned} x' &= \sigma(y - x), \\ y' &= rx - y - xz, \\ z' &= xy - bz, \end{aligned} \tag{6.4.119}$$

which is a minimalist model of thermal convection in the atmosphere. Here $r, b, \sigma > 0$ are constants. For some parameters (e.g., $\sigma = 10, b = \frac{8}{3}, r = 28$ originally chosen by Lorenz), the system displays complicated dynamics (strange, or Lorenz, attractor) and was the starting point for the development of chaos theory; see [48, Section 12.6]. For other values of parameters, however, the behaviour of solutions to (6.4.119) is quite regular.

Use the candidate Lyapunov function

$$V(x, y, z = \alpha x^2 + \beta y^2 + \gamma z^2,$$

where α, β, γ are constants to be determined to show that $(0, 0, 0)$ is an asymptotically stable equilibrium if $r < 1$.

Exercise 6.4.3 Show that $(1, 0)$ is an equilibrium for

$$\begin{aligned} x' &= 1 - 3x + 3x^2 + 2y^2 - x^3 - 2xy^2, \\ y' &= y - 2xy + x^2y - y^3. \end{aligned} \tag{6.4.120}$$

Use the transformation (6.4.103) to transform (6.4.120) to a system having an equilibrium at $(0, 0)$. Then, use the candidate Lyapunov function

$$V(x, y) = \alpha x^2 + \beta x^2y^2 + \gamma y^4, \quad (x, y) \in \mathbb{R}^2,$$

where α, β and γ are positive constants to be determined, to show that $(0, 0)$ is asymptotically stable.

In the previous examples, the origin is asymptotically stable, but there is no basis to say it is globally asymptotically stable. For instance, in Example 6.54, V is a strict Lyapunov function only in $\Omega \neq \mathbb{R}^2$. However, even if $\boldsymbol{x}^* = 0$ is the only equilibrium in $\mathbb{R}^n$ and V is a strict Lyapunov function on the whole space, $\boldsymbol{x}^*$ need not be globally asymptotically stable, as we will see in the following example.

Example 6.55 Consider

$$\begin{aligned} x' &= -\frac{6x}{\left(1+x^2\right)^2} + 2y, \\ y' &= -\frac{2(x+y)}{\left(1+x^2\right)^2}. \end{aligned} \tag{6.4.121}$$

First, we find the equilibria. The second equation gives $y = -x$ so that the first equation yields

$$x\left(6 + 2\left(1+x^2\right)^2\right) = 0,$$

whose unique solution is $x = 0$. Hence (6.4.121) has exactly one equilibrium $(0, 0)$.

Denote

$$f(x, y) = \left(-\frac{6x}{\left(1+x^2\right)^2} + 2y, -\frac{2(x+y)}{\left(1+x^2\right)^2}\right)$$

and consider

$$V(x, y) = \frac{x^2}{1+x^2} + y^2. \tag{6.4.122}$$

This function is defined on $\mathbb{R}^2$ and satisfies $V(0,0) = 0$ and $V(x,y) > 0$ for $(x,y) \neq (0,0)$. Since

$$\nabla V(x,y) = \left(\frac{2x}{\left(1+x^2\right)^2}, 2y \right),$$

$$\begin{aligned} V'(x,y) &= \langle \nabla V(x,y), \boldsymbol{f}(x,y) \rangle \\ &= \left\langle \left(\frac{2x}{\left(1+x^2\right)^2}, 2y \right), \left(-\frac{6x}{\left(1+x^2\right)^2} + 2y, -\frac{2(x+y)}{\left(1+x^2\right)^2} \right) \right\rangle \\ &= \frac{12x^2}{\left(1+x^2\right)^4} + \frac{4xy}{\left(1+x^2\right)^2} - \frac{4xy}{\left(1+x^2\right)^2} - \frac{4y^2}{\left(1+x^2\right)^2} \\ &= -\frac{4}{\left(1+x^2\right)^2} \left(\frac{3x^2}{\left(1+x^2\right)^2} + 4y^2 \right). \end{aligned}$$

Hence, $V'(0,0) = 0$, $V'(x,y) < 0$ for $(x,y) \neq (0,0)$ and the assumptions of Theorem 6.51 are satisfied on $\mathbb{R}^2$. Therefore, $(0,0)$ is asymptotically stable.

Next, we show that $(0,0)$ is not globally stable even though V is a strict Lyapunov function on $\mathbb{R}^2$. Consider the right branch of the hyperbola $y = \frac{2}{x-\sqrt{2}}$, that is, the branch in the half-plane $\left\{(x,y) \in \mathbb{R}^2 : x > \sqrt{2}\right\}$. Then

$$\frac{dy}{dx} = -\frac{2}{\left(x-\sqrt{2}\right)^2}.$$

Let us find the expression $\frac{y'(t)}{x'(t)}$ along this hyperbola. Using (6.4.121), we find

$$\frac{y'}{x'} = \frac{\frac{2(x+y)}{\left(1+x^2\right)^2}}{\frac{-6x}{\left(1+x^2\right)^2} + 2y} = \frac{x+y}{3x - y\left(1+x^2\right)^2}.$$

Substituting $y = \frac{2}{x-\sqrt{2}}$, we get

$$\begin{aligned} \frac{y'}{x'} &= \frac{x + \frac{2}{x-\sqrt{2}}}{3x - \frac{2}{x-\sqrt{2}}\left(1+x^2\right)^2} = \frac{x(x-\sqrt{2})+2}{3x(x-\sqrt{2}) - 2\left(1+x^2\right)^2} \\ &= \frac{-x^2+\sqrt{2}x-2}{2x^4+x^2+3\sqrt{2}x+2} = -\frac{\left(x-\frac{\sqrt{2}}{2}\right)^2 + \frac{3}{2}}{2x^4+x^2+3\sqrt{2}x+2}. \end{aligned}$$

The expression above is negative since $x > \sqrt{2} > 0$. The ratio above gives the tangent of the angle between the axis Ox and the vector field f. Now we will show that

$$\frac{y'}{x'} = \frac{-x^2 + \sqrt{2}x - 2}{2x^4 + x^2 + 3\sqrt{2}x + 2} > -\frac{2}{\left(x - \sqrt{2}\right)^2} = \frac{dy}{dx},$$

where the ratio on the right-hand side is the tangent along the hyperbola. After some algebra, we find

$$\frac{(-x^2 + \sqrt{2}x - 2)(x - \sqrt{2})^2 + 2(2x^4 + x^2 + 3\sqrt{2}x + 2)}{(2x^4 + x^2 + 3\sqrt{2}x + 2)(x - \sqrt{2})^2} > 0.$$

Again, the denominator is positive since $x > \sqrt{2} > 0$. It suffices to show that also the numerator is positive, that is, after some simplifications, that

$$x(x^3 + \sqrt{2}x^2 - 2x + 6\sqrt{2}) > 0. \tag{6.4.123}$$

Denote $h(x) = x^3 + \sqrt{2}x^2 - 2x + 6\sqrt{2}$. Then

$$h'(x) = 3x^2 + 2\sqrt{2}x - 2.$$

Since $h(\sqrt{2}) = 8\sqrt{2} > 0$ and $h'(x) > 0$ for $x > \sqrt{2}$, (6.4.123) is positive for $x > \sqrt{2}$. Finally

$$\frac{y'}{x'} > \frac{dy}{dx}.$$

Since both $\frac{x'}{y'}$ and $\frac{dy}{dx}$ are negative, the angle between f and the positive semi-axis Ox is greater than the angle between the tangent to the right branch of hyperbola $y = \frac{2}{x-\sqrt{2}}$ and this semi-axis. Thus, the vector field f points inside the set $\left\{(x, y) \in \mathbb{R}^2 : y > \frac{2}{x-\sqrt{2}}\right\}$ (see Fig. 6.29), which means, by Lemma B.20, that the trajectories originating inside this set cannot cross the hyperbola. Hence, we have trajectories that do not converge to $(0, 0)$; therefore, it is not a globally asymptotically stable equilibrium. This is illustrated in Fig. 6.29, where we see that trajectories starting to the right of the hyperbola at the top and the bottom will not converge $(0, 0)$.

Let us reflect on where the problem lies. We know that the trajectories cannot cross the level curves of the Lyapunov function in the increasing direction. This condition is certainly satisfied, but the crux of the proof of Theorem 6.39 (and hence Theorem 6.51) was that for any ball B_r we could find a ball B_δ, possibly very small, the trajectories originating from which were bounded. This, combined with the monotonicity of V along trajectories, ensured the convergence of the trajectories

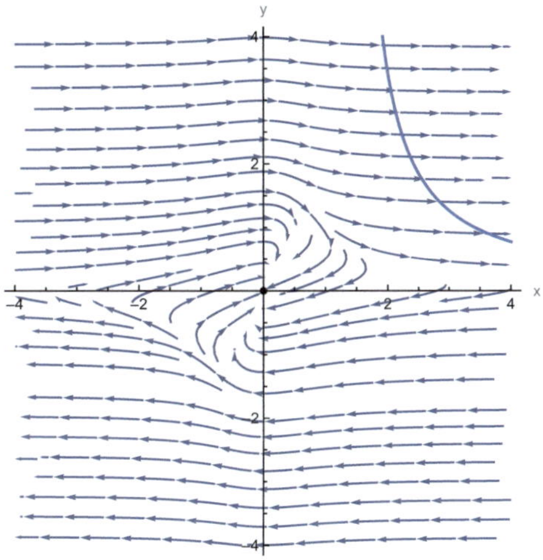

Fig. 6.29 Streamlines of (6.4.121) and the hyperbola $y = \frac{2}{x-\sqrt{2}}$, $x > \sqrt{2}$

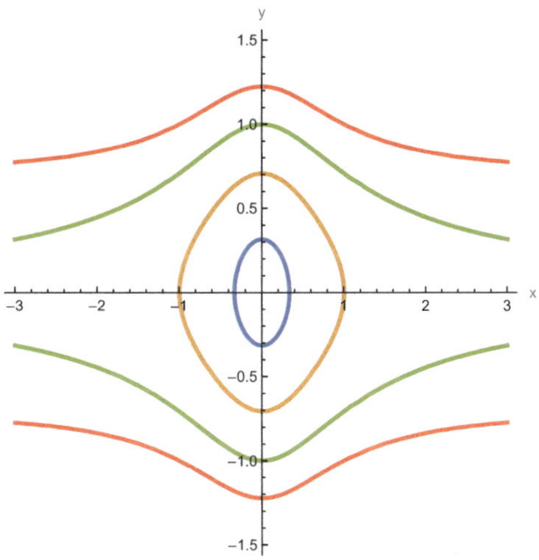

Fig. 6.30 Level curves of the Lyapunov function V given by (6.4.122)

starting from B_δ to 0. The above property, however, does not guarantee that any trajectory is bounded. For instance, the ball B_δ can be the same for all sufficiently large r. In this example, we see that the level curves $\{(x, y) : V(x, y) = c\}$ are not closed for large c, see Fig. 6.30, and the trajectories, even moving in the direction of the decreasing values of c, can escape to infinity, as seen in Fig. 6.31.

Thus, for a strict Lyapunov function V to ensure the global stability of a unique equilibrium in an invariant set, it will be sufficient to ensure that all level curves of V are closed. This leads us to the following result.

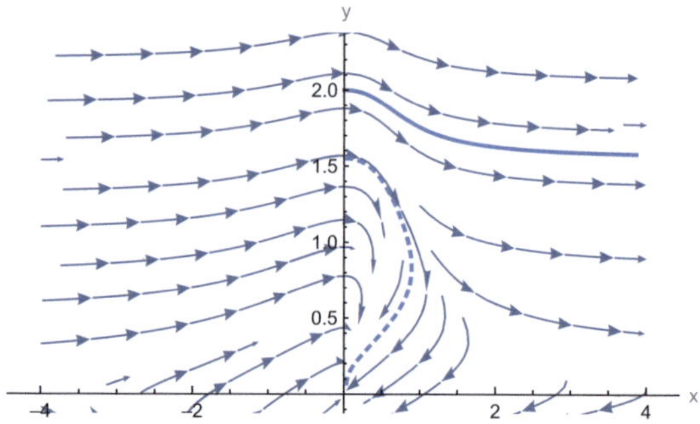

Fig. 6.31 Trajectories of (6.4.121) converging to the equilibrium (0, 0) (dashes) and moving away from it (solid)

Theorem 6.56 *Let* 0 *be a unique equilibrium of* (6.2.17) *in* $\mathbb{R}^d$. *Assume that* $V : \mathbb{R}^d \to [0, \infty)$ *is a strict Lyapunov function for* (6.2.17) *additionally satisfying*

$$V(\boldsymbol{x}) \to \infty \quad \textit{if} \quad \|\boldsymbol{x}\| \to \infty. \tag{6.4.124}$$

Then, the zero equilibrium is globally asymptotically stable.

Proof Let $\boldsymbol{x} \in \mathbb{R}^d$ be arbitrary. From the properties of the Lyapunov function, we have

$$V(\phi(t, \boldsymbol{x})) \leq V(\boldsymbol{x}), \quad \boldsymbol{x} \in \mathbb{R}^d,$$

for all $t \geq 0$, hence (6.4.124) implies that $\phi(t, \boldsymbol{x})$ is bounded and thus the solution exists for all $t \geq 0$. Since the solution stays in a bounded set, we can apply the proof of Theorem 6.51 to show that it converges to 0. Since $\boldsymbol{x}$ is arbitrary, 0 is globally asymptotically stable. □

Remark 6.57 As in Remarks 6.42 and 6.52, Theorem 6.56 works if $\boldsymbol{x}^*$ is at the boundary of a closed invariant region Ω of (6.2.17), with the obvious modification that (6.4.124) is satisfied for $\boldsymbol{x} \in \Omega$; see again [51, Theorem 2.1] or [19, Podrozdział 5.4].

Exercise 6.4.4 Show that the Lyapunov function (6.4.122) does not satisfy (6.4.124).

Example 6.58 In Sect. 6.2.4, we considered the system

$$\begin{aligned} x' &= ax - bxy - ex^2, \\ y' &= -cy + dxy - fy^2, \end{aligned} \tag{6.4.125}$$

where a, b, e, c, d, f are positive, describing predator-prey interactions in an environment with limited resources. We have shown that the predator extinction equilibrium

$$\left(\frac{a}{e}, 0\right)$$

is globally asymptotically stable in $\overline{\mathbb{R}}_{+}^{2}$ if

$$ad - ce > 0.$$

Here, we consider the case when

$$ce - ad < 0. \tag{6.4.126}$$

Then, there exists a co-existence equilibrium

$$(x^*, y^*) = \left(\frac{af + cb}{bd + ef}, \frac{ad - ce}{bd + ef}\right). \tag{6.4.127}$$

To simplify calculations, we move the equilibrium to the point (1, 1) by defining new variables

$$\xi = \frac{x}{x^*}, \qquad \eta = \frac{y}{y^*}.$$

Then

$$\begin{aligned} \xi' &= \frac{x'}{x^*} = a\xi \left(1 - \frac{by^*}{a}\eta - \frac{ex^*}{a}\xi\right), \\ \eta' &= \frac{y'}{y^*} = -c\eta \left(1 - \frac{dx^*}{c}\xi + \frac{fy^*}{c}\eta\right). \end{aligned} \tag{6.4.128}$$

Since $(\xi, \eta) = (1, 1)$ is the equilibrium, substituting these values into (6.4.128) makes the left-hand sides equal to zero, that is,

$$\begin{aligned} \frac{by^*}{a} + \frac{ex^*}{a} &= 1, \\ \frac{dx^*}{c} - \frac{fy^*}{c} &= 1. \end{aligned}$$

Hence, defining

$$g = \frac{ex^*}{a}, \qquad h = \frac{fy^*}{c},$$

we write (6.4.128) in the form

$$\begin{aligned} \xi' &= a\xi \left(1 - g\xi - (1 - g)\eta\right), \\ \eta' &= -c\eta \left(1 - (1 + h)\xi + h\eta\right). \end{aligned} \tag{6.4.129}$$

Since (6.4.125) is a perturbation of the Lotka–Volterra system, it makes sense to consider the Lyapunov function for the latter,

$$V(\xi,\eta) = \frac{A}{a} f(\xi) + \frac{B}{c} f(\eta) = \frac{A}{a}(\xi - \ln \xi - 1) + \frac{B}{c}(\eta - \ln \eta - 1), \qquad (6.4.130)$$

as the candidate for the Lyapunov function for (6.4.129), where we took into account that $(1, 1)$ is the equilibrium; A and B are positive constants to be determined. Using Exercise 6.4.1, we see that $V(\xi,\eta) > 0$ for $(\xi,\eta) \in \mathbb{R}^2_+$. Since $f'(\xi) = 1 - \frac{1}{\xi}$, $\xi f'(\xi) = \xi - 1$, and hence

$$\begin{aligned} V'(\xi,\eta) = \langle \nabla V(\xi,\eta), (\xi',\eta')\rangle &= A(\xi - 1)(1 - g\xi - (1-g)\eta) \\ &\quad - B(\eta - 1)(1 - (1+h)\xi + h\eta). \end{aligned}$$

The first step is to eliminate the mixed terms, that is, we must have

$$A(1-g) = B(1+h),$$

so we can take

$$A = 1, \qquad B = \frac{1-g}{1+h}.$$

Then, we have

$$\begin{aligned} V'(\xi,\eta) &= (\xi - 1) - g\xi(\xi - 1) + (1-g)\eta - \frac{1-g}{1+h}(\eta - 1) \\ &\quad - (1-g)\xi + \frac{h(1-g)}{1+h}(\eta - 1)\eta \\ &= -g(\xi - 1)^2 - 1 + g - \frac{h(1-g)}{1+h}\left((\eta - 1)^2 + 1 + \frac{1}{h}\right) \\ &= -g(\xi - 1)^2 - \frac{h(1-g)}{1+h}(\eta - 1)^2, \end{aligned}$$

and $V' < 0$ unless $(\xi,\eta) = (1, 1)$. Thus, $(1, 1)$ is asymptotically stable for (6.4.129) and hence (x^*, y^*) is asymptotically stable for (6.4.125). Since V is radially unbounded, we see that it is also globally asymptotically stable in $\mathbb{R}^2_+$, see Fig. 6.32.

Exercise 6.4.5 Use the candidate Lyapunov function

$$V(x, y) = Ax - B\ln x + Cy$$

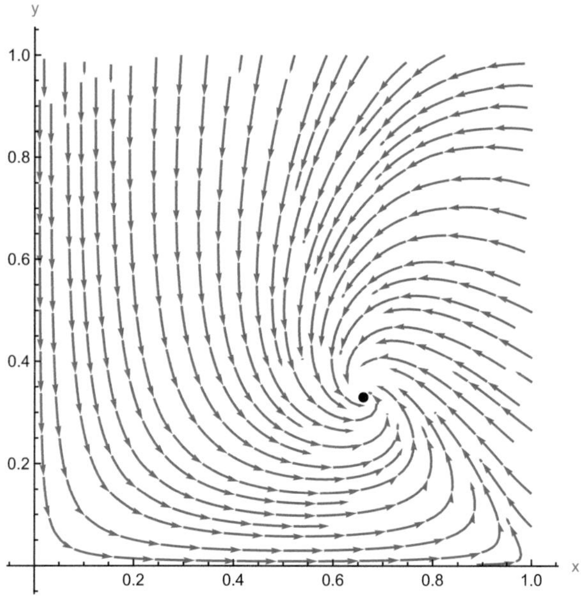

Fig. 6.32 Streamlines of (6.4.125) with $a = b = c = e = f = 1$ and $d = 2$, so that $(x^*, y^*) = \left(\frac{2}{3}, \frac{1}{3}\right)$

to show that the predator extinction equilibrium

$$\left(\frac{a}{e}, 0\right)$$

is asymptotically stable if

$$ad - ce \geq 0.$$

Example 6.59 (SIR Model with Immigration) In this example, we discuss extending a known Lyapunov function for a simpler dynamical system to conclude about a system that is close to it. A general result in this direction, based on the inverse Lyapunov theorem, Theorem 6.75, can be found in [58, Section 5.2].

$$\begin{aligned} S' &= \Lambda - \mu S - \beta SI, \\ I' &= \beta SI - (\mu + \gamma)I + I_{im}, \\ R' &= \gamma I - \mu R. \end{aligned} \tag{6.4.131}$$

If $I_{im} = 0$, the DFE is given by $(S^*, 0, 0) = \left(\frac{\Lambda}{\mu}, 0, 0\right)$ and $R_0 = \frac{\beta S^*}{\mu+\gamma}$. We define $Z = S - S^*$ and, dropping the R equation,

$$\begin{aligned} Z' &= -\mu Z - \beta ZI - \beta S^* I, \\ I' &= \beta ZI + \beta S^* I - (\mu + \gamma)I + I_{im}, \end{aligned} \tag{6.4.132}$$

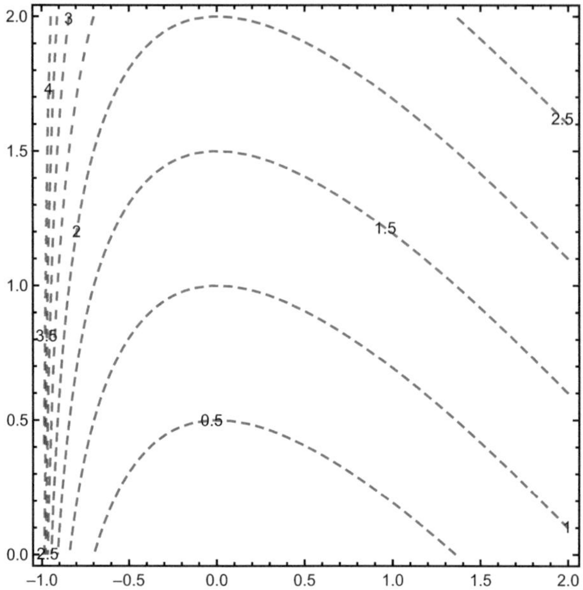

Fig. 6.33 Level curves of V

with the immigration-free DFE equal to $(0, 0)$. The immigration-free Lyapunov function is

$$V(Z, I) = Z - S^* \ln\left(\frac{Z}{S^*} + 1\right) + I,$$

with the level curves illustrated in Fig. 6.33. By Example 6.4.1, we have $V(Z, I) > 0$ if $(Z, I) \neq 0$ and

$$V'(Z, I) = -\mu\frac{Z^2}{Z + S^*} - I(\mu + \gamma)(1 - R_0) < 0$$

as long as $R_0 < 1$ since $Z + S^* = S > 0$; the level curves of V' are shown in Fig. 6.34.

Recall the identity

$$(\mu + \gamma)(1 - R_0) = \mu + \gamma - \beta S^*,$$

which we will use below. Next, we apply V to the model with migration. Then

$$V'(Z, I) = -\mu\frac{Z^2}{Z + S^*} - I(\mu + \gamma)(1 - R_0) + I_{im},$$

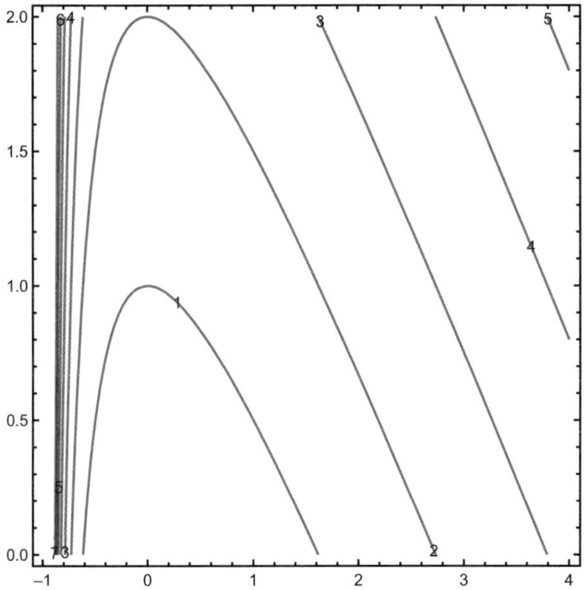

Fig. 6.34 Level curves of V'

and V' is negative as long as $(Z(t), I(t))$ satisfies

$$\mu \frac{Z^2(t)}{Z(t) + S^*} + I(t)(\mu + \gamma)(1 - R_0) > I_{im},$$

or

$$I(t) > -\frac{\mu}{(\mu + \gamma)(1 - R_0)} \frac{Z^2(t)}{Z(t) + S^*} + \frac{1}{(\mu + \gamma)(1 - R_0)} I_{im}. \tag{6.4.133}$$

Define

$$I(Z) := -\frac{\mu}{(\mu + \gamma)(1 - R_0)} \frac{Z^2}{Z + S^*} + \frac{1}{(\mu + \gamma)(1 - R_0)} I_{im}, \tag{6.4.134}$$

for $Z > -S^*$. It is positive for $Z < \frac{I_{im} + \sqrt{I_{im}^2 + 4 I_{im} \Lambda}}{2\mu}$ and reaches maximum $\frac{1}{(\mu+\gamma)(1-R_0)} I_{im}$ at $Z = 0$. Keeping the assumption $R_0 < 1$, we consider the case

$$\beta S^* \geq \gamma. \tag{6.4.135}$$

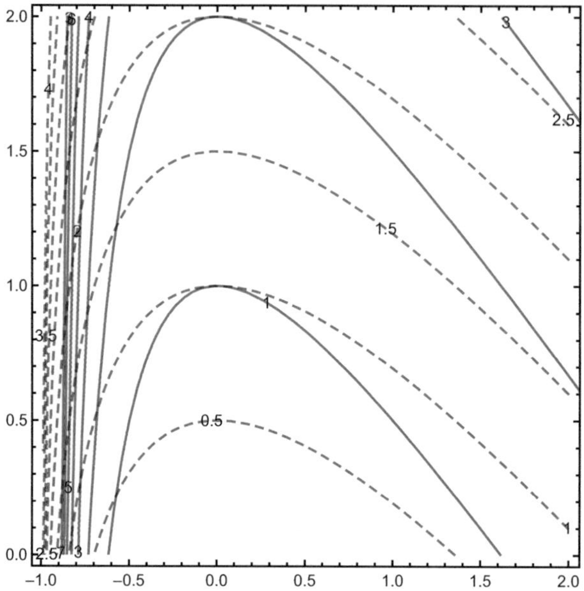

Fig. 6.35 Level curves of V' (solid) and V (dashed) when (6.4.135) is satisfied

As long as $I(t)$ satisfies (6.4.133), it strictly decreases as V is a Lyapunov function in this region. Then, any such trajectory must reach the contour line of V',

$$I = -\frac{\mu}{(\mu+\gamma)(1-R_0)}\frac{Z^2}{Z+S^*} + \frac{1}{(\mu+\gamma)(1-R_0)}I_{im}, \tag{6.4.136}$$

at some t_1 and, at this time, it is inside the contour line of V:

$$Z - S^* \ln\left(\frac{Z}{S^*}+1\right) + I = \frac{1}{(\mu+\gamma)(1-R_0)}I_{im}, \tag{6.4.137}$$

see Fig. 6.35, and, by the invariance, will stay there for all future times. So, in particular, we have

$$I(t) \le \frac{1}{(\mu+\gamma)(1-R_0)}I_{im}$$

as $t \to \infty$.

Exercise 6.4.6 Justify the claim that if a trajectory $(I(t), Z(t))$ satisfies (6.4.133) for some t_0, then it must reach the level curve (6.4.134) for some $t_1 > t_0$. Hint. You can follow the argument of the proof of Theorem 6.51.

Exercise 6.4.7

(i) Show that

$$f(x) = a\frac{x^2}{x+1} - x + \ln(x+1) \geq 0, \quad x > -1,$$

provided $a \geq 1$.

(ii) Use this result to show that (6.4.136) implies (6.4.137) provided (6.4.135) is satisfied.

(iii) Can you determine a sublevel set of V containing (6.4.136) if (6.4.135) is not satisfied, and hence estimate $I(t)$ in terms of I_{im}?

Let us return to Example 6.43 to motivate a further refinement of a Lyapunov theory.

Example 6.60 Consider again the equation

$$u'' + cu' + g(u) = 0,$$

which, in the context of mechanics, is called the Liénard equation. We adopt the same assumptions as in Example 6.43. Writing the equation as a system, we get

$$\begin{aligned} x_1' &= x_2, \\ x_2' &= -g(x_1) - cx_2. \end{aligned} \tag{6.4.138}$$

As we have shown,

$$V(x_1, x_2) = \frac{1}{2}x_2^2 + \int\limits_0^{x_1} g(\sigma)d\sigma$$

is the Lyapunov function in

$$\Omega = \{(x_1, x_2);\ |x_1| < k, x_2 \in \mathbb{R}\},$$

satisfying

$$V'(x_1, x_2) = -x_2x_2' + g(x_1)x_1' = -cx_2^2 - g(x_1)x_2 + g(x_1)x_2 = -cx_2^2.$$

Hence, V is a Lyapunov function for (6.4.138), ensuring that $(0, 0)$ is stable. However, it fails to be a strict Lyapunov function as there is no neighbourhood of $(0, 0)$ on which V' is strictly positive. Now, if we look closer at this example, we see that we should be able to prove something more. Namely, if we can ensure that a trajectory stays away from $L = \{(x_1, x_2) :\ x_2 = 0, x_1 \neq 0\}$ then, following the proof of Theorem 6.39, we obtain that it must converge to $(0, 0)$. On the other hand, at any point of L, the vector field is transversal to L, so the trajectory cannot stay on L, and the field would have to be tangent to L. Thus, it is to be expected that the trajectory must eventually reach $(0, 0)$. We shall provide a rigorous and more general result of this type below.

Theorem 6.61 (LaSalle Invariance Principle) *Let $\boldsymbol{x}^* = 0$ be a stationary point of (6.2.17) and let V be a Lyapunov function on some neighbourhood $\Omega \ni 0$. If, for $\boldsymbol{x} \in \Omega$, the (forward) trajectory $\Gamma_{\boldsymbol{x}}$ is bounded with limit points in Ω and M is the largest invariant subset of*

$$E := \{\boldsymbol{x} \in \Omega : V'(\boldsymbol{x}) = 0\}, \tag{6.4.139}$$

then

$$\phi(t, \boldsymbol{x}) \to M \quad as \quad t \to \infty. \tag{6.4.140}$$

Remark 6.62 The convergence of the solution to the set M is understood in the sense explained in Lemma 6.49, point 6.

Proof By assumptions, Lemma 6.49, point 1., implies that for any $\boldsymbol{x} \in \Omega$ satisfying the assumption of the theorem, $\emptyset \neq \omega(\Gamma_{\boldsymbol{x}}) \subset \Omega$. Since V is a Lyapunov function, $V(\phi(t, \boldsymbol{x}))$ is a non-increasing function of t which is bounded below by zero. Hence, there is $c \geq 0$ such that

$$\lim_{t\to\infty} V(\phi(t, \boldsymbol{x})) = c.$$

Now let $\boldsymbol{y} \in \omega(\Gamma_{\boldsymbol{x}})$. Then, $\phi(t_n, \boldsymbol{x}) \to \boldsymbol{y}$ for some $(t_n)_{n\in\mathbb{N}}$ converging to ∞ as $n \to \infty$. On the other hand, by the continuity of V, we have

$$\lim_{t\to\infty} V(\phi(t_n, \boldsymbol{x})) = c.$$

Consequently, V is constant on $\omega(\Gamma_{\boldsymbol{x}})$.

Now, by Lemma 6.49, point 3., $\omega(\Gamma_{\boldsymbol{x}})$ is invariant so that if $\boldsymbol{y} \in \omega(\Gamma_{\boldsymbol{x}})$, then $\phi(t, \boldsymbol{y}) \in \omega(\Gamma_{\boldsymbol{x}})$ for all t. Thus, $V(\phi(t, \boldsymbol{y})) = c$ for all t and $\boldsymbol{y} \in \omega(\Gamma_{\boldsymbol{x}})$ and hence

$$V'(\boldsymbol{y}) = 0 \quad \text{for} \quad \boldsymbol{y} \in \omega(\Gamma_{\boldsymbol{x}}).$$

So

$$\omega(\Gamma_{\boldsymbol{x}}) \subset M \subset E.$$

But $\phi(t, \boldsymbol{x}) \to \omega(\Gamma_{\boldsymbol{x}})$ as $t \to \infty$, so $\phi(t, z) \to M$ as $t \to \infty$ for all $z \in \Omega$ satisfying $\omega(\Gamma_z) \subset \Omega$. □

Remark 6.63 The value of this result lies in the fact that it provides a bound for the ω-limit set, which may be very complicated, by a set which is easier to find.

Remark 6.64 From the proof of the LaSalle principle, we see that V doesn't need to satisfy the first condition of the Lyapunov function in Definition 6.38; it is sufficient if V is bounded below. Also, as previously, the set E can intersect the boundary of an invariant region.

Remark 6.65 As we know, examples of open forward invariant sets are the sublevel sets of V,

$$\Omega_k = \{\boldsymbol{x} : \; V(\boldsymbol{x}) < k\}$$

and they can be used as the set Ω in the LaSalle principle as long as they are bounded.

Remark 6.66 La Salle principle gives immediate proof of Theorem 6.39. Indeed, in the domain of applicability of this theorem, $M = E = \{0\}$.

Combining Theorems 6.56 and 6.61, we obtain

Corollary 6.67 *Assume that there is a Lyapunov function for (6.2.17) defined on the whole* $\mathbb{R}^d$ *which satisfies additionally* $V(\boldsymbol{y}) \to \infty$ *as* $\|\boldsymbol{y}\| \to \infty$. *If* 0 *is the only invariant set of* $E = \{\boldsymbol{x} \in \mathbb{R}^d \; : \; V'(\boldsymbol{x}) = 0\}$, *then* 0 *is globally asymptotically stable.*

Example 6.68 (Global Stability of SEIR Equilibria) Let us continue with the SEIR model of Example 6.35 and determine the global stability of the equilibria, which we found there, namely the disease-free one

$$\boldsymbol{E}_0 = \left(\frac{\Lambda}{\mu}, 0, 0, 0\right), \tag{6.4.141}$$

and the endemic one

$$\begin{aligned} \boldsymbol{E}_1 &= (S^*, E^*, I^*, R^*) \\ &= \left(\frac{(\mu + \gamma)(\mu + \nu)}{\gamma\beta}, \frac{\mu+\nu}{\gamma}\frac{\mu}{\beta}(R_0 - 1), \frac{\mu}{\beta}(R_0 - 1), \frac{\nu}{\beta}(R_0 - 1)\right), \end{aligned} \tag{6.4.142}$$

which exists in $\mathbb{R}^4_+$ if

$$R_0 = \frac{\Lambda\beta\gamma}{(\mu+\gamma)(\mu+\nu)\mu} > 1. \tag{6.4.143}$$

Proposition 6.69 *If* $R_0 < 1$, *then the disease-free equilibrium* $\boldsymbol{E}_0$ *is globally asymptotically stable (in* $\mathbb{R}^4_+$*).*

Proof Since in (6.3.84) the first three equations are independent of the last one, if $\left(\frac{\Lambda}{\mu}, 0, 0\right)$ is attracting in $\mathbb{R}^3_+$, then also $R(t) \to 0$ as $t \to \infty$ for any nonnegative $R(0)$. This follows from

$$R(t) = e^{-\mu t}R(0) + \nu e^{-\mu t}\int_0^t e^{\mu s}I(s)ds$$

and the l'Hôspital rule. Hence, consider

$$\begin{aligned} S' &= \Lambda - \beta SI - \mu S, \\ E' &= \beta SI - (\mu + \gamma)E, \\ I' &= \gamma E - (\mu + \nu)I. \end{aligned} \tag{6.4.144}$$

The first two equations have a similar structure to the Lotka–Volterra equation, and we are looking at the equilibrium of the boundary equilibrium, so we can try the function introduced in Exercise 6.4.5, which, in the current setting, takes the form

$$V(S, E, I) = \alpha\left(S - S^* - S^* \ln\frac{S}{S^*}\right) + \frac{1}{\mu+\gamma}E + \frac{1}{\gamma}I, \tag{6.4.145}$$

where α is to be determined, and $S^* = \frac{\Lambda}{\mu}$.

Using Exercise 6.4.1, we establish that $V > 0$ for $(S, E, I) \neq \left(\frac{\Lambda}{\mu}, 0, 0\right)$. Next,

$$\begin{aligned} V'(S, E, I) &= \alpha\left(1 - \frac{S^*}{S}\right)(\Lambda - \beta SI - \mu S) \\ &\quad + \frac{\beta SI - (\mu+\gamma)E}{\mu+\gamma} + \frac{\gamma E - (\mu+\nu)I}{\gamma} \\ &= 2\alpha\Lambda - \beta\alpha SI - \alpha\mu S - \frac{\Lambda^2\gamma}{\mu S} + \frac{\Lambda\beta\alpha}{\mu}I + \frac{\beta}{\mu+\gamma}SI - \frac{\mu+\nu}{\gamma}I. \end{aligned}$$

Choosing $\alpha = \frac{1}{\mu+\gamma}$ eliminates the cross product SI, giving in the last line

$$2\alpha\Lambda - \alpha\mu S - \frac{\Lambda^2\alpha}{\mu S} + \frac{\Lambda\beta\alpha}{\mu}I - \frac{\mu+\nu}{\gamma}I = -\gamma\Lambda\left(\frac{\Lambda}{\mu S} + \frac{\mu S}{\Lambda} - 2\right) + \frac{\mu+\nu}{\gamma}(R_0 - 1)I.$$

Since the last term is negative due to $R_0 < 1$, we focus on the first term. Denoting $x = \frac{\Lambda}{\mu S}$, the term inside the bracket can be transformed as

$$x + \frac{1}{x} - 2 = \frac{(x-1)^2}{x} > 0, \quad x \neq 1, x > 0.$$

Since $x = 1$ if and only if $S = \frac{\Lambda}{\mu} = S^*$, we see that $V' < 0$ if $(S, E, I) \neq (S^*, 0, 0)$. As V is radially unbounded in the $\mathbb{R}^3_+$, we find that $(S^*, 0, 0)$ is globally stable. □

Next, we turn our attention to the endemic equilibrium. We have

Proposition 6.70 *If $R_0 > 1$, then the endemic equilibrium*

$$\begin{aligned} \boldsymbol{E}_1 &= (S^*, E^*, I^*, R^*) \\ &= \left(\frac{(\mu+\gamma)(\mu+\nu)}{\gamma\beta}, \frac{\mu+\nu}{\gamma}\frac{\mu}{\beta}(R_0 - 1), \frac{\mu}{\beta}(R_0 - 1), \frac{\nu}{\beta}(R_0 - 1) \right) \end{aligned} \tag{6.4.146}$$

is globally asymptotically stable (in $\mathbb{R}^4_+$).

Proof As before, it is sufficient to consider the first three components, (S, E, I). We know they stay in the positive orthant $\mathbb{R}^3_+$ if they start in it. Arguing as in the DFE case and noting that we are looking at an interior equilibrium, we consider the candidate Lyapunov function based on (6.4.130), which in the current notation takes the form

$$\begin{aligned} V(S, E, I) = \kappa_1 \left(S - S^* - S^* \ln \frac{S}{S^*} \right) + \kappa_2 \left(E - E^* - E^* \ln \frac{E}{E^*} \right) \\ + \kappa_3 \left(I - I^* - I^* \ln \frac{I}{I^*} \right), \end{aligned} \tag{6.4.147}$$

where κ_i, $i = 1, 2, 3$, are constants to be determined. As in the previous proof, $V(S^*, E^*, I^*) = 0$ and $V > 0$ otherwise. Moreover, V is radially unbounded in $\mathbb{R}^3_+$. So, it remains to determine the sign of V'. We have

$$\begin{aligned} V'(S, E, I) &= \kappa_1 \left(1 - \frac{S^*}{S} \right) S' + \kappa_2 \left(1 - \frac{E^*}{E} \right) E' + \kappa_3 \left(1 - \frac{I^*}{I} \right) I' \\ &= \kappa_1 \left(1 - \frac{S^*}{S} \right) (\Lambda - \beta S I - \mu S) \\ &\quad + \kappa_2 \left(1 - \frac{E^*}{E} \right) (\beta S I - (\mu + \gamma) E) \\ &\quad + \kappa_3 \left(1 - \frac{I^*}{I} \right) (\gamma E - (\mu + \nu) I). \end{aligned}$$

First, we observe that (6.3.90) yields

$$\Lambda = \beta S^* I^* + \mu S^*$$

and thus

$$\kappa_1\left(1-\frac{S^*}{S}\right)(\Lambda-\beta SI-\mu S)=-\kappa_1\mu\frac{(S-S^*)^2}{S}+\kappa_1\beta S^*I^*-\kappa_1\beta SI\\ -\kappa_1\beta\frac{S^{*2}I^*}{S}+\kappa_1\beta S^*I.$$

Multiplying out the other brackets, we obtain

$$\begin{aligned}V'(S,E,I)=&-\kappa_1\mu\frac{(S-S^*)^2}{S}+\kappa_1\beta S^*I^*-\kappa_1\beta SI-\kappa_1\beta\frac{S^{*2}I^*}{S}+\kappa_1\beta S^*I\\ &+\kappa_2\beta SI-\kappa_2(\mu+\gamma)E-\kappa_2\beta\frac{E^*}{E}SI+\kappa_2E^*(\mu+\gamma)\\ &+\kappa_3\gamma E-\kappa_3(\mu+\nu)I-\kappa_3\gamma\frac{I^*E}{I}+\kappa_3I^*(\mu+\nu).\end{aligned}$$

First, we observe that by taking $\kappa_1=\kappa_2=1$, we cancel the terms containing SI. Next, we group together terms containing linear terms. We have

$$\begin{aligned}&E(\gamma\kappa_3-(\mu+\gamma))\\ &I(\beta S^*-\kappa_3(\mu+\nu))=I(\mu+\nu)\left(\frac{\mu+\gamma}{\gamma}-\kappa_3\right),\end{aligned}$$

where we used (6.4.146) to express S^*. Hence, putting

$$\kappa_3=\frac{\mu+\gamma}{\gamma}$$

eliminates the linear terms, leaving

$$\begin{aligned}V'(S,E,I)=&-\mu\frac{(S-S^*)^2}{S}+\beta S^*I^*-\beta\frac{S^{*2}I^*}{S}-\beta\frac{E^*}{E}SI\\ &+E^*(\mu+\gamma)-(\mu+\gamma)\frac{I^*E}{I}+\frac{\mu+\gamma}{\gamma}I^*(\mu+\nu).\end{aligned}$$

Now, we use (6.3.88) and (6.3.89) to get

$$\beta S^*I^*=(\mu+\gamma)E^*,$$

and again the formula for S^* to get

$$V'(S,E,I)=-\mu\frac{(S-S^*)^2}{S}+\beta S^*I^*\left(3-\frac{S^*}{S}-\frac{E^*SI}{ES^*I^*}-\frac{I^*E}{IE^*}\right),$$

where again we used

$$\mu + \gamma = \frac{\beta S^* I^*}{E^*}.$$

Let us denote

$$x_1 = \frac{S^*}{S}, \quad x_2 = \frac{E^* S I}{E S^* I^*}, \quad x_3 = \frac{I^* E}{I E^*},$$

and observe that

$$x_1 x_2 x_3 = 1.$$

What is the maximum of $x_1 + x_2 + x_3$ under this constraint? We can use the relation between the harmonic and geometric means:

$$\frac{x_1 + \ldots + x_n}{n} \geq \sqrt[n]{x_1 \ldots x_n}, \quad x_1, \ldots, x_n \geq 0, \tag{6.4.148}$$

to ascertain that $x_1 + x_2 + x_3 \geq 3$ with the equality attained if $x_1 = x_2 = x_3 = 1$; see Exercise 6.4.8. Hence, we obtain

$$V'(S, E, I) \leq 0$$

with

$$V'(S, E, I) = 0$$

on the set

$$\left\{(S, E, I) : \; S = S^*, I = \frac{I^*}{E^*} E\right\}.$$

Now, along the line $I = \frac{I^*}{E^*} E$, $S = S^*$, the first two components of the field are given by $\Lambda - \beta I S^* - \mu S^*$ and 0, whereas the line is parallel to $(0, I^*, E^*)$. Thus, the only invariant set on the line is the equilibrium (S^*, E^*, I^*) and, by the LaSalle principle, the ω-limit set of any trajectory originating from initial conditions not on the S axis consists of the endemic equilibrium. □

Exercise 6.4.8 Prove (6.4.148). Hint. Let

$$\alpha := \frac{x_1 + \ldots + x_n}{n}, \quad \sqrt[n]{x_1 \ldots x_n} =: \beta.$$

One of the simplest proofs starts with the observation that (6.4.148) obviously holds if $x_1 = x_2 = \ldots = x_n$. If not, we must have $x_i < \alpha < x_j$ for some i, j. Show that replacing x_j and x_j by $x_i+x_j-\alpha$ and α, respectively, does not change the arithmetic mean but increases the geometric mean. Continue the process if necessary until all terms on both sides are equal, and then deduce (6.4.148).

Not always does the LaSalle principle give a definitive answer about the stability of equilibria, and concluding it requires additional considerations.

Example 6.71 Consider the following SIS model with balanced Malthusian growth and disease-induced deaths

$$\begin{aligned} S' &= \mu N + \gamma I - \beta\frac{SI}{N} - \mu S = (\mu+\gamma)I - \beta\frac{SI}{N}, \\ I' &= \beta\frac{SI}{N} - (\gamma+\mu+\delta)I, \end{aligned} \tag{6.4.149}$$

where $N = S + I$.

Exercise 6.4.9 Show that all solutions to (6.4.149) originating from nonnegative initial conditions $S(0)$, $I(0)$ satisfying $S(0) + I(0) > 0$, are nonnegative, bounded and globally defined.

First, we observe that any point $(S, 0)$, $S > 0$ is an equilibrium so that the equilibria are nonisolated. Hence, the linearization techniques and classical Lyapunov theorems are not available.

We can use

$$N = S + I,$$

as the Lyapunov function. Here, we have

$$N' = -\delta I \le 0,$$

so that, by the LaSalle principle, $(S(t), I(t)) \to \{(S, I) : \ S \ge 0, I = 0\}$. In particular, $\lim_{t\to\infty} I(t) = 0$. So, the disease will be eliminated. However, we do not know whether this is due to the elimination of the whole population (that is, whether also $N(t) \to 0$) or whether there are survivors.

Let us consider the nullclines of (6.4.149). We have

- $S' = 0$ if and only if $I = 0$ or

$$I(\mu+\gamma) - (\beta-\mu-\gamma)S = 0 \quad\Rightarrow\quad I = \left(\frac{\beta}{\mu+\gamma} - 1\right)S,$$

and

- $I' = 0$ if and only if $I = 0$ or

$$I(\mu+\gamma+\delta) - (\beta-\mu-\gamma-\delta)S = 0 \quad\Rightarrow\quad I = \left(\frac{\beta}{\mu+\gamma+\delta} - 1\right)S.$$

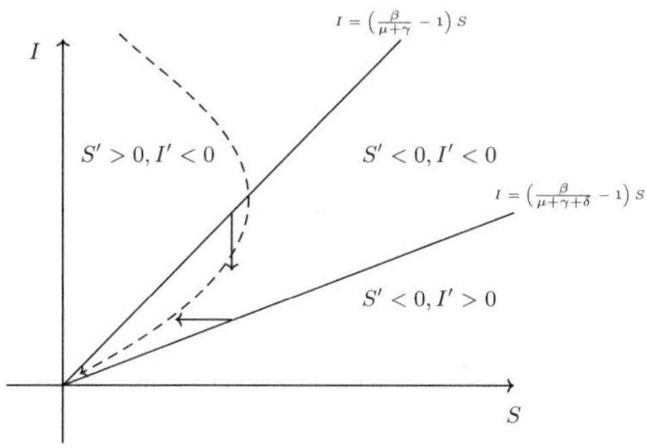

Fig. 6.36 Nullclines of (6.4.149) with $\frac{\beta}{\mu+\gamma} > \frac{\beta}{\mu+\gamma+\delta} > 1$

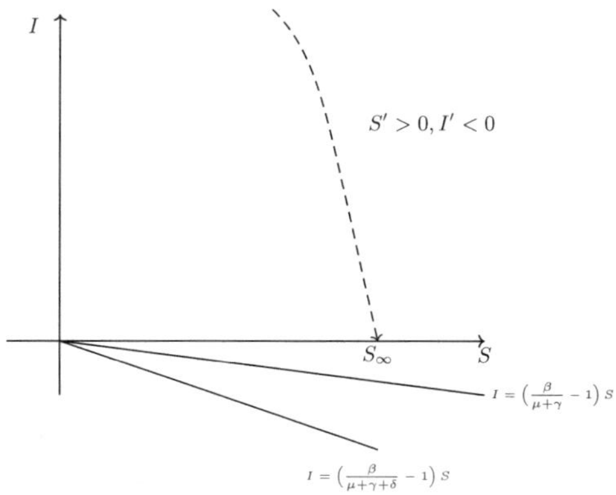

Fig. 6.37 Nullclines of (6.4.149) with $\frac{\beta}{\mu+\gamma+\delta} < \frac{\beta}{\mu+\gamma} < 1$

Based on Fig. 6.36, we see that as long as $\beta > \mu + \gamma$, $S(t) \to 0$, so that the elimination of the disease is achieved by the elimination of the population. On the other hand, Fig. 6.37 shows that if $\beta \le \mu+\gamma$, then $S(t)$ converges to a nonzero limit, which depends on the initial condition. In other words, the population with disease-induced mortality will survive only if the population with no disease-induced deaths has $R_0 < 1$.

Exercise 6.4.10 Consider the equation of the form

$$x'' + f(x)x' + g(x) = 0. \tag{6.4.150}$$

Show, that by defining $x_1 = x$ and $x_2 = x_1' + F(x_1)$, where $F(x) = \int^x f(\xi)d\xi$, we can write (6.4.150) as the system

$$x_1' = x_2 - F(x_1),$$
$$x_2' = -g(x_1).$$

Apply this approach to the so-called van der Pol equation

$$x'' - ax'(x^2 - 1) + x = 0,$$

where $a > 0$ is a constant. This equation describes damped oscillations and is a prototype of the FitzHugh–Nagumo system describing excitable systems such as neurons.

Apply the above change of variables to transform (9.1.2) into the system

$$\begin{aligned} x_1' &= x_2 + a\left(\frac{1}{3}x_1^3 - x_1\right), \\ x_2' &= -x_1. \end{aligned} \tag{6.4.151}$$

For the Lyapunov function

$$V(x_1, x_2) = \frac{1}{2}(x_1^2 + x_2^2),$$

find V' and determine the region where $V' \leq 0$. Then, find the largest set of the form

$$V_k = \{(x_1, x_2) : \ V(x_1, x_2) < k\}$$

contained in the region where $V' \leq 0$; see Remark 6.40. Hence, use the LaSalle principle to conclude that $(0, 0)$ is an asymptotically stable equilibrium for (6.4.151). What is its basin of attraction?

Exercise 6.4.11 Consider the equation

$$x'' + 2ax' + x + x^3 = 0, \tag{6.4.152}$$

where a is a constant satisfying $0 < a < 1$. Write this equation as a system and find its equilibria.

Compare (6.4.152) with Example 6.30 and hence use the first integral obtained there as a candidate Lyapunov function at $(0, 0)$. Then, use Corollary 6.67 to show that $(0, 0)$ is globally asymptotically stable on $\mathbb{R}^2$.

Exercise 6.4.12 Consider in a domain of $\mathbb{R}^d$ the system

$$\boldsymbol{y}' = \boldsymbol{f}(\boldsymbol{y}), \tag{6.4.153}$$

where $\boldsymbol{f}$ has continuous first partial derivatives in some neighbourhood of an equilibrium $\boldsymbol{x}^*$, and its linearization

$$\boldsymbol{x}' = \mathcal{A}\boldsymbol{x} + o(\|\boldsymbol{x}\|). \tag{6.4.154}$$

where $\boldsymbol{x} = \boldsymbol{y} - \boldsymbol{x}^*$; see Lemma 6.33. Assume that $\mathcal{A}$ has distinct eigenvalues $\{\lambda_1, \ldots, \lambda_d\}$ with the corresponding eigenvectors $\{\boldsymbol{e}_1, \ldots, \boldsymbol{e}_d\}$ and let $\{\boldsymbol{f}_1, \ldots, \boldsymbol{f}_d\}$ be the eigenvectors of $\mathcal{A}^T$, see Theorem 5.21. We assume that the eigenvectors are normalized eigenvectors so that

$$\langle \boldsymbol{f}_i, \boldsymbol{e}_j \rangle = \delta_{ij},$$

(that is, 1 for $i = 1$ and 0 for $i \neq j$). Then, using (5.3.85), we can write any $\boldsymbol{x}$ as

$$\boldsymbol{x} = \sum_{i=1}^{d} \langle \boldsymbol{f}_i, \boldsymbol{x} \rangle \boldsymbol{e}_i.$$

Show that

$$\frac{d}{dt} \langle \boldsymbol{f}_i, \boldsymbol{x} \rangle = \lambda_i \langle \boldsymbol{f}_i, \boldsymbol{x} \rangle + o(\|\boldsymbol{x}\|), \quad i = 1, \ldots, d.$$

Use this fact to show that for any positive $\alpha_1, \ldots, \alpha_d$,

$$V(\boldsymbol{x}) = \sum_{i=1}^{d} \alpha_i \overline{\langle \boldsymbol{f}_i, \boldsymbol{x} \rangle} \langle \boldsymbol{f}_i, \boldsymbol{x} \rangle$$

is a Lyapunov function for (6.4.154).

6.5 Inverse Lyapunov Theorem and Its Applications

Lyapunov functions are useful tools allowing for proving stability or asymptotic stability of equilibria of (6.2.17), so it would seem that if we know that an equilibrium is stable, looking for a Lyapunov function at this point no longer makes sense. It follows, however, that knowing that such a function exists, we can prove several interesting results. For instance, all stability definitions are qualitative, whereas knowing a Lyapunov function furnishes quantitative information, such as estimating the rate of convergence and the basin of attraction to equilibrium. Also, if we have a Lyapunov function for a simple system, we can stretch it to work for systems close to it, as seen in Example 6.59.

The construction of a Lyapunov function at an equilibrium depends on the rate of convergence of solutions to it. We restrict our construction to the so-called exponentially stable equilibria, defined below; for other types of stability, see, e.g., [51, 58].

Definition 6.72 We say that the equilibrium $\boldsymbol{x}^* = 0$ of (6.2.17) is exponentially stable if there are constants $k, \gamma, \delta > 0$, such that for any τ and $\boldsymbol{x}(\tau) = \mathring{\boldsymbol{x}} \in B_\delta$ and $t \geq \tau$ we have

$$\boldsymbol{x}(t) \leq k\|\mathring{\boldsymbol{x}}\|e^{-\gamma(t-\tau)}. \tag{6.5.155}$$

If the estimate is valid for any $\mathring{\boldsymbol{x}}$ (in the domain of the equation), we say that $\boldsymbol{x}^* = 0$ is globally exponentially stable.

Exercise 6.5.1 Show that an equilibrium satisfying assumptions of point 1. of Theorem 6.34 is exponentially stable.

Exercise 6.5.2 Show that $x^* = 0$ is an asymptotically stable but not exponentially stable equilibrium of

$$x' = -x^3.$$

We will precede the formulation and proof of the main result by two auxiliary lemmas. Let $\boldsymbol{x}(t) = \phi(t, \tau, \boldsymbol{y})(= \phi(t - \tau, \boldsymbol{y}))$ be the maximally defined solution to

$$\boldsymbol{x}' = \boldsymbol{f}(\boldsymbol{x}), \quad \boldsymbol{x}(\tau) = \boldsymbol{y} \in \Omega, \tag{6.5.156}$$

where $\Omega \subset \mathbb{R}^d$ is convex and open, and $\boldsymbol{f}$ is a differentiable function on $\overline{\Omega}$. By assumption,

$$\|\mathcal{J}_{\boldsymbol{f}}\| \leq L \quad \text{on} \quad \overline{\Omega}$$

for some constant L. We use the matrix norm, that is, the operator norm consistent with the norm in $\mathbb{R}^d$, see Sect. A.2.

Lemma 6.73 *Assume that $\Omega \ni \boldsymbol{x}^* = 0$ is an equilibrium for* (6.5.156). *Then, on the interval of existence of $\phi(t, \tau, \boldsymbol{y})$ we have*

$$\|\boldsymbol{y}\|_2 e^{-L(t-\tau)} \leq \|\boldsymbol{x}(t)\|_2 \leq \|\boldsymbol{y}\|_2 e^{L(t-\tau)}, \tag{6.5.157}$$

where we use the euclidean norm $\|\cdot\|_2$, see Sect. A.1.

Proof Since Ω is convex, Taylor's formula and $\boldsymbol{f}(0) = 0$ yields

$$\|\boldsymbol{f}(\boldsymbol{x})\|_2 \leq L\|\boldsymbol{x}\|_2.$$

Further, denoting $\zeta(t) = \|x(t)\|^2$, we get

$$\zeta'(t) = \frac{d}{dt}\langle \boldsymbol{x}(t), \boldsymbol{x}(t)\rangle = 2\langle \boldsymbol{x}(t), \boldsymbol{f}(\boldsymbol{x}(t))\rangle, \tag{6.5.158}$$

and hence

$$\left|\frac{d}{dt}\|\boldsymbol{x}(t)\|_2^2\right| \leq 2L\|\boldsymbol{x}(t)\|_2^2.$$

Thus

$$-2L\zeta(t) \leq \zeta'(t) \leq 2L\zeta(t),$$

the integration of which with the initial condition $\zeta(\tau) = \|\boldsymbol{y}\|_2^2$, yields (6.5.157). □

Lemma 6.74 *On the interval of existence of* $\boldsymbol{x}(t) = \phi(t, \tau, \boldsymbol{y})$ *we have*

$$\frac{\partial}{\partial \tau}\phi(t, \tau, \boldsymbol{y}) + \nabla_{\boldsymbol{y}}\phi(t, \tau, \boldsymbol{y})\boldsymbol{f}(\boldsymbol{y}) = 0. \tag{6.5.159}$$

Proof Using Theorem B.18 (with Remark B.19) and, in particular, (B.4.57), we have

$$\begin{aligned} \frac{\partial}{\partial \tau}\phi(t, \tau, \boldsymbol{y}) &= -\boldsymbol{f}(\boldsymbol{y}) + \int_\tau^t \mathcal{J}_f(\phi(\sigma, \tau, \boldsymbol{y}))\frac{\partial}{\partial \tau}\phi(\sigma, \tau, \boldsymbol{y})d\sigma, \\ \nabla_{\boldsymbol{y}}\phi(t, \tau, \boldsymbol{y}) &= \mathcal{I} + \int_\tau^t \mathcal{J}_f(\phi(\sigma, \tau, \boldsymbol{y}))\nabla_{\boldsymbol{y}}\phi(\sigma, \tau, \boldsymbol{y})d\sigma, \end{aligned} \tag{6.5.160}$$

where we used notation (B.4.58). Hence

$$\frac{\partial}{\partial \tau}\phi(t, \tau, \boldsymbol{y})|_{t=\tau} = -\boldsymbol{f}(\boldsymbol{y}) \quad \text{and} \quad \nabla_{\boldsymbol{y}}\phi(t, \tau, \boldsymbol{y})|_{t=\tau} = \mathcal{I}. \tag{6.5.161}$$

Differentiating (6.5.160) with respect to t, we obtain

$$\begin{aligned} \frac{\partial}{\partial t}\frac{\partial}{\partial \tau}\phi(t, \tau, \boldsymbol{y}) &= \mathcal{J}_f(\phi(t, \tau, \boldsymbol{y}))\frac{\partial}{\partial \tau}\phi(t, \tau, \boldsymbol{y}), \\ \frac{\partial}{\partial t}\nabla_{\boldsymbol{y}}\phi(t, \tau, \boldsymbol{y}) &= \mathcal{J}_f(\phi(t, \tau, \boldsymbol{y}))\nabla_{\boldsymbol{y}}\phi(t, \tau, \boldsymbol{y}). \end{aligned} \tag{6.5.162}$$

Multiplying the second equation of (6.5.162) by $\boldsymbol{f}(\boldsymbol{y})$ and adding to the first, we obtain a linear equation for $\frac{\partial}{\partial \tau}\phi(t, \tau, \boldsymbol{y}) + \nabla_{\boldsymbol{y}}\phi(t, \tau, \boldsymbol{y})\boldsymbol{f}(\boldsymbol{y})$,

$$\begin{aligned} &\frac{\partial}{\partial t}\left(\frac{\partial}{\partial \tau}\phi(t, \tau, \boldsymbol{y}) + \nabla_{\boldsymbol{y}}\phi(t, \tau, \boldsymbol{y})\boldsymbol{f}(\boldsymbol{y})\right) \\ &\qquad = \mathcal{J}_f(\phi(t, \tau, \boldsymbol{y}))\left(\frac{\partial}{\partial \iota}\phi(t, \tau, \boldsymbol{y}) + \nabla_{\boldsymbol{y}}\phi(t, \tau, \boldsymbol{y})\boldsymbol{f}(\boldsymbol{y})\right) \end{aligned}$$

with the initial condition

$$\left.\frac{\partial}{\partial \tau}\phi(t,\tau,\boldsymbol{y}) + \nabla_{\boldsymbol{y}}\phi(t,\tau,\boldsymbol{y})\boldsymbol{f}(\boldsymbol{y})\right|_{t=\tau} = -\boldsymbol{f}(\boldsymbol{y}) + \mathcal{J}\boldsymbol{f}(\boldsymbol{y}) = 0,$$

where we used (6.5.161). Then, the uniqueness of solutions yields (6.5.159). □

Theorem 6.75 *Assume that the function* $\boldsymbol{f}$ *in* (6.2.17) *is continuously differentiable on* $\Omega = B_r$, $r > 0$, *and its Jacobi matrix* $\mathcal{J}_{\boldsymbol{f}}$ *is bounded there. If* $\boldsymbol{x}^* = 0$ *is an exponentially stable equilibrium of* (6.2.17) *in some set* $\Omega_0 \subset \Omega$, *then there is a continuously differentiable function* $V : \Omega_0 \to \mathbb{R}$ *satisfying*

$$c_1\|\boldsymbol{x}\|^2 \le V(\boldsymbol{x}) \le c_2\|\boldsymbol{x}\|^2, \tag{6.5.163}$$

$$\|\nabla V(\boldsymbol{x})\| \le c_4\|\boldsymbol{x}\|, \tag{6.5.164}$$

$$V'(\boldsymbol{x}) = \langle \nabla V(\boldsymbol{x}), \boldsymbol{f}(\boldsymbol{x})\rangle \le -c_3\|\boldsymbol{x}\|^2, \tag{6.5.165}$$

for some positive constants c_1, c_2, c_3, c_4. *Furthermore, if* $r = \infty$ *and* $\boldsymbol{x}^* = 0$ *is globally exponentially stable, then* V *is defined on* $\mathbb{R}^d$ *and satisfies* (6.5.163)–(6.5.165) *there.*

Proof We will prove (6.5.163)–(6.5.165) for the Euclidean norm, but they are valid for any norm on $\mathbb{R}^d$, see Theorem A.3.

Let $\max_{y\in\Omega_0}\|\mathcal{J}_f(y)\| \le L$ and consider the solution $\boldsymbol{y}(t) = \phi(t,0,\boldsymbol{x}) = \phi(t,\boldsymbol{x})$ of the problem

$$\boldsymbol{y}' = \boldsymbol{f}(\boldsymbol{y}), \quad \boldsymbol{y}(0) = \boldsymbol{x} \in \Omega_0,$$

for which

$$\|\boldsymbol{y}(t)\|_2 \le k\|\boldsymbol{x}\|_2 e^{-\xi t}, \tag{6.5.166}$$

for some $\xi > 0$. We shall show that

$$V(\boldsymbol{x}) = \int_0^\delta \langle \phi(t,\boldsymbol{x}), \phi(t,\boldsymbol{x})\rangle dt,$$

with appropriate $\delta > 0$, is a Lyapunov function satisfying (6.5.163)–(6.5.165) with constants

$$c_1 = \frac{1-e^{-2L\delta}}{2L}, \quad c_2 = \frac{k^2(1-e^{-2\xi\delta})}{2\xi},$$

$$c_3 = 1 - k^2e^{-2\xi\delta}, \quad c_4 = \frac{2k}{L-\xi}(e^{(L-\xi)\delta} - 1).$$

Before we begin the proof, we observe that for the estimates to make sense, we must have $L \neq \xi$, $Lk^2 > \xi$, and that δ has to be large enough to ensure that $c_3 > 0, c_4 > 0$ and $c_1 \leq c_2$. This is possible, as we can always enlarge L and k to satisfy these requirements.

Using (6.5.166), we get (6.5.163) by

$$V(\boldsymbol{x}) = \int_0^\delta \|\phi(t,x)\|_2^2 dt \leq k^2 \|x\|_2^2 \int_0^\delta e^{-2\xi t} dt = \frac{k^2(1-e^{-2\delta\xi})}{2\xi}\|x\|_2^2.$$

For the reverse inequality, (6.5.157) with $\tau = 0$ yields

$$V(\boldsymbol{x}) = \int_0^\delta \|\phi(t,x)\|_2^2 dt \geq \|x\|_2^2 \int_0^\delta e^{-2Lt} dt = \frac{1-e^{-2\delta L}}{2L}\|x\|_2^2.$$

Next, similarly to (6.5.158), we obtain

$$\frac{\partial}{\partial x_i}\langle \phi(t,\boldsymbol{x}), \phi(t,\boldsymbol{x})\rangle = 2\left\langle \phi(t,\boldsymbol{x}), \frac{\partial}{\partial x_i}\phi(t,\boldsymbol{x})\right\rangle,$$

hence

$$\nabla_{\boldsymbol{x}}\langle \phi(t,\boldsymbol{x}), \phi(t,\boldsymbol{x})\rangle = 2\nabla_{\boldsymbol{x}}^T \phi(t,\boldsymbol{x})\phi(t,x). \tag{6.5.167}$$

Hence, by (6.5.159) with $\frac{\partial}{\partial t}\phi = -\frac{\partial}{\partial \tau}\phi$ and (6.5.166), we get

$$\begin{aligned} V'(\boldsymbol{x}) &= \left\langle \nabla_{\boldsymbol{x}} \int_0^\delta \langle \phi(t,\boldsymbol{x}), \phi(t,\boldsymbol{x})\rangle dt, \boldsymbol{f}(\boldsymbol{x})\right\rangle = 2\int_0^\delta \langle \phi(t,\boldsymbol{x}), \nabla_{\boldsymbol{x}}\phi(t,\boldsymbol{x})\boldsymbol{f}(\boldsymbol{x})\rangle dt \\ &= \int_0^\delta \frac{\partial}{\partial t}\langle \phi(t,\boldsymbol{x}), \phi(t,\boldsymbol{x})\rangle dt = \|\phi(\delta,\boldsymbol{x})\|_2^2 - \|\boldsymbol{x}\|_2^2 \leq -(1-k^2e^{-2\xi\delta})\|\boldsymbol{x}\|_2^2, \end{aligned}$$

where we need to select δ so that $c_3 = (1-k^2e^{-2\xi\delta}) > 0$. Finally, using (6.5.167), the second equation in (6.5.162) with the initial condition $\nabla_{\boldsymbol{x}}\phi(0,\boldsymbol{x}) = \mathcal{I}$ and the bound for $\mathcal{I}_f$, we get

$$\begin{aligned} \|\nabla V_{\boldsymbol{x}}(\boldsymbol{x})\|_2 &\leq 2\int_0^\delta \|\nabla_{\boldsymbol{x}}^T\phi(t,\boldsymbol{x})\phi(t,x)\|_2 dt \\ &\leq 2\int_0^\delta \|\phi(t,\boldsymbol{x})\|_2 \|\nabla_{\boldsymbol{x}}\phi(t,\boldsymbol{x})\| dt \leq 2k\|\boldsymbol{x}\|_2 \int_0^\delta e^{(L-\xi)t} dt \\ &= \frac{2k}{L-\xi}(e^{(L-\xi)\delta}-1)\|\boldsymbol{x}\|_2. \end{aligned}$$

□

Remark 6.76 From the proof, it is easy to see that the result is valid if $\boldsymbol{x}^* = 0$ is an equilibrium at the boundary of an invariant region Ω_0 as V is constructed using trajectories starting from $\boldsymbol{x}$ and staying inside Ω_0.

Hierarchic Systems and the Vidyasagar Theorem
In many applications, we face so-called hierarchical (triangular) systems

$$\begin{aligned} \boldsymbol{x}' &= \boldsymbol{f}(\boldsymbol{x}), \\ \boldsymbol{y}' &= \boldsymbol{g}(\boldsymbol{x}, \boldsymbol{y}). \end{aligned} \tag{6.5.168}$$

We can also consider systems consisting of several systems equations of the same structure. To simplify considerations, here we assume that $\boldsymbol{f} : \mathbb{R}^n \to \mathbb{R}^n$ and $\boldsymbol{g} : \mathbb{R}^n \times \mathbb{R}^m \to \mathbb{R}^m$ are differentiable functions and the relevant equilibria are exponentially stable. Still, the result below is valid in a more general context, [98, Section 5.8.4].

Theorem 6.77 *Consider* (6.5.168). *Assume that* $\boldsymbol{x}^* = 0 \in \mathbb{R}^n$ *is exponentially stable for* $\boldsymbol{x}' = \boldsymbol{f}(\boldsymbol{x})$ *and* $\boldsymbol{y}^* = 0 \in \mathbb{R}^m$ *is exponentially stable for the system* $\boldsymbol{y}' = \boldsymbol{g}(0, \boldsymbol{y})$. *Then* $(\boldsymbol{x}^*, \boldsymbol{y}^*) = (0, 0) \in \mathbb{R}^n \times \mathbb{R}^m$ *is exponentially stable. If both equilibria are globally asymptotically stable and all forward trajectories of* (6.5.168) *are bounded, then* $(0, 0)$ *is globally exponentially stable.*

Proof Theorem 6.75 ensures that there are Lyapunov functions $V_1(\boldsymbol{x})$ for $\boldsymbol{x}' = \boldsymbol{f}(\boldsymbol{x})$ and $V_2(\boldsymbol{y})$ for $\boldsymbol{y}' = \boldsymbol{g}(0, \boldsymbol{y})$, which satisfy

$$\begin{aligned} &c_1\|\boldsymbol{x}\|_2^2 \le V_1(\boldsymbol{x}) \le c_2\|\boldsymbol{x}\|_2^2, \quad \|\nabla V_1(\boldsymbol{x})\| \le c_4\|\boldsymbol{x}\|_2, \\ &V_1'(\boldsymbol{x}) = \langle \nabla V_1(\boldsymbol{x}), \boldsymbol{f}(\boldsymbol{x})\rangle \le -c_3\|\boldsymbol{x}\|_2^2 \end{aligned} \tag{6.5.169}$$

and

$$\begin{aligned} &d_1\|\boldsymbol{y}\|_2^2 \le V_2(\boldsymbol{y}) \le d_2\|\boldsymbol{y}\|_2^2, \quad \|\nabla V_2(\boldsymbol{y})\| \le d_4\|\boldsymbol{y}\|_2, \\ &V_2'(\boldsymbol{y}) = \langle \nabla V_2(\boldsymbol{y}), \boldsymbol{g}(0, \boldsymbol{y})\rangle \le -d_3\|\boldsymbol{y}\|_2^2. \end{aligned} \tag{6.5.170}$$

First, we prove the stability of $(0, 0)$. Consider a neighbourhood U_ϵ of $(0, 0)$ given by

$$U_\epsilon = \{(\boldsymbol{x}, \boldsymbol{y}) : \ \|\boldsymbol{x}\|_2 < \epsilon, \|\boldsymbol{y}\|_2 < \epsilon\}$$

and let

$$L := \max_{(\boldsymbol{x}, \boldsymbol{y}) \in U_\epsilon} \|\mathcal{J}_{\boldsymbol{g}}(\boldsymbol{x}, \boldsymbol{y})\|.$$

Then, evaluating V_2' along the solution $(\boldsymbol{x}(t), \boldsymbol{y}(t))$ of the full system (6.5.168), we get

$$\begin{aligned} V_2'(\boldsymbol{y}) &= \langle \nabla V_2(\boldsymbol{y}), g(\boldsymbol{x}, \boldsymbol{y})\rangle = \langle \nabla V_2(\boldsymbol{y}), \boldsymbol{g}(0, \boldsymbol{y})\rangle + \langle \nabla V_2(\boldsymbol{y}), \boldsymbol{g}(\boldsymbol{x}, \boldsymbol{y}) - \boldsymbol{g}(0, \boldsymbol{y})\rangle \\ &\le -d_4\|\boldsymbol{y}\|_2^2 + d_3\|\boldsymbol{g}(\boldsymbol{x}, \boldsymbol{y}) - \boldsymbol{g}(0, \boldsymbol{y})\|_2 \le -d_4\|\boldsymbol{y}\|_2^2 + d_3L\|\boldsymbol{x}\|_2. \end{aligned} \tag{6.5.171}$$

Consider now some $\delta_1 < \epsilon$. Since $\boldsymbol{x}(t)$ does not depend on $\boldsymbol{y}$ and $\boldsymbol{x}^* = 0$ is stable, we can find $\delta_2 < \epsilon$ such that if $\|\mathring{\boldsymbol{x}}\|_2 < \delta_2$, then $\|\boldsymbol{x}(t)\|_2 < \frac{d_4\delta_1^2}{d_3L} =: \delta_3$. Take a solution $(\boldsymbol{x}(t), \boldsymbol{y}(t))$ with $\|\mathring{\boldsymbol{x}}\|_2 < \delta_2$. We know that then $\|\boldsymbol{x}(t)\|_2 < \delta_3$ and assume that $\|\boldsymbol{y}(t)\|_2 > \delta_1$. Then, for such t, $V_2(\boldsymbol{y}(t))$ is decreasing. Now, the first inequalities in (6.5.170) imply that for any t,

$$\begin{aligned} \|\boldsymbol{y}(t)\|_2^2 &\le \frac{1}{d_1}V_2(\boldsymbol{y}(t)) \le \frac{1}{d_1}\max\left\{\max_{\|\boldsymbol{y}\|_2\le\delta_1} V_2(\boldsymbol{y}), V_2(\mathring{\boldsymbol{y}})\right\} \\ &\le \frac{d_2}{d_1}\max\{\delta_1^2, \|\mathring{\boldsymbol{y}}\|_2^2\} = \frac{d_2\delta_1^2}{d_1} =: \delta_4 \end{aligned}$$

provided $\|\mathring{\boldsymbol{y}}\|_2^2 < \delta_1$. We note that δ_4 can be arbitrarily small. Hence, $\boldsymbol{y}(t)$ cannot wander off 0. In summary, for any $\delta > 0$ we can select δ_1 such that $\delta_3, \delta_4 < \delta$, and δ_2 for which

$$\|\boldsymbol{x}(t)\|_2 < \delta, \quad \|\boldsymbol{y}(t)\|_2 < \delta,$$

provided $\|\boldsymbol{x}\|_2 < \delta_2$ and $\|\boldsymbol{y}\|_2 < \delta_1$. Hence, we proved the stability of (0, 0).

To show local asymptotic (exponential) stability, we first note that $\boldsymbol{x}(t) \to 0$ as $t \to \infty$. To prove that also $\boldsymbol{y}(t) \to 0$ as $t \to \infty$, we note that from the stability of solutions, the estimate (6.5.171) is satisfied on any solution originating close to (0, 0), since the solutions stay inside U_ϵ and the constant L is uniform on U_ϵ. Hence, by (6.5.170), we have

$$\frac{d}{dt}V_2(y(t)) \le -\frac{d_4}{d_1}V_2(y(t)) + d_3L\|\boldsymbol{x}(t)\|_2. \tag{6.5.172}$$

Now, from (6.5.169),

$$\frac{d}{dt}V_1(\boldsymbol{x}(t)) \le -\frac{c_3}{c_2}V_1(\boldsymbol{x}(t)).$$

Hence,

$$V_1(\boldsymbol{x}(t)) \le V_1(\mathring{\boldsymbol{x}})e^{-\frac{c_3}{c_2}t}$$

and, using again (6.5.169), we get

$$\|\boldsymbol{x}(t)\| \leq \left(\frac{V(x(t))}{c_1}\right)^{\frac{1}{2}} \leq \left(\frac{c_2}{c_1}\right)^{\frac{1}{2}} \|\mathring{\boldsymbol{x}}\| e^{-\frac{c_3}{2c_2}t}.$$

Thus, (6.5.172) takes the form

$$\frac{d}{dt}V_2(y(t)) \leq -\frac{d_4}{d_1}V_2(y(t)) + d_3 L\sqrt{\frac{c_2}{c_1}}\|x_{10}\|_2 e^{-\frac{c_3}{2c_2}t}.$$

It is a scalar differential inequality; hence, V_2 is dominated by the solution of the equation

$$\psi' = -\alpha\psi + \beta e^{-\xi t},$$

given by

$$\psi(t) = \psi(0)e^{-\alpha t} + \frac{\beta}{\alpha - \xi}\left(e^{-\xi t} - e^{-\alpha t}\right)$$

(we can always modify the constants so that $\alpha \neq \xi$). Thus, $\psi(t) \to 0$ as $t \to \infty$, hence

$$\|\boldsymbol{y}(t)\|_2^2 \leq \frac{1}{d_1}V_2(\boldsymbol{y}(t)) \to 0, \quad t \to \infty,$$

and (0, 0) is locally asymptotically stable.

Finally, if all forward trajectories are bounded, their ω-limit sets are nonempty by Lemma 6.49, point 1. The assumption that $\boldsymbol{x}^* = 0$ is globally asymptotically stable for $\boldsymbol{x}' = \boldsymbol{f}(\boldsymbol{x})$ implies the convergence of $\boldsymbol{x}(t)$ to 0 for any initial condition. Hence, for any initial condition $(\mathring{\boldsymbol{x}}, \mathring{\boldsymbol{y}})$, $\omega(\Gamma_{(\mathring{\boldsymbol{x}},\mathring{\boldsymbol{y}})}) \subset \{0\} \times \mathbb{R}^m$. On the other hand, for any $(0, \mathring{\boldsymbol{y}})$ we have $\omega(\Gamma_{(0,\mathring{\boldsymbol{y}})}) = (0, 0)$. From Lemma 6.49, point 5., we get $(0, 0) \in \omega(\Gamma_{(\mathring{\boldsymbol{x}},\mathring{\boldsymbol{y}})})$. Hence, for any trajectory, there are points arbitrarily close to (0, 0). In other words, there is t' we have $\phi(t, \phi(t', (\mathring{\boldsymbol{x}}, \mathring{\boldsymbol{y}}))) \to (0, 0)$ as $t \to \infty$. Using the flow property

$$\phi(t, \phi(t', (\mathring{\boldsymbol{x}}, \mathring{\boldsymbol{y}})) = \phi(t + t', (\mathring{\boldsymbol{x}}, \mathring{\boldsymbol{y}})) = \phi(\tau, (\mathring{\boldsymbol{x}}, \mathring{\boldsymbol{y}})) \to (0, 0), \quad \tau \to \infty,$$

and therefore (0, 0) is globally asymptotically stable. □

Remark 6.78 Using Remark 6.76, we see that the result is valid if $\mathbb{R}^n \times \mathbb{R}^m$ is replaced by $(\boldsymbol{x}^*, \boldsymbol{y}^*) \in \Omega = \Omega_n \times \Omega_m$ with Ω, Ω_n and $\{\boldsymbol{x}^*\} \times \Omega_m$ being invariant sets of respective dynamical systems and $(\boldsymbol{x}^*, \boldsymbol{y}^*)$ is possibly a boundary equilibrium. Moreover, the proof of the result on global stability allows for the existence of repelling equilibria if we modify the definition of the globally asymptotically stable equilibrium to exclude constant trajectories of the other equilibria.

Example 6.79 Consider an SI system with demography given by the simplified logistic model

$$\begin{aligned} S' &= \Lambda - \beta SI - \mu S, \\ I' &= \beta SI - \mu I, \end{aligned} \tag{6.5.173}$$

with positive parameters Λ, β, μ.

Exercise 6.5.3 Show that (6.5.173) has global nonnegative solutions for nonnegative initial conditions. Show that

$$R_0 = \frac{\beta\Lambda}{\mu^2} = \frac{\beta N^*}{\mu}.$$

Introducing the total population $N = S + I$, we can write (6.5.173) as

$$\begin{aligned} N' &= \Lambda - \mu N, \\ I' &= \beta(N - I)I - \mu I. \end{aligned} \tag{6.5.174}$$

This system is hierarchical. The first equation has a unique globally exponentially stable equilibrium (show this)

$$N^* = \frac{\Lambda}{\mu},$$

and the second equation gives two stationary points

$$I_1^* = 0, \quad I_2^* = \frac{\Lambda}{\mu} - \frac{\mu}{\beta},$$

with I_2^* being positive if and only if $R_0 > 1$. To use the Vidyasagar theorem, we consider the equation

$$I' = f(I) := \beta(N^* - I)I - \mu I = -\beta I^2 + (\beta N^* - \mu)I; \tag{6.5.175}$$

the shape of f for different values of R_0 is presented in Fig. 6.38. We see that for $R_0 \leq 1$, the DFE equilibrium $I_1^* = 0$ is globally asymptotically stable (though not exponentially stable if $R_0 = 1$) in $I \geq 0$. Hence, the DFE $\left(\frac{\Lambda}{\mu}, 0\right)$ for (6.5.173) is globally stable (the case $R_0 = 1$ requires a more general version of the Vidyasagar theorem, [98, Section 5.8.4], dealing with equilibria which are globally asymptotically stable but not exponentially stable. If $R_0 > 1$, then $I_1^* = 0$ becomes a repelling equilibrium and hence $I_2^* = \frac{\Lambda}{\mu} - \frac{\mu}{\beta}$ becomes a globally exponentially stable equilibrium in $I > 0$ and thus, by Remark 6.78, the endemic equilibrium $\left(\frac{\mu}{\beta}, \frac{\Lambda}{\mu} - \frac{\mu}{\beta}\right)$ is globally asymptotically stable in $\mathbb{R}^2_+$.

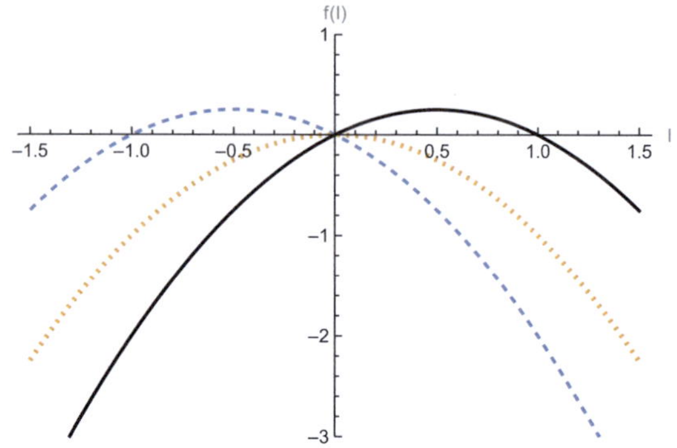

Fig. 6.38 The graphs of $f(I)$ for $R_0 < 1$ (dashed line), $R_0 = 1$ (dotted line) and $R_0 > 1$ (solid line)

Exercise 6.5.4 If the assumption that forward trajectories are bounded is not satisfied, the equilibrium need not be globally stable. Consider

$$\begin{aligned} x' &= -x, \\ y' &= y\,(x^2y^2 - 1). \end{aligned} \tag{6.5.176}$$

Show that

1. Show that $x = 0$ is globally asymptotically stable for the first equation and $y = 0$ is globally asymptotically stable for the second equation with $x = 0$.
2. Show that the first quadrant is invariant for (6.5.176).
3. Show that $z(t) = x(t)y(t) \geq 0$ satisfies

$$z' = z^3 - 2z.$$

4. Show that $z = 0$ is an attractive equilibrium and $z = \sqrt{2}$ is a repeller. Conclude that if $\mathring{x}\mathring{y} > \sqrt{2}$, then $(x(t), y(t))$ cannot converge to $(0, 0)$ and hence $(0, 0)$ is not globally asymptotically stable.
5. Show that $x(t) = \mathring{x}e^{-t}$ and $y(t) = \frac{\sqrt{2}}{\mathring{x}}e^{t}$, $\mathring{x} > 0$, is a solution to (6.5.176) and hence the assumption of boundedness of positive trajectories is not satisfied.

6.6 Existence (and Non-existence) of Periodic Trajectories

The linearization theorem, Theorem 6.34, may suggest that locally nonlinear dynamics is the same as linear. This is, however, false even in 2 dimensions, as shown in the proof of point 3. of that theorem. The Poincaré-Bendixon theory, which we shall discuss in this section, provides a complete description of two-dimensional dynamics by fully classifying possible limit sets for planar autonomous systems. The reason for this restriction is the application of the Jordan theorem, which states that every simple closed curve divides the plane into an interior region bounded by the curve and an exterior region containing all other points on the plane. While intuitively simple, its proof is not trivial. The interested reader is referred to [44, 94] for more accessible proofs; here, we will use it as a tool.

6.6.1 Preliminaries

Let $\phi(t, \boldsymbol{x})$ be the flow generated by the planar version of (6.2.17),

$$\begin{aligned} x_1' &= f_1(x_1, x_2), \\ x_2' &= f_2(x_1, x_2), \end{aligned} \tag{6.6.177}$$

where $\boldsymbol{x} = (x_1, x_2) \in \mathbb{R}^2$ and $\boldsymbol{f} = (f_1, f_2)$. Throughout this chapter, we assume that $\boldsymbol{f}$ is a differentiable function. By a *local transversal* to ϕ, we mean a line segment L, which all trajectories cross from one side. In other words, the field $\boldsymbol{f}$ always forms either an acute or an obtuse angle with the normal to L.

Lemma 6.80 *If $\boldsymbol{x}_0$ is not a stationary point of (6.6.177), then it is always possible to construct a local transversal in a neighbourhood of $\boldsymbol{x}_0$.*

If $\boldsymbol{x}_0$ is not stationary point, then $\boldsymbol{v} = \boldsymbol{f}(\boldsymbol{x}_0) \neq 0$. Take a coordinate system with the origin at $\boldsymbol{x}_0$, and axes parallel to $\boldsymbol{v}$ and take $\boldsymbol{w}$ perpendicular to $\boldsymbol{v}$. Let (v, w) be the coordinates of a point in this system. Then (6.6.177) can be written as

$$\begin{aligned} v' &= a + O(\|(v, w) - \boldsymbol{x}_0\|), \\ w' &= O(\|(v, w) - \boldsymbol{x}_0\|). \end{aligned}$$

Here $a \neq 0$ is the norm of $\boldsymbol{v}$. Now we can choose line L through $\boldsymbol{x}_0$ along $\boldsymbol{w}$. As long as we are close to $\boldsymbol{x}_0$, $v' \neq 0$ and the trajectories cross L in the same direction (positive if $a > 0$ and negative otherwise).

It follows that the times at which trajectories cross the local transversal change smoothly with the initial points. This fact has several implications, as shown in the proof of the Poincaré-Bendixon theorem below and Example 6.83.

Consider a trajectory originating from $\boldsymbol{x}$ reaches a local transversal at time t_0 (see Fig. 6.39).

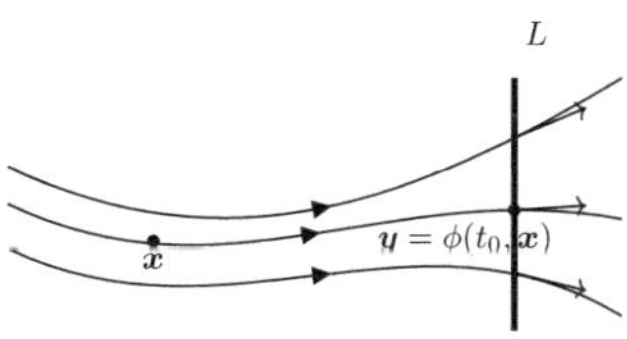

Fig. 6.39 Local transversal of the flow ϕ

Lemma 6.81 *Let V be a local transversal for $\{\phi(t,\cdot)\}_{t\in\mathbb{R}}$ and let $\phi(t_0,\boldsymbol{x}) = \boldsymbol{y}$. Then there is an open neighbourhood U of $\boldsymbol{x}$ in $\mathbb{R}^2$ and a unique function $\tau \in C^1(U)$, such that $\tau(\boldsymbol{x}) = t_0$ and*

$$\phi(\tau(z), z) \in L,$$

for any $z \in U$.

Proof Let $\boldsymbol{v} = f(\boldsymbol{y}) \neq 0$ and take as a local transversal L at $\boldsymbol{y}$ an appropriate segment of the line perpendicular to $\boldsymbol{v}$, as in the proof of Lemma 6.80, . For the solution $\phi(t, z)$ we define

$$g(t, z) = \langle \boldsymbol{v}, \phi(t, z)\rangle - \langle \boldsymbol{v}, \boldsymbol{y}\rangle$$

defined on some neighbourhood of $(0, \boldsymbol{x})$. We have

$$g(t_0, \boldsymbol{x}) = 0,$$

and, since the solution is differentiable in t, we have

$$\frac{\partial}{\partial t} g(t, z)|_{(t,z)=(t_0,\boldsymbol{x})} = \langle \boldsymbol{v}, f(\boldsymbol{y})\rangle \neq 0.$$

From the Implicit Function Theorem, see [6, Theorem VII.8.2], there is a neighbourhood U of $\boldsymbol{x}$ on which there is a unique differentiable function $U \ni z \mapsto t(z)$ such that $t(\boldsymbol{x}) = t_0$ and

$$g(t(z), z) \equiv 0$$

for $z \in U$. Furthermore,

$$0 = g(t(z), z) = \langle \boldsymbol{v}, \phi(t(z), z) - \boldsymbol{y}\rangle,$$

which shows that $\phi(t(z), z) - \boldsymbol{y}$ is orthogonal to $\boldsymbol{v}$. Hence $\phi(t(z), z) \in L$ as $\boldsymbol{y} \in L$. □

Remark 6.82 With an appropriate modification of the definition of the local transversal, the above result is valid in $\mathbb{R}^n$.

The usefulness of the concept of local transversal and of Lemma 6.81 is further illustrated in the following example.

Example 6.83 Let us recall the Lotka–Volterra equation

$$\begin{aligned} x_1' &= rx_1 - \alpha x_1 x_2, \\ x_2' &= -sx_2 + \beta x_1 x_2. \end{aligned} \tag{6.6.178}$$

In Sect. 6.2.4, we have shown that the solution of (6.6.178) are periodic, but the phase-plane methods cannot deliver any information about their period. Since we can find periods of linear models, we linearize (6.6.178) in the neighbourhood of the positive equilibrium (x_1^*, x_2^*)

$$x_1^* = \frac{s}{\beta}, \quad x_2^* = \frac{r}{\alpha}.$$

Introducing new variables $\boldsymbol{u} = (u_1, u_2)$ by

$$u_1 = x_1 - \frac{s}{\beta}, \quad u_2 = x_2 - \frac{r}{\alpha},$$

we write (6.6.178) as

$$\begin{aligned} u_1' &= \left(u_1 + \frac{s}{\beta}\right)\left(r - \alpha\left(u_2 + \frac{r}{\alpha}\right)\right) = -\frac{\alpha s}{\beta}u_2 - \alpha u_1 u_2, \\ u_2' &= \left(u_2 + \frac{r}{\alpha}\right)\left(\beta\left(u_1 + \frac{s}{\beta}\right) - s\right) = \frac{\beta r}{\alpha}u_1 + \beta u_1 u_2. \end{aligned} \tag{6.6.179}$$

Hence, the linearization of (6.6.178) is given by

$$\begin{aligned} v_1' &= -\frac{\alpha s}{\beta}v_2, \\ v_2' &= \frac{\beta r}{\alpha}v_1, \end{aligned} \tag{6.6.180}$$

with $\boldsymbol{v} = (v_1, v_2)$. Consider now the initial conditions $\boldsymbol{u}(0) = \boldsymbol{v}(0) = \mathring{\boldsymbol{u}}$ satisfying $\|\mathring{\boldsymbol{u}}\| \leq \delta$ for some small $\delta > 0$ so that the original initial conditions are close to the equilibrium. Then, we re-scale the problem according to

$$\underline{\boldsymbol{u}} - \frac{\boldsymbol{u}}{\delta}, \quad \underline{\boldsymbol{v}} = \frac{\boldsymbol{v}}{\delta}. \tag{6.6.181}$$

This leads to

$$\begin{aligned} \underline{u}_1' &= -\frac{\alpha s}{\beta}\underline{u}_2 - \delta\alpha\underline{u}_1\underline{u}_2 \\ \underline{u}_2' &= \frac{\beta r}{\alpha}\underline{u}_1 + \delta\beta\underline{u}_1\underline{u}_2, \end{aligned} \tag{6.6.182}$$

and

$$\begin{aligned}\underline{v}_1' &= -\frac{\alpha s}{\beta}\underline{v}_2,\\ \underline{v}_2' &= \frac{\beta r}{\alpha}\underline{v}_1,\end{aligned} \tag{6.6.183}$$

with initial condition $\underline{\boldsymbol{u}}(0) = \underline{\boldsymbol{v}}(0) = \mathring{\underline{\boldsymbol{u}}}$ satisfying $\|\mathring{\underline{\boldsymbol{u}}}\| \leq 1$.

Exercise 6.6.1 Use Example 5.3.62 to show that the period T of solution to (6.6.183) is $T = \frac{2\pi}{\sqrt{sr}}$.

Using Theorem B.16 to the dynamical systems $\underline{\phi}$ and $\underline{\psi}$ generated, respectively, by (6.6.182) i (6.6.182), we see that for any $\epsilon > 0$ and any time τ, there is $\delta_0 > 0$ such that for any $0 < \delta < \delta_0$ we have

$$\|\underline{\phi}(t, \mathring{\underline{u}}, \delta) - \underline{\psi}(t, \mathring{\underline{u}}, \delta)\| \leq \epsilon$$

on $[0, \tau]$. Denoting by ϕ and ψ dynamical systems generated by, respectively, (6.6.179) i (6.6.180), we have

$$\phi(t, \mathring{\boldsymbol{u}}) = \delta\underline{\phi}(t, \delta^{-1}\mathring{\boldsymbol{u}}, \delta) = \delta\underline{\phi}(t, \mathring{\underline{\boldsymbol{u}}}, \delta), \quad \psi(t, \mathring{\boldsymbol{u}}) = \delta\underline{\psi}(t, \delta^{-1}\mathring{\boldsymbol{u}}, \delta) = \delta\underline{\psi}(t, \mathring{\underline{\boldsymbol{u}}}, \delta),$$

Hence,

$$\|\phi(t, \mathring{\boldsymbol{u}}) - \psi(t, \mathring{\boldsymbol{u}})\| = \|\delta\underline{\phi}(t, \mathring{\underline{\boldsymbol{u}}}, \delta) - \delta\underline{\psi}(t, \mathring{\underline{\boldsymbol{u}}}, \delta)\| \leq \delta\epsilon.$$

Let us reflect on the obtained result. We know that the solutions to (6.6.180) are periodic; therefore, solutions to (6.6.179) are arbitrarily close to periodic solutions originating from the same initial conditions provided they are sufficiently small, on any finite $[0, \tau]$. Since solutions to (6.6.182) are also periodic, both solutions will stay close to each other at all t. It remains to show that the periods are also close to each other.

Let us write (6.6.182) as

$$\underline{\boldsymbol{u}}' = f(\underline{\boldsymbol{u}}, \delta), \quad \underline{\boldsymbol{u}}(0) = \mathring{\underline{\boldsymbol{u}}}, \tag{6.6.184}$$

and note that (6.6.184) with $\delta = 0$ gives (6.6.183). The solutions of (6.6.184) with $\delta > 0$ and $\delta = 0$ are, respectively, $\underline{\phi}(t, \mathring{\underline{\boldsymbol{u}}})$ and $\underline{\psi}(t, \mathring{\underline{\boldsymbol{u}}})$, $t \in \mathbb{R}$. From Problem 6.6.1 we know that the period of $\underline{\psi}$ (and ψ) is given by $T = \frac{2\pi}{\sqrt{sr}}$. Take a local transversal L in the neighbourhood of $\mathring{\underline{\boldsymbol{u}}}$ (for $\boldsymbol{f}(\cdot, 0)$, as well as $\boldsymbol{f}(\cdot, \delta)$ for small $|\delta|$). We assume L is a segment perpendicular to a vector $\boldsymbol{v} = \boldsymbol{f}(\mathring{\underline{\boldsymbol{u}}}, 0)$. The trajectory $\gamma_{\mathring{\underline{\boldsymbol{u}}}}$ of the flow $\underline{\psi}(t, \mathring{\underline{\boldsymbol{u}}}, \delta)$ returns to $\mathring{\underline{\boldsymbol{u}}}$ at $t = T$. Since the trajectory $\Gamma_{\mathring{\underline{\boldsymbol{u}}}}$ of the flow $\underline{\phi}(t, \mathring{\underline{\boldsymbol{u}}}, \delta)$ stays close to $\gamma_{\mathring{\underline{\boldsymbol{u}}}}$ on an interval containing $[0, T]$ and $\gamma_{\mathring{\underline{\boldsymbol{u}}}}$ crosses L, $\Gamma_{\mathring{\underline{\boldsymbol{u}}}}$ also crosses L at $\boldsymbol{w} := \underline{\phi}(t_1, \mathring{\boldsymbol{u}}, \delta)$ for some t_1; we assume that $t_1 > 0$ is the first crossing time. Let

$\boldsymbol{w} \neq \mathring{\underline{\boldsymbol{u}}}$. Since $\Gamma_{\mathring{\underline{\boldsymbol{u}}}}$ is a closed curve, we must have $\mathring{\underline{\boldsymbol{u}}} = \underline{\phi}(T', \mathring{\underline{\boldsymbol{u}}}, \delta)$ for some T'. Clearly, since $\mathring{\underline{\boldsymbol{u}}} \in L$ and $\boldsymbol{w} \neq \mathring{\underline{\boldsymbol{u}}}$, we have $T' > t_1$. Consider curve C composed of $\Gamma_{\mathring{\underline{\boldsymbol{u}}}}$ between $\mathring{\underline{\boldsymbol{u}}}$ and $\boldsymbol{w}$ and the part of L between $\mathring{\underline{\boldsymbol{u}}}$ i $\boldsymbol{w}$. It is a simple curve; hence, Jordan's theorem implies that C divides $\mathbb{R}^2$ into two open sets, bounded D_1 and unbounded D_2. The trajectories of $\underline{\phi}$ can only cross C at the points of L and we can assume that they move from D_1 to D_2 (the other case is analogous). The definition of local transversal implies that for some $T' > t_2 > t_1$, we have $\underline{\phi}(t_2, \mathring{\underline{\boldsymbol{u}}}, \delta) \in D_2$. The periodicity yields $\mathring{\underline{\boldsymbol{u}}} = \underline{\phi}(T', \mathring{\underline{\boldsymbol{u}}}, \delta)$, that is, the trajectory would have to return to L from D_2. This, however, is impossible since the vector field on L is directed into D_2, and the trajectory cannot cross $C \setminus L$ to return D_1. Thus $\boldsymbol{w} = \mathring{\underline{\boldsymbol{u}}} = \underline{\phi}(t_1, \mathring{\underline{\boldsymbol{u}}}, \delta) = \underline{\phi}(T', \mathring{\underline{\boldsymbol{u}}}, \delta)$.

The estimate of t_1 is similar to the proof of Lemma 6.81, only here we consider the dependence on δ. Thus, we consider

$$g(t, \delta) = \langle \boldsymbol{v}, \underline{\phi}(t, \mathring{\underline{\boldsymbol{u}}}, \delta)\rangle - \langle \boldsymbol{v}, \mathring{\underline{\boldsymbol{u}}}\rangle,$$

defined in some neighbourhood of $(T, 0)$; we have

$$g(T, 0) = 0.$$

Next,

$$\frac{\partial}{\partial t} g(t, \delta)|_{(t,\delta)=(T,0)} = \langle \boldsymbol{v}, \boldsymbol{f}(\mathring{\underline{\boldsymbol{u}}}, 0)\rangle \neq 0.$$

As before, the Implicit Function Theorem implies the existence of an interval $I = (-\underline{\delta}, \underline{\delta})$ and a unique function $\delta \mapsto t(\delta)$ differentiable on I satisfying $t(0) = T$ and

$$g(t(\delta), \delta) = 0$$

dla $\delta \in I$. Further, since

$$0 = g(t(\delta), \delta) = \langle \boldsymbol{v}, \underline{\phi}(t(\delta), \mathring{\underline{\boldsymbol{u}}}, \delta) - \mathring{\underline{\boldsymbol{u}}}\rangle,$$

we have $\underline{\phi}(t(\delta), \mathring{\underline{\boldsymbol{u}}}, \delta) \in L$. By the previous part, $\underline{\phi}(t(\delta), \mathring{\underline{\boldsymbol{u}}}, \delta) = \mathring{\underline{\boldsymbol{u}}}$, that is, $t(\delta) = T'$. In other words, for small $|\delta|$ (and the initial conditions from the unit ball B_1), the period of solutions $t \mapsto \underline{\phi}(t, \mathring{\underline{\boldsymbol{u}}}, \delta)$ of (6.6.182) is close to $T = \frac{2\pi}{\sqrt{sr}}$. It means, by (6.6.181), that the solutions to (6.6.179) with the initial conditions from B_δ, and hence the solutions of the Lotka–Volterra equation (6.6.178) with initial conditions close to the equilibrium, also have periods close to T.

6.6.2 The Poincaré–Bendixon Theory

Having found a local transversal L at $\boldsymbol{x}$, we can construct a *flow box* around $\boldsymbol{x}$ by taking a collection of trajectories with starting points on L and t running from $-\delta$ to δ with δ small enough that all trajectories exist over this time interval, that is,

$$V_{\boldsymbol{x}} = \{\boldsymbol{y} : \ \boldsymbol{y} = \phi(t, \boldsymbol{z}), \ \boldsymbol{z} \in L, t \in (-\delta, \delta)\}.$$

An important property of the flow box is that if a trajectory starts close enough to $\boldsymbol{x}$, it crosses L (either forward or backward). Precisely speaking, we have

Lemma 6.84 *Let L be a transversal at* $\boldsymbol{x}$*. There is a neighbourhood D of* $\boldsymbol{x}$ *such that for any* $\boldsymbol{y} \in D$ *there are* $t_1 < t_2$ *such that* $\Gamma_{t_1,t_2} := \{\boldsymbol{z} : \ \boldsymbol{z} = \phi(t, \boldsymbol{y}), t \in (t_1, t_2)\} \subset D$ *and* $\Gamma_{t_1,t_2} \cap L \neq \emptyset$.

Proof We chose coordinates (z_1, z_2) as in the proof of Lemma 6.80 so that $\boldsymbol{x} = (0, 0)$ and $\boldsymbol{f}(\boldsymbol{x}) = (a, 0)$ where we select $a > 0$. Then, L is on the x_2 axis. Let D_δ be the ball $\|(z_1, z_2)\| = |z_1| + |z_2| < \delta$ (that is, a square with diagonals on L and $\boldsymbol{f}(\boldsymbol{x})$. We can choose δ small enough for the following conditions to hold:

(a) The slope of $\boldsymbol{f}(\boldsymbol{z})$ is strictly between -1 and 1, that is, $|f_2(\boldsymbol{z})/f_1(\boldsymbol{z})| < 1$,
(b) $f_1(\boldsymbol{z}) > a/2$,

for $\boldsymbol{z} \in D_\delta$. This is possible by the continuity of $\boldsymbol{f}$ in a neighbourhood of $\boldsymbol{x}$. Notice, in particular, that $\boldsymbol{f}$ is always transversal to the sides of D_δ and pointing outward at the right-hand side and inward at the left-hand side of it. Hence, a trajectory can only leave D_δ at the right and enter at the left. Now, clearly, for any solution $\boldsymbol{z}(t)$,

$$z_1(t) - z_1(0) = \int\limits_0^t f_1(\boldsymbol{z}(s))ds > \frac{a}{2}t.$$

Since the maximum value of $z_1(t) - z_1(0)$ is 2δ, the solution starting from any point $\boldsymbol{z}_0 \in D_\delta$ must leave it in time shorter than $4\delta/a$. Let t_2 be the smallest value at which $\phi(t, \boldsymbol{z}_0)$ intersects the right-hand side of D_δ. Similarly, $\phi(t, \boldsymbol{z}_0)$ intersects the left-hand side of D_δ at some time $t \in (-4\delta/a, 0)$ and hence there is $t_1 < 0$ at which this happens for the first time (going backwards). Hence the segment $\Gamma_{t_1,t_2} = \{\boldsymbol{z} : \boldsymbol{z} = \phi(t, \boldsymbol{y}), t \in (t_1, t_2)\} \subset D_\delta$ and also $\Gamma_{t_1,t_2} \cap L \neq \emptyset$. □

Lemma 6.85 *If a trajectory* $\Gamma_{\boldsymbol{x}}$ *intersects a local transversal several times, the successive crossing points move monotonically along the transversal, see Fig. 6.40.*

Proof Consider two successive crossings $\boldsymbol{y}_1 = \phi(t_1, \boldsymbol{x})$ and $\boldsymbol{y}_2 = \phi(t_2, \boldsymbol{x})$ with, say, $t_1 < t_2$ and a closed curve S composed of the piece Γ' of the trajectory between $\boldsymbol{y}_1$ and $\boldsymbol{y}_2$ and the piece L' of the transversal between these two points. By Jordan's theorem, S divides $\mathbb{R}^2$ into two disjoint open sets with one, say D_1, bounded the other, say D_2, unbounded. We can assume that the flow through L' is from D_1 into

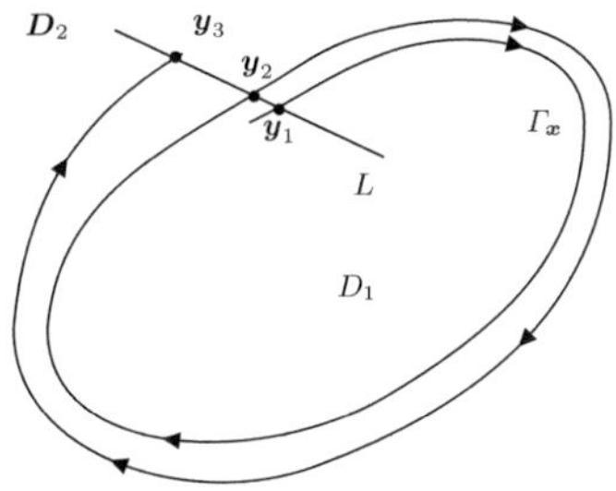

Fig. 6.40 Monotonic character of the subsequent intersections of the flow with the local transversal

D_2. Consider $\boldsymbol{y}_3 = \phi(t_3, \boldsymbol{x}) \in L$ with $t_3 > t_2$. Taking $t' = t_2 + \epsilon$ with ϵ sufficiently small, we can be sure that $\phi(t', \boldsymbol{x})$ is outside D_1. If we assume that $\boldsymbol{y}_3 \in L'$, then there is ϵ' such that $\phi(t_3 - \epsilon', \boldsymbol{x}) \in D_1$. Hence, the trajectory between t' and $t_3 - \epsilon'$ joins points of D_1 and D_2. However, it cannot cross Γ' as trajectories cannot cross; also, it cannot enter D_1 through L' as the flow is in the opposite direction. By a similar argument, $\boldsymbol{y}_3$ cannot belong to the sub-segment of L, which is inside D_1 (we note that this sub-segment cannot stick outside D_1 as this would require that a point moves along a piece of trajectory in two directions at once). Thus, it must belong to the subsegment outside D_1 with $\boldsymbol{y}_2$ as the end-point. □

An important corollary is:

Corollary 6.86 *If $\boldsymbol{x} \in \omega(\Gamma_{\boldsymbol{x}_0})$ is not a stationary point and $\boldsymbol{x} \in \Gamma_{\boldsymbol{x}_0}$, then $\Gamma_{\boldsymbol{x}} (= \Gamma_{\boldsymbol{x}_0})$ is a closed curve.*

Proof Since $\boldsymbol{x} \in \Gamma_{\boldsymbol{x}_0}$, $\omega(\Gamma_{\boldsymbol{x}_0}) = \omega(\Gamma_{\boldsymbol{x}})$ (the limit set depends on the trajectory and not on the initial point). Hence, $\boldsymbol{x} \in \omega(\Gamma_{\boldsymbol{x}})$. Choose L as a local transversal at $\boldsymbol{x}$. By Lemma 6.84, for sufficiently large T we have an increasing sequence $t_i > t_{i-1} \geq T$ such that $\phi(t_i, \boldsymbol{x}) \to \boldsymbol{x}$ as $t_i \to \infty$ and $\phi(t_i, \boldsymbol{x}) \in L$ (see Lemma 6.84). Also, $\phi(0, \boldsymbol{x}) = \boldsymbol{x}$. Suppose $\phi(t_i, \boldsymbol{x}) \neq \boldsymbol{x}$, $t_i > T$, then successive intercepts are bounded away from $\boldsymbol{x}$ which contradicts the fact that $\boldsymbol{x} \in \omega(\Gamma_{\boldsymbol{x}})$. Hence $\phi(t_1, \boldsymbol{x}) = \boldsymbol{x}$ for some t and, by Theorem 6.27, the solution is periodic. □

Remark 6.87 Corollary 6.86 formalizes an intuitively clear observation that no point on an orbit $\Gamma_{\boldsymbol{x}}$ can be the limit of other points $\phi(t, \boldsymbol{x})$ on this orbit with $t \to \infty$ (so no orbit can approach its point even asymptotically) unless it is closed. Then, of course, we have $\phi(t, \boldsymbol{x}) = \phi(t + nT, \boldsymbol{x})$ with $t + nT \to \infty$ as $n \to \infty$, where T is a period of the solution.

Theorem 6.88 *If an orbit $\Gamma_{\boldsymbol{x}_0}$ enters and does not leave a closed bounded domain Ω, which contains no equilibrium points, then $\omega(\Gamma_{\boldsymbol{x}_0})$ is a closed orbit.*

Proof First we prove that $\omega(\Gamma_{\boldsymbol{x}_0})$ contains a closed orbit. Let $\boldsymbol{x} \in \omega(\Gamma_{\boldsymbol{x}_0})$. There are two possibilities.

(i) If $\boldsymbol{x} \in \Gamma_{\boldsymbol{x}_0}$, then, by Corollary 6.86, $\Gamma_{\boldsymbol{x}_0}$ is a closed orbit.
(ii) If $\boldsymbol{x} \notin \Gamma_{\boldsymbol{x}_0}$, then since $\boldsymbol{x} \in \omega(\Gamma_{\boldsymbol{x}_0})$, the orbit $\Gamma_{\boldsymbol{x}} \subset \omega(\Gamma_{\boldsymbol{x}_0})$ by Lemma 6.49 (3) and, because $\omega(\Gamma_{\boldsymbol{x}_0})$ is closed, $\omega(\Gamma_{\boldsymbol{x}}) \subset \omega(\Gamma_{\boldsymbol{x}_0})$. Let $\boldsymbol{x}^* \in \omega(\Gamma_{\boldsymbol{x}}) \subset \omega(\Gamma_{\boldsymbol{x}_0})$.

If $\boldsymbol{x}^* \in \Gamma_{\boldsymbol{x}}$ then, again by Corollary 6.86, $\Gamma_{\boldsymbol{x}} \subset \omega(\Gamma_{\boldsymbol{x}_0})$ is a closed orbit. This leaves the only possibility that $\boldsymbol{x}^* \notin \Gamma_{\boldsymbol{x}}$. Consider a local transversal L at $\boldsymbol{x}^*$, which exists as $\boldsymbol{x}^*$ is not an equilibrium. Arguing as in the proof of Corollary 6.86, we have a sequence $(\boldsymbol{p}_i)_{i\in\mathbb{N}}$, $\boldsymbol{p}_i = \phi(t_i, \boldsymbol{x}) \in L$ with $\boldsymbol{p}_i \to \boldsymbol{x}^*$ as $t_i \to \infty$ in a monotonic way. On the other hand, $\boldsymbol{p}_i \in \omega(\Gamma_{\boldsymbol{x}_0})$ and L is also a local transversal at each $\boldsymbol{p}_i$. This means that there are sequences on the trajectory $\Gamma_{\boldsymbol{x}_0}$ converging monotonically along L to, say, $\boldsymbol{p}_i$ and $\boldsymbol{p}_{i+1}$. Assume $\|\boldsymbol{p}_i - \boldsymbol{p}_{i+1}\| > \epsilon$. We have, say, $\|\phi(t_1, \boldsymbol{x}_0) - \boldsymbol{p}_i\| < \epsilon/4$, then there must be $t_2 > t_1$ such that $\|\phi(t_2, \boldsymbol{x}_0) - \boldsymbol{p}_{i+1}\| < \epsilon/4$ but then, at the next t_3 at which $\Gamma_{\boldsymbol{x}_0}$ intersects L, $\phi(t_3, \boldsymbol{x}_0)$ must be closer than $\epsilon/4$ to both $\boldsymbol{p}_i$ and $\boldsymbol{p}_{i+1}$, by Lemma 6.85. This is a contradiction, which shows that $\boldsymbol{x}^* \in \Gamma_{\boldsymbol{x}}$ and thus $\omega(\Gamma_{\boldsymbol{x}_0})$ contains a closed orbit.

The final step of the proof is to show that this orbit, say, $\Gamma \subset \omega(\Gamma_{\boldsymbol{x}_0})$ is equal to $\omega(\Gamma_{\boldsymbol{x}_0})$. We will show that $\omega(\Gamma_{\boldsymbol{x}_0}) \subset \Gamma$, that is, that $\phi(t, \boldsymbol{x}_0) \to \Gamma$ as $t \to \infty$ in the sense of Lemma (6.49) 6. In other words, we will show that for each $\epsilon > 0$ there is $t > 0$ such that for every $t' > t$ there is $\boldsymbol{p} \in \Gamma$ (possibly depending on t) which satisfies $\|\phi(t, \boldsymbol{x}_0) - \boldsymbol{p}\| < \epsilon$. The argument again uses the properties of a local transversal. Let $z \in \Gamma \subset \omega(\Gamma_{\boldsymbol{x}_0})$ and consider a local transversal L at z. Using Lemmas 6.84 and 6.85, we find a sequence $t_0 < t_1 < \ldots t_n \to \infty$, such that $\boldsymbol{x}_n = \phi(t_k, \boldsymbol{x}_0) \in L$, $\boldsymbol{x}_n \to z$ (in a monotonic way along L). Moreover, we select a sequence $(t_n)_{n\in\mathbb{N}}$ to correspond to subsequent intercepts of $\Gamma_{\boldsymbol{x}_0}$ with L so that $\phi(t, \boldsymbol{x}_0) \notin L$ for $t_n < t < t_{n+1}$. By Lemma 6.81 we know that for sufficiently large n, the times $t_{n+1} - t_n$ between subsequent crossings of the local transversal depend continuously on x_n, that is, the times between subsequent crossings are bounded uniformly in n by, say, T_1 (and T_1 can be made close to the period T of Γ). Using Theorem B.16, we see that for any ϵ there is n_0 such that for any $n \geq n_0$ and $0 \leq t' \leq T_1$

$$\|\phi(t', \boldsymbol{x}_n) - \phi(t', z)\| < \epsilon.$$

Hence, let us fix an arbitrary ϵ and corresponding n_0. We take arbitrary $t > t_{n_0}$ and select $n > n_0$ such that $t_n < t < t_{n+1}$. Then

$$\|\phi(t, \boldsymbol{x}) - \phi(t - t_n, z)\| = \|\phi(t - t_n, \boldsymbol{x}_n) - \phi(t - t_n, z)\| = \|\phi(t', \boldsymbol{x}_n) - \phi(t', z)\| < \epsilon,$$

since $0 < t' < T_1$, independently of n. Since $\phi(t - t_n, z) \in \Gamma$ for any n and t, we see that $\phi(t, \boldsymbol{x}) \to \Gamma$. Hence, by Lemma 6.49 6., $\omega(\Gamma_{\boldsymbol{x}_0}) \subset \Gamma$, and thus $\Gamma = \omega(\Gamma_{\boldsymbol{x}_0})$. □

Remark 6.89 The proof of Theorem 6.88 actually shows that a stronger result is valid. Namely, if $\boldsymbol{x} \in \omega(\Gamma_{\boldsymbol{x}_0})$ is such that $\omega(\Gamma_{\boldsymbol{x}})$ contains a non-stationary point, then $\omega(\Gamma_{\boldsymbol{x}})$ is a closed orbit since we can draw a local transversal through this non-stationary point. This fact will be used to prove the following corollary.

Corollary 6.90 *Let $\Gamma_{\boldsymbol{x}}$ be a bounded trajectory of* (6.6.177) *with a differentiable field $\boldsymbol{f}$ for which equilibria are isolated. Then, there are only three possibilities.*

1. *$\omega(\Gamma_{\boldsymbol{x}})$ is an equilibrium,*
2. *$\omega(\Gamma_{\boldsymbol{x}})$ is a periodic orbit,*
3. *$\omega(\Gamma_{\boldsymbol{x}})$ consists of a finite number of equilibria $\boldsymbol{p}_1, \ldots, \boldsymbol{p}_k$, connected by trajectories Γ with $\alpha(\Gamma) = \boldsymbol{p}_i$ and $\omega(\Gamma) = \boldsymbol{p}_j$.*

Proof By Lemma 6.49 1., the set $\omega(\Gamma_{\boldsymbol{x}})$ is bounded and closed and thus can contain only a finite number of equilibria. If it contains only equilibria, then it contains only one of them, since $\omega(\Gamma_{\boldsymbol{x}})$ is connected by Lemma 6.49 4., which covers the first possibility. If this is not the case, then $\omega(\Gamma_{\boldsymbol{x}})$ contains non-equilibria and, by Lemma 6.49 3. (invariance), it contains trajectories through these points. Let $\boldsymbol{u} \in \omega(\Gamma_{\boldsymbol{x}})$ be an arbitrary non-equilibrium. If $\omega(\Gamma_{\boldsymbol{u}})$ and $\alpha(\Gamma_{\boldsymbol{u}})$ are equilibria, then case 3. holds. Assume then that $\omega(\Gamma_{\boldsymbol{u}})$ contains a non-equilibrium point, say, $\boldsymbol{z}$. Then, arguing as in the proof of Theorem 6.88 (which is possible as $\boldsymbol{z}$ is not an equilibrium), $\Gamma_{\boldsymbol{u}}$ is a periodic orbit, which means that $\omega(\Gamma_{\boldsymbol{x}})$ contains a periodic orbit. But then, by the second part of the proof of Theorem 6.88, the whole $\omega(\Gamma_{\boldsymbol{x}})$ is a periodic orbit. The same argument applies if $\alpha(\Gamma_{\boldsymbol{u}})$ contains a non-equilibrium point. □

Remark 6.91 Case 3 of the previous corollary can be fine-tuned. Namely, if $\boldsymbol{p}_1$ and $\boldsymbol{p}_2$ are two equilibria in $\omega(\Gamma_{\boldsymbol{x}})$, then there is at most one trajectory $\Gamma \subset \omega(\Gamma_{\boldsymbol{x}})$ with $\alpha(\Gamma) = \boldsymbol{p}_1$ and $\omega(\Gamma) = \boldsymbol{p}_2$. Indeed, assume to the contrary that we have two trajectories Γ_1, Γ_2 with this property and take points $\boldsymbol{q}_1 \in \Gamma_1$ close to $\boldsymbol{p}_1$ and $\boldsymbol{q}_2 \in \Gamma_2$ close to $\boldsymbol{p}_2$. Since $\boldsymbol{q}_i, i = 1, 2$, are not equilibria, there are local transversals L_1 and L_2 at these points. Since $\Gamma_i, i = 1, 2$, are in $\omega(\Gamma_{\boldsymbol{x}})$ the trajectory $\Gamma_{\boldsymbol{x}}$ crosses L_1 in the same direction as Γ_1 and L_2 in the direction of Γ_2. Since the direction of the field along the transversals is the same as of the above trajectories, the region bounded by L_1, the segment of $\Gamma_{\boldsymbol{x}}$ between L_1 and L_2, L_2, and corresponding segments of Γ_2 and Γ_1, is positively invariant (that is the trajectory $\Gamma_{\boldsymbol{x}}$ must stay inside). But this is a contradiction as the segments of trajectories Γ_1 and Γ_2 outside this region form a part of the limit set for $\Gamma_{\boldsymbol{x}}$.

Hence, if $\omega(\Gamma_{\boldsymbol{x}})$ contains two equilibria, then they must be joined either by a single trajectory or two trajectories running in opposite directions (and thus forming, together with the equilibria, a loop).

Definition 6.92 We say that a closed orbit Γ is a limit cycle if for some $\boldsymbol{x} \notin \Gamma$ we have $\Gamma \subset \omega(\Gamma_{\boldsymbol{x}})$ (ω-limit cycle) or $\Gamma \subset \alpha(\Gamma_{\boldsymbol{x}})$ (α-limit cycle).

In the proof of Theorem 6.88, we showed that a limit cycle Γ has the following property: there is $\boldsymbol{x} \notin \Gamma$ such that

$$\phi(t, \boldsymbol{x}) \to \Gamma, \qquad t \to \infty.$$

Geometrically, it means that some trajectory spirals towards Γ as $t \to \infty$ (or $t \to -\infty$). It follows that limit cycles have a certain stability property.

Proposition 6.93 *Let Γ be an ω-limit cycle such that $\Gamma \subset \omega(\Gamma_{\boldsymbol{x}})$, $\boldsymbol{x} \notin \Gamma$. Then there is a neighbourhood V of $\boldsymbol{x}$ such that for any $\boldsymbol{y} \in V$ we have $\Gamma \in \omega(\Gamma_{\boldsymbol{y}})$.*

Proof Let Γ be an ω-limit cycle and let $\phi(t, \boldsymbol{x})$ spiral toward Γ as $t \to \infty$. Take $z \in \Gamma$ and a transversal L at z. Take $t_0 < t_1$ such that $\boldsymbol{x}_0 = \phi(t_0, \boldsymbol{x}), \boldsymbol{x}_1 = \phi(t_1, \boldsymbol{x}) \in L$ with $\phi(t, \boldsymbol{x}) \notin L$ for $t_0 < t < t_1$. For sufficiently large t_0 the segment $L_{\boldsymbol{x}_0,\boldsymbol{x}_1}$ of L between $\boldsymbol{x}_0$ and $\boldsymbol{x}_1$ does not intersect Γ. Then the region A bounded by Γ, the part of $\Gamma_{\boldsymbol{x}}$ between $\boldsymbol{x}_0$ and $\boldsymbol{x}_1$ and $L_{\boldsymbol{x}_0,\boldsymbol{x}_1}$ is forward invariant, as is the set $B = A \setminus \Gamma$. For sufficiently large $t > 0$, $\phi(t, \boldsymbol{x})$ is in the interior of A. But than, the same is true for $\phi(t, \boldsymbol{y})$ for $\boldsymbol{y}$ sufficiently close to $\boldsymbol{x}$. Such $\phi(t, \boldsymbol{y})$ stays in A and, by the Poincareé–Bendixon theorem must spiral towards Γ. □

Corollary 6.94 *Let Γ be a closed orbit enclosing an open set Ω (contained in the domain of $\boldsymbol{f}$). Then Ω contains an equilibrium.*

Proof Assume Ω contains no equilibrium. Our first step is to show that there must be the 'smallest' closed orbit. We know that orbits don't intersect, so if we have a collection of closed orbits Γ_n, the enclosed regions Ω_n form a nested sequence with decreasing areas A_n. We also note that if we have a sequence $\boldsymbol{x}_n$ of points on Γ_n converging to $\boldsymbol{x} \in \Omega$, then $\boldsymbol{x}$ also is on a closed orbit, say Γ_0. Otherwise, by Poincaré–Bendixon theorem, $\phi(t, \boldsymbol{x})$ would spiral towards a limit cycle, but then, by Proposition 6.93, this would be true for $\phi(t, \boldsymbol{x}_n)$ for sufficiently large n, which contradicts the assumption that $\boldsymbol{x}_n$ is on a closed orbit.

Let $0 \leq A$ be an infimum of all areas of regions enclosed by closed orbits. Then $A = \lim\limits_{n\to\infty} A_n$ for some sequence of Γ_n. Let $\boldsymbol{x}_n \in \Gamma_n$. Since $\Omega \cup \Gamma$ is compact, we may assume $\boldsymbol{x}_n \to \boldsymbol{x}$ and, by the previous part, $\boldsymbol{x}$ belongs to a closed orbit Γ_0. Using, for instance, Lemma 6.81 and Theorem B.16, that orbits Γ_n get arbitrarily close to Γ_0, so A is the area enclosed by Γ_0. Since Γ_0 does not reduce to a point, $A > 0$. But then the region enclosed by Γ_0 contains neither equilibria nor closed orbit, which contradicts the Poincaré-Bendixon theorem. □

A difficult part in applying the Poincaré–Bendixon theorem is to find a *trapping region*, that is, a closed and bounded region which contains no equilibria and which trajectories cannot leave. Quite often, this is an annular region with the equilibrium (source) in the hole so that the trajectories can enter through the inner boundary and the outer boundary is chosen in such a way that the vector field always points inside the region. We illustrate this in the following example.

Example 6.95 We show that the second-order equation

$$z'' + (z^2 + 2(z')^2 - 1)z' + z = 0$$

has a nontrivial periodic solution. Let us start with writing this equation as the system

$$\begin{aligned} x' &= y, \\ y' &= -x + y(x^2 + 2y^2 - 1) \end{aligned} \tag{6.6.185}$$

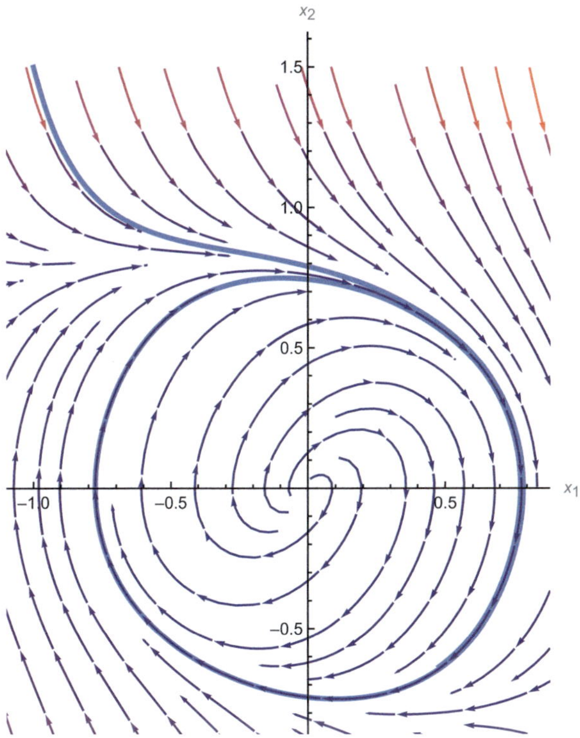

Fig. 6.41 Streamlines and a trajectory converging to the limit cycle of (6.6.185)

Clearly, (0, 0) is an equilibrium and, by linearization we see that this is an unstable equilibrium; hence, there is a chance that the flow in a small neighbourhood of the origin will be outward. Thus, let us write the equation in polar coordinates. We get

$$\frac{d}{dt}(x^2 + y^2) = 2xx' + 2yy' = 2y^2(1 - x^2 - 2y^2).$$

We observe that $1 - x^2 - 2y^2 > 0$ for $x^2 + y^2 < \frac{1}{2}$ and $1 - x^2 - 2y^2 < 0$ for $x^2 + y^2 > 1$. Hence, any solution which starts in the annulus $\frac{1}{2} < x^2 + y^2 < 1$ must stay there. Since the annulus does not contain an equilibrium, a periodic orbit must be inside it, see Fig 6.41.

Exercise 6.6.2 Show that $S = \left\{(x, y) : \frac{1}{4} \leq x^2 + y^2 \leq 4\right\}$ is a trapping region for the system

$$x' = -y + x(1 - x^2 - y^2),$$
$$y' = x + y(1 - x^2 - y^2).$$

Hence, it has a periodic trajectory.

Let us consider a more sophisticated example pertaining to the population balance equation at a subcellular level. Here, following [90], we consider a model of glycolysis.

Example 6.96 Glycolysis is a fundamental biochemical process in which living cells obtain energy by breaking down sugar. In many cells, such as yeast cells, glycolysis can proceed in an oscillatory fashion.

A simple model of glycolysis presented by Sel'kov is given by

$$\begin{aligned} x' &= -x + ay + x^2y, \\ y' &= b - ay - x^2y, \end{aligned} \tag{6.6.186}$$

where x and y are, respectively, concentrations of ADP (adenosine diphosphate) and F6P (fructose-6-phosphate), and $a, b > 0$ are kinetic parameters. We shall show that under certain assumptions, there are, indeed, periodic solutions to the system.

First, we find nullclines. We obtain that $x' = 0$ along the curve $y = x/(a + x^2)$ and $y' = 0$ along $y = b/(a+x^2)$. The nullclines are sketched in Fig. 6.42, along with some representative vectors. Our first step is to construct a trapping region. First, we observe that no trajectory can escape across any coordinate axis. Also, since $y' < 0$ above both nullclines, no trajectory can cross a horizontal line in this region going upward. This leaves only the question of closing this region from the right. We note that $x' + y' = b - x < 0$ as long as $x > b$. Thus, in this region $y' < -x'$ and, since $x' > 0$, we have $y'/x' < -1$, which geometrically means that the slope is

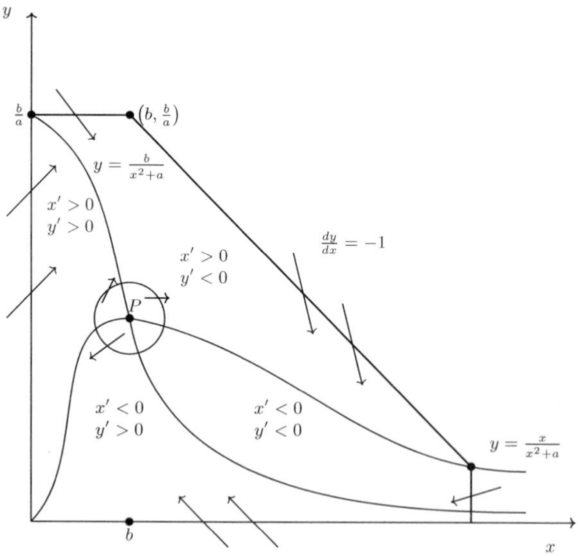

Fig. 6.42 Nullclines, equilibrium and the trapping region for (6.6.186)

steeper than any straight line with slope -1. In other words, any trajectory in this region will cross a line with the slope -1 from above-right to below-left. Hence, if we draw such a line crossing the horizontal line $y = \frac{b}{a}$ at $x = b$, then trajectories from the left cannot escape through it. Finally, we continue the line till it intersects the nullcline $y = \frac{x}{a+x^2}$ and note that below this nullcline $x' < 0$ so that we can close the region by a vertical segment to obtain a trapping region; see Fig. 6.42.

Can we conclude that there is a periodic trajectory? Well, no, as there is an equilibrium $\left(b, \frac{b}{a+b^2}\right)$ inside the region. However, this is not necessarily bad news. If the equilibrium is a repeller, then there is a neighbourhood V of the equilibrium such that any trajectory starting from V will eventually reach a circle of a fixed radius (see the proof of point 2 of Theorem 6.34). Thus, the boundary of V can be taken as the inner boundary of the required set. So, we need to prove that $\left(b, \frac{b}{a+b^2}\right)$ is a repeller. The Jacobi matrix for (6.6.186) at (x_0, y_0) is

$$\mathcal{J} = \begin{pmatrix} -1 + 2x_0y_0 & a + x_0 \\ -2x_0y_0 & -(a + x_0^2) \end{pmatrix}.$$

The Jacobian at the equilibrium is $a + b^2$ and the trace $tr\,\mathcal{A}$ is

$$tr\,\mathcal{A} = -\frac{b^2 + (2a-1)b^2 + (a+a^2)}{a+b^2}.$$

The dividing line $tr\,\mathcal{A} = 0$ occurs when

$$b^2 = \frac{1}{2}(1 - 2a \pm \sqrt{1-4a}).$$

Hence, by Remark 6.12, the equilibrium is unstable $tr\,\mathcal{A} > 0$, and asymptotically stable if $tr\,\mathcal{A} < 0$, see Fig. 6.43. The system has a closed orbit for parameters in the region corresponding to $tr\,\mathcal{A} > 0$, shown on the left of Fig. 6.44.

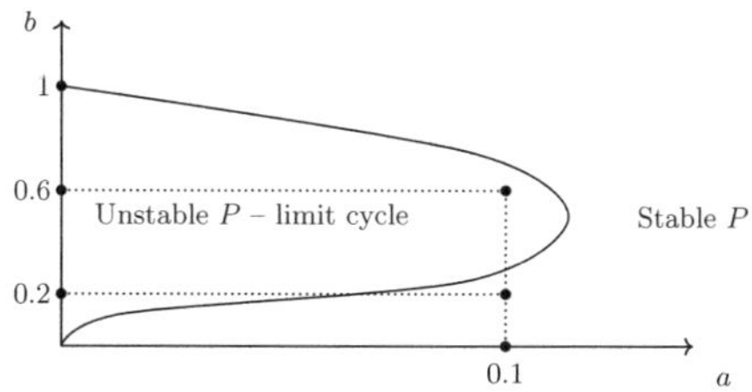

Fig. 6.43 Region of parameters a i b, where $\tau > 0$

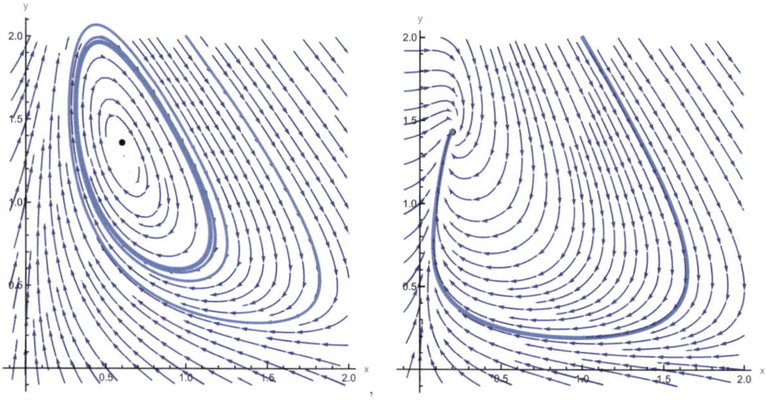

Fig. 6.44 Limit cycle of (6.6.186) for $b = 0.6, a = 0.1$ and asymptotically stable equilibrium with $b = 0.2, a = 0.1$

6.6.3 *Other Criteria for Existence and Non-existence of Periodic Orbits*

When discussing he Lyapunov functions, we observed that when V was constant along trajectories, they were closed so that the solutions were periodic. While not always true (as level sets of V can be unbounded), there is a general result pertaining to situations when the system is conservative. To be more precise, let us consider the system

$$\boldsymbol{x}' = \boldsymbol{f}(\boldsymbol{x}), \quad \boldsymbol{x}(0) = \mathring{\boldsymbol{x}}. \tag{6.6.187}$$

We say that there is a conserved quantity for this system if there exists a scalar continuous function $E(\boldsymbol{x})$ which is constant along trajectories, that is, $\frac{d}{dt}E(\phi(t, \mathring{\boldsymbol{x}})) = 0$. We also require $E(\boldsymbol{x})$ to be nonconstant on every open set to avoid trivial examples. We have the following result.

Proposition 6.97 *Assume that there is a conserved quantity of the system (6.6.187), where $\boldsymbol{x} \in \mathbb{R}^2$, and that $\boldsymbol{x}^*$ is an isolated fixed point of (6.6.187). If $\boldsymbol{x}^*$ is a local minimum of E, then there is a neighbourhood of $\boldsymbol{x}^*$, where all trajectories are closed.*

Proof Let $\bar{B}_{\boldsymbol{x}^*,\delta}$ be a closed ball centred at $\boldsymbol{x}^*$ where E attains absolute minimum. We can assume $E(\boldsymbol{x}^*) = 0$. Since E is constant on trajectories, each trajectory is contained in a level set of E,

$$E_c = \{(x_1, x_2) : E(x_1, x_2) = c\} \cap B_\delta.$$

The set E_c is nonempty, bounded and closed for all sufficiently small $0 < c \leq c_0$. We use Theorem 6.88. For each c, there are two possibilities: either the trajectory stays in E_c, in which case also its ω-limit set is in E_c, and thus it is a closed orbit or the trajectory leaves E_c so that there is a point $\boldsymbol{y}$ on the trajectory where $E(\boldsymbol{y}) = c$ and $\boldsymbol{y} = \delta$.

Assume that the second case happens for every $c > 0$; we can consider a sequence $c_n \to 0$. Then we have a sequence $\boldsymbol{y}_n$ with $\|\boldsymbol{y}_n\| = \delta$ for which $E(\boldsymbol{y}_n) = c_n \to 0$. But then, by the continuity of E, we find a point $\boldsymbol{y}_0$ with $\|\boldsymbol{y}_0\| = \delta$ and $E(\boldsymbol{y}_0) = 0$, contradicting the assumption that $\boldsymbol{y}^*$ is an absolute minimum in B_δ. □

Sometimes, it is equally important to show that certain regions have no periodic orbits. We have already seen one such criterion related to Lyapunov functions: if there is a strict Lyapunov function in a certain region surrounding an equilibrium, then there are no periodic orbits in this region. Here, we consider two more "negative" criteria, the first of which is the Dulac-Bendixon criterion. Recall that for a differentiable function $\boldsymbol{f} : \Omega \to \mathbb{R}^d$, its divergence is defined as

$$\operatorname{div} \boldsymbol{f} = \sum_{i=1}^{d} \frac{\partial f_i}{\partial x_i}.$$

Proposition 6.98 *Let Ω be a simply connected (with no holes) open region in $\mathbb{R}^2$. Consider a planar system $\boldsymbol{x}' = \boldsymbol{f}(\boldsymbol{x})$ with $\boldsymbol{f}$ continuously differentiable in Ω and suppose that there is a scalar differentiable function $g : \Omega \to \mathbb{R}$ such that $div(g\boldsymbol{f}) \neq 0$ on Ω. Then there are no periodic orbits of this system in Ω.*

Proof Suppose $\Gamma \subset \Omega$ is a periodic orbit. Using Green's theorem, [32, Section 5.1], and the assumption,

$$0 \neq \int\!\!\int_{\Omega} \operatorname{div}(g\boldsymbol{f}) dx_1 dx_2 = \oint_{\Gamma} g\langle \boldsymbol{f}, \boldsymbol{n}\rangle d\sigma,$$

where $\boldsymbol{n}$ is the normal to Γ and $d\sigma$ the line element along Γ. However, since $\boldsymbol{n}$ is normal to Γ, we have $\boldsymbol{n} \cdot \boldsymbol{f} = 0$ as $\boldsymbol{f}$ is tangent to any trajectory. Thus, the left-hand side is zero, and we obtain a contradiction. □

Exercise 6.6.3 Using the function $g(x, y) = \frac{1}{xy}$, show that the variant of Lotka–Volterra model,

$$x' = x(1 - a_1 x + b_1 y),$$
$$y' = y(1 - a_2 y + b_2 x),$$

with $a_i > 0$ (to model the effect of overcrowding), does not have periodic orbits in the first quadrant.

Another criterion for the non-existence of periodic orbits, which, contrary to the Dulac–Bendixon criterion, works in arbitrary dimension, pertains to gradient systems. A system $\boldsymbol{x}' = \boldsymbol{f}(\boldsymbol{x})$ is called a gradient systems with potential function V if $\boldsymbol{f} = -\nabla V$ in some region Ω for a scalar function V. Recall that in Example 6.37, we showed that one-dimensional equations are always gradient systems. We have

Proposition 6.99 *Closed orbits are impossible in gradient systems provided* $V \neq const$.

Proof Suppose that there is a closed orbit Γ corresponding to a periodic solution with period T. Define $\Delta V(t_2, t_1)$ to be the change of V along Γ between $\phi(t_2, \mathring{\boldsymbol{x}})$ and $\phi(t_1, \mathring{\boldsymbol{x}})$. Clearly, $\Delta V(T, 0) = 0$ as the orbit is closed and V is continuous. On the other hand

$$0 = \Delta V(T, 0) = \int_0^T \frac{dV(\phi(t, \mathring{\boldsymbol{x}})}{dt} dt = \int_0^T \langle \nabla V(\phi(t, \mathring{\boldsymbol{x}}), \boldsymbol{x}' \rangle dt$$

$$= -\int_0^T \|\boldsymbol{x}'(t)\|^2 dt < 0,$$

as the trajectory is not an equilibrium due to $V \neq const$. This contradiction proves the statement. □

Exercise 6.6.4 Show that

$$x' = \sin y,$$
$$y' = x \cos y,$$

is a gradient system, so it does not have periodic solutions.

Exercise 6.6.5 Consider $\boldsymbol{x}' = \boldsymbol{f}(\boldsymbol{x})$. Show that there are no periodic trajectories in any region where $\langle \boldsymbol{f}(x), x \rangle \neq 0$. Hint. Show that

$$\frac{d}{dt}\|\boldsymbol{x}(t)\|^2 = 2\langle \boldsymbol{f}(\boldsymbol{x}(t)), \boldsymbol{x}(t) \rangle$$

and then use the argument of Proposition 6.99 to the increment $\boldsymbol{x}(T) - \boldsymbol{x}(0)$, where T is a period of the assumed periodic solution. Give a geometric interpretation of this result.

Example 6.100 Let us extend the analysis of (6.2.45) to cover the case $R_0 > 1$. Let us recall that we work with the non-dimensional version of (6.2.45),

$$\begin{aligned} x' &= \rho(1 - x) - R_0 x y, \\ y' &= (R_0 x - 1) y, \end{aligned} \tag{6.6.188}$$

where

$$\rho = \frac{\mu}{\nu + \mu}.$$

When $R_0 > 1$, we have two equilibria—the disease free equilibrium (DFE)

$$(x_0^*, y_0^*) = (1, 0)$$

and the endemic equilibrium (EE)

$$(x_1^*, y_1^*) = \left(\frac{1}{R_0}, \rho\left(1 - \frac{1}{R_0}\right)\right).$$

We found that

$$V = \{(x, y) : \ x \geq 0, y \geq 0, x + y \leq 1\} \tag{6.6.189}$$

is an invariant region for (6.6.188), see (6.2.51) and Fig. 6.45. Clearly, DFE cannot be globally stable. We proceed by analysing the local stability of these equilibria. The Jacobi matrix of (6.6.188) at any equilibrium (x^*, y^*) is

$$\mathcal{J}(x^*, y^*) = \begin{pmatrix} -\rho - R_0 y^* & -R_0 x^* \\ R_0 y^* & R_0 x^* - 1 \end{pmatrix}. \tag{6.6.190}$$

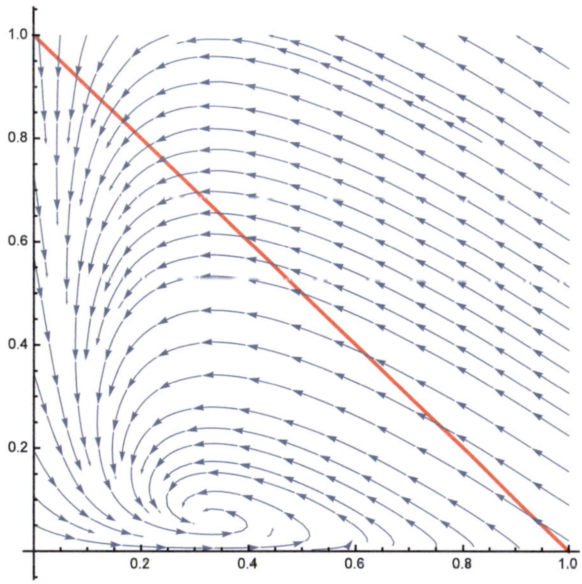

Fig. 6.45 Streamlines and V as the basin of attraction for (6.6.188) with $R_0 > 1$

Hence, at DFE, we have

$$\mathcal{J}(1,0) = \begin{pmatrix} -\rho & -R_0 \\ 0 & R_0 - 1 \end{pmatrix} \tag{6.6.191}$$

and, since the matrix is upper triangular, we immediately get eigenvalues

$$\lambda_1 = -\rho, \quad \lambda_2 = R_0 - 1.$$

The first eigenvalue is negative, while the second is negative if $R_0 < 1$. As we observed earlier, there is no other equilibrium (in the biologically feasible domain), and we proved that DFE is globally asymptotically stable. The same was proved true if $R_0 = 1$, so we only need to consider the case $R_0 > 1$. In this case, DFE is a saddle, and we move to investigate the endemic equilibrium. We have

$$\mathcal{J}(x_1^*, y_1^*) = \begin{pmatrix} -\rho - R_0 y_1^* & -R_0 x_1^* \\ R_0 y_1^* & R_0 x_1^* - 1 \end{pmatrix}. \tag{6.6.192}$$

From the equation for EE, we have $R_0 x_1^* - 1 = 0$, so

$$\mathcal{J}(x_1^*, y_1^*) = \begin{pmatrix} -\rho - R_0 y_1^* & -1 \\ R_0 y_1^* & 0 \end{pmatrix}, \tag{6.6.193}$$

and we obtain the characteristic equation as

$$\lambda^2 + (\rho + R_0 y_1^*)\lambda + R_0 y_1^* = 0. \tag{6.6.194}$$

Using $y_1^* = \rho\left(1 - \frac{1}{R_0}\right)$, we find

$$\rho + R_0 y_1^* = \rho + R_0 \rho \left(1 - \frac{1}{R_0}\right) = \rho R_0,$$

$$R_0 y_1^* = R_0 \rho \left(1 - \frac{1}{R_0}\right) = \rho(R_0 - 1),$$

so that (6.6.194) becomes

$$\lambda^2 + \rho R_0 \lambda + \rho(R_0 - 1) = 0. \tag{6.6.195}$$

Using Remark 6.12, we conclude that the endemic equilibrium is locally asymptotically stable. To provide a more detailed description of the character of the equilibrium, we find the roots of the characteristic equation. We have

$$\lambda_{1,2} = \frac{-\rho R_0 \pm \sqrt{\Delta}}{2}, \qquad \Delta = (\rho R_0)^2 - 4\rho(R_0 - 1).$$

Hence, if $\Delta > 0$, the characteristic equation has two negative real roots, and therefore the endemic equilibrium is a stable node. If $\Delta < 0$, the characteristic equation has two complex conjugate roots with negative real parts; hence, the endemic equilibrium is a stable focus. see Fig. 6.45.

Next, we consider the global stability of the endemic equilibrium. We prove the following proposition.

Theorem 6.101 *Assume* $R_0 > 1$. *If* $y_0 > 0$, *then the solution* $\phi(t, (x_0, y_0))$ *of* (6.6.188) *satisfies*

$$\lim_{t\to\infty} \phi(t, (x_0, y_0)) = (x_1^*, y_1^*) = \left(\frac{1}{R_0}, \rho\left(1 - \frac{1}{R_0}\right)\right).$$

In other words

$$\omega(\Gamma_{(x_0,y_0)}) = (x_1^*, y_1^*)$$

provided $y_0 > 0$.

Proof We know from the previous considerations that

$$\omega(\Gamma_{(x_0,y_0)}) \subset V = \{(x, y) : x \geq 0, y \geq 0, x + y \leq 1\}.$$

Moreover, for the initial condition of the form $(x_0, y_0) = (x_0, 0)$, system (6.2.49) can be reduced to the single equation

$$x' = \rho(1 - x), \quad x(0) = x_0,$$

with the solution $x(t) = 1 + (x_0 - 1)e^{-\rho t}$ and hence the solution to (6.2.49) is $(1 + (x_0 - 1)e^{-\rho t}, 0)$. Therefore the whole line $y = 0$ is a trajectory and $\omega(\Gamma_{(x_0,0)}) = (1, 0)$. First, we rule out periodic trajectories in V using the Dulac–Bendixon criterion, Proposition 6.98. We observe that taking $g = 1/y$ simplifies the coordinates of the vector field and we get

$$\frac{\partial}{\partial x}\frac{\rho(1 - x) - R_0 xy}{y} + \frac{\partial}{\partial y}\frac{y(R_0 x - 1)}{y} = -\frac{\rho}{y} - R_0 < 0.$$

Hence, there are no periodic orbits for $y > 0$. However, if there was a periodic orbit with $y = 0$, then this orbit and the orbit $\{y = 0\}$ would intersect, which is impossible.

Thus, we only have to prove that $(1,0) \notin \omega(\Gamma_{(x_0,y_0)})$ with $y_0 > 0$. The problem here is that the structure of the vector field allows, in principle, the existence of trajectories that unwind, getting closer to $(1,0)$ at each turn. Assume the contrary, that is, that $(1,0) \in \omega(\Gamma_{(x_0,y_0)})$. We aim to show that then $\{y = 0\} \subset \omega(\Gamma_{(x_0,y_0)})$. So, if $(1,0) \in \omega(\Gamma_{(x_0,y_0)})$, then there is a sequence $(t_n)_{n\in\mathbb{N}}$ with $t_n \to \infty$ as $n \to \infty$ such that $(x(t_n), y(t_n)) \to (1,0)$. We can assume that $x(t_n) > 1/R_0$ for all n. Then we have $y'(t_n) > 0$ for any n. Let us take arbitrary n. Since $y(t_n) > 0$ and $y(t_n) \to 0$ as $n \to \infty$, we cannot have $y'(t) \geq 0$ for $t \geq t_n$. Thus, there is $\bar{t}_n > t_n$ with $y'(\bar{t}_n) < 0$. Then there must be an index $\bar{n} > n$ such that $y'(t_{\bar{n}}) > 0$. To avoid working with subsequences, we select from $(t_n)_{n\in\mathbb{N}}$ a subsequence (without changing notation) in such a way that $t_n < \bar{t}_n < t_{n+1}$ (that is, we throw away from $(t_n)_{n\in\mathbb{N}}$ all elements with indices between n and $\bar{n}$ and rename $\bar{n}$ to be $n+1$). Then there must be $\hat{t}_n \in (\bar{t}_n, t_{n+1})$, where $y'(\hat{t}_n) = 0$. Again, we select $\hat{t}_n$ to be the largest t in (t_n', t_{n+1}) with this property. Then we have

$$0 \leq y(\hat{t}_n) \leq y(t_n), \quad x(\hat{t}_n) = 1/R_0$$

as $(x(\hat{t}_n), y(\hat{t}_n))$ is on the $y'-$nullcline and $y'(t) > 0$ for $t \in (\hat{t}_n, t_{n+1})$. Thus, by the Sandwich Theorem, $y(\hat{t}_n) \to 0$ as $n \to \infty$ and thus $\left(\frac{1}{R_0}, 0\right) \in \omega(\Gamma_{(x_0,y_0)})$. But then, by Lemma 6.49 3., $\{y = 0\} \in \omega(\Gamma_{(x_0,y_0)})$ as $\{y = 0\}$ is an orbit. On the other hand, as before,

$$(x + y)' = x' + y' = \rho(1 - x) - y$$

hence

$$(x + y)' \geq -(x + y) + \rho,$$

that is,

$$(x + y)(t) \geq \rho + (x_0 + y_0 - 1)e^{-t}.$$

Thus, if $(x, y) \in \omega(\Gamma_{(x_0,y_0)})$, $x + y \geq \rho$, and hence only points $(x, 0)$ with $x \geq \rho$ could potentially be in $\omega(\Gamma_{(x_0,y_0)})$. This contradicts the above result that $\{y = 0\} \subset \omega(\Gamma_{(x_0,y_0)})$. Hence DFE is not in $\omega(\Gamma_{(x_0,y_0)})$ and the theorem is proved. □

Example 6.102 Let us consider a simplified version of a vector-borne disease without the exposed classes and disease-induced death (6.1.10)

$$\begin{aligned}
S_h' &= B_h(N_h) - \lambda_h S_h + \gamma_h R_h - \mu_h S_h,\\
I_h' &= \lambda_h S_h - (\nu_h + \mu_h) I_h,\\
R_h' &= \nu_h I_h - (\gamma_h + \mu_h) R_h,\\
S_v' &= B_v(N_v) - \lambda_v S_v - \mu_v S_v,\\
I_v' &= \lambda_v S_v - \mu_v I_v.
\end{aligned} \tag{6.6.196}$$

For the infection force (6.1.15) corresponding to a small number of mosquitoes relative to humans, and using the total populations $N_h = S_h + I_h + R_h$ and $N_v = S_v + I_v$, we re-write (6.6.196) as

$$\begin{aligned} N_h' &= B_h(N_h) - \mu_h N_h, \\ I_h' &= \beta_h \frac{I_v S_h}{N_h} - (\nu_h + \mu_h) I_h, \\ R_h' &= \nu_h I_h - (\gamma_h + \mu_h) R_h, \\ N_v' &= B_v(N_v) - \mu_v N_v, \\ I_v' &= \beta_v \frac{I_h S_v}{N_h} - \mu_v I_v. \end{aligned} \tag{6.6.197}$$

We assume that the equations for the human and vector total populations have unique globally exponentially stable equilibria N_h^* and N_v^*, respectively. Since the mosquito population evolves much faster than humans, it is plausible to assume that from the point of view of the human population, the mosquito population is at equilibrium depending, however, on the state of the former. In this case, (6.6.196) can be approximated uniformly in time by

$$N_h{}' = B_h(N_h) - \mu_h N_h, \tag{6.6.198a}$$

and

$$\begin{aligned} I_h{}' &= f(I_h, R_h) := \beta_v \beta_h N_v^* \frac{N_h - I_h - R_h}{N_h} \frac{I_h}{\mu_v N_h + \beta_v I_h} - (\nu_h + \mu_h) I_h, \\ R_h{}' &= g(I_h, R_h) := \nu_h I_h - (\gamma_h + \mu_h) R_h. \end{aligned} \tag{6.6.198b}$$

A rigorous proof that such an approximation is valid can be found in [16]. Here, we shall focus on analysing (6.6.198).

First, writing the infection term as

$$\beta_v \beta_h N_v^* \frac{N_h - I_h - R_h}{N_h} \frac{I_h}{\mu_v N_h + \beta_v I_h} = \beta_v \beta_h N_v^* \frac{N_h - I_h - R_h}{N_h(\mu_v N_h + \beta_v I_h)} I_h,$$

we see that the number of infections due to one infective in a fully susceptible population at equilibrium ($I_h = R_h = 0$) is equal to $\frac{\beta_v \beta_h N_v^*}{\mu_v N_h^*}$. Then, the average time of sojourn in the infective state is $\frac{1}{\nu_h + \mu_h}$ and hence

$$R_0 = \frac{\beta_v \beta_h N_v^*}{\mu_v N_h^* (\nu_h + \mu_h)}.$$

To get endemic equilibria of (6.6.198), first we find

$$R_h^* = \frac{\nu_h}{\gamma_h + \mu_h} I_h^*, \tag{6.6.199}$$

and then the nontrivial I_h^* is given by

$$I_h^* = N_h^*(\mu_h + \gamma_h)\frac{\beta_v \beta_h N_v^* - \mu_v(\mu_h + \nu_h)N_h^*}{\beta_v \beta_h N_v^*(\mu_h + \gamma_h + \nu_h) + \beta_h(\mu_h + \gamma_h)(\mu_h + \nu_h)N_h^*}, \tag{6.6.200}$$

and we see that I^* is positive for $R_0 > 1$.

Next, we use the Vidyasagar theorem, Theorem 6.77, to establish the global stability of (N_h^*, I_h^*, R_h^*). For this, we consider (6.6.198b) with N_h replaced by the fixed equilibrium N_h^* of (6.6.198a). Then, at any equilibrium point (I_h^*, R_h^*), the Jacobi matrix of (6.6.198b) is given by

$$\mathcal{J}_{(f,g)}(I_h^*, R_h^*) = \begin{pmatrix} -\frac{\beta_v I_h^*}{\mu_v N_h^* + \beta_v I_h^*}\left(\frac{\beta_h N_h^*}{N_h^*} + (\nu_h + \mu_h)\right) & -I_h^* \frac{\beta_h \beta_v N_v^*}{N^*(\mu_v N_h^* + \beta_v I_h^*)} \\ \nu_h & -(\gamma_h + \mu_h) \end{pmatrix}.$$

If $R_0 > 1$, we have a unique $I_h^* > 0$ and the trace of $J_{(f,g)}(I_h^*, R_h^*)$ is negative, while its determinant is positive, thus (I_h^*, R_h^*) is locally asymptotically stable. Next, consider the Dulac function

$$\phi(I_h, R_h) = \frac{\mu_v N_h^* + \beta_v I_h}{I_h},$$

well-defined and differentiable in the open first quadrant $\mathbb{R}_+^2$.

Then, we have

$$\begin{aligned}
&\frac{\partial}{\partial I_h}(\phi(I_h, R_h) f(I_h, R_h)) + \frac{\partial}{\partial R_h}(\phi(I_h, R_h) f(I_h, R_h)) \\
&= \frac{\partial}{\partial I_h}\left(\beta_v \beta_h N_v^* \frac{N_h^* - I_h - R_h}{N_h^*} - (\nu_h + \mu_h)(\mu_v N_h^* + \beta_v I_h)\right) \\
&+ \frac{\partial}{\partial R_h}\left(\nu_h(\mu_v N_h^* + \beta_v I_h) - (\gamma_h + \mu_h)R_h \frac{\mu_v N_h^* + \beta_v I_h}{I_h}\right) \\
&= -\frac{\beta_v \beta_h N_v^*}{N_h^*} - (\gamma_h + \mu_h)\beta_v - (\gamma_h + \mu_h)\frac{\mu_v N_h^* + \beta_v I_h}{I_h} < 0,
\end{aligned}$$

whenever $I_h > 0$. Thus, there are no closed orbits in $\mathbb{R}_+^2$. No ω-limit set of any point $(\mathring{I}_h, \mathring{R}_h) \in \mathbb{R}_+^2$ intersects the semi-axes $\{I_h = 0, R_h > 0\}$ or $\{R_h = 0, I_h > 0\}$. Indeed, in the latter case, the field points inside $\mathbb{R}_+^2$, while in the former case, the whole semi-axis is a trajectory with its ω-limit point being $(0, 0)$, which, by invariance, would belong to $\omega(\Gamma_{(\mathring{I},\mathring{R})})$. Since $I_h' > 0$ in the positive vicinity of $(0, 0)$, no internal trajectory can approach $\{I_h = 0, R_h > 0\}$ close to $(0, 0)$. Hence, this part, belonging to $\omega(\Gamma_{(\mathring{I},\mathring{R})})$, must be a part of an internal trajectory, which, however, would violate the uniqueness. Since the trajectories are bounded, their ω-limit sets are compact and contained in $\mathbb{R}^2_+$. Hence, using Corollary 6.90, the ω-limit set of any orbit coincides with (I_h^*, R_h^*) and therefore (I_h^*, R_h^*) is globally exponentially stable in $\mathbb{R}^2_+$. By the Vidyasagar theorem, (N_h^*, I_h^*, R_h^*) is globally asymptotically stable.

Chapter 7
Discrete Time Non-linear Models for Interacting Species and Structured Populations

In the previous chapter, we discussed at length methods for analysing systems of differential equations that occur in epidemiology, ecology, and other fields pertaining to population dynamics. However, the interactions often occur in discrete, equally spaced moments of time, for instance, during the short breeding seasons. We also have seen systems in age-structured one-species models. Then, it is more convenient to model such populations using systems of difference equations. So far, we have dealt only with linear models, but they can be easily generalized to non-linear ones, e.g., by introducing density-dependent coefficients (such as logistic growth). While we do not intend to give a comprehensive account of non-linear discrete systems, see, e.g., [34, 35, 42], we will discuss two models to indicate how the methods used in continuous time can be adapted in the discrete scenario and used for their stability analysis.

7.1 Models

7.1.1 Host-Parasitoid System

Discrete difference equation models apply most readily to groups such as insect populations with a natural time division into discrete generations. A model which has received considerable attention from experimental and theoretical biologists is the *host-parasitoid* system. Let us begin by introducing a definition of a parasitoid, [27, 49]. Predators kill their prey, typically for food. Parasites live in or on a host and often draw food, shelter, or other requirements from that host without killing it. Female parasitoids, in turn, typically search and kill but do not consume their hosts. Rather, they *oviposit* (deposit eggs) on, in, or near the host and use it as a source

J. Banasiak, *Introduction to Mathematical Methods in Population Theory*, Springer Undergraduate Mathematics Series,
https://doi.org/10.1007/978-3-031-65491-6_7

of food and shelter for the developing young. There are around 50,000 species of wasp-like parasitoids, 15,000 fly-type parasitoids and 3000 species of other orders.

Typical insect species, both host and parasitoid, have several life stages, including eggs, larvae, pupae and adults. In most cases, eggs are attached to the outer surface of the host during its larval or pupal stage or injected into the host's flesh. The larval parasitoids develop and grow at the expense of their host, consuming it and eventually killing it before they pupate.

A simple model for this system has the following set of assumptions:

1. Hosts that have been parasitised will give rise to the next generation of parasitoids.
2. Hosts not parasitised will give rise to their own prodigy.
3. The fraction of parasitised hosts depends on the rate of encounter of the two species; in general, this fraction may depend on the densities of one or both species.

It is instructive to consider this minimal set of interactions first and examine their consequences. We define:

- $N(k)$—density (number) of host species in generation k,
- $P(k)$—density (number) of parasitoid in generation k,
- $f = f(N, P)$—the number of encounters in unit time (rate),
- λ—host reproductive rate,
- c—average number of viable eggs laid by the parasitoid on a single host.

Then our assumptions 1.–3. lead to the following system of equations

$$\begin{aligned} N(k+1) &= \lambda N(k) f(N(k), P(k)), \\ P(k+1) &= cN(k)(1 - f(N(k), P(k))), \end{aligned} \tag{7.1.1}$$

$k = 0, 1, \ldots$. To proceed, we have to specify the rate of encounters f. One of the earliest models is the Nicholson-Bailey model; see [27, Section 2.2].

The Nicholson-Bailey Model

Nicholson and Bailey added two assumptions to the list 1.–3., namely:

4. Encounters occur randomly. The number of encounters N_e of the host with the parasitoid is, therefore, proportional to the product of their (number) densities,

$$N_e = \alpha N P,$$

where α is a constant, which represents the searching efficiency of the parasitoid. (This kind of assumption presupposing random encounters is known as the *law of mass action*.)

5. Only the first encounter between a host and a parasitoid is significant (once the host has been parasitised, it gives rise to exactly c parasitoid progeny; any other encounter with an egg-laying parasitoid will not increase or decrease this number.

Based on the latter assumption, we have to distinguish only between those hosts that have had no encounters and those that had n encounters, $n \geq 1$. Because the encounters are random, one can represent the probability of r encounters by some distribution based on the average number of encounters that take place per unit time. Poisson distribution is one of the simplest distributions used in such a context. We introduced it in Sect. 2.1.5 as a limiting case of the binomial distribution, giving the probability of exactly r events in n independent trails if the probability of one occurrence is p. If we let the number of trials n go to infinity in such a way that the average number of events $\mu = np$ stays constant (so p must tend to zero), then the probability of exactly r events is given by

$$p(r) = \frac{e^{-\mu}\mu^r}{r!},$$

which is the Poisson distribution.

In the case of host-parasitoid interaction, the average number of encounters per host per unit time is

$$\mu = \frac{N_e}{N},$$

that is, by 4.,

$$\mu = aP.$$

Hence, the probability of a host not having any encounter with parasitoid is

$$p(0) = e^{-aP}.$$

Assuming that the parasitoids search independently and that their searching efficiency is constant a, from (7.1.1) we obtain the Nicholson-Bailey system

$$\begin{aligned} N(k+1) &= \lambda N(k)e^{-aP(k)}, \\ P(k+1) &= cN(k)\lambda(1 - e^{-aP(k)}), \end{aligned} \tag{7.1.2}$$

$k = 0, 1, \ldots$.

7.1.2 Nonlinear Age-Structured Model

Consider a single species population with two age classes: juveniles and adults. Let $X(k)$ be the number of juveniles at the time k and let $Y(k)$ be the number of adults. We assume that the fertility rate for adults is b, c is the survival rate of juveniles, that is, a fraction c of juveniles present at time k become adults at $k+1$, and the others die. In each time period, only the density-dependent fraction $s - DY(k)$ of the adult population survives. These assumptions lead to the system

$$\begin{aligned} X(k+1) &= bY(k), \\ Y(k+1) &= cX(k) + Y(k)(s - DY(k)). \end{aligned} \tag{7.1.3}$$

We re-write this equation in a form which is more convenient for analysis by introducing new unknowns $X(k) = b\hat{X}(k)/D$ and $Y(k) = \hat{Y}(k)/D$, which converts (7.1.3) into

$$\begin{aligned} \hat{X}(k+1) &= \hat{Y}(k), \\ \hat{X}(k+1) &= a\hat{X}(k) + \hat{Y}(k)(s - \hat{Y}(k)), \end{aligned} \tag{7.1.4}$$

where $a = cb > 0$.

7.1.3 Discrete SIR Model

A large class of nonlinear models is provided by mathematical epidemiology. Here, the models describe the movement of individuals between subpopulations characterized by different infection statuses due to infection, recovery, vaccination, loss of immunity, etc.

Extending [23, Section 2.9], we consider the population divided into three classes: susceptibles S, infectives I and removed (immune or dead) R. We do not consider any births in the process. Within one cycle from time k to time $k+1$, the probability of an infective meeting someone is α' and thus meeting a susceptible is $\alpha' S/N$, where N is the size of the population at time k; further, a fraction α'' of these encounters result in an infection. We denote $\alpha = \alpha'\alpha''$. Moreover, we assume that a fraction β of individuals (except class S) can become susceptible (could be reinfected) and a fraction γ of infectives move to R. This results in the system

$$\begin{aligned} S(k+1) &= S(k) - \frac{\alpha}{N(k)} I(k)S(k) + \beta(I(k) + R(k)), \\ I(k+1) &= I(k) + \frac{\alpha}{N(k)} I(k)S(k) - \gamma I(k) - \beta I(k), \\ R(k+1) &= R(k) - \beta R(k) + \gamma I(k). \end{aligned} \tag{7.1.5}$$

We observe that

$$S(k+1) + I(k+1) + R(k+1) = S(k) + I(k) + R(k) = const = N,$$

so that the total population does not change in time.

This observation can be used to reduce the (7.1.5) to a two dimensional system

$$\begin{aligned} S(k+1) &= S(k) - \frac{\alpha}{N} I(k)S(k) + \beta(N - S(k)), \\ I(k+1) &= I(k)(1 - \gamma - \beta) + \frac{\alpha}{N} I(k)S(k). \end{aligned} \tag{7.1.6}$$

The modelling indicates that we need to assume

$$0 < \gamma + \beta < 1 \quad \text{and} \quad 0 < \alpha < 1. \tag{7.1.7}$$

7.2 Stability Analysis

As in the scalar case, our main interest is finding and determining the stability of the equilibria. For this, however, we have to do some mathematics.

We shall be concerned with autonomous systems of difference equations

$$\boldsymbol{x}(k+1) = \boldsymbol{f}(\boldsymbol{x}(k)), \tag{7.2.8}$$

where $\boldsymbol{x}(0) = \mathring{\boldsymbol{x}}$ is given. Here, $\boldsymbol{x} = (x_1, \ldots, x_d) \in \mathbb{R}^d$ and $\boldsymbol{f}(\boldsymbol{x}) = (f_1(\boldsymbol{x}), \ldots, f_d(\boldsymbol{x}))$ is a continuous function from $\mathbb{R}^d$ into $\mathbb{R}^d$. In what follows, $\|\cdot\|$ is any norm on $\mathbb{R}^d$, unless specified otherwise, see Chap. A.

The main concepts are exactly as in Sect. 4.2, but specified to the vector case. Thus, $\boldsymbol{x}^* \in \mathbb{R}^d$ is called an equilibrium point of (7.2.8) if

$$\boldsymbol{x}^* = \boldsymbol{f}(\boldsymbol{x}^*) \tag{7.2.9}$$

and the definition of stability reads

Definition 7.1

(a) The equilibrium $\boldsymbol{x}^*$ is stable if for given $\epsilon > 0$ there is $\delta > 0$ such that for any $\boldsymbol{x}$ and for any $k > 0$, $\|\mathring{\boldsymbol{x}} - \boldsymbol{x}^*\| < \delta$ implies $\|\boldsymbol{f}^k(\boldsymbol{x}) - \boldsymbol{x}^*\| < \epsilon$ for all $k > 0$. If $\boldsymbol{x}^*$ is not stable, then it is called unstable (that is, $\boldsymbol{x}^*$ is unstable if there is $\epsilon >$ such that for any $\delta > 0$ there are $\boldsymbol{x}$ and k such that $\|\boldsymbol{x} - \boldsymbol{x}^*\| < \delta$ and $\|\boldsymbol{f}^k(\boldsymbol{x}) - \boldsymbol{x}^*\| \geq \epsilon$.)

(b) The point $\boldsymbol{x}^*$ is called attracting if there is $\eta > 0$ such that

$$\|\boldsymbol{x}(0) - x^*\| < \eta \text{ implies } \lim_{k\to\infty} \boldsymbol{x}(k) = \boldsymbol{x}^*.$$

If $\eta = \infty$, then $\boldsymbol{x}^*$ is called a global attractor or globally attracting.

(c) The point $\boldsymbol{x}^*$ is called an asymptotically stable equilibrium if it is stable and attracting. If $\eta = \infty$, ten $\boldsymbol{x}^*$ is said to be a globally asymptotically stable equilibrium.

Though the attracting and asymptotically stable equilibria seem similar, the distinction is necessary as (in dimensions higher than 2) we may have unstable and attracting equilibria, as demonstrated in the following example.

Example 7.2 Let us consider a sequence of points in $\mathbb{R}^2$, in polar coordinates (r, ϕ) defined by the system

$$r(k+1) = \sqrt{r(k)}, \quad r(0) = \mathring{r},$$
$$\phi(k+1) = \sqrt{2\pi\phi(k)}, \quad \phi(0) = \mathring{\phi},$$

$k = 0, 1, \ldots, r_0 > 0, 0 \le \phi_0 \le 2\pi$. Direct calculations show that

$$r(k) = r_0^{2^{-k}},$$
$$\phi(k) = (2\pi)^{1-2^{-k}}\theta_0^{2^{-k}}$$

$k = 0, 1, \ldots$. Thus, the equilibria are $(0, 0)$ and $(1, 0)$. We see that for the initial condition $(r_0, 0)$ with $r_0 > 0$, we have $r(k) \to 1, \theta(k) \to 0$ is a monotonic way. On the other hand, let us take $r_0 > 0$ and $\phi_0 = \delta\pi$ with r_0 close to 1 and δ close to zero, so that (r_0, ϕ_0) is in a neighbourhood of $(1, 0)$. Then, while $r(k)$ tends monotonically to 1, we have $\phi(k) = 2\pi\left(\frac{\delta}{2}\right)^{2^{-k}}$ and, for $0 < \delta < 2$, $\phi(k)$ tends to monotonically to 2π. This means that $(r(k), \phi(k))$ first will move away from $(1, 0)$ and then, getting closer to the unit circle, will converge to $(1, 0)$ from the other side of the x-axis. This shows that $(1, 0)$ is attracting but not stable.

7.2.1 *Stability of Linear Systems*

We consider the linear autonomous system

$$\boldsymbol{x}(k+1) = \mathcal{A}\boldsymbol{x}(k), \quad \boldsymbol{x}(0) = \mathring{\boldsymbol{x}}, \quad k = 0, 1, 2, \ldots. \tag{7.2.10}$$

From Sect. 5.3, we know that the solution is given by $\boldsymbol{x}(k) = \mathcal{A}^k\mathring{\boldsymbol{x}}, k = 1, 2, \ldots$. The long-term behaviour of solutions was discussed in Sect. 5.3.8, where, however, we focused on the case when there was a simple dominant eigenvalue of $\mathcal{A}$. In this section, we interpret those considerations in the context of the stability concepts and extend them to the case of general eigenvalues.

Assume that $\mathcal{A}$ is non-singular to simplify considerations. Then the origin 0 is always a unique equilibrium point of (7.2.10). Otherwise, we would have higher dimensional subspaces consisting of equilibria.

We have the following result.

Theorem 7.3 *The following statements hold:*

1. *The zero solution of (7.2.10) is stable if and only if the spectral radius of $\mathcal{A}$ satisfies $\rho(\mathcal{A}) \leq 1$ and the eigenvalues of unit modulus are semi-simple;*
2. *The zero solution is asymptotically stable if and only if $\rho(\mathcal{A}) < 1$.*

Proof Let $\lambda_i, i = 1, \ldots, \lambda_p$, be distinct eigenvalues of $\mathcal{A}$, each with algebraic multiplicity n_i so that $n_1 + \ldots + n_p = d$. We assume that $|\lambda_1| \geq |\lambda_2| \geq \ldots \geq |\lambda_p| > 0$. For each $1 \leq r \leq p$, let $\boldsymbol{v}_r^1, \ldots, \boldsymbol{v}_r^{n_r}$ be the set of eigenvectors and associated eigenvectors belonging to λ_r. Following the considerations of Sect. 5.3.3, each $\boldsymbol{v}_r^j$ is a solution to

$$(\mathcal{A} - \lambda_r \mathcal{I})^{m_j^r} \boldsymbol{v}_r^j = 0$$

for some $1 \leq m_j^r \leq n_r$, where some (even all) js may correspond to the same m_j^r. Then, we can write the solution as

$$\mathcal{A}^k \mathring{\boldsymbol{x}} = \sum_{r=1}^{p} \left(\sum_{j=1}^{n_r} c_r^j \mathcal{A}^k \boldsymbol{v}_r^j \right), \tag{7.2.11}$$

where c_r^j are coefficients of the expansion of $\mathring{\boldsymbol{x}}$ in the basis consisting of $\boldsymbol{v}_r^j$, $1 \leq r \leq p$, and $1 \leq j \leq n_r$. Hence, as in (5.3.72),

$$\begin{aligned} \mathcal{A}^k \boldsymbol{v}_r^j &= \Big(\lambda_r^k \mathcal{I} + k \lambda_r^{k-1} (\mathcal{A} - \lambda_r \mathcal{I}) + \ldots \\ &+ \frac{k!}{(m_j^r - 1)!(k - m_j^r + 1)!} \lambda_r^{k - m_j^r + 1} (\mathcal{A} - \lambda_r \mathcal{I})^{m_j^r - 1} \Big) \boldsymbol{v}_r^j. \end{aligned} \tag{7.2.12}$$

We emphasize again that (7.2.12) is a finite sum for any k consisting of at most m_j^r terms. Using the triangle inequality for norms, we obtain

$$\begin{aligned} \|\mathcal{A}^k \boldsymbol{v}_r^j\| &\leq |\lambda_r|^k \Big(1 + k |\lambda_r|^{-1} (\|\mathcal{A}\| + |\lambda_r|) + \ldots \\ &+ P_{m_j^r - 1}(k) |\lambda_r|^{-m_j^r + 1} (|\mathcal{A}| + |\lambda_r|)^{m_j^r - 1} \Big) \|\boldsymbol{v}_r^j\| \\ &= |\lambda_r|^k k^{m_j^r - 1} \Big(k^{-m_j^r + 1} + k^{m_j^r - 2} |\lambda_r|^{-1} (\|\mathcal{A}\| + |\lambda_r|) + \ldots \\ &+ \frac{P_{m_j^r - 1}(k)}{k^{m_j^r - 1}} |\lambda_r|^{-m_j^r + 1} (\|\mathcal{A}\| + |\lambda_r|)^{m_j^r - 1} \Big) \|\boldsymbol{v}_r^j\| \\ &\leq C_j^r |\lambda_r|^k k^{m_j^r - 1} \leq C_j^r |\lambda_r|^k k^{n_r - 1}, \end{aligned} \tag{7.2.13}$$

where the constant C^r_j does not depend on d, $P_{m^r_j-1}$ are polynomial in k of order $m^r_j - 1$, and we used $m^r_j \leq n_r$. Next, as in (5.3.42), we observe that the vector $\boldsymbol{c}$ consisting of constants c^j_r is given by

$$\boldsymbol{c} = \begin{pmatrix} | & \dots & | \\ \boldsymbol{v}^1_1 & \dots & \boldsymbol{v}^{n_p}_p \\ | & \dots & | \end{pmatrix}^{-1} \mathring{\boldsymbol{x}},$$

and thus, for some constant M,

$$\|\boldsymbol{c}\| \leq M\|\mathring{\boldsymbol{x}}\|. \tag{7.2.14}$$

Assume now that $\rho(\mathcal{A}) < 1$, that is, all eigenvalues have absolute values smaller than 1. Then

$$\|\mathcal{A}^k\mathring{\boldsymbol{x}}\| \leq \sum_{r=1}^{p} |\lambda_r|^k k^{n_r-1} \left(\sum_{j=1}^{n_r} |c^j_r| C^r_j\right) \leq M'\|\mathring{\boldsymbol{x}}\| \sum_{r=1}^{p} |\lambda_r|^k k^{n_r-1},$$

where

$$M' = M \max_{1\leq r\leq p} \sum_{j=1}^{n_r} C^r_j,$$

and we used the fact that in (7.2.14) we can use $\|\boldsymbol{c}\| = \max\{|c^j_r|\}$. Since $1 > \rho(\mathcal{A}) = |\lambda_1| \geq |\lambda_2| \geq \ldots \geq |\lambda_p|$, for arbitrary $1 > \eta > \rho(\mathcal{A})$ we have

$$\|\mathcal{A}^k\mathring{\boldsymbol{x}}\| \leq M'p\|\mathring{\boldsymbol{x}}\|\eta^k \left(\frac{\rho(\mathcal{A})}{\eta}\right)^k k^{d-1}.$$

Now, for any $a < 1$ and $d > 0$, we have $\lim\limits_{k\to\infty} a^k k^d = 0$, hence there is a constant L such that $\left(\frac{\rho(\mathcal{A})}{\eta}\right)^k k^{d-1} \leq L$ for all k. Thus, there is a constant $K > 0$ such that

$$\|\mathcal{A}^k\mathring{\boldsymbol{x}}\| \leq K\|\mathring{\boldsymbol{x}}\|\eta^k, \tag{7.2.15}$$

and the zero solution is asymptotically stable.

If $\rho(\mathcal{A}) = 1$, that is, there are eigenvalues of unit modulus, but they are semi-simple, then the expansions (7.2.13) reduce to the first term ($j_1 = 1$ in each case) so that in such a case

$$\|\mathcal{A}^n \boldsymbol{v}^j_r\| \leq C^r_j,$$

and the solution is stable (but not asymptotically stable).

If an eigenvalue λ_r with $|\lambda_r| = 1$ is not semi-simple, then some m^r_j is bigger than one, and we have polynomial entries in (7.2.12). Consider an associated eigenvector $\boldsymbol{v}^j_r$ corresponding to this case. Then

$$\begin{aligned}\|\mathcal{A}^k \boldsymbol{v}^j_r\| &= \|\boldsymbol{v}^j_r + k\lambda_r^{-1}(\mathcal{A} - \lambda_r)\boldsymbol{v}^j_r + \ldots + P_{m^r_j - 1}(k)\lambda_r^{-m^r_j+1}(\mathcal{A} - \lambda_r)^{m^r_j - 1}\boldsymbol{v}^j_r\| \\ &\geq k^{m^r_j - 1}\Big| k^{-m^r_j+1}|P_{m^r_j-1}(k)|\|(\mathcal{A} - \lambda_r)^{m^r_j-1}\boldsymbol{v}^j_r\| \\ &\quad - k^{-1}\|k^{-m^r_j+2}\boldsymbol{v}^j_r + k^{-m^r_j+3}\lambda_r^{-1}(\mathcal{A} - \lambda_r)\boldsymbol{v}^j_r + \ldots\|\Big|.\end{aligned}$$

The coefficient $\|(\mathcal{A} - \lambda_r)^{m^r_j-1}\boldsymbol{v}^j_r\|$ is non-zero and the first term inside the absolute value bars converges to a finite limit $\frac{1}{(m^r_j-1)!}$ and the second to zero, hence $\|\mathcal{A}^n\boldsymbol{v}^j_r\|$ diverges to infinity. Thus, taking initial conditions of the form $\epsilon\boldsymbol{v}^j_r$, we see that we can take an arbitrarily small initial condition, producing an unbounded solution; thus, the zero solution is unstable.

Finally, if $\rho(\mathcal{A}) > 1$, then the argument as above gives instability of the zero solution. □

7.2.2 Stability by Linearization

Let us return to (7.2.8). We assume that f is a function with continuous first-order partial derivatives in some open set $\Omega \subset \mathbb{R}^n$. In Lemma 6.33 we proved that, in some neighbourhood of $\boldsymbol{y}_0 \in \Omega$,

$$\boldsymbol{f}(\boldsymbol{x} + \boldsymbol{y}_0) = \boldsymbol{f}(\boldsymbol{y}_0) + \mathcal{A}\boldsymbol{x} + \boldsymbol{g}(\boldsymbol{x}), \tag{7.2.16}$$

where

$$\mathcal{A} = \begin{pmatrix} \frac{\partial f_1}{\partial x_1}(\mathring{\boldsymbol{y}}) & \ldots & \frac{\partial f_1}{\partial x_d}(\mathring{\boldsymbol{y}}) \\ \vdots & \ddots & \vdots \\ \frac{\partial f_d}{\partial x_1}(\mathring{\boldsymbol{y}}) & \ldots & \frac{\partial f_d}{\partial x_d}(\mathring{\boldsymbol{y}}) \end{pmatrix}$$

is the Jacobi matrix of $\boldsymbol{f}$, and $\boldsymbol{g}(\boldsymbol{x})/\|\boldsymbol{x}\|$ is a continuous function in some neighbourhood of $\boldsymbol{y}_0$ and vanishes at $\boldsymbol{x} = 0$.

Let $\boldsymbol{y}^0 = \boldsymbol{x}^*$ be an isolated equilibrium of (7.2.8) and $\mathcal{A}$ be the matrix defined above. We say that

$$\boldsymbol{x}(k+1) = \mathcal{A}\boldsymbol{x}(k) \tag{7.2.17}$$

is the linearization of (7.2.8) around an equilibrium $\boldsymbol{x}^*$.

Further, mimicking the derivation of (2.1.49), the solution of the nonhomogeneous system of equations

$$\boldsymbol{x}(k+1) = \mathcal{A}\boldsymbol{x}(k) + \boldsymbol{g}(k), \tag{7.2.18}$$

where $\boldsymbol{x} = (x_1, \ldots, x_d)$, $\boldsymbol{g} = (g_1, \ldots, g_d)$ and $\mathcal{A} = \{a_{ij}\}_{1\leq i,j\leq d}$, with the initial condition $\boldsymbol{x}(0) = \mathring{\boldsymbol{x}}$, is given by the formula

$$\boldsymbol{x}(k) = \mathcal{A}^k\mathring{\boldsymbol{x}} + \sum_{r=0}^{k-1} \mathcal{A}^{k-r-1}\boldsymbol{g}(r). \tag{7.2.19}$$

We shall need a discrete version of Gronwall's lemma (see Lemma B.11 for the continuous version) .

Lemma 7.4 *Let* $(z(k))_{k\in\mathbb{N}}$ *and* $(h(k))_{k\in\mathbb{N}}$ *be two sequences of real numbers and let* $h(k) \geq 0$ *for all* $k \geq k_0 \geq 1$. *If for all such* k,

$$z(k) \leq M\left(z(k_0) + \sum_{j=k_0}^{k-1} h(j)z(j)\right) \tag{7.2.20}$$

for some $M > 0$, *then*

$$z(k) \leq z(k_0)\prod_{j=k_0}^{k-1}(1 + Mh(j)) \tag{7.2.21}$$

and

$$z(k) \leq z(k_0)\exp\sum_{j=k_0}^{k-1} Mh(j). \tag{7.2.22}$$

Proof Consider the equation

$$u(k) = M\left(u(k_0) + \sum_{j=k_0}^{k-1} h(j)u(j)\right), \quad u(k_0) = z(k_0).$$

From non-negativity, by induction, we obtain $z(k) \leq u(k)$ for $k \geq k_0$. Hence

$$u(k+1) - u(k) = Mh(k)u(k)$$

or, equivalently,

$$u(k+1)) = (1 + Mh(k))u(k).$$

Hence,

$$u(k) = u(k_0) \prod_{j=k_0}^{k-1} (1 + Mh(j)),$$

which proves (7.2.21). The second follows from the formula $1 + Mh(j) \le e^{Mh(j)}$. □

Theorem 7.5 *Assume that $\boldsymbol{f}$ is a C^1 function and $\boldsymbol{x}^*$ is an equilibrium point. If the zero solution of the linearized system (7.2.17) at $\boldsymbol{x}^*$ is asymptotically stable, then the equilibrium $\boldsymbol{x}^*$ is asymptotically stable.*

Proof We have

$$\boldsymbol{x}(k+1) = \boldsymbol{f}(\boldsymbol{x}(k)) = \boldsymbol{f}(\boldsymbol{x}(k)) - \boldsymbol{f}(\boldsymbol{x}^*) + \boldsymbol{x}^*.$$

Denoting $\boldsymbol{y}(k) = \boldsymbol{x}(k) - \boldsymbol{x}^*$ and using Lemma 6.33, we obtain

$$\boldsymbol{y}(k+1) = \mathcal{A}\boldsymbol{y}(k) + \boldsymbol{g}(\boldsymbol{y}(k)),$$

so that, by (7.2.19),

$$\boldsymbol{y}(k) = \mathcal{A}^k \boldsymbol{y}(0) + \sum_{r=0}^{k-1} \mathcal{A}^{k-r-1}\boldsymbol{g}(\boldsymbol{y}(r)).$$

By Theorem 7.3, the asymptotic stability of the linearized system is equivalent to (7.2.15), we get

$$\|\boldsymbol{y}(k)\| \le K\eta^k \|\boldsymbol{y}(0)\| + K\eta^{-1} \sum_{r=0}^{k-1} \eta^{k-r} \|\boldsymbol{g}(\boldsymbol{y}(r))\|.$$

For a given $\epsilon > 0$, there is $\delta > 0$ such that $\|\boldsymbol{g}(\boldsymbol{y})\| < \epsilon\|\boldsymbol{y}\|$ whenever $\|\boldsymbol{y}\| < \delta$. So, as long as we can keep $\|\boldsymbol{y}(r)\| < \delta$ for $r \le k-1$,

$$\eta^{-k}\|\boldsymbol{y}(k)\| \le K\|\boldsymbol{y}(0)\| + K\epsilon \sum_{r=0}^{k-1} \eta^{-r-1}\|\boldsymbol{y}(r)\|.$$

Applying the Gronwall inequality for $z(k) = \eta^{-k}\|\boldsymbol{y}(k)\|$, we obtain

$$\eta^{-k}\|\boldsymbol{y}(k)\| \le \|\boldsymbol{y}(0)\| \prod_{j=0}^{k-1} (1 + K\epsilon\eta^{-1}).$$

Thus

$$\|\boldsymbol{y}(k)\| \leq \|\boldsymbol{y}(0)\|(\eta + K\epsilon)^k.$$

Let us choose $\epsilon < (1-\eta)/K$ so that $\eta + K\eta < 1$. Thus, by induction, $\|\boldsymbol{y}(0)\| < \delta$, we have $\|\boldsymbol{y}(k)\| < \|\boldsymbol{y}(0)\| < \delta$ and the equilibrium is asymptotically stable. □

We can also prove that if $\rho(\mathcal{A}) > 1$, then the equilibrium is unstable, but the proof is more involved; see Theorem 6.34, point 2., where we show an analogous result for systems with continuous time.

The stability of the equilibria for nonlinear systems when $\rho(\mathcal{A}) = 1$ cannot be decided based on linearization; see the analogous results for continuous-time systems in Theorem 6.34, point 3.

7.3 Stability Analysis of Models

7.3.1 SIR Model

Let us start with finding the equilibria of (7.1.6). These are solutions of

$$\begin{aligned} S &= F_1(S, I) = S - \frac{\alpha}{N} IS + \beta(N - S), \\ I &= F_2(S, I) = I(1 - \gamma - \beta) + \frac{\alpha}{N} IS. \end{aligned} \tag{7.3.23}$$

The pair $(S, I) = (N, 0)$ is a solution of this system, giving a disease-free equilibrium. If $I \neq 0$, then, dividing the second equation by I and denoting $\delta = \gamma + \beta$, we find $S = N\delta/\alpha$, which yields $I = \beta N(\alpha - \delta)/\alpha\delta$; this is an endemic equilibrium. Thus

$$\boldsymbol{E}_1^* = (N, 0), \qquad \boldsymbol{E}_2^* = \left(\frac{N\delta}{\alpha}, \frac{\beta N(\alpha - \delta)}{\alpha\delta}\right).$$

To find the linearization, we find the Jacobi matrix of (F_1, F_2) by calculating

$$\begin{aligned} F_{1,S}(S, I) &= 1 - \frac{\alpha}{N} I - \beta, \quad F_{1,I}(S, I) = -\frac{\alpha}{N}, \\ F_{2,S}(S, I) &= \frac{\alpha}{N} I, \quad F_{2,I}(S, I) = 1 - \delta + \frac{\alpha}{N} S; \end{aligned}$$

thus we have

$$\mathcal{J}(\boldsymbol{E}_1^*) = \begin{pmatrix} 1-\beta & -\alpha \\ 0 & 1-\delta+\alpha \end{pmatrix}$$

and

$$\mathcal{J}(\boldsymbol{E}_2^*) = \begin{pmatrix} 1 - \frac{\alpha\beta}{\delta} & -\delta \\ \frac{\alpha\beta}{\delta} - \beta & 1 \end{pmatrix}.$$

To determine whether the magnitude of the eigenvalues is smaller or larger than one, we could find the eigenvalues and directly compute their magnitude, but this is generally time-consuming and not always informative. There are other, more straightforward methods.

Interlude: How to Determine Whether the Eigenvalues of a 2×2 Matrix Have a Magnitude Less Than One Without Solving the Quadratic Equation—The Jury Criterion

Consider the equation

$$\lambda^2 - B\lambda + A = 0,$$

where B and A are real coefficients. The roots are given by

$$\lambda_{1,2} = \frac{B \pm \sqrt{B^2 - 4A}}{2}.$$

We consider two cases. First, let $B^2 - 4A > 0$ so that the roots are real. Then $|\lambda_i| < 1, i = 1, 2,$ if and only if

$$-2 - B < \sqrt{B^2 - 4A} < 2 - B$$

and

$$-2 - B < -\sqrt{B^2 - 4A} < 2 - B.$$

Squaring the second inequality in the first expression, we obtain

$$1 - B + A > 0.$$

Similarly, squaring the first inequality in the second expression, we get

$$1 + B + A > 0.$$

Next, we get $2 - B > 0$ from the first and $-2 - B < 0$ from the second inequality, hence $|B| < 2$ and, since $B^2 - 4A \geq 0$, we have $A < 1$. Combining, we can write

$$|B| < 1 + A < 2. \tag{7.3.24}$$

Conversely, from (7.3.24) we get $-1 < B/2 < 1$ so that the midpoint between the roots is indeed inside $(-1, 1)$. Now, if $B > 0$, then we must only make sure that

$$\frac{B}{2} + \frac{\sqrt{B^2 - 4A}}{2} < 1.$$

This is equivalent to the following chain of inequalities (as $1 - B/2 > 0$)

$$\frac{\sqrt{B^2 - 4A}}{2} < 1 - \frac{B}{2} \Longleftrightarrow \frac{B^2 - 4A}{4} < 1 - B + \frac{B^2}{4} \Longleftrightarrow B < 1 + A.$$

Similarly, if $B < 0$, then we must only make sure that

$$\frac{B}{2} - \frac{\sqrt{B^2 - 4A}}{2} > -1.$$

This is equivalent to the following chain of inequalities (as $1 + B/2 > 0$)

$$\frac{\sqrt{B^2 - 4A}}{2} < 1 + \frac{B}{2} \Longleftrightarrow \frac{B^2 - 4A}{4} < 1 + B + \frac{B^2}{4} \Longleftrightarrow -B < 1 + A.$$

Hence, (7.3.24) is sufficient.

Assume now that $4A - B^2 > 0$ so that the roots are complex conjugate. Since absolute values of complex conjugate numbers are equal, and A is the product of the roots, we must have $A < 1$ (in fact, $0 < A < 1$ from the condition on the discriminant). We must prove that

$$1 - |B| + A > 0. \tag{7.3.25}$$

But in this case, $|B| < 2\sqrt{A}$ so that if

$$1 - 2\sqrt{A} + A > 0$$

holds on $(0, 1)$, than (7.3.25) holds as well. But the former is nothing but $(1 - \sqrt{A})^2 > 0$ on this open interval. Hence, (7.3.25) is proved.

Conversely, assume (7.3.24) holds. In the first part, we already proved that it yields the desired result if $4A - B^2 < 0$, we can assume that $4A - B^2 > 0$. This yields $A > 0$ and hence $0 < A < 1$ yields $\lambda\bar{\lambda} = |\lambda|^2 = A < 1$.

For a matrix

$$\mathcal{A} = \begin{pmatrix} a_{11} & a_{12} \\ a_{21} & a_{22} \end{pmatrix}$$

its eigenvalues are determined by solving

$$
\begin{aligned}
0 = det \begin{pmatrix} a_{11} - \lambda & a_{12} \\ a_{21} & a_{22} - \lambda \end{pmatrix} &= \begin{pmatrix} a_{11} & a_{12} \\ a_{21} & a_{22} \end{pmatrix} - \begin{pmatrix} \lambda & a_{12} \\ 0 & a_{22} \end{pmatrix} - \begin{pmatrix} a_{11} & 0 \\ a_{21} & \lambda \end{pmatrix} \\
&= \lambda^2 - \lambda(a_{11} + a_{22}) + det\mathcal{A} \\
&= \lambda^2 - \lambda tr\,\mathcal{A} + det\mathcal{A},
\end{aligned}
$$

where, recall, the trace tr is defined $tr\mathcal{A} = a_{11} + a_{22}$. Hence, the condition for stability can be expressed as

$$|tr\,\mathcal{A}| < 1 + det\,\mathcal{A} < 2 \tag{7.3.26}$$

Returning to our model, we find

$$|tr\,J_{E_1^*}| = |2 - \beta - \delta + \alpha| = 2 - \beta - \delta + \alpha,$$

where we can drop the absolute value bars by (7.1.7) and

$$det\,J_{E_1^*} = 1 - \delta + \alpha - \beta(1 - \delta + \alpha)$$

so that condition (7.3.26) can be written as

$$2 - \beta - \delta + \alpha < 2 - \delta + \alpha - \beta(1 - \delta + \alpha) < 2$$

Subtracting $2 - \beta - \delta + \alpha$ from each term we obtain

$$0 < \beta(\delta - \alpha) < \delta - \alpha + \beta.$$

This gives $\delta - \alpha > 0$ while the second condition is automatically satisfied as $0 < \beta < 1$ and $(\delta - \alpha) > 0$. Hence, the equilibrium $(N, 0)$ is asymptotically stable if

$$\delta = \beta + \gamma > \alpha.$$

Consider now the endemic equilibrium $\boldsymbol{E}_2^*$. Here we have

$$|tr\mathcal{J}(\boldsymbol{E}_2^*)| = \left|2 - \frac{\beta\alpha}{\delta}\right| = 2 - \frac{\beta\alpha}{\delta},$$

as $2(\gamma + \beta) - \beta\alpha = 2\gamma + \beta(2 - \alpha) > 0$, and

$$det\mathcal{J}(\boldsymbol{E}_2^*) = 1 - \frac{\beta\alpha}{\delta} + \alpha\beta - \delta\beta,$$

so that condition (7.3.26) can be written as

$$2 - \frac{\beta\alpha}{\delta} < 2 - \frac{\beta\alpha}{\delta} + \alpha\beta - \delta\beta < 2.$$

Subtracting $2 - \frac{\beta\alpha}{\delta}$ from both sides, we get

$$0 < \beta(\alpha - \delta) < \frac{\beta\alpha}{\delta},$$

from where $\alpha - \delta > 0$. The second condition is equivalent to $(\alpha - \delta) < \alpha/\delta$, that is, $\delta(\alpha - \delta) < \alpha$ but this is always satisfied as $\delta = \beta + \gamma < 1$ by (7.1.7). Hence, the endemic disease equilibrium

$$\left(\frac{N\delta}{\alpha}, \frac{\beta N(\alpha - \delta)}{\alpha\delta}\right)$$

is asymptotically stable provided

$$\alpha > \gamma + \beta.$$

We note that these conditions are consistent with the modelling process. The disease-free equilibrium is stable if the infection rate is smaller than the removal rate of infected individuals. On the other hand, in the opposite case, we have an endemic disease.

7.3.2 Nicholson–Bailey Model

Recall that the model is given by

$$\begin{aligned} N(k+1) &= \lambda N(k) e^{-aP(k)}, \\ P(k+1) &= cN(k)(1 - e^{-aP(k)}). \end{aligned} \tag{7.3.27}$$

The equilibria of (7.3.27) are obtained by solving

$$\begin{aligned} N &= \lambda N e^{-aP}, \\ P &= cN(1 - e^{-aP}). \end{aligned}$$

This gives either trivial equilibrium $\boldsymbol{E}_1^* = (N, P) = (0, 0)$ or, using

$$\lambda = e^{a\bar{P}},$$

we find $\boldsymbol{E}_2^* = (\bar{N}, \bar{P})$, where

$$\bar{P} = \frac{\ln \lambda}{a}, \tag{7.3.28}$$

and hence

$$\bar{N} = \frac{\lambda \ln \lambda}{(\lambda - 1)ac}. \tag{7.3.29}$$

Clearly, $\lambda > 1$ for $\bar{N}$ to be positive. To analyse the stability of the equilibria, first we define

$$F(N, P) = Ne^{-aP}, \quad G(N, P) = cN(1 - e^{-aP}).$$

Then, $F_{,N}(N, P) = e^{-aP}$, $F_{,P}(N, P) = -aNe^{-aP}$ and $G_{,N}(N, P) = c(1 - e^{-aP})$, $G_{,P}(N, P) = cNe^{-aP}$ and we obtain the Jacobi matrix at $\boldsymbol{E}_1^*$ as

$$\mathcal{J}(\boldsymbol{E}_1^*) = \begin{pmatrix} 1 & 0 \\ 0 & 0 \end{pmatrix},$$

and at $\boldsymbol{E}_2^*$ as

$$\mathcal{J}(\boldsymbol{E}_2^*) = \begin{pmatrix} 1 & -a\bar{N} \\ c\left(1 - \frac{1}{\lambda}\right) & \frac{ca\bar{N}}{\lambda} \end{pmatrix}.$$

We shall focus on the nontrivial equilibrium and use (7.3.26) in the subsequent considerations. At $(\bar{N}, \bar{P})$ we obtain

$$tr\, \mathcal{J}(\boldsymbol{E}_2^*) = 1 + \frac{\lambda}{\lambda - 1},$$

$$det\, \mathcal{J}(\boldsymbol{E}_2^*) = \frac{ca\bar{N}}{\lambda} + ca\bar{N}\left(1 - \frac{1}{\lambda}\right) = ca\bar{N} = \frac{\lambda \ln \lambda}{\lambda - 1}.$$

We know that $\lambda > 1$. Consider the function

$$S(\lambda) = \lambda - 1 - \lambda \ln \lambda,$$

see Exercise 6.4.1. We have $S(1) = 0$, $S'(\lambda) = -\ln \lambda$ so that $S'(\lambda) < 0$ for $\lambda > 1$. Thus, $S(\lambda) < 0$ for $\lambda > 1$ and hence

$$\lambda \ln \lambda > \lambda - 1 \text{ for } \lambda > 1.$$

Consequently,

$$det\,\mathcal{J}(\boldsymbol{E}_2^*) > 1$$

for any $\lambda > 1$; hence, the nontrivial equilibrium is unstable.

Most natural parasitoid-host systems in nature are more stable than the Nicholson-Bailey seems to indicate, and thus the model is not a satisfactory representation of real systems. We shall try to improve the system by modifying some parameters to see whether this could introduce stabilizing factors. We shall discuss the following modification:

> In the absence of parasitoids, the host population grows to some limited density (determined by the carrying capacity K of the environment).

Thus, the original system (7.3.27) would be amended as follows:

$$\begin{aligned} N(k+1) &= \lambda(N(k))N(k)e^{-aP(k)}, \\ P(k+1) &= cN(k)(1-e^{-aP(k)}), \end{aligned} \tag{7.3.30}$$

where for $\lambda(N(k))$ we might adopt the Ricker formula (2.1.25),

$$\lambda(N(k)) = \exp r\left(1 - \frac{N(k)}{K}\right),$$

where $r > 0$. With this choice, we obtain a modified Nicholson–Bailey system

$$\begin{aligned} N(k+1) &= N(k)\exp\left(r\left(1-\frac{N(k)}{K}\right) - aP(k)\right), \\ P(k+1) &= cN(k)(1-\exp(-aP(k))). \end{aligned} \tag{7.3.31}$$

We simplify this system by introducing $n(k) = N(k)/K$ and $p(k) = aP(k)$. This converts (7.3.31) into

$$\begin{aligned} n(k+1) &= n(k)\exp\left(r(1-n(k)) - p(k)\right), \\ \partial(k+1) &= caKn(k)(1-\exp(-p(k))), \end{aligned} \tag{7.3.32}$$

and, in what follows, we denote $caK = C$.

The equilibria are obtained by solving

$$\begin{aligned} n &= n\exp\left(r\,(1-n) - p\right), \\ p &= Cn(1-\exp(-p)). \end{aligned}$$

We discard the trivial equilibrium (0, 0) so that we are left with

$$1 = \exp\left(r\left(1-n\right)-p\right), \tag{7.3.33}$$

$$p = Cn(1-\exp(-p)). \tag{7.3.34}$$

Using the monotonicity of the exponential function, we obtain

$$p = r\left(1-n\right), \tag{7.3.35}$$

$$Cn = \frac{p}{1-\exp(-p)}. \tag{7.3.36}$$

It is clear that one non-trivial equilibrium point is given by $\bar{n}_1 = 1$ ($\bar{N}_1 = K$), $\bar{p}_1 = \bar{P}_1 = 0$. Is there any other equilibrium point? To answer this question, we use (7.3.33) to re-write (7.3.36) as

$$p = Cn\left(1-\exp\left(-r\left(1-n\right)\right)\right)$$

so that, using (7.3.35), n satisfies

$$\frac{r\left(1-n\right)}{Cn} = 1-\exp\left(-r\left(1-n\right)\right).$$

Define the following functions,

$$\begin{aligned} f_1(n) &= \frac{r\left(1-n\right)}{Cn} = \frac{r}{C}\left(\frac{1}{n}-1\right), \\ f_2(n) &= 1-\exp\left(-r\left(1-n\right)\right). \end{aligned} \tag{7.3.37}$$

First, we observe that, indeed, $f_1(1) = f_2(1) = 0$, which gives the equilibrium obtained above. Next,

$$f_1'(n) = -\frac{r}{Cn^2}, \qquad f_2'(n) = -r\exp\left(-r\left(1-n\right)\right);$$

hence, both functions are decreasing for $n > 0$. Furthermore, $f_1'(1) = -\frac{r}{C}$ and $f_2'(1) = -r$. If we assume $C \geq 1$, then the graph of f_1 is below the graph of f_2 for n smaller than and close to $n = 1$. Furthermore, $f_2(0) = 1-\exp(-r)$ and $f_1(n) \to +\infty$ as $n \to 0^+$. This implies the existence of at least one more equilibrium $(\bar{n}_2, \bar{p}_2)$. To show that there are no other equilibria, we find

$$f_1''(n) = \frac{2r}{Cn^3}, \qquad f_2''(n) = -r^2\exp\left(-r\left(1-n\right)\right),$$

so that f_1 is convex down and f_2 is convex down. In other words, $g(n) = f_1(n) - f_2(n)$ satisfies $g''(n) > 0$ which means that $g'(n)$ is strictly increasing and thus $g(n)$ can have at most two zeros. Thus, we found all possible equilibria of the system.

Let us focus on the last equilibrium describing the coexistence of the parasitoid and the host. Let us consider the stability of this equilibrium. The first step is to linearize the system around the equilibrium. To this end, we return to (7.3.31) and define

$$F(n,p) = n\exp\left(r\left(1-n\right)-p\right), \quad G(n,p) = Cn(1-\exp(-p)).$$

Thus,

$$\begin{aligned}
F_{,n}(n,p) &= (1-rn)\exp\left(r\left(1-n\right)-p\right),\\
F_{,p}(n,p) &= -n\exp\left(r\left(1-n\right)-p\right),\\
G_{,n}(n,p) &= C(1-\exp(-p)),\\
G_{,p}(n,p) &= Cn\exp(-p).
\end{aligned}$$

At $(\bar{n}_2, \bar{p}_2)$ we find, by (7.3.33),

$$\begin{aligned}
F_{,n}(\bar{n}_2,\bar{p}_2) &= 1-r\bar{n}_2,\\
F_{,p}(\bar{n}_2,\bar{p}_2) &= -\bar{n}_2.
\end{aligned}$$

For the other two derivatives, we find, by (7.3.34) and (7.3.33), that

$$1-\exp(-\bar{p}_2) = \frac{\bar{p}_2}{C\bar{n}_2} = \frac{r\left(1-\bar{n}_2\right)}{C\bar{n}_2}$$

and thus

$$\exp(-\bar{p}_2) = \frac{C\bar{n}_2 - r(1-\bar{n}_2)}{C\bar{n}_2}.$$

Hence

$$\begin{aligned}
G_{,n}(\bar{n}_2,\bar{p}_2) &= \frac{r\left(1-\bar{n}_2\right)}{C\bar{n}_2},\\
G_{,p}(\bar{n}_2,\bar{p}_2) &= C\bar{n}_2 - r(1-\bar{n}_2).
\end{aligned}$$

Thus, the Jacobi matrix is given by

$$\mathcal{J}(\bar{n}_2,\bar{p}_2) = \begin{pmatrix} 1-r\bar{n}_2 & -\bar{n}_2 \\ \frac{r(1-\bar{n}_2)}{\bar{n}_2} & C\bar{n}_2 - r(1-\bar{n}_2) \end{pmatrix}.$$

The trace of the matrix $\mathcal{J}(\bar{n}_2, \bar{p}_2)$ is given by $1 - \bar{n}_2 + C\bar{n}_2$ and the determinant is given by

$$\det \mathcal{J}(\bar{n}_2, \bar{p}_2) = \bar{n}_2(C(1 - r\bar{n}_2) + r^2(1 - \bar{n}_2)).$$

Thus, the Jury condition for stability is

$$|1 - r + C\bar{n}_2| < \bar{n}_2(C(1 - r\bar{n}_2) + r^2(1 - \bar{n}_2)) + 1 < 2.$$

Through computer simulations, it can be found that there is a range of parameters for which this equilibrium is stable.

Example 7.6 Consider (7.3.32) with $r = 0.075$ and $C = 1.5$. Then, the value of n at the intersection of graphs of f_1 and f_2,

$$\frac{1}{20}\left(\frac{1}{n} - 1\right) = 1 - \exp\left(-0.075\,(1 - n)\right),$$

gives the co-existence equilibrium

$$(\bar{n}_2, \bar{p}_2) \approx (0.675, 0.024),$$

see Fig. 7.1. Then,

$$\bar{n}_2(C(1 - r\bar{n}_2) + r^2(1 - \bar{n}_2)) + 1 \approx 1.962$$

and

$$1 - r + C\bar{n}_2 \approx 0.0243,$$

so that $(\bar{n}_2, \bar{p}_2)$ is asymptotically stable, as seen in Fig. 7.2.

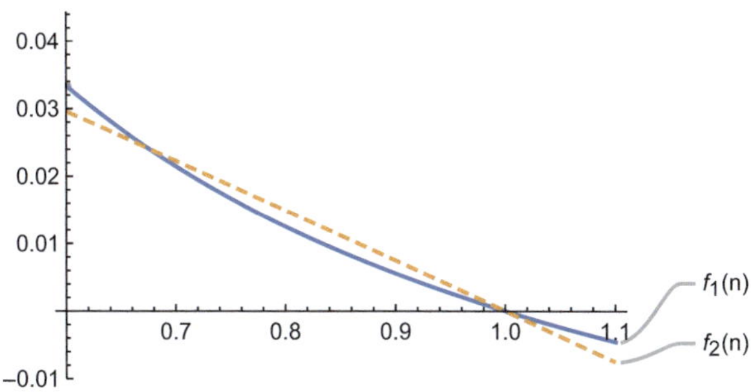

Fig. 7.1 Values of n at the equilibria as the abscissae of the intersection points of the graphs of f_1 and f_2

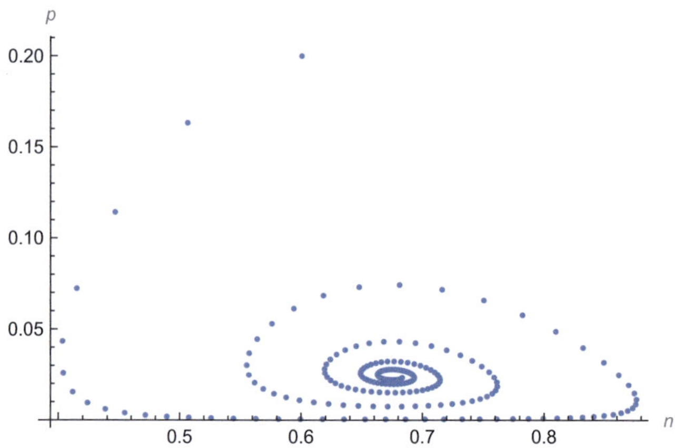

Fig. 7.2 Iterates defined by (7.3.32) with $r = 0.075$ and $C = 1.5$, originating from $(n(0), p(0)) = (0.6, 0.2)$

Exercise 7.3.1 Give other examples of parameters r and C for which the endemic equilibrium is stable. Also, give an example of the parameters C and r for which the endemic equilibrium is unstable.

Chapter 8
Positivity in Natural Science Models

8.1 Next Generation Matrix Method in Epidemiology

As we have seen, the basic reproduction number R_0, introduced in Definition 6.1, is one of the most critical parameters in epidemiological (and not only) modelling, giving the threshold value at which the disease-free equilibrium DFE becomes unstable. We have derived R_0 directly from the definition for relatively simple models considered so far. This approach becomes, however, cumbersome for more involved cases. A systematic approach was proposed in [39], where the authors defined R_0 as the spectral radius of the so-called next-generation operator. This definition was specified for systems of ordinary differential equations in [96], see also [24, Chapter 6], which is the approach we present in this section. We note an alternative approach for determining the threshold value at which DFE loses stability given in [56].

8.1.1 Additional Properties of Metzler Matrices

Metzler matrices, introduced in Remark 5.31 and discussed in Sect. C.3.2, play a crucial role in analysing epidemiological and ecological models. Several important results necessary for developing the theory in this chapter can be proved using dynamical systems methods, as we demonstrate now.

J. Banasiak, *Introduction to Mathematical Methods in Population Theory*, Springer Undergraduate Mathematics Series,
https://doi.org/10.1007/978-3-031-65491-6_8

Theorem 8.1 *Let $\mathcal{A}$ be a Metzler matrix. The following conditions are equivalent.*

(*i*) $s(\mathcal{A}) < 0$;
(*ii*) *there is* $\boldsymbol{v} \gg 0$ *satisfying* $-\mathcal{A}\boldsymbol{v} \gg 0$;
(*iii*) $\mathcal{A}$ *is invertible with* $-\mathcal{A}^{-1} \geq 0$.

Proof Assumption (i) implies the existence of $\mathcal{A}^{-1}$ and, by (5.3.86) and Proposition 5.30, there follows the existence of $\int_0^\infty e^{t\mathcal{A}} dt \geq 0$. Then, by (5.3.59),

$$\begin{aligned}\int_0^t e^{t\mathcal{A}} dt &= \sum_{j=0}^{\infty} \frac{t^{j+1}\mathcal{A}^j}{(j+1)!} = \mathcal{A}^{-1} \sum_{j=0}^{\infty} \frac{t^{j+1}\mathcal{A}^{j+1}}{(j+1)!} \\ &= \mathcal{A}^{-1}\left(-\mathcal{I} + \sum_{j=0}^{\infty} \frac{t^j \mathcal{A}^j}{j!}\right) = \mathcal{A}^{-1}\left(-\mathcal{I} + e^{t\mathcal{A}}\right). \end{aligned} \tag{8.1.1}$$

Using again (5.3.86),

$$\int_0^\infty e^{t\mathcal{A}} dt = \lim_{t\to\infty} \mathcal{A}^{-1}\left(-\mathcal{I} + e^{t\mathcal{A}}\right) = -\mathcal{A}^{-1} \geq 0,$$

hence (i) implies (iii).

Next, we show that (iii) implies (ii). Since $\mathcal{A}^{-1}$ is a non-zero matrix, (iii) implies that $-\mathcal{A}^{-1}$ has at least one positive entry, say, $\hat{a}_{ij} > 0$. Then, for instance, for $\boldsymbol{v} = -(1, \ldots, 1)$, $\boldsymbol{c}_0 := \mathcal{A}^{-1}\boldsymbol{v} = (\ldots, \hat{a}_{ij}, \ldots) > 0$, (the other coordinates are non-negative due to (iii)) and $-\mathcal{A}\boldsymbol{c}_0 = -\boldsymbol{v} \gg 0$. If we define $\boldsymbol{c} := \boldsymbol{c}_0 - \epsilon\boldsymbol{v}$, then $\boldsymbol{c} \gg 0$ for $\epsilon > 0$. Then $-\mathcal{A}\boldsymbol{c} = -\boldsymbol{v}(1-\epsilon) \gg 0$ for $\epsilon < 1$.

Finally, we show that (ii) implies (i). Consider on $\overline{\mathbb{R}}^d_+$ the system

$$\boldsymbol{x}'(t) = \mathcal{A}^T\boldsymbol{x}, \quad \boldsymbol{x}(0) = \mathring{\boldsymbol{x}}, \tag{8.1.2}$$

where the transposed matrix $\mathcal{A}^T$ is also a Metzler matrix. Using (iii), there is $\boldsymbol{c} \gg 0$ satisfying $-\mathcal{A}\boldsymbol{c} \gg 0$. For such a $\boldsymbol{c}$, we define

$$V(\boldsymbol{x}) = \langle \boldsymbol{c}, \boldsymbol{x} \rangle$$

on $\overline{\mathbb{R}}^n_+$. Clearly, $V(\boldsymbol{x}) > 0$ unless $\boldsymbol{x} = 0$. Further,

$$V' = \langle \boldsymbol{c}, \boldsymbol{x}' \rangle = \langle \boldsymbol{c}, \mathcal{A}^T\boldsymbol{x} \rangle = \langle \mathcal{A}\boldsymbol{c}, \boldsymbol{x} \rangle < 0,$$

as long as $\boldsymbol{x} \neq 0$. Using Theorem 6.51 and Remark 6.52, we see that $\boldsymbol{x} = 0$ is asymptotically stable relative to $\overline{\mathbb{R}}^d_+$. Then $s(\mathcal{A}) = s(\mathcal{A}^T) < 0$. Indeed, otherwise, the dominant eigenvalue would be non-negative, and so would be the left and right eigenvectors belonging to it, by Proposition C.12 and Theorem C.19. Then, for $\mathring{\boldsymbol{x}} \gg$

0, the coefficient of the leading term of (5.3.86) would be positive, contradicting the convergence to zero of solutions to (8.1.2). □

Definition 8.2 A Metzler matrix $\mathcal{A}$ satisfying $s(\mathcal{A}) < 0$ is called a Metzler stable or a Hurwitz matrix.

In the forthcoming considerations, we will be dealing with Metzler matrices in the form

$$\mathcal{A} = \mathcal{F} - \mathcal{V}. \tag{8.1.3}$$

It follows that the stability of $\mathcal{A}$ is to some extent independent of the way $\mathcal{A}$ is split into $\mathcal{F}$ and $\mathcal{V}$. To make it precise, we introduce the following definition, [56, 97].

Definition 8.3 We say that (8.1.3) is a regular splitting of $\mathcal{A}$ if $-\mathcal{V}$ is a stable Metzler matrix and $\mathcal{F} \geq 0$.

We need the following lemma.

Lemma 8.4 *Let $\mathcal{A}$ be a non-negative matrix. Then $(\mathcal{I} - \mathcal{A})^{-1}$ exists, is given by $(\mathcal{I} - \mathcal{A})^{-1} = \sum_{k=0}^{\infty} \mathcal{A}^k$ and is non-negative if and only if $r(\mathcal{A}) < 1$.*

Proof If $r(\mathcal{A}) < 1$, then $\max\{|\lambda| : \lambda \in \sigma(\mathcal{A})\} < 1$. As in the derivation of the Neumann series, see, e.g., [71, Section 3.8], we have

$$\mathcal{I} - (\mathcal{I} - \mathcal{A})\sum_{k=0}^{m} \mathcal{A}^k = \mathcal{A}^{m+1}.$$

By (5.3.85) and (5.3.88a), $\|\mathcal{A}^{m+1}\|$ is dominated by $cr^{m+1}(\mathcal{A})$ for some constant c; hence $\mathcal{A}^{m+1}$ converges to zero exponentially fast, so the series

$$\sum_{k=0}^{\infty} \mathcal{A}^k$$

converges absolutely. By the continuity of $\mathcal{I} - \mathcal{A}$,

$$(\mathcal{I} - \mathcal{A})\sum_{k=0}^{\infty} \mathcal{A}^k = \mathcal{I}$$

and, since, by assumption, $\mathcal{I} - \mathcal{A}$ is invertible ($1 \notin \sigma(\mathcal{A})$), we obtain

$$(\mathcal{I} - \mathcal{A})^{-1} = \sum_{k=0}^{\infty} \mathcal{A}^k. \tag{8.1.4}$$

The non-negativity of $\mathcal{A}$ implies $(\mathcal{I} - \mathcal{A})^{-1} \geq 0$.

Conversely, if $r(\mathcal{A}) \geq 1$, then, using Theorems C.10 and C.12, there is $\boldsymbol{e} > 0$ such that

$$\mathcal{A}\boldsymbol{e} = r(\mathcal{A})\boldsymbol{e}.$$

Then

$$\sum_{k=0}^{\infty} \mathcal{A}^k \boldsymbol{e} = (\sum_{k=0}^{\infty} r^k(\mathcal{A}))\boldsymbol{e},$$

and the series diverges.

Now, take again the eigenvector $\boldsymbol{e} > 0$ satisfying

$$(\mathcal{I} - \mathcal{A})\boldsymbol{e} = (1 - r(\mathcal{A}))\boldsymbol{e}.$$

If $r(\mathcal{A}) = 1$, $(\mathcal{I} - \mathcal{A})$ is not injective and hence not invertible. If $r(\mathcal{A}) > 1, \boldsymbol{x} = -\boldsymbol{e}$ is a solution to

$$(\mathcal{I} - \mathcal{A})\boldsymbol{x} = (r(\mathcal{A}) - 1)\boldsymbol{e} \geq 0,$$

and hence $(\mathcal{I} - \mathcal{A})^{-1}$, should it exist, is not non-negative. □

Note that this result does not preclude the existence of $(\mathcal{I} - \mathcal{A})^{-1}$, but only states that if $(\mathcal{I} - \mathcal{A})^{-1}$ exists, it cannot be given by (8.1.4).

Exercise 8.1.1 For the matrix

$$\mathcal{A} = \begin{pmatrix} 3 & 2 \\ 4 & 1 \end{pmatrix}$$

find $r(\mathcal{A})$ and $(\mathcal{I} - \mathcal{A})^{-1}$. Explain, why this does this contradict Proposition 8.4.

The main result is

Theorem 8.5 ([97, Theorem 3.29]) *If $\mathcal{A} = \mathcal{B} - \mathcal{C}$ is a regular splitting of a Metzler matrix $\mathcal{A}$, then $\mathcal{A}$ is Metzler stable if and only if $r(\mathcal{B}\mathcal{C}^{-1}) < 1$.*

Proof Assume that $\mathcal{A}$ is Metzler stable and hence $-\mathcal{B}\mathcal{A}^{-1}$; also $\mathcal{B}\mathcal{C}^{-1} \geq 0$. Then, from

$$\mathcal{A} = \mathcal{B} - \mathcal{C},$$

we obtain

$$-\mathcal{B}\mathcal{A}^{-1} = \mathcal{B}(\mathcal{C} - \mathcal{B})^{-1} = \mathcal{B}\mathcal{C}^{-1}\mathcal{C}(\mathcal{C} - \mathcal{B})^{-1} = \mathcal{B}\mathcal{C}^{-1}(\mathcal{I} - \mathcal{B}\mathcal{C}^{-1})^{-1},$$

which, in particular, ensures that $\lambda = 1 \notin \sigma(\mathcal{B}\mathcal{C}^{-1})$, by Lemma 8.4. Let $\mu \in \sigma(-\mathcal{B}\mathcal{A}^{-1})$. Then, for some $\boldsymbol{e} \neq 0$,

$$\mathcal{B}\mathcal{C}^{-1}(\mathcal{I} - \mathcal{B}\mathcal{C}^{-1})^{-1}\boldsymbol{e} = \mu\boldsymbol{e},$$

and, using the commutativity of $\mathcal{B}\mathcal{C}^{-1}$ and $(\mathcal{I} - \mathcal{B}\mathcal{C}^{-1})^{-1}$,

$$\mathcal{B}\mathcal{C}^{-1}\boldsymbol{e} = \mu(\mathcal{I} - \mathcal{B}\mathcal{C}^{-1})\boldsymbol{e},$$

hence

$$\mathcal{B}\mathcal{C}^{-1}\boldsymbol{e} = \frac{\mu}{1+\mu}\boldsymbol{e}. \tag{8.1.5}$$

We can run the calculations in the reverse direction, and thus we see that $\mu \in \sigma(-\mathcal{B}\mathcal{A}^{-1})$ if and only if $\lambda = \frac{\mu}{1+\mu} \in \sigma(\mathcal{B}\mathcal{C}^{-1})$. We are interested in estimating $\lambda = r(\mathcal{B}\mathcal{C}^{-1}) = \frac{\mu}{1+\mu}$ for some $\mu \in \sigma(-\mathcal{B}\mathcal{A}^{-1})$, which is real and non-negative. Since

$$\frac{\mu}{1+\mu} = \frac{\mu(1+\bar{\mu})}{|1+\mu|^2} = \frac{\mu+|\mu|^2}{|1+\mu|^2},$$

λ is real if and only if μ is real. Further, to $r(\mathcal{B}\mathcal{C}^{-1})$, there belongs a nonnegative eigenvector so that we can restrict our attention to $\boldsymbol{e} \geq 0$ in (8.1.5). Since $\boldsymbol{e}$ is also the eigenvector of $-\mathcal{B}\mathcal{A}^{-1}$ belonging to μ, μ must be non-negative. Hence, to calculate $r(\mathcal{B}\mathcal{C}^{-1})$ it suffices to use real nonnegative μ. But the function $\lambda(\mu) = \frac{\mu}{1+\mu}$ is increasing on $[0, \infty)$ and satisfy $0 \leq \lambda(\mu) < 1$ there. Thus

$$r(\mathcal{B}\mathcal{C}^{-1}) = \frac{r(-\mathcal{B}\mathcal{A}^{-1})}{1+r(-\mathcal{B}\mathcal{A}^{-1})} < 1. \tag{8.1.6}$$

To prove the converse, let $r(\mathcal{B}\mathcal{C}^{-1}) < 1$. Since $\mathcal{B}\mathcal{C}^{-1}$ in non-negative, $-\mathcal{I} + \mathcal{B}\mathcal{C}^{-1}$ is a Metzler matrix. Further, $r(\mathcal{B}\mathcal{C}^{-1}) < 1$ implies $s(-\mathcal{I} + \mathcal{B}\mathcal{C}^{-1}) < 0$ and hence it is Metzler stable. Then, by Theorem 8.1, we have

$$-(-\mathcal{I} + \mathcal{B}\mathcal{C}^{-1})^{-1} \geq 0$$

and thus, by $\mathcal{C}^{-1} \geq 0$,

$$0 \leq -(-\mathcal{I} + \mathcal{B}\mathcal{C}^{-1})^{-1}\mathcal{C}^{-1} = -(\mathcal{B} - \mathcal{C})^{-1}.$$

This means that $\mathcal{A} = \mathcal{B} - \mathcal{C}$ is invertible, so that $-\mathcal{A}^{-1} \geq 0$, and hence $\mathcal{A}$ is Metzler stable. □

Corollary 8.6 *Let the assumptions of Theorem 8.5 be satisfied. Then $s(\mathcal{A}) \leq 0$ if and only if $r(\mathcal{B}\mathcal{C}^{-1}) \leq 1$.*

Proof We take arbitrary $\varepsilon > 0$ and consider $\mathcal{A}_\varepsilon = \mathcal{B} - \mathcal{C} - \varepsilon\mathcal{I}$, which also is a Metzler matrix. Let $\lambda \in \sigma(\mathcal{B} - \mathcal{C})$ and $\boldsymbol{x}$ be an eigenvector belonging to it. Then

$$\mathcal{A}_\varepsilon \boldsymbol{x} = (\mathcal{B} - \mathcal{C} - \varepsilon\mathcal{I})\boldsymbol{x} = (\lambda - \varepsilon)\boldsymbol{x},$$

and $\Re(\lambda - \varepsilon) < 0$ or, in other words, $s(\mathcal{A}_\varepsilon) < 0$. Hence, $\mathcal{A}_\varepsilon$ is an invertible Metzler matrix. Then Theorem 8.5 applied to $\mathcal{B}$ and $\mathcal{C} + \varepsilon\mathcal{I}$ implies $r(\mathcal{B}(\mathcal{C} + \varepsilon\mathcal{I})^{-1}) < 1$ if and only if $s(\mathcal{A}_\varepsilon) < 0$. Since $\mathcal{C}^{-1}$ exists, $(\mathcal{C} + \varepsilon I)^{-1}$ converges to $\mathcal{C}^{-1}$ as $\varepsilon \to 0$. Further, both the spectral radius and the spectral bound of a matrix are given by continuous functions of eigenvalues, and the eigenvalues depend continuously on the entries of the respective matrices, see [74, Sections 3.1.1 & 3.1.2]. Hence, as in (8.1.6),

$$r(\mathcal{B}(\mathcal{C} + \epsilon\mathcal{I})^{-1}) = \frac{r(-\mathcal{B}\mathcal{A}_\varepsilon^{-1})}{1 + r(-\mathcal{B}\mathcal{A}_\varepsilon^{-1})} \leq 1$$

for arbitrary ε, which implies that if $s(\mathcal{A}) \leq 0$, then $r(\mathcal{B}\mathcal{C}^{-1}) \leq 1$. Conversely, if $r(\mathcal{B}\mathcal{C}^{-1}) \leq 1$, then we consider the approximation of $\mathcal{C}_\epsilon$ given by

$$\mathcal{C}_\varepsilon := (1 + \varepsilon)\mathcal{C}.$$

Let $\lambda, \boldsymbol{e}$ be an eigenpair of $\mathcal{B}\mathcal{C}^{-1}$, that is,

$$\mathcal{B}\mathcal{C}^{-1}\boldsymbol{e} = \lambda\boldsymbol{e}.$$

Then

$$\mathcal{B}\mathcal{C}_\varepsilon^{-1}\boldsymbol{e} = \mathcal{B}((1 + \varepsilon)\mathcal{C})^{-1}\boldsymbol{e} = \frac{1}{1 + \varepsilon}\mathcal{B}\mathcal{C}^{-1}\boldsymbol{e} = \frac{\lambda}{1 + \varepsilon}\boldsymbol{e}.$$

Consequently, $r(\mathcal{B}\mathcal{C}_\varepsilon^{-1}) < 1$. Then, by Theorem 8.5, $s(\mathcal{A} - \mathcal{C} - \varepsilon\mathcal{C}) < 0$, so all eigenvalues of $\mathcal{A} - \mathcal{C} - \varepsilon\mathcal{C}$ have negative real parts and the continuity of eigenvalues with respect to ε precludes the existence of an eigenvalue of $\mathcal{A}$ with positive real part. □

Exercise 8.1.2 Let $\mathcal{B}$ and $\mathcal{C}$ be $d \times d$ matrices and $\mathcal{C}$ be invertible. Show that $r(\mathcal{B}\mathcal{C}) = r(\mathcal{C}\mathcal{B})$.

Exercise 8.1.3 Let

$$\mathcal{A} = \begin{pmatrix} -4 & 1 & 0 & 0 \\ 1 & -4 & 1 & 0 \\ 0 & 1 & -4 & 1 \\ 0 & 0 & 1 & -4 \end{pmatrix}$$

and

$$\mathcal{C}_1 = \begin{pmatrix} 4 & 0 & 0 & 0 \\ 0 & 4 & 0 & 0 \\ 0 & 0 & 4 & 0 \\ 0 & 0 & 0 & 4 \end{pmatrix}, \quad \mathcal{C}_2 = \begin{pmatrix} 4 & -1 & 0 & 0 \\ -1 & 4 & 0 & 0 \\ 0 & 0 & 4 & 0 \\ 0 & 0 & 0 & 4 \end{pmatrix}, \quad \mathcal{C}_3 = \begin{pmatrix} 4 & -1 & 0 & 0 \\ -1 & 4 & -1 & 0 \\ 0 & -1 & 4 & 0 \\ 0 & 0 & 0 & 4 \end{pmatrix}.$$

Find $s(\mathcal{A})$, regular splits $\mathcal{A} = \mathcal{B}_i - \mathcal{C}_i$ and $r(\mathcal{B}_i \mathcal{C}_i^{-1})$ for $i = 1, 2, 3$.

8.1.2 Next Generation Matrix

This section introduces a systematic method of finding the basic reproduction number R_0. Our presentation is based on [24, Chapter 6].

Let us subdivide a given population into n groups of infective individuals and m groups free from infection. Each group is characterized by different states of infection or susceptibility, e.g., different strains, infectivity or immunity levels, etc. Let $\boldsymbol{x} \in \mathbb{R}^n$ and $\boldsymbol{y} \in \mathbb{R}^m$ represent sizes or densities of each group and let

- F_i—the rate at which the secondary infections increase the size of the ith group;
- V_i—the rate at which the progression of the disease, recovery or death decreases the size of the ith group;
- g_j—the rate of change of the size of the jth uninfected population due to demographic processes (unrelated to infections).

Then we can write a general epidemiological model as

$$x_i' = F_i(\boldsymbol{x}, \boldsymbol{y}) - V_i(\boldsymbol{x}, \boldsymbol{y}), \quad i = 1, \ldots, n, \tag{8.1.7a}$$

$$y_j' = g_j(\boldsymbol{x}, \boldsymbol{y}), \quad j = 1, \ldots, m. \tag{8.1.7b}$$

Functions F_i and V_i need not be uniquely defined and may depend on how a concrete model is interpreted. However, by Theorem 8.5, reasonable choices lead to the same threshold value for R_0.

The disease-free equilibria (DFE) of (8.1.7) are of the form $(0, \boldsymbol{y})$.

The method of finding R_0 for (8.1.7) uses the linearization of (8.1.7) at DFE. To ensure the existence of this equilibrium and the well-posedness of the model, we assume:

(A1) $F_i(0, \boldsymbol{y}) = 0$ and $V_i(0, \boldsymbol{y}) = 0$ for all $\boldsymbol{y} \geq 0, i = 1, \ldots, n$.
This assumption ensures that all new infections are internal, that is, there is no migration of infectives from outside.

(A2) $F_i(\boldsymbol{x}, \boldsymbol{y}) \geq 0$ for all $\boldsymbol{x}, \boldsymbol{y} \geq 0, i = 1, \ldots, n$.
This assumption reflects that $\boldsymbol{F} = (F_1, \ldots, F_n)$ gives new infections and thus cannot be negative.

(A3) $V_i(\boldsymbol{x}, \boldsymbol{y}) \leq 0$ for $x_i = 0, i = 1, \ldots, n$.
The interpretation of this assumption is that since V_i represents the outflow of infectives from the ith group, it must be zero (or negative) if there are no infectives in this group.

(A4) $\sum_{i=1}^{n} V_i(\boldsymbol{x}, \boldsymbol{y}) \geq 0$ for all $\boldsymbol{x}, \boldsymbol{y} \geq 0$.
Since the above sum represents the total outflow from all infected compartments, if negative, at least one compartment would contain new infections. This is, however, impossible as all new infections are accounted for in $\boldsymbol{F}$.

(A5) We assume that the disease-free system

$$\boldsymbol{y}' = \boldsymbol{g}(0, \boldsymbol{y}) \tag{8.1.8}$$

has a unique equilibrium $\boldsymbol{y}^*$, which is globally asymptotically and exponentially stable on $\overline{\mathbb{R}}^m_+$, that is, $s(\mathcal{J}_{\boldsymbol{g}}(0, \boldsymbol{y}^*)) < 0$.

Assumption (A1) ensures that the disease-free space $(0, \boldsymbol{y})$ is invariant. Indeed, (A1) implies $\boldsymbol{x}' = 0$ if $x_i = 0$ for $i = 1, \ldots, n$.

According to Definition 6.1 of R_0, we assume that we introduce a single infective individual into a fully susceptible population ($\boldsymbol{x} = 0$). Thus, the initial spread of the disease is determined by the linearization of (8.1.7a) at $(0, y^*)$ and hence, from (A1), we get

$$\frac{\partial F_i}{\partial y_j}(0, \boldsymbol{y}^*) = \frac{\partial V_i}{\partial y_j}(0, y^*) = 0, \quad i = 1, \ldots, n, \ j = 1, \ldots, m.$$

This means that the rows of the Jacobi matrices of $\boldsymbol{F}$ and $\boldsymbol{V}$ corresponding to the partial derivatives with respect to y_j, evaluated at DFE, consist of zeros. Thus the linearization of (8.1.7a) is independent of $\boldsymbol{y}$, and hence

$$\boldsymbol{x}' = (\mathcal{F} - \mathcal{V})\boldsymbol{x}, \tag{8.1.9}$$

where $\mathcal{F}$ i $\mathcal{V}$ are given by

$$\begin{aligned} \mathcal{F} = (f_{ik})_{1 \leq i,k \leq n} &= \left(\frac{\partial F_i}{\partial x_k}(0, \boldsymbol{y}^*)\right)_{1 \leq i,k \leq n}, \\ \mathcal{V} = (v_{ik})_{1 \leq i,k \leq n} &= \left(\frac{\partial V_i}{\partial x_k}(0, \boldsymbol{y}^*)\right)_{1 \leq i,k \leq n}. \end{aligned} \tag{8.1.10}$$

In other words, the Jacobi matrices of $\boldsymbol{F}$ and $\boldsymbol{V}$ with respect to the variables representing the infectives, evaluated at DFE, are given by (8.1.10). Hence, the stability of DFE for (8.1.7) is determined by (8.1.9).

Since $\boldsymbol{V}$ describes the outflow of infectives, the infected compartments should be depleted as $t \to \infty$. Hence the solution of

$$\boldsymbol{x}' = -\mathcal{V}\boldsymbol{x}, \quad \boldsymbol{x}(0) = \mathring{\boldsymbol{x}}, \tag{8.1.11}$$

that is, the solution to (8.1.9) when $\boldsymbol{F} = 0$ (no secondary infections) must converge to 0 as $t \to \infty$. In particular, the only equilibrium of (8.1.11) is $\boldsymbol{x}^* = 0$. Therefore, we adopt one more assumption

(A6) The matrix $\mathcal{V}$ is invertible

In the next section, we shall show that (A1), (A4) and (A6) imply that $-\mathcal{V}$ is a stable Metzler matrix, and thus the calculations below are justified. As in the calculations of R_0 based on Definition 6.1, the number of secondary infections caused by one infective is given by the product of the expected duration of the illness and the rate at which the infective infects susceptible. Here, these values must be calculated for each of the n infected groups. The expected time an infected individual spends in a given group is given by

$$\int_0^\infty \varphi(t, \mathring{\boldsymbol{x}})dt,$$

where $\varphi(t, \mathring{\boldsymbol{x}}) = e^{-\mathcal{V}t}\mathring{\boldsymbol{x}}$ is the solution to (8.1.11) with the initial condition of the form $\mathring{\boldsymbol{x}} = (0, \ldots, 0, 1, 0, \ldots, 0)$. The solution describes the path taken by an average infective introduced at $t = 0$ into the jth group, $j = 1, \ldots, n$, from the first contact to the recovery or death. In other words, we can interpret $\varphi_j(t, x_0)$, $j = 1, \ldots, n$, as the density of the distribution of individuals.

Thus, by (5.3.59),

$$\int_0^\infty \varphi(t, \mathring{\boldsymbol{x}})dt = \int_0^\infty e^{-\mathcal{V}t}\mathring{\boldsymbol{x}}dt = -\mathcal{V}^{-1}\left(\lim_{t\to\infty} e^{-\mathcal{V}t} - \mathcal{I}\right)\mathring{\boldsymbol{x}} = \mathcal{V}^{-1}\mathring{\boldsymbol{x}}.$$

The non-negative entries (i, j) of $\mathcal{V}^{-1}$ can be interpreted as the expected time spent in group i by an individual from group j. The same entry in $\mathcal{F}$ represents the rate at which secondary infections due to an individual in group j appear in group i. Thus, the expected number of secondary infections is given by

$$\mathcal{F}\int_0^\infty e^{-\mathcal{V}t}\mathring{\boldsymbol{x}}dt = \mathcal{F}\mathcal{V}^{-1}\mathring{\boldsymbol{x}}.$$

The matrix $\mathcal{F}\mathcal{V}^{-1}$ is called the next-generation matrix. The (i, j) entry of this matrix expresses the expected number of secondary infections in group i caused by an infective introduced into group j (under the assumption that the environment remains constant for the duration of the disease).

8.1.3 Definition of R_0 by the Next Generation Matrix Method

We will show that $\mathcal{F}\mathcal{V}^{-1} \geq 0$. First, we prove that $\mathcal{F} \geq 0$. Indeed, $(A1)$ and $(A2)$ imply

$$\begin{aligned} f_{ij} &= \frac{\partial F_i}{\partial x_j}(0, \mathbf{y}^*) = \lim_{x_j \to 0+} \frac{F_i(0, \dots, 0, x_j, 0, \dots, 0, \mathbf{y}^*) - F_i(0, \dots, 0, \mathbf{y}^*)}{x_j} \\ &= \lim_{x_j \to 0+} \frac{F_i(0, \dots, 0, x_j, 0, \dots, 0, \mathbf{y}^*)}{x_j} \geq 0, \quad i, j = 1, \dots, n. \end{aligned}$$

Next, we show that $-\mathcal{V}$ is a stable Metzler matrix. For any $1 \leq i, j \leq n$ and $j \neq i$, using again $(A1)$ and $(A3)$, we obtain

$$\begin{aligned} v_{ij} &= \frac{\partial V_i}{\partial x_j}(0, \mathbf{y}^*) = \lim_{x_j \to 0+} \frac{V_i(0, \dots, 0, x_j, 0, \dots, 0, \mathbf{y}^*) - V_i(0, \dots, 0, \mathbf{y}^*)}{x_j} \\ &= \lim_{x_j \to 0+} \frac{V_i(0, \dots, 0, x_j, 0, \dots, 0, \mathbf{y}^*)}{x_j} \leq 0, \end{aligned}$$

so $-\mathcal{V}$ is a Metzler matrix. Similarly, fixing $j \in \{1, \dots, n\}$, $(A1)$ and $(A4)$ imply

$$\begin{aligned} \sum_{i=1}^{n} v_{ij} &= \sum_{i=1}^{n} \frac{\partial V_i}{\partial x_j}(0, \mathbf{y}^*) \\ &= \sum_{i=1}^{n} \lim_{x_j \to 0+} \frac{V_i(0, \dots, 0, x_j, 0, \dots, 0, \mathbf{y}^*) - V_i(0, \dots, 0, \mathbf{y}^*)}{x_j} \\ &= \lim_{x_j \to 0+} \frac{\sum_{i=0}^{n} V_i(0, \dots, 0, x_j, 0, \dots, 0, \mathbf{y}^*)}{x_j} \geq 0. \end{aligned}$$

System (8.1.11) in coordinates takes the form

$$\begin{aligned} x_1'(t) &= -v_{11}x_1(t) - \dots - v_{1n}x_n(t), \\ x_2'(t) &= -v_{21}x_1(t) - \dots - v_{2n}x_n(t), \\ &\vdots \\ x_n'(t) &= -v_{n1}x_n(t) - \dots - v_{nn}x_n(t). \end{aligned} \tag{8.1.12}$$

Since $-\mathcal{V}$ is a Metzler matrix, the solutions of (8.1.12) for $\mathring{\boldsymbol{x}} \geq 0$ are non-negative. Adding up the equations of (8.1.12), (8.1.3) and the non-negativity of solutions yield

$$\frac{d}{dt}||\boldsymbol{x}(t)||_1 = \frac{d}{dt}\sum_{i=1}^{n} x_i(t) = -\sum_{j=1}^{n}\sum_{i=1}^{n} v_{ij}x_j(t) \leq 0.$$

Hence,

$$\|\boldsymbol{x}(t)\|_1 \leq \|\mathring{\boldsymbol{x}}\|_1,$$

and therefore, the zero equilibrium of (8.1.11) is stable. Using (5.3.86), we see that the real parts of the eigenvalues of $-\mathcal{V}$ are non-positive. As seen in Fig. C.1 in Sect. C.3.2, there are no imaginary eigenvalues, and the case $\lambda_{\max} = 0$ is ruled out by the invertibility of $\mathcal{V}$. Hence,

$$s(-\mathcal{V}) < 0. \tag{8.1.13}$$

and thus, by Theorem 8.1, $\mathcal{V}^{-1} \geq 0$.

We have proved that $\mathcal{F}\mathcal{V}^{-1} \geq 0$ and hence, using Proposition C.10, we know that the spectral radius of $\mathcal{F}\mathcal{V}^{-1}$ is its (nonnegative) eigenvalue. We define

$$R_0 := r(\mathcal{F}\mathcal{V}^{-1}). \tag{8.1.14}$$

8.1.4 R_0 and Local and Global Stability of DFE

According to Definition 6.1, we expect that $R_0 < 1$ implies that the disease disappears, while it spreads if $R_0 > 1$. Mathematically, DFE should be stable if $R_0 < 1$ and unstable $R_0 > 1$. The theorem below confirms that (8.1.14) is consistent with this expectation.

Theorem 8.7 ([96, Theorem 2]) *Consider the epidemiological model* (8.1.7). *Then,*

(*i*) *if $R_0 < 1$, then DFE is locally asymptotically stable;*
(*ii*) *if $R_0 > 1$, then DFE is unstable.*

Proof

(*i*) Let $\mathcal{G}_{21}$ and $\mathcal{G}_{22}$ be the Jacobi matrices if $\boldsymbol{g}$ with respect to $\boldsymbol{x}$ and $\boldsymbol{y}$ evaluated at $(0, \boldsymbol{y}^*)$. Then the Jacobi matrix of (8.1.7) at $(0, \boldsymbol{y}^*)$ is the block matrix

$$\mathcal{G} = \begin{pmatrix} \mathcal{F} - \mathcal{V} & 0 \\ \mathcal{G}_{21} & \mathcal{G}_{22} \end{pmatrix}.$$

The structure of $\mathcal{J}$ implies that the eigenvalues of $\mathcal{J}$ are the eigenvalues of $\mathcal{F}-\mathcal{V}$ and $\mathcal{J}_{22}$. By Assumption (A5), the eigenvalues of $\mathcal{J}_{22}$ have negative real parts. Next, using the previous section, $\mathcal{F}-\mathcal{V}$ is a regular splitting, so from (8.1.13) and Theorem 8.5, $s(\mathcal{F}-\mathcal{V})<0$. Hence, $s(\mathcal{J})<0$ and the thesis follows from Theorem 6.34.

To prove (ii), we again use Theorem 6.34. For this, we show that if $R_0>1$, there is an eigenvalue of $\mathcal{J}$ (and hence of $\mathcal{F}-\mathcal{V}$) with a positive real part. This, however, is equivalent to proving that $s(\mathcal{F}-\mathcal{V})\le 0$ implies $R_0=r(\mathcal{F}\mathcal{V}^{-1})\le 1$, but the latter follows from Corollary 8.6. □

Example 8.8 Let us return to the SEIR model (6.3.84) introduced in Example 6.35,

$$\begin{aligned}
S' &= \Lambda - \mu S - \beta S I,\\
E' &= \beta S I - (\mu+\gamma)E,\\
I' &= \gamma E - (\mu+\nu)I,\\
R' &= \nu I - \mu R,
\end{aligned} \tag{8.1.15}$$

where, as before,

- Λ is the total birth rate;
- β is the infection transmission rate;
- μ is the natural death rate;
- γ is the disease incubation rate;
- ν is the recovery rate,

and all parameters are positive.

The vector field of (8.1.15) is given by

$$f(S,E,I,R) = (\Lambda-\mu S-\beta SI, \beta SI-(\mu+\gamma)E, \gamma E-(\mu+\nu)I, \nu I-\mu R).$$

The existence of global (on $(0,\infty)$) nonnegative solutions follows from Example 6.3.4. The disease-free equilibrium is given by

$$DFE = (S^0, E^0, I^0, R^0) = \left(\frac{\Lambda}{\mu}, 0, 0, 0\right).$$

The endemic equilibrium EE is given by

$$\begin{aligned}
S^* &= \frac{(\mu+\gamma)(\mu+\nu)}{\beta\gamma},\\
I^* &= \frac{\beta\Lambda\gamma-\mu(\mu+\gamma)(\mu+\nu)}{\beta(\mu+\gamma)(\mu+\nu)},
\end{aligned}$$

$$E^* = \frac{\beta \Lambda \gamma - \mu(\mu+\gamma)(\mu+\nu)}{\gamma\beta(\mu+\gamma)},$$
$$R^* = \frac{\nu\left(\beta \Lambda \gamma - \mu(\mu+\gamma)(\mu+\nu)\right)}{\mu\beta(\mu+\gamma)(\mu+\nu)},$$

which has biological meaning if

$$\beta \Lambda \gamma - \mu(\mu+\gamma)(\mu+\nu) > 0, \tag{8.1.16}$$

that is, if R_0, defined in (6.3.87), is bigger than 1. Next, we use (8.1.14) to define the basic reproduction number. It is clear that the infected compartments are E and I. According to the definitions of the functions $\boldsymbol{F}$ and $\boldsymbol{V}$

$$\boldsymbol{F}((E,I),(S,R)) = \begin{pmatrix} \beta S I \\ 0 \end{pmatrix}, \quad \boldsymbol{V}((E,I),(S,R)) = \begin{pmatrix} (\mu+\gamma)E \\ -\gamma E + (\mu+\nu)I \end{pmatrix}.$$

We note that though the term γE increases the size of I, these are not secondary infections; thus, it is not included in $\boldsymbol{F}$.

Let us check that assumptions (A1)–(A6) are satisfied. Since

$$F_1((0,0),(S,R)) = F_2((0,0),(S,R)) = V_1((0,0),(S,R))$$
$$= V_2((0,0),(S,R)) = 0,$$

so (A1) holds. Next, observe that

$$F_1((E,I),(S,R)) = \beta S I \geq 0 \text{ provided } S, I \geq 0,$$

and

$$F_2((E,I),(S,R)) = 0,$$

so (A2) is satisfied. Moreover,

$$V_1((0,I),(S,R)) = 0$$

and

$$V_2((E,0),(S,R)) = -\gamma E \leq 0 \text{ if } E \geq 0,$$

yielding (A3). Since

$$\sum_{i=1}^{2} V_i((E,I),(S,R)) = \mu E + (\mu+\nu)I \geq 0 \text{ provided } E, I \geq 0,$$

(A4) holds. Consider (A5). The disease-free system takes the form

$$S' = \Lambda - \mu S,$$
$$R' = -\mu R,$$

whose unique equilibrium is $\left(\frac{\Lambda}{\mu}, 0\right)$. The system consists of two independent equations which can be analyzed separately. The first equation is solved by

$$S(t) = C_1 e^{-\mu t} + \frac{\Lambda}{\mu}, \quad C_1 \in \mathbb{R},$$

with

$$\lim_{t\to\infty} S(t) = \frac{\Lambda}{\mu},$$

independently of C_1. Similarly,

$$R(t) = C_2 e^{-\mu t}, \quad C_2 \in \mathbb{R}$$

and

$$\lim_{t\to\infty} R(t) = 0.$$

Hence, $\left(\frac{\Lambda}{\mu}, 0\right)$ is globally exponentially stable for the disease-free system.

Thus, (8.1.15) satisfies assumptions (A1)–(A5).

Next, according to the definition of the matrices $\mathcal{F}$ and $\mathcal{V}$, we obtain

$$\mathcal{F} = \begin{pmatrix} 0 & \frac{\beta\Lambda}{\mu} \\ 0 & 0 \end{pmatrix} \quad \text{and} \quad \mathcal{V} = \begin{pmatrix} \mu+\gamma & 0 \\ -\gamma & \mu+\nu \end{pmatrix}$$

and we see that $-\mathcal{V}$ is Metzler stable so (A6) is satisfied. Since

$$\mathcal{V}^{-1} = \begin{pmatrix} \frac{1}{\mu+\gamma} & 0 \\ \frac{\gamma}{(\mu+\gamma)(\mu+\nu)} & \frac{1}{\mu+\nu} \end{pmatrix},$$

the next-generation matrix takes the form

$$\mathcal{F}\mathcal{V}^{-1} = \begin{pmatrix} \frac{\beta\gamma\Lambda}{\mu(\mu+\gamma)(\mu+\nu)} & \frac{\beta\Lambda}{\mu(\mu+\nu)} \\ 0 & 0 \end{pmatrix}.$$

The eigenvalues of $\mathcal{F}\mathcal{V}^{-1}$ are given by

$$\lambda_1 = \frac{\beta\gamma\Lambda}{\mu(\mu+\gamma)(\mu+\nu)}, \qquad \lambda_2 = 0,$$

and hence

$$R_0 = \frac{\beta\gamma\Lambda}{\mu(\mu+\gamma)(\mu+\nu)},$$

which is consistent with (6.3.87).

By Theorem 8.7, DFE is locally asymptotically stable if $R_0 < 1$, and unstable $R_0 > 1$.

We note a significant reduction in the complexity of the problem thanks to the next-generation matrix method. While a direct approach to determine the stability of DFE for (6.3.87) requires finding eigenvalues of the 4×4 Jacobi matrix (6.3.93), here it suffices to consider just a 2×2 next generation matrix.

Exercise 8.1.4 Consider a modification of (6.4.131) in which the exposed class is mildly infective

$$\begin{aligned} S' &= \Lambda - \mu S - \beta SI - \beta_1 SE, \\ E' &= \beta SI + \beta_1 SE - (\mu+\gamma)E, \\ I' &= \gamma E - (\mu+\nu)I, \\ R' &= \nu I - \mu R, \end{aligned} \tag{8.1.17}$$

where $\beta_1 < \beta$. Show that

$$\mathcal{F}\mathcal{V}^{-1} = \begin{pmatrix} \frac{\beta_1\Lambda}{\mu(\mu+\gamma)} + \frac{\beta\gamma\Lambda}{\mu(\mu+\gamma)(\mu+\nu)} & \frac{\beta\Lambda}{\mu(\mu+\nu)} \\ 0 & 0 \end{pmatrix}.$$

and hence

$$R_0 = \frac{\beta_1\Lambda}{\mu(\mu+\gamma)} + \frac{\beta\gamma\Lambda}{\mu(\mu+\gamma)(\mu+\nu)}.$$

Example 8.9 To illustrate the use of the next-generation matrix method in a more involved case, let us return to a version of the malaria model (6.1.10) with the demography of both humans and mosquitoes modelled by the simplified logistic law, the infection force given by the mass action law and without disease-induced death,

$$
\begin{aligned}
S_h' &= \Lambda_h - \beta_h I_v S_h + \gamma_h R_h - \mu_h S_h,\\
E_h' &= \beta_h I_v S_h - \delta_h E_h - \mu_h E_h,\\
I_h' &= \delta_h E_h - \nu_h I_h - \mu_h I_h,\\
R_h' &= \nu_h I_h - \gamma_h R_h - \mu_h R_h,\\
S_v' &= \Lambda_v - \beta_v I_h S_v - \mu_v S_v,\\
E_v' &= \beta_v I_h S_v - \delta_v E_v - \mu_v E_v,\\
I_v' &= \delta_v E_v - \mu_v I_v.
\end{aligned}
\tag{8.1.18}
$$

We see that $\boldsymbol{x} = (E_h, I_h, E_v, I_v)$ are the infected classes and $\boldsymbol{y} = (S_h, R_h, S_v)$ are the disease-free classes. According to the adopted definitions, we have

$$
\boldsymbol{F}(\boldsymbol{x}, \boldsymbol{y}) = \begin{pmatrix} \beta_h S_h I_v \\ 0 \\ \beta_v S_v I_h \\ 0 \end{pmatrix},
$$

whereas

$$
\boldsymbol{V}(\boldsymbol{x}, \boldsymbol{y}) = \begin{pmatrix} (\delta_h + \mu_h) E_h \\ (\nu_h + \mu_h) I_h - \delta_h E_h \\ (\delta_v + \mu_v) E_v \\ \mu_v I_v - \delta_v E_m \end{pmatrix}.
$$

Let us check whether assumptions (A1)–(A6) are satisfied. First, observe that for E_h, I_h, E_v and $I_v = 0$, we have

$$
F_i((0, 0, 0, 0), \boldsymbol{y}) = 0 \ \text{for} \ i = 1, 2, 3, 4,
$$

$$
V_i((0, 0, 0, 0), \boldsymbol{y}) = 0 \ \text{for} \ i = 1, 2, 3, 4,
$$

and hence (A1) is satisfied.

Assumption (A2) follows from

$$
\begin{aligned}
&F_1(\boldsymbol{x}, \boldsymbol{y}) = \beta_h S_h I_v \geq 0 \ \text{for} \ \beta_h, S_h, I_v \geq 0,\\
&F_2(\boldsymbol{x}, \boldsymbol{y}) = 0,\\
&F_3(\boldsymbol{x}, \boldsymbol{y}) = \beta_v S_v I_h \geq 0 \ \text{for} \ \beta_v, S_v, I_h \geq 0,\\
&F_4(\boldsymbol{x}, \boldsymbol{y}) = 0.
\end{aligned}
$$

Assumption (A3) is also satisfied. Indeed,

$$\begin{aligned}
&V_1((0, I_h, E_v, I_v), \mathbf{y}) = 0, \\
&V_2((E_h, 0, E_v, I_v), \mathbf{y}) = -\delta_h E_h \leq 0 \text{ for } \delta_h, E_h \geq 0, \\
&V_3((E_h, I_h, 0, I_v), \mathbf{y}) = 0, \\
&V_4((E_h, I_h, E_v, 0), \mathbf{y}) = -\delta_v E_v \leq 0 \text{ for } \delta_v, E_v \geq 0.
\end{aligned}$$

Further, for $\mu_h, E_h, \nu_h, I_h, \mu_v, E_v, I_v \geq 0$, we obtain

$$\begin{aligned}
\sum_{i=1}^{4} V_i(\boldsymbol{x}, \mathbf{y}) &= (\delta_h + \mu_h)E_h + (\nu_h + \mu_h)I_h - \delta_h E_h + (\delta_v + \mu_v)E_v + \mu_v I_v - \delta_v E_v \\
&= \mu_h E_h + (\nu_h + \mu_h)I_h + \mu_v E_v + \mu_v I_v \geq 0,
\end{aligned}$$

so (A4) is satisfied.

Next, we look at assumption (A5). The disease-free system takes the form

$$\begin{aligned}
S_h' &= \Lambda_h - \mu_h S_h + \gamma_h R_h, \\
R_h' &= -(\mu_h + \gamma_h)R_h, \\
S_v' &= \Lambda_v - \mu_v S_v,
\end{aligned} \tag{8.1.19}$$

whose unique equilibrium is

$$\mathbf{y}^* = \left(\frac{\Lambda_h}{\mu_h}, 0, \frac{\Lambda_v}{\mu_v}\right).$$

System (8.1.19) can be solved explicitly. We immediately obtain the solution

$$R_h(t) = R_h(0)e^{-(\mu_h+\gamma_h)t}, \tag{8.1.20}$$

satisfying

$$\lim_{t\to\infty} R_h(t) = 0.$$

Then, the equation for S_h takes the form

$$S_h' = -\mu_h S_h + \Lambda_h + \gamma_h R_h(0)e^{-(\mu_h+\gamma_h)t}, \tag{8.1.21}$$

and its solution is

$$S_h(t) = \left(R_h(t) + S_h(t) - \frac{\Lambda_h}{\mu_h}\right)e^{-\mu_h t} - R_h(0)e^{-(\mu_h+\gamma_h)t} + \frac{\Lambda_h}{\mu_h}.$$

Since all parameters are positive,

$$\lim_{t\to\infty} S_h(t) = \frac{\Lambda_h}{\mu_h}.$$

Finally, in a similar way, the solution

$$S_v(t) = \left(S_v(0) - \frac{\Lambda_v}{\mu_v} \right) e^{-\mu_v t} + \frac{\Lambda_v}{\mu_v},$$

satisfies

$$\lim_{t\to\infty} S_v(t) = \frac{\Lambda_v}{\mu_v}.$$

Thus, $\boldsymbol{y}^*$ is globally asymptotically stable for the disease-free system, and it is easy to see that it is also exponentially stable.

Hence, (A1)–(A5) are satisfied.

Next, the Jacobi matrices of $\boldsymbol{F}$ and $\boldsymbol{V}$ at DFE, see (8.1.10), are given by

$$\mathcal{F} = \begin{pmatrix} 0 & 0 & 0 & \frac{\beta_h \Lambda_h}{\mu_h} \\ 0 & 0 & 0 & 0 \\ 0 & \frac{\beta_v \Lambda_v}{\mu_v} & 0 & 0 \\ 0 & 0 & 0 & 0 \end{pmatrix}$$

and

$$\mathcal{V} = \begin{pmatrix} \delta_h + \mu_h & 0 & 0 & 0 \\ -\delta_h & \nu_h + \mu_h & 0 & 0 \\ 0 & 0 & \delta_v + \mu_v & 0 \\ 0 & 0 & -\delta_v & \mu_v \end{pmatrix}.$$

Matrix $\mathcal{V}$ is block-diagonal, with each block being a tridiagonal matrix. Thus

$$\mathcal{V}^{-1} = \begin{pmatrix} \frac{1}{\delta_h + \mu_h} & 0 & 0 & 0 \\ \frac{\delta_h}{(\delta_h + \mu_h)(\nu_h + \mu_h)} & \frac{1}{\nu_h + \mu_h} & 0 & 0 \\ 0 & 0 & \frac{1}{\delta_v + \mu_v} & 0 \\ 0 & 0 & \frac{\delta_v}{\mu_v(\delta_v + \mu_v)} & \frac{1}{\mu_v} \end{pmatrix},$$

and therefore,

$$\mathcal{F}\mathcal{V}^{-1} = \begin{pmatrix} 0 & 0 & \frac{\beta_h \Lambda_h \delta_v}{\mu_h \mu_v (\delta_v + \mu_v)} & \frac{\beta_h \Lambda_h}{\mu_h \mu_v} \\ 0 & 0 & 0 & 0 \\ \frac{\beta_v \Lambda_v \delta_h}{\mu_v (\delta_h + \mu_h)(\nu_h + \mu_h)} & \frac{\beta_v \Lambda_v}{\mu_v (\nu_h + \mu_h)} & 0 & 0 \\ 0 & 0 & 0 & 0 \end{pmatrix}.$$

We see that the characteristic equation of $\mathcal{F}\mathcal{V}^{-1}$ is given by

$$det(\mathcal{F}\mathcal{V}^{-1} - \lambda \mathcal{I}) = \lambda^2 \left(\lambda^2 - \frac{\beta_h \Lambda_h \delta_v}{\mu_h \mu_v (\delta_v + \mu_v)} \frac{\beta_v \Lambda_v \delta_h}{\mu_v (\delta_h + \mu_h)(\nu_h + \mu_h)} \right) = 0,$$

whose roots are

$$\lambda_{1,2} = 0, \quad \lambda_{2,3} = \pm \sqrt{\frac{\delta_h \beta_h \Lambda_h \delta_v \beta_v \Lambda_v}{\mu_h (\delta_h + \mu_h)(\nu_h + \mu_h) \mu_v^2 (\delta_v + \mu_v)}}.$$

Therefore,

$$R_0 = \sqrt{\frac{b^2 \delta_h \beta_h \Lambda_h \delta_v \beta_v \Lambda_v}{\mu_h (\delta_h + \mu_h)(\nu_h + \mu_h) \mu_v^2 (\delta_v + \mu_v)}}. \tag{8.1.22}$$

Thus, the basic reproduction number of the DFE for the malaria model (8.1.18) is given by

$$R_0 = \sqrt{R_h R_v}, \tag{8.1.23}$$

where

$$R_h = \frac{\delta_h \beta_h \Lambda_h}{\mu_h (\delta_h + \mu_h)(\nu_h + \mu_h)}, \tag{8.1.24}$$

$$R_v = \frac{\delta_v \beta_v \Lambda_v}{\mu_v^2 (\delta_v + \mu_v)}. \tag{8.1.25}$$

In (8.1.22), the expression $\frac{\delta_h}{\delta_h + \mu_h}$ is the probability that a human survives the exposed period and becomes infective and $\frac{1}{\nu_h + \mu_h}$ is the average time a human is infective until he/she dies or recovers. Similarly, $\frac{\delta_v}{\delta_v + \mu_v}$ is the probability that a mosquito survives the exposed period and becomes infective, and $\frac{1}{\mu_v}$ is the time a mosquito stays infective until it dies.

Thus, we can write (8.1.22) as (8.1.23), where R_h, R_v are given by (8.1.24) and (8.1.25), respectively. Here, R_h gives the number of humans infected by a single mosquito in a unit time in a completely susceptible human population, and R_v is the number of mosquitoes infected by one infective human in a unit time in a completely susceptible population, so R_0 is the number of infected humans in a unit time caused by introducing one infected human (or one infected mosquito) is the environment consisting of completely susceptible humans and mosquitoes.

Exercise 8.1.5 (Influenza—A Dynamics in the Emergence of Drug Resistance, [57]) We consider the spread of the flu, where the flu virus can mutate from a wild to a drug-resistant strain in a human population subject to a vaccination campaign. Thus, we divide the population into five groups: susceptibles (S), vaccinated (V), infected with a wild strain of the virus (I_W), infected with a drug-resistant virus strain (I_R) and recovered (R).

We use standard assumptions for compartmental models. Additionally, we assume that an individual can be infected with only one virus strain, and the vaccination is imperfect. This leads to the following model

$$
\begin{aligned}
S' &= \Lambda + \vartheta R - (\phi + \mu + \beta_W I_W + \beta_R(1+b^2)I_R)S,\\
V' &= \phi S - (\varepsilon\beta_W I_W + \varepsilon\beta_R(1+b^2)I_R + \mu)V,\\
I_W' &= \beta_W I_W S + \varepsilon\beta_W I_W V - (b + \mu + a_W + \gamma_W)I_W,\\
I_R' &= \beta_R(1+b^2)I_R S + \varepsilon\beta_R(1+b^2)I_R V + bI_W - (\mu + \gamma_R + a_R)I_R,\\
R' &= \gamma_W I_W + \gamma_R I_R - (\vartheta + \mu)R,
\end{aligned}
\tag{8.1.26}
$$

where

- β_W—transmission rate of the wild strain;
- β_R—transmission rate of the drug-resistant strain;
- ε—the probability of vaccine's failure;
- ϕ—vaccination rate;
- b—the virus' mutation rate;
- γ_W—recovery rate in the I_W group;
- γ_R—recovery rate in the I_R group;
- ϑ—immunity loss rate;
- μ_W—disease-induced death rate due to the wild strain infection;
- μ_R—disease-induced death rate due to the drug-resistant strain infection;
- μ—natural death rate;
- Λ—total growth rate of the population.

The flow chart of the model is presented in Fig. 8.1.

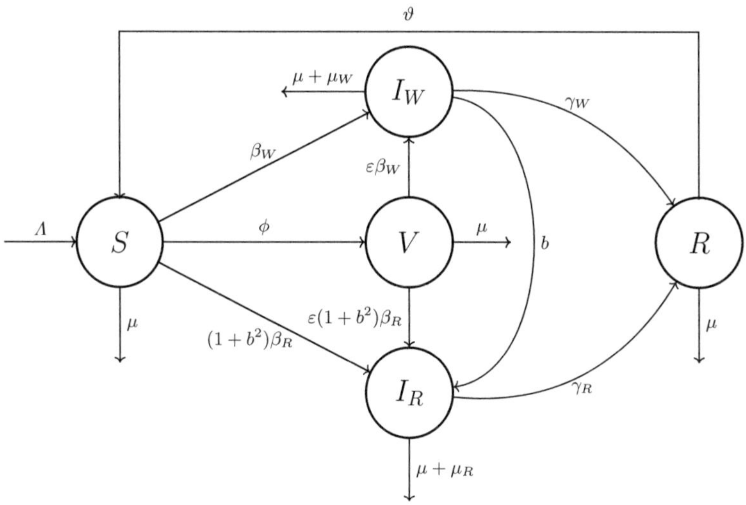

Fig. 8.1 The flowchart of (8.1.26)

1. Identify the groups of infectives and susceptibles, find DFE and show that $\boldsymbol{F}$ and $\boldsymbol{V}$, given by

$$\boldsymbol{F}((I_W, I_R), (S, V, R)) = \begin{pmatrix} \beta_W I_W S + \varepsilon \beta_W I_W V \\ (1+b^2)\beta_R I_R S + \varepsilon(1+b^2)\beta_R I_R V \end{pmatrix},$$

and

$$\boldsymbol{V}((I_W, I_R), (S, V, R)) = \begin{pmatrix} (b + \mu + a_W + \gamma_W) I_W \\ (\mu + \gamma_R + a_R) I_R - b I_W \end{pmatrix},$$

satisfy assumptions (A1)–(A5).
2. Find $\mathcal{F}$ and $\mathcal{V}$ and show that $-\mathcal{V}$ is Metzler stable.
3. Consider the model without vaccinations ($\phi = 0$). Show that, in this case,

$$R_0 = \max\{R_w, R_r\}$$

is the basic reproduction number of (8.1.26), where

$$R_w = \frac{\Lambda \beta_W \mu}{\mu^2(b + \mu + \mu_W + \gamma_W)}$$

is the expected number of secondary infections caused by one individual infected with the wild strain of the virus in a totally susceptible population, and

$$R_r = \frac{\Lambda(1+b^2)\beta_R\mu}{\mu^2(\mu+\gamma_R+\mu_R)}$$

is the expected number of secondary infections caused by one individual infected with the drug-resistant strain of the virus.

4. Consider now (8.1.26) with $\varphi > 0$. Show that the basic reproduction number in this case is

$$R_0^V = \max\{R_w^V, R_r^V\},$$

where

$$R_w^V = \frac{\Lambda\beta_W(\phi\varepsilon+\mu)}{\mu(\mu+\phi)(b+\mu+\mu_W+\gamma_W)}$$

and

$$R_r^V = \frac{\Lambda(1+b^2)\beta_R(\phi\varepsilon+\mu)}{\mu(\mu+\phi)(\mu+\gamma_R+\mu_R)},$$

with analogous interpretation of R_w^V and R_r^V.

Assumptions (A1)–(A6) allow, in certain cases, to show the global asymptotic stability of DFE. We have

Theorem 8.10 ([24, Theorem 2 in Chapter 6])

(*i*) *Assume that* (8.1.7) *can be written as*

$$\begin{aligned} \boldsymbol{x}' &= \mathcal{A}\boldsymbol{x} - \boldsymbol{f}(\boldsymbol{x},\boldsymbol{y}), \\ \boldsymbol{y}' &= \boldsymbol{g}(\boldsymbol{x},\boldsymbol{y}). \end{aligned} \tag{8.1.27}$$

(*ii*) *Assumption (A5) is satisfied and* $R_0 < 1$ *for* (8.1.27)*, that is,* $DFE = (0, \boldsymbol{y}^*)$ *is a locally asymptotically stable equilibrium of* (8.1.27)*;*

(*iii*) $\overline{\mathbb{R}}_+^{m+n}$ *is invariant for* (8.1.27)*;*

(*iv*) *the trajectories of* (8.1.27) *are bounded. If* $\mathcal{A}$ *is a Metzler stable matrix and* $\boldsymbol{f} \geq 0$ *on* $\overline{\mathbb{R}}_+^{m+n}$*, then DFE is globally asymptotically stable.*

Proof Let $(\boldsymbol{x}(t), \boldsymbol{y}(t)) = \phi(t, (\mathring{\boldsymbol{x}}, \mathring{\boldsymbol{y}}))$ be the solution to (8.1.27) with the initial condition $(\mathring{\boldsymbol{x}}, \mathring{\boldsymbol{y}})$. First, we show $\boldsymbol{x}(t) \to 0$ as $t \to \infty$. Consider

$$\boldsymbol{x}' = \mathcal{A}\boldsymbol{x} - \boldsymbol{f}(\boldsymbol{x},\boldsymbol{y}).$$

Using the variation of constants formula (5.3.40), we get

$$\boldsymbol{x}(t) = e^{t\mathcal{A}}\mathring{\boldsymbol{x}} - \int\limits_0^t e^{(t-s)\mathcal{A}} \boldsymbol{f}(\boldsymbol{x}(s), \boldsymbol{y}(s))ds.$$

Since $\mathcal{A}$ is a Metzler matrix, Theorem 5.30 implies $e^{(t-s)\mathcal{A}} \geq 0$ for $t \geq s$. Since $\boldsymbol{f} \geq 0$,

$$\boldsymbol{x}(t) = e^{t\mathcal{A}}\mathring{\boldsymbol{x}} - \int\limits_0^t e^{-(t-s)\mathcal{A}} \boldsymbol{f}(\boldsymbol{x}(s), \boldsymbol{y}(s))ds \leq e^{t\mathcal{A}}\mathring{\boldsymbol{x}},$$

thus

$$0 \leq \boldsymbol{x}(t) \leq e^{t\mathcal{A}}\mathring{\boldsymbol{x}},$$

and hence, since $\mathcal{A}$ is Metzler stable,

$$0 \leq \lim_{t\to\infty} \boldsymbol{x}(t) \leq \lim_{t\to\infty} e^{t\mathcal{A}}\mathring{\boldsymbol{x}} = 0.$$

We have to prove that the component $\boldsymbol{y}(t)$ of the solution $\phi(t, (\mathring{\boldsymbol{x}}, \mathring{\boldsymbol{y}}))$ converges to DFE for any $(\mathring{\boldsymbol{x}}, \mathring{\boldsymbol{y}})$. Assumption (A5) implies that this is true if $(\mathring{\boldsymbol{x}}, \mathring{\boldsymbol{y}}) = (0, \mathring{\boldsymbol{y}})$. Consider an arbitrary sequence $(t_n)_{n\in\mathbb{N}}$ with $t_n \to \infty$ as $n \to \infty$. Since the trajectory $\Gamma_{(\mathring{x},\mathring{y})}$ is bounded, point 1. of Lemma 6.49 ensures $\omega(\Gamma_{(\mathring{x},\mathring{y})}) \neq \emptyset$ and, by the above, any point in $\omega(\Gamma_{(\mathring{x},\mathring{y})})$ is of the form $(0, \boldsymbol{y})$ and, by (A5),

$$\lim_{t\to\infty} \phi(t, (0, \boldsymbol{y})) = (0, \boldsymbol{y}^*),$$

that is, $\omega(\Gamma_{(0,\boldsymbol{y})}) = \{(0, \boldsymbol{y}^*)\}$. From point 5. of Lemma 6.49 we infer $\omega(\Gamma_{(0,\boldsymbol{y})}) \subset \omega(\Gamma_{(\mathring{x},\mathring{y})})$. Hence $(0, \boldsymbol{y}^*) \in \omega(\omega(\Gamma_{(\mathring{x},\mathring{y})}))$ and thus there is a sequence $(t_n)_{n\in\mathbb{N}}$ converging to ∞ for which

$$\lim_{n\to\infty} \phi(t_n, (\mathring{\boldsymbol{x}}, \mathring{\boldsymbol{y}})) = (0, \boldsymbol{y}^*).$$

Since $(0, \boldsymbol{y}^*)$ is locally asymptotically stable, there is $\delta > 0$ such that if

$$\|(\boldsymbol{\xi}, \boldsymbol{\eta}) - (0, \boldsymbol{y}^*)\| < \delta,$$

then

$$\lim_{t\to\infty} \phi(t, (\boldsymbol{\xi}, \boldsymbol{\eta})) = (0, \boldsymbol{y}^*).$$

Since $\lim_{n\to\infty} \phi(t_n, (\mathring{\boldsymbol{x}}, \mathring{\boldsymbol{y}})) = (0, \boldsymbol{y}^*)$, there exists $n_0 \in \mathbb{N}$ for which

$$\|(0, \boldsymbol{y}^*) - \phi(t_{n_0}, (\mathring{\boldsymbol{x}}, \mathring{\boldsymbol{y}}))\| < \delta.$$

Hence

$$\lim_{t\to\infty} \phi(t, (\mathring{\boldsymbol{x}}, \mathring{\boldsymbol{y}}))) = (0, \boldsymbol{y}^*).$$

□

In the notation of this section, we typically use $\mathcal{F} - \mathcal{V}$ as $\mathcal{A}$, which is Metzler stable provided (A6) is satisfied and $R_0 < 1$ by Theorem 8.5.

The main role in the proof of Theorem 8.10 is played by the convergence of the sizes of the infective classes to 0. Note that assumption (A5), that is, global stability of DFE equilibrium in the disease-free environment combined with its local asymptotic stability in the environment with disease, gives global asymptotic stability of DFE. This can be summed up as

Corollary 8.11 *Assume that* (8.1.7) *satisfies* *(A5). If* $\boldsymbol{x}(t) \to 0$ *for any initial condition and DFE is locally asymptotically stable, then it is globally asymptotically stable.*

This result can sometimes be used as an alternative to the LaSalle principle, Theorem 6.61, as we will illustrate in Example 8.12. As often happens, the derivative of the Lyapunov function vanishes on the disease-free set, say M, see, e.g., Example 6.71. To show the global stability of DFE using the LaSalle principle, we need to prove that {DFE} is the only invariant subset of M. However, sometimes it is easier to show that assumption (A5) is satisfied. Then Corollary 8.11 ensures the global stability of DFE.

Example 8.12 Let us return to the SEIR model (6.4.131). We proved the global stability of its DFE using the LaSalle principle in Example 6.68. Here, we use Theorem 8.10 for this purpose. Let $R_0 < 1$. We can write (6.4.131) as

$$\begin{aligned}
\boldsymbol{x}' &= (\mathcal{F} - \mathcal{V})\boldsymbol{x} - \begin{pmatrix} \beta(S^0 - S)I \\ 0 \end{pmatrix} \\
S' &= \Lambda - \mu S - \beta S I \\
R' &= \alpha I - \mu R,
\end{aligned}$$

where $\boldsymbol{x} = (E, I)$, $S^0 = \dfrac{\Lambda}{\mu}$, so assumption (i) of Theorem 8.10 is satisfied with

$$\mathcal{A} = \mathcal{F} - \mathcal{V}, \quad \boldsymbol{f}(x, y) = \boldsymbol{f}(S, E, I, R) = \begin{pmatrix} \beta(S^0 - S)I \\ 0 \end{pmatrix}.$$

We know that the trajectories of (6.4.131) are non-negative and bounded, so assumptions (iii) and (iv) are satisfied. Since $R_0 < 1$, Theorem 8.7 implies local asymptotic (and exponential) stability of DFE, so assumption (ii) holds. Further, from Theorem 8.5, it follows that $\mathcal{F} - \mathcal{V}$ is a nonsingular Metzler matrix. We are left with proving $f \geq 0$, for which, due to the non-negativity of solutions and parameters of the model, it suffices to show $S \leq S^0$. Let us recall that the size of the population N satisfies

$$N' = \Lambda - \mu N,$$

whose unique equilibrium is $N^* = S^0 = \frac{\Lambda}{\mu}$, and

$$N(t) = \left(N(0) - \frac{\Lambda}{\mu}\right)e^{-\mu t} + \frac{\Lambda}{\mu}$$

is its solution originating from $N(0)$. Consider the following two cases.

Case 1. If $N(0) \leq \frac{\Lambda}{\mu}$, then $\left(N(0) - \frac{\Lambda}{\mu}\right) \leq 0$ and hence

$$S(t) \leq N(t) \leq \frac{\Lambda}{\mu} = S^0,$$

hence Theorem 8.10 implies that DFE attracts all trajectories satisfying $N(0) \leq \frac{\Lambda}{\mu}$.

Case 2. If $N(0) > \frac{\Lambda}{\mu}$, then we can have two subcases.

(a) If $S(t) \geq S^0$ for all $t \geq 0$, than, since $S(t) \leq N(t) \to S^0$, we get the convergence of $S(t)$ to S^0. Hence, all other components of the solution converge to 0.
(b) There is $T > 0$ such that $S(T) \leq S^0$. Then, $S(t) \leq S^0$ for all $t \geq T$ and we can consider (6.4.131) with the initial condition imposed at $t = T$, so this subcase is reduced Case 1.

Combining Case 1 and Case 2, we obtain the global stability of DFE.

8.2 Cooperative Systems

There is a class of differential equations systems describing population interactions, which generate relatively simple dynamics. Returning to the general model (6.1.1) and the terminology introduced there, these are mutualistic, also called cooperative, systems in which each population positively impacts the others. This induces monotonicity of solutions and, in turn, regular long-term behaviour. We shall present a brief introduction to the theory of cooperative systems based on [88] and lecture notes [82]. Several further examples can be found in [43, 87].

8.2.1 Differential Inequalities

A version of the property that ensures the positivity of solutions to system (6.2.17),

$$\boldsymbol{x}' = \boldsymbol{f}(t, \boldsymbol{x}), \quad \boldsymbol{x}(t_0) = \mathring{\boldsymbol{x}}, \tag{8.2.28}$$

given in Theorem B.21, has further applications; thus, we shall formalize it.

Definition 8.13 A vector function $\boldsymbol{f} = (f_1, \ldots, f_d) : [t_0, t_1) \times D \to \mathbb{R}^d$, $-\infty < t_0 < t_1 \leq \infty$, $D \subset \mathbb{R}^d$, or, equivalently, system (8.2.28), is said to be of type K (or to be quasimonotone increasing) if for each $i = 1, \ldots, d$, we have $f_i(t, \boldsymbol{x}) \leq f_i(t, \boldsymbol{y})$ for any $\boldsymbol{x}, \boldsymbol{y} \in D$ with $x_j \leq y_j$ for $j \neq i$ and $x_i = y_i$.

Exercise 8.2.1 Show that if $\boldsymbol{f}$ is of type K and $\boldsymbol{f}(0) \geq 0$, then solutions to (8.2.28) with $\mathring{\boldsymbol{x}}$ are nonnegative for $t \geq 0$.

Exercise 8.2.2 Consider the system

$$\begin{aligned} x_1' &= f(x_1) + \phi(x_1, x_2), \\ x_2 &= g(x_2) + \psi(x_1, x_2). \end{aligned} \tag{8.2.29}$$

Find conditions on f, g, ϕ and ψ ensuring that

(i) $\overline{\mathbb{R}}_+^d$ is invariant under the flow generated by (8.2.29),
(ii) System (8.2.29) is of type K.

Give examples of systems of type K but not preserving $\overline{\mathbb{R}}_+^d$, and conversely.

When there is no need to specify the time domain, we will use

$$\Omega := [t_0, t_1) \times D.$$

As the first application of K type functions, we prove the monotonicity of solutions to differential equations with respect to their right-hand sides. Precisely, we compare solutions to system (8.2.28) with functions satisfying differential inequalities

$$z' \leq \boldsymbol{f}(t, z), \quad z(t_0) = \mathring{z}, \tag{8.2.30}$$

or

$$\boldsymbol{y}' \geq \boldsymbol{f}(t, \boldsymbol{y}), \quad \boldsymbol{y}(t_0) = \mathring{\boldsymbol{y}}. \tag{8.2.31}$$

Theorem 8.14 *Suppose f satisfies the assumptions of the Picard–Lindelöf theorem, Theorem B.10, on Ω and is of type K there. If z is a differentiable function satisfying* (8.2.30) *on (t_0, t_1) with $\mathring{z} \leq \mathring{x}$, then $z(t) \leq x(t)$ on $[t_0, t_1]$. If y is a differentiable function satisfying* (8.2.31) *on (t_0, t_1) with $\mathring{y} \geq \mathring{x}$, then $y(t) \geq x(t)$ on $[t_0, t_1)$.*

Proof First, we modify (8.2.28) by

$$x'_\epsilon = f_\epsilon(t, x_\epsilon) = f(t, x_\epsilon) + \epsilon e, \quad x_\epsilon(t_0) = \mathring{x} + \epsilon e \geq 0, \tag{8.2.32}$$

where $e = (1, \ldots, 1)$ and $\epsilon > 0$. By Theorem B.16, $x_\epsilon(t)$ is defined on (t_0, t_1) for sufficiently small ϵ. Then we have

$$z' < f_\epsilon(t, z), \quad z(t_0) = \mathring{z}. \tag{8.2.33}$$

Let us fix ϵ. Since $\mathring{z} \leq \mathring{x} < x_\epsilon(t_0)$ and both z and x_ϵ are continuous, we have $z(t) < x_\epsilon(t)$ for $t \in [t_0, t_\delta)$ for some $t_\delta > 0$. Arguing as in Theorem B.21, we see that if the thesis does not hold, then there is $k \in \{1, \ldots, d\}$ such that t_k is the smallest time for which $z(t_k) \leq x(t_k)$, $z_k(t) < x_{\epsilon,k}(t)$ for $t \in [t_0, t_k)$ and $z_k(t_k) = x_{\epsilon,k}(t_k)$. Therefore,

$$z'_k(t_k) = \lim_{h\to 0} \frac{z_k(t_k) - z_k(t_k - h)}{h} \geq \lim_{h\to 0} \frac{x_{\epsilon,k}(t_k) - x_{\epsilon,k}(t_k - h)}{h} = x'_{\epsilon,k}(t_k)$$

and

$$f_k(t_k, z(t_k)) \geq z'_k(t_k) \geq x'_{\epsilon,k}(t_k) = f_{\epsilon,k}(t_k, x_\epsilon)(t_k) > f_k(t_k, x_\epsilon(t_k)),$$

which contradicts the K condition, since $z(t_k) \leq x(t_k)$ and $z_k(t_k) = x_{\epsilon,k}(t_k)$. Using again Theorem B.16,

$$x_\epsilon(t) \to x(t),$$

on $[t_0, t_1)$, hence $z(t) \leq x(t)$ there. The second case is proved in the same way. □

Remark 8.15 A more general version of this result, which involves weaker notions of the differentiability, can be found in [30].

If D is a convex set, the K condition can be replaced by a condition on the Jacobi matrix of f.

Corollary 8.16 *Let D be convex. Let f be a continuous function of $(t, x) \in [t_0, t_1] \times D$ and continuously differentiable there with respect to x, and $\mathcal{J}_f(t, x)$ be a Metzler matrix for each $(t, x) \in \Omega$. If $\phi(t, t_0, \xi_1)$ and $\phi(t, t_0, \xi_2)$ are solutions to* (8.2.28) *with $\xi_1 \leq \xi_2$, then*

$$\phi(t, t_0, \xi_1) \leq \phi(t, t_0, \xi_2) \tag{8.2.34}$$

for $t \in [t_0, t_1)$.

Proof If we take $\boldsymbol{u}, \boldsymbol{v} \in D$ with $\boldsymbol{u} \leq \boldsymbol{v}$ with $u_i = v_i$ for some i, then, using the convexity of D, Taylor's formula gives

$$f_i(t, \boldsymbol{v}) - f_i(t, \boldsymbol{u}) = \int_0^1 \sum_{j=1, j\neq i}^{d} \frac{\partial f_i}{\partial x_j}(t, \boldsymbol{u} + r(\boldsymbol{v} - \boldsymbol{u}))(v_j - u_j)dr \geq 0,$$

by the definition of a Metzler matrix. Thus, $\boldsymbol{f}$ is of type K, and the result follows from Theorem 8.14, where we consider only the part concerning the inequality of the initial conditions.

Corollary 8.17 *Assume that $\boldsymbol{f}$ satisfies assumptions of Corollary 8.16 and suppose that for some $\boldsymbol{\xi} \in D$, $\boldsymbol{f}(t, \boldsymbol{\xi}) \geq 0$ for all $t \geq t_0$. Then the solution $\phi(t, t_0, \boldsymbol{\xi})$ of* (8.2.28) *satisfies $\phi(t, t_0, \boldsymbol{\xi}) \geq \boldsymbol{\xi}$ for all $t \geq t_0$. An analogous conclusion holds if the inequalities are reversed.*

□

Proof Let us introduce the change of variables $\boldsymbol{y} = \boldsymbol{x} - \boldsymbol{\xi}$. Then

$$\boldsymbol{y}'(t) = \boldsymbol{x}'(t) = \boldsymbol{f}(t, \boldsymbol{y}(t) + \boldsymbol{\xi}) =: \boldsymbol{g}(t, \boldsymbol{y}(t)), \quad \boldsymbol{y}(t_0) = \boldsymbol{x}(t_0) - \boldsymbol{\xi} = 0.$$

Since $\mathcal{J}_{\boldsymbol{g}}$ is a Metzler matrix, $\boldsymbol{g}$ is of type K. Take a vector $\boldsymbol{y} \geq 0$ with $y_i = 0$ and $\boldsymbol{x} = 0$ in Definition 8.13. Hence, by this definition and the assumption,

$$f_i(t, \boldsymbol{\xi} + \boldsymbol{y}) = g_i(t, \boldsymbol{y}) \geq g_i(t, 0) = f_i(t, \boldsymbol{\xi}) \geq 0.$$

Using Theorem B.21, we see that $\boldsymbol{y}(t) \geq 0$, which implies $\boldsymbol{x}(t) \geq \boldsymbol{\xi}$. □

Definition 8.18 System (8.2.28) is called cooperative in Ω if $\mathcal{J}_{\boldsymbol{f}}(t, \boldsymbol{x})$ is a Metzler matrix for any $(t, \boldsymbol{x}) \in \Omega$. If $\mathcal{J}_{\boldsymbol{f}}(t, \boldsymbol{x})$ is additionally an irreducible matrix for any $(t, \boldsymbol{x}) \in \Omega$, then we say that (8.2.28) is a cooperative and irreducible system.

Remark 8.19 Linear system

$$\boldsymbol{y}'(t) = \mathcal{A}(t)\boldsymbol{y}(t),$$

where $\mathcal{A}(t)$ is a continuous matrix function, and Metzler for any t, is a cooperative system. It is clearly cooperative and irreducible if $\mathcal{A}(t)$ is an irreducible Metzler matrix for any t.

Consider (8.2.28) with sufficiently smooth $\boldsymbol{f}$, see Remark B.19, and let $\boldsymbol{x}(t) = \phi(t, t_0, \boldsymbol{\xi})$ be its solution satisfying $\boldsymbol{x}(t_0) = \boldsymbol{\xi}$. By Theorem B.18, the derivatives $\frac{\partial \phi}{\partial \xi_i}$ exist and satisfy

$$\frac{\partial}{\partial t}\left(\frac{\partial \phi(t, t_0, \boldsymbol{\xi})}{\partial \xi_i}\right) = \mathcal{J}_{\boldsymbol{f}}(t, \phi(t, t_0, \boldsymbol{\xi}))\frac{\partial \phi(t, \boldsymbol{\xi})}{\partial \xi_i}, \qquad \frac{\partial \phi(t_0, t_0, \boldsymbol{\xi})}{\partial xi_i} = \boldsymbol{e}_i, \tag{8.2.35}$$

$i = 1, \dots, d$, where

$$\mathcal{J}_f(t, \phi(t, t_0, \boldsymbol{\xi})) = \nabla_{\boldsymbol{x}} \boldsymbol{f}(t, \boldsymbol{x})|_{\boldsymbol{x}=\phi(t,t_0,\boldsymbol{\xi})}. \tag{8.2.36}$$

Now, if (8.2.28) is cooperative, then $\mathcal{J}_f(t, \phi(t, t_0, \boldsymbol{\xi}))$ is a Metzler matrix and hence (8.2.35) is also a cooperative system. Since $\boldsymbol{e}_i \geq 0$, $\frac{\partial \phi(t,\boldsymbol{\xi})}{\partial \xi_i} \geq 0$ for $i = 1, \dots, d$. In compact form, we can write

$$\mathcal{J}_{\phi,\boldsymbol{\xi}}(t, t_0, \boldsymbol{\xi}) = \begin{pmatrix} \frac{\partial \phi_1(t,t_0,\boldsymbol{\xi})}{\partial \xi_1} & \dots & \frac{\partial \phi_1(t,t_0,\boldsymbol{\xi})}{\partial \xi_d} \\ \vdots & \ddots & \vdots \\ \frac{\partial \phi_d(t,t_0,\boldsymbol{\xi})}{\partial \xi_1} & \dots & \frac{\partial \phi_d(t,t_0,\boldsymbol{\xi})}{\partial \xi_d} \end{pmatrix} \geq 0. \tag{8.2.37}$$

We can get a stronger result if we additionally assume that $\mathcal{J}_f$ is irreducible.

Theorem 8.20 *Assume that* (8.2.28) *is cooperative and irreducible on* Ω. *Then*

$$\mathcal{J}_{\phi,\boldsymbol{\xi}}(t, t_0, \boldsymbol{\xi}) \gg 0 \tag{8.2.38}$$

for $t > t_0$. *Furthermore, if* $\boldsymbol{\xi}_1, \boldsymbol{\xi}_2 \in D$ *satisfy* $\boldsymbol{\xi}_1 < \boldsymbol{\xi}_2$, *then*

$$\phi(t, t_0, \boldsymbol{\xi}_1) \ll \phi(t, t_0, \boldsymbol{\xi}_2) \tag{8.2.39}$$

for $t > t_0$.

Proof To shorten notation, we introduce

$$\mathcal{J}_{\phi,\boldsymbol{\xi}}(t, t_0, \boldsymbol{\xi}) =: (\varphi_{ij}(t))_{1\leq i,j\leq d}, \quad \mathcal{J}_f(t, \phi(t, t_0, \boldsymbol{\xi})) =: (f_{ij}(t))_{1\leq i,j\leq d}.$$

We observe that $\mathcal{J}_{\phi,\boldsymbol{\xi}}(t, t_0, \boldsymbol{\xi})$ is a fundamental matrix for (8.2.35), and hence it is an invertible matrix for each $t \in [t_0, t_1)$, see Definition 5.7 and Proposition 5.8.

In component form, we have

$$\varphi'_{ij}(t) = f_{i1}(t)\varphi_{1j}(t) + \dots + f_{ii}(t)\varphi_{ij}(t) + \dots + f_{id}(t)\varphi_{dj}(t), \quad i = 1, \dots, d.$$

Since $\mathcal{J}_f(t, \phi(t, t_0, \boldsymbol{\xi}))$ is a Metzler matrix, $f_{ij} \geq 0$ for $i \neq j$, and (8.2.37) implies $\varphi_{ij} \geq 0$ for all $i, j = 1, \dots, d$. Hence,

$$\varphi'_{ij}(t) \geq f_{ii}(t)\varphi_{ij}(t),$$

thus if $\varphi_{ij}(\tau) > 0$ for some $\tau > t_0$, then $\varphi_{ij}(t) > 0$ for $t > \tau$. Define

$$Z_{ij} = \{t > t_0 : \varphi_{ij}(t) = 0\}.$$

Then Z_{ij} is either empty, or is of the form $(t_0, \tau_{ij}]$. Clearly, $Z_{ii} = \emptyset$ since $\varphi_{ii}(t_0) = 1$ for any $i = 1, \ldots, d$.

Let us assume that $Z_{ij} \neq \emptyset$ for some i, j and define $S_j = \{k : \varphi_{kj}(t) = 0 \text{ for all } t \in (t_0, \tau_{ij}]\}$. We have $S_j \neq \emptyset$ since $i \in S_j$. Also, $S_j \neq \{1, \ldots, d\}$ as otherwise the whole column $\varphi_{1j}(t), \ldots, \varphi_{dj}(t)$ would be zero on $(t_0, \tau_{ij}]$, contradicting the invertibility of $\mathcal{J}_{\phi,\xi}(t, t_0, \xi)$. Let $\emptyset \neq S_j^c$ be the complement of S_j in $\{1, \ldots, d\}$. Thus, for $t \in (t_0, \tau_{ij}]$ and $k \in S_j$, we have

$$0 = \varphi'_{kj}(t) = \sum_{l=1}^{d} a_{kl}(t)\varphi_{lj}(t) = \sum_{l \in S_j^c} a_{kl}(t)\varphi_{lj}(t). \tag{8.2.40}$$

Since $k \in S_j$, the term with $a_{kk}(t)$ does not appear in the sum on the right-hand side. Further, by the definition of S_j, for any $l \in S_j^c$, there is $\tau_l < \tau_{ij}$ for which $\varphi_{lj}(\tau_l) > 0$, but, by the first part of the proof, $\varphi_{lj}(t) > 0$ on $(\tau_l, \tau_{ij}]$. Thus, taking $\tau = \max\{\tau_l : l \in S_j^c\}$, we have $\varphi_{lj}(t) > 0$ on (τ, τ_{ij}) for any $l \in S_j^c$. Therefore, (8.2.40) implies that $a_{kl}(t) = 0$ on (τ, τ_{ij}) for any $l \in S_j^c$. Since $k \in S_j$ was arbitrary, $a_{kl}(t) = 0$ on (τ, τ_{ij}) for any $k \in S_j, l \in S_j^c$. Consider a matrix $\mathcal{A}$ with such entries,

$$\mathcal{A} = \begin{pmatrix} a_{11} & \cdots & a_{1l_1} & \cdots & a_{1l_r} & \cdots & a_{1d} \\ \vdots & \ddots & \vdots & \ddots & \vdots & \ddots & \vdots \\ a_{k_1 1} & \cdots & 0 & \cdots & 0 & \cdots & a_{k_1 d} \\ \vdots & \ddots & \vdots & \ddots & \vdots & \ddots & \vdots \\ a_{k1} & \cdots & a_{kl_1} & \cdots & a_{kl_r} & \cdots & a_{kd} \\ \vdots & \ddots & \vdots & \ddots & \vdots & \ddots & \vdots \\ a_{k_q 1} & \cdots & 0 & \cdots & 0 & \cdots & a_{k_q d} \\ \vdots & \ddots & \vdots & \ddots & \vdots & \ddots & \vdots \\ a_{d1} & \cdots & a_{dl_1} & \cdots & a_{dl_r} & \cdots & a_{dd} \end{pmatrix},$$

where $k_1, \ldots, k_q \in S_j$ and $l_1, \ldots, l_r \in S_j^c$, and a vector of the form $\boldsymbol{\alpha} = (0, \ldots, 0, \alpha_{l_1}, \ldots, \alpha_{l_r}, 0, \ldots, 0)$, so the zeroes are exactly at the places with indices from S_j. Any row of $\mathcal{A}$ with an index belonging to S_j has zero entries at the positions with indices from S_j^c; thus, the multiplication of such a row and $\boldsymbol{\alpha}$ gives zero. Therefore, $\mathcal{A}\boldsymbol{\alpha}$ is a vector of the same form; hence, $\mathcal{A}$ cannot be irreducible.

Finally,

$$\phi(t, t_0, \boldsymbol{\xi}_1) - \phi(t, t_0, \boldsymbol{\xi}_2) = \int_0^1 \mathcal{J}_{\phi,\xi}(t, t_0, \boldsymbol{\xi}_1 + r(\boldsymbol{\xi}_2 - \boldsymbol{\xi}_1))(\boldsymbol{\xi}_2 - \boldsymbol{\xi}_1)dr \gg 0,$$

since $\mathcal{J}_{\phi,\xi} \gg 0$ and $\boldsymbol{\xi}_2 - \boldsymbol{\xi}_1 > 0$. □

Exercise 8.2.3 Consider the system

$$\begin{aligned} x' &= f(t, x, y), \\ y' &= g(t, x, y), \end{aligned} \tag{8.2.41}$$

where $\boldsymbol{F} = (f, g)$ is defined on $\mathbb{R} \times D$, $D \subset R^2$. Assume that $f(t, x_1, y_1) \geq f(t, x_2, y_2)$ and $g(t, x_1, y_1) \leq g(t, x_2, y_2)$ if $x_1 \geq x_2$ and $y_1 \leq y_2$, $t \in [t_0, t_1]$.

1. Show that the change of variables $x = u$, $y = -v$ transforms $\boldsymbol{F}$ into a type K function.
2. Consider (8.2.41) as a system in $\mathbb{R}^p \times \mathbb{R}^q$,

$$\begin{aligned} \boldsymbol{x}' &= \boldsymbol{f}(t, \boldsymbol{x}, \boldsymbol{y}), \\ \boldsymbol{y}' &= \boldsymbol{g}(t, \boldsymbol{x}, \boldsymbol{y}). \end{aligned} \tag{8.2.42}$$

   Formulate conditions on $\boldsymbol{f}$ and $\boldsymbol{g}$, which transform $\boldsymbol{F} = (\boldsymbol{f}, \boldsymbol{g})$ into a type K function.
3. Reformulate the results of this section for such systems.

Exercise 8.2.4 Consider system

$$\begin{aligned} x_1' &= x_1(1 - x_1), \\ x_2' &= \lambda x_2(1 - x_2 - \gamma x_1), \end{aligned} \tag{8.2.43}$$

$\lambda, \gamma > 0$, which is the space-homogeneous version of the system of tree–grass interactions, [18].

(i) Determine whether (8.2.43) is monotone.
(ii) Show that the box $[0, 1] \times [0, 1] \in \overline{\mathbb{R}}^2_+$ is invariant. Hint. You can use Theorem 8.14 to show that $x_2 \leq 1$. Conclude that the solutions to (8.2.43) exist for all $t \geq 0$ if $(\mathring{x}_1, \mathring{x}_2) \in [0, 1] \times [0, 1]$.
(iii) Show that the transformation

$$u_1 = x_1, \quad u_2 = 1 - x_2,$$

transforms (8.2.43) into a monotone system in $[0, 1] \times [0, 1]$.

8.2.2 *Monotone Systems*

We observe that (8.2.34) and (8.2.39) express that the cooperativity of a system of differential equations implies the monotonicity of its solutions with respect to the initial conditions. This property has far-reaching implications and thus deserves a more formal definition. As our interest lies in the existence of equilibria and convergence of trajectories to them, we shall focus on autonomous systems as in (6.2.17).

Definition 8.21 Let $\phi(t, \boldsymbol{\xi})$, $t \geq 0$, be the flow generated by the autonomous Cauchy problem

$$\boldsymbol{x}' = \boldsymbol{f}(\boldsymbol{x}), \quad \boldsymbol{x}(0) = \boldsymbol{\xi} \tag{8.2.44}$$

in a convex open domain $D \subset \mathbb{R}^d$. We say that ϕ is a monotone flow if

$$\boldsymbol{\xi}_1 \leq \boldsymbol{\xi}_2 \quad \text{implies} \quad \phi(t, \boldsymbol{\xi}_1) \leq \phi(t, \boldsymbol{\xi}_2). \tag{8.2.45}$$

The flow ϕ is called strongly monotone, if

$$\boldsymbol{\xi}_1 < \boldsymbol{\xi}_2 \quad \text{implies} \quad \phi(t, \boldsymbol{\xi}_1) \ll \phi(t, \boldsymbol{\xi}_2). \tag{8.2.46}$$

Remark 8.22 If this does not create any misunderstanding, systems of equations (8.2.44) generating monotone (resp. strongly monotone) dynamical systems will also be called monotone (resp. strongly monotone).

With this new terminology, Corollary 8.16 and Theorem 8.20 can be reformulated as

Theorem 8.23 *If* (8.2.44) *is cooperative in D, then ϕ is monotone in D. If* (8.2.44) *is cooperative and irreducible in D, then ϕ is strongly monotone in D.*

Theorem 8.24 *Let $\Gamma_{\boldsymbol{\xi}}$ be a (forward) trajectory of a monotone system* (8.2.44)*, with a compact closure in D. If either*

1. $0 \leq \boldsymbol{f}(\boldsymbol{\xi})$ (or $0 \geq \boldsymbol{f}(\boldsymbol{\xi})$), or
2. $\boldsymbol{\xi} \ll \phi(T, \boldsymbol{\xi})$ (or $\boldsymbol{\xi} \gg \phi(T, \boldsymbol{\xi})$) for some $T > 0$,

then $\omega(\Gamma_{\boldsymbol{\xi}})$ is an equilibrium.

Proof

1. Using Corollary 8.17, we have

$$\boldsymbol{\xi} \leq \phi(t, \boldsymbol{\xi})$$

and, using the semigroup property of the flow and the monotonicity

$$\phi(s, \boldsymbol{\xi}) \leq \phi(s, \phi(t, \boldsymbol{\xi})) = \phi(s + t, \boldsymbol{\xi}), \quad t, s \geq 0.$$

Thus, $\phi(t, \boldsymbol{\xi})$ is nondecreasing and, since $\Gamma_{\boldsymbol{\xi}}$ has compact closure, there is a limit

$$\lim_{t \to \infty} \phi(t, \boldsymbol{\xi}) = \boldsymbol{e},$$

and, by Corollary 6.32, $\boldsymbol{e}$ is an equilibrium.

2. The assumption $\boldsymbol{\xi} \ll \phi(T, \boldsymbol{\xi})$ implies, as above,

$$\phi(nT, \boldsymbol{\xi}) \leq \phi((n+1)T, \boldsymbol{\xi}), \quad n \in \mathbb{N}.$$

Hence, there is $\boldsymbol{p}$ such that

$$\lim_{n\to\infty} \phi(nT, \boldsymbol{\xi}) = \boldsymbol{p}.$$

Further, the continuity of the flow with respect to the initial conditions, Theorem B.16, and $\phi((n+1)T, \boldsymbol{\xi}) = \phi(T, \phi(nT, \boldsymbol{\xi}))$ imply

$$\begin{aligned}\boldsymbol{p} &= \lim_{n\to\infty} \phi((n+1)T, \boldsymbol{\xi}) = \lim_{n\to\infty} \phi(T, \phi(nT, \boldsymbol{\xi}))\\ &= \phi(T, \lim_{n\to\infty} \phi(nT, \boldsymbol{\xi})) = \phi(T, \boldsymbol{p}).\end{aligned}$$

Hence, the trajectory through $\boldsymbol{p} \in \omega(\Gamma_{\boldsymbol{\xi}})$ is periodic of period T. By Lemma 6.49, point 3., $\omega(\Gamma_{\xi})$ contains the periodic trajectory $\{\phi(t, \boldsymbol{p}) : t \in \mathbb{R}\}$. Let us show that $\omega(\Gamma_{\boldsymbol{\xi}}) = \{\phi(t, \boldsymbol{p}) : 0 \leq t \leq T\}$. Let $\boldsymbol{z} \in \omega(\Gamma_{\boldsymbol{\xi}})$. Then $\boldsymbol{z} = \lim_{t_p\to\infty} \phi(t_p, \boldsymbol{\xi})$ for a sequence $(t_p)_{p\in\mathbb{N}}$; we can select this sequence to be increasing and $t_{p+1} - t_p > T$. Hence, we can find a subsequence $(n_p)_{p\in\mathbb{N}}$ such that

$$n_pT \leq t_p < (n_p + 1)T.$$

Since $0 \leq t_p - n_pT < T$, there is a subsequence, which we denote in the same way, such that

$$\lim_{p\to\infty} (t_p - n_pT) = \tau$$

for some finite τ. Then we have

$$\phi(t_p, \boldsymbol{\xi}) = \phi(n_pT, \phi(t_p - n_pT, \boldsymbol{\xi})).$$

Since $\phi(t, \boldsymbol{\xi})$ is a solution of (8.2.44), $\frac{\partial\phi}{\partial t}(t, \boldsymbol{\xi})$ is bounded on any compact subset of $(0, \infty) \times D$. Hence, the Mean Value Theorem implies that ϕ is continuous in t uniformly with respect to $\boldsymbol{\xi}$ in any compact subset of D. Thus

$$\boldsymbol{z} = \lim_{p\to\infty} \phi(t_p, \boldsymbol{\xi}) = \lim_{p\to\infty} \phi(t_p - n_pT, \phi(n_pT, \boldsymbol{\xi})) = \phi(\tau, \boldsymbol{y}),$$

which means that $\boldsymbol{z} \in \Gamma_{\boldsymbol{y}}$. Hence, $\omega(\Gamma_{\boldsymbol{\xi}}) \subset \Gamma_{\boldsymbol{y}}$, and thus $\omega(\Gamma_{\boldsymbol{\xi}})$ is a periodic orbit of period T. Let us find the minimal period of $\phi(t, \boldsymbol{p})$. For this, we define $\Pi = \{\tau \in \mathbb{R} : \phi(t + \tau, \boldsymbol{p}) = \phi(t, \boldsymbol{p})\}$ to be the set of all possible periods. By the continuity of ϕ, Π is a closed set. Also, if $\tau, \sigma \in \Pi$, then

$$\phi(t + \tau + \sigma, \boldsymbol{p}) = \phi(t + \tau, \phi(\sigma, \boldsymbol{p})) = \phi(t + \tau, \boldsymbol{p}) = \phi(t, \boldsymbol{p}).$$

Similarly, since $\boldsymbol{p} = \phi(\sigma - \sigma, \boldsymbol{p}) = \phi(\sigma, \phi(-\sigma, \boldsymbol{p})) = \phi(-\sigma, \boldsymbol{p})$,

$$\phi(t + \tau - \sigma, \boldsymbol{p}) = \phi(t + \tau, \phi(-\sigma, \boldsymbol{p})) = \phi(t, \phi(-\sigma, \boldsymbol{p})) = \phi(t, \boldsymbol{p})$$

so Π is a closed subgroup of additive group $(\mathbb{R}, +)$, which contains $nT, n \in \mathbb{Z}$. Now, since $\boldsymbol{\xi} \ll \phi(T, \boldsymbol{\xi})$, the continuity of ϕ implies that $\boldsymbol{\xi} \ll \phi(T + \sigma, \boldsymbol{\xi})$ for $\sigma \in (-\delta, \delta)$, where $\delta > 0$ is sufficiently small. Repeating the argument of the previous part, $(T - \delta, T + \delta) \subset \Pi$ and hence, by the subgroup property, $(-\delta, \delta) \subset \Pi$. But, again by the subgroup property, for any $\tau \in \Pi$, $(\tau - \delta, \tau + \delta) \subset \Pi$ and therefore Π is an open set. Being also closed, $\Pi = \mathbb{R}$ and hence $\omega(\Gamma_{\boldsymbol{\xi}})$ is an equilibrium.

□

Corollary 8.25 *If* (8.2.44) *is strongly monotone, then the conclusion of Theorem 8.24, point 2., holds if* $\boldsymbol{\xi} < \phi(T, \boldsymbol{\xi})$ *for some* $T > 0$.

Proof By (8.2.46), for $\boldsymbol{\eta} = \phi(T, \boldsymbol{\xi})$,

$$\boldsymbol{\eta} \ll \phi(T, \boldsymbol{\eta}),$$

so, since $\Gamma_{\boldsymbol{\xi}} = \Gamma_{\boldsymbol{\eta}}$, we can repeat the proof of Theorem 8.24, point 2., for $\boldsymbol{\eta}$. □

The above results restrict the possible long-term behaviour of solutions to cooperative systems.

Corollary 8.26 *A monotone dynamical system cannot have a nontrivial (not an equilibrium) stable limit cycle.*

Proof Assume that Γ is an attracting periodic orbit and let $\boldsymbol{y} \in \Gamma$. Then there is a ball $B_{\boldsymbol{y},\rho}$ such that for any $\boldsymbol{x} \in B_{\boldsymbol{y},\rho}$, $\phi(t, \boldsymbol{x})$ converges to Γ. Let $\boldsymbol{z} \ll \boldsymbol{y}$. There is $\delta > 0$ such that $\boldsymbol{z} + \lambda(\boldsymbol{y} - \boldsymbol{z}) \in B_{\boldsymbol{y},\rho}$ for $\lambda \in [1 - \delta, 1 + \delta]$. Define $\boldsymbol{\xi}_\delta = \boldsymbol{y} - \delta(\boldsymbol{y} - \boldsymbol{z})$ and $\boldsymbol{\eta}_\delta = \boldsymbol{y} + \delta(\boldsymbol{y} - \boldsymbol{z})$. Then $\boldsymbol{\xi} \ll \boldsymbol{y} \ll \boldsymbol{\eta}$. Taking $0 < \theta < \delta$, we have $\boldsymbol{\xi}_\delta \ll \boldsymbol{\xi}_\theta \ll \boldsymbol{y} \ll \boldsymbol{\eta}_\theta \ll \boldsymbol{\eta}_\delta$ and $\{\boldsymbol{x} : \boldsymbol{\xi}_\theta \ll \boldsymbol{x} \ll \boldsymbol{\eta}_\theta\}$ is an open neighbourhood of $\boldsymbol{y}$. But, since Γ is attracting, there must be T such that $\boldsymbol{\xi}_\theta \ll \phi(T, \boldsymbol{\xi}_\delta) \ll \boldsymbol{\eta}_\theta$. Thus, we have $\boldsymbol{\xi}_\delta \ll \phi(T, \boldsymbol{\xi}_\delta)$ and, by Theorem 8.24, $\phi(t, \boldsymbol{\xi}_\delta)$ converges to equilibrium, contradicting the assumption that it converges to Γ. □

This result can be strengthened for planar dynamical systems.

Corollary 8.27 *A monotone two-dimensional dynamical system cannot have a periodic orbit.*

Proof Let Γ be a periodic orbit and $\boldsymbol{p} \in \Gamma$. Then $\boldsymbol{f}(\boldsymbol{p})$ is tangent to Γ at $\boldsymbol{p}$. Let $\theta(p)$ be the angle between the x_1-axis and $\boldsymbol{f}(\boldsymbol{p})$. When we move $\boldsymbol{p}$ around $\boldsymbol{\Gamma}$, then, by continuity, $\theta(\boldsymbol{p})$ will change by 2π, so there must be $\boldsymbol{p}$ with $\theta(\boldsymbol{p}) \in \left[0, \frac{\pi}{2}\right]$, hence $\boldsymbol{f}(\boldsymbol{p}) \geq 0$ and $\Gamma = \omega(\Gamma)$ is an equilibrium. □

Exercise 8.2.5 Show that

$$\begin{aligned} x_1' &= x_1(1 - x_1 - \alpha x_2), \\ x_2' &= \lambda x_2(1 - x_2 - \gamma x_1), \end{aligned} \tag{8.2.47}$$

where $\alpha > 0, \lambda > 0, \gamma > 0$, does not have periodic trajectories in $[0, 1] \times [0, 1]$. Hint. Show the invariance $[0, 1] \times [0, 1]$ and use Exercise 8.2.4.

It follows that monotone flows preserve any 'order' introduced in Sect. A.3.1. Let $\prec$ denote any relation $\leq$, $<$ or $\ll$.

Proposition 8.28 *Let* (8.2.44) *be a differentiable monotone system on D. If $\boldsymbol{x} \prec \boldsymbol{y}$, then $\phi(t, \boldsymbol{x}) \prec \phi(t, \boldsymbol{y}), t \geq 0, \boldsymbol{x}, \boldsymbol{y} \in D$.*

Proof That $\boldsymbol{x} \leq \boldsymbol{y}$ implies $\phi(t, \boldsymbol{x}) \leq \phi(t, \boldsymbol{y})$ follows from the definition. To prove the other two implications, we observe that since (8.2.37) is a fundamental matrix, it is invertible and thus $\phi(t, \cdot)$ is a diffeomorphism for any t so, in particular, it is one-to-one. Thus, since $\boldsymbol{x} < \boldsymbol{y}$ implies $\boldsymbol{x} \leq \boldsymbol{y}$ and $\boldsymbol{x} \neq \boldsymbol{y}$, $\phi(t, \boldsymbol{x}) \leq \phi(t, \boldsymbol{y})$ by monotonicity and $\phi(t, \boldsymbol{x}) \neq \phi(t, \boldsymbol{y})$ by injectivity, implies $\phi(t, \boldsymbol{x}) < \phi(t, \boldsymbol{y})$. Finally, by monotonicity, the image of the order interval $[\boldsymbol{x}, \boldsymbol{y}]$ is $[\phi(t, \boldsymbol{x}), \phi(t, \boldsymbol{y})]$. Now, if $\boldsymbol{x} \ll \boldsymbol{y}$, the interior of $[\boldsymbol{x}, \boldsymbol{y}]$, $\prod_{i=1}^{d} (x_i, y_i)$, is non-empty. Since $\phi(t, \cdot)$ is a diffeomorphism, the interior of its image, $[\phi(t, \boldsymbol{x}), \phi(t, \boldsymbol{y})]$, also is non-empty. This means that $\phi(t, \boldsymbol{x}) \ll \phi(t, \boldsymbol{y})$ as otherwise we would have $\phi_i(t, \boldsymbol{x}) = \phi_i(t, \boldsymbol{y})$ for some i which would imply that the interior of $[\phi(t, \boldsymbol{x}), \phi(t, \boldsymbol{y})]$ is empty. □

We need one more auxiliary result, an extension of Corollary 8.17.

Proposition 8.29 *Let* (8.2.44) *be a differentiable monotone system on a positively invariant convex set D. Then*

$$D_+ := \{\boldsymbol{x} \in D : \ \boldsymbol{f}(\boldsymbol{x}) \succ 0\} \tag{8.2.48}$$

is positively invariant, and the flow is nondecreasing there and

$$D_- := \{\boldsymbol{x} \in D : \ \boldsymbol{f}(\boldsymbol{x}) \prec 0\} \tag{8.2.49}$$

is also positively invariant, and the flow is nonincreasing there.

Proof Let $\phi(t, \boldsymbol{\xi})$ be the flow generated by (8.2.44). Define $z(t) = \boldsymbol{f}(\phi(t, \boldsymbol{\xi}))$. Then, as in Theorem B.18, we have

$$z'(t) = \mathcal{J}_{\boldsymbol{f}}(\phi(t, \boldsymbol{\xi}))\phi'(t, \boldsymbol{\xi}) = \mathcal{J}_{\boldsymbol{f}}(\phi(t, \boldsymbol{\xi}))z(t),$$

with the initial condition $z(0) = \boldsymbol{f}(\phi(0, \boldsymbol{\xi})) = \boldsymbol{f}(\boldsymbol{\xi})$. Let $z(t) = \psi(t, 0, \boldsymbol{\xi})$ be the flow generated by the above equation. From the linearity of ψ, we see that $\psi(t, 0, 0) = 0$. Since $\mathcal{J}_{\boldsymbol{f}}(\phi(t, \boldsymbol{\xi}))$ is a Metzler matrix, Proposition 5.30 implies that ψ leaves $\overline{\mathbb{R}}^d_+$ invariant. Now, by Proposition 5.8, the fundamental matrix of the above system is nonsingular and hence $\psi(t, 0, \boldsymbol{\xi})$ is an isomorphism for any t. Hence, the thesis of Proposition 8.28 is valid and we have $\psi(t, 0, \boldsymbol{\xi}) > \psi(t, 0, 0) = 0$ if $\boldsymbol{\xi} > 0$ and $\psi(t, 0, \boldsymbol{\xi}) \gg \psi(t, 0, 0) = 0$ if $\boldsymbol{\xi} \gg 0$. Therefore, both $\overline{\mathbb{R}}^d_+ \setminus \{0\}$ and $\mathbb{R}^d_+$ are also invariant. Let E denote any of these sets. If $\boldsymbol{\xi} \in D_+$, then $\boldsymbol{f}(\boldsymbol{\xi})$ belongs to some E. From the invariance, $z(t) = \boldsymbol{f}(\phi(t, \boldsymbol{\xi}))$ stays in E, which means that $\phi(t, \boldsymbol{\xi}) \in D_+$. This proves the invariance of D_+. Now, from the invariance, $\boldsymbol{f}(\phi(t, \boldsymbol{\xi})) \succ 0$ along the trajectory and hence $\frac{\partial \phi}{\partial t}(t, \boldsymbol{\xi}) \succ 0$, which shows that $\phi(t, \boldsymbol{\xi})$ is nondecreasing.

The proof for D_- is analogous. □

Theorem 8.30 *Let* (8.2.44) *be a differentiable dynamical system on an open set* $\mathbb{O} \subset \mathbb{R}^d$ *and monotone on a convex invariant set D such that* $\overline{D} \subset \mathbb{O}$. *Assume that* $\boldsymbol{\xi}, \boldsymbol{\eta} \in D$ *satisfy* $\boldsymbol{\xi} \ll \boldsymbol{\eta}$ *and* $\boldsymbol{f}(\boldsymbol{\xi}) \geq 0$ *and* $\boldsymbol{f}(\boldsymbol{\eta}) \leq 0$. *Then,*

1. $[\boldsymbol{\xi}, \boldsymbol{\eta}] \cap D$ *is invariant under the flow* ϕ,
2. *if there is a unique equilibrium* $\boldsymbol{e}$ *in* $[\boldsymbol{\xi}, \boldsymbol{\eta}] \cap \overline{D}$, *then it is globally asymptotically stable there.*

Proof Consider $\boldsymbol{x} \in [\boldsymbol{\xi}, \boldsymbol{\eta}] \cap D$. Since $\boldsymbol{f}(\boldsymbol{\xi}) \geq 0$, then, by Proposition 8.29, $\phi(t, \boldsymbol{\xi})$ is nondecreasing, so $\boldsymbol{\xi} \leq \phi(t, \boldsymbol{\xi})$. Monotonicity implies $\phi(t, \boldsymbol{\xi}) \leq \phi(t, \boldsymbol{x})$, so $\boldsymbol{\xi} \leq \phi(t, \boldsymbol{\xi}) \leq \phi(t, \boldsymbol{x}), t \geq 0$. Similarly, $\phi(t, \boldsymbol{x}) \leq \phi(t, \boldsymbol{\eta}) \leq \boldsymbol{\eta}, t \geq 0$. This shows that $[\boldsymbol{\xi}, \boldsymbol{\eta}] \cap D$ is invariant under the flow and, by Theorem 8.24, point 1., $\phi(t, \boldsymbol{\xi})$ and $\phi(t, \boldsymbol{\eta})$ converge as $t \to \infty$ and the limits are equilibria in $[\boldsymbol{\xi}, \boldsymbol{\eta}] \cap \overline{D}$. If there is a unique equilibrium $\boldsymbol{e}$, we must have

$$\lim_{t\to\infty} \phi(t, \boldsymbol{\xi}) = \lim_{t\to\infty} \phi(t, \boldsymbol{\eta}) = \boldsymbol{e}.$$

Since

$$\boldsymbol{\xi} \leq \phi(t, \boldsymbol{\xi}) \leq \phi(t, \boldsymbol{x}) \leq \phi(t, \boldsymbol{\eta}) \leq \boldsymbol{\eta}$$

for any $\boldsymbol{x} \in [\boldsymbol{\xi}, \boldsymbol{\eta}]$, any trajectory $\phi(t, \boldsymbol{x})$ originating in $[\boldsymbol{\xi}, \boldsymbol{\eta}]$, converges to $\boldsymbol{e}$, so $\boldsymbol{e}$ is globally attracting in $[\boldsymbol{\xi}, \boldsymbol{\eta}]$. The stability follows from the fact that the order interval $[\phi(t, \boldsymbol{\xi}), \phi(t, \boldsymbol{\eta})]$ containing $\boldsymbol{e}$ is invariant and can be made arbitrarily small by taking sufficiently large t. □

Example 8.31 Consider an SIS system

$$\begin{aligned} S' &= \Lambda - \lambda SI - \mu S + \gamma I =: f_1(S, I), \\ I' &= \lambda SI - \mu I - \gamma I =: f_2(S, I), \end{aligned} \tag{8.2.50}$$

describing the situation in which the recovery does not induce immunity, and thus, the recovered infectives return to susceptible at the rate γ; as before, Λ is the total population births rate, μ is the per capita death rate and β is the force of infection.

Since $\frac{\partial f_1}{\partial I} = -S < 0$ for $S > 0$, (8.2.50) is not cooperative on $\overline{\mathbb{R}}^2_+$.

However, re-writing (8.2.50) in terms of the total population $N = S + I$ and I,

$$\begin{aligned} N' &= \Lambda - \mu N =: g_1(N, I), \\ I' &= \lambda I \left(N - I - \frac{\mu + \gamma}{\lambda} \right) =: g_2(N, I), \end{aligned} \tag{8.2.51}$$

we have

$$\frac{\partial g_1}{\partial I} = 0, \quad \frac{\partial g_2}{\partial N} = \lambda I \geq 0.$$

System (8.2.51) is biologically well-posed in

$$D = \{(N, I) \in \overline{\mathbb{R}}_+^2 : I \leq N\}.$$

We see that $\{(N, I) \in \overline{\mathbb{R}}_+^2 : I = 0\}$ and have a disease-free equilibrium,

$$\boldsymbol{E}_0^* = \left(\frac{\Lambda}{\mu}, 0 \right),$$

and an endemic equilibrium,

$$\boldsymbol{E}_1^* = \left(\frac{\Lambda}{\mu}, \frac{\Lambda}{\mu} - \frac{\mu + \gamma}{\lambda} \right), \tag{8.2.52}$$

which makes epidemiological sense if

$$R_0 = \frac{\lambda \Lambda}{\mu(\mu + \gamma)} > 1.$$

Exercise 8.2.6 Prove that $\boldsymbol{E}_0^*$ is globally asymptotically stable if $R_0 \leq 1$.

In what follows, we assume $R_0 > 1$. We see $\boldsymbol{g}(N_0, I_0) \geq 0$ as long as

$$I_0 + \frac{\mu + \gamma}{\lambda} < N_0 < \frac{\Lambda}{\mu} \text{ and } I_0 > 0, \tag{8.2.53}$$

and $\boldsymbol{g}(N_1, I_1) \leq 0$ as long as

$$\frac{\Lambda}{\mu} < N_1 < I_1 + \frac{\mu + \gamma}{\lambda}. \tag{8.2.54}$$

We observe that $\boldsymbol{E}_1^* \in [(N_0, I_0), (N_1, I_1)]$. The reason for taking $I_0 > 0$ is that otherwise $[\left(\frac{\mu+\gamma}{\lambda}, 0\right), (N_1, I_1)]$ would contain also $\boldsymbol{E}_1^* = \left(\frac{\Lambda}{\mu}, 0\right)$. Using

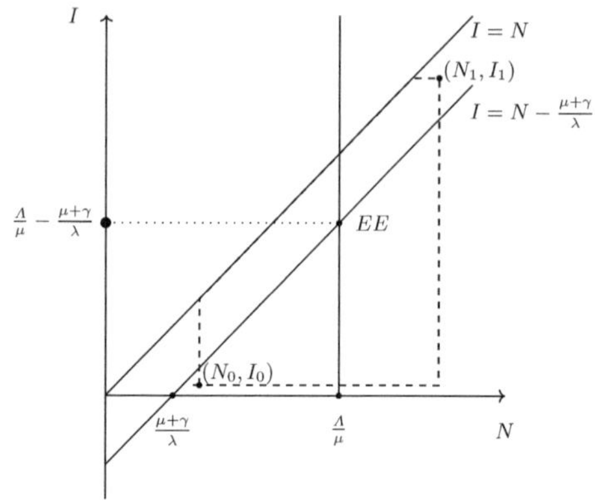

Fig. 8.2 The intersection of $[(N_0, I_0), (N_1, I_1)]$ (dashed line boundary) and D (the sector between $I = 0$ and $I = N$) in Example 8.31

Theorem 8.30, we can claim that the endemic equilibrium (8.2.52) is globally asymptotically stable in any set

$$[(N_0, I_0), (N_1, I_1)] \cap D$$

for any (N_0, I_0) and (N_1, I_1) satisfying (8.2.53) and (8.2.54), see Fig. 8.2.

Exercise 8.2.7 Using the Vidyasagar theorem, or otherwise, show that if $R_0 > 1$, then $\boldsymbol{E}_1^*$ is globally asymptotically stable in $D \cap \{(N, I) \in \overline{\mathbb{R}}_+^2 : I_0 > 0\}$.

Example 8.32 Consider the system

$$x_i' = x_i \left(r_i - a_{ii}x_i + \sum_{\substack{j=1 \\ j \neq i}}^{d} a_{ij}x_j \right), \qquad i = 1, \ldots, d, \tag{8.2.55}$$

or, in compact notation,

$$\boldsymbol{x}' = \mathcal{diag}\, \boldsymbol{r}\boldsymbol{x} + \mathcal{diag}\, \boldsymbol{x}\, \mathcal{A}\boldsymbol{x} =: \boldsymbol{F}(\boldsymbol{x}), \tag{8.2.56}$$

where, for a given vector $\boldsymbol{a} \in \mathbb{R}^d$, we denote

$$\mathcal{diag}\, \boldsymbol{a} = \begin{pmatrix} a_1 & 0 & \cdots & 0 \\ 0 & a_2 & \cdots & 0 \\ \vdots & \vdots & \ddots & \vdots \\ 0 & 0 & \cdots & a_k \end{pmatrix}. \tag{8.2.57}$$

We assume that $\mathcal{A}$ is a Metzler stable matrix, so that (8.2.55) is cooperative, and $\boldsymbol{r} \gg 0$. Clearly, $\boldsymbol{x} = 0$ is the extinction equilibrium, and we have a positive coexistence equilibrium,

$$\boldsymbol{e} = -\mathcal{A}^{-1}\boldsymbol{r} \gg 0.$$

Indeed, $\boldsymbol{e} \geq 0$ by Theorem 8.1 and, since $\mathcal{A}^{-1}$ is invertible, it cannot have a zero row, and thus strict positivity follows from $\boldsymbol{r} \gg 0$. Now, let us take arbitrary $\boldsymbol{b} \gg \boldsymbol{r}$. Then,

$$\boldsymbol{f} := -\mathcal{A}^{-1}\boldsymbol{b} \gg \boldsymbol{e},$$

where, again, the strict inequality follows from the non-negativity and invertibility of $-\mathcal{A}^{-1}$. Moreover, for $k > 1$, we have

$$k\boldsymbol{f} = -\mathcal{A}^{-1}k\boldsymbol{b},$$

and

$$\boldsymbol{r} + \mathcal{A}(-\mathcal{A}^{-1}(k\boldsymbol{b})) = \boldsymbol{r} - k\boldsymbol{b} \ll \boldsymbol{r} - \boldsymbol{b} \ll 0,$$

thus $\boldsymbol{F}(k\boldsymbol{b}) \ll 0$ since $k\boldsymbol{b} \gg 0$. Further, since $\mathcal{A}0 = 0$ and $\boldsymbol{r} \gg 0$, by continuity of the matrix multiplication, we can find $\rho > 0$ such that $\boldsymbol{r} + \mathcal{A}\boldsymbol{c} \gg 0$ for any $\|\boldsymbol{c}\| \leq \rho$. Thus, $\boldsymbol{F}(l\boldsymbol{b}) \gg 0$ for any $0 < l \leq \dfrac{\rho}{\|\boldsymbol{b}\|}$. We have $l\boldsymbol{b} \ll k\boldsymbol{b}$ for all sufficiently small l and large k, $\boldsymbol{F}(l\boldsymbol{b}) \geq 0$ and $\boldsymbol{F}(k\boldsymbol{b}) \leq 0$, we can apply Theorem 8.30 to claim that $\boldsymbol{e}$ is the unique equilibrium in $\mathbb{R}^d_+$, which is globally stable there.

Let us note that the above result does not preclude the existence of equilibria on the faces of the positive orthant $\mathbb{R}^d_+$, but they cannot be stable in $\overline{\mathbb{R}}^d_+$, as illustrated in the following exercise.

Exercise 8.2.8 Consider the cooperative system

$$\begin{aligned} x_1' &= F_1(x_1, x_2) = x_1(1 - x_1) + x_1x_2, \\ x_2' &= F_2(x_1, x_2) = x_2(1 - x_2) + \frac{1}{2}x_1x_2, \end{aligned} \tag{8.2.58}$$

and let $\boldsymbol{F} = (F_1, F_2)$.

(i) Find all equilibria of (8.2.58) and determine their local asymptotic stability.
(ii) Are there any invariant sets where the boundary equilibria are asymptotically stable?
(iii) Find points $\boldsymbol{a}$ and $\boldsymbol{b}$ such that $\boldsymbol{a} \ll \boldsymbol{b}$ and $\boldsymbol{F}(\boldsymbol{a}) \geq 0$ and $\boldsymbol{F}(\boldsymbol{b}) \leq 0$. Show that they can be found to be, respectively, arbitrarily close to 0 and arbitrarily large.

Conclude that the interior equilibrium is globally asymptotically stable on $\mathbb{R}^2_+$ but not globally asymptotically stable on $\overline{\mathbb{R}}^2_+$.

(iv) By considering

$$x_1' = x_1(1 - x_1) + x_1x_2,$$
$$x_2' = x_2(1 - x_2) + x_1x_2,$$

show that the assumption that $\mathcal{A}$ is a Metzler stable matrix is essential for the conclusion of Example 8.32 to hold.

The results proved above are helpful, provided we know the equilibria. In the next section, we shall show that by imposing stronger assumptions on the dynamical system, we can prove the existence of a nontrivial globally stable equilibrium in a cooperative system.

8.2.3 Sublinearity and the Existence of Equilibria

The ideas here are based on results in [61] about the number of solutions to operator equations in positive cones; they were used in epidemiological context in [53] and [82], the presentation below follows the latter. The additional assumption required here was called strong concavity in [61], but later, it was referred to as sublinearity.

Definition 8.33 Let $\boldsymbol{f}$ be a function $\boldsymbol{f} : \overline{\mathbb{R}}^d_+ \to \overline{\mathbb{R}}^d_+$. We say that $\boldsymbol{f}$ is sublinear if for any $\boldsymbol{x} > 0$ and $0 < \lambda < 1$ we have

$$\lambda \boldsymbol{f}(\boldsymbol{x}) \leq \boldsymbol{f}(\lambda \boldsymbol{x}). \tag{8.2.59}$$

We say that $\boldsymbol{f}$ is strictly sublinear if for any $\boldsymbol{x} \gg 0$ and $0 < \lambda < 1$ we have

$$\lambda \boldsymbol{f}(\boldsymbol{x}) < \boldsymbol{f}(\lambda \boldsymbol{x}). \tag{8.2.60}$$

Finally, $\boldsymbol{f}$ is strongly sublinear if for any $\boldsymbol{x} \gg 0$ and $0 < \lambda < 1$ we have

$$\lambda \boldsymbol{f}(\boldsymbol{x}) \ll \boldsymbol{f}(\lambda \boldsymbol{x}). \tag{8.2.61}$$

Example 8.34 A linear operator is sublinear but not strictly sublinear.

Example 8.35 For scalar functions, strict and strong sublinearity are the same. To give a geometric interpretation of the definition, we observe that, for a given $x_0 > 0$, $\{(\lambda x_0, \lambda f(x_0));\ 0 < \lambda < 1\}$ is an open segment connecting $(0, 0)$ and $(x_0, f(x_0))$ and thus strict sublinearity of f means that the graph of f lies above the segment for all $0 < x < x_0$, see Fig. 8.3. We see that strict concavity of f implies strict sublinearity. However, the reverse is not true, as in the definition of sublinearity, we

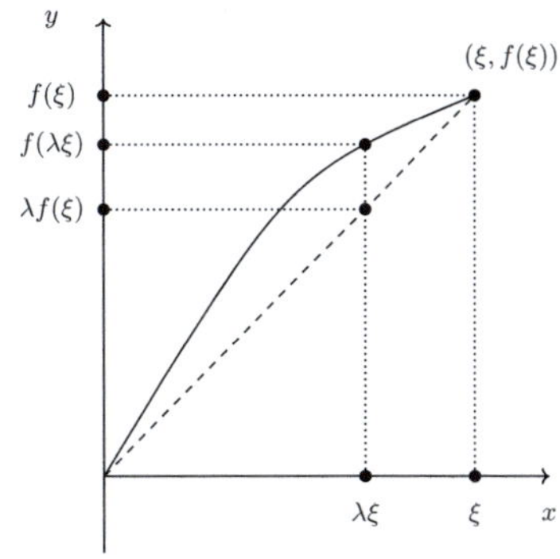

Fig. 8.3 A sublinear function of one variable

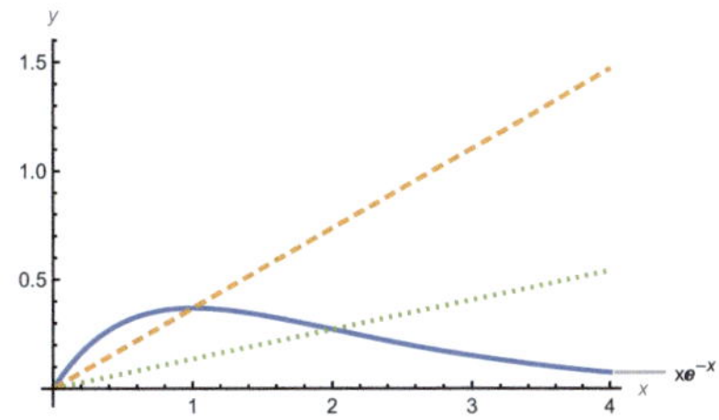

Fig. 8.4 Strict sublinearity of a nonconcave function $f(x) = xe^{-x}$

only consider 0 as the left-end point. For instance, $f(x) = xe^{-x}$ is not concave, but it is sublinear. Indeed, for $x > 0, 0 < \lambda < 1$ we have $e^{x(1-\lambda)} > 1$, hence $e^{-\lambda x} > e^{-x}$ and, finally,

$$f(\lambda x) = \lambda x e^{-\lambda x} > \lambda x e^{-x} = \lambda f(x).$$

This situation is illustrated in Fig. 8.4.

Exercise 8.2.9 Show that

(i) the Holling II function

$$f(x) = \frac{x}{k + x},$$

$k > 0$, and

(ii) logistic function

$$f(x) = x(1 - x),$$

are strictly sublinear on $[0, \infty)$.

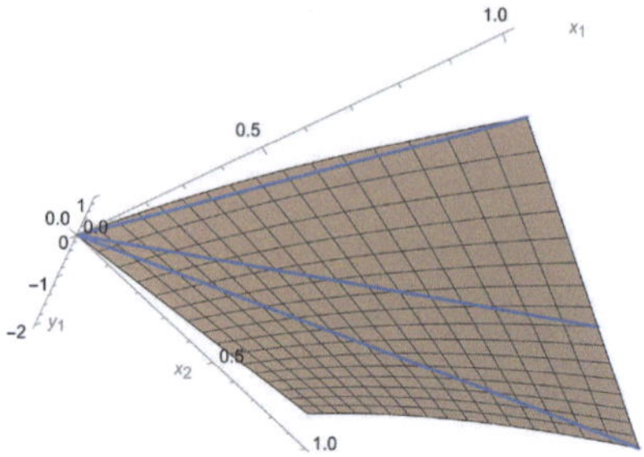

Fig. 8.5 An illustration of inequality (8.2.63) for $y_1 = (1 - \lambda x_1) \sum_{j=1}^{2} x_j - 2x_1$: the segments $\lambda(\boldsymbol{x}, f_1())$, $0 < \lambda < 1$, lie below the graph of $f_1(\boldsymbol{x})$

Example 8.36 Consider $\boldsymbol{y} = \boldsymbol{f}(\boldsymbol{x})$ given by

$$\begin{aligned} y_1 &= f_1(x_1, x_2) = (1 - x_1) \sum_{j=1}^{2} \beta_{1j} x_j - \alpha_1 x_1, \\ y_2 &= f_2(x_1, x_2) = (1 - x_2) \sum_{j=1}^{2} \beta_{2j} x_j - \alpha_1 x_2, \end{aligned} \tag{8.2.62}$$

where $\beta_{ij} > 0$, $1 \le i, j \le 2$ and $\alpha_i > 0$, $1 \le i \le 2$.

Let $x_1 > 0$, $x_2 > 0$ and $\lambda \in (0, 1)$. Since $1 - \lambda x_i > 1 - x_i$, for $i = 1, 2$, we have

$$f_i(\lambda x_1, \lambda x_2) = (1 - \lambda x_i) \sum_{j=1}^{2} \beta_{1j} \lambda x_j - \alpha_1 \lambda x_1 > \lambda (1 - x_i) \sum_{j=1}^{2} \beta_{1j} x_j - \alpha_1 x_1, \tag{8.2.63}$$

and hence $\boldsymbol{f}$ is strongly sublinear, see Fig. 8.5. On the other hand, $\boldsymbol{y} = \boldsymbol{f}(\boldsymbol{x})$ given by

$$\begin{aligned} y_1 &= f_1(x_1, x_2) = (1 - x_1) \sum_{j=1}^{2} \beta_{1j} x_j - \alpha_1 x_1, \\ y_2 &= f_2(x_1, x_2) = \sum_{j=1}^{2} \beta_{2j} x_j - \alpha_1 x_2, \end{aligned} \tag{8.2.64}$$

is strictly, but not strongly, sublinear. This follows from Example 8.34, as even if we have $x_1 > 0$, $x_2 > 0$, then

$$f_2(\lambda x_1, \lambda x_2) = \sum_{j=1}^{2} \beta_{2j} \lambda x_j - \alpha_1 \lambda x_2 = \lambda \left(\sum_{j=1}^{2} \beta_{2j} x_j - \alpha_1 x_2 \right) = \lambda f_2(x_1, x_2).$$

Proposition 8.37 *Assume that* (8.2.44) *is a cooperative system with strictly sublinear* $\boldsymbol{f}$*, with the flow* ϕ *preserving* $\overline{\mathbb{R}}^d_+$*. Then,* ϕ *is monotone and strictly sublinear. If, in addition,* (8.2.44) *is irreducible, then* ϕ *is strongly monotone and strongly sublinear.*

Proof The statements concerning the monotonicity and strong monotonicity of the flow follow from Theorem 8.23. As far as the sublinearity is concerned, let us consider $\boldsymbol{x} \gg 0$ and $\lambda \in (0, 1)$ and define

$$\boldsymbol{y}(t) = \phi(t, \lambda\boldsymbol{x}) - \lambda\phi(t, \boldsymbol{x}).$$

Then

$$\begin{aligned}\boldsymbol{y}'(t) &= \boldsymbol{f}(\phi(t, \lambda\boldsymbol{x})) - \lambda\boldsymbol{f}(\phi(t, \boldsymbol{x}))\\ &= (\boldsymbol{f}(\phi(t, \lambda\boldsymbol{x}))) - \boldsymbol{f}(\lambda\phi(t, \boldsymbol{x})) + (\boldsymbol{f}(\lambda\phi(t, \boldsymbol{x})) - \lambda\boldsymbol{f}(\phi(t, \boldsymbol{x})))\\ &= \mathcal{A}(t)\boldsymbol{y}(t) + \mathcal{B}(t),\end{aligned}$$

with

$$\boldsymbol{y}(0) = 0,$$

where

$$\mathcal{A}(t) = \int_0^1 \mathcal{J}_{\boldsymbol{f}}(\lambda\phi(t, \boldsymbol{x}) + s(\phi(t, \lambda\boldsymbol{x}) - \lambda\phi(t, \boldsymbol{x})))ds$$

and

$$\mathcal{B}(t) = \boldsymbol{f}(\lambda\phi(t, \boldsymbol{x})) - \lambda\boldsymbol{f}(\phi(t, \boldsymbol{x})).$$

Since (8.2.44) is cooperative, Corollary 8.16 implies that f is of type K. Further, by assumption, $\phi(t, 0) \geq 0$ and then, since $\boldsymbol{x} \gg 0$, the monotonicity of ϕ implies $0 \leq \phi(t, 0) \ll \phi(t, \boldsymbol{x})$ by Proposition 8.28.

Denote by $\mathcal{Y}(t, s)$ the fundamental matrix of

$$\mathcal{Y}(t) = \mathcal{A}(t)\mathcal{Y}(t), \quad \mathcal{Y}(s) = \mathcal{I}_d.$$

Then, by the variation of constants formula (5.3.40), $\boldsymbol{y}$ is given by

$$\boldsymbol{y}(t) = \int_0^t \mathcal{Y}(t, s)\mathcal{B}(s)ds.$$

Since $\mathcal{J}_{\boldsymbol{f}}(\boldsymbol{x})$ is a Metzler matrix for any $\boldsymbol{x}$, we see, as in the proof of Corollary 8.16, that $\mathcal{A}(t)$ is also a Metzler matrix for any t. Thus $\mathcal{Y}(t, s) \geq 0$ for $t \geq s$ by

Proposition 5.30, and, since it is nonsingular, from $\mathcal{B}(t) > 0$ we infer $\mathcal{Y}(t,s)\mathcal{B}(s) > 0$. Hence $\mathbf{y}(t) > 0$ and ϕ is strictly sublinear.

The second assertion from the fact that since $\mathcal{J}_f(\mathbf{x})$ is an irreducible Metzler matrix, we prove, as in Theorem 8.20, that $\mathcal{Y}(t,s) \gg 0$ for $t > s$ and hence, since $\mathcal{B}(t) > 0$, $\mathcal{Y}(t,s)\mathcal{B}(s) \gg 0$. This shows that ϕ is strongly sublinear. □

The main application of sublinearity, first noticed in [61, Theorem 6.3], is the uniqueness of positive equilibria. We have

Lemma 8.38 *Let* (8.2.44) *be cooperative, irreducible and strongly sublinear on* $\mathbb{R}^d_+$ *and let its flow preserve* $\overline{\mathbb{R}}^d_+$. *Then* (8.2.44) *cannot have two distinct equilibria in* $\mathbb{R}^d_+$.

Proof Consider the flow ϕ of (8.2.44). Since (8.2.44) is cooperative irreducible, it is strongly monotone and sublinear by Proposition 8.37. Assume that $\mathbf{p} = (p_1, \ldots, p_d) \gg 0$ and $\mathbf{q} = (q_1, \ldots, q_d) \gg 0$ be two different equilibria and $p_{i_r} > q_{i_r}, r = 1, \ldots, u$, $p_{j_r} = q_{j_r}, r = 1, \ldots, v$ and $p_{k_r} < q_{k_r}, r = 1, \ldots, w$, with $u + v + w = d$. Then, defining $s_p := \min\limits_{1 \le r \le u} \frac{q_{i_r}}{p_{i_r}}$, we have $0 < s_p < 1$ if $\{i_1, \ldots, i_u\} \neq \emptyset$ (otherwise we set $s_p = 1$), and

$$s_p p_{i_s} = p_{i_s} \min_{1 \le r \le u} \frac{q_{i_r}}{p_{i_r}} \le q_{i_s}.$$

This implies

$$s_p \mathbf{p} \le q$$

and s_p is the largest number with this property (since the equality above is attained if the minimum is attained at $\frac{q_{i_s}}{p_{i_s}}$). Similarly, defining $s_q := \min\limits_{1 \le r \le w} \frac{q_{k_r}}{p_{k_r}}$, we get

$$s_q \mathbf{q} \le \mathbf{q},$$

hence, defining $s = \min\{s_p, s_q\}$ we have $0 < s < 1$ (as one of the sets $\{i_1, \ldots, i_u\}$ and $\{i_1, \ldots, i_w\}$ is nonempty; otherwise $\mathbf{p} = \mathbf{q}$) and s is the largest scalar for which simultaneously

$$s\mathbf{p} \le \mathbf{p}, \quad s\mathbf{q} \le \mathbf{q},$$

see Fig. 8.6. Now, strong sublinearity and monotonicity imply

$$\mathbf{p} = \phi(t, \mathbf{p}) \ge \phi(t, s\mathbf{q}) \gg s\phi(t, \mathbf{q}) = s\mathbf{q}$$

and

$$\mathbf{q} = \phi(t, \mathbf{q}) \ge \phi(t, s\mathbf{p}) \gg s\phi(t, \mathbf{p}) = s\mathbf{p},$$

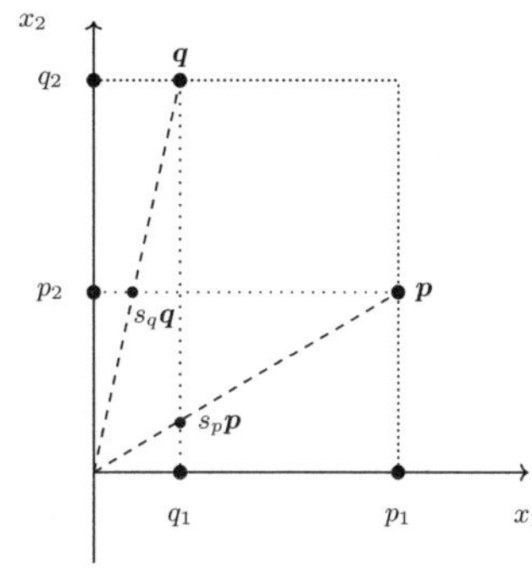

Fig. 8.6 The construction of s_p and s_q

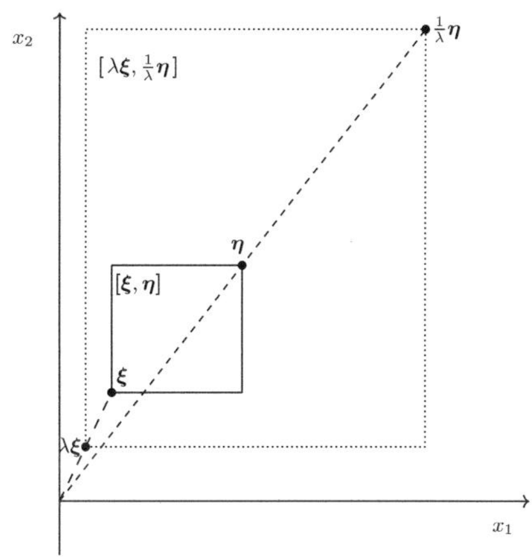

Fig. 8.7 Order intervals $[\boldsymbol{\xi}, \boldsymbol{\eta}]$ and $[\lambda\boldsymbol{\xi}, \frac{1}{\lambda}\boldsymbol{\eta}]$ for $\lambda = \frac{1}{2}$

which shows that s cannot be the largest number satisfying both $s\boldsymbol{p} \leq \boldsymbol{q}$ and $s\boldsymbol{q} \leq \boldsymbol{p}$ (we noted above that in one case, we have the equality at least at one position). □

Proposition 8.39 *Let* (8.2.44) *be cooperative, irreducible and strictly sublinear in* $\mathbb{R}^d_+$*, whose flow* ϕ *preserves* $\overline{\mathbb{R}}^d_+$*. If there are* $0 \ll \boldsymbol{\xi} \ll \boldsymbol{\eta}$ *such that* $\boldsymbol{f}(\boldsymbol{\xi}) > 0$ *and* $\boldsymbol{f}(\boldsymbol{\eta}) \leq 0$*, then the positive orthant* $\mathbb{R}^d_+$ *is positively invariant and there is a unique positive equilibrium* $\boldsymbol{e} \in [\boldsymbol{\xi}, \boldsymbol{\eta}]$*, which is globally attracting in* $\mathbb{R}^d_+$*.*

Proof Theorem 8.30 implies that $[\boldsymbol{\xi}, \boldsymbol{\eta}]$ is positively invariant. Next, strong sublinearity implies that for $0 < \lambda < 1$ we obtain $\boldsymbol{f}(\lambda\boldsymbol{\xi}) > \lambda\boldsymbol{f}(\boldsymbol{\xi}) \geq 0$ and $\lambda\boldsymbol{f}\left(\frac{1}{\lambda}\boldsymbol{\eta}\right) < \boldsymbol{f}(\boldsymbol{\eta}) \leq 0$ and so $\boldsymbol{f}\left(\frac{1}{\lambda}\boldsymbol{\eta}\right) \ll 0$. Using Theorem 8.30 again, we see that $[\lambda\boldsymbol{\xi}, \frac{1}{\lambda}\boldsymbol{\eta}]$ is also invariant. Letting $\lambda \to 0^+$, we see that the open orthant $\mathbb{R}^d_+$ is invariant under the flow and also all forward trajectories are bounded, see Fig. 8.7. Using Theorem 8.30 yet again, $\phi(t, \boldsymbol{\xi})$ converges to a strictly positive equilibrium, say, $\boldsymbol{e}$, which, by Lemma 8.38 is unique. Since $[\boldsymbol{\xi}, \boldsymbol{\eta}]$ is invariant, any trajectory starting in $[\boldsymbol{\xi}, \boldsymbol{\eta}]$ stays there, and hence the equilibrium belongs to $[\boldsymbol{\xi}, \boldsymbol{\eta}]$ by its closedness. To show the stability of $\boldsymbol{e}$, we consider $[\lambda\boldsymbol{e}, \frac{1}{\lambda}\boldsymbol{e}]$, which is invariant under the flow, as in the previous part of the proof, and can be made as small as we wish by taking $\lambda \to 1$. □

The next theorem is a refinement of the above results.

Theorem 8.40 *Let (8.2.44) be a cooperative and strictly sublinear system in $\overline{\mathbb{R}}^d_+$, whose flow ϕ preserves $\overline{\mathbb{R}}^d_+$ and assume that all trajectories in $\overline{\mathbb{R}}^d_+$ are bounded. Then*

1. *If $f(0) = 0$, that is, the origin is an equilibrium, then either it is globally asymptotically stable, or there is an equilibrium $e \gg 0$, which is globally asymptotically stable in $\overline{\mathbb{R}}^d_+ \setminus \{0\}$.*
2. *If $f(0) > 0$, there is a unique equilibrium $e \gg 0$, which is globally asymptotically stable on $\overline{\mathbb{R}}^d_+$.*

Proof First, let us assume that there is an equilibrium $e \gg 0$. By Lemma 8.38, it is the only equilibrium in the open orthant $\mathbb{R}^d_+$. Further, by the strong monotonicity, if $x > 0$, then $\phi(t, x) \gg \phi(t, 0) \geq 0$. Thus, a trajectory starting at any face of $\overline{\mathbb{R}}^d_+ \setminus \{0\}$ immediately enters $\mathbb{R}^d_+$, and thus, there are no other equilibria except possibly at 0. Then we can repeat the proof of Proposition 8.39 for the pair λe and $\frac{1}{\lambda}e$, $0 < \lambda < 1$. Indeed, $\lambda e \ll \frac{1}{\lambda}e$, $f(e) = 0$ and, by strong sublinearity, $f(\lambda e) \gg \lambda f(e) = 0$ and $\lambda f\left(\frac{1}{\lambda}e\right) \ll f(e)$. Hence, any trajectory starting in $\overline{\mathbb{R}}^d_+ \setminus \{0\}$ converges to e, and its stability follows as in Proposition 8.39.

So, we are left with proving that if there is $\mathring{x} \in \overline{\mathbb{R}}^d_+$ such that $\phi(t, \mathring{x})$ does not converge to 0, then there is a strictly positive equilibrium. Since, by assumption, $\Gamma_{\mathring{x}}$, and hence $\omega(\Gamma_{\mathring{x}})$, are bounded, there exists a least upper bound $y \in \overline{\mathbb{R}}^d_+ \setminus \{0\}$ (with respect to the order $\leq$ on $\mathbb{R}^d$). Let $z \in \omega(\Gamma_{\mathring{x}})$. By Lemma 6.49, point 3., $\omega(\Gamma_{\mathring{x}})$ is invariant, so $\phi(-t, z) \in \omega(\Gamma_{\mathring{x}})$ for any $t \geq 0$. By the monotonicity, from $\phi(-t, z) \leq y$ we obtain $z \leq \phi(t, y)$, $t \geq 0$. This shows that $\phi(t, y)$ is an upper bound of $\omega(\Gamma_{\mathring{x}})$, hence $y \leq \phi(t, y)$. If y is an equilibrium, then, since $y > 0$, $y = \phi(t, y) \gg 0$ by the first part of the proof, and the proof is complete. If not, we have $y < \phi(t, y)$ and, by the strong monotonicity,

$$\phi(t, y) \ll \phi(t, \phi(t, y)) = \phi(2t, y)$$

and hence

$$y \ll \phi(2t, y).$$

By assumption, the trajectory $\phi(t, y)$ is bounded and thus the assumptions of Theorem 8.24, point 2., are satisfied and the limit $\lim_{t\to\infty} \phi(t, y)$ exists. By

Corollary 6.32, the limit is an equilibrium, which is strictly positive by the first part of the proof; hence, it must be $\boldsymbol{e}$. □

8.2.4 Applications of Theorem 8.40

Example 8.41 (Ross–Macdonald Model with Migrations, [43, 82]) In this example, we follow [82]. An approach that avoids directly using the concept of sublinearity can be found in [43]. Let us now assume that humans stay at d different physical locations, called patches, and the size of the population occupying patch i has size N_i, $i = 1, \ldots, d$. We neglect the demography in each patch; hence, the only changes of N_i are due to inter-patch migrations. The dynamics of human migrations is thus described by

$$N_i' = \sum_{j=1, j\neq i}^{d} m_{ij} N_j - N_i \sum_{j=1, j\neq i}^{d} m_{ji}, \tag{8.2.65}$$

or, in compact form,

$$\boldsymbol{N}' = \mathcal{M}\boldsymbol{N}, \tag{8.2.66}$$

where $\mathcal{M} = (m_{ij})_{1\le i,j\le d}$ and we defined $m_{ii} = -\sum\limits_{j=1, j\neq i}^{d} m_{ji}$, see Sect. 5.1.2.

We consider the Ross–Macdonald model (6.2.59) on d patches, denoting, for $i = 1, \ldots, d$, by $I_{h,i}$ the number of infective humans on patch i and by $I_{v,i}$ and $N_{v,i}$ the number of infective mosquitos and their total population on patch i, respectively. We assume that infective mosquitos reside on p patches, which we number so that they are the first ones, and thus, we assume $I_{v,i} = 0$ if $i > p$. To simplify analysis, we assume that the parameters μ_v, β_v, γ_h and β_h of (6.2.59) are independent of the patch. It is reasonable to assume that there are no migrations of mosquitos between patches, and thus $N_{v,i}$ is constant for each $i = 1, \ldots, d$. Hence, for $1 \le i \le p$, we have

$$\begin{aligned}
I_{i,h}' &= \frac{\beta_h}{N_{h,i}} I_{v,i}(N_{h,i} - I_{h,i}) - \gamma_h I_{h,i} + \sum_{j=1, j\neq i}^{d} m_{ij} I_{h,j} - I_{h,i} \sum_{j=1, j\neq i}^{d} m_{ji},\\
I_{v,i}' &= \frac{\beta_v}{N_{h,i}} I_{h,i}(N_{v,i} - I_{v,i}) - \mu_v I_{v,i},\\
I_{h,i}(0) &= \mathring{I}_{h,i}, \quad I_{v,i}(0) = \mathring{I}_{v,i},
\end{aligned} \tag{8.2.67}$$

and for $i > p$ the only infections occurring in patch i are due to migrations, and there are no mosquitoes; hence we have

$$
\begin{aligned}
I'_{i,h} &= -\gamma_h I_{h,i} + \sum_{j=1, j\neq i}^{d} m_{ij} I_{h,j} - I_{h,i} \sum_{j=1, j\neq i}^{d} m_{ji}, \\
I_{h,i}(0) &= \mathring{I}_{h,i}.
\end{aligned}
\tag{8.2.68}
$$

Complemented by (8.2.66), we have a $2d + p$ dimensional system of equations.

To write the system in a compact form, we introduce the notation

$$
\begin{aligned}
\boldsymbol{I}_h &= (I_{h,1}, \ldots, I_{h,d}), \quad \boldsymbol{I}_v = (I_{v,1}, \ldots, I_{v,p}, 0, \ldots, 0), \\
\boldsymbol{N}_v &= (N_{v,1}, \ldots, N_{v,p}, 0, \ldots, 0).
\end{aligned}
$$

Further, for a given vector $\boldsymbol{X} \in \mathbb{R}^d$, we introduce the canonical projection $\pi : \mathbb{R}^d \to \mathbb{R}^p$ by

$$\pi(X_1, \ldots, X_d) = (X_1, \ldots, X_p).$$

Then, (8.2.66)–(8.2.68), can be written as

$$\boldsymbol{N}'_h = \mathcal{M}\boldsymbol{N}_h, \tag{8.2.69a}$$

$$\boldsymbol{I}'_h = \beta_h (\mathcal{diag}\boldsymbol{N}_h)^{-1} \mathcal{diag}(\boldsymbol{N}_h - \boldsymbol{I}_h)\boldsymbol{I}_v - \gamma_h \boldsymbol{I}_h + \mathcal{M}\boldsymbol{I}_h, \tag{8.2.69b}$$

$$\pi \boldsymbol{I}'_v = \beta_v (\mathcal{diag}\, \pi \boldsymbol{N}_h)^{-1} \mathcal{diag}(\pi \boldsymbol{N}_v - \pi \boldsymbol{I}_v)\pi \boldsymbol{I}_h - \mu_v \pi \boldsymbol{I}_v, \tag{8.2.69c}$$

where the notation $\mathcal{diag}$ was introduced in (8.2.57). Due to the structure of the right-hand-side of (8.2.69), the initial conditions for $\boldsymbol{N}_h$ must satisfy $\boldsymbol{N}_h(0) = \mathring{\boldsymbol{N}}_h \gg 0$.

Due to the properties of $\mathcal{M}$, the evolution occurs in the affine hyperplane in $\mathbb{R}^{2d+p}$ defined by

$$\sum_{i=1}^{d} N_{h,i} = \sum_{i=1}^{d} \mathring{N}_{h,i} =: N = \mathring{N}_h, \tag{8.2.70}$$

where N is the total human population, equal to the initial human population. Denote $\boldsymbol{1} = (1, \ldots, 1)$ and $\boldsymbol{N} = N\boldsymbol{1} = (N, \ldots, N)$.

It is natural to assume that the initial condition for the infective humans satisfies

$$0 \leq \mathring{\boldsymbol{I}}_h \leq \mathring{\boldsymbol{N}}_h, \tag{8.2.71}$$

and for the infective vectors

$$0 \le \mathring{\boldsymbol{I}}_v \le \boldsymbol{N}_v. \tag{8.2.72}$$

Then we have

Proposition 8.42 *Let the initial conditions* $(\mathring{\boldsymbol{N}}_h, \mathring{\boldsymbol{I}}_h, \mathring{\boldsymbol{N}}_v)$ *for* (8.2.69) *satisfy* (8.2.70)–(8.2.72). *Then, all forward trajectories stay in the set*

$$\mathbb{A} = \left\{ (\boldsymbol{N}_h, \boldsymbol{I}_h, \boldsymbol{I}_v) \in \mathbb{R}_+^{2d+p} : \sum_{i=1}^{d} N_{h,i} = N, \boldsymbol{I}_h \le \boldsymbol{N}_h, \boldsymbol{I}_v \le \boldsymbol{N}_v \right\}. \tag{8.2.73}$$

Proof Since (8.2.69a) is independent of the other unknowns and $\mathcal{M}$ is a Metzler matrix, Proposition 5.30 implies that $0 \le \boldsymbol{N}_h(t)$. Hence, by (8.2.70), $\sum_{i=1}^{d} N_{h,i}(t) = N$. Theorem B.21 implies that also $\boldsymbol{I}_h(t) \ge 0$ and $\boldsymbol{I}_v(t) \ge 0$ on the maximal interval of existence of the solution. Next, we see that for each $0 \le i \le p$, we have

$$I'_{v,i}|_{I_{v,i}=N_{v,i}} = -\mu_v N_{v,i} < 0$$

since $N_{v,i} = \mathring{N}_{v,i} > 0$ for $i = 1, \ldots, p$, by definition. Hence, $\boldsymbol{I}_v \le \boldsymbol{N}_v$. To show the statement for $\boldsymbol{I}_h$, we rewrite (8.2.69a), (8.2.69b) in the form involving susceptible humans by defining $\boldsymbol{S}_h = \boldsymbol{N}_h - \boldsymbol{I}_h$, as in (6.2.58). We have

$$\boldsymbol{S}'_h = -\beta_h (\mathcal{diag}\boldsymbol{N}_h)^{-1} \mathcal{diag}\boldsymbol{S}_h \boldsymbol{I}_v + \gamma_h \boldsymbol{I}_h + \mathcal{M}\boldsymbol{S}_h, \tag{8.2.74a}$$

$$\boldsymbol{I}'_h = \beta_h (\mathcal{diag}\boldsymbol{N}_h)^{-1} \mathcal{diag}\boldsymbol{S}_h \boldsymbol{I}_v - \gamma_h \boldsymbol{I}_h + \mathcal{M}\boldsymbol{I}_h. \tag{8.2.74b}$$

Using again Theorem B.21, we see that $\boldsymbol{S}_h \ge 0$ and hence

$$\boldsymbol{I}_h \le \boldsymbol{N}_h. \tag{8.2.75}$$

□

In what follows, we assume that $\mathcal{M}$ is an irreducible Metzler matrix. Then, by Theorem C.21, there is a strictly positive eigenvector $\boldsymbol{e} \gg 0$ belonging to a simple dominant eigenvalue equal to the spectral bound $s(\mathcal{M})$ (and it is the only strictly positive eigenvector of $\mathcal{M}$). By (8.2.65), $\boldsymbol{1}\mathcal{M} = \mathcal{M}^T\boldsymbol{1} = 0$. Since $N = N(t) = \sum_{i=1}^{d} N_{h,i}(t)$, each $N_{h,i}, i = 1, \ldots, d$, are bounded. Thus, there are no eigenvalues with positive real parts; therefore, $s(\mathcal{M}) = 0$. We normalize $\boldsymbol{e}$ to satisfy

$$\langle \boldsymbol{e}, \boldsymbol{1} \rangle = 1.$$

Using (5.3.86), we see that for any $\mathring{\boldsymbol{N}}_{\boldsymbol{h}} \in \mathbb{R}^d_+$

$$e^{t\mathcal{M}}\mathring{\boldsymbol{N}}_h = \langle \mathbf{1}, \mathring{\boldsymbol{N}}_h\rangle \boldsymbol{e} + \boldsymbol{\psi}(t) = N\boldsymbol{e} + \boldsymbol{\psi}(t), \tag{8.2.76}$$

where $\langle \boldsymbol{\psi}(t), \mathbf{1}\rangle = 0$ for all t and $\boldsymbol{\psi}(t) \to 0$ exponentially as $t \to \infty$. Thus, for a fixed total human population N, for any initial condition $\mathring{\boldsymbol{N}}_{\boldsymbol{h}}$ satisfying $\sum\limits_{i=1}^{d} \mathring{N}_{h,i} = N$, the evolution of $e^{t\mathcal{M}}\mathring{\boldsymbol{N}}_{\boldsymbol{h}}$ occurs in the affine hyperplane $N\boldsymbol{e}+W_0$, where $W_0 \perp \mathbf{1}$. In other words, for a given N, the hyperplane $W_N := N\boldsymbol{e}+W_0$ is invariant under $e^{t\mathcal{M}}$ and $\boldsymbol{N}^* := N\boldsymbol{e}$ is a unique globally asymptotically stable equilibrium of (8.2.65) in W_N. It is easy to see that $W_N = \{\boldsymbol{N}_h \in \mathbb{R}^d : \sum\limits_{i=1}^{d} N_{h,i} = N\}$ and

$$\mathbb{A} = W_N \times \left\{(\boldsymbol{I}_h, \boldsymbol{I}_v) \in \mathbb{R}^{d+p}_+ : \boldsymbol{I}_h \leq \boldsymbol{N}_h, \boldsymbol{I}_v \leq \boldsymbol{N}_v\right\}. \tag{8.2.77}$$

Thus, we will try to use the Vidyasagar theorem, Theorem 6.77, to reduce the analysis of the stability of (8.2.69) to that of a smaller $d + p$ system

$$\boldsymbol{I}_h' = \beta_h(\mathcal{diag}\boldsymbol{N}^*)^{-1}\mathcal{diag}(\boldsymbol{N}^* - \boldsymbol{I}_h)\boldsymbol{I}_v - \gamma_h\boldsymbol{I}_h + \mathcal{M}\boldsymbol{I}_h, \tag{8.2.78a}$$

$$\pi\boldsymbol{I}_v' = \beta_v(\mathcal{diag}\,\pi\boldsymbol{N}^*)^{-1}\mathcal{diag}(\pi\boldsymbol{N}_v - \pi\boldsymbol{I}_v)\pi\boldsymbol{I}_h - \mu_v\pi\boldsymbol{I}_v. \tag{8.2.78b}$$

With constant total populations, we can normalize the variables by introducing $x_i = \frac{I_{h,i}}{N^*_{h,i}}$, $y_i = \frac{I_{v,i}}{N_{v,i}}$ and $n_i = \frac{N_{v,i}}{N^*_{h,i}}$; we have $n_i = 0, i = p+1, \ldots, d$. Then

$$\boldsymbol{x}' = \beta_h\mathcal{diag}\,\boldsymbol{n}\,\mathcal{diag}(\mathbf{1} - \boldsymbol{x})\,\boldsymbol{y} - \gamma_h\boldsymbol{x} + \mathcal{D}\boldsymbol{x}, \tag{8.2.79a}$$

$$\pi\boldsymbol{y}' = \beta_v\mathcal{diag}(\pi\mathbf{1} - \pi\boldsymbol{y})\pi\boldsymbol{x} - \mu_v\pi\boldsymbol{y}, \tag{8.2.79b}$$

where $\mathcal{D} = (d_{ij})_{1\leq i,j\leq d}$, with $d_{ij} = \frac{N^*_{h,j}m_{ij}}{N^*_{h,i}}$ for $i \neq j$ and $d_{ii} = m_{ii} = -\sum\limits_{\substack{j\neq i\\ j=1}}^{d} m_{ji}$,

or, in other words,

$$\mathcal{D} = (\mathcal{diag}\boldsymbol{N}^*)^{-1}\mathcal{M}\,\mathcal{diag}\boldsymbol{N}^*. \tag{8.2.80}$$

Hence

$$\mathcal{D}\mathbf{1} = (\mathcal{diag}\boldsymbol{N}^*)^{-1}\mathcal{M}\,\mathcal{diag}\boldsymbol{N}^*\,\mathbf{1} = (\mathcal{diag}\,\boldsymbol{e})^{-1}\mathcal{M}\,\mathcal{diag}\,\boldsymbol{e}\,\mathbf{1} = (\mathcal{diag}\,\boldsymbol{e})^{-1}\mathcal{M}\boldsymbol{e} = 0,$$

that is,

$$d_{ii} + \sum_{\substack{j\neq i\\ j=1}}^{d} d_{ij} = 0, \quad i = 1, \ldots, d. \tag{8.2.81}$$

We have

Proposition 8.43 *The order interval* $\mathbb{I}_1 = \{(\boldsymbol{x}, \boldsymbol{y}) : (\boldsymbol{x}, \boldsymbol{y}) \in [0, \mathbf{1}]\} = [0, 1]^{d+p}$ *is invariant for* (8.2.79).

Proof The nonnegativity follows from Theorem B.21. The fact that $\boldsymbol{y}(t) \leq 1$ is proved as $\boldsymbol{I}_v \leq \boldsymbol{N}_v$ in Proposition 8.42. Finally, consider $x_i = 1$ and $0 \leq x_j \leq 1$, $j \neq i$. Then, by (8.2.81), y and hence, no trajectory crosses any face of $\mathbb{I}_1$. □

We will make the notation still more compact by introducing $\boldsymbol{X} := (\boldsymbol{x}, \boldsymbol{y}) \in \mathbb{R}^{d+p}$ and block matrices

$$\mathcal{B} = \begin{pmatrix} 0 & 0 & \mathcal{I}_p \\ 0 & \mathcal{I}_{d-p} & 0 \\ \mathcal{I}_p & 0 & 0 \end{pmatrix},$$

$$\mathcal{V} = \begin{pmatrix} \mathcal{D} - \gamma_h \mathcal{I}_d & 0 \\ 0 & -\mu_v \mathcal{I}_p \end{pmatrix} = \begin{pmatrix} \mathcal{D}_{11} - \gamma_h \mathcal{I}_p & \mathcal{D}_{12} & 0 \\ \mathcal{D}_{21} & \mathcal{D}_{22} - \gamma_h \mathcal{I}_{d-p} & 0 \\ 0 & 0 & -\mu_v \mathcal{I}_p \end{pmatrix},$$

$$\mathcal{L} = \begin{pmatrix} \beta_h \mathcal{diag}\, \pi \boldsymbol{n} & 0 & 0 \\ 0 & 0 & 0 \\ 0 & 0 & \beta_v \mathcal{I}_p \end{pmatrix},$$

where, for a matrix $\mathcal{A} = (a_{ij})_{1 \leq i, j \leq d}$, we denoted

$$\mathcal{A} := \begin{pmatrix} \mathcal{A}_{11} & \mathcal{A}_{12} \\ \mathcal{A}_{21} & \mathcal{A}_{22} \end{pmatrix}, \tag{8.2.82}$$

so $\mathcal{A}_{11} = (a_{ij})_{1 \leq i, j \leq p}$, $\mathcal{A}_{12} = (a_{ij})_{1 \leq i \leq p, p+1 \leq j \leq d}$, $\mathcal{A}_{22} = (a_{ij})_{p+1 \leq i, j \leq d}$, $\mathcal{A}_{21} = (a_{ij})_{p+1 \leq i \leq d, 1 \leq j \leq p}$, and $\mathcal{I}_k$ is the $k \times k$ unit matrix. We observe that $\mathcal{B}$ and $\mathcal{L}$ are nonnegative matrices. Further, $\mathcal{F}$ is block-diagonal, $s(-\mu_v \mathcal{I}_p) = -\mu_v < 0$, and, since $s(\mathcal{D}) = 0$, we see that $s(\mathcal{V}) < 0$. Hence, $\mathcal{V}$ is a stable Metzler matrix. Using this notation, (8.2.79) can be written as

$$\boldsymbol{X}' = \mathcal{L} \mathcal{diag}(\mathbf{1} - \boldsymbol{X}) \mathcal{B} \boldsymbol{X} + \mathcal{V} \boldsymbol{X} = (\mathcal{L} \mathcal{diag}(\mathbf{1} - \boldsymbol{X}) \mathcal{B} + \mathcal{V}) \boldsymbol{X}. \tag{8.2.83}$$

We have

Proposition 8.44 *System* (8.2.79) *is cooperative in* $\mathbb{I}_1$ *and strongly monotone in its interior.*

Proof The Jacobi matrix of (8.2.83) at $\boldsymbol{X}$ is given by

$$\mathcal{J}(\boldsymbol{X}) = \mathcal{L} \mathcal{diag}(\mathbf{1} - \boldsymbol{X}) \mathcal{B} + \mathcal{V} - \mathcal{L} \mathcal{diag}(\mathcal{B} \boldsymbol{X}). \tag{8.2.84}$$

Since $\mathcal{L}\mathcal{diag}(\mathcal{B}\boldsymbol{X})$ is diagonal, $\mathcal{J}(\boldsymbol{X})$ is a Metzler matrix for $0 \leq \boldsymbol{X} \leq \mathbf{1}$ and hence (8.2.83) is a cooperative system on $\mathbb{I}_1$. To prove the strong monotonicity of (8.2.83), it is enough to show that $\mathcal{J}(\boldsymbol{X})$ is irreducible on $(0, 1)^{d+p}$. For this, we investigate the structure of $\mathcal{J}(\boldsymbol{X})$ in detail. We have

$$\mathcal{J}(\boldsymbol{X})$$
$$= \begin{pmatrix} \mathcal{D}_{11} - \gamma_h \mathcal{I}_p - \beta_h \mathcal{diag}\, \pi \boldsymbol{n}\, \mathcal{diag}\, \pi \boldsymbol{y} & \mathcal{D}_{12} & \beta_h \mathcal{diag}\, \pi \boldsymbol{n}\, \mathcal{diag}\, \pi (\mathbf{1} - \boldsymbol{x}) \\ \mathcal{D}_{21} & \mathcal{D}_{22} - \gamma_h \mathcal{I}_{d-p} & 0 \\ \beta_v \mathcal{diag}\, \pi (\mathbf{1} - \boldsymbol{y}) & 0 & -\mu_v \mathcal{I}_p - \beta_v \mathcal{diag}\, \pi \boldsymbol{x} \end{pmatrix}.$$

We know that the diagonal does not impact irreducibility, and the irreducibility of $\mathcal{D}$ implies that the states $x_i, i = 1, \ldots, d$, are interconnected, that is, we can pass from any state x_i to any other state x_j. Now, if $\boldsymbol{y} \ll \mathbf{1}$, then the entries $(d+1, 1), \ldots, (d+p, p)$ in the bottom left block of $\mathcal{J}(\boldsymbol{X})$ are positive. Thus, there is a connection from $x_i, i = 1, \ldots, p$, to y_i. But, since any x_i is connected to any x_j, there is a path from any $x_i, i = 1, \ldots, d$, to any $y_j, j = 1, \ldots, p$. Similarly, if $\boldsymbol{x} \ll 1$, then the entries $(1, d+1), \ldots, (p, d+p)$ in the top right block of $\mathcal{J}(\boldsymbol{X})$ are positive, and hence any $y_i, i = 1, \ldots, p$, is connected $x_i, i = 1, \ldots, p$, and thus, by the irreducibility of $\mathcal{D}$, to any other $x_j, j = 1, \ldots, d$. □

Remark 8.45 We emphasize that we proved irreducibility of $\mathcal{J}(\boldsymbol{X})$ for $\boldsymbol{X} \in (0, 1)^{d+p}$ and at $\boldsymbol{X} = 0$. We note that if the trajectory starts at the boundary of $(0, 1)^{d+p}$, away from $\boldsymbol{X} = 0$, then, by Proposition 8.43, it immediately enters $(0, 1)^{d+p}$. Thus, its long-term behaviour is determined by the properties of the field of (8.2.78) inside $\mathbb{I}_1$.

Proposition 8.46 *The origin is the disease-free equilibrium of* (8.2.79) *and*

$$R_0^2 = \frac{\beta_h \beta_v}{\mu_v} r(-\mathcal{diag}(\pi n)(\mathcal{D} - \gamma_h \mathcal{I})_{11}^{-1}). \tag{8.2.85}$$

Proof The Jacobi matrix $\mathcal{J}(\boldsymbol{X})$ of (8.2.79), given by (8.2.84), evaluated at DFE $= 0$ is given by

$$\mathcal{J}(0) = \mathcal{L}\mathcal{B} + \mathcal{V} = \begin{pmatrix} 0 & 0 & \beta_h \mathcal{diag}\, \pi \boldsymbol{n} \\ 0 & 0 & 0 \\ \beta_v \mathcal{I}_p & 0 & 0 \end{pmatrix}$$
$$+ \begin{pmatrix} \mathcal{D}_{11} - \gamma_h \mathcal{I}_p & \mathcal{D}_{12} & 0 \\ \mathcal{D}_{21} & \mathcal{D}_{22} - \gamma_h \mathcal{I}_{d-p} & 0 \\ 0 & 0 & -\mu_v \mathcal{I}_p \end{pmatrix}. \tag{8.2.86}$$

By the definitions of the matrices, $\mathcal{L}\mathcal{B}$ is non-negative, and $\mathcal{V}$ is Metzler stable, so (8.2.86) is a regular splitting, see Definition 8.3. We see that $\mathcal{L}\mathcal{B}$ represents secondary infections and $\mathcal{V}$ are transfers, hence $R_0 = r(-\mathcal{L}\mathcal{B}\mathcal{V}^{-1})$. Due to the block structure of $\mathcal{V}$, we have,

$$\mathcal{V}^{-1} = \begin{pmatrix} (\mathcal{D} - \gamma_h \mathcal{I}_d)^{-1}_{11} & (\mathcal{D} - \gamma_h \mathcal{I}_d)^{-1}_{12} & 0 \\ (\mathcal{D} - \gamma_h \mathcal{I}_d)^{-1}_{21} & (\mathcal{D} - \gamma_h \mathcal{I}_d)^{-1}_{22} & 0 \\ 0 & 0 & -\frac{1}{\mu_v}\mathcal{I}_p \end{pmatrix}.$$

Thus,

$$-\mathcal{L}\mathcal{B}\mathcal{V}^{-1} = \begin{pmatrix} 0 & 0 & \frac{\beta_h}{\mu_v}\mathit{diag}\,\pi \boldsymbol{n} \\ 0 & 0 & 0 \\ -\beta_v(\mathcal{D} - \gamma_h \mathcal{I}_d)^{-1}_{11} & 0 & 0 \end{pmatrix},$$

and its characteristic equation is

$$\begin{aligned} 0 = det(\lambda \mathcal{I}_{d+p} + \mathcal{L}\mathcal{B}\mathcal{V}^{-1}) &= \begin{pmatrix} \lambda \mathcal{I}_p & 0 & \frac{\beta_h}{\mu_v}\mathit{diag}\,\pi \boldsymbol{n} \\ 0 & \lambda \mathcal{I}_{d-p} & 0 \\ -\beta_v(\mathcal{D} - \gamma_h \mathcal{I}_d)^{-1}_{11} & 0 & \lambda \mathcal{I}_p \end{pmatrix} \\ &= \lambda^{d-p} det \begin{pmatrix} \lambda \mathcal{I}_p & \frac{\beta_h}{\mu_v}\mathit{diag}\,\pi \boldsymbol{n} \\ -\beta_v(\mathcal{D} - \gamma_h \mathcal{I}_d)^{-1}_{11} & \lambda \mathcal{I}_p \end{pmatrix}, \end{aligned}$$

so $0 \in \sigma(-\mathcal{L}\mathcal{B}\mathcal{V}^{-1})$ is an eigenvalue of multiplicity $d-p$ and the other eigenvalues are obtained by looking at the eigenvalue problem

$$\lambda^2 \boldsymbol{v} + \frac{\beta_h \beta_v}{\mu_v}\mathit{diag}\,\pi \boldsymbol{n}(\mathcal{D} - \gamma_h \mathcal{I}_d)^{-1}_{11} \boldsymbol{v} = 0,$$

equivalent to finding zeroes of the remaining determinant. Hence,

$$R_0^2 = r\left(-\frac{\beta_h \beta_v}{\mu_v}\mathit{diag}\,\pi \boldsymbol{n}(\mathcal{D} - \gamma_h \mathcal{I}_d)^{-1}_{11}\right).$$

□

Remark 8.47 In the scalar case ($\mathcal{D} = 0$, $\boldsymbol{n} = \frac{N_v}{N_h} = m$), we have

$$R_0^2 = m\frac{\beta_h \beta_v}{\mu_v \gamma_h},$$

as in (6.2.61).

Remark 8.48 Returning to the original notation, using (8.2.80) and the commutativity of diagonal matrices, we have

$$
\begin{aligned}
(\mathcal{D} - \gamma_h \mathcal{I}_d)_{11}^{-1} &= \left((\mathcal{diag}\mathbf{N}^*)^{-1}\mathcal{M}\,\mathcal{diag}\mathbf{N}^* - \gamma_h \mathcal{I}_d\right)_{11}^{-1} \\
&= \left((\mathcal{diag}\mathbf{N}^*)^{-1}\mathcal{M}\,\mathcal{diag}\mathbf{N}^* - \gamma_h (\mathcal{diag}\mathbf{N}^*)^{-1}\mathcal{diag}\mathbf{N}^*\right)_{11}^{-1} \\
&= \left(\mathcal{diag}\,\pi\,\mathbf{N}^*)^{-1}(\mathcal{M} - \gamma_h \mathcal{I}_d\right)_{11}^{-1} \mathcal{diag}\,\pi\,\mathbf{N}^*,
\end{aligned}
$$

so

$$
R_0^2 = r\left(-\frac{\beta_h\beta_v}{\mu_v}\mathcal{diag}\,\pi\,\mathbf{N}_v \left(\mathcal{diag}\,\pi\,\mathbf{N}^*)^{-2}(\mathcal{M} - \gamma_h \mathcal{I}_d\right)_{11}^{-1} \mathcal{diag}\,\pi\,\mathbf{N}^*\right). \tag{8.2.87}
$$

Now, using Theorem 8.7, we have

Corollary 8.49 *Let* R_0 *be defined by* (8.2.87). *The disease-free equilibrium of* (8.2.78) *is locally asymptotically stable if* $R_0 < 1$ *and unstable if* $R_0 > 1$.

Next, we pass to the global stability of equilibria.

Theorem 8.50 *Consider system* (8.2.79). *Then*

1. *If* $R_0 \leq 1$, *then the disease-free equilibrium is globally asymptotically stable on* $\mathbb{I}_1 = [0, 1]^{d+p}$.
2. *If* $R_0 > 1$, *then there exists a unique positive equilibrium* $0 \ll (\mathbf{x}^*, \mathbf{y}^*) \ll 1$, *which is globally asymptotically stable on* $[0, 1]^{d+p} \setminus \{0\}$.

Proof To prove 1., recall that the Jacobi matrix of (8.2.79) at DFE is given by

$$
\mathcal{J}(0) = \mathcal{L}\mathcal{B} + \mathcal{V}, \tag{8.2.88}
$$

and assume $R_0^2 = r(-\mathcal{L}\mathcal{B}\mathcal{V}^{-1}) \leq 1$. As we noted in the proof of Proposition 8.46, $\mathcal{L}\mathcal{B} \geq 0$ and $\mathcal{V}$ is Metzler stable, so, by Corollary 8.6, $s(\mathcal{J}(0)) \leq 0$ if and only if $r(-\mathcal{L}\mathcal{B}\mathcal{V}^{-1}) \leq 1$. By Remark 8.45, $\mathcal{J}(0)$ is an irreducible Metzler matrix; hence, there exists a left eigenvector $\mathbf{e} \gg 0$ satisfying

$$
\mathbf{e}\mathcal{J}(0) = s(\mathcal{J}(0))\mathbf{e} \leq 0.
$$

Let us consider the candidate Lyapunov function

$$
V(\mathbf{X}) = \langle \mathbf{e}, \mathbf{X} \rangle, \quad \mathbf{X} \in \overline{\mathbb{R}}_+^{d+p}. \tag{8.2.89}
$$

Since $\boldsymbol{e} \gg 0$, we see $V(\boldsymbol{X}) \geq 0$ with $V(\boldsymbol{X}) = 0$ only at $\boldsymbol{X} = 0$. Further,

$$V'(\boldsymbol{X}) = \langle \boldsymbol{e}, (\mathcal{L}\mathcal{d}iag(\mathbf{1} - \boldsymbol{X})\mathcal{B} + \mathcal{V})\boldsymbol{X} \rangle = \langle \boldsymbol{e}, \mathcal{J}(0)\boldsymbol{X} \rangle - \langle \boldsymbol{e}, \mathcal{L}\mathcal{d}iag\boldsymbol{X}\mathcal{B}\boldsymbol{X} \rangle.$$

If $R_0 < 1$, then $s(\mathcal{J}(0)) < 0$ and

$$\langle \boldsymbol{e}, \mathcal{J}(0)\boldsymbol{X} \rangle = s(\mathcal{J}(0))\langle \boldsymbol{e}, \boldsymbol{X} \rangle < 0$$

and V is a strict Lyapunov function. If $R_0 = 1$, then $s(\mathcal{J}(0)) = 0$ and

$$\langle \boldsymbol{e}, \mathcal{J}(0)\boldsymbol{X} \rangle = s(\mathcal{J}(0))\langle \boldsymbol{e}, \boldsymbol{X} \rangle = 0,$$

so we only have

$$V'(\boldsymbol{X}) = -\langle \boldsymbol{e}, \mathcal{L}\mathcal{d}iag\boldsymbol{X}\mathcal{B}\boldsymbol{X} \rangle;$$

$V'(\boldsymbol{X})$ vanishes if $\mathcal{L}\mathcal{d}iag\boldsymbol{X}\mathcal{B}\boldsymbol{X}$ is orthogonal to $\boldsymbol{e}$. Now, similarly to (8.2.82), let us write $\boldsymbol{X} = (\boldsymbol{x}_1, \boldsymbol{x}_2, \boldsymbol{y})$, with $\boldsymbol{x}_1 = (x_1, \ldots, x_p)$, $\boldsymbol{x}_2 = (x_{p+1}, \ldots, x_d)$, and, for simplicity, we wrote $\boldsymbol{y} = \pi\,\boldsymbol{y}$. Further, for vectors $\boldsymbol{a}$ and $\boldsymbol{b}$ of the same dimension q, $\boldsymbol{ab} := (a_1b_1, \ldots, a_qb_q)$. Then

$$\begin{aligned}
\mathcal{L}\mathcal{d}iag\boldsymbol{X}\mathcal{B}\boldsymbol{X} &= \begin{pmatrix} \beta_h \mathcal{d}iag\,\pi\boldsymbol{n} & 0 & 0 \\ 0 & 0 & 0 \\ 0 & 0 & \beta_v\mathcal{I}_p \end{pmatrix} \mathcal{d}iag\boldsymbol{X} \begin{pmatrix} 0 & 0 & \mathcal{I}_p \\ 0 & \mathcal{I}_{d-p} & 0 \\ \mathcal{I}_p & 0 & 0 \end{pmatrix} \boldsymbol{X} \\
&= \begin{pmatrix} \beta_h \mathcal{d}iag\,\pi\boldsymbol{n} & 0 & 0 \\ 0 & 0 & 0 \\ 0 & 0 & \beta_v\mathcal{I}_p \end{pmatrix} \begin{pmatrix} \boldsymbol{x}_1\boldsymbol{y} \\ \boldsymbol{x}_2\boldsymbol{x}_2 \\ \boldsymbol{x}_1\boldsymbol{y} \end{pmatrix} \\
&= \begin{pmatrix} \beta_h \pi\boldsymbol{n}\boldsymbol{x}_1\boldsymbol{y} \\ 0 \\ \beta_v \boldsymbol{x}_1\boldsymbol{y} \end{pmatrix},
\end{aligned}$$

and, since $\boldsymbol{e} \gg 0$,

$$\begin{aligned}
M &= \{\boldsymbol{X} \in \overline{\mathbb{R}}_+^{d+p} : V'(\boldsymbol{X}) = 0\} \\
&= \{(\boldsymbol{x}_1, \boldsymbol{x}_2, \boldsymbol{y}) \in \overline{\mathbb{R}}_+^{d+p} : x_i y_i = 0, \text{ for } i = 1, \ldots, p\}.
\end{aligned}$$

By the LaSalle principle, we must show that the largest invariant subset of M is $\{0\}$. Let us assume $x_i = 0$ for $i \in I_1 = \{i_1, \ldots i_r\}$ and $y_j = 0$ for $j \in I_2 = \{j_1, \ldots j_s\}$, $I_1 \cup I_2 = \{1, \ldots, p\}$.

Let us consider $x_i = 0$ for some $i \in I_1$. By (8.2.79a), the ith component of the vector field on the hyperplane $x_i = 0$ is given by

$$\beta_h n_i y_i + \sum_{j=1, j\neq i}^{d} d_{ij} x_j,$$

so the only way the trajectory can stay on this hyperplane is if $y_i = 0$ and all x_j for which $d_{ij} > 0$ satisfy $x_j = 0$. Note that this includes also x_js with indices j satisfying $p+1 \leq j \leq d$. Running this argument for all $i \in I_1$, we get $y_i = 0$ for $i \in I_1$. Similarly, if $y_j = 0$ for $j \in I_2$, using (8.2.79b), we see that the vector field on this hyperplane is given by

$$\beta_v x_j;$$

hence, the hyperplane is invariant only if $x_j = 0$, which gives $x_j = 0$ for $j \in I_2$.

Finally, since $\mathcal{D}$ is irreducible, there is no invariant subspace in $M_0 := (0, \ldots, 0, x_{p+1}, \ldots x_d, 0, \ldots 0)$. Indeed, in M_0, (8.2.79) takes the form

$$\begin{aligned} \boldsymbol{x}_1' &= \mathcal{D}_{12}\boldsymbol{x}_2, \\ \boldsymbol{x}_2' &= \mathcal{D}_{22}\boldsymbol{x}_2. \end{aligned}$$

and $\mathcal{D}_{12} \neq 0$ by irreducibility. Consider $x_j > 0$ for some $j \in J_2 := \{p+1, \ldots, p\}$. If there is $d_{ij} > 0$ for some $i \in J_1 := \{1, \ldots, p\}$, then immediately $x_i' > 0$ and hence M_0 is not invariant. Assume then that $d_{ij} = 0$ for all $i \in J_1$. By irreducibility of $\mathcal{D}$, there is a path from $j \in J_2$ to $i \in J_1$. In other words, see (5.4.115), we must have

$$d_{ii_i} \cdot \ldots \cdot d_{rs} \cdot d_{sj_l} \cdot \ldots \cdot d_{j_1 j} > 0,$$

where $i, i_1, \ldots r \in J_1$ and $s, j_l, \ldots j_1, j \in J_2$. By (5.4.114), we see that for $r = \max_{j\in J_2} |d_{jj}|$, we find that $\tilde{\mathcal{D}}_{22} = r\mathcal{I} + \mathcal{D}_{22} \geq 0$ and by (5.3.59)

$$e^{t\mathcal{D}_{22}} = e^{-rt} e^{t\tilde{\mathcal{D}}_{22}}.$$

Using the fact that $\tilde{\mathcal{D}}_{22} \geq 0$, (5.4.115) and (5.3.59), we find

$$x_s(t) = \left[e^{t\tilde{\mathcal{D}}_{22}} \begin{pmatrix} 0 \\ \vdots \\ x_j \\ \vdots \\ 0 \end{pmatrix} \right]_s \geq e^{-rt} \frac{d_{sj_l} \cdot \ldots \cdot d_{j_1 j} x_j}{(l+1)!} t^{l+1} > 0.$$

Note that the diagonal entries do not affect irreducibility; hence, the relevant entries in $\mathcal{D}_{22}$ and $\tilde{\mathcal{D}}_{22}$ are the same. This contradicts the earlier result that for the invariance of the hyperplane $x_r = 0$, we need $x_j = 0$ for all j for which $d_{rj} > 0$ so, in particular, $x_s = 0$.

Finally, let us consider $R_0 > 1$. Then, DFE is unstable. Take $\boldsymbol{X} = (\boldsymbol{x}_1, \boldsymbol{x}_2, \boldsymbol{y}) \gg 0$ and $\lambda \in (0, 1)$. Since $\mathcal{B}$ and $\mathcal{V}$ are linear and we have

$$\begin{aligned}
&\mathcal{L}\mathcal{d}iag(\mathbf{1} - \lambda\boldsymbol{X})\mathcal{B}(\lambda\boldsymbol{X})\\
&= \lambda \begin{pmatrix} \beta_h \mathcal{d}iag\, \pi \boldsymbol{n} & 0 & 0 \\ 0 & 0 & 0 \\ 0 & 0 & \beta_v \mathcal{I}_d \end{pmatrix} \begin{pmatrix} \mathbf{1} - \lambda\boldsymbol{x}_1 & 0 & 0 \\ 0 & \mathbf{1} - \lambda\boldsymbol{x}_2 & \\ 0 & 0 & \mathbf{1} - \lambda\boldsymbol{y} \end{pmatrix} \begin{pmatrix} \boldsymbol{y} \\ \boldsymbol{x}_2 \\ \boldsymbol{x}_1 \end{pmatrix}\\
&= \lambda \begin{pmatrix} \beta_h \mathcal{d}iag\, \pi \boldsymbol{n}(\mathbf{1} - \lambda\boldsymbol{x}_1)\boldsymbol{y} & 0 & 0 \\ 0 & 0 & 0 \\ 0 & 0 & \beta_v(\mathbf{1} - \lambda\boldsymbol{y})\boldsymbol{x}_1 \end{pmatrix}\\
&> \lambda \begin{pmatrix} \beta_h \mathcal{d}iag\, \pi \boldsymbol{n}(\mathbf{1} - \boldsymbol{x}_1)\boldsymbol{y} & 0 & 0 \\ 0 & 0 & 0 \\ 0 & 0 & \beta_v(\mathbf{1} - \boldsymbol{y})\boldsymbol{x} \end{pmatrix},
\end{aligned}$$

as long as $\lambda \in (0, 1)$ and $0 \ll \boldsymbol{X} \leq \mathbf{1}$. Hence,

$$\mathcal{L}\mathcal{d}iag(\mathbf{1} - \lambda\boldsymbol{X})\mathcal{B}(\lambda\boldsymbol{X}) + \mathcal{V}(\lambda\boldsymbol{X}) > \lambda(\mathcal{L}\mathcal{d}iag(\mathbf{1} - \lambda\boldsymbol{X})\mathcal{B}\boldsymbol{X} + \mathcal{V}\boldsymbol{X}),$$

and the system is strictly sublinear. Since, by Proposition 8.44, the flow is strongly monotone in the interior of $\mathcal{I}_1$, Proposition 8.37 implies that the flow is strongly sublinear. We will apply Theorem 8.40. It is clear that it can be used in the invariant box $\mathbb{I}_1 = [0, 1]^{d+p}$. We observe, however, that Theorem 8.40 requires strong monotonicity to hold on the $\mathbb{I}_1$, while Proposition 8.44 gives it only on $(0, 1)^{d+p} \cup \{0\}$. Returning to the proof of Theorem 8.40, we see that the strong monotonicity on the faces of $\mathbb{I}_1$ contained in the faces of the non-negative orthant is used only to eliminate possible boundary equilibria. Here, using Remark 8.45, we see that the flow direction on these faces rules out such equilibria. Since the equilibria on other faces are also impossible, see Proposition 8.43, point 2. of the present theorem has been proved. □

Using the Vidyasagar theorem, Theorem 6.77 with Remark 6.78, we have

Corollary 8.51 *Consider* (8.2.69) *with a given total human population* $N > 0$ *and let* $\boldsymbol{N}^*$ *be the equilibrium of* (8.2.69a) *in* W_N. *Then*

1. *If* $R_0 \leq 1$, *then DFE* = $(\boldsymbol{N}^*, 0, 0)$ *is globally asymptotically stable in* $\mathbb{A}$.
2. *If* $R_0 > 1$, *then there exist* $0 \ll \boldsymbol{I}_h^* \ll N^*, 0 \ll \boldsymbol{I}_v^* \ll N_v$ *such that* $(\boldsymbol{N}^*, \boldsymbol{I}_h, \boldsymbol{I}_v)$ *is globally asymptotically stable in* $\mathbb{A} \setminus \{\boldsymbol{I}_h = 0, \boldsymbol{I}_v = 0\}$

where $\mathbb{A}$ *is defined in* (8.2.73) *(see also* (8.2.77).

Proof The proof requires a small modification of the proof of the Vidyasagar theorem. Let us fix N and denote by $\boldsymbol{E}$ either DFE or the positive endemic equilibrium of (8.2.78). First, we observe that we have local asymptotic stability of either equilibrium $(\boldsymbol{N}^*, \boldsymbol{E})$ of (8.2.69) in $\mathbb{A}$. Consider any initial condition $\mathring{\boldsymbol{X}} := (\mathring{\boldsymbol{N}}_h, \mathring{\boldsymbol{I}}_h, \mathring{\boldsymbol{I}}_v)$ satisfying (8.2.70)–(8.2.72) for some fixed mosquitoes populations $\boldsymbol{N}_v$. Then, by (8.2.75), the ω-limit set of the trajectory $\Gamma_{\mathring{\boldsymbol{X}}}$ satisfies

$$\omega(\Gamma_{\mathring{\boldsymbol{X}}}) \subset \{\boldsymbol{N}^*\} \times \{(\boldsymbol{I}_h, \boldsymbol{I}_v) : 0 \le \boldsymbol{I}_h \le \boldsymbol{N}^*, 0 \le \boldsymbol{I}_v \le \boldsymbol{N}_v\},$$

which is invariant, by Proposition 8.43 written in terms of the original variables.

Now, from Theorem 8.50, the respective equilibrium $\boldsymbol{E}$ of (8.2.78) is globally asymptotically stable in $\{(\boldsymbol{I}_h, \boldsymbol{I}_v) : 0 \le \boldsymbol{I}_h \le \boldsymbol{N}^*, 0 \le \boldsymbol{I}_v \le \boldsymbol{N}_v\}$ and hence $\boldsymbol{E} \in \omega(\Gamma_{\mathring{\boldsymbol{X}}})$. Its local stability in $\mathbb{A}$ implies $\omega(\Gamma_{\mathring{\boldsymbol{X}}}) = \boldsymbol{E}$. □

Exercise 8.2.10 (Population Dynamics of Anopheles Mosquito, [7]) We consider a mosquito population, which we divide into two classes: aquatic (eggs, larvae and pupae) and terrestrial (adult females and males). For simplicity, here we group aquatic stages into one class whose size will be denoted by A, though a model involving all aquatic stages can be considered, [82]. In the adult stage, we consider the populations: Y of females not laying eggs, F of fertilized females, and M of males. The life cycle is visualised in Fig. 8.8, which leads to the system

$$\begin{aligned} A' &= \phi F - (\gamma + \mu_1 + \mu_2 A)A, \\ Y' &= r\gamma A - (\beta + \mu_Y)Y, \\ F' &= \beta Y - \mu_F F, \\ M' &= (1 - r)\gamma A - \mu_M M. \end{aligned} \tag{8.2.90}$$

Here, ϕ is the rate of oviposition per adult female (that is, the rate at which an adult female lays eggs), γ is the hatching rate, μ_1 and $\mu_2 A$ are, respectively, the natural and density-dependent death rates in the aquatic stage, r is the fraction of young females in the emerging population, β is the mating rate and μ_Y, μ_F and μ_M are

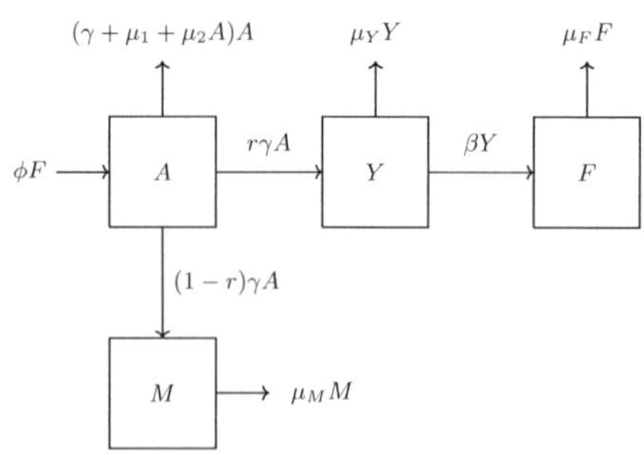

Fig. 8.8 Life cycle of a mosquito

the death rates among, respectively, young females, fertilized females and males. All parameters are assumed positive with $0 < r < 1$.

1. Prove that the system (8.2.90) is well posed in $\overline{\mathbb{R}}^4_+$.
2. Show that if $\lim_{t\to\infty} A(t) = A^*$, then $\lim_{t\to\infty} M(t) = \dfrac{(1-r)\gamma A^*}{\mu_M}$. Conclude that, since the first three equations in (8.2.90) do not involve M, the stability analysis of (8.2.90) can be reduced to that of

$$\begin{aligned} A' &= f_1(A, Y, F) := \phi F - (\gamma + \mu_1 + \mu_2 A)A, \\ Y' &= f_2(A, Y, F) := r\gamma A - (\beta + \mu_Y)Y, \\ F' &= f_3(A, Y, F) := \beta Y - \mu_F F. \end{aligned} \tag{8.2.91}$$

3. Show that the Jacobi matrix of (8.2.91) is given by

$$\mathcal{J}((A, Y, F)) = \begin{pmatrix} -(\gamma + \mu_1 + \mu_2 A) & 0 & \phi \\ r\gamma & -(\beta + \mu_Y) & 0 \\ 0 & \beta & -\mu_F \end{pmatrix}. \tag{8.2.92}$$

4. Check that $(0, 0, 0)$ is an equilibrium and find a regular splitting of $\mathcal{J}((0, 0, 0))$. Hence, show that $(0, 0, 0)$ is locally asymptotically stable if

$$R_0 := \sqrt[3]{\frac{\phi r\gamma\beta}{\mu_F(\gamma + \mu_1)(\beta + \mu_Y)}} < 1 \tag{8.2.93}$$

and unstable if $R_0 > 1$.
5. Show that $f = (f_1, f_2, f_3)$ is cooperative and irreducible; hence it generates a strongly monotone flow on $\overline{\mathbb{R}}^3_+$. Then, show that f is strictly sublinear; hence the flow is strongly sublinear. Use Theorem 8.40 to show that

   - DFE $= (0, 0, 0)$ is globally asymptotically stable in $\overline{\mathbb{R}}^3_+$ if $R_0 < 1$.
   - If $R_0 > 1$, there is a strictly positive equilibrium $(A^*, Y^*, F^*) \gg 0$ which is globally asymptotically stable in $\overline{\mathbb{R}}^3_+ \setminus \{0\}$.

6. Show that the positive equilibrium is given by

$$\begin{aligned} A^* &= \frac{1}{\mu_2}\left(\frac{r\gamma\beta\phi}{(\beta + \mu_Y)\mu_F} - \gamma - \mu_1\right), \\ Y^* &= \frac{r\gamma}{\mu_2(\beta + \mu_Y)}\left(\frac{r\gamma\beta\phi}{(\beta + \mu_Y)\mu_F} - \gamma - \mu_1\right), \\ F^* &= \frac{r\gamma\beta}{\mu_2\mu_F(\beta + \mu_Y)}\left(\frac{r\gamma\beta\phi}{(\beta + \mu_Y)\mu_F} - \gamma - \mu_1\right). \end{aligned}$$

By expressing (A^*, Y^*, F^*) in terms of R_0, show that a positive equilibrium exists only if $R_0 > 1$.

7. Show that for any fixed F, $Y = \frac{\mu_F}{\beta} F$ and $A = \frac{\mu_F(\beta + \mu_Y)}{r\beta\gamma} F$, we have

$$\begin{aligned}
f_1(A, Y, F) &= F\frac{\mu_F(\beta+\mu_Y)(\gamma+\mu_1)}{r\beta\gamma}\left(R_0^3 - 1 - \frac{\mu_2\mu_F(\beta+\mu_Y)}{r\beta\gamma(\gamma+\mu_1)}F\right),\\
f_2(A, Y, F) &= 0,\\
f_3(A, Y, F) &= 0.
\end{aligned}$$

Hence, if $R_0 = 1$, $f\left(\frac{\mu_F(\beta+\mu_Y)}{r\beta\gamma}F, \frac{\mu_F}{\beta}F, F\right) < 0$ for any $F > 0$. Since $f(0) \geq 0$, and DFE is the only equilibrium in $\overline{\mathbb{R}}^3_+$ if $R_0 = 1$, use Theorem 8.30 to conclude that DFE is globally asymptotically stable there also in this case.

8. Can this argument be used to show global asymptotic stability of DFE if $R_0 < 1$?

Exercise 8.2.11 (Multi-Group Infection Model, [63, 82]) Let us consider a population divided into r groups, $N_1, \ldots, N_r$, differing from each other by, say, a specific individual's attribute such as age group, immunity level, etc. We assume that each group has susceptibles S_j and infectives I_j, an infective from each group I_j can infect a susceptible in any other group S_i, and the infected remains in its group (the age of an individual does not change upon infection). A specific example of such a situation is given in [63]. If we consider a SIRS version of such a scenario with a short recovery so that the recovered individual remains in the same state through the illness and recovery, we arrive at the following system

$$\begin{aligned}
S_i' &= \Lambda_i - S_i\sum_{j=1}^{r}\lambda_{ij}\frac{I_j}{N_j} - \mu S_i + \gamma_i R_i, \quad i = 1, \ldots, r,\\
I_i' &= S_i\sum_{j=1}^{r}\lambda_{ij}\frac{I_j}{N_j} - \mu I_i - \nu_i I_i, \quad i = 1, \ldots, r, \qquad (8.2.94)\\
R_i' &= \nu_i I_i - \mu R_i - \gamma_i R_i, \quad i = 1, \ldots, r,
\end{aligned}$$

where $N_i = S_i + I_i + R_i$ and, as before, positive parameters $\Lambda_i, \mu, \nu_i, \gamma_i$, $i = 1, \ldots, r$, are, respectively, the total birth rate in the ith subpopulation, the death rate (the same in each subpopulation), per capita recovery and immunity loss rates. The non-negative matrix $\mathcal{L} = (\lambda_{ij})_{1\leq i,j\leq r}$ gives the forces of infection between the ith class of susceptibles and the jth class of susceptibles.

1. Show that (8.2.94) can be written as

$$N_i' = \Lambda_i - \mu N_i, \quad i = 1, \ldots, r,$$
$$I_i' = (N_i - I_i)\sum_{j=1}^{r}\lambda_{ij}\frac{I_j}{N_j} - \mu I_i - \nu_i I_i, \quad i = 1, \ldots, r, \tag{8.2.95}$$
$$R_i' = \nu_i I_i - \mu R_i - \gamma_i R_i, \quad i = 1, \ldots, r,$$

2. Show that if $\lim_{t\to\infty} I_i(t) = I_i^*$, then $\lim_{t\to\infty} R_i(t) = \frac{\nu_i I_i^*}{\mu + \gamma_1}$. Conclude that, since the equations for N_i do not involve R_i, the equations for R_i can be discarded in the stability analysis of (8.2.95).
3. Justify the use of the Vidyasagar theorem for (8.2.95) and show that the stability analysis of (8.2.95) can be rewritten as

$$x_i' = (1 - x_i)\sum_{j=1}^{r}\lambda_{ij}x_j - \alpha_i x_i, \quad i = 1, \ldots, r, \tag{8.2.96}$$

where $\alpha_i = \mu - \nu_i$ and $x_i = \frac{I_i}{N_i}$, $i = 1, \ldots, r$, or, in short,

$$\boldsymbol{x}' = \mathcal{diag}(\mathbf{1} - \boldsymbol{x})\mathcal{L}\boldsymbol{x} - \mathcal{diag}\,\boldsymbol{\alpha x}, \tag{8.2.97}$$

where $\boldsymbol{\alpha} = (\alpha_1, \ldots, \alpha_r)$.
4. Show that $\mathbb{I}_r = [0, 1]^r$ is invariant under the flow of (8.2.96) and hence the solution exist for all $t \geq 0$. Further, show that no face $F_i = \{\boldsymbol{x} \in \mathbb{I}_r : x_i = 1\}$, $i = 1, \ldots, r$, is invariant under the flow of (8.2.97).
5. Show that the Jacobi matrix of $\mathcal{diag}\,(\mathbf{1} - \boldsymbol{x})\mathcal{L}\boldsymbol{x} - \mathcal{diag}\,\boldsymbol{\alpha x}$ is given by

$$\mathcal{J}(\boldsymbol{x}) = \mathcal{diag}(\mathbf{1} - \boldsymbol{x})\mathcal{L} - \mathcal{diag}\mathcal{L}\boldsymbol{x} - \mathcal{diag}\,\boldsymbol{\alpha} \tag{8.2.98}$$

6. Show that $\mathcal{J}(\boldsymbol{x})$ is a Metzler matrix for any $\boldsymbol{x} \in \mathbb{I}_r$.
7. Show that if $\mathcal{L} = (\lambda_{ij})_{1\leq i,j\leq r}$ is irreducible, then $\mathcal{J}(\boldsymbol{x})$ is irreducible for $0 \leq \boldsymbol{x} \ll \mathbf{1}$. Conclude that (8.2.96) is strongly monotone as long as $\boldsymbol{x} \ll \mathbf{1}$.
8. Using the fact that an irreducible matrix cannot have a zero row (why?), show that $\mathcal{J}(\boldsymbol{x})$ is strongly sublinear for $0 \ll \boldsymbol{x} \ll \mathbf{1}$ and strictly sublinear for $0 \ll \boldsymbol{x} < \mathbf{1}$.
9. The origin $\boldsymbol{x} = 0$ is DFE for (8.2.96) with $\mathcal{J}(0) = \mathcal{L} - \mathcal{diag}\,\boldsymbol{\alpha}$. Show that this is a regular splitting of $\mathcal{J}(0)$ and hence $s(\mathcal{J}(0)) < 0$ if and only if

$$r(\mathcal{L}(\mathcal{diag}\,\boldsymbol{\alpha})^{-1}) = r\left(\left(\begin{matrix} \frac{\lambda_{11}}{\alpha_1} & \cdots & \frac{\lambda_{1r}}{\alpha_r} \\ \vdots & \ddots & \vdots \\ \frac{\lambda_{r1}}{\alpha_1} & \cdots & \frac{\lambda_{rr}}{\alpha_r} \end{matrix}\right)\right) < 1.$$

10. Use Theorem 8.40 to conclude that

   - If $r(\mathcal{L}(\mathscr{diag}\,\boldsymbol{\alpha})^{-1}) < 1$, then DFE is globally asymptotically stable for (8.2.96) in $\mathbb{I}_r$.
   - If $r(\mathcal{L}(\mathscr{diag}\,\boldsymbol{\alpha})^{-1}) > 1$, then there is an equilibrium $0 \ll e \ll \mathbf{1}$ which is globally asymptotically stable for (8.2.96) in $\mathbb{I}_r \setminus \{0\}$.

11. By considering the Lyapunov function (8.2.89) and following the argument in the proof of Theorem 8.50, show that DFE is globally asymptotically stable in $\mathbb{I}_r$ also if $R_0 = 1$, see [43].
12. Show that the application of the Vidyasagar theorem was justified. Translate these results for the full system (8.2.94).

Part III
Models with Continuous Age Structure

Chapter 9
McKendrick–von Foerster Model

9.1 From Discrete Leslie Model to Continuous McKendrick–von Foerster Problem

In the classical Leslie model, the census is taken in equal intervals, which, for convenience, we also assumed to be the unit of time. If the time between censuses and the length of each age class is instead taken to be $h > 0$, then, starting from some time t, the Leslie model will take the form

$$\begin{pmatrix} x_0(t+h) \\ x_h(t+h) \\ x_{2h}(t+h) \\ \vdots \\ x_{(n-1)h}(t+h) \end{pmatrix} = \begin{pmatrix} f_0(h) & f_h(h) & \cdots & f_{(n-2)h}(h) & f_{(n-1)h}(h) \\ s_0(h) & 0 & \cdots & 0 & 0 \\ 0 & s_h(h) & \cdots & 0 & 0 \\ \vdots & \vdots & \cdots & \vdots & \vdots \\ 0 & 0 & \cdots & s_{(n-2)h}(h) & 0 \end{pmatrix} \begin{pmatrix} x_0(t) \\ x_h(t) \\ x_{2h}(t) \\ \vdots \\ x_{(n-1)h}(t) \end{pmatrix}. \tag{9.1.1}$$

The maximal age of individuals $\omega = nh$ is thus divided into n age intervals $[0, h), [h, 2h) \ldots [(n-1)h, nh)$ with the convention that if the age a of an individual falls into $[kh, (k+1)h)$, it is considered to be kh. In this definition, as in the discrete case, nobody reaches the age ω. Thus, $x_a(t)$ denotes the number of individuals of age a, $s_a = \frac{l_{a+h}}{l_a}$ is the probability of survival to the age of $a + h$ conditioned upon surviving up to age a with $l_0 = 1$ and $f_a = m_{a+h}s_h$ is the effective fecundity with m_{a+h} being the average fertility of females of age $a + h$. We note that $1 - s_a$ is the

J. Banasiak, *Introduction to Mathematical Methods in Population Theory*, Springer Undergraduate Mathematics Series,
https://doi.org/10.1007/978-3-031-65491-6_9

number of individuals who do not survive from a to $a+h$. We make the following assumptions and notation: for any $a \geq 0$,

$$\lim_{h\to 0^+} s_a(h) = s_a(0) = 1, \tag{9.1.2}$$

$$\lim_{h\to 0^+} \frac{1-s_a(h)}{h} =: \mu(a), \tag{9.1.3}$$

$$\lim_{h\to 0^+} \frac{f_a(h)}{h} =: \beta(a), \tag{9.1.4}$$

where $f_a(h)$, $\mu(a)$ and $\beta(a)$ are continuous functions. To explain this notation, we note that the probability of survival over a very short period of time should be close to 1, as in Eq. (9.1.2). Further, using Sect. 2.1.3, we note that if the death rate μ is constant, then the probability of surviving over a short time interval h approximately is $s_a(h) = 1 - \mu h$ for any a and thus the limit in (9.1.3) can serve as a more general definition of the age dependant death rate. Similarly, if the average number of births per female over the unit time is a constant β, then the number of births over h will be βh. Hence, the last equation gives the general definition of the age-dependent birth rate, which, moreover, is independent of the survival rate by (9.1.2).

Finally, we assume that there is a continuous density function $n(t, a)$ such that

$$x_a(t) = \int_a^{a+h} n(t,\alpha)d\alpha. \tag{9.1.5}$$

We are going to derive a differential equation for n. Consider a fixed age $a = ih > 0$. From (9.1.1), we see that

$$x_{a+h}(t+h) = s_a(h)x_a(t), \qquad a = 0, h, \ldots, (n-2)h. \tag{9.1.6}$$

Using (9.1.5),

$$x_{a+h}(t+h) = \int_{a+h}^{a+2h} n(t+h,\alpha)d\alpha = \int_a^{a+h} n(t+h,\alpha+h)d\alpha,$$

thus (9.1.6) can be written as

$$\int_a^{a+h} n(t+h,\alpha+h)d\alpha = s_a(h)\int_a^{a+h} n(t,\alpha)d\alpha.$$

We re-write it as

$$\frac{1}{h^2}\left(\int_a^{a+h} n(t+h,\alpha+h)d\alpha - \int_a^{a+h} n(t,\alpha)d\alpha\right) = -\frac{1-s_a(h)}{h^2}\int_a^{a+h} n(t,\alpha)d\alpha.$$

We assume that the directional derivative along the vector $\mathbf{1} = (1, 1)$,

$$D_{\mathbf{1}}n(t,a) = \lim_{h\to 0^+}\frac{n(t+h,a+h)-n(t,a)}{h},$$

exists at each (t, a) and the limit is locally uniform in the sense that

$$n(t+h,a+h)-n(t,a) = D_{\mathbf{1}}n(t,a)h + R(t,a,h),$$

where each (t, a) has a neighbourhood such that

$$\frac{R(\tau,\alpha,h)}{h} \le c(h)$$

for some $c(h)$ independent of (τ, α) in this neighbourhood and $c(h) \to 0$. Then, we can write

$$\frac{1}{h^2}\left(\int_a^{a+h} n(t+h,\alpha+h)d\alpha - \int_a^{a+h} n(t,\alpha)d\alpha\right)$$
$$= \frac{1}{h}\int_a^{a+h} D_{\mathbf{1}}n(t,\alpha)d\alpha + \frac{1}{h}\int_a^{a+h} R(t,\alpha,h)d\alpha.$$

If we assume that $D_{\mathbf{1}}n(t, \alpha)$ is continuous at $\alpha = a$, then the Fundamental Theorem of Calculus yields

$$\lim_{h\to 0^+}\frac{1}{h^2}\left(\int_a^{a+h} n(t+h,\alpha+h)d\alpha - \int_a^{a+h} n(t,\alpha)d\alpha\right) = D_{\mathbf{1}}n(t,a).$$

Similarly, since we assumed n to be continuous, using (9.1.3) we obtain

$$\lim_{h\to 0^+}\frac{1-s_a(h)}{h^2}\int_a^{a+h} n(t,\alpha)d\alpha = \mu(a)n(t,a).$$

Combining, we obtain that at every point of continuity of n and $D_{\mathbf{1}}n$, n satisfies

$$D_{\mathbf{1}}n(t,a) = -\mu(a)n(t,a), \qquad a > 0, t > 0.$$

Assuming that the partial derivatives $\frac{\partial n}{\partial t}, \frac{\partial n}{\partial a}$ at (t, a) exist, we can further transform the last equation to

$$\frac{\partial n}{\partial t}(t, a) + \frac{\partial n}{\partial a}n(t, a) = -\mu(a)n(t, a), \qquad a > 0, t > 0.$$

This is the most commonly used form of the equation for n. However, as we shall see later, it is not the best form for its analysis, and it is incorrect in many cases as the differentiability assumptions are often not satisfied.

Next, consider the class of neonates in the Leslie formulation,

$$x_0(t + h) = \sum_{j=0}^{n-1} f_{jh}(h)x_{jh}(t),$$

which can be rewritten as

$$\frac{1}{h}x_0(t + h) = \sum_{j=0}^{n-1} \frac{1}{h} f_{jh}(h)\frac{1}{h}x_{jh}(t)h.$$

If n is continuous and f is differentiable at 0, then

$$\frac{1}{h}x_{jh}(t) = \frac{1}{h}\int_{jh}^{(j+1)h} n(t, \alpha)d\alpha = n(t, jh + \theta_j h)$$

and

$$\frac{f_{jh}(h)}{h} = \beta(jh + \theta'_j h)$$

for some $0 < \theta_j, \theta'_j < 1$. Thus

$$n(t, \theta_j h) = \sum_{j=0}^{n-1} n(t, jh + \theta_j h)\beta(jh + \theta'_j h)h.$$

Since we assumed that β is a continuous function, the right-hand side is the Riemann sum, and we can pass to the limit as $h \to 0^+$, getting

$$n(t, 0) = \int_0^{\omega} \beta(\alpha)n(t, \alpha)d\alpha.$$

Thus, we arrived at the classical formulation of the McKendrick–von Foerster model

$$\frac{\partial n}{\partial t}(t,a) + \frac{\partial n}{\partial a}(t,a) = -\mu(a)n(t,a), \quad a > 0, t > 0, \tag{9.1.7a}$$

$$n(t,0) = \int_0^\omega \beta(\alpha)n(t,\alpha)d\alpha, \quad t > 0, \tag{9.1.7b}$$

$$n(0,a) = \mathring{n}(a), \quad a > 0, \tag{9.1.7c}$$

where the last equation provides the initial distribution of the population.

Remark 9.1 Assumption (9.1.5) is not strictly necessary as it is possible to show the existence of n as the limit (in a certain sense) of $x_a(t)$, see [95].

If $\omega < +\infty$, then we have to ensure that $n(t,a) = 0$ for $t \geq 0, a \geq \omega$, which can be done either by imposing an additional boundary condition on n or by introducing assumptions on the coefficients, which ensure that no individual survives beyond ω. If $\omega = \infty$, then, instead of such an additional condition, we impose some requirements on the behaviour of the solution at ∞, e.g., that they are integrable over $[0,\infty)$.

9.2 Linear Constant Coefficient Case

Before we embark on a more advanced analysis of (9.1.7a)–(9.1.7c), let us get a taste of the structure of the problem by solving the simplest case with $\mu(a) = \mu$ and $\beta(a) = \beta$,

$$\frac{\partial n}{\partial t}(t,a) + \frac{\partial n}{\partial a}(t,a) = -\mu n(t,a). \tag{9.2.1a}$$

coupled with the boundary condition

$$n(t,0) = \beta \int_0^\infty n(t,a)da, \tag{9.2.1b}$$

and the initial condition

$$n(0,a) = \mathring{n}(a). \tag{9.2.1c}$$

First, let us simplify (9.2.1a) by introducing the integrating factor $e^{\mu a}$ so that

$$\frac{\partial}{\partial t}(e^{\mu a}n(t,a)) = -\frac{\partial a}{\partial a}(e^{\mu a}n(t,a)),$$

and denote $u(t,a) = e^{\mu a}n(t,a)$. Then,

$$u(t,0) = n(t,0) = \beta \int_0^\infty e^{-\mu a}u(t,a)da$$

with $u(0,a) = e^{\mu a}\mathring{n}(a) =: u_0(a)$. Now, if we knew $\psi(t) = u(t,0)$, then, by Example D.2,

$$u(t,a) = \begin{cases} u_0(a-t), & t < a, \\ \psi(t-a), & a < t. \end{cases} \tag{9.2.2}$$

The boundary condition can be rewritten as

$$\psi(t) = \beta \int_0^\infty e^{-\mu a}u(t,a)da = \beta \int_0^t e^{-\mu a}\psi(t-a)da + \beta \int_t^\infty e^{-\mu a}u_0(a-t)da$$

$$= \beta e^{-\mu t}\int_0^t e^{\mu\sigma}\psi(\sigma)d\sigma + \beta e^{-\mu t}\int_0^\infty e^{-\mu r}u_0(r)dr,$$

which, upon denoting $\phi(t) = \psi(t)e^{\mu t}$ and using the original initial value, takes the form

$$\phi(t) = \beta \int_0^t \phi(\sigma)d\sigma + \beta \int_0^\infty \mathring{n}(r)dr. \tag{9.2.3}$$

Now, if we differentiate both sides, we get

$$\phi' = \beta\phi,$$

which is just a first-order linear equation. Letting $t = 0$ in (9.2.3), we obtain the initial value for ϕ: $\phi(0) = \beta \int_0^\infty \mathring{n}(r)dr$. Then

$$\phi(t) = \beta e^{\beta t}\int_0^\infty \mathring{n}(r)dr$$

and

$$\psi(t) = \beta e^{(\beta-\mu)t} \int_0^\infty \mathring{n}(r)dr.$$

Hence,

$$n(t,a) = e^{-\mu a}u(t,a) = e^{-\mu t} \begin{cases} \mathring{n}(a-t), & t < a, \\ \beta e^{\beta(t-a)} \int_0^\infty \mathring{n}(r)dr, & a < t. \end{cases}$$

Observe that

$$\lim_{a\to t^+} n(t,a) = e^{-\mu t}\mathring{n}(0)$$

and

$$\lim_{a\to t^-} n(t,a) = \beta e^{-\mu t} \int_0^\infty \mathring{n}(r),$$

so that the solution is continuous, not alone differentiable, only if the initial condition satisfies the following compatibility condition

$$\mathring{n}(0) = \beta \int_0^\infty \mathring{n}(r)dr, \tag{9.2.4}$$

see Figs. 9.1 and 9.2. Thus, as we noted earlier, we must be very careful with using (9.1.7a)–(9.1.7c) in the differential form and interpreting the solution.

Exercise 9.2.1 Consider a population in which individuals can reach arbitrary age, with a constant death rate μ but which can reproduce at a constant rate β but only for age $a \leq 10$. Write down the McKendrick–von Foerster model for such a population and set up the integral equation to solve it.

Exercise 9.2.2 Find the solution to (9.2.1a) with

(a) $\mathring{n}(a) = e^{-a}$, and

- $\beta = 3, \mu = 1,$
- $\beta = 1, \mu = 0.5;$

(b) $\mathring{n}(a) = \dfrac{1}{1+a^2}$, and $\beta = 3, \mu = 1;$

(c) In case (b), find the value of β for which the solution is continuous.

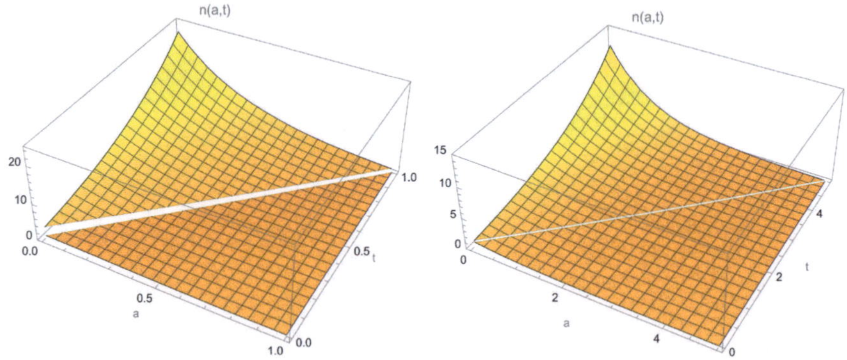

Fig. 9.1 Discontinuity of the population density $n(t, a)$ when (9.2.4) is not satisfied (left) and continuous $n(t, a)$ when (9.2.4) holds—$\mathring{n}(a) = e^{-a}$ in both cases, but $\beta = 3$, $\mu = 1$ in the former and $\beta = 1$, $\mu = 0.5$ in the latter, see Exercise 9.2.2

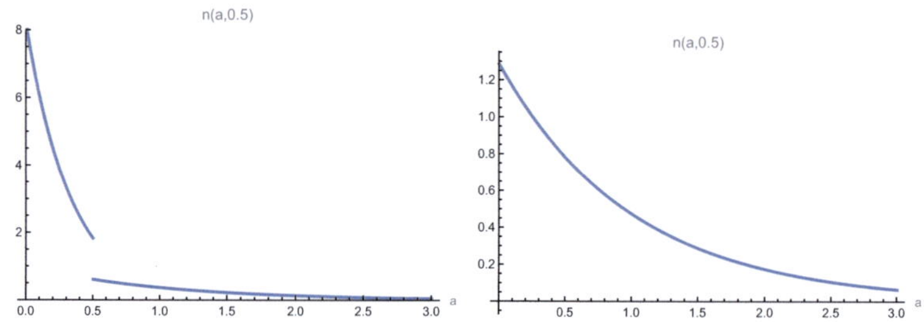

Fig. 9.2 Snapshots of the population densities $n(a, 0.5)$ shown in Fig. 9.1

9.3 General Linear McKendrick–von Foerster Problem

The ideas used to solve the McKendrick–von Foerster equation in the constant coefficient case can also be used in more general situations, but, unfortunately, the resulting integral equation (9.2.3) cannot be explicitly solved. Before, however, we discuss the solvability of more general cases, let us introduce certain functions related to (9.1.7a)–(9.1.7c), which are relevant to the population dynamics.

9.3.1 Demographic Parameters of the McKendrick–von Foerster Problem

Consider again the general McKendrick–von Foerster problem (9.1.7),

$$\frac{\partial n}{\partial t}(t, a) + \frac{\partial n}{\partial a}(t, a) = -\mu(a)n(t, a), \quad a > 0, t > 0,$$

$$n(t,0) = \int_0^{\omega} \beta(\alpha)n(t,\alpha)d\alpha, \quad t > 0,$$

$$n(0,a) = \mathring{n}(a), \quad a > 0.$$

We recall that $\beta(a)$ is the *age-specific fertility* which can be defined as the number of newborns, in one unit of time, coming from a single individual whose age is in the short time age interval $[a, a + \Delta a)$. So, the number of births coming from all individuals in the population aged between a_1 and a_2 in a unit of time is

$$\int_{a_1}^{a_2} \beta(\alpha)n(t,\alpha)d\alpha$$

and we can define the *total birth rate* as

$$B(t) = \int_0^{\omega} \beta(\alpha)n(t,\alpha)d\alpha,$$

which gives the total number of newborns in a unit time (ω is the maximum age in the population).

Let us consider the death rate $\mu(a)$, which is the average number of deaths per unit time and per unit of the population aged a. We can relate $\mu(a)$ to several vital population characteristics. Similarly to the discrete case, we introduce the *survival probability* $S(a)$ as the proportion of the initial population surviving to age a. We can relate μ and S by the following argument. Consider a population beginning with $\mathring{n}$ individuals of age 0. Then $\mathring{n}(a)S(a)(= n(a))$ is the average number of individuals that survived to age a. The decline in the population over a short age period $[a, a + \Delta a]$ is $\mathring{n}(a)S(a) - \mathring{n}(a)S(a + \Delta a)$. On the other hand, this decline can only be attributed to deaths: if the death rate is μ, then in this age interval, we will have approximately $\mathring{n}(a)S(a)\mu(a)\Delta a$ deaths. Equating and passing to the limit as $\Delta a \to 0$ yields

$$\frac{dS}{da} = -S\mu,$$

that is,

$$S(a) = S(0)e^{-\int_0^a \mu(\sigma)d\sigma}$$

(compare (2.1.6)). Since, however, the probability of surviving to age 0 is 1, we have

$$S(a) = e^{-\int_0^a \mu(\sigma)d\sigma}. \tag{9.3.6}$$

We note that if no individuals can survive beyond ω, we must have $S(\omega) = 0$ or, equivalently,

$$\int_0^\omega \mu(\sigma)d\sigma = \infty. \tag{9.3.7}$$

Exercise 9.3.1 Find the possible age range $[0, \omega)$ of individuals in a population with the death rate given by: (i) $\mu(a) = e^a$; (ii) $\mu(a) = \frac{1}{70 - a}$; (iii) $\mu(a) = \frac{1}{\sqrt{|a - 64|}}$; iv) $\mu(a) = |a - 64|^{-3/2}$.

Similarly to Sect. 2.1.3, these considerations can be used to find the average lifespan of individuals in the population. In fact, the average lifespan is the mean value of the length of life in the population, which can be expressed as

$$L = \int_0^\omega ap(a)da,$$

where $p(a)$ is the probability (density) of an individual dying at age a. If we approximate the integral as the Riemann sum

$$L \approx \sum_i a_i p(a_i)\Delta a_i,$$

where $p(a_i)$ is the probability that an individual survived till the age a_i and then died, then

$$p(a_i) = S(a_i)\mu(a_i).$$

Exercise 9.3.2 Show that $S(a)\mu(a)$ is a probability density (that is, its integral over $[0, \omega)$ is one).

Using the result of the exercise, integration by parts and $S(\omega) = 0$,

$$L = \int_0^\omega a\mu(a)e^{-\int_0^a \mu(s)ds}da = -\int_0^\omega a\frac{d}{da}e^{-\int_0^a \mu(s)ds}da = \int_0^\omega S(a)da.$$

Exercise 9.3.3 Find the expected lifespan in the populations with death rates introduced in Exercise 9.3.1.

Further, we introduce

$$K(a) = \beta(a)S(a), \tag{9.3.8}$$

which is called the maternity function and describes the rate of birth relative to the surviving fraction of the population and is the continuous equivalent to the coefficients $f_0, f_1, \ldots, f_{n-1}$. Finally, we define

$$R = \int\limits_0^{\omega} \beta(a)S(a)da, \tag{9.3.9}$$

and call it net reproduction rate of the population. It is the expected number of offspring an individual produces during her reproductive life.

9.3.2 *Solution of the Linear McKendrick–von Foerster Problem*

One of the easiest ways of analysing the general McKendrick–von Foerster model (9.1.7) is to reduce it to an integral equation like in Sect. 9.2. However, the technicalities are slightly more involved due to the age dependence of mortality and maternity functions. First, we simplify (9.1.7a) by introducing the integrating factor

$$\frac{\partial}{\partial t}\left(e^{\int\limits_0^a \mu(\sigma)d\sigma} n(t,a)\right) = -\frac{\partial}{\partial a}\left(e^{\int\limits_0^a \mu(\sigma)d\sigma} n(t,a)\right), \tag{9.3.10}$$

and denote $u(t,a) = e^{\int\limits_0^a \mu(\sigma)d\sigma} n(t,a)$. Then

$$u(t,0) = n(t,0) = \int\limits_0^{\omega} \beta(a) e^{-\int\limits_0^a \mu(\sigma)d\sigma} u(t,a)da = \int\limits_0^{\omega} K(a)u(t,a)da,$$

where we recognized that the kernel in the integral above is the maternity function introduced in (9.3.8). Further, $u(0,a) = e^{\int\limits_0^a \mu(s)ds} \mathring{n}(a) =: u_0(a)$. Also, the right-hand side defines the total birth rate $B(t)$.

As before, if we knew $B(t) = u(t, 0)$, then we would have

$$u(t,a) = \begin{cases} u_0(a-t), & t < a, \\ B(t-a), & a < t. \end{cases} \tag{9.3.11}$$

The boundary condition can be rewritten as

$$\begin{aligned} B(t) &= \int_0^\infty \beta(a) e^{-\int_0^a \mu(\sigma)d\sigma} u(t,a)da \\ &= \int_0^t \beta(a) e^{-\int_0^a \mu(\sigma)d\sigma} B(t-a)da + \int_t^\infty \beta(a) e^{-\int_0^a \mu(\sigma)d\sigma} u_0(a-t)da \\ &= \int_0^t K(t-a)B(a)da + \int_0^\infty \beta(a+t) e^{-\int_0^{a+t} \mu(\sigma)d\sigma} e^{\int_0^a \mu(s)ds} \mathring{n}(a)da, \end{aligned} \tag{9.3.12}$$

where we extended coefficients by zero beyond $a = \omega$ to shorten the notation. Summarizing, we arrived at the integral equation for the total birth rate

$$B(t) = \int_0^t K(t-a)B(a)da + G(t), \tag{9.3.13}$$

where

$$G(t) = \int_0^\infty \beta(a+t)\frac{S(a+t)}{S(a)}\mathring{n}(a)da \tag{9.3.14}$$

is a known function. Explicitly, we have

$$\begin{aligned} B(t) &= \int_0^t K(t-a)B(a)da + \int_0^{\omega-t} \beta(a+t)\frac{S(a+t)}{S(a)}\mathring{n}(a)da \\ &= \int_0^t K(t-a)B(a)da + \int_t^{\omega} \beta(a)\frac{S(a)}{S(a-t)}\mathring{n}(a-t)da \end{aligned}$$

for $0 \leq t \leq \omega$, and

$$B(t) = \int_0^\omega K(t-a)B(a)da$$

for $t > \omega$.

Unlike in Sect. 9.2, this equation cannot be solved explicitly, and we have to use a more abstract approach. The basis for this is the Picard iteration process used in Theorem B.10, but the details are different as, on the one hand, the problem is linear, and, on the other, we deal with multivariable functions. The latter makes the problem infinite-dimensional. To explain this, Theorem B.10 deals with functions of time, taking values in $\mathbb{R}^d$, that is, functions which for each t are finite-dimensional vectors. However, here, the function $u(t,a)$ for each t is a function of a, and the space of functions of one variable is infinite-dimensional (reflect on this). We must introduce a proper mathematical framework to deal with such functions. As in the discrete finite-dimensional case (A.1.4), here the natural space is $L_1([0,\omega))$ with the norm

$$\|n\|_1 = \int_0^\omega |n(\alpha)|d\alpha;$$

see Sect. B.2.1. Since the integral is an extension of summation, analogously, with $n \geq 0$ being the density of the population distribution with respect to age, $\|n\|_1$ is the total population.

Let us observe that for the spaces of integrable (but not for locally integrable, see Remark B.2) functions, it does not matter whether they are defined on open, half-closed or closed intervals. Thus, we introduce the notation

$$I_\omega = (0,\omega)$$

and use the notation $[0,\omega)$ only when the behaviour of a function close to zero is important. Therefore the state space is the space $X = L_1(I_\omega)$ of (Lebesgue) integrable functions on I_ω, discussed in Sect. B.2.1. As noted above, we often consider $(t,a) \to n(t,a)$ as a function $t \to n(t,\cdot)$, that is, the value of this function for each t is a function with a as its argument. For such functions, we consider the space $C([0,T], L_1(I_\omega))$ of continuous $L_1(I_\omega)$-valued continuous functions. We make the following assumptions:

(i)

$$0 \leq \beta \in L_\infty(I_\omega), \tag{9.3.15}$$

(ii)

$$0 \leq \mu \in L_{1,loc}([0,\omega)) \tag{9.3.16}$$

with

$$\int_0^\omega \mu(\alpha)d\alpha = \infty, \tag{9.3.17}$$

(iii)

$$0 \leq \mathring{n} \in L_1(I_\omega); \tag{9.3.18}$$

see Sect. B.2.1. Now, if (9.3.15)–(9.3.18) are satisfied, then we can show that K is a non-negative bounded function, which is zero for $t \geq \omega$ and G is a continuous function which also is zero for $t \geq \omega$. If, additionally

$$\mathring{n} \in W_1^1(I_\omega) \quad \text{and} \quad \mu\mathring{n} \in L_1(I_\omega), \tag{9.3.19}$$

(recall that W_1^1 denotes the Sobolev space of functions from L_1 with generalized derivatives also belonging to L_1, see Sect. B.2.2), then G is differentiable with bounded derivative. Indeed, let us look at G for $t < \omega$,

$$G(t) = \int_t^\omega \beta(a)\frac{S(a)}{S(a-t)}\mathring{n}(a-t)da = \int_t^\omega \beta(a)e^{-\int_{a-t}^{a}\mu(s)ds}\mathring{n}(a-t)da.$$

If we formally differentiate using the Leibniz rule, we get

$$G'(t) = -\beta(t)S(t)\mathring{n}(0) + \int_t^\omega \beta(a)e^{-\int_{a-t}^{a}\mu(s)ds}\mu(a-t)\mathring{n}(a-t)da$$
$$+\int_t^\omega \beta(a)e^{-\int_{a-t}^{a}\mu(s)ds}\mu(a-t)\mathring{n}'(a-t)da,$$

so we see that for the existence of the integrals, we need the integrability of $\mu\mathring{n}$ and differentiability of $\mathring{n}$. Then we can prove the main result

Theorem 9.2 *If (9.3.15)–(9.3.18) are satisfied, then (9.3.13) has a unique continuous and non-negative solution. If, additionally, (9.3.19) is satisfied, then B is differentiable with B' bounded on bounded intervals.*

Proof Similarly to Theorem B.10, we define Picard iterates

$$B^0(t) = G(t),$$
$$B^{k+1}(t) = G(t) + \int_0^t K(t-s)B^k(s)ds, \quad k = 1, 2, \ldots, . \tag{9.3.20}$$

Take $T > 0$. Then, for any $t \in [0, T]$ we have

$$|B^1(t) - B^0(t)| = \int_0^t |K(t-s)G(s)|ds \le tK_mG_m,$$

where $K_m = \sup_{0\le t\le T} |K(s)|$ and $G_m = \sup_{0\le t\le T} |G(s)|$. Then

$$|B^2(t) - B^1(t)| \le K_m \int_0^t |B^1(s) - B^0(s)|ds \le \frac{K_m^2 G_m}{2} t^2,$$

and, by induction,

$$|B^{k+1}(t) - B^k(t)| \le K_m \int_0^t |B^k(s) - B^{k-1}(s)|ds \le \frac{K_m^{k+1} G_m}{(k+1)!} t^{k+1}. \tag{9.3.21}$$

Further,

$$\lim_{k\to\infty} B^{k+1}(t) = G(t) + \lim_{k\to\infty} \sum_{i=0}^{k} (B^{i+1}(t) - B^i(t))$$

with

$$\sup_{0\le t\le T} \left| \sum_{i=0}^{k} (B^{i+1}(t) - B^i(t)) \right| \le \sum_{i=0}^{k} \sup_{0\le t\le T} \left| B^{i+1}(t) - B^i(t) \right| \le F_m \sum_{i=0}^{k} \frac{(TK_m)^{k+1}}{(k+1)!}.$$

The series on the right-hand side converges to $F_m e^{TK_m}$ and thus $(B^k(t))_{k\ge0}$ converges uniformly to a continuous solution B of (9.3.13). The uniqueness follows by the Gronwall inequality, Lemma B.11.

If, in addition, (9.3.19) is satisfied, then B^k can be differentiated with respect to t and the iterates

$$V^k := \frac{d}{dt} B^k, \quad k = 1, 2, \ldots,$$

satisfy the recurrence

$$V^{k+1}(t) = F'(t) + K(t)F(0) + \int_0^t K(t-s)V^k(s)ds,$$

which converges uniformly to some continuous function V, which, by the theorem of uniform convergence of derivatives, [31, Section7.4], must be the derivative of B. □

Once we have B, we can recover n by (9.3.11) and back substitution

$$n(t,a) = e^{-\int_0^a \mu(\sigma)d\sigma} u(t,a) = \begin{cases} \frac{S(a)}{S(a-t)}\mathring{n}(a-t), & t < a, \\ S(a)B(t-a), & a < t. \end{cases} \tag{9.3.22}$$

Thus, if (9.3.19) is satisfied in addition to (9.3.15)–(9.3.18), then it is easy to see that n defined above satisfies Eq. (9.1.7a) everywhere except the line $a = t$. Along this line, we have, as before,

$$\lim_{a\to t^+} n(t,a) = S(0)\mathring{n}(0) = \mathring{n}(0)$$

and

$$\lim_{a\to t^-} n(t,a) = S(0)B(0) = \int_0^\omega \beta(a)\mathring{n}(a)da.$$

Hence, to ensure at least the continuity of the solution, we need to assume the compatibility condition

$$\mathring{n}(0) = \int_0^\omega \beta(a)\mathring{n}(a)da. \tag{9.3.23}$$

We note that if a function is continuous at a point and differentiable in both one-sided neighbourhoods, then it is a Lipschitz function, which is, in fact, differentiable almost everywhere (since Lipschitz continuous functions are absolutely continuous and the latter are differentiable almost everywhere, [77, Section 5.4]). On the other hand, if a function has a jump at a point, its derivative at this point is the Dirac delta function (compare Example B.5). Thus, we can state that if (9.3.23) is satisfied, then the solution is continuous and satisfies (9.1.7a) almost everywhere. If we do not assume (9.3.23), then we can still claim that the solution satisfies

$$D_{\mathbf{1}}n(t,a) = \lim_{h\to 0^+} \frac{n(t+h,a+h) - n(t,a)}{h} = -\mu(a)n(t,a), \qquad a > 0, t > 0.$$

Furthermore, both the birth rate B and the solution n grow at most exponentially. Indeed, consider again (9.3.13),

$$B(t) = \int_0^t K(t-a)B(a)da + G(t),$$

with G given by (9.3.14), where

$$S(a) = e^{-\int_0^a \mu(\sigma)d\sigma}$$

and $K(a) = \beta(a)S(a)$. We see that $K(t) \leq \|\beta\|_\infty$ and $G(t) \leq \|\beta\|_\infty \|\mathring{n}\|_1$ so that

$$B(t) \leq \max_{0\leq a\leq \omega} \beta(a) \int_0^t B(s)ds + \max_{0\leq a\leq \omega} \beta(a) \int_0^\omega \mathring{n}(s)ds$$

$$=: \|\beta\|_\infty \int_0^t B(s)ds + \|\beta\|_\infty \|\mathring{n}\|_1,$$

which, by Gronwall's inequality, yields

$$B(t) \leq \|\beta\|_\infty \|\mathring{n}\|_1 e^{t\|\beta\|_\infty}. \tag{9.3.24}$$

This gives

$$\|n(t,\cdot)\|_1 \leq \int_0^t B(t-s)S(s)ds + \int_t^\infty \frac{S(s)}{S(s-t)}\mathring{n}(s-t)ds$$

$$\leq \|\beta\|_\infty \|\mathring{n}\|_1 \left(\int_0^t e^{(t-s)\|\beta\|_\infty} ds + 1 \right),$$

where we used $S(s)/S(s-t) \leq 1$. Then, integrating,

$$\|n(\cdot,t)\|_1 \leq \|\mathring{n}\|_1 + \|\mathring{n}\|_1 e^{t\|\beta\|_\infty}(1 - e^{-t\|\beta\|_\infty}) = \|\mathring{n}\|_1 e^{t\|\beta\|_\infty}. \tag{9.3.25}$$

Remark 9.3 Note, that (9.3.25) is not optimal. We shall provide a better estimate below.

9.4 Long Time Behaviour of Solutions

As in the case of systems of ordinary differential equations, the central question in population dynamics is determining the long-term behaviour of the solutions. We shall now proceed in this direction for the McKendrick–von Foerster model.

9.4.1 *Warm Up—The Constant Coefficients Case*

Consider the constant coefficient problem (9.2.1) again. It has a solution given by

$$n(t,a) = e^{-\mu t}\begin{cases} \mathring{n}(a-t), & t < a, \\ \beta e^{\beta(t-a)} \int\limits_0^\infty \mathring{n}(r)dr, & a < t. \end{cases} \tag{9.4.26}$$

What happens with the population when $t \to \infty$? If t is large, we can consider only the second part of the solution,

$$n(t,a) = \beta \mathring{N} e^{-\beta a} e^{t(\beta-\mu)}, \quad a < t,$$

where $\mathring{N} = \int\limits_0^\infty \mathring{n}(r)dr$. Denote by $r = \beta - \mu$ the net growth rate. We see that if $r = 0$, we have

$$n(t,a) = \beta \mathring{N} e^{-\mu a}, \quad a < t,$$

and one can surmise that

$$n(t,a) \approx \beta \mathring{N} e^{-\beta a},$$

for large t and all $a > 0$. If we assume that $\mathring{n}$ is bounded, this can be easily checked. Indeed

$$\begin{aligned} n(t,a) &= \begin{cases} e^{-\mu t}\mathring{n}(a-t), & t < a, \\ \beta e^{-\mu a}\mathring{N}, & a < t, \end{cases} \\ &= \beta e^{-\mu a}\mathring{N} + \begin{cases} -\beta e^{-\mu a}\mathring{N} + e^{-\mu t}\mathring{n}(a-t), & t < a, \\ 0, & a < t, \end{cases} \end{aligned}$$

and, since for $t < a$, $e^{-\mu a} < e^{-\mu t}$, we have

$$|-\beta e^{-\mu a}\mathring{N} + e^{-\mu t}\mathring{n}(a-t)| \leq Ce^{-\mu t},$$

where $C = \max\{\beta \mathring{N}, \|\mathring{n}\|_\infty\}$. In other words,

$$n(t,a) = \beta e^{-\mu a}\mathring{N} + O(e^{-\mu t}). \tag{9.4.27}$$

So we see that for large t, the solution has the shape of $\beta e^{-\mu a}$, independent of the initial data, multiplied by the scalar $\mathring{N}$. Thus, the initial age distribution practically has no effect on the solution's shape. In other words, the population's age distribution after a long time is independent of the initial age distribution. If $r \neq 0$, we can write

$$e^{-rt}n(t,a) = \beta \mathring{N} e^{-\beta a}, \quad a < t,$$

and, as before,

$$\begin{aligned} e^{-rt}n(t,a) &= \begin{cases} e^{-\beta t}\mathring{n}(a-t), & t < a, \\ \beta e^{-\beta a}\mathring{N}, & a < t, \end{cases} \\ &= \beta e^{-\mu a}\mathring{N} + \begin{cases} -\beta e^{-\beta a}\mathring{N} + e^{-\beta t}\mathring{n}(a-t), & t < a, \\ 0, & a < t, \end{cases} \end{aligned}$$

and, since $e^{-\beta a} < e^{-\beta t}$ for $t < a$, we have

$$|-\beta e^{-\beta a}\mathring{N} + e^{-\beta t}\mathring{n}(a-t)| \leq Ce^{-\beta t},$$

where $C = \max\{\beta \mathring{N}, \|\mathring{n}\|_\infty\}$. In other words,

$$n(t,a) = \mathring{N}e^{rt}\beta e^{-\beta a} + O(e^{-\mu t}), \tag{9.4.28}$$

where we used $e^{rt}e^{-\beta t} = e^{(\beta-\mu)t}e^{-\beta t} = e^{-\mu t}$. Hence, the population is described by the Malthusian part $\mathring{N}e^{rt}$, which is independent of the population's age profile, multiplied by the age profile $\beta e^{-\beta a}$, see Fig. 9.3. The profile is called the stable age distribution, and the property described above is the same as the asynchronous exponential growth property (AEG) for models with discrete age structure, described in Sect. 5.2.1.

In what follows, we shall prove that this property holds for the general McKendrick–von Foerster model. First, however, bearing in mind that the AEG property for models with discrete age structure is related to the spectral properties of the problem, we demonstrate that also, in this case, the stable age distribution can be obtained by considering the eigenvalue problem for (9.2.1). To this end, let us solve

$$\begin{aligned} \lambda n(a) + n'(a) &= -\mu n(a), \\ n(0) &= \beta \int_0^\infty n(s)ds. \end{aligned} \tag{9.4.29}$$

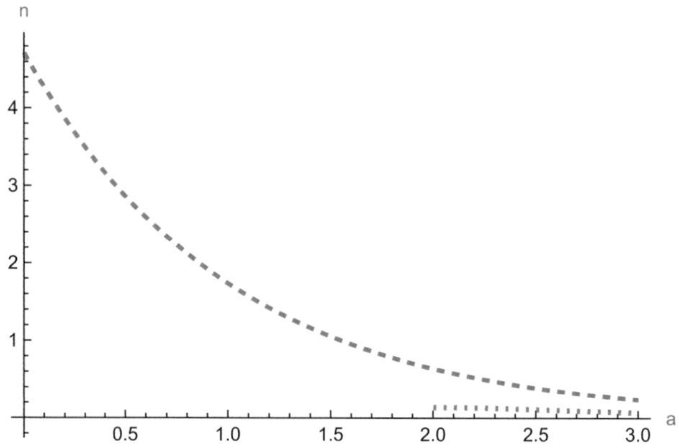

Fig. 9.3 Comparison of the solution $n(2, a)$ given by (9.4.26) (dotted line) and the stable profile $\beta \mathring{N} e^{-\beta a}$ for $\mu = \beta = 1$, $\mathring{n}(a) = \frac{1}{1+a^2}$ (dashed line); we see that up to $a = t(= 2)$, the solutions coincide, and for $a > t$ the difference is smaller than the exponential $\beta \mathring{N} e^{-\beta a} \leq \beta \mathring{N} e^{-\beta t}$

The first equation is simply a linear equation with the general solution

$$n(a) = Ce^{-(\mu+\lambda)a},$$

while the nonlocal initial condition yields

$$1 = \beta \int_0^\infty e^{-(\mu+\lambda)a} da,$$

where we cancelled the constant C. This is an example of the so-called Lotka renewal equation, which we will encounter later. In our case, we can solve it explicitly. By integration, we get

$$1 = \frac{\beta}{\mu + \lambda}, \tag{9.4.30}$$

that is,

$$\lambda = \beta - \mu = r,$$

and hence,

$$n(a) = Ce^{-\beta a}.$$

So, the unique eigenvalue of (9.4.29) is (in this case) precisely the net growth rate. This eigenvalue is simple, and the corresponding eigenvector is, up to a constant, the stable age distribution. As we shall see, this is not a coincidence.

9.4.2 *Long Time Behaviour—General Case*

In this section, we assume that $\omega < \infty$.

Inequality (9.3.24) shows that B is exponentially bounded, see (B.3.16), and hence we can apply the Laplace transform for the analysis of (9.3.13). The basic properties of this transform are collected in Sect. B.3. In the case of B, (9.3.24) shows that $\hat{B}(\lambda)$ is analytic in $\Re\lambda > \omega := \|\beta\|_\infty \|\mathring{n}\|_1$.

Using (B.3.21), (9.3.13) yields

$$\hat{B}(\lambda) = \hat{B}(\lambda)\hat{K}(\lambda) + \hat{G}(\lambda). \tag{9.4.31}$$

Hence,

$$\hat{B}(\lambda) = \frac{\hat{G}(\lambda)}{1 - \hat{K}(\lambda)} = \hat{G}(\lambda) + \frac{\hat{G}(\lambda)\hat{K}(\lambda)}{1 - \hat{K}(\lambda)}. \tag{9.4.32}$$

By (9.3.8) and (9.3.14), both K and G are zero for $t > \omega$ so, as noted in Sect. B.3, $\hat{G}$ and $\hat{K}$ are entire functions. Hence, the only singularities of $\hat{B}$ are due to the zeroes of $1 - \hat{K}$. Since $\hat{K}$ is an entire function, these zeroes are isolated of finite order (thus giving rise to poles of $\hat{B}$ and with no finite accumulation point). However, there may be infinitely many of them, requiring some care when handling the inverse.

Let us look closer at the equation

$$\hat{K}(\lambda) = 1, \tag{9.4.33}$$

or, explicitly,

$$\int_0^\infty \beta(a) e^{-\lambda a - \int_0^a \mu(\sigma) d\sigma} da = 1, \tag{9.4.34}$$

where $\lambda \in \mathbb{C}$.

Remark 9.4 We observe that (9.4.34) is a continuous copy of the discrete renewal equation (5.4.124) if one replaces λ of the latter by e^λ. However, as we shall see below, the continuous case does not admit any cyclic behaviour.

Theorem 9.5 *Equation (9.4.33) has exactly one real root, say $\lambda = \lambda_*$, of algebraic multiplicity 1. All other roots λ_j of (9.4.33) occur as complex conjugates (the real* *root is its own conjugate). Moreover, $\Re\lambda_j < \lambda_*$ for any j, there could be only denumerable number of them and, in each strip $a < \Re\lambda < b$, there there is at most a finite number of them.*

Proof We introduce the real function

$$\psi(\lambda) = \int_0^\infty e^{-\lambda a} K(a)da$$

for $\lambda \in \mathbb{R}$. We note that this function is well defined on $\mathbb{R}$ since K is nonzero only on a finite interval. Also, because of this, it is continuous and differentiable, see Remark 9.6 below. Then

$$\lim_{\lambda\to-\infty} \psi(\lambda) = \infty,$$
$$\lim_{\lambda\to\infty} \psi(\lambda) = 0.$$

Moreover,

$$\psi'(\lambda) = -\int_0^\infty ae^{-\lambda a} K(a)da < 0,$$
$$\psi''(\lambda) = \int_0^\infty a^2e^{-\lambda a} K(a)da > 0,$$

so that ψ is a strictly decreasing and convex function. Since it is continuous, it takes on every positive value exactly once. Thus, in particular, there is exactly one real value λ_* satisfying (9.4.33).

Suppose $\lambda = u + iv$ is a root of (9.4.33). Then

$$1 = \int_0^\infty e^{-va}(\cos(-ua) + i\sin(-ua))K(a)da,$$

and, taking the real and imaginary part,

$$\int_0^\infty e^{-va} K(a)\cos ua\, da = 1,$$
$$\int_0^\infty e^{-va} K(a)\sin ua\, da = 0.$$

We observe that these two equations are invariant under the change $v \to -v$ so that $\bar{\lambda} = u - iv$ also satisfies (9.4.33).

To prove the second part, we note that since the variable a is continuous, there must be a range of a, say, $[\alpha, \beta]$ over which $\cos ua < 1$. Thus,

$$\int_0^\infty e^{-va} K(a)da > \int_0^\infty e^{-va} K(a) \cos ua\, da = 1.$$

However

$$\int_\alpha^\beta e^{-\lambda_* a} K(a)da = 1,$$

and a direct comparison of these two integrals yields $\lambda_* > v = \Re\lambda$.

The last part follows from the fact that since $\hat{K} - 1$ is an entire function, by the principle of isolated zeros, it can have only finitely many zeros in each bounded set. Thus, there could be no more than a denumerable amount of them in $\mathbb{C}$. Finally, since $\hat{K} \to 0$ as $|\lambda| \to \infty$ in any strip (see Sect. B.3), we also see that there can be only finitely many of them in any vertical strip. □

Remark 9.6 In the proof above, the continuity of ψ is a consequence of the boundedness of the support of the definition of K. In general, if we allow K to be nonzero on $[0, \infty)$, then the above statement is not true. Consider $K(a) = c(1 + a^2)^{-1}$ with $c < 2/\pi$. Then,

$$\psi(\lambda) = c \int_0^\infty \frac{e^{-\lambda a}}{1 + a^2} da,$$

and, by direct calculations, $\psi(\lambda) < 1$ for $\lambda \geq 0$ and $\psi(\lambda) = \infty$ for $\lambda < 0$. Hence, Eq. (9.4.33) has no real solution.

In general, if we want to extend Theorem 9.5 to the case $\omega = \infty$, we have to prove that the range of ψ contains 1. For instance, in the constant coefficient case, ψ is given by (9.4.30),

$$\psi(\lambda) = \frac{m}{\lambda + \mu},$$

and though it is discontinuous at $\lambda = -\mu$, its range for $\lambda \in (-\mu, \infty)$ is $\mathbb{R}$ and the statement of Theorem 9.5 holds.

Exercise 9.4.1 Find $S(a)$ if $\mu(a) = \dfrac{1}{70-a}$, $0 \leq a < 70$. Write down (9.4.33) if

(i) $\beta(a) = \beta_0 > 0$,
(ii) $\beta(a) = e^{-a}$,
(iii) $\beta(a) = \beta_0 > 0$ for $15 \leq a \leq 60$ and 0 otherwise.

In each case, determine whether λ^* is positive, negative or zero and try to estimate its value (you can use some fixed value of β_0).

Exercise 9.4.2 Find $S(a)$ if $\mu(a) = \dfrac{1}{\sqrt{|64-a|}}$. Write down (9.4.33) if $\beta(a) = \beta_0 > 0$ and determine whether (9.4.33) can have a real solution λ^* for some/all values of β_0.

Observe that the function ψ crosses the vertical axis at

$$R := \psi(0) = \int_0^\infty K(a)da, \tag{9.4.35}$$

which is precisely the net reproductive rate. R must exceed 1 for λ_* to be positive, $R = 1$ if and only if $\lambda_* = 0$ and, finally, $R < 1$ if and only if $\lambda_* < 0$.

Next, we shall show that the sign of λ_* indeed determines the long-time behaviour of the population.

Let us consider the second term in the last formula of (9.4.32),

$$\hat{H}(\lambda) = \frac{\hat{G}(\lambda)\hat{K}(\lambda)}{1-\hat{K}(\lambda)}.$$

We noted that $\hat{G}(\lambda)$ and $\hat{K}(\lambda)$ tend to zero as $|\lambda| \to \infty$ in any half plane $\Re\lambda > \delta$, $\delta \in \mathbb{R}$. Furthermore, on any line $\{\sigma + iy;\ y \in \mathbb{R}\}$, which does not meet any root of (9.4.33), we have $\inf_{y\in\mathbb{R}} |1 - \hat{K}(\sigma+iy)| =: m > 0$ and

$$\begin{aligned}
\int_{-\infty}^{\infty} \left|\frac{\hat{G}(\sigma+iy)\hat{K}(\sigma+iy)}{1-\hat{K}(\sigma+iy)}\right| dy &\leq \frac{1}{m}\int_{-\infty}^{\infty} \left|\hat{G}(\sigma+iy)\hat{K}(\sigma+iy)\right| dy \\
&\leq \frac{1}{m}\int_{-\infty}^{\infty} \left|\int_0^\infty e^{-iyt}[e^{-\sigma t}G(t)]dt\right. \\
&\quad \cdot \left.\int_0^\infty e^{-iyt}[e^{-\sigma t}K(t)]dt\right| dy \\
&= \frac{1}{m}\int_{-\infty}^{\infty} \left|\mathcal{F}[\psi_{\sigma,G}](y)\cdot\mathcal{F}[\psi_{\sigma,K}](y)\right| dy < \infty,
\end{aligned} \tag{9.4.36}$$

where $\mathcal{F}$ is the Fourier transform and the functions $\psi_{\sigma,G}$ and $\psi_{\sigma,K}$ are the extensions of the respective functions by 0, as defined in (B.3.22). Since F and G have bounded supports in $\mathbb{R}_+$, multiplied by $e^{-\sigma t}$ they are $L_2(\mathbb{R})$ functions, and thus, the finiteness of the above integral follows from the Plancherel theorem, [99, Section 6.6 (c)], [79, Theorem 9.13]. Inverting $\hat{H}(\lambda)$, we have

$$H(t) = \frac{1}{2\pi i}\int\limits_{\sigma-i\infty}^{\sigma+i\infty} \frac{\hat{G}(\sigma+iy)\hat{K}(\sigma+iy)}{1-\hat{K}(\sigma+iy)} e^{(\sigma+iy)t}dy$$

for any $\sigma > \lambda_*$. Hence

$$B(t) = G(t) + H(t).$$

To estimate $H(t)$ we note that, by properties of $\hat{H}$, we can shift the line of integration to $\{\sigma_1 + iy : \; y \in \mathbb{R}\}$, where $\Re\lambda_1 < \sigma_1 < \lambda_*$ and λ_1 is the eigenvalue with the largest real part less than λ_*, see Fig. 9.4. Then, the Cauchy residue theorem gives

$$H(t) = H_1(t) + H_2(t),$$

where

$$\begin{aligned} H_1(t) &= res_{\lambda=\lambda_*}\frac{e^{\lambda t}\hat{G}(\lambda)\hat{K}(\lambda)}{1-\hat{K}(\lambda)} = \lim_{\lambda\to\lambda_*}(\lambda-\lambda_*)\frac{e^{\lambda t}\hat{G}(\lambda)\hat{K}(\lambda)}{1-\hat{K}(\lambda)} \\ &= \lim_{\lambda\to\lambda_*}(\lambda-\lambda_*)\frac{e^{\lambda t}\hat{G}(\lambda)\hat{K}(\lambda)}{\hat{K}(\lambda_*)-\hat{K}(\lambda)} \\ &= e^{\lambda_* t}\frac{\hat{G}(\lambda_*)\hat{K}(\lambda_*)}{-\hat{K}'_\lambda(\lambda_*)} = B_0 e^{\lambda_* t}, \end{aligned}$$

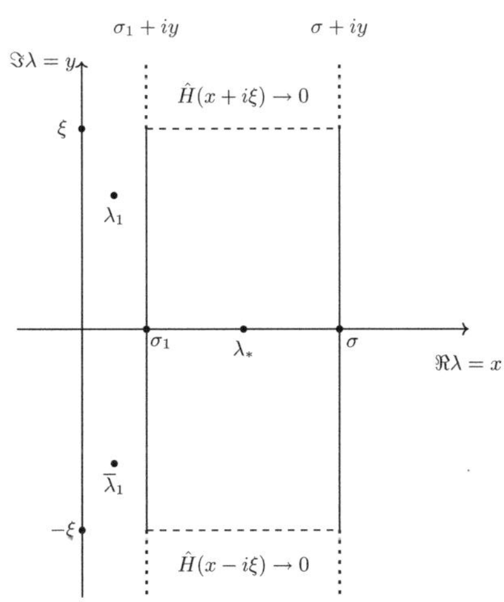

Fig. 9.4 Change of the integration line allowed since $\hat{H}(x \pm i\xi) \to 0$ as $\xi \to \pm\infty$

with

$$B_0 := \frac{\int_0^\infty e^{-\lambda_* a} G(a) da}{\int_0^\infty a e^{-\lambda_* a} K(a) da}$$

and

$$H_2(t) = \frac{1}{2\pi i} \int_{\sigma_1 - i\infty}^{\sigma_1 + i\infty} \frac{\hat{G}(\sigma_1 + iy)\hat{K}(\sigma_1 + iy)}{1 - \hat{K}(\sigma_1 + iy)} e^{(\sigma_1 + iy)t} dy.$$

The latter satisfies the estimate

$$|H_2(t)| \leq \frac{e^{\sigma_1 t}}{2\pi} \int_{\sigma_1 - i\infty}^{\sigma_1 + i\infty} \left| \frac{\hat{G}(\sigma_1 + iy)\hat{K}(\sigma_1 + iy)}{1 - \hat{K}(\sigma_1 + iy)} \right| dy = B_1 e^{\sigma_1 t},$$

where B_1 is a constant. Thus, we arrived at the representation

$$B(t) = G(t) + e^{\lambda_* t} B_0 + e^{\sigma_1 t} B_1.$$

However, remembering that $G(t) = 0$ for $t \geq \omega$ and assuming that $B_0 \neq 0$, we can write

$$B(t) = B_0 e^{\lambda_* t} \left(1 + \frac{e^{-\lambda_* t} G(t)}{B_0} + e^{(\sigma_1 - \lambda_*)t} \frac{B_1}{B_0}\right) = B_0 e^{\lambda_* t} \left(1 + \Omega(t)\right), \tag{9.4.37}$$

where $\Omega(t) \to 0$ as $t \to \infty$.

Now, $B_0 = 0$ if and only if $G(t) = 0$ for all $t \geq 0$ but then, from uniqueness, $B(t) = 0$ for all t. Let us interpret this condition. We have

$$0 = G(t) = \int_0^\infty \beta(a + t) \frac{S(a + t)}{S(a)} \mathring{n}(a) da,$$

which, by nonnegativity of $\mathring{n}$, is possible only if

$$\beta(a + t)\mathring{n}(a) = 0$$

for $a \in [0, \omega)$ and $t \geq 0$. This only occurs if the support of β is to the left of the support of $\mathring{n}$ (as the support of $\beta(\cdot + t)$ moves to the left as t increases). In other words, this case occurs only if the original population is too old to become fertile.

In this case,

$$n(t,a) = \begin{cases} \mathring{n}(a-t)\frac{S(a)}{S(a-t)}, & t < a, \\ 0, & a < t. \end{cases} \tag{9.4.38}$$

Otherwise, we can write

$$n(t,a) = \begin{cases} \mathring{n}(a-t)\frac{S(a)}{S(a-t)}, & t < a, \\ B_0 e^{\lambda_*(t-a)}\left(1+\Omega(t-a)\right) S(a), & a < t. \end{cases} \tag{9.4.39}$$

Now, since $\omega < +\infty$, we see that for $t \geq \omega$ we have

$$n(t,a) = B_0 e^{\lambda_*(t-a)}\left(1+\Omega(t-a)\right) S(a),$$

and we identify the stable age distribution

$$n_\infty(a) = e^{-\lambda_* a - \int\limits_0^a \mu(s)ds}.$$

so that

$$\lim_{t\to\infty} e^{-\lambda_* t} n(t,a) = e^{-\lambda_* a - \int\limits_0^a \mu(s)ds}$$

on $[0,\omega)$ (provided the supports of $\mathring{n}$ and β meet).

Finally, we noted in (9.4.35) that $\lambda_* > 0$, $\lambda_* = 0$ and $\lambda_* < 0$ if and only if, respectively, $R > 1$, $R = 1$ and $R < 1$. Thus, the population is growing if $R > 1$, it is stable if $R = 1$ and it decays if $R < 1$ (again if supports of $\mathring{n}$ and β meet), following the interpretation of the parameter R.

Exercise 9.4.3

(i) Show that asymptotically, the proportion of individuals $n_\infty(a), a \in I_\omega$, to newborns $n_\infty(0)$ is given by

$$\frac{n_\infty(a)}{n_\infty(0)} = e^{-\lambda_* a} S(a).$$

(ii) A census of a certain population revealed that there were 1000 individuals of age 40 and 800 individuals of age 50. Further, it is known that out of a large cohort of N, 600 individuals reach age 40, and 400 reach age 50. Assuming that the population reached the stage of stable growth, estimate λ_*. How much time will it take for the population to double?

Exercise 9.4.4 Assume that in a certain population that has reached the stage of stable growth, a male marries at the age of 40 years with two brides of age 20 years each. Further, assume that the male-to-female ratio at birth is one and all individuals in the population die at a constant (independent of age) death rate $\mu(a) = 0.015$. Find the growth rate of the community for it to be viable. Hint. There must be twice as many females of 20 as males of 40.

Chapter 10
Basic Nonlinear Models

10.1 Nonlinear Extensions of the McKendrick–von Foerster Model

10.1.1 Density-Dependent Demographic Coefficients

In Sect. 9.2, we introduced the age-structured McKendrick–von Foerster population model

$$\frac{\partial}{\partial t}n(t,a) + \frac{\partial}{\partial a}n(t,a) = -\mu(t,a)n(t,a), \qquad t, a > 0,$$

$$n(t,0) = \int\limits_0^{\omega} n(t,a)\beta(t,a)da, \quad t > 0,$$

$$n(0,a) = \mathring{n}(a), \quad a > 0,$$

where μ is the death rate, β is the maternity rate and $\omega \leq \infty$ is the maximum age of individuals in the population.

In many cases, the assumption that μ and β do not depend on the population is a serious oversimplification. A more realistic is the system

$$\begin{aligned}&\frac{\partial}{\partial t}n(t,a) + \frac{\partial}{\partial a}n(t,a) = -\mu(a,t,N_{\gamma_1})n(t,a), \qquad t, a > 0,\\ &n(t,0) = \int\limits_0^{\omega} n(t,a)\beta(a,t,N_{\gamma_2})da, \quad t > 0, \\ &n(0,a) = \mathring{n}(a), \quad a > 0,\end{aligned} \tag{10.1.1}$$

J. Banasiak, *Introduction to Mathematical Methods in Population Theory*, Springer Undergraduate Mathematics Series,
https://doi.org/10.1007/978-3-031-65491-6_10

where

$$N_{\gamma_i}(t) := \int_0^\infty \gamma_i(a)n(t,a)da, \qquad i = 1, 2,$$

are weighted total populations at time t. The weights γ_i account for the fact that the death and maternity rates may be more sensitive to the density of the population at particular ages.

These assumptions make (10.1.1) a nonlinear equation that can be solved explicitly only in exceptional cases by using a technique often referred to as the linear chain trick. Later, we shall describe some cases yielding to this approach, but first, we will discuss another nonlinear extension.

10.1.2 An Epidemiological System with Age Structure

Standard epidemiological models, introduced in Sect. 6.1.2, treat the population as homogeneous apart from the differences due to the infection status. Here, we confine ourselves to consider SIR models when, recall, the population is divided into three main classes: susceptibles S (individuals who are not sick and can be infected), infectives I (individuals who have the disease and can infect others) and removed R (individuals who were infective but recovered and are now immune, dead or isolated). Then, the total population at time t is given by

$$N(t) = S(t) + I(t) + R(t).$$

In many models, it is assumed that the population size is constant, disregarding thus the vital dynamics due to births and deaths. Then, the SIR model, introduced in (6.1.7), is given by

$$\begin{aligned} S' &= -\lambda S + \gamma I, \\ I' &= \lambda S - (\gamma + \nu)I, \qquad (10.1.2) \\ R' &= \nu I, \end{aligned}$$

with $S(0) = \mathring{S}$, $I(0) = \mathring{I}$, $R(0) = \mathring{R}$ and $\mathring{S} + \mathring{I} + \mathring{R} = N$. As before, the parameter λ is the force of infection, γ is the recovery rate without immunity and ν is the recovery/removal rate. While ν and γ are usually taken to be constant, the force of infection requires a constitutive law, which we discussed in Sect. 6.1.2. However, in the case of constant total population, the mass action law and the standard incidence are mathematically the same and can be written as

$$\lambda = c\phi\frac{I}{N}, \tag{10.1.3}$$

where c is the contact rate (the number of contacts that a single individual has with other individuals in the population per unit time, ϕ is the ineffectiveness, that is, the probability that a contact with an infective will result in infection, and $\frac{I}{N}$ is the probability that the contacted individual is infectious. In what follows, we shall denote $\beta = \frac{c\phi}{N}$.

There are many other assumptions underlying (10.1.2): apart from the population being homogenous, we assume that no multiple infections are possible, an infected individual immediately becomes infective, etc.

Concerning the nature of the diseases, the basic distinction is between those which are not lethal and do not impart immunity (influenza, common cold) and those which could be caught only once (leading to death or immunity), such as measles or AIDS. In the first case, $\nu = 0$ and the model is referred to as an SIS model and in the second $\gamma = 0$, and the model is called an SIR model.

In many cases, the infection rate significantly varies with age; thus, it is important to consider the population's age structure. Thus, we expect the interaction of the vital dynamics and the infection mechanism to produce a nontrivial behaviour.

To introduce the model, we note again that, in the absence of the disease, the age-dependent density of the population $n(t, a)$ would be the solution of the linear model introduced in (9.1.7a)–(9.1.7c). However, because of the epidemics, the population, as before, is partitioned into three classes: susceptibles, infectives, and removed, represented by their respective age-specific densities $s(t, a), i(t, a)$ and $r(t, a)$. Now, if we look at the population of susceptibles, then we see that it is losing individuals at the rate $\lambda(t, a)s(t, a)$ and gaining at the rate $\delta(a)i(t, a)$, where we have taken into account that the infection force and the cure rate are age dependent. Similarly, the source terms for the other two classes are given by the (age-dependent) terms of the (6.1.7) model. This leads to the system

$$\begin{aligned}
&\frac{\partial}{\partial t}s(t, a) + \frac{\partial}{\partial a}s(t, a) + \mu(a)s(t, a) = -\lambda(t, a)s(t, a) + \gamma(a)i(t, a),\\
&\frac{\partial}{\partial t}i(t, a) + \frac{\partial}{\partial a}i(t, a) + \mu(a)i(t, a) = \lambda(t, a)s(t, a) - (\nu(a) + \gamma(a))i(t, a),\\
&\frac{\partial}{\partial t}r(t, a) + \frac{\partial}{\partial a}r(t, a) + \mu(a)r(t, a) = \nu(a)i(t, a).
\end{aligned} \tag{10.1.4}$$

$$s(t,0) = \int_0^\omega \beta(a)(s(t,a) + (1-q)i(t,a) + (1-w)r(t,a))da,$$

$$i(t,0) = q\int_0^\omega \beta(a)i(t,a)da, \tag{10.1.5}$$

$$r(t,0) = w\int_0^\omega \beta(a)r(t,a)da,$$

where $q \in [0, 1]$ and $w \in [0, 1]$ are the vertical transmission coefficients of infectiveness and immunity, respectively. The system is complemented by initial conditions $s(0,a) = \mathring{s}(a), i(0,a) = \mathring{i}(a)$ and $r(0,a) = \mathring{r}(a)$. Here, we assumed that the disease did not significantly affect the death and birth coefficients. In particular, if we assume that the solution of (10.1.4) exists in such a way that all terms are separately well defined, then adding the equations together, we obtain that the total population density $n(t,a) = s(t,a) + i(t,a) + r(t,a)$ satisfies

$$\frac{\partial}{\partial t}n(t,a) + \frac{\partial}{\partial a}n(t,a) = -\mu(a)n(t,a),$$

$$n(t,0) = \int_0^\omega \beta(a)n(t,a)da,$$

$$n(0,a) = \mathring{n}(a) = \mathring{s}(a) + \mathring{i}(a) + \mathring{r}(a),$$

that is, as expected from the model, the disease does not change the global picture of the evolution of the population.

Finally, we must specify a constitutive relation for the force of infection λ. This is usually given by given by the equation

$$\lambda(t,a) = K_0(a)i(t,a) + \int_0^\omega K(a,s)i(t,s)ds, \tag{10.1.6}$$

where K_0 and K are known functions. The two terms on the right-hand side are the intracohort and intercohort terms, respectively. The intracohort term describes the situation in which individuals can only be infected by those of their own age. In contrast, the intercohort term describes cases in which individuals of any age can infect each other.

Exercise 10.1.1 Give examples of functions K which describe

(i) a disease in which only children can infect each other;
(ii) a disease in which children can affect each other and adults, but adults cannot infect children,
(iii) a disease in which adults can infect each other and children, but the latter do not infect anybody.

Analysis of this problem requires abstract theory, which will be developed later.

10.1.3 *An Exactly Solvable Nonlinear Model—A Linear Chain Trick*

Consider

$$\begin{aligned} \frac{\partial}{\partial t}n(t,a) + \frac{\partial}{\partial a}n(t,a) &= -\mu(N(t))n(t,a), \qquad t > 0, 0 < a < \omega, \\ n(t,0) &= \int\limits_0^\omega \beta(N(t))n(t,a)da, \quad t > 0, \\ n(0,a) &= \mathring{n}(a), \quad 0 < a < \omega. \end{aligned} \tag{10.1.7}$$

We assume that there is a nonnegative classical solution n to this problem defined for $t \in [0, t_{max})$ such that

$$\int_0^\omega \frac{\partial}{\partial t}n(t,a)da = \frac{\partial}{\partial t}\int_0^\omega n(t,a)da,$$

and $n(\omega, t) = 0$ for all $t \in [0, t_{max})$. Then, by these assumptions,

$$\begin{aligned} \int\limits_0^\omega \frac{\partial}{\partial a}n(t,a)da &= n(\omega,t) - n(t,0) = 0 - \beta(N(t))\int\limits_0^\omega n(t,a)da \\ &= -\beta(N(t))N(t), \end{aligned} \tag{10.1.8}$$

and we obtain an ordinary differential equation for the total population

$$\begin{aligned} N' &= N(\beta(N) - \mu(N)), \\ N(0) &= \mathring{N} = \int\limits_0^\omega \mathring{n}(a)da. \end{aligned} \tag{10.1.9}$$

This simple separable equation can be solved explicitly or in quadratures, depending on the form of μ and β. An example allowing for explicit solutions will be discussed later. Here, we shall address two more theoretical issues.

First, let us ask ourselves whether this procedure solves the original problem. To wit, suppose N is a the solution to (10.1.9) defined on an interval $[0, t_{max})$. Substituting this (known) N into (10.1.7), we obtain a linear boundary value problem which can be explicitly solved, yielding a solution, say, u. Is this a solution to the original nonlinear problem (10.1.7)? We observe that u solves

$$
\begin{aligned}
\frac{\partial}{\partial t}u(t,a) + \frac{\partial}{\partial a}u(t,a) &= -\mu(N(t))n(t,a), \qquad t > 0, 0 < a < \omega,\\
u(t,0) &= \int_0^\omega \beta(N(t))u(t,a)da, \quad t > 0, \qquad (10.1.10)\\
u(0,a) &= \mathring{n}(a), \quad 0 < a < \omega,
\end{aligned}
$$

where N is a given function, whereas n is a solution to

$$
\begin{aligned}
\frac{\partial}{\partial t}n(t,a) + \frac{\partial}{\partial a}n(t,a) &= -\mu\left(\int_0^\omega n(t,a)da\right)n(t,a), \qquad t > 0, 0 < a < \omega,\\
n(t,0) &= \int_0^\omega n(t,a)\beta\left(\int_0^\omega n(t,a)da\right)da, \quad t > 0,\\
n(0,a) &= \mathring{n}(a), \quad 0 < a < \omega.
\end{aligned} \qquad (10.1.11)
$$

Integrating (10.1.10) we find that $U(t) = \int_0^\omega u(t,a)da$ satisfies

$$
\begin{aligned}
\frac{dU}{dt} &= U(\beta(N) - \mu(N)),\\
U(0) &= \mathring{N}.
\end{aligned} \qquad (10.1.12)
$$

Using the fact that N is a given solution to (10.1.9), the factor $(\beta(N) - \mu(N))$ in the latter and in (10.1.12) is the same, $U - N$ satisfies

$$
\begin{aligned}
\frac{d(U-N)}{dt} &= (U-N)(\beta(N) - \mu(N)),\\
U(0) - N(0) &= 0,
\end{aligned}
$$

which is a linear equation and thus uniquely solvable. Hence, $U(t) = N(t)$ on $[0, t_{max})$. Thus u satisfies

$$\frac{\partial}{\partial t}u(t,a) + \frac{\partial}{\partial a}u(t,a) = -\mu\left(\int_0^{\omega} u(t,a)da\right)n(t,a), \qquad t > 0, 0 < a < \omega$$

$$u(t,0) = \int_0^{\omega} \beta\left(\int_0^{\omega} u(t,a)da\right)u(t,a)da, \quad t > 0,$$

$$u(0,a) = \mathring{n}(a), \quad 0 < a < \omega,$$

and hence $u(t,a) = n(t,a)$, provided (10.1.7) is uniquely solvable (which is a separate issue.)

The second issue we consider here is the long-time behaviour of solutions. Assume that there is a unique nontrivial equilibrium for (10.1.9) denoted by N^*. This can be ensured if, e.g., $\beta(N)$ is a decreasing and $\mu(N)$ an increasing function with $\mu(0) < \beta(0)$ or if $\mu(0) = \beta(0)$ and $\beta'' - \mu'' < 0$. Under these assumptions, N^* is the global attractor, that is, $N(t) \to N^*$ as $t \to \infty$. Is there a corresponding stationary density $n^*(a)$ such that

$$N^* = \int_0^{\infty} n^*(a)da?$$

A stationary density is a solution to

$$\begin{aligned} \frac{\partial}{\partial a}n(a) &= -\mu\left(\int_0^{\omega} n(a)da\right)n(a), \qquad 0 < a < \omega, \\ n(0) &= \int_0^{\omega} n(a)\beta\left(\int_0^{\omega} n(a)da\right). \end{aligned} \tag{10.1.13}$$

We can follow the argument for the time-dependent solution. Assuming $n(\omega) = 0$ and integrating the first equation over I_ω, we find

$$-N\beta(N) = -\mu(N)N,$$

so that the total population at the stationary state is the equilibrium population. Further, again arguing as in the time-dependent case, we see that the nonlinear, nonlocal equation (10.1.13) for the stationary state n^* can be replaced by the linear one

$$\frac{\partial}{\partial a}n(a) = -\mu\left(N^*\right)n(a), \qquad 0 < a < \omega$$
$$n(0) = N^*\beta(N^*),$$

which can be explicitly solved, yielding

$$n^*(a) = N^*\beta(N^*)e^{-\mu(N^*)}. \tag{10.1.14}$$

This can be considered as the stable age profile of the population. However, the question whether $n(t,a) \to n^*(a)$ as $t \to \infty$ remains to be addressed.

Example 10.1 Consider

$$\frac{\partial}{\partial t}n(t,a) + \frac{\partial}{\partial a}n(t,a) = -\mu_0 N(t)n(t,a), \qquad t > 0,\ 0 < a < \infty,$$
$$n(t,0) = \beta_0 \int_0^\infty n(t,a)N(t)da, \quad t > 0, \tag{10.1.15}$$
$$n(0,a) = \mathring{n}(a), \quad 0 < a < \infty,$$

where μ_0, β_0 are positive constants.

Let us integrate the first and the second equation with respect to a. Using the considerations from the beginning of the subsection, we obtain

$$\frac{dN}{dt} = N^2(\beta_0 - \mu_0),$$
$$N(0) = \mathring{N} = \int_0^\infty \mathring{n}(a)da. \tag{10.1.16}$$

There are two cases to consider.

(a) $\mu_0 = \beta_0$.
In this case, $N(t) = \mathring{N}$ for all $0 \le t \le t_{max}$ and (10.1.15) turns into

$$\frac{\partial}{\partial t}n(t,a) + \frac{\partial}{\partial a}n(t,a) = -\mu_0\mathring{N}n(t,a), \qquad t > 0, 0 < a < \infty,$$
$$n(t,0) = \beta_0\mathring{N}\int_0^\infty n(t,a)da, \quad t > 0,$$
$$n(0,a) = \mathring{n}(a), \quad a > 0, \tag{10.1.17}$$

which can be transformed to the form (9.3.11) with the boundary condition given explicitly. We could use the formula which we derived there, but here, we can use the fact that

$$n(t,0) = \int_0^\infty n(t,a)\beta_0 \mathring{N} da = \beta_0 \mathring{N} \int_0^\infty n(t,a) da = \beta_0 \mathring{N} N(t) = \beta_0 \mathring{N}^2,$$

and hence, we deal with a straightforward initial boundary value problem

$$\frac{\partial n(t,a)}{\partial t} + \frac{\partial n(t,a)}{\partial a} = -\mu_0 \mathring{N} n(t,a), \qquad t, a > 0,$$
$$n(t,0) = \beta_0 \mathring{N}^2, \quad t > 0,$$
$$n(0,a) = \mathring{n}(a), \quad a > 0,$$

the solution of which is, by (9.3.11),

$$n(t,a) = e^{-\mu_0 \mathring{N} a} u(t,a) = e^{-\mu_0 \mathring{N} a} \begin{cases} e^{\mu_0 \mathring{N}(a-t)} \mathring{n}(a-t), & t < a, \\ \beta_0 \mathring{N}^2, & a < t, \end{cases}$$
$$= \begin{cases} e^{-\mu_0 \mathring{N} t} \mathring{n}(a-t), & t < a, \\ e^{-\mu_0 \mathring{N} a} \beta_0 \mathring{N}^2, & a < t. \end{cases}$$

(b) $\mu_0 \neq \beta_0$.
In this case, the solution of (10.1.16) is given by

$$\frac{1}{\mathring{N}} - \frac{1}{N(t)} = (\beta_0 - \mu_0)t,$$

that is,

$$N(t) = \frac{\mathring{N}}{1 - (\beta_0 - \mu_0)\mathring{N} t}. \tag{10.1.18}$$

At this moment, we note that the evolution of the population is determined by the sign of $r_0 = \beta_0 - \mu_0$, which plays the role of the net growth rate. If $r_0 > 0$ (that is, if the maternity coefficient exceeds the mortality rate), the population will explode at $t_{max} = \frac{1}{r_0 \mathring{N}}$. On the other hand, if $r_0 < 0$ (that is, if the maternity coefficient is lower than the mortality rate), the population will exist for all time, gradually decaying to 0. However, in both cases, we can give explicit formulae for $n(t,a)$. Indeed, similarly to (a), we arrive at the straightforward initial boundary value problem

$$\begin{aligned}\frac{\partial}{\partial t}n(t,a)+\frac{\partial}{\partial a}n(t,a) &= -\mu_0 N(t)n(t,a), \qquad t,a>0,\\ n(t,0) &= \beta_0 N^2(t), \quad t>0,\\ n(0,a) &= \mathring{n}(a), \quad a>0,\end{aligned} \tag{10.1.19}$$

where $N(t)$ is given (10.1.18). However, contrary to (10.1.18) (and (9.2.1a)), the coefficient μ is dependent on time, and thus the reduction must be made with more care. We begin by noting that

$$\frac{d}{dt}e^{\mu_0\int\limits_0^t N(s)ds} = \mu_0 N(t)e^{\mu_0\int\limits_0^t N(s)ds}$$

and thus the first equation of (10.1.19) can be written as

$$\frac{\partial}{\partial t}\left(n(t,a)e^{\mu_0\int\limits_0^t N(s)ds}\right)+\frac{\partial}{\partial a}\left(n(t,a)e^{\mu_0\int\limits_0^t N(s)ds}\right)=0.$$

If we denote

$$u(t,a)=n(t,a)e^{\mu_0\int\limits_0^t N(s)ds},$$

then

$$u(0,a)=n(0,a)e^{\mu_0\int\limits_0^0 N(s)ds}=\mathring{n}(a)$$

and

$$u(t,0)=n(t,0)e^{\mu_0\int\limits_0^t N(s)ds}=\beta_0 N^2(t)e^{\mu_0\int\limits_0^t N(s)ds},$$

so that, using again (9.3.11), we obtain

$$\begin{aligned}n(t,a) &= e^{-\mu_0\int\limits_0^t N(s)ds}u(t,a)\\ &= e^{-\mu_0\int\limits_0^t N(s)ds}\begin{cases}\mathring{n}(a-t), & t<a,\\ \beta_0 e^{\mu_0\int\limits_0^{t-a} N(s)ds}N^2(t-a), & a<t,\end{cases}\end{aligned}$$

$$= \begin{cases} e^{-\mu_0 \int\limits_0^t N(s)ds} \mathring{n}(a-t), & t < a, \\ \beta_0 e^{-\mu_0 \int\limits_{t-a}^t N(s)ds} N^2(t-a), & a < t, \end{cases}$$

as long as N is defined. For our particular case, we have

$$\int N(s)ds = \mathring{N} \int \frac{ds}{1 - r_0 \mathring{N} t} = \ln \left(\frac{1}{1 - \mathring{N} r_0 t} \right)^{\frac{1}{r_0}} + C$$

for $0 \leq t < \frac{1}{r_0 \mathring{N}}$ if $r_0 > 0$, and for $0 \leq t < \infty$ otherwise. Hence, for such t,

$$e^{-\mu_0 \int\limits_0^t N(s)ds} = (1 - r_0 \mathring{N} t)^{\frac{\mu_0}{r_0}}$$

and

$$e^{-\mu_0 \int\limits_{t-a}^t N(s)ds} = \left(\frac{1 - r_0 \mathring{N} t}{1 - r_0 \mathring{N}(t-a)} \right)^{\frac{\mu_0}{r_0}}.$$

Thus

$$n(t,a) = \begin{cases} (1 - r_0 \mathring{N} t)^{\frac{\mu_0}{r_0}} \mathring{n}(a-t), & t < a, \\ \frac{\beta_0 \mathring{N}^2}{(1 - r_0 \mathring{N}(t-a))^2} \left(\frac{1 - r_0 \mathring{N} t}{1 - r_0 \mathring{N}(t-a)} \right)^{\frac{\mu_0}{r_0}}, & a < t, \end{cases}$$

where $t < t_{max}$ if $r_0 > 0$. We note that the assumption (10.1.8) is satisfied if $\lim_{a \to \infty} \mathring{n}(a) = 0$, so the procedure is justified for such initial conditions.

In this example, we cannot talk about a stable age profile, as there is no equilibrium of the total population, or every positive number is an equilibrium (if $\mu_0 = \beta_0$).

Exercise 10.1.2 Use the above method to solve the constant coefficient McKendrick–von Foerster problem

$$n_t(a,t) + n_a(a,t) = -\mu n(a,t). \tag{10.1.20}$$

with

$$n(0,t) = \beta \int\limits_0^\infty n(a,t)da, \qquad n(a,0) = \mathring{n}(a),$$

that is, first find the equation for the evolution of the total population $N(t) = \int_0^\infty n(a,t)da$ (assuming $\lim_{a\to\infty} n(a,t) = 0$ for any t).

10.1.4 Another Example of Linear Chain Trickery, [60, Chapter 24]

Consider a class of organisms of potentially infinite longevity which can reproduce immediately after birth and whose age-specific birth rate decreases exponentially with age. This results in the system

$$\begin{aligned} &\frac{\partial}{\partial t}n(t,a) + \frac{\partial}{\partial a}n(t,a) = -\mu(N(t))n(t,a), \qquad t>0, 0<a<\infty,\\ &n(t,0) = \beta_0 \int_0^\infty e^{-\gamma a}n(t,a)da, \quad t>0,\\ &n(0,a) = \mathring{n}(a), \quad 0<a<\infty, \end{aligned} \tag{10.1.21}$$

where $\gamma, \beta_0 > 0$ are constants and μ is a monotonic function. Assuming, as before, that there is a nonnegative classical solution n to this problem defined for $t \in [0,\infty)$ such that

$$\int_0^\infty \frac{\partial}{\partial t}n(t,a)da = \frac{\partial}{\partial t}\int_0^\infty n(t,a)da$$

and $\lim_{a\to\infty} n(t,a) = 0$ for all $t \in [0,\infty)$, we integrate the equation with respect to a over $[0,\infty)$,

$$\int_0^\infty \frac{\partial}{\partial t}n(t,a) + \lim_{a\to\infty} n(t,a) - n(t,0) = -\mu(N(t))\int_0^\infty n(t,a)da,$$

to obtain

$$N' = B - \mu(N)N, \tag{10.1.22}$$

where

$$B(t) = n(t,0) = \beta_0 \int_0^\infty e^{-\gamma a}n(t,a)da. \tag{10.1.23}$$

To close the system, we multiply the equation by the maternity function $\beta_0 e^{-\gamma a}$,

$$\frac{\partial}{\partial t}\left(\beta_0 e^{-\gamma a} n(t,a)\right) + \beta_0 e^{-\gamma a}\frac{\partial}{\partial a} n(t,a) = -\mu(N(t))\beta_0 e^{-\gamma a} n(t,a),$$

and integrate over $[0, \infty)$ so that

$$B'(t) + \int_0^\infty \beta_0 e^{-\gamma a}\frac{\partial}{\partial a} n(t,a)da = -\mu(N(t))B(t). \tag{10.1.24}$$

The integral can be transformed by integration by parts

$$\int_0^\infty e^{-\gamma a}\frac{\partial}{\partial a} n(t,a)da = -n(t,0) + \gamma\int_0^\infty e^{-\gamma a} n(t,a)da = -\beta_0 B(t) + \gamma B(t),$$

so that (10.1.24) can be written as

$$B'(t) - \beta_0 B(t) + \gamma B(t) = -\mu(N(t))B(t). \tag{10.1.25}$$

Hence, this time, we have obtained a system of differential equations,

$$\begin{aligned} N' &= B - \mu(N)N, \\ B' &= (\beta_0 - \gamma - \mu(N))B. \end{aligned} \tag{10.1.26}$$

This system can be solved in some special cases; otherwise, we must use a phase-plane analysis.

Constant Death Rate

We begin with a simple case when $\mu(N) = \mu$ is a constant. Then, we obtain the decoupled system

$$\begin{aligned} N' &= B - \mu N, \\ B' &= (\beta_0 - \gamma - \mu)B. \end{aligned}$$

This gives

$$B(t) = B_0 e^{(\beta_0 - \gamma - \mu)t},$$

where

$$B_0 = B(0) = \beta_0\int_0^\infty e^{-\gamma a}\mathring{n}(a)da.$$

Then, the first equation becomes a linear nonhomogeneous equation

$$N' = -\mu N + B_0 e^{(\beta_0 - \gamma - \mu)t}.$$

Using the integrating factor, we have

$$(N e^{\mu t})' = B_0 e^{(\beta_0 - \gamma)t},$$

that is,

$$N(t) = \mathring{N} e^{-\mu t} + \frac{B_0}{\beta_0 - \gamma}\left(e^{(\beta_0-\gamma-\mu)t} - e^{-\mu t}\right),$$

provided $\beta_0 \neq \gamma$. Note that finding N is not necessary for finding n. Indeed, in our case (10.1.21) takes the form

$$\begin{aligned} \frac{\partial}{\partial t}n(t,a) + \frac{\partial}{\partial a}n(t,a) &= -\mu n(t,a), \qquad t>0,\ 0<a<\infty,\\ n(t,0) = B(t) &= B_0 e^{(\beta_0-\gamma-\mu)t}, \quad t>0,\\ n(0,a) &= \mathring{n}(a), \quad 0<a<\infty. \end{aligned} \tag{10.1.27}$$

Then, as in (9.3.11), we get

$$n(t,a) = e^{-\mu t}\begin{cases} \mathring{n}(a-t), & t<a,\\ \beta_0 e^{(\beta_0-\gamma)(t-a)} \int\limits_0^\infty e^{-\gamma r}\mathring{n}(r)dr, & a<t. \end{cases}$$

As before, the procedure is justified if $\lim_{a\to\infty}\mathring{n}(a) = 0$.

General Death Rate

Here, we assume that μ is a strictly increasing function.

In general, the system (10.1.21) cannot be explicitly solved, and the best thing we can hope for is to determine the asymptotic behaviour of the solution. We shall use phase plane analysis to do this. The nullclines are given by

$$\begin{aligned} B &= \mu(N)N,\\ 0 &= (\beta_0 - \gamma - \mu(N))B, \end{aligned} \tag{10.1.28}$$

and the stationary points are at the intersection of these nullclines. We immediately see that we have trivial equilibrium $(0, 0)$, and the nontrivial ones are given as the solutions of

$$\mu(N) = \beta_0 - \gamma, \qquad B = \mu(N)N.$$

A nontrivial equilibrium exists if $\beta_0 - \gamma$ is in the range of μ (so, in particular, $\beta_0 - \gamma > 0$), and then it is unique by monotonicity of μ. Assume this is the case and denote the equilibrium by (N^*, B^*). As usual, we begin by looking at the stability of the equilibrium. The Jacobi matrix is given by

$$\mathcal{J}(N, B) = \begin{pmatrix} -\mu'(N)N - \mu(N) & 1 \\ -\mu'(N)B & (\beta_0 - \gamma - \mu(N)) \end{pmatrix},$$

and

$$\mathcal{J}(N^*, B^*) = \begin{pmatrix} -\mu'(N^*)N^* - \mu(N^*) & 1 \\ -\mu'(N^*)B^* & 0 \end{pmatrix},$$

with the characteristic equation

$$\lambda^2 + (\mu'(N^*)N^* + \mu(N^*))\lambda + \mu'(N^*)B^* = 0.$$

Under the natural biological assumption adopted above, $\mu' > 0$, so, using Vieta's formulae, we see that if there are two real roots, then they are of the same sign with their sum negative, which yields that they are both negative. If they are complex, they are complex conjugates; thus, the equal real parts are negative. Hence, the equilibrium is asymptotically stable.

We can better understand the dynamics by looking for an invariant manifold (line) $B = \phi(N)$ for (10.1.26), where ϕ is the unknown function such that

$$N' = \phi(N) - \mu(N)N.$$

Further, by (10.1.23), $N = 0$ implies $B = 0$, so $\phi(0) = 0$. For it to determine an invariant manifold, the second equation must be identically satisfied, which yields

$$\phi'(N)N' = \phi'(N)(\phi(N) - \mu(N)N) = B' = (\beta_0 - \gamma - \mu(N))\phi(N).$$

Considering

$$\phi(N) = \phi_1 N + \ldots,$$

we obtain

$$(\phi_1 + \ldots)(\phi_1 N + \ldots - \mu(N)N) = (\beta_0 - \gamma - \mu(N))(\phi_1 N + \ldots). \qquad (10.1.29)$$

Hence, we see that

$$B = (\beta_0 - \gamma)N$$

is an exact solution to (10.1.29) and hence the invariant line (manifold). Thus, if (B, N) is on the invariant manifold, it will stay there. Moreover, if $x(t) := B(t) - (\beta_0 - \gamma)N(t) \neq 0$, then

$$x'(t) = B'(t) - (\beta_0 - \gamma)N'(t) = -\mu(N(t))x(t),$$

and hence, it will decrease to 0 monotonically (even exponentially, provided $\mu(N)$ is separated from zero). On the manifold, the equation reduces to

$$N' = ((\beta_0 - \gamma) - \mu(N))N,$$

which produces a logistic-like behaviour. It is also clear that there cannot be closed orbits or damped oscillations around the nontrivial equilibrium, as such orbits would have to cut the invariant manifold, which is impossible. We can get a global phase portrait if we note that since $\mu' > 0$, for the existence of the stationary state N^*, we must have $\mu(0) < \beta_0 - \gamma$. This shows that

$$\frac{d}{dN}(N\mu(N))|_{N=0} = \mu(0) < \beta_0 - \gamma$$

and thus the nullcline $B = \mu(N)N$ enters the first quadrant below the invariant manifold $B = (\beta_0 - \gamma)N$. Furthermore, since along this nullcline, the vector field is of the form $(0, (\beta_0 - \gamma - \mu(N))B)$ and along the invariant manifold it is given by $(\beta_0 - \gamma - \mu(N))N(1, (\beta_0 - \gamma))$ we see that the only intersection of the nullcline and the invariant manifold can occur at the equilibrium (N^*, B^*). Hence, the nullcline and the invariant manifold must be situated with respect to each other as in Fig. 10.1. From the picture, we see that the sectors between the nullcline and the invariant manifold are trapping regions and any trajectory starting outside them must eventually get there by monotonicity, the fact that $B = 0$ is also an invariant manifold and the fact that any finite limit point of a trajectory must be

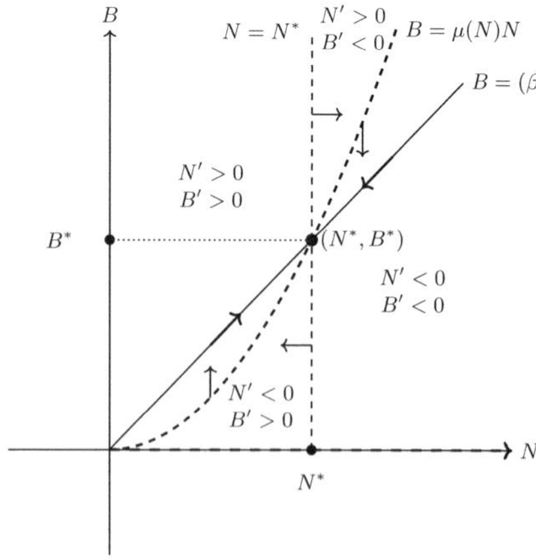

Fig. 10.1 Nullclines of (10.1.26) (dashed) and the invariant line (solid)

Fig. 10.2 Streamlines of (10.1.26) with $\mu(N) = N$, $\beta_0 = 2$ and $\gamma = 1$, the invariant manifold $B = N$ and the globally stable equilibrium (1, 1)

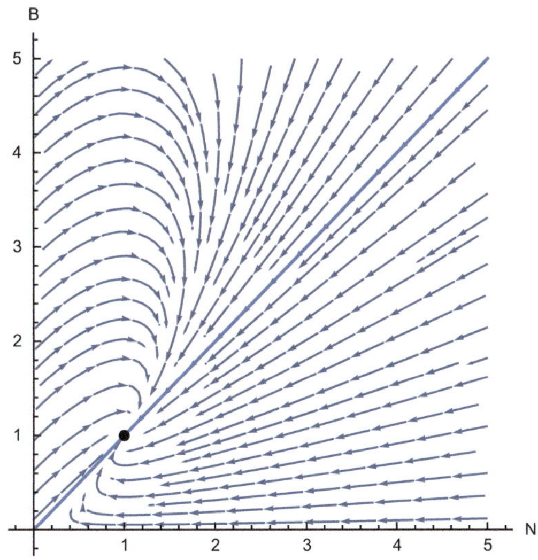

an equilibrium, see Fig. 10.2. Finally, using Eq. (10.1.14), we find that the stable age profile in this case is

$$n^*(a) = (\beta_0 - \gamma)N^* e^{-(\beta_0-\gamma)a}.$$

As explained earlier, the total population $N(t)$ converges to N^*, the total population of the stationary density n^*, but proving that the densities converge requires a separate theory.

10.2 Solvability of the SIR/SIS System with Age Structure

Let us recall the system

$$\begin{aligned}
&\frac{\partial}{\partial t}s(t,a) + \frac{\partial}{\partial a}s(t,a) + \mu(a)s(t,a) = -\lambda(t,a)s(t,a) + \gamma(a)i(t,a),\\
&\frac{\partial}{\partial t}i(t,a) + \frac{\partial}{\partial a}i(t,a) + \mu(a)i(t,a) = \lambda(t,a)s(t,a) - (\nu(a) + \gamma(a))i(t,a),\\
&\frac{\partial}{\partial t}r(t,a) + \frac{\partial}{\partial a}r(t,a) + \mu(a)r(t,a) = \nu(a)i(t,a),
\end{aligned} \tag{10.2.30}$$

with the boundary conditions

$$
\begin{aligned}
s(t,0) &= \int_0^{\omega} \beta(a)(s(t,a) + (1-q)i(t,a) + (1-w)r(t,a))da,\\
i(t,0) &= q\int_0^{\omega} \beta(a)i(t,a)da, \qquad\qquad (10.2.31)\\
r(t,0) &= w\int_0^{\omega} \beta(a)r(t,a)da,
\end{aligned}
$$

where $q \in [0, 1]$ and $w \in [0, 1]$ are the vertical transmission coefficients of infectivity and immunity, respectively, and with the initial conditions

$$
s(0,a) = \mathring{s}(a), \quad i(0,a) = \mathring{i}(a), \quad r(0,a) = \mathring{r}(a). \qquad (10.2.32)
$$

To deal with the solvability of this system, we have to provide a more general framework for the original McKendrick–von Foerster problem.

10.2.1 Solution of the McKendrick–von Foerster Problem as a Semi-dynamical System

Let us recall that in Sect. 9.3.2, we considered the problem

$$
\begin{aligned}
\frac{\partial}{\partial t}n(t,a) + \frac{\partial}{\partial a}n(t,a) &= -\mu(a)n(t,a),\\
n(t,0) &= \int_0^{\omega} \beta(a)n(t,a)da, \qquad\qquad (10.2.33)\\
n(0,a) &= \mathring{n}(a),
\end{aligned}
$$

and proved that, under assumptions (9.3.15)–(9.3.18), it has a unique solution

$$
n(t,a) = \begin{cases} \frac{S(a)}{S(a-t)}\mathring{n}(a-t), & t < a,\\ S(a)B(t-a), & a < t, \end{cases} \qquad (10.2.34)
$$

where B is the solution of the Volterra equation

$$
\begin{aligned}
B(t) &= \int_0^{t} K(t-a)B(a)da + \int_0^{\omega - t} \beta(a+t)\frac{S(a+t)}{S(a)}\mathring{n}(a)da\\
&= \int_0^{t} K(t-a)B(a)da + \int_t^{\omega} \beta(a)\frac{S(a)}{S(a-t)}\mathring{n}(a-t)da
\end{aligned} \qquad (10.2.35)
$$

for $0 \le t \le \omega$, and

$$B(t) = \int_0^\omega K(t-a)B(a)da \tag{10.2.36}$$

for $t > \omega$. The solution satisfies the estimate

$$\|n(t,\cdot)\|_{L_1(I_\omega)} \le \|\mathring{n}\|_{L_1(I_\omega)} e^{t\|\beta\|_\infty}. \tag{10.2.37}$$

From now on, we assume that $\omega < \infty$. The case of a potentially infinite lifespan $\omega = \infty$ is similar but requires slightly different proof techniques.

The unique solvability of (10.2.33) in the sense of formula (10.2.34) allows for the dynamical systems approach to the problem. This amounts to looking at the solution not through individual trajectories but more globally as a flow, that is, a family of mappings $\phi(t, \boldsymbol{x})$ of the state space into itself, parametrized by time, see Sect. 6.2, where the flow was introduced in the finite-dimensional setting. Let us recall that our state space is $X = L_1(I_\omega)$ of all population densities with finite total population. To emphasize that in the current context, the flow is linear with respect to the initial condition, we introduce the following notation

$$[T(t)\mathring{n}](a) := \phi(t, \mathring{n})(a) = n(t,a), \quad a \in I_\omega, t \ge 0, \tag{10.2.38}$$

where n is the solution defined by (10.2.34). From (10.2.37), we see that $(T(t))_{t\ge 0}$ is a family of linear bounded operators on X with at most exponential growth in time, often called a semigroup (of linear operators). In what follows, we investigate the properties of $(T(t))_{t\ge 0}$.

Earlier, we observed that, in general, the solution (10.2.34) is discontinuous along the characteristic $a = t$. Here, we prove that condition (9.3.23) is essential for the solution's regularity.

Theorem 10.2 *Let us assume that (9.3.15)–(9.3.18) are satisfied. Further, let*

$$\mathring{n} \in W_1^1(I_\omega), \quad \mu\mathring{n} \in L_1(I_\omega), \tag{10.2.39}$$

and

$$\mathring{n}(0) = \int_0^\omega \beta(a)\mathring{n}(a)da. \tag{10.2.40}$$

Then $n \in C([0,\infty)\times[0,\omega])$, $\mu(\cdot)n(t,\cdot) \in L_1(I_\omega)$ *for any* $t > 0$, $n \in W_1^1(\mathbb{R}_+ \times I_\omega)$, *that is,* $\frac{\partial n}{\partial t}, \frac{\partial n}{\partial a}$ *exist almost everywhere on* $\mathbb{R}_+ \times I_\omega$ *and (10.2.33) is satisfied almost everywhere.*

Proof First we note that, by Theorem B.8, functions from $W_1^1((0,\omega))$ are (absolutely) continuous and thus, by Theorem 9.2 and formula (10.2.34), n is continuous outside the characteristic $t = a$. Using argument as in (9.3.23) and the relevant assumption we find that n is continuous on $[0, \infty) \times [0, \omega]$. Similarly, if (10.2.39) is satisfied, then, again by Theorem 9.2, B is differentiable almost everywhere, and thus n is differentiable almost everywhere away from the characteristic $t = a$. Then the continuity along $t = a$ shows that $n \in W_1^1(\mathbb{R}_+ \times I_\omega)$. To simplify the argument, let as consider $\omega = \infty$, $\Delta_1 = \{(t, a);\ t > a\}$ and $\Delta_2 = \{(t, a);\ t < a\}$ and functions $f_i \in W_1^1(\Delta_i)$, $i = 1, 2$, satisfying $f_1(t, t) = f_2(t, t)$. Define F by $F|_{\Delta_1} = f_1$ and $F|_{\Delta_2} = f_2$. Let $\phi \in C_0^\infty(\mathbb{R}_+^2)$. Then

$$
\begin{aligned}
\int_{\mathbb{R}_+^2} F\phi_a dadt &= \int_{\Delta_1} f_1\phi_a dadt + \int_{\Delta_2} f_2\phi_a dadt \\
&= \int_0^\infty \left(\int_0^t f_1\phi_a da\right) dt + \int_0^\infty \left(\int_t^\infty f_2\phi_a da\right) dt \\
&= -\int_0^\infty \left(\int_0^t f_{1,a}\phi da\right) dt + \int_0^\infty \left(\int_t^\infty f_{2,a}\phi da\right) dt \\
&\quad + \int_0^\infty (f_1(t, t) - f_2(t, t))\phi(t, t)dt \\
&= -\int_{\mathbb{R}_+^2} \tilde{F}\phi dadt,
\end{aligned}
$$

where $\tilde{F}$ is given by $\tilde{F}|_{\Delta_1} = f_{1,a}$ and $\tilde{F}|_{\Delta_2} = f_{2,a}$. Thus, $\tilde{F} = F_a$ is the generalized derivative of F and is integrable since its components are. The analogous argument holds for the derivative with respect to t.

Finally, we have

$$
\begin{aligned}
\int_0^\omega \mu(a)n(t, a)da &= \int_0^{\min\{t,\omega\}} \mu(a)B(t - a)S(a)da \\
&\quad + \int_{\min\{t,\omega\}}^\omega \mu(a)\frac{S(a)}{S(a - t)}\mathring{n}(t - a)da \\
&\leq \max_{s\in[0,\omega]} |B(s)| \int_0^{\min\{t,\omega\}} \mu(a)S(a)da \\
&\quad + e^{\int_0^\omega \mu(s)ds} \max_{s\in[0,\omega]} |\mathring{n}(s)| \int_{\min\{t,\omega\}}^\omega \mu(a)S(a)da \\
&\leq \max_{s\in[0,\omega]} |B(s)| + e^{\int_0^\omega \mu(s)ds} \max_{s\in[0,\omega]} |\mathring{n}(s)|,
\end{aligned}
$$

where we used the fact that μS is a probability density. □

Lemma 10.3 *The set*

$$
\mathcal{D} = \left\{\psi \in W_1^1(I_\omega) :\ \mu\psi \in L_1(I_\omega), \psi(0) = \int_0^\omega \beta(a)\psi(a)da\right\} \tag{10.2.41}
$$

is dense in $L_1(I_\omega)$.

Proof Let $f \in L_1(I_\omega)$. By (B.2.7), for any fixed $\epsilon > 0$ we can find $\psi \in C_0^\infty(I_\omega)$ such that

$$\|f - \psi\|_{L_1(I_\omega)} \leq \epsilon. \tag{10.2.42}$$

Since ψ is of compact support, $\mu\psi \in L_1(I_\omega)$ is automatically satisfied. We have to construct ψ_ϵ satisfying (10.2.40) and close to ψ in L_1. For a given $\delta > 0$, let

$$\psi_\delta(a) := \psi(a) + \alpha_\delta e^{-a/\delta},$$

where α_δ is a constant to be determined. We require

$$\psi_\delta(0) = \alpha_\delta = \int_0^\omega \beta(a)\psi(a)da + \alpha_\delta \int_0^\omega e^{-a/\delta}\beta(a)da. \tag{10.2.43}$$

Now,

$$0 < \int_0^\omega e^{-a/\delta}\beta(a)da \leq \delta \sup_{a\in[0,\omega]} \beta(a) \int_0^{\omega/\delta} e^{-s}ds \leq \delta\|\beta\|_\infty,$$

and therefore, for sufficiently small δ, (10.2.43) can be solved for α_δ, yielding

$$\alpha_\delta = \frac{\int_0^\omega \beta(a)\psi(a)da}{1 - \int_0^\omega e^{-a/\delta}\beta(a)da},$$

and hence

$$|\alpha_\delta| \leq \frac{\sup\limits_{a\in[0,\omega]} |\beta(a)|\|\psi\|_{L_1(I_\omega)}}{1 - \delta\|\beta\|_\infty} \leq C\|\psi\|_{L_1(I_\omega)},$$

where C is independent of δ for sufficiently small δ. Thus,

$$\|\psi - \psi_\delta\|_{L_1(I_\omega)} \leq |\alpha_\delta| \int_0^\omega e^{-a/\delta}da \leq \delta C\|\psi\|_{L_1((I_\omega)}.$$

Thus, for a given $\psi \in C_0^\infty((0,\omega))$ satisfying (10.2.42), we can choose $\delta = \epsilon/C\|\psi\|_{L_1(I_\omega)}$ so that

$$\|f - \psi_\delta\|_{L_1(I_\omega)} \leq 2\epsilon$$

with $\phi_\delta \in \mathcal{D}$. $\square$

The flow ϕ in the finite-dimensional case is differentiable in time and continuous with respect to the initial condition. While the continuity with respect to the initial condition of each operator $T(t)$ of the semigroup in the infinite-dimensional situation follows from (10.2.37), in general, we can only expect the continuity of the semigroup with respect to time for each fixed $\mathring{n}$. This is called the strong continuity of $(T(t))_{t\geq 0}$.

Proposition 10.4 *If (9.3.15)–(9.3.18) hold, then the solution n defined by (10.2.34) satisfies*

$$n \in C([0,\infty), L_1(I_\omega)). \tag{10.2.44}$$

In other words, the family $(T(t))_{t\geq 0}$ is strongly continuous.

Proof Let us fix any $t_0 \in (0,\infty)$. Clearly, if $\psi \in C([0,t_0] \times [0,\omega])$, then $\psi \in C([0,t_0], L_1([0,\omega])$ by the uniform continuity of ψ and the boundedness of the domain of integration. Let n be the solution with initial condition $\mathring{n} \in L_1(I_\omega)$ and consider a sequence $\psi_k \in \mathcal{D}$ such that $\psi_k \to \mathring{n}$ in L_1. Let further n_k be the solution with initial condition ψ_k. Then, by (10.2.37),

$$\sup_{t\in[0,t_0]} \|n(\cdot,t) - n_k(\cdot,t)\|_{L_1(I_\omega)} \leq e^{t_0\|\beta\|_\infty}\|\mathring{n} - \phi_k\|_{L_1(I_\omega)},$$

which shows that n is a uniform in time limit of continuous functions (by Theorem 10.2) and hence $n \in C([0,t_0], L_1(I_\omega))$. □

Proposition 10.5 *The family $(T(t))_{t\geq 0}$ has the semigroup property, that is, for any $\mathring{n} \in X$ and $t, \tau \geq 0$, we have*

$$T(t+\tau)\mathring{n} = T(t)T(\tau)\mathring{n}. \tag{10.2.45}$$

Proof We defined

$$[T(t)\mathring{n}](a) = n(t,a),$$

where $n(t,a)$ is a unique solution to (9.3.13) with initial condition $\mathring{n} \in L_1([0,\infty))$. Since $T(t+\tau)\mathring{n}$ and $T(t)[T(\tau)\mathring{n}]$ are solutions of the same equation taking the same value $T(\tau)\mathring{n}$ at $t=0$, the uniqueness implies (10.2.45). □

The previous results show that $(T(t))_{t\geq 0}$ is a strongly continuous dynamical system, also called a strongly continuous semigroup, that is, it is a family of bounded linear operators satisfying, for any $n \in X$,

1. $T(t+\tau)n = T(t)T(\tau)n, \quad t,\tau \geq 0;$
2. $T(0)n = n;$
3. $\lim_{t\to 0^+} T(t)n = n.$

Remark 10.6 A prototype of a semigroup of operators is the exponential function of a matrix, $e^{t\mathcal{A}}, t \geq 0$, in $\mathbb{R}^d$, see (A.3.11). The semigroup property follows directly from (A.3.11) as $t\mathcal{A}$ and $s\mathcal{A}$ commute for any $t, s \in \mathbb{R}$.

Exercise 10.2.1 Prove that

$$\lim_{t \to 0^+} e^{t\mathcal{A}} \boldsymbol{x} = \boldsymbol{x}$$

for any $\boldsymbol{x} \in \mathbb{R}^d$.

Thus, semigroups of operators can be considered as generalizations of exponentials of operators. However, the involved operators need not be bounded, which is the case here, where we deal with differential operators. Then, due to the difficulties with proving the convergence of the series of unbounded operators, (A.3.11) is not a good way to define the semigroup generated by $\mathcal{A}$, and we need to find alternative ways to do it.

The reader interested in the theory of semigroups of strongly continuous semigroups is referred to the monographs [8, 75] or the survey in [10]. Here, we shall briefly recall the basics needed in the sequel.

As in the case of finite-dimensional flows generated by systems of differential equations, where

$$\frac{\partial}{\partial t}\phi(t, \boldsymbol{x}) = \boldsymbol{f}(\phi(t, \boldsymbol{x})),$$

we expect a similar relation between the semigroup of operators and a differential equation. However, the situation in infinite-dimensional state spaces is not so simple as $T(t)n$ need not be differentiable for each n. Thus, we introduce the (infinitesimal) generator $\mathcal{A}$ of $(T(t))_{t\geq 0}$ by

$$\mathcal{A}n = \lim_{h \to 0^+} \frac{T(h)n - n}{h}, \tag{10.2.46}$$

for any n for which the limit exists in X. The set of such $n \in X$ is called the domain of $\mathcal{A}$ and denoted $D(\mathcal{A})$. Typically, $\mathcal{A}$ is an unbounded operator and $D(\mathcal{A}) \neq X$. It follows that if $\mathring{n} \in D(\mathcal{A})$, then $T(t)\mathring{n}$ is differentiable with respect to t, $T(t) \in D(\mathcal{A})$ for all $t \geq 0$ and

$$\frac{d}{dt}T(t)\mathring{n} = \mathcal{A}T(t)\mathring{n}, \qquad T(0)\mathring{n} = \mathring{n}, \tag{10.2.47}$$

that is, $(T(t))_{t\geq 0}$ gives solutions to the Cauchy problem (10.2.47), and $T(t)\mathring{n}$ is the (semi)flow generated by this equation due to the semigroup property 3. above. We emphasize that contrary to finite dimensional problems, the solution need not be defined for $t < 0$.

In the case of the McKendrick–von Foerster problem, it can be proved that $\mathcal{A} = -\frac{\partial}{\partial a} - \mu$ with $D(\mathcal{A}) = \mathcal{D}$, but the proof is outside the scope of these notes, see [10].

Finally, we note the following fact, which will be important in analysing the nonlinear problem.

Proposition 10.7 *Under assumptions (9.3.15)–(9.3.18), if $\mathring{n}$ is bounded, then $n(t, a)$ is bounded on $[0, t_0] \times [0, \omega)$ for any $0 \leq t_0 < \infty$.*

Proof From the formula (9.3.22),

$$n(t, a) = \begin{cases} \frac{S(a)}{S(a-t)}\mathring{n}(a-t), & t < a, \\ S(a)B(t-a), & a < t, \end{cases}$$

we see that since $S(a)/S(a-t) \leq 1$, $S(a) \leq 1$, the part in $t < a$ is bounded by the boundedness of $\mathring{n}$. Also, we have proved that the iterates (9.3.20) defining B converge uniformly on each bounded time interval and thus $B(t-a)$ is bounded on $[0, t_0] \times [0, \omega)$. □

10.2.2 Linear Systems

To address the solvability of the McKendrick–von Foerster epidemiological system, we first look at its linear part

$$\begin{aligned} \frac{\partial}{\partial t}s(t, a) + \frac{\partial}{\partial a}s(t, a) + \mu(a)s(t, a) - \gamma(a)i(t, a) &= 0, \\ \frac{\partial}{\partial t}i(t, a) + \frac{\partial}{\partial a}i(t, a) + \mu(a)i(t, a) + (\nu(a) + \gamma(a))i(t, a) &= 0, \\ \frac{\partial}{\partial t}r(t, a) + \frac{\partial}{\partial a}r(t, a) + \mu(a)r(t, a) - \nu(a)i(t, a) &= 0, \end{aligned} \tag{10.2.48}$$

with the boundary conditions

$$\begin{aligned} s(t, 0) &= \int_0^\omega \beta(a)(s(t, a) + (1-q)i(t, a) + (1-w)r(t, a))da, \\ i(t, 0) &= q\int_0^\omega \beta(a)i(t, a)da, \\ r(t, 0) &= w\int_0^\omega \beta(a)r(t, a)da, \end{aligned} \tag{10.2.49}$$

and the initial conditions

$$s(0,a) = \mathring{s}(a), \quad i(0,a) = \mathring{i}(a), \quad r(0,a) = \mathring{r}(a). \tag{10.2.50}$$

The problem is an example of a more general vector McKendrick–von Foerster system

$$\frac{d}{dt}\boldsymbol{n} = \mathcal{S}\boldsymbol{n} + \mathcal{M}\boldsymbol{n}, \tag{10.2.51}$$

where $\boldsymbol{n}(t,a) = (n_1(t,a), \dots, n_d(t,a))$ and $n_i(t,a)$ is the population density at time t of individuals in patch i and being of age a (in the discussed model, $\boldsymbol{n} = (s,i,r)$). Further,

$$\mathcal{S}\boldsymbol{n} = -\frac{\partial}{\partial a}\boldsymbol{n} = \left(-\frac{\partial}{\partial a}n_1, \dots, -\frac{\partial}{\partial a}n_d\right) \tag{10.2.52}$$

describes ageing, $\mathcal{M}(a) = (\mu_{ij}(a))_{1\le i,j\le d}$ is the mortality/projection matrix, which, in our case, is

$$\mathcal{M} = \begin{pmatrix} -\mu & \delta & 0 \\ 0 & -(\mu+\delta+\gamma) & 0 \\ 0 & \gamma & -\mu \end{pmatrix} = \begin{pmatrix} -\mu & 0 & 0 \\ 0 & -\mu & 0 \\ 0 & 0 & -\mu \end{pmatrix} + \begin{pmatrix} 0 & \delta & 0 \\ 0 & -(\delta+\gamma) & 0 \\ 0 & \gamma & 0 \end{pmatrix}. \tag{10.2.53}$$

Here, the first matrix describes deaths, which is an intrapatch phenomenon, and the second refers to migrations between patches thus, it is a Kolmogorov matrix, see Sect. 5.1.2. We require that general $\mathcal{M}$ considered below has the same structure, that is, it can be written as

$$\mathcal{M} = -\mathcal{D}iag\,\mu + \mathcal{Q},$$

where $\boldsymbol{\mu} = (\mu_1, \dots, \mu_d)$, $\mathcal{Q}$ is a bounded Kolmogorov matrix, that is, it is nonnegative off-diagonal, the sum of entries in each column is 0, and each entry is a bounded function of a. This structure has important consequences as far as the asymptotic properties are concerned.

We supplement this system with the McKendrick–von Foerster boundary condition

$$\boldsymbol{n}|_{a=0} = \boldsymbol{n}(t,0) = [\boldsymbol{B}\boldsymbol{n}](t) = \int_0^\omega \mathcal{B}(a)\boldsymbol{n}(t,a)da, \tag{10.2.54}$$

and $\mathcal{B}(a) = (\beta_{ij}(a))_{1\le i,j\le d}$ is the fertility matrix. We note that births may be interpatch phenomena, like in our case,

$$\mathcal{B} = \begin{pmatrix} \beta & \beta(1-q) & \beta(1-w) \\ 0 & \beta q & 0 \\ 0 & 0 & \beta w \end{pmatrix}. \tag{10.2.55}$$

The initial condition is given by

$$\boldsymbol{n}|_{t=0} = \boldsymbol{n}(0,a) = \mathring{\boldsymbol{n}}(a). \tag{10.2.56}$$

The natural phase space for the problem is $X = (L_1(I_\omega))^d = \underbrace{L_1(I_\omega) \times \cdots \times L_1(I_\omega)}_{d\ times}$.

We denote by X_+ the subset of X consisting of vectors $\boldsymbol{n}$, which are coordinate-wise nonnegative almost everywhere. Further, $X_\infty = (L_\infty(I_\omega))^d$.

By adapting our earlier consideration concerning the scalar case, we can prove the solvability result as follows. Let us denote by $\mathcal{V}_{\mathcal{M}}(a,b)$ the fundamental matrix of the equation

$$z_a'(a) = \mathcal{M}(a)z(a),$$

that is, $z(a) = \mathcal{V}_{\mathcal{M}}(a,b)\mathring{z}$ satisfies the above equation with $z(b) = \mathring{z}$, see Definition 5.7 (the matrix $\mathcal{V}_{\mathcal{M}}$ plays the role of the integrating factor in the scalar case). The inverse $\mathcal{V}_{\mathcal{M}}^{-1}(a,b) = \mathcal{V}_{\mathcal{M}}(b,a)$, $a,b \in \mathbb{R}$, of a fundamental matrix always exists and thus $\mathcal{V}_{\mathcal{M}}(a,b) = \mathcal{V}_{\mathcal{M}}(a,0)\mathcal{V}_{\mathcal{M}}^{-1}(b,0)$. With this, we can write the solution to (10.2.51) with (10.2.54) and (10.2.56) as

$$\boldsymbol{n}(t,a) = \begin{cases} \mathcal{V}_{\mathcal{M}}(a,a-t)\mathring{\boldsymbol{n}}(a-t), & a > t, \\ \mathcal{V}_{\mathcal{M}}(0,a)\boldsymbol{\psi}(t-a), & a < t, \end{cases} \tag{10.2.57}$$

with $a \in [0,\omega)$, where $\boldsymbol{\psi}$ satisfies the Volterra equation

$$\boldsymbol{\psi}(t) = \int_0^t \mathcal{B}(a)\mathcal{V}_{\mathcal{M}}(a,0)\boldsymbol{\psi}(t-a)da + \int_t^\omega \mathcal{B}(a)\mathcal{V}_{\mathcal{M}}(a,a-t)\mathring{\boldsymbol{n}}(a-t)da,$$

with the same convention for $t > \omega$ as in (9.3.12). As in the scalar case, we can prove

Theorem 10.8 *Solutions* (10.2.57) *define a strongly continuous semigroup* $(\mathcal{T}(t))_{t\ge 0}$ *on X satisfying*

$$||\mathcal{T}(t)|| \le e^{\bar{b}t},$$

where $\bar{b} := \sup_{a \in I_\omega} ||B(a)||$. *Furthermore, if* $\mathring{\boldsymbol{n}} \in X_+$, *then* $\boldsymbol{n}(t, \cdot) \in X_+$ *and if* $\mathring{\boldsymbol{n}} \in X_\infty \cap X$, *then* $\boldsymbol{n}(t, \cdot) \in X_\infty \cap X$.

If, additionally,

$$\mathring{\boldsymbol{n}} \in (W_1^1(I_\omega))^d, \quad \mathcal{M}\mathring{\boldsymbol{n}} \in (L_1(I_\omega))^d, \tag{10.2.58}$$

and

$$\mathring{\boldsymbol{n}}(0) = \int_0^\omega \mathcal{B}(a)\mathring{\boldsymbol{n}}(a)da, \tag{10.2.59}$$

then $\boldsymbol{n} \in (W_1^1(\mathbb{R}_+ \times I_\omega))^d$ *and* (10.2.51) *is satisfied almost everywhere.*

Remark 10.9 The above result can be made more precise using the methods of the semigroup theory, [10]. It follows that if we define $\mathcal{A}$ to be $-\mathcal{diag}\frac{\partial}{\partial a} - \mathcal{M}$ on the domain

$$D_{\mathcal{A}} = \{\boldsymbol{n} \in (W_1^1(I_\omega))^d, \boldsymbol{n}(0) = \boldsymbol{B}\boldsymbol{n}\}, \tag{10.2.60}$$

$\mathcal{A}$ generates the strongly continuous semigroup $(\mathcal{T}(t)(t))_{t\geq 0_{t\geq 0}}$ on X satisfying

$$||\mathcal{T}(t)|| \leq e^{(\bar{b}-\underline{\mu})t},$$

where $\underline{\mu} := \inf_{a\in[0,\omega), 1\leq j\leq d} \mu_j(a)$.

10.2.3 The Nonlinear System

With the notation of the previous section, the problem (10.2.30)–(10.2.32) can be written in a compact form

$$\begin{aligned} \frac{d}{dt}\boldsymbol{n} &- \mathcal{A}\boldsymbol{n} + \mathfrak{F}(\boldsymbol{n}), \quad t > 0, \\ \boldsymbol{n}|_{a=0} &= \boldsymbol{B}\boldsymbol{n}, \\ \boldsymbol{n}|_{t=0} &= \mathring{\boldsymbol{n}}, \end{aligned} \tag{10.2.61}$$

where $\boldsymbol{n} = (s, i, r)$, $\mathcal{A} = \mathcal{S} + \mathcal{M}$ with $\mathcal{S}$ and $\mathcal{M}$ defined by (10.2.52) and (10.2.53). Further, $\boldsymbol{B}$ is defined by (10.2.55) and $\mathfrak{F}$ is a nonlinear perturbation,

$$\mathfrak{F}((s, i, r)) = \begin{pmatrix} -\lambda & 0 & 0 \\ \lambda & 0 & 0 \\ 0 & 0 & 0 \end{pmatrix} \begin{pmatrix} s \\ i \\ r \end{pmatrix}, \tag{10.2.62}$$

where the force of infection depends on the solution through the formula (10.1.6),

$$\lambda(t,a) = K_0(a)i(t,a) + \int\limits_0^{\omega} K(a,\alpha)i(t,\alpha)d\alpha, \tag{10.2.63}$$

where $K_0(a)$ and $K(a,\alpha)$ are known bounded functions. How do we deal with such problems? First, we note that the boundary condition is a part of the definition of the domain of $\mathcal{A}$, see (10.2.60), and thus, if we find a solution to (10.2.61), then it must satisfy $\boldsymbol{n} \in D(\mathcal{A})$ and automatically satisfies the boundary condition. Thus, for now, we shall discard it in the notation.

The main idea is to extend the variation of constants formula (5.3.64) to nonlinear problems. To wit, if we know the solution $\boldsymbol{n}$ to (10.2.61), then $\boldsymbol{f}(t) := \mathfrak{F}(\boldsymbol{n}(t))$ is a known function and hence, using (5.3.64), $\boldsymbol{n}$ should be given as the solution of the integral equation

$$\boldsymbol{n}(t) = \mathcal{T}(t)\mathring{\boldsymbol{n}} + \int\limits_0^t \mathcal{T}(t-\sigma)\mathfrak{F}(\boldsymbol{n}(\sigma))d\sigma. \tag{10.2.64}$$

Since, in general, the semigroup is not defined for $t < 0$ and the operators it consists of may not be invertible, the applicability of (5.3.64) must be justified. Let $\boldsymbol{n}$ be the solution to

$$\begin{aligned} \frac{d}{dt}\boldsymbol{n} &= \mathcal{A}\boldsymbol{n} + \mathfrak{F}, \quad t > 0, \\ \boldsymbol{n}|_{t=0} &= \mathring{\boldsymbol{n}}, \end{aligned} \tag{10.2.65}$$

where we assume that $\mathfrak{F}$ is known. Then, in particular, $\boldsymbol{n} \in D(\mathcal{A})$. Let us fix arbitrary $t > 0$, and define $\boldsymbol{u}(\sigma) = \mathcal{T}(t-\sigma)\boldsymbol{n}(\sigma)$ for $0 < \sigma < t$. Then $\boldsymbol{u}$ is differentiable and

$$\begin{aligned} \boldsymbol{u}'(\sigma) &= -\mathcal{T}(t-\sigma)\mathcal{A}\boldsymbol{n}(\sigma) + \mathcal{T}(t-\sigma)\boldsymbol{n}'(\sigma) \\ &= -\mathcal{T}(t-\sigma)\mathcal{A}\boldsymbol{n}(\sigma) + \mathcal{T}(t-\sigma)\left(\mathcal{A}\boldsymbol{n}(\sigma) + \mathfrak{F}(\sigma)\right) = \mathcal{T}(t-\sigma)\mathfrak{F}(\sigma), \end{aligned}$$

which yields (10.2.64) by integration from 0 to t, and the application of the semigroup property 2.

Even if we do not know $\boldsymbol{n}$, (10.2.64) offers a simplification of (10.2.61) by not involving unbounded operator $\mathcal{A}$. Of course, a solution to (10.2.64) is not necessarily a solution to (10.2.61), but at least it seems to be a step in the right direction in the sense that any solution to (10.2.61) must be a solution to (10.2.64).

To avoid any ambiguity, solutions to (10.2.64) are called mild, or integral, solutions to (10.2.61).

The problem (10.2.64) can be solved by Picard iterations, similar to the linear case (9.3.20),

$$\boldsymbol{n}_0(t) = \mathring{\boldsymbol{n}},$$
$$\boldsymbol{n}_{k+1}(t) = \mathcal{T}(t)\mathring{\boldsymbol{n}} + \int\limits_0^t \mathcal{T}(t-\sigma)\mathfrak{F}(\boldsymbol{n}_k(\sigma))d\sigma, \tag{10.2.66}$$

however, handling the nonlinearity $\mathfrak{F}$ requires more care, as in the Picard–Lindelöf theorem, (B.4.34).

Let us recall that by $Y = C([t_0, t_1], X)$ we denoted the space of continuous functions $[t_0, t_1] \ni t \to \boldsymbol{u}(t) \in X$, where $-\infty < t_0 < t_1 < \infty$. With the norm

$$\|\boldsymbol{u}(\cdot)\|_Y = \sup_{t\in[t_0,t_1]} \|\boldsymbol{u}(t)\|_X,$$

Y becomes a Banach space. First, assume that $\mathfrak{F}$ satisfies the global Lipschitz condition on X, that is, there is $L > 0$ such that for any $\boldsymbol{u}, \boldsymbol{v} \in X$

$$\|\mathfrak{F}(\boldsymbol{u}) - \mathfrak{F}(\boldsymbol{v})\|_X \leq L\|\boldsymbol{u} - \boldsymbol{v}\|, \tag{10.2.67}$$

see Example B.15. In this case, we can let $t_0 = 0$. Then, returning to (10.2.66), we obtain, for any $t \in [0, t_1]$,

$$\begin{aligned}\|\boldsymbol{n}_{k+1}(t) - \boldsymbol{n}_k(t)\|_X &\leq \int\limits_0^t \|\mathcal{T}(t-\sigma)(\mathfrak{F}(\boldsymbol{n}_k(\sigma)) - \mathfrak{F}(\boldsymbol{n}_{k-1}(\sigma)))\|_X d\sigma \\ &\leq Me^{\bar{b}t_1}Lt\|\boldsymbol{n}_k - \boldsymbol{n}_{k-1}\|_Y,\end{aligned} \tag{10.2.68}$$

which, by induction as in (9.3.21), yields

$$\sup_{t\in[0,t_1]} \|\boldsymbol{n}_{k+1}(t) - \boldsymbol{n}_k(t)\|_X \leq \frac{(Me^{\bar{b}t_1}L)^k}{k!}\|\boldsymbol{n}_1 - \mathring{\boldsymbol{n}}\|_Y$$

and, as for (9.3.20), this shows that $(\boldsymbol{n})_{k\geq 0}$ converges to a continuous solution to (10.2.64) defined on the whole interval $[0, t_1]$ for any $t_1 < \infty$. Such solutions are called global.

If we consider another solution $\boldsymbol{v}$ to (10.2.64) with the initial condition $\mathring{\boldsymbol{v}}$, then

$$\|\boldsymbol{n}(t) - \boldsymbol{v}(t)\|_X \leq Me^{\bar{b}t_1}\|\mathring{\boldsymbol{n}} - \mathring{\boldsymbol{v}}\|_X + \int\limits_0^t \|\mathcal{T}(t-\sigma)(\mathfrak{F}(\boldsymbol{n}(\sigma)) - \mathfrak{F}(\boldsymbol{v}(\sigma)))\|_X d\sigma$$

$$\leq Me^{\bar{b}t_1}\|\mathring{\boldsymbol{n}} - \mathring{\boldsymbol{v}}\|_X + Me^{\bar{b}t_1}L\int\limits_0^t \|\boldsymbol{n}(\sigma) - \boldsymbol{v}(\sigma)\|_X d\sigma,$$

and Gronwall's inequality gives

$$\|\boldsymbol{n}(t) - \boldsymbol{v}(t)\|_X \leq Me^{\bar{b}t_1}e^{MLt_1e^{\bar{b}t_1}}\|\mathring{\boldsymbol{n}} - \mathring{\boldsymbol{v}}\|_X, \quad 0 \leq t \leq t_1,$$

so that we obtain that the solution is (Lipschitz) continuous with respect to the initial data and is unique (by putting $\mathring{\boldsymbol{n}} = \mathring{\boldsymbol{v}}$).

However, as discussed in Sect. B.4, even for a simple scalar nonlinearity such as $\mathfrak{F}(u) = u^2$ we have

$$|\mathfrak{F}(u) - \mathfrak{F}(v)| = |u + v||u - v|$$

and thus $\mathfrak{F}$ is Lipschitz continuous as long as we restrict u and v to satisfy $|u|, |v| \leq c$ for some constant c. Thus, a quadratic nonlinearity is Lipschitz continuous, but not globally, as the Lipschitz constant depends on bounds for u and v. Recall that such functions are called locally Lipschitz continuous. Precisely, $\mathfrak{F}$ is said to satisfy a local Lipschitz condition on X if for any $c > 0$ there is $L = L_c$ such that

$$\|\mathfrak{F}(\boldsymbol{u}) - \mathfrak{F}(\boldsymbol{v})\|_X \leq L_c\|\boldsymbol{u} - \boldsymbol{v}\|_X, \tag{10.2.69}$$

whenever $\|u\|_X, \|v\|_X \leq c$.

In such a case, we cannot use directly (10.2.68) as the constant L changes with the iterates and can grow to infinity. Then, similarly to the proof of Theorem B.10, we modify this argument to ensure that all the iterates will stay in a fixed bounded set of X. Since, however, we operate in an infinite-dimensional space X, we use a slightly different approach.

Theorem 10.10 *Let $\mathfrak{F} : X \to X$ be a locally Lipschitz continuous function. If $\mathcal{A}$ is the generator of a semigroup $(\mathcal{T}(t))_{t\geq 0}$, then for every $\mathring{\boldsymbol{n}} \in X$ and every $t_0 \in \mathbb{R}$ there is $t_{max} > t_0$ such that the Cauchy problem*

$$\begin{aligned} \frac{d}{dt}\boldsymbol{n} &= \mathcal{A}\boldsymbol{n} + \mathfrak{F}(\boldsymbol{n}), \quad t > t_0, \\ \boldsymbol{n}|_{t=t_0} &= \mathring{\boldsymbol{n}}, \end{aligned} \tag{10.2.70}$$

has a unique mild solution $\boldsymbol{n}$ *on* $[t_0, t_{max})$. *Moreover, if* $t_{max} < \infty$, *then*

$$\lim_{t \to t_{max}} \|\boldsymbol{n}(t)\|_X = \infty.$$

Proof The proof consists of finding $\delta > 0$ and a closed subset $C([t_0, t_0 + \delta], X)$ of $C([t_0, t_1], X)$ such that the iterates (10.2.66) stay in this subset or, in other words, that the mapping

$$[\boldsymbol{F}(\boldsymbol{n})](t) = \mathcal{T}(t - t_0)\mathring{\boldsymbol{n}} + \int_{t_0}^{t} \mathcal{T}(t - \sigma)\mathfrak{F}(\boldsymbol{n}(\sigma))d\sigma,$$

is a self-map of this set. Hence, let us take some $0 < \delta < 1$ and denote $M_0 = \sup_{t_0 \le t \le t_0+1} \|\mathcal{T}(t)\|$, $r_0 > M_0\|\boldsymbol{n}_0\|_X + 1$. Then, for $t_0 \le t \le t_0 + \delta$ and $\boldsymbol{n} \in B_{r_0} \subset C([t_0, t_0 + \delta], X)$ (recall, B_{r_0} is the open ball centred at 0 and with radius r_0),

$$\begin{aligned}\|[\boldsymbol{F}(\boldsymbol{n})](t)\|_X &\le M_0\|\boldsymbol{n}_0\|_X + M_0 \int_{t_0}^{t} \|\mathfrak{F}(\boldsymbol{n}(\sigma)) - \mathfrak{F}(0)\|_X d\sigma + M_0(t - t_0)\|\mathfrak{F}(0)\|_X \\ &\le M_0\|\boldsymbol{n}_0\|_X + \delta M_0 L_{r_0} K_0 + M_0\delta\|\mathfrak{F}(0)\|_X \\ &\le M_0(\|\boldsymbol{n}_0\|_X + \delta(L_{r_0} r_0 + \|\mathfrak{F}(0)\|_X)).\end{aligned}$$

Hence, we see that if we take the length of the time interval δ to be

$$\delta = \min\left\{1, \frac{1}{M_0(L_{r_0} r_0 + \|\mathfrak{F}(0)\|_X)}\right\}, \tag{10.2.71}$$

then

$$\|[\boldsymbol{F}(\boldsymbol{n})](t)\|_X < r_0$$

and the $k + 1$ iterate is in B_{r_0}, provided the kth one is. Then on B_{r_0}, the function $\mathfrak{F}$ is globally Lipschitz, and the iterates converge to the mild solution of (10.2.70).

From what we just proved, it follows that if $\boldsymbol{n}$ is a mild solution on a closed interval $[t_0, t_0+\tau]$, then it can be extended onto $[t_0, t_0+\tau+\delta]$ with $\delta > 0$ defined by (10.2.71) with t_0 replaced by $t_0+\tau$. Of course, δ can decrease with every step as M_0 and L_{r_0} can increase. Let $[t_0, t_{max})$ be the maximal interval to which the solution can be extended by such a procedure. If $t_{max} < \infty$, then $\|\boldsymbol{n}(t)\|_X \to \infty$ as $t \to t_{max}$. Indeed, otherwise we would have a sequence $t_n \to t_{max}$ such that $\|\boldsymbol{n}(t_n)\|_X \le C$ for some constant C. However, at each t_n, we can extend the solution to the interval $[t_n, t_n + \delta]$ with δ independent of t_n since δ depends only on C (L_{r_0} also depends on the norm of $\boldsymbol{n}$) and the linear semigroup). For t_n sufficiently close to t_{max}, we would then have $t_n + \delta > t_{max}$, which contradicts the definition of t_{max}. □

The above theorem does not address the question of whether our mild solution is the solution to (10.2.70), that is, whether it can be differentiated and whether it belongs to $D(\mathcal{A})$ and thus satisfies the boundary conditions. A full discussion of regularity is beyond the scope of these lectures; see [75, Theorem 6.1.5]. We only note that for $\boldsymbol{n}$ to be a classical solution to (10.2.70), it suffices that $\mathring{\boldsymbol{n}} \in D(\mathcal{A})$ and $\boldsymbol{n} \to \mathfrak{F}(\boldsymbol{n})$ is continuously differentiable.

Let us return to the epidemiological problem (10.2.61) with $\mathfrak{F}$ given by (10.2.62)–(10.2.63). Our state space is $X = (L_1(I_\omega))^3$ and

$$\|\boldsymbol{n}\|_X = \|(s,i,r)\|_X = \|s\|_{L_1(I_\omega)} + \|i\|_{L_1(I_\omega)} + \|r\|_{L_1(I_\omega)}.$$

To simplify the discussion, we first consider the intercohort infection and disregard r, which does not impact the other variables. Then, for $\boldsymbol{n}_j = (s_j, i_j)$, $j = 1, 2$,

$$\mathfrak{F}(\boldsymbol{n}_1) - \mathfrak{F}(\boldsymbol{n}_2) = \begin{pmatrix} -\int_0^\omega K i_1 d\alpha & 0 \\ \int_0^\omega K i_1 d\alpha & 0 \end{pmatrix} \begin{pmatrix} s_1 \\ i_1 \end{pmatrix} - \begin{pmatrix} -\int_0^\omega K i_2 d\alpha & 0 \\ \int_0^\omega K i_2 d\alpha & 0 \end{pmatrix} \begin{pmatrix} s_2 \\ i_2 \end{pmatrix}$$

$$= \begin{pmatrix} -s_1 \int_0^\omega K i_1 d\alpha + s_2 \int_0^\omega K i_2 d\alpha \\ s_1 \int_0^\omega K i_1 d\alpha - s_2 \int_0^\omega K i_2 d\alpha \end{pmatrix}.$$

It is now easy to see that $\mathfrak{F}$ is locally Lipschitz continuous as

$$\begin{aligned} \|\mathfrak{F}(\boldsymbol{n}_1) - \mathfrak{F}(\boldsymbol{n}_2)\|_X &\le 2 \int_0^\omega \Bigg| s_1(a) \int_0^\omega K(a,\alpha) i_1(\alpha) d\alpha \\ &\quad - s_2(a) \int_0^\omega K(a,\alpha) i_2(\alpha) d\alpha \Bigg| da \\ &\le 2\Bigg(\int_0^\omega |s_1(a) - s_2(a)| \int_0^\omega K(a,\alpha) |i_1(\alpha)| d\alpha da \\ &\quad + \int_0^\omega |s_2(a)| \int_0^\omega K(a,\alpha) |i_1(\alpha) - i_2(\alpha)| d\alpha da \Bigg) \\ &\le C \|(s_1 - s_2, i_1 - i_2)\|_X, \end{aligned}$$

where

$$C = 2 \sup_{a,\alpha\in\mathbb{R}_+} K(a,\alpha) \max\{\|i_1\|_{L_1(I_\omega)}, \|s_2\|_{L_1(I_\omega)}\}.$$

Hence, the problem has a unique solution defined at least on some interval $[0, \delta]$.

We note that the case with intracohort infection is slightly different as the product is of two integrable functions is not necessarily integrable; see [15]. However, from Proposition 10.7, we know that if $\mathring{\boldsymbol{n}} = (\mathring{s}, \mathring{i}) \in X_{1,\infty} := L_1(I_\omega) \cap L_\infty(I_\omega)$, then also $(i(t), s(t)) := \mathcal{F}(t)\mathring{\boldsymbol{n}} \in X_{1,\infty}$ and so does the product is. Thus, we have

$$\begin{aligned}\|\mathfrak{F}(\boldsymbol{n}_1) - \mathfrak{F}(\boldsymbol{n}_2)\|_{X_{1,\infty}} &\leq 2\int_0^\omega |K_0(a)(i_1(a)s_1(a) - s_2(a)i_2(a))|\, da\\ &\leq 2 \sup_{a\in\mathbb{R}_+} K_0(a) \int_0^\omega (|s_1(a) - s_2(a)||i_1(a)|da + |s_2(a)||i_1(a) - i_2(a)|)\, da\\ &\leq C\|(s_1 - s_2, i_1 - i_2)\|_{X_{1,\infty}},\end{aligned}$$

where

$$C = 2 \sup_{a\in\mathbb{R}_+} K_0(a) \max\{\|i_1\|_{L_\infty(I_\omega)}, \|s_2\|_{L_\infty(I_\omega)}\},$$

and, as above, there exists a mild solution to (10.2.61) on some interval $[0, \delta]$.

Can this solution be extended to $[0, \infty)$? We observe that if $i, s \geq 0$ then in, say, intercohort case,

$$\|\boldsymbol{n}(t)\|_X = \int_0^\omega (|i(t,a)| + |s(t,a)|)da = \int_0^\omega (i(t,a) + s(t,a))da = \int_0^\omega n(t,a)da,$$

where n is the solution of the equation obtained by adding together (10.1.4) and (10.1.5). Since we know that n exists for all t, $\|\boldsymbol{n}(t)\|_X$ would be bounded for any finite t and thus would be extendable to $[0, \infty)$. Let's look at the iterates (10.2.66). We see that if $\mathring{\boldsymbol{n}} \geq 0$ and, since the linear semigroup $(\mathcal{T}(t))_{t\geq 0}$ preserves positivity, the iterates, and thus the solution, will be positive if $\mathfrak{F}(\boldsymbol{n}) \geq 0$ for $\boldsymbol{n} \geq 0$. However, $\mathfrak{F}$ is not positive.

To solve this problem, we observe that the iterations in Theorem 10.10 are performed on a fixed ball in X (or $X_{1,\infty}$). This means that in the iterations, we can always assume that the argument $\boldsymbol{n}$ of $\mathfrak{F}$ satisfies, with the respective norm, $\|\boldsymbol{n}\| \leq C$ for some constant C.

Then we observe that the problem (10.2.70) is equivalent to

$$\frac{d}{dt}\boldsymbol{n} = (\mathcal{A}\boldsymbol{n} - \rho\boldsymbol{n}) + (\rho\boldsymbol{n} + \mathfrak{F}(\boldsymbol{n})) = \mathcal{A}_\rho\boldsymbol{n} + \mathfrak{F}_\rho(\boldsymbol{n}),$$

for any $\rho \in \mathbb{R}$. It is easy to see that the semigroup generated by $\mathcal{A}_\rho$ is $(\mathcal{T}_\rho(t))_{t\geq 0} := (e^{-\rho t}\mathcal{T}(t))_{t\geq 0}$. The semigroup $(\mathcal{T}_\rho(t))_{t\geq 0}$ also preserves positivity. Therefore $\boldsymbol{n}$ is the mild solution to

$$\boldsymbol{n}(t) = \mathcal{T}_\rho(t)\mathring{\boldsymbol{n}} + \int_0^t \mathcal{T}_\rho(t-\sigma)\mathfrak{F}_\rho(\boldsymbol{n}(\sigma))\,d\sigma, \quad 0 \leq t < \delta. \tag{10.2.72}$$

In our case, we have

$$\begin{aligned}\mathfrak{F}_\rho(\boldsymbol{n}) &= \begin{pmatrix} -\int\limits_0^\omega K(a,\alpha)i(\alpha)d\alpha & 0 \\ \int\limits_0^\omega K(a,\alpha)i(\alpha)d\alpha & 0 \end{pmatrix}\begin{pmatrix} s \\ i \end{pmatrix} + \rho\begin{pmatrix} s \\ i \end{pmatrix} \\ &= \begin{pmatrix} -s(a)\int\limits_0^\omega K(a,\alpha)i(\alpha)d\alpha + \rho s(a) \\ s(a)\int\limits_0^\omega K(a,\alpha)i(\alpha)d\alpha + \rho i(a) \end{pmatrix}\end{aligned}$$

and we see that if we take $\rho > C\sup_{a,\alpha\in I_\omega} K(a,\alpha)$, then $\mathfrak{F}_\rho(\boldsymbol{n}) \geq 0$ for any $\boldsymbol{n} \geq 0$ satisfying $\|\boldsymbol{n}\| \leq C$. Thus, all iterates are nonnegative, and hence the solution is nonnegative. Based on the earlier argument, we have global solvability for the age-structured epidemiological problem.

Appendix A
Spaces, Norms and Order

We begin by introducing relevant notation and mathematical concepts. By $\mathbb{N}$, we denote the set of natural numbers, and $\mathbb{N}_0 := \mathbb{N} \cup \{0\}$; $\mathbb{Z}$ is the set of all integers. By $\mathbb{R}_+$ we understand $(0, \infty)$ and $\overline{\mathbb{R}}_+ := [0, \infty)$.

The state space X in any process in population dynamics should consist of real entries. In this book, we mostly work in finite-dimensional state space $X = \mathbb{R}^d$. However, it is often convenient to use complex numbers to write calculations in a more compact form. Thus, we formulate the relevant results for the complex d-dimensional space $X = \mathbb{C}^d$. Vectors in $\mathbb{R}^d$ (or $\mathbb{C}^d$) are denoted by boldface letters, $\boldsymbol{x} = (x_1, \ldots, x_d)$. Occasionally, in Chaps. 9 and 10, the state spaces are spaces of functions, which are infinite-dimensional; thus, we shall briefly mention them below. In this book, we will only use linear, or vector, state spaces, that is, spaces in which the addition of elements and their multiplication by numbers, obeying usual algebraic rules, are defined.

A.1 Vector Norms

As we saw earlier, e.g., the solution to (5.3.22) is given by the sequence $(\mathcal{A}^k)_{k\geq 1}$ of iterates operating in the state space X that, in this case, is $\mathbb{C}^d$. Then, in fact, (5.2.17) is a statement about the limit of $\mathcal{A}^k \mathring{\boldsymbol{y}}$ as $k \to \infty$, so we must introduce a way of measuring distances in X, which we call a metric in X. Usually, we want to make the metric consistent with the linear structure of X, that is, with the vectors' addition and multiplication of them by scalars. Such metrics are defined by norms. A norm is a function $\|\cdot\| : X \to \overline{\mathbb{R}}_+$ satisfying, for any $\boldsymbol{x}, \boldsymbol{y} \in X, \alpha \in \mathbb{C}$,

$$\|\boldsymbol{x}\| = 0 \quad \text{iff} \quad \boldsymbol{x} = 0, \qquad \|\alpha \boldsymbol{x}\| = |\alpha|\|\boldsymbol{x}\|, \qquad \|\boldsymbol{x} + \boldsymbol{y}\| \leq \|\boldsymbol{x}\| + \|\boldsymbol{y}\|. \tag{A.1.1}$$

J. Banasiak, *Introduction to Mathematical Methods in Population Theory*, Springer Undergraduate Mathematics Series,
https://doi.org/10.1007/978-3-031-65491-6

In plain language, a norm is a parameter we use to describe the state of a system. A vector space equipped with a norm is called a normed space

In normed spaces, the distance between two points $\boldsymbol{x}$ and $\boldsymbol{y}$ is given by $\|\boldsymbol{x}-\boldsymbol{y}\|$. We say that a sequence $(\boldsymbol{x}(k))_{k\geq 0}$ converges to $\boldsymbol{x}$ in norm $\|\cdot\|$, and write

$$\lim_{k\to\infty} \boldsymbol{x}(k) = \boldsymbol{x},$$

if

$$\lim_{k\to\infty} \|\boldsymbol{x}(k)-\boldsymbol{x}\| = 0.$$

When we do not know the limit element of a sequence, we first check whether $(\boldsymbol{x}(k))_{k\geq 0}$ is a Cauchy sequence, that is, whether

$$\lim_{k,m\to\infty} \|\boldsymbol{x}(k)-\boldsymbol{x}(m)\| = 0.$$

Not every Cauchy sequence in a normed space must converge. If in a normed space X every Cauchy sequence converges (to an element of X), we say that X is complete and call X a Banach space.

As we mentioned earlier, norms define the distance between points in X, and thus, they allow the introduction of several geometric objects in X. For instance, we define an open ball centred at $\boldsymbol{x}$ with radius r by

$$B_{\boldsymbol{x},r} = \{\boldsymbol{y}\in X :\ \|\boldsymbol{y}-\boldsymbol{x}\| < r\}.$$

The closed ball $\overline{B}_{\boldsymbol{x},r}$ is

$$\overline{B}_{\boldsymbol{x},r} = \{\boldsymbol{y}\in X :\ \|\boldsymbol{y}-\boldsymbol{x}\| \leq r\},$$

while the sphere we denote by

$$S_{\boldsymbol{x},r} = \{\boldsymbol{y}\in X :\ \|\boldsymbol{y}-\boldsymbol{x}\| = r\}.$$

If $\boldsymbol{x} = 0$, then we shall use the notation $B_{0,r} = B_r$, $\overline{B}_{0,r} = \overline{B}_r$ and $S_{0,r} = S_r$. We say that $\Omega\subset X$ is open if for any $\boldsymbol{x}\in\Omega$ there is $B_{\boldsymbol{x},r}\subset\Omega$. By $\overline{\Omega}$, $\operatorname{Int}\Omega$ and $\partial\Omega$ we denote, respectively, its closure (Ω together with all limits of sequences from Ω), interior (all points $\boldsymbol{x}$ which belong to Ω together with some $B_{\boldsymbol{x},r}$), and the boundary defined as $\overline{\Omega}\setminus\operatorname{Int}\Omega$. We say that Ω is closed if it contains all limits of sequences from Ω and that it is compact if from any sequence of elements of Ω, one can select a subsequence converging to an element of Ω. In finite-dimensional normed spaces, compact sets are exactly bounded and closed sets; see [5, Thorem III 3.5].

In the analysis of ω-limit sets, see Sect. 6.4.2, we encountered the concept of the convergence to a set. Let us make it more precise. If $\boldsymbol{x}\notin\Omega$, then we define the distance from $\boldsymbol{x}$ to Ω by

$$d(\boldsymbol{x}, \Omega) := \inf_{z \in \Omega} \|\boldsymbol{x} - z\|. \tag{A.1.2}$$

We say that $(\boldsymbol{x}(k))_{k\geq 0}$ converges to Ω, if

$$\lim_{k\to\infty} d(\boldsymbol{x}(k), \Omega) = 0. \tag{A.1.3}$$

We see that if $\boldsymbol{x}(k) \to \Omega$ as $k \to \infty$, then there exists a sequence $(\boldsymbol{y}(k))_{k\geq 0} \subset \Omega$ such that

$$\lim_{k\to\infty} (\boldsymbol{x}(k) - \boldsymbol{y}(k)) = 0.$$

Remark A.1 If, in particular, Ω is compact, then there is a subsequence of $(\boldsymbol{y}(k))_{k\geq 0}$ that converges to an element $\boldsymbol{y} \in \Omega$; hence, there exists a subsequence of $(\boldsymbol{x}(k))_{k\geq 0}$ convergent to $\boldsymbol{y} \in \Omega$. In general, from any subsequence of a sequence converging to Ω, we can select a subsequence converging to an element of Ω. On the other hand, since

$$d(\boldsymbol{x}, \Omega) \leq \|\boldsymbol{x} - \boldsymbol{y}\|$$

for any $\boldsymbol{y} \in \Omega$ (by the definition of inf), we see that if $\boldsymbol{x}(k) \to \boldsymbol{y}$ as $k \to \infty$ for some $\boldsymbol{y} \in \Omega$, then $\boldsymbol{x}(k) \to \Omega$.

There is a variety of state spaces and norms we can introduce in them. Infinite-dimensional function spaces will be discussed in more detail in Sect. B.2. Now, we will focus on finite-dimensional cases. Here, we also can introduce several norms, which, nevertheless, define the same notion of convergence and reduce any finite-dimensional case to $\mathbb{C}^d$. To explain this, let $\{\boldsymbol{e}_1, \ldots, \boldsymbol{e}_d\}$ be a basis in X with $\boldsymbol{x} = (x_1, \ldots, x_d)$ with respect to this basis. In this way, we can identify X with $\mathbb{C}^d$. The best-known norm in $\mathbb{C}^d$ is the Euclidean norm

$$\|\boldsymbol{x}\|_2 = \sqrt{\sum_{i=1}^{d} |x_i|^2},$$

representing the geometric distance of $\boldsymbol{x}$ from the origin $\boldsymbol{0}$ in an Euclidean space. However, the Euclidean norm is often not convenient for particular applications. In the context of the population theory, we will use the so-called l^1-norm,

$$\|\boldsymbol{x}\|_1 = \sum_{i=1}^{d} |x_i|, \tag{A.1.4}$$

and the l^∞-norm

$$\|\boldsymbol{x}\|_\infty = \max_{1\le i\le d} |x_i|. \tag{A.1.5}$$

The l^∞-norm is used when we are interested in controlling the highest values of the solutions, e.g., the highest concentration of the reacting substances. We observe that

$$\|\boldsymbol{x}\|_\infty = \max_{1\le i\le d} |x_i| \le \sum_{i=1}^{d} |x_i| = \|\boldsymbol{x}\|_1 = \sum_{i=1}^{d} |x_i| \le \sum_{i=1}^{d} \max_{1\le i\le d} |x_i| = d\|\boldsymbol{x}\|_\infty. \tag{A.1.6}$$

The most familiar concept of the convergence of a sequence of vectors $(\boldsymbol{x}(k))_{k\ge 0}$ in $\mathbb{C}^d$ is the coordinate-wise convergence. We say that $(\boldsymbol{x}(k))_{k\ge 0}$ converges coordinate-wise to $\boldsymbol{x}$ if for each $i = 1, \dots, d$,

$$\lim_{k\to\infty} x_i(k) = x_i$$

or, equivalently,

$$\lim_{k\to\infty} (x_i(k) - x_i) = 0.$$

We have

Lemma A.2 *The following are equivalent:*

1. $(\boldsymbol{x}(k))_{k\ge 0}$ *converges coordinate-wise to* 0;
2. $(\boldsymbol{x}(k))_{k\ge 0}$ *converges to* 0 *in either norm* $\|\cdot\|_1, \|\cdot\|_\infty$.

Proof The statements follow directly from the definition of the limit and the fact the X is finite-dimensional. Indeed, 1. implies that for any $\epsilon > 0$ and any $i = 1, \dots, d$ there is k_i such that for all $k \ge k_i$ we have $|x_i(k) - x_i| < \epsilon$. But then, taking $K = \max\{k_1, \dots, k_d\}$ we see that for any $\epsilon > 0$ there is K such that for all $k \ge K$ and all $i = 1, \dots, d$, we have $|x_i(k) - x_i| < \epsilon$. But this implies that $\|\boldsymbol{x}(k) - \boldsymbol{x}\|_\infty < \epsilon$ and $\|\boldsymbol{x}(k) - \boldsymbol{x}\|_1 < d\epsilon$ for $k \ge K$ which shows 2.

Conversely, if for any $\epsilon > 0$ there is K such that for all $k \ge K$ we have $\|\boldsymbol{x}(k) - \boldsymbol{x}\|_\infty < \epsilon$, then the definition of $\|\cdot\|_\infty$ implies that for all $k \ge K$ and $i = 1, \dots, d$ we have $|x_i(k) - x_i| < \epsilon$ and hence $(\boldsymbol{x}(k))_{k\ge 0}$ converges coordinate-wise. An analogous argument can be applied for $\|\cdot\|_1$ as, if the sum of nonnegative terms is smaller than ϵ, so must be each term. □

Inequalities (A.1.6) are an example of a more general property of norms in arbitrary normed spaces. Namely, any two norms $\|\cdot\|_a, \|\cdot\|_b$ for which there exist $c_a, C_a > 0$ such that for all $\boldsymbol{x} \in X$,

$$c_a\|\boldsymbol{x}\|_a \le \|\boldsymbol{x}\|_b \le C_a\|\boldsymbol{x}\|_a,$$

are called equivalent. Thus, (A.1.6) says that $\|\cdot\|_1$ and $\|\cdot\|_\infty$ are equivalent. However, the disadvantage of using particular norms is that they depend on the basis in X. Also, as we see later, it is advantageous to use norms defined without referring to any particular basis in some cases. For this, we have

Theorem A.3 *Any two norms on $\mathbb{C}^d$, $0 < d < \infty$, are equivalent.*

Proof It is enough to prove that any norm is equivalent to $\|\cdot\|_\infty$ in $\mathbb{C}^d$ with the canonical basis $\{\boldsymbol{e}_1, \ldots, \boldsymbol{e}_d\}$. First, we have

$$\|\boldsymbol{x}\| = \left\|\sum_{i=1}^{d} x_i \boldsymbol{e}_i\right\| \leq \sum_{i=1}^{d} |x_i|\|\boldsymbol{e}_i\| \leq C\|\boldsymbol{x}\|_\infty,$$

where $C = \sum_{i=1}^{d} \|\boldsymbol{e}_i\|$. Then, using the reverse triangle inequality, for any $\boldsymbol{x}, \boldsymbol{y} \in \mathbb{C}^d$ and any norm on $\mathbb{C}^d$ we have

$$|\|\boldsymbol{x}\| - \|\boldsymbol{y}\|| \leq \|\boldsymbol{x} - \boldsymbol{y}\| \leq C\|\boldsymbol{x} - \boldsymbol{y}\|_\infty,$$

and hence $\boldsymbol{x} \mapsto \|\boldsymbol{x}\|$ as a function from X with the norm $\|\cdot\|_\infty$ into $\overline{\mathbb{R}}_+$ is continuous (more precisely, Lipschitz continuous). Let us consider the set

$$\Omega = \{\boldsymbol{x} \in X : \|\boldsymbol{x}\|_\infty = 1\}.$$

This set is the unit sphere in the $\|\cdot\|_\infty$ norm; hence, it is bounded and closed. By e.g., [31, Chapter 1, Supplement S.2] or [5, Corollary III.3.7], any continuous real function on a bounded and closed set of $\mathbb{C}^d$ attains its maximum and minimum (note that since $\mathbb{C}^d$ can be identified with $\mathbb{R}^{2d}$, we can use here the standard version of the Weierstrass theorem on real spaces). Thus, there are $\boldsymbol{y}, \boldsymbol{z} \in \Omega$ such that for any $\boldsymbol{v} \in \Omega$ we have

$$0 < \|\boldsymbol{y}\| \leq \|\boldsymbol{v}\| \leq \|\boldsymbol{z}\|,$$

where the LHS inequality follows from the first axiom of the norm, $\|\boldsymbol{y}\| = 0$ if and only if $\boldsymbol{y} = \boldsymbol{0}$, which would contradict $\|\boldsymbol{y}\|_\infty = 1$. Since for any $\boldsymbol{x} \in X$, $\boldsymbol{v} = \boldsymbol{x}/\|\boldsymbol{x}\|_\infty \in \Omega$ and hence we obtain

$$m\|\boldsymbol{x}\|_\infty \leq \|\boldsymbol{x}\| \leq M\|\boldsymbol{x}\|_\infty,$$

where $m = \|\boldsymbol{y}\|$ and $M = \|\boldsymbol{z}\|$. This proves the statement. □

Remark A.4 Theorem A.3 is not valid in infinite-dimensional spaces.

Corollary A.5 *Let $(\boldsymbol{x}(k))_{k\geq 0}$ be a sequence in X and let $\|\cdot\|$ be any norm on X. Then, the following are equivalent.*

1. *$(\boldsymbol{x}(k))_{k\geq 0}$ converges to $\boldsymbol{x}$ in the norm $\|\cdot\|$.*
2. *For any basis $(\boldsymbol{f}_1, \ldots, \boldsymbol{f}_d)$ in X, if $\boldsymbol{x}(k) = \sum_{i=1}^{d} \xi_i(k)\boldsymbol{f}_i$, then for each $i = 1, \ldots, d$,*

$$\lim_{k\to\infty} \xi_i(k) = \xi_i,$$

where $\boldsymbol{x} = \sum_{i=1}^{d} \xi_i \boldsymbol{f}_i$.

Proof The statement follows since $\|\boldsymbol{x}\|_\infty = \max_{1\leq i\leq d} |\xi_1|$ is a norm for any basis $(\boldsymbol{f}_1, \ldots, \boldsymbol{f}_d)$ and, by Lemma A.2, coordinate-wise convergence in any basis is equivalent to the convergence in $\|\cdot\|_\infty$ with respect to this basis which, in turn, is equivalent to the convergence in any norm by Theorem A.3. □

Since any Cauchy sequence in $\mathbb{R}$ converges, Corollary A.5 yields

Corollary A.6 *Any finite-dimensional normed space is a Banach space.*

A.2 Operator and Matrix Norms

Let X and Y be normed spaces and $\mathcal{A} : X \to Y$ be a linear operator between them. In other words, $\mathcal{A}$ is a function defined on X with values in Y satisfying

$$\mathcal{A}(\alpha\boldsymbol{x} + \beta\boldsymbol{y}) = \alpha\mathcal{A}\boldsymbol{x} + \beta\mathcal{A}\boldsymbol{y}, \quad \boldsymbol{x}, \boldsymbol{y} \in X, \alpha, \beta \in \mathbb{C}.$$

If $X = \mathbb{C}^d$, we identify linear operators with matrices. Consider the set $\mathbb{M}_d$ of all $d \times d$ matrices. Since any $\mathcal{A} \in \mathbb{M}_d$ can be identified with a vector in $\mathbb{C}^{d\times d}$, $\mathbb{M}_d$ is a linear space equipped with any vector norm. This approach, however, would deprive us of many important properties of $\mathbb{M}_d$, which sets it apart from $\mathbb{C}^{d\times d}$—the possibility of evaluating $\mathcal{A}\boldsymbol{x}$ or multiplying $\mathcal{A}$ by another $\mathcal{B} \in \mathbb{M}_d$. With this in mind, we introduce the so-called operator norm on $\mathbb{M}_d$. To motivate the definition, let us consider the multiplication $X \ni \boldsymbol{x} \mapsto \mathcal{A}\boldsymbol{x} \in X$, where we use the same norm $\|\cdot\|_1$ to measure the sizes of both input and output. We have

$$\|\mathcal{A}\boldsymbol{x}\|_1 = \sum_{i=1}^{d} \left| \sum_{j=1}^{d} a_{ij}x_j \right| \leq \sum_{j=1}^{d} |x_j| \sum_{i=1}^{d} |a_{ij}| \leq \|\boldsymbol{x}\|_1 \max_{1\leq j\leq d} \sum_{i=1}^{d} |a_{ij}| = L\|\boldsymbol{x}\|_1, \tag{A.2.7}$$

where

$$L = \max_{1 \le j \le d} \sum_{i=1}^{d} |a_{ij}|$$

is a scalar controlling the change of the magnitude of $\boldsymbol{x}$ after the multiplication by $\mathcal{A}$.

Following the above idea, it is natural to define, for any linear operator $\mathcal{A}$ between normed spaces X and Y, the operator norm

$$\|\mathcal{A}\| = \inf\{L : \|\mathcal{A}\boldsymbol{x}\| \le L\|\boldsymbol{x}\| \text{ for any } \boldsymbol{x} \in X\}. \tag{A.2.8}$$

The operators $\mathcal{A}$ for which such an L can be found are called bounded operators. Any bounded operator is a continuous function from X to Y by linearity. We have

Proposition A.7

1. *Equivalently to* (A.2.8), *we have*

$$\|\mathcal{A}\| = \sup_{x \ne 0} \frac{\|\mathcal{A}\boldsymbol{x}\|}{\|\boldsymbol{x}\|} = \sup_{\|x\|=1} \|\mathcal{A}\boldsymbol{x}\| = \sup_{\|x\| \le 1} \|\mathcal{A}\boldsymbol{x}\|. \tag{A.2.9}$$

2. *The space of bounded linear operators is a normed space with the norm* $\mathcal{A} \mapsto \|\mathcal{A}\|$.
3. *If* $X = \mathbb{C}^d$, *then all suprema and infima in* (A.2.8) *and* (A.2.9) *are attained.*

Proof

1. Let

$$|||\mathcal{A}||| = \sup_{x \ne 0} \frac{\|\mathcal{A}\boldsymbol{x}\|}{\|\boldsymbol{x}\|}.$$

From the definition of supremum, we have

$$|||\mathcal{A}|||\|\boldsymbol{x}\| \ge \|\mathcal{A}\boldsymbol{x}\|, \qquad \boldsymbol{x} \subset X,$$

(for $\boldsymbol{x} = 0$, the inequality is obvious) and hence $\|\mathcal{A}\| \le |||\mathcal{A}|||$. On the other hand, for any L such that $\|\mathcal{A}\boldsymbol{x}\| \le L\|\boldsymbol{x}\|$ we have

$$|||\mathcal{A}||| = \sup_{x \ne 0} \frac{\|\mathcal{A}\boldsymbol{x}\|}{\|\boldsymbol{x}\|} \le L,$$

hence, by taking infimum, we get $|||\mathcal{A}||| \le \|\mathcal{A}\|$, so $|||\mathcal{A}||| = \|\mathcal{A}\|$. Next,

$$\sup_{x\neq 0} \frac{\|\mathcal{A}\boldsymbol{x}\|}{\|\boldsymbol{x}\|} = \sup_{x\neq 0} \left\| \mathcal{A}\frac{\boldsymbol{x}}{\|\boldsymbol{x}\|} \right\| = \sup_{\|\boldsymbol{y}\|=1} \|\mathcal{A}\boldsymbol{y}\|.$$

For the final equality, we clearly have $\|\mathcal{A}\| \le \sup_{\|\boldsymbol{x}\|\le 1} \|\mathcal{A}\boldsymbol{x}\|$ as the set over which we take supremum on the right-hand side is larger. On the other hand, $\sup_{\|\boldsymbol{x}\|\le 1} \|\mathcal{A}\boldsymbol{x}\| \le \sup_{\|\boldsymbol{x}\|\le 1} \|\mathcal{A}\|\|\boldsymbol{x}\| \le \|\mathcal{A}\|$.

2. The linearity of the space of operators is clear. To prove the norm property, by 1., we can use $\|\mathcal{A}\| = \sup_{\|\boldsymbol{x}\|=1} \|\mathcal{A}\boldsymbol{x}\|$. The first two axioms of (A.1.1) are obvious. The triangle inequality can be obtained from

$$\begin{aligned}\|\mathcal{A}+\mathcal{B}\| &= \sup_{\|\boldsymbol{x}\|=1} \|\mathcal{A}\boldsymbol{x}+\mathcal{B}\boldsymbol{x}\| \le \sup_{\|\boldsymbol{x}\|=1} \{\|\mathcal{A}\boldsymbol{x}\| + \|\mathcal{B}\boldsymbol{x}\|\}\\ &\le \sup_{\|\boldsymbol{x}\|=1} \|\mathcal{A}\boldsymbol{x}\| + \sup_{\|\boldsymbol{x}\|=1} \|\mathcal{B}\boldsymbol{x}\| = \|\mathcal{A}\| + \|\mathcal{B}\|,\end{aligned}$$

where the last inequality follows as summing separate suprema of $\|\mathcal{A}\boldsymbol{x}\|$ and $\|\mathcal{B}\boldsymbol{x}\|$ cannot give a smaller value than the supremum of $\|\mathcal{A}\boldsymbol{x}\| + \|\mathcal{B}\boldsymbol{x}\|$.

3. Since, by

$$|\|\mathcal{A}\boldsymbol{x}\| - \|\mathcal{A}\boldsymbol{y}\|| \le \|\mathcal{A}(\boldsymbol{x}-\boldsymbol{y})\| \le \|\mathcal{A}\|\|\boldsymbol{x}-\boldsymbol{y}\|, \qquad \boldsymbol{x}, \boldsymbol{y} \in X,$$

$\boldsymbol{x} \mapsto \|\mathcal{A}\boldsymbol{x}\|$ is a continuous function and $\{\boldsymbol{x} \in X;\ \|\boldsymbol{x}\| = 1\}$ is closed and bounded, and hence compact, subset of $X = \mathbb{C}^d$, $\boldsymbol{x} \mapsto \|\mathcal{A}\boldsymbol{x}\|$ attains its maximum on $\{\boldsymbol{x} \in X;\ \|\boldsymbol{x}\| = 1\}$. This shows that each supremum in (A.2.9) is attained. For (A.2.8), we see that $|||\mathcal{A}||| = \|\mathcal{A}\|$ and $\|\mathcal{A}\|$ belongs to the set on the right-hand side, so the infimum is attained.

□

Example A.8 Let us find the operator norm of $\mathcal{A}$ acting on the space $\mathbb{C}^d$ with norm $\|\cdot\|_1$. In (A.2.7) we have found that $\|\mathcal{A}\| \le \max_{1\le j\le d} \sum_{i=1}^{d} |a_{ij}|$. On the other hand, let

$$\sum_{i=1}^{d} |a_{ij_0}| = \max_{1\le j\le d} \sum_{i=1}^{d} |a_{ij}|,$$

and consider $\boldsymbol{x}_0 = (0, \ldots, 0, 1, 0, \ldots, 0)$ with 1 at the j_0th place. Then $\|\boldsymbol{x}_0\|_1 = 1$ and

$$\|\mathcal{A}\boldsymbol{x}_0\|_1 = \sum_{i=1}^{d} \left| \sum_{j=1}^{d} a_{ij}x_j \right| = \sum_{i=1}^{d} |a_{ij_0}|.$$

Hence,

$$\|\mathcal{A}\| = \sup_{\|\boldsymbol{x}\|_1=1} \|\mathcal{A}\boldsymbol{x}\|_1 \geq \|\mathcal{A}\boldsymbol{x}_0\|_1 = \sum_{i=1}^{d} |a_{ij_0}| = \max_{1\leq j\leq d} \sum_{i=1}^{d} |a_{ij}|,$$

so

$$\|\mathcal{A}\| = \max_{1\leq j\leq d} \sum_{i=1}^{d} |a_{ij}|.$$

We have a result corresponding to Corollary A.5.

Corollary A.9 *Let $(\mathcal{A}(k))_{k\geq 0}$ be a sequence of matrices in $X = \mathbb{C}^d$. Then, the following are equivalent.*

1. *$(\mathcal{A}(k))_{k\geq 0}$ converges in the operator norm in $\mathbb{M}_d$.*
2. *For any $\boldsymbol{x} \in X$, $(\mathcal{A}(k)\boldsymbol{x})_{k\geq 0}$ converges in any norm in X.*
3. *For any basis in X, if $\mathcal{A}(k) = (a_{ij}(k))_{1\leq i,j\leq d}$ is the matrix representation of $\mathcal{A}_k$ in this basis, then the coordinate sequences $(a_{ij}(k))_{k\geq 0}$ converge for any $1 \leq i, j \leq d$ as $k \to \infty$.*

Proof Considering $\mathcal{A}$ as a d^2 dimensional vector, we see, by Theorem A.3, that the operator norm is equivalent to any other norm on $\mathbb{C}^{d\times d}$. Hence, the equivalence of 1. and 3. follows from Corollary A.5.

Let $\|\mathcal{A}(k) - \mathcal{A}\| \to 0$ as $k \to \infty$. Then,

$$\|\mathcal{A}(k)\boldsymbol{x} - \mathcal{A}\boldsymbol{x}\| = \|(\mathcal{A}(k) - \mathcal{A})\boldsymbol{x}\| \leq \|\mathcal{A}(k) - \mathcal{A}\|\|\boldsymbol{x}\|,$$

showing that 1. implies 2. Now, if 2. holds, then choosing $\boldsymbol{x} = \boldsymbol{e}_j = (0, \ldots, 0, 1, 0 \ldots, 0)$ in the chosen basis, we have

$$\mathcal{A}(k)\boldsymbol{e}_j = (a_{1j}(k), \ldots, a_{dj}(k)),$$

and, by assumption, the vector $(a_{1j}(k), \ldots, a_{dj}(k))$ converges in the norm $\|\cdot\|$. However, again by Corollary A.5, this implies the coordinate-wise convergence of any $(a_{ij}(k))_{k\geq 0}$ for any $1 \leq i, j \leq d$; hence statement 1. follows from the first part of the proof. □

A.3 Functions of Matrices

Let us recall that we are interested in solving systems of difference (5.3.20) and differential (5.3.21) equations,

$$\boldsymbol{y}(k+1) = \mathcal{A}\boldsymbol{y}(k), \quad \boldsymbol{y}(0) = \mathring{\boldsymbol{y}},$$

and

$$\boldsymbol{y}' = \mathcal{A}\boldsymbol{y}, \quad \boldsymbol{y}(0) = \mathring{\boldsymbol{y}}.$$

While the solution in the first case is given by the iterations

$$\boldsymbol{y}(k) = \mathcal{A}^k \mathring{\boldsymbol{y}},$$

where all elements are well-defined since the multiplication of matrices is allowed, expressing the solution of the system of differential equations in terms of $\mathcal{A}$ is not so straightforward.

Passing to the second equation, recall that for a single equation $y' = ay$, $y(0) = \mathring{y}$, where a is a scalar constant, the general solution is given by $y(t) = e^{ta}\mathring{y}$. Similarly, we would like to say that the general solution to (5.3.56) is $\boldsymbol{y}(t) = e^{t\mathcal{A}}\mathring{\boldsymbol{y}}$. The problem is that we do not know what it means to evaluate the exponential of a matrix. However, if we reflect on the fact that the exponential of a number can be evaluated as the power (Maclaurin) series

$$e^{ta} = \sum_{k=0}^{\infty} \frac{t^k a^k}{k!},$$

where the only operations on the argument a involve additions, scalar multiplications and taking integer powers, we conclude that the above expression can also be written for a matrix, that is, we can define

$$e^{t\mathcal{A}} = \mathcal{I} + t\mathcal{A} + \frac{1}{2}t^2\mathcal{A}^2 + \frac{1}{3!}t^3\mathcal{A}^3 + \cdots + \frac{1}{k!}t^k\mathcal{A}^k + \cdots, \tag{A.3.10}$$

where $\mathcal{I}$ denotes the identity matrix. Each term of the sum is well defined by the properties of matrices and, having in mind that in $\mathbb{M}_d$ we have a norm, and thus the notion of convergence of a sequence/series is well defined, we can establish the existence of the infinite sum

$$e^{t\mathcal{A}} = \lim_{n\to\infty} \sum_{k=0}^{n} \frac{t^k\mathcal{A}^k}{k!} = \sum_{k=0}^{\infty} \frac{t^k\mathcal{A}^k}{k!}. \tag{A.3.11}$$

Indeed, denoting $S_n(t) = \sum\limits_{k=0}^{n} \frac{t^k\mathcal{A}^k}{k!}$, for any $n > m \in \mathbb{N}$ and $t \in [-T, T]$, $0 < T < \infty$, we have

$$\|S_n(t) - S_m(t)\| = \left\| \sum_{k=m+1}^{n} \frac{t^k\mathcal{A}^k}{k!} \right\| \leq \sum_{k=m+1}^{n} \frac{T^k\|\mathcal{A}\|^k}{k!},$$

and on the right-hand side, we recognize partial sums of the Maclaurin expansion of the scalar exponent $e^{T\|\mathcal{A}\|}$, which converge to zero as $n, m \to \infty$. Thus, we may say that the series of matrices in (A.3.11) converges absolutely and uniformly on any interval $[-T, T]$ and thus $\mathbb{R} \ni t \mapsto e^{t\mathcal{A}} \in \mathbb{M}_d$ is a well-defined and continuous function of $t \in \mathbb{R}$, see [31, Section 7.4]. It is actually a power series in t (and in $\mathcal{A}$, what will be used later), and thus, like a scalar power series, it can be differentiated term-wise with respect to t any number of times. In particular,

$$\begin{aligned}\frac{d}{dt}e^{t\mathcal{A}} &= \mathcal{A} + t\mathcal{A}^2 + \frac{t^2}{2!}\mathcal{A}^3 + \cdots + \frac{t^{k-1}}{(k-1)!}\mathcal{A}^k + \cdots \\ &= \mathcal{A}\left(\mathcal{I} + t\mathcal{A} + \frac{t^2}{2!}\mathcal{A}^2 + \cdots + \frac{t^{k-1}}{(k-1)!}\mathcal{A}^{k-1} + \cdots\right) = \mathcal{A}e^{t\mathcal{A}} = e^{t\mathcal{A}}\mathcal{A},\end{aligned} \tag{A.3.12}$$

where the fact that we can factor out $\mathcal{A}$ from the infinite converging series follows from the fact that $\boldsymbol{x} \mapsto \mathcal{A}\boldsymbol{x}$ is a continuous function, as shown in the proof of point 3. of Proposition A.7.

Thus, $e^{\mathcal{A}}$ behaves similarly to a scalar exponential. There are, however, some differences. While for $a, b \in \mathbb{C}$ we have $e^{a+b} = e^a e^b$, matrix exponentials have this property if $\mathcal{A}$ and $\mathcal{B}$ commute, that is, if $\mathcal{A}\mathcal{B} = \mathcal{B}\mathcal{A}$. Then,

$$\begin{aligned}e^{\mathcal{A}}e^{\mathcal{B}} &= \lim_{n\to\infty}\left(\sum_{k=0}^{n}\frac{\mathcal{A}^k}{k!}\right)\left(\sum_{k=0}^{n}\frac{\mathcal{B}^k}{k!}\right) \\ &= \lim_{n\to\infty}\left(I + (\mathcal{A}+\mathcal{B}) + \frac{1}{2!}(\mathcal{A}^2 + 2\mathcal{A}\mathcal{B} + \mathcal{B}^2) + \cdots\right. \\ &\qquad \left. + \frac{1}{n!}(\mathcal{A}+\mathcal{B})^n + R_n(\mathcal{A}, \mathcal{B})\right) \\ &= e^{\mathcal{A}+\mathcal{B}},\end{aligned} \tag{A.3.13}$$

where $R_n(\mathcal{A}, \mathcal{B})$ contains all terms of the form $\frac{\mathcal{A}^j\mathcal{B}^i}{i!j!}$ with $0 < i, j \le n$ and $j + i > n$; naturally, $i + j \le 2n$. If we fix $n < r \le 2n$, then each term $\frac{\mathcal{A}^j\mathcal{B}^i}{i!j!}$ with $i + j = r$ appears only once, and the number of terms of this form will be smaller than the number of such terms with $i + j = r, 0 \le i, j \le r$. Thus

$$\begin{aligned}\left\|\sum_{1\le i,j\le n, i+j=r}\frac{\mathcal{A}^j\mathcal{B}^i}{i!j!}\right\| &\le \sum_{0\le i,j\le n, i+j=r}\frac{\|\mathcal{A}\|^j\|\mathcal{B}\|^i}{i!j!} \\ &\le \frac{1}{r!}\sum_{0\le i,j\le r, i+j=r}\frac{r!}{i!j!}\|\mathcal{A}\|^j\|\mathcal{B}\|^i = \frac{1}{r!}(\|\mathcal{A}\| + \|\mathcal{B}\|)^r.\end{aligned}$$

Hence,

$$\lim_{n\to\infty} \|R_n(\mathcal{A}, \mathcal{B})\| \leq \lim_{n\to\infty} \sum_{k=n+1}^{2n} \frac{(\|\mathcal{A}\| + \|\mathcal{B}\|)^k}{k!} = 0,$$

as the latter sum is a partial sum of the exponent $e^{\|\mathcal{A}\|+\|\mathcal{B}\|}$.

Using the above method, we can define $f(\mathcal{A})$ for any analytic function f. Namely, if f has an expansion

$$f(z) = \sum_{n=0}^{\infty} a_n z^n,$$

$a_n \in \mathbb{C}, n = 0, 1, \ldots$, convergent in a disc $\{z \in \mathbb{C} : |z| < r\}$ for some $r > 0$, then we can define

$$f(\mathcal{A}) = \sum_{n=0}^{\infty} a_n \mathcal{A}^n,$$

as long as $\|\mathcal{A}\| < r$. The same applies if f is expanded around any point z_0 in its domain of analyticity and also if z_0 is an isolated singularity of f, but then the Taylor expansion should be replaced by the Laurent expansion

$$f(z) = \sum_{n=-\infty}^{\infty} b_n (z - z_0)^n.$$

A.3.1 *Positivity in Vector Spaces*

First, we note that, in the context of population theory, if a given difference equation/system of equations describes a population's evolution, that is, if the solution is the population size or density, then clearly solutions emanating from non-negative data must stay non-negative. Thus, we have to extend the notion of positivity to vectors. We say that a vector $\boldsymbol{x} = (x_1, \ldots, x_d)$ is non-negative, if for all $i = 1, \ldots, d$, $x_i \geq 0$ and write $\boldsymbol{x} \geq 0$. If $0 \neq \boldsymbol{x} \geq 0$ (that is, at least one $x_i > 0$), then we write $\boldsymbol{x} > 0$ and say that $\boldsymbol{x}$ is positive. Finally, when $x_i > 0$ for all i, we use the notation $\boldsymbol{x} \gg 0$ and say that $\boldsymbol{x}$ is strictly positive. Then, we define

$$X_+ = \overline{\mathbb{R}}^d_+ = \{\boldsymbol{x} \in \mathbb{R}^d : \boldsymbol{x} \geq 0\},$$

and by $\mathbb{R}^d_+$ we denote the interior of X_+, that is,

$$\mathbb{R}^d_+ = \{\boldsymbol{x} \in \mathbb{R}^d : \boldsymbol{x} \gg 0\}.$$

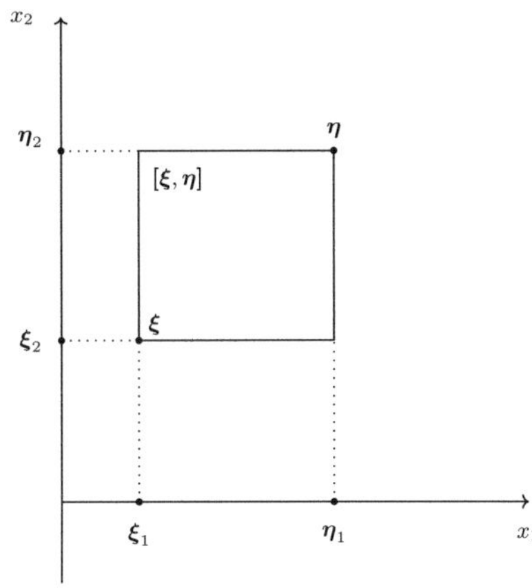

Fig. A.1 Order interval $[\boldsymbol{\xi}, \boldsymbol{\eta}]$ in $\mathbb{R}^2$

Similarly, we say that a matrix $\mathcal{A} = (a_{ij})_{1 \le i,j \le d}$ is non-negative (resp. positive or strictly positive) and write $\mathcal{A} \geq 0$ (resp. $\mathcal{A} > 0$ or $\mathcal{A} \gg 0$) if $a_{ij} \geq 0$ for all $i, j = 1, \ldots, d$, (resp. if additionally $a_{ij} > 0$ for some i, j, or $a_{ij} > 0$ for all $i, j = 1, \ldots, d$.)

An important role in Chap. 8 is played by the order interval

$$[\boldsymbol{\xi}, \boldsymbol{\eta}] = \{\boldsymbol{x} \in \mathbb{R}^d : \xi_i \leq x_i \leq \eta_i, \ i = 1, \ldots, d\}. \tag{A.3.14}$$

As illustrated in Fig. A.1, an order interval in $\mathbb{R}^d$ is a closed box and, in particular, with $\mathbf{1} = (1, \ldots, 1)$, $[-\mathbf{1}, \mathbf{1}]$ equals the closed unit ball in the l^∞-norm. Thus, we see that if $\boldsymbol{\xi} \ll \eta$, then $[\boldsymbol{\xi}, \boldsymbol{\eta}]$ has a nonempty interior.

Appendix B
Systems of Differential Equations

B.1 Basic Notation

Introducing a universal notation for derivatives that would not become awkward in some situations is difficult. Thus, we will use several symbols depending on the context. For a scalar function of a single variable, we use

$$f', f'', \ldots, f^{(n)}, \ldots,$$

or

$$\frac{df}{dx}, \frac{d^2 f}{dx^2}, \ldots, \frac{d^n f}{dx^n}, \ldots,$$

if we need to indicate the variable with respect to which we differentiate.

Partial derivatives of the first order of a scalar function of several variables $\boldsymbol{x} = (x_1, \ldots, x_d) \in \mathbb{R}^d$ will be denoted by

$$\frac{\partial f}{\partial x_i}, \ \partial_{x_i} f \quad \text{or} \quad f_{x_i}, \quad i = 1, \ldots, d.$$

Then, the derivative of f with respect to $\boldsymbol{x} \in \mathbb{R}^d$,

$$\partial_{\boldsymbol{x}} f = \nabla_{\boldsymbol{x}} f = \nabla f,$$

J. Banasiak, *Introduction to Mathematical Methods in Population Theory*, Springer Undergraduate Mathematics Series,
https://doi.org/10.1007/978-3-031-65491-6

that is, the gradient of f is a vector. If $\boldsymbol{f}$ is vector valued, $\boldsymbol{f}(\boldsymbol{x}) \in \mathbb{R}^m$ for $\boldsymbol{x} \in \mathbb{R}^d$, then we use the notation

$$\frac{\partial f_j}{\partial x_i} \quad \text{or} \quad \partial_{x_i} f_j, \quad 1 \le i \le d, 1 \le j \le m.$$

In this case, the derivative of $\boldsymbol{f}$,

$$\partial_{\boldsymbol{x}} \boldsymbol{f} = \nabla_{\boldsymbol{x}} \boldsymbol{f} = \mathcal{J}_{\boldsymbol{f}}, \tag{B.1.1}$$

where the last option is used when the variable of differentiation is clear, is a matrix called the Jacobi matrix

$$\begin{pmatrix} \frac{\partial f_1}{\partial x_1} & \cdots & \frac{\partial f_1}{\partial x_d} \\ \vdots & \ddots & \vdots \\ \frac{\partial f_m}{\partial x_1} & \cdots & \frac{\partial f_m}{\partial x_d} \end{pmatrix}.$$

If we are only interested in their order, we use the multi-index notation for higher-order derivatives. By a multi-index we understand a d-tuple $\beta := (\beta_1, \ldots, \beta_d)$, $\beta_i \in \mathbb{N}_0$, and define $|\beta| := \beta_1 + \cdots + \beta_d$. Then,

$$\partial_{\boldsymbol{x}}^{\beta} f := \frac{\partial^{\beta} f}{\partial_{x_1}^{\beta_1} \cdots \partial_{x_d}^{\beta_n}}$$

is any derivative of f of order $|\beta|$. Thus, $\sum\limits_{|\beta|=0}^{k} \partial^{\beta} f$ means the sum of all derivatives of f of the order less than, or equal to, k.

B.2 Spaces of Functions

We deal with basic partial differential equations in Chaps. 9 and 10. Thus, it is convenient to introduce several concepts and notations related to their integrability and differentiability properties. We will skip details of the theory for the lack of space, referring the interested reader to, e.g., [2, 26, 77].

B.2.1 Spaces of Continuous and Integrable Functions

Let f be a function defined on a set $\Omega \subset \mathbb{R}^d$ and let $\boldsymbol{x} \in \Omega$.

If $\Omega \subset \mathbb{R}^d$ is an open set, then, for $k \in \mathbb{N}_0$, the symbol $C^k(\Omega)$ denotes the set of k-times continuously differentiable functions in Ω. In particular, for $k = 0$, we denote by $C(\Omega) := C^0(\Omega)$ the set of all continuous functions in Ω. Functions from $C^k(\Omega)$ need not be bounded in Ω. If they are required to be bounded together with their derivatives up to the order k, then the corresponding set is denoted by $C^k(\overline{\Omega})$.

For a continuous function f defined on Ω, we define the support of f by

$$\operatorname{supp} f = \overline{\{\boldsymbol{x} \in \Omega : f(x) \neq 0\}}.$$

The set of all functions with compact support in Ω which have continuous derivatives of order smaller than, or equal to, k is denoted by $C_0^k(\Omega)$. As above, $C_0(\Omega) := C_0^0(\Omega)$ is the set of all continuous functions with compact support in Ω and

$$C_0^\infty(\Omega) := \bigcap_{k=0}^{\infty} C_0^k(\Omega).$$

Example B.1 The function

$$\omega(\boldsymbol{x}) = \begin{cases} \exp\left(\frac{1}{\|\boldsymbol{x}\|^2-1}\right) & \text{for } \|\boldsymbol{x}\| < 1, \\ 0 & \text{for } \|\boldsymbol{x}\| \geq 1, \end{cases} \tag{B.2.2}$$

is in $C_0^\infty(\mathbb{R}^d)$, with support in the unit ball B_1 of $\mathbb{R}^d$.

Another important class of spaces is the class of spaces $L_p(\Omega)$, $1 \leq p < \infty$, of functions that are Lebesgue integrable with power p. Their proper introduction requires the definition of the Lebesgue measure $m(\Omega)$, which is a natural extension of the volume of a set (note, however, that not to every set we can associate its Lebesgue measure). A Lebesgue measurable function is such that the preimage of any interval is Lebesgue measurable. Any Lebesgue measurable function can be approximated by continuous functions (Łuzin's theorem, [2, Theorem 4.2]) or by piecewise constant functions [77, Section 4.2] and the Lebesgue integral can be thus defined as the limit of the Riemann integrals of the approximating functions. For our applications, it is sufficient to know that continuous or piecewise continuous functions are Lebesgue measurable and their Lebesgue integrals are the same as their Riemann integrals. Then the spaces $L_p(\Omega)$ are defined as the spaces of functions for which

$$\|f\|_p := \|f\|_{L_p(\Omega)} = \left(\int_\Omega |f(\boldsymbol{x})|^p d\boldsymbol{x}\right)^{1/p} < \infty. \tag{B.2.3}$$

The function $f \mapsto \|f\|_p$ is a norm in the space $L_p(\Omega)$ provided we identify functions which differ only on a set of measure zero. We say that such functions are equal almost everywhere. The space $L_p(\Omega)$ with the norm (B.2.3) is complete, and hence it is a Banach space, [2, Theorem 2.16].

It is customary to finish the scale of L_p spaces by the space $L_\infty(\Omega)$ defined to be the space of all Lebesgue measurable functions, which are bounded almost everywhere in Ω, that is, bounded everywhere except possibly on a set of measure zero. The corresponding norm is defined by

$$\|f\|_\infty := \|f\|_{L_\infty(\Omega)} := \inf\{M;\ m(\{\mathbf{x} \in \Omega : |f(\mathbf{x})| > M\}) = 0\}. \tag{B.2.4}$$

The expression on the right-hand side of (B.2.4) is frequently referred to as the *essential supremum* of f over Ω.

For a bounded set Ω and $1 \le p \le p' \le \infty$, we have

$$L_{p'}(\Omega) \subset L_p(\Omega) \tag{B.2.5}$$

and, for $f \in L_\infty(\Omega)$,

$$\|f\|_\infty = \lim_{p\to\infty} \|f\|_p, \tag{B.2.6}$$

which justifies the notation. However,

$$\bigcap_{1\le p<\infty} L_p(\Omega) \ne L_\infty(\Omega),$$

as demonstrated by the function $f(x) = \ln x,\ x \in (0, 1]$. If $m(\Omega) = \infty$, then neither (B.2.5) nor (B.2.6) hold.

Occasionally, we need functions that are integrable with power p only over bounded and closed subsets of Ω. Spaces of such functions are denoted by $L_{p,loc}(\Omega)$. A function $f \in L_{1,loc}(\Omega)$ is called *locally integrable* (in Ω).

Remark B.2 Observe that, in contrast to $L_p(\Omega)$ spaces, here it matters whether Ω is open, closed or contains a part of its boundary. For instance, $L_{1,loc}(\mathbb{R}_+)$ contains all functions integrable on any interval $[a, b]$, $0 < a < b < \infty$, whereas functions from $L_{1,loc}(\overline{\mathbb{R}}_+) = L_{1,loc}([0, \infty))$ are integrable on $[a, b]$ with $0 \le a < b < \infty$. Hence, if $f(x) = \frac{1}{x}$, then $f \in L_{1,loc}(\mathbb{R}_+)$ but $f \notin L_{1,loc}(\overline{\mathbb{R}}_+)$.

Let $\Omega \subset \mathbb{R}^d$ be an open set. It is clear that

$$C_0^\infty(\Omega) \subset L_p(\Omega)$$

for $1 \le p \le \infty$. If $p \in [1, \infty)$, then $C_0^\infty(\Omega)$ is dense in $L_p(\Omega)$, that is,

$$\overline{C_0^\infty(\Omega)} = L_p(\Omega), \tag{B.2.7}$$

where the closure is taken in the L_p-norm, [2, Theorem 2.19]. The proof of this fact uses regularizations of f defined by

$$f_\epsilon(\boldsymbol{x}) := \int_{\mathbb{R}^d} f(\boldsymbol{x}-\boldsymbol{y})\omega_\epsilon(\boldsymbol{y})d\boldsymbol{y} = \int_{\mathbb{R}^d} f(\boldsymbol{y})\omega_\epsilon(\boldsymbol{x}-\boldsymbol{y})d\boldsymbol{y}, \tag{B.2.8}$$

where

$$\omega_\epsilon(\boldsymbol{x}) := C_\epsilon \omega\left(\frac{\boldsymbol{x}}{\epsilon}\right),$$

and the constants C_ϵ are chosen so that $\int_{\mathbb{R}^d} \omega_\epsilon(\boldsymbol{x})d\boldsymbol{x} = 1$. Functions ω_ϵ are also $C_0^\infty(\mathbb{R}^d)$ with supports equal to B_ϵ. Because the effective domain of integration in the second integral of (B.2.8) is $B_{\boldsymbol{x},\epsilon}$, f_ϵ is well defined whenever f is a locally integrable function. Similarly, if the support of f is bounded, then $\operatorname{supp} f_\epsilon$ is also bounded, and it is contained in the ϵ-neighbourhood of $\operatorname{supp} f$. The functions f_ϵ are infinitely differentiable with

$$\partial_{\boldsymbol{x}}^\beta f(\boldsymbol{x}) = \int_{\mathbb{R}^d} f(\boldsymbol{y})\partial_{\boldsymbol{x}}^\beta \omega_\epsilon(\boldsymbol{x}-\boldsymbol{y})d\boldsymbol{y} \tag{B.2.9}$$

for any β. It can be proved that if $f \in L_p(\mathbb{R}^n)$, $1 \le p < \infty$, then

$$f = \lim_{\epsilon\to 0^+} f_\epsilon \tag{B.2.10}$$

in $L_p(\mathbb{R}^d)$. Because any function $f \in L_p(\mathbb{R}^d)$ can be approximated by L_p-functions with supports in B_N, $N \to \infty$, (B.2.7) follows. Since the latter can be accomplished by simply restricting f to B_N and extending it by zero outside, formula (B.2.8) also shows that non-negative C_0^∞ functions approximate non-negative L_p functions.

B.2.2 Generalized Derivatives and Sobolev Spaces

In Chap. 9, we encountered discontinuous functions, which potentially should be solutions of differential equations. This indicates that we should revisit the concept of the derivative. The following considerations are based on the du Bois-Reymond lemma; see [26, Corollary 4.24].

Lemma B.3 *Let $\Omega \subset \mathbb{R}^d$ be an open set and let $f \in L_{1,loc}(\Omega)$ be such that*

$$\int_\Omega f(\boldsymbol{x})\phi(\boldsymbol{x})d\boldsymbol{x} = 0$$

for any $\phi \in C_0(\Omega)$. Then $f = 0$ is almost everywhere on Ω.

Then, we can formulate

Definition B.4 We say that a function $g \in L_{1,loc}(\Omega)$ is the generalized (or distributional) derivative of $f \in L_{1,loc}(\Omega)$ of order α, denoted by $\partial_{\boldsymbol{x}}^{\alpha} f$, if

$$\int_{\Omega} g(\boldsymbol{x})\phi(\boldsymbol{x})d\boldsymbol{x} = (-1)^{|\alpha|}\int_{\Omega} f(\boldsymbol{x})\partial_{\boldsymbol{x}}^{\alpha}\phi(\boldsymbol{x})d\boldsymbol{x} \tag{B.2.11}$$

for any $\phi \in C_0^{\infty}(\Omega)$.

Note that g is well-defined by the du Bois-Reymond lemma.

We observe that for a k-times continuously differentiable function f, g is just the classical kth derivative of f and (B.2.11) is the integration by parts formula.

Example B.5 Let us find the generalized derivative of $f(x) = |x|$ on $\mathbb{R}$. We have

$$\begin{aligned}
\int_{-\infty}^{\infty} |x|\phi'(x)dx &= -\int_{-\infty}^{0} x\phi'(x)dx + \int_{0}^{\infty} x\phi'(x)dx \\
&= -x\phi(x)|_{-\infty}^{0} + \int_{-\infty}^{0}\phi(x)dx + x\phi(x)|_{0}^{\infty} - \int_{0}^{\infty}\phi(x)dx \\
&= \int_{-\infty}^{0}\phi(x)dx - \int_{0}^{\infty}\phi(x)dx = -\int_{-\infty}^{\infty}\operatorname{sign} x\phi(x)dx.
\end{aligned}$$

Since $g(x) = \operatorname{sign} x$ is a locally integrable function on $\mathbb{R}$, it is the generalized derivative of $f(x) = |x|$.

Next, let us try to find the generalized derivative of $f(x) = \operatorname{sign} x$. Similarly, we obtain

$$\begin{aligned}
\int_{-\infty}^{\infty} \operatorname{sign} x\phi'(x)dx &= -\int_{-\infty}^{0}\phi'(x)dx + \int_{0}^{\infty}\phi'(x)dx \\
&= -\phi(x)|_{-\infty}^{0} + \phi(x)|_{0}^{\infty} = -2\phi(0).
\end{aligned}$$

Is there a locally integrable function g such that

$$\int_{-\infty}^{\infty} g(x)\phi(x)dx = 2\phi(0)$$

for any $\phi \in C_0^{\infty}(\mathbb{R})$? If we take any $\phi \in C_0^{\infty}(\mathbb{R}_+)$, then we can extend it by zero to $\mathbb{R}_-$ as a C^{∞} function and hence

$$\int_{-\infty}^{\infty} g(x)\phi(x)dx = 0$$

for any $\phi \in C_0^\infty(\mathbb{R}_+)$. Therefore, by the du Bois-Reymond lemma, $g = 0$ almost everywhere on $\mathbb{R}_+$. An analogous argument gives $g = 0$ on $\mathbb{R}_-$. But, for the integration, the value of g at 0 is irrelevant, and hence

$$\int_{-\infty}^{\infty} g(x)\phi(x)dx = 0$$

for any $\phi \in C_0^\infty(\mathbb{R})$, which gives a contradiction.

This does not mean that $x \mapsto \operatorname{sign} x$ cannot be differentiated. Its derivative, however, is not a function but a measure, or a distribution, called the Dirac distribution or Dirac delta function.

The concept of generalized derivative allows for defining spaces of differentiable functions in the framework of integrable functions.

Definition B.6 Let $\Omega \subseteq \mathbb{R}^d$ be an open set. The Sobolev space $W_p^m(\Omega)$, $1 \leq p \leq \infty$, $m \in \mathbb{N}_0$, is defined as

$$W_p^m(\Omega) := \{u \in L_p(\Omega) : \partial^\alpha u \in L_p(\Omega),\ 0 \leq |\alpha| \leq m\}, \tag{B.2.12}$$

where, recall, $\partial^\alpha = \partial_{x_1}^{\alpha_1} \ldots \partial_{x_n}^{\alpha_n}$, $|\alpha| = \alpha_1 + \ldots \alpha_n$, is the distributional derivative of order $|\alpha|$.

The function

$$\|u\|_{m,p} := \|u\|_{W_p^m(\Omega)} := \left(\sum_{0 \leq |\alpha| \leq m} \|\partial^\alpha u\|_{L_p(\Omega)}^p \right)^{1/p}. \tag{B.2.13}$$

is a norm on $W_p^m(\Omega)$ making it a Banach space, [2, Theorem 3.3].

In Chaps. 9 and 10, we mainly use the space $W_1^1(\mathbb{R}_+)$. Sobolev spaces of functions of a single variable have very nice properties, which we will discuss now. First, observe that, in contrast to the classical derivative, we cannot calculate the distributional derivative at a point. So, in what follows, we need to differentiate between the distributional ∂f and the classical f' derivatives of f, provided the latter exists.

Lemma B.7 *If $f \in L_{1,loc}(\mathbb{R})$ and $\partial f = 0$, then $f = const$ almost everywhere.*

Proof First, we observe that if $f \in L_{1,loc}(\mathbb{R})$ satisfies $\partial f = 0$, then

$$\int_{\mathbb{R}} f(x)\phi'(x)dx = 0 \tag{B.2.14}$$

for any $\phi \in C_0^\infty(\mathbb{R})$. We note that clearly $\phi' \in C_0^\infty(\mathbb{R})$ but not every $\varphi \in C_0^\infty(\mathbb{R})$ can be expressed as $\varphi = \phi'$ for some $\phi \in C_0^\infty(\mathbb{R})$, since such a φ must satisfy

$$\int_{-\infty}^{\infty} \varphi(x)dx = \int_{-\infty}^{\infty} \phi'(x)dx = \lim_{x\to\infty} \phi(x) - \lim_{x\to-\infty} \phi(x) = 0$$

on account of the compact support of ϕ. Conversely, if $\varphi \in C_0^\infty(\mathbb{R})$ is such that $\int_{-\infty}^{\infty} \varphi(x)dx = 0$, then

$$\phi(x) = \int_{-\infty}^{x} \varphi(s)ds,$$

$\phi' = \varphi$ and $\phi \in C_0^\infty(\mathbb{R})$. To prove the latter, we observe that $\phi(x) = 0$ for all $x < \min\{\operatorname{supp}\varphi\}$ and also ϕ is constant for $x > \max\{\operatorname{supp}\varphi\}$ and thus must be zero there.

Because of this, we cannot use the du Bois-Reymond lemma directly in (B.2.14), but we need to express ϕ' in terms of an arbitrary $\omega \in C_0^\infty(\mathbb{R})$. Using the discussion above, we observe that if $\psi \in C_0^\infty(\mathbb{R})$ satisfies $\int_{\mathbb{R}} \psi dx = 1$, then for any $\omega \in C_0^\infty(\mathbb{R})$, the function

$$\varphi(x) = \omega(x) - \psi(x) \int_{-\infty}^{\infty} \omega(s)ds$$

satisfies

$$\int_{-\infty}^{\infty} \varphi(s)ds = 0.$$

Hence, there is $\phi \in C_0^\infty(\mathbb{R})$ satisfying

$$\phi'(x) = \omega(x) - \psi(x) \int_{-\infty}^{\infty} \omega(s)ds.$$

Thus, for any $\omega \in C_0^\infty(\mathbb{R})$,

$$\int_{-\infty}^{\infty} f(x)\phi'(x)dx = \int_{-\infty}^{\infty} f(x)(\omega(x) - \psi(x) \int_{-\infty}^{\infty} \omega(s)ds)dx = 0,$$

or

$$\int_{-\infty}^{\infty} \left(f(x) - \int_{-\infty}^{\infty} f(s)\psi(s)ds \right) \omega(x)ds = 0$$

and thus $f = const$ almost everywhere. □

Theorem B.8 *Assume that $u, \partial u \in L_{1,loc}(\mathbb{R})$. Then there is a continuous function U equal to u almost everywhere such that for any $x, y \in \mathbb{R}$ we have*

$$U(x) = U(y) + \int_y^x \partial u(s)ds \tag{B.2.15}$$

and $U' = \partial U = \partial u$ almost everywhere.

Proof First we observe that if $v(x) = \int_y^x f(s)ds$ for $f \in L_{1,loc}(\mathbb{R})$ and any fixed y, then v is continuous and $\partial v = f$. Indeed,

$$\begin{aligned}\int_{-\infty}^{\infty} v(x)\phi'(x)dx &= \int_y^{\infty}\left(\int_y^x f(s)ds\right)\phi'(x)dx - \int_{-\infty}^{y}\left(\int_x^y f(s)ds\right)\phi'(x)dx\\ &= \int_y^{\infty} f(s)\left(\int_s^{\infty}\phi'(x)dx\right)ds\\ &\quad -\int_{-\infty}^{y} f(s)\left(\int_{-\infty}^{s}\phi'(x)dx\right)ds = -\int_{-\infty}^{\infty} f(s)\phi(s)ds.\end{aligned}$$

With this result, let $u, \partial u \in L_{1,loc}(\mathbb{R})$ and set $\bar{u}(x) = \int_0^x \partial u(s)ds$. Then, $\partial\bar{u} = \partial u$ almost everywhere and, by Lemma B.7, $\bar{u} + C = u$ almost everywhere. Defining $U = \bar{u} + C$, we see that U is continuous and, by

$$U(y) = \int_0^y \partial u(s)ds + C,$$

we obtain

$$U(x) - U(y) = \int_0^x \partial u(s)ds - \int_0^y \partial u(s)ds = \int_y^x \partial u(s)ds.$$

□

Remark B.9 Functions satisfying (B.2.15) are called absolutely continuous.

B.3 The Laplace Transform

As in the previous section, we restrict ourselves to providing basic facts and notation from the theory of Laplace transforms, which are needed in Chap. 9. More comprehensive sources are [9, 40, 84].

Let a function $f \in L_{1,loc}([0, \infty))$ be exponentially bounded, that is, there are $M \in \mathbb{R}_+, \sigma \in \mathbb{R}$,

$$|f(t)| \leq Me^{\sigma t}, \quad t \geq 0. \tag{B.3.16}$$

The Laplace transform of f is defined by

$$\hat{f}(\lambda) := [\mathcal{L}f](\lambda) := \int_0^\infty e^{-\lambda t} f(t)dt, \tag{B.3.17}$$

where the improper integral in (B.3.17) is absolutely and uniformly convergent in any half-plane $\overline{S}_{\sigma'} := \{\lambda \in \mathbb{C} : \Re\lambda \geq \sigma'\} \subset \mathbb{C}, \sigma' > \sigma$, and hence $\hat{f}$ is defined and analytic in the open half-plane $S_\sigma = \{\lambda \in \mathbb{C} : \Re\lambda > \sigma\}$. For our applications, it is also important to note that if the support of f is contained in a finite interval (a, b), then its Laplace transform is defined and analytic everywhere in $\mathbb{C}$—this follows as in such a case, we can let the constant σ tend to $-\infty$. We recall that functions analytic in $\mathbb{C}$ are called entire. The Laplace transform's power lies in transforming differential and integral operations into algebraic ones. In particular, if f satisfies (B.3.16) with $\sigma \geq 0$ and

$$F(t) := \int_0^t f(s)ds, \quad t \geq 0,$$

then

$$\widehat{F}(\lambda) = \frac{\hat{f}(\lambda)}{\lambda}, \quad \Re\lambda > \sigma. \tag{B.3.18}$$

Further, if f is differentiable (it suffices that it is absolutely continuous) for $t \geq 0$ and f' satisfies (B.3.16) with some $\sigma \geq 0$, then $\hat{f}(\lambda)$ exists for $\Re\lambda > \sigma$, and

$$\widehat{f'}(\lambda) = \lambda\hat{f}(\lambda) - f(0), \quad \Re\lambda > \sigma. \tag{B.3.19}$$

Finally, let $f, g \in L_{1,loc}([0, \infty))$ satisfy (B.3.16). We define the convolution of f and g by

$$(f * g)(t) = \int_0^t f(t-s)g(s)ds = \int_0^t f(s)g(t-s)ds; \tag{B.3.20}$$

it can be shown that it exists for almost any t, [99, Section 4.1]. Then, using the definition of the Laplace transform and changing the order of integration, we obtain

$$[\mathcal{L}(f * g)](\lambda) = [\mathcal{L}f](\lambda) \cdot [\mathcal{L}g](\lambda), \quad \Re\lambda > \sigma. \tag{B.3.21}$$

Moreover, we also use the property that $\hat{f}(\lambda) \to 0$ as $|\lambda| \to \infty$ in any closed strip contained in the domain of analyticity of $\hat{f}$.

This result is easy to prove if f is differentiable, and both f and f' satisfy (B.3.16). Indeed, then (B.3.19) yields

$$\hat{f}(\lambda) = \frac{\widehat{f'}(\lambda)}{\lambda} + \frac{f(0)}{\lambda}.$$

In general, denoting $\lambda = x + iy$, we write, see [84, Chapter 4],

$$[\mathcal{L}f](\lambda) = \int_0^\infty e^{-iyt}[e^{-xt}f(t)]dt = \int_0^\infty e^{-iyt}\phi_{x,f}(t)dt = \mathcal{F}[\phi_{x,f}](y),$$

where $\mathcal{F}$ is the Fourier transform and

$$\phi_{x,f}(t) = \begin{cases} e^{-xt}f(t) & \text{for } t \geq 0, \\ 0 & \text{for } t < 0. \end{cases} \tag{B.3.22}$$

Then, the result follows from the Riemann-Lebesgue lemma, e.g., [78, Theorem 9.6].

While using the Laplace transform converts the calculus operation of functions to algebraic operations on their transforms, greatly simplifying the analysis, the crucial part of the process is returning to the original variables. There are several inversion theorems, essentially based on the inversion of the Fourier transform, see [40, Chapter 24] or [84, Chapter 4]. A sufficient version given in the latter source states that if f is a continuous function with a piecewise continuous derivative f' such that f and f' satisfy (B.3.16), then

$$f(t) = \frac{1}{2\pi i}\int\limits_{c-i\infty}^{c+i\infty} e^{\lambda t}\hat{f}(\lambda)d\lambda := \frac{1}{2\pi i}\lim_{T\to\infty}\int\limits_{c-iT}^{c+iT} e^{\lambda t}\hat{f}(\lambda)d\lambda, \tag{B.3.23}$$

where the integration is carried along any vertical line in the domain of analyticity of $\hat{f}$, that is, with $c > \sigma$.

B.4 Cauchy Problem for Ordinary Differential Equations

Ordinary differential equations describe processes that depend only on one independent variable. As explained in Chap. 1, a single equation

$$y' = f(t, y)$$

describes the evolution of a quantity y which does not interact with its environment, and the rate of change of y is fully determined by its state (and the state of the environment independent of y) at time t. There are two immediate generalizations of such a first-order scalar equation. One is a differential equation of a higher-order

$$y^{(n)} = F(t, y', y'', \ldots, y^{(n-1)}) = 0, \tag{B.4.24}$$

(where, for simplicity, we consider only equations solved with respect to the highest derivative), and the other is a system of first-order equations, that is,

$$\boldsymbol{y}' = \boldsymbol{f}(t, \boldsymbol{y}), \tag{B.4.25}$$

where

$$\boldsymbol{y}(t) = (y_1(t), \dots, y_d(t)),$$

and, in general,

$$\boldsymbol{f}(t, \boldsymbol{y}) = (f_1(t, y_1, \dots, y_d), \dots, f_d(t, y_1, \dots, y_d))$$

is a nonlinear function of t and $\boldsymbol{y}$, describing the way the components of the system and the environment affect each other, see Fig. 1.3. As explained in Remark 6.2, at least from the theoretical point of view, there is no need to consider these two extensions separately, as any equation of higher order can be always written as a first-order system (the converse, in general, is not true). Therefore, from now on, we will be concerned with systems of differential equations (B.4.25).

B.4.1 Local Existence and Basic Properties

Let us consider the Cauchy problem for (B.4.25):

$$\begin{aligned} \boldsymbol{y}' &= \boldsymbol{f}(t, \boldsymbol{y}), \\ \boldsymbol{y}(t_0) &= \mathring{\boldsymbol{y}}. \end{aligned} \tag{B.4.26}$$

We shall present an elementary, though long, proof of the solvability of (B.4.26). On the one hand, a glimpse at how the solution is obtained is necessary to analyse its properties, and, on the other, the proof can be repeated almost without any change in more complex situations, such as described in Theorems 9.2 and 10.10.

Before we formulate and discuss the main results of this chapter, we shall recall a few notions from vector analysis. Let $\boldsymbol{y}(t)$ be a vector-valued function with values in $X = \mathbb{C}^d$. We have seen in Corollary A.5, that all notions of norm convergence in X are equivalent to coordinate-wise convergence. Then we can say that $\boldsymbol{y}(t)$ is continuous if each of its coordinates is continuous. Similarly, we define the derivative $\boldsymbol{y}'(t) = (y_1'(t), \dots, y_d'(t))$ and the integral

$$\int_a^b \boldsymbol{y}(t)dt = \left(\int_a^b y_1(t)dt, \dots, \int_a^b y_d(t)dt \right).$$

Using the definition of the Riemann integral as the limit of sums, the triangle inequality and the continuity of the norm, we can establish that

$$\left\| \int_a^b \boldsymbol{y}(t)dt \right\| \leq \int_a^b \|\boldsymbol{y}(t)\|dt. \tag{B.4.27}$$

We can now formulate and prove the Picard–Lindelöf theorem.

Theorem B.10 *If the function $\boldsymbol{f}$ is continuous in a cylinder $\Xi := \{(t, \boldsymbol{x}) : |t - t_0| \leq a, \|\boldsymbol{y} - \mathring{\boldsymbol{y}}\| \leq b\}$, where $a, b > 0$, and there is $L > 0$ such that for all $(t, \boldsymbol{y}_1), (t, \boldsymbol{y}_2) \in \Xi$*

$$\|\boldsymbol{f}(t, \boldsymbol{y}_1) - \boldsymbol{f}(t, \boldsymbol{y}_2)\| \leq L\|\boldsymbol{y}_1 - \boldsymbol{y}_2\|, \tag{B.4.28}$$

then the initial value problem (B.4.26) has one and only one solution $\boldsymbol{y}(t) = (y_1(t), \ldots, y_d(t))$ defined at least on the interval $|t - t_0| \leq \alpha$, where $\alpha = \min\{a, b/M\}$ and $M = \max_{(t,y)\in\Xi} \|f(t, y)\|$.

A function $\boldsymbol{f}$ satisfying (B.4.28) is called Lipschitz continuous (precisely, Lipschitz continuous in $\boldsymbol{y}$, uniformly in t). In particular, if $\boldsymbol{f}$ has continuous partial derivatives in Ξ with respect to $\boldsymbol{y}$, then it is there Lipschitz continuous, since by Taylor's expansion

$$\|\boldsymbol{f}(t, \boldsymbol{y}_1) - \boldsymbol{f}(t, \boldsymbol{y}_2)\| = \left\|\mathcal{J}_f(t, \theta\boldsymbol{y}_1 + (1-\theta)\boldsymbol{y}_2)(\boldsymbol{y}_1 - \boldsymbol{y}_2)\right\| \leq L\|\boldsymbol{y}_1 - \boldsymbol{y}_2\|, \tag{B.4.29}$$

where $\mathcal{J}_f = \left(\frac{\partial f_j}{\partial y_i}\right)_{1\leq i,j\leq d}$ is the Jacobi matrix of $\boldsymbol{f}$, $\theta \in [0, 1]$ so that $(t, \theta\boldsymbol{y}_1 + (1-\theta)\boldsymbol{y}_2)$ is a point in Ξ and L is the maximum of $\|\mathcal{J}_f\|$ over Ξ. However, the scalar function $f(y) = |y|$ is Lipschitz continuous but not continuously differentiable.

We split the proof of Theorem B.10 into a few steps. First, we shall prove a fundamental result known as Gronwall's lemma.

Lemma B.11 *If $f(t)$, $g(t)$ are continuous and nonnegative for $t \in [t_0, t_0 + \alpha]$, $\alpha > 0$, and $c > 0$, then*

$$f(t) \leq c + \int_{t_0}^t f(s)g(s)ds \tag{B.4.30}$$

on $[t_0, t_0 + \alpha]$ implies

$$f(t) \leq c\exp\left(\int_{t_0}^t g(s)ds\right) \tag{B.4.31}$$

for all $[t_0, t_0 + \alpha]$.

If f satisfies (B.4.30) with $c = 0$, then $f(t) = 0$ on $[t_0, t_0 + \alpha]$.

Proof Let $c > 0$ and define $F(t) = c + \int_{t_0}^{t} f(s)g(s)ds$ on $[t_0, t_0 + \alpha]$. Then $F(t) \geq f(t)$ and $F(t) > 0$ on this interval. Differentiating, we get $F'(t) = f(t)g(t)$ and therefore

$$\frac{F'(t)}{F(t)} = \frac{f(t)g(t)}{F(t)} \leq \frac{f(t)g(t)}{f(t)} = g(t).$$

However, the left-hand side is equal to $\frac{d}{dt} \ln F(t)$ so that

$$\frac{d}{dt} \ln F(t) \leq g(t),$$

or, integrating,

$$\ln F(t) - \ln F(t_0) \leq \int_{t_0}^{t} g(s)ds.$$

Since $F(t_0) = c$, we obtain

$$F(t) \leq c \exp\left(\int_{t_0}^{t} g(s)ds\right)$$

but since, as we observed, $f(t) \leq F(t)$, we have

$$f(t) \leq c \exp\left(\int_{t_0}^{t} g(s)ds\right),$$

which proves the first part. If $c = 0$, then we cannot use the above argument directly, as it would involve taking a logarithm of zero. However, if

$$f(t) \leq \int_{t_0}^{t} f(s)g(s)ds,$$

then

$$f(t) \leq c + \int_{t_0}^{t} f(s)g(s)ds$$

for any $c > 0$, and so

$$0 \le f(t) \le c \exp\left(\int_{t_0}^{t} g(s)ds\right)$$

for any $c > 0$. This yields $f(t) = 0$ for all t in $[t_0, t_0 + \alpha]$. □

Proof of Theorem B.10 The uniqueness of solutions to (B.4.26) can be proved using Gronwall's lemma. Let $\mathbf{y}_1(t)$ and $\mathbf{y}_2(t)$ be two solutions of the Cauchy problem (B.4.26) in Ξ with the same initial condition $\mathring{\mathbf{y}}$, that is,

$$\mathbf{y}_1'(t) = \mathbf{f}(t, \mathbf{y}_1(t)), \qquad \mathbf{y}_1(t_0) = \mathring{\mathbf{y}}$$

and

$$\mathbf{y}_2'(t) = \mathbf{f}(t, \mathbf{y}_2(t)), \qquad \mathbf{y}_2(t_0) = \mathring{\mathbf{y}}.$$

Then, integrating, subtracting the second equation from the first, and using the equality at t_0, we see that

$$\mathbf{y}_1(t) - \mathbf{y}_2(t) = \int_0^t (\mathbf{f}(t, \mathbf{y}_1(s)) - \mathbf{f}_2(t, \mathbf{y}_2(s)))ds.$$

Thus, by (B.4.27),

$$\begin{aligned} \|\mathbf{y}_1(t) - \mathbf{y}_2(t)\| &= \left\| \int_{t_0}^t (\mathbf{f}(t, \mathbf{y}_1(s)) - \mathbf{f}_2(t, \mathbf{y}_2(s)))ds \right\| \\ &\le \int_{t_0}^t \|\mathbf{f}(t, \mathbf{y}_1(s)) - \mathbf{f}_2(t, \mathbf{y}_2(s))\|ds \\ &\le L \int_{t_0}^t \|\mathbf{y}_1(s)) - \mathbf{y}_2(s)\|ds, \end{aligned} \tag{B.4.32}$$

and we can use the second part of Gronwall's lemma to claim that $\|\mathbf{y}_1(t) - \mathbf{y}_2(t)\| = 0$. Hence, by the axioms of the norm, $\mathbf{y}_1(t) = \mathbf{y}_2(t)$ for all t satisfying $|t - t_0| < \alpha$.

The proof of the existence is much more complicated. Firstly, we convert the Cauchy problem to an integral equation by integrating both sides of the equation in (B.4.26) and using the initial condition. This yields

$$\boldsymbol{y}(t) = \mathring{\boldsymbol{y}} + \int_{t_0}^{t} \boldsymbol{f}(s, \boldsymbol{y}(s))ds. \tag{B.4.33}$$

If $\boldsymbol{y}(t)$ is a differentiable solution to (B.4.26), then of course (B.4.33) is satisfied. On the other hand, if $\boldsymbol{y}(t)$ is a continuous solution to (B.4.33), then it is also differentiable by the fundamental theorem of calculus, and we see that we obtain the solution of (B.4.26) by differentiating (B.4.33). Thus, we shall concentrate on finding continuous solutions to (B.4.33). Our approach is to define the so-called Picard's iterates by

$$\begin{aligned} \boldsymbol{y}_0(t) &= \mathring{\boldsymbol{y}}, \\ \boldsymbol{y}_n(t) &= \mathring{\boldsymbol{y}} + \int_{t_0}^{t} \boldsymbol{f}(s, \boldsymbol{y}_{n-1}(s))ds, \end{aligned} \tag{B.4.34}$$

and proving that they converge to the solution.

As a first step, we shall show that the iterates remain in Ξ. Namely, we show that if M, a, b, α are defined as in the formulation of Picard's theorem and $\boldsymbol{y}_n$ is defined as in (B.4.34), then for any n

$$\|\boldsymbol{y}_n(t) - \mathring{\boldsymbol{y}}\| \le M|t - t_0| \tag{B.4.35}$$

for $|t - t_0| \le \alpha$ and hence, by the definition of α, $\boldsymbol{y}_n(t) \in \Xi$.

To prove (B.4.35), we note that for $n = 0$ the estimate (B.4.35) is obvious, so to proceed with induction, we shall assume that it is valid for some $n > 0$ and, taking $n + 1$, we have

$$\|\boldsymbol{y}_{n+1}(t) - \mathring{\boldsymbol{y}}\| = \left\| \int_{t_0}^{t} \boldsymbol{f}(s, \boldsymbol{y}_n(s))ds \right\| \le \int_{t_0}^{t} \|\boldsymbol{f}(s, \boldsymbol{y}_n(s))\|ds.$$

However, by the remark above and the induction assumption, $\boldsymbol{y}_n(s)$ is in Ξ as long as $|t - t_0| \le \alpha$ and thus $\|\boldsymbol{f}(s, \boldsymbol{y}_n(s))\| \le M$. Hence,

$$\|\boldsymbol{y}_{n+1}(t) - \mathring{\boldsymbol{y}}\| \le M|t - t_0|.$$

In the next step, we shall show that the sequence of Picard's iterates converges. To make things simpler, we shall convert the sequence into a series, the convergence of which is easier to establish. To this end, writing

$$\boldsymbol{y}_n(t) = \mathring{\boldsymbol{y}} + (\boldsymbol{y}_1(t) - \boldsymbol{y}_0) + (\boldsymbol{y}_2(t) - \boldsymbol{y}_1(t)) + \ldots + (\boldsymbol{y}_n(t) - \boldsymbol{y}_{n-1}(t)), \qquad \text{(B.4.36)}$$

we will try to show that

$$\sum_{n=0}^{\infty} \|\boldsymbol{y}_{n+1}(t) - \boldsymbol{y}_n(t)\| < +\infty$$

for any $|t - t_0| \leq \alpha$, which would give the convergence of the series.

We use induction again. Assume that $t > t_0$, the analysis for $t < t_0$ being analogous. Firstly, proceeding as in (B.4.32), we observe that for $n > 1$

$$\begin{aligned} \|\boldsymbol{y}_n(t) - \boldsymbol{y}_{n-1}(t)\| &\leq \int_{t_0}^{t} \|\boldsymbol{f}(s, \boldsymbol{y}_{n-1}(s)) - \boldsymbol{f}(s, \boldsymbol{y}_{n-2}(s))\| ds \\ &\leq L \int_{t_0}^{t} \|\boldsymbol{y}_{n-1}(s) - \boldsymbol{y}_{n-2}(s)\| ds, \end{aligned} \qquad \text{(B.4.37)}$$

where we used the fact that each iterate is in Ξ by (B.4.35). Now, for $n = 1$ we obtain

$$\|\boldsymbol{y}_1(t) - \boldsymbol{y}_0\| \leq M(t - t_0),$$

while for $n = 2$

$$\begin{aligned} \|\boldsymbol{y}_2(t) - \boldsymbol{y}_1(t)\| &\leq \int_{t_0}^{t} \|\boldsymbol{f}(s, \boldsymbol{y}_1(s)) - \boldsymbol{f}(s, \boldsymbol{y}_0)\| ds \leq L \int_{t_0}^{t} \|\boldsymbol{y}_1(s) - \boldsymbol{y}_0\| ds \\ &\leq ML \int_{t_0}^{t} (s - t_0) ds = \frac{ML}{2}(t - t_0)^2. \end{aligned}$$

This motivate the induction assumption

$$\|\boldsymbol{y}_n(t) - \boldsymbol{y}_{n-1}(t)\| \leq \frac{ML^{n-1}}{n!}(t - t_0)^n,$$

and further, by (B.4.37),

$$\begin{aligned}\|\boldsymbol{y}_{n+1}(t)-\boldsymbol{y}_n(t)\| &\le \int_{t_0}^{t} \|\boldsymbol{f}(s,\boldsymbol{y}_n(s))-\boldsymbol{f}(s,\boldsymbol{y}_{n-1}(s))\|ds\\ &\le L\int_{t_0}^{t}\|\boldsymbol{y}_n(s)-\boldsymbol{y}_{n-1}(s)\|ds\\ &\le \frac{ML^n}{n!}\int_{t_0}^{t}(s-t_0)^n = \frac{ML^n}{(n+1)!}(t-t_0)^{n+1}.\end{aligned}$$

Now, because $|t-t_0|<\alpha$, we see that

$$\|\boldsymbol{y}_n(t)-\boldsymbol{y}_{n-1}(t)\| \le \frac{ML^{n-1}}{n!}\alpha^n,$$

so

$$\sum_{n=0}^{\infty}\|\boldsymbol{y}_{n+1}(t)-\boldsymbol{y}_n(t)\| \le \sum_{n=1}^{\infty}\frac{ML^{n-1}}{n!}\alpha^n = \frac{M}{L}(e^{\alpha L}-1),$$

which is finite. Thus, the sequence $\boldsymbol{y}_n(t)$ converges for any t satisfying $|t-t_0|\le\alpha$. Let us denote the limit by $\boldsymbol{y}(t)$. By (B.4.36), we obtain

$$\begin{aligned}\boldsymbol{y}(t) &= \boldsymbol{y}_0 + (\boldsymbol{y}_1(t)-\boldsymbol{y}_0)+\ldots+(\boldsymbol{y}_n(t)-\boldsymbol{y}_{n-1}(t))+\ldots\\ &= \boldsymbol{y}_0+\sum_{n=0}^{\infty}(\boldsymbol{y}_{n+1}(t)-\boldsymbol{y}_n(t))\end{aligned} \tag{B.4.38}$$

and so

$$\begin{aligned}\|\boldsymbol{y}(t)-\boldsymbol{y}_n(t)\| &= \left\|\sum_{j=n}^{\infty}(\boldsymbol{y}_{j+1}(t)-\boldsymbol{y}_j(t))\right\| \le M\sum_{j=n}^{\infty}L^j\frac{(t-t_0)^{j+1}}{(j+1)!}\\ &\le \frac{M}{L}\sum_{j=n}^{\infty}\frac{L^{j+1}\alpha^{j+1}}{(j+1)!},\end{aligned} \tag{B.4.39}$$

where the tail of the series is convergent to zero. This shows that the series (B.4.38) is absolutely and uniformly convergent for $|t-t_0|\le\alpha$. We use it to show the continuity of the limit function $\boldsymbol{y}(t)$, in the same way as in the proof of continuity of the limit of uniformly convergent series of continuous function, see, e.g., [31,

Section 7.4]. We provide proof for the completeness of the argument. Firstly, we observe that, by induction, $\boldsymbol{y}_n(t)$ is continuous if $\boldsymbol{y}_{n-1}(t)$ is. In fact, we have

$$\begin{aligned}\|\boldsymbol{y}_n(t_1) - \boldsymbol{y}_n(t_2)\| &\leq \left\| \int_{t_0}^{t_1} \boldsymbol{f}(s, \boldsymbol{y}_n(s))ds - \int_{t_0}^{t_2} \boldsymbol{f}(s, \boldsymbol{y}_n(s))ds \right\| \\ &\leq \left\| \int_{t_1}^{t_2} \boldsymbol{f}(s, \boldsymbol{y}_n(s))ds \right\| \leq M|t_2 - t_1|.\end{aligned}$$

Next, let t_i be arbitrary numbers satisfying $|t_i - t_0| \leq \alpha$ for $i = 1, 2$. By (B.4.39), we can find n so large that

$$\frac{M}{L}\sum_{j=n}^{\infty}\frac{L^{j+1}\alpha^{j+1}}{(j+1)!} < \epsilon/3,$$

so that

$$\|\boldsymbol{y}(t_i) - \boldsymbol{y}_n(t_i)\| < \epsilon/3, \qquad i = 1, 2.$$

Since $\boldsymbol{y}_n(t)$ is continuous, we can find $\delta > 0$ such that if $|t_1 - t_2| < \delta$, then

$$\|\boldsymbol{y}_n(t_1) - \boldsymbol{y}_n(t_2)\| < \epsilon/3.$$

Combining, we see that whenever $|t_1 - t_2| < \delta$ and $|t_i - t_0| < \alpha$ for $i = 1, 2$, we have

$$\|\boldsymbol{y}(t_1) - \boldsymbol{y}(t_2)\| \leq \|\boldsymbol{y}(t_1) - \boldsymbol{y}_n(t_1)\| + \|\boldsymbol{y}_n(t_1) - \boldsymbol{y}_n(t_2)\| + \|\boldsymbol{y}_n(t_2) - \boldsymbol{y}(t_2)\| \leq \epsilon,$$

so that $\boldsymbol{y}(t)$ is continuous.

The last step is to prove that the obtained function is indeed the solution of the Cauchy problem in the integral form (B.4.33),

$$\boldsymbol{y}(t) = \mathring{\boldsymbol{y}} + \int_{t_0}^{t} \boldsymbol{f}(s, \boldsymbol{y}(s))ds.$$

Since, by construction,

$$\boldsymbol{y}_{n+1}(t) = \boldsymbol{y}_0 + \int_{t_0}^{t} \boldsymbol{f}(s, \boldsymbol{y}_n(s))ds,$$

we obtain

$$\boldsymbol{y}(t) = \lim_{n\to\infty} \boldsymbol{y}_{n+1}(t) = \mathring{\boldsymbol{y}} + \lim_{n\to\infty} \int_{t_0}^{t} \boldsymbol{f}(s, \boldsymbol{y}_n(s))ds.$$

Hence, we have to prove that

$$\lim_{n\to\infty} \int_{t_0}^{t} \boldsymbol{f}(s, \boldsymbol{y}_n(s))ds = \int_{t_0}^{t} \boldsymbol{f}(s, \boldsymbol{y}(s))ds. \tag{B.4.40}$$

Firstly, note that the right-hand side is well-defined as both $\boldsymbol{y}(t)$ and $\boldsymbol{f}$ are continuous functions so that the composition $\boldsymbol{f}(s, \boldsymbol{y}(s))$ is continuous and the integral is well-defined. Thus we can write, by (B.4.39),

$$\begin{aligned}\left\| \int_{t_0}^{t} \boldsymbol{f}(s, \boldsymbol{y}_n(s))ds - \int_{t_0}^{t} \boldsymbol{f}(s, \boldsymbol{y}(s))ds \right\| &\leq \int_{t_0}^{t} \|\boldsymbol{f}(s, \boldsymbol{y}_n(s)) - \boldsymbol{f}(s, \boldsymbol{y}(s))\|ds \\ &\leq L \int_{t_0}^{t} \|\boldsymbol{y}_n(s) - \boldsymbol{y}(s)\|ds \\ &\leq L\frac{M}{L}\sum_{j=n}^{\infty} \frac{L^{j+1}\alpha^{j+1}}{(j+1)!} \int_{t_0}^{t} ds \\ &\leq M\alpha \sum_{j=n}^{\infty} \frac{L^{j+1}\alpha^{j+1}}{(j+1)!}.\end{aligned}$$

As before, the sum above approaches zero as $n \to \infty$ and therefore (B.4.40), and hence the whole theorem, are proved. □

Let us illustrate the significance of the assumptions of the Picard–Lindelöf theorem.

Example B.12 It is easy to see that there are at least two solutions to the Cauchy problem

$$\begin{aligned} y' &= t\sqrt{|y|}, \quad t \in \mathbb{R}, \\ y(0) &= 0, \end{aligned}$$

namely, $y(t) \equiv 0$ and $z(t) \equiv \frac{1}{16}t^4$. In this case, $f(t, y) = t\sqrt{|y|}$ is not Lipschitz continuous in any rectangle $|t| \leq a$, $|y| \leq b$ as for $y > 0$

$$|\sqrt{y} - \sqrt{0}| = \frac{1}{\sqrt{y}}|y - 0|,$$

and $\frac{1}{\sqrt{y}}$ is unbounded in any right neighbourhood of $(0, 0)$.

Another important aspect of the Picard–Lindelöf theorem is that it is local, that is, it ensures the existence of a solution only in a possibly very small neighbourhood of the initial data. While, in general, the solutions to (B.4.26) are not defined for all $t \in \mathbb{R}$, the interval $|t - t_0| \leq \alpha$ usually is not the best possible, as illustrated below.

Example B.13 The solution to the initial value problem

$$y' = 1 + y^2,$$
$$y(0) = 0,$$

is given by $y(t) = \tan t$. This solution is defined only for $-\pi/2 < t < \pi/2$. Let us check this equation against the Picard–Lindelöf theorem. We have $f(t, y) = 1 + y^2$ and $f_y(t, y) = 2y$, and both functions are continuous on the whole plane. Let Ξ be the rectangle $|t| \leq a$, $|y| \leq b$, then

$$M = \max_{(t,y)\in\Xi} |f(t, y)| = 1 + b^2,$$

and the solution exists for

$$|t| \leq \alpha = \min\left\{a, \frac{b}{1 + b^2}\right\}.$$

Since a can be arbitrary, the maximal interval of existence predicted by the Picard–Lindelöf theorem is determined by the maximum of $\frac{b}{1+b^2}$, which is equal to $\frac{1}{2}$, considerably smaller than the true maximal of existence.

To address the problem indicated in the previous example, we introduce the concept of the continuation of the solution. Precisely, according to the Picard–Lindelöf theorem, for any point $(t_0, \mathring{\mathbf{y}})$ around which its assumptions are satisfied, there is an interval on which there is only one solution of (B.4.26). However, in principle, there could be two solutions $\mathbf{y}^{(1)}(t)$ and $\mathbf{y}^{(2)}(t)$, which coincide over the interval of uniqueness mentioned in the theorem, but branching for larger times. Assuming that around any point of the space $\mathbb{R} \times \mathbb{R}^d$ the assumptions of the Picard–Lindelöf theorem are satisfied, such a scenario is impossible. In fact, if $\mathbf{y}^{(1)}(t) = \mathbf{y}^{(2)}(t)$ over some interval I then, by the continuity of solutions, there is the largest t, say t_1, having this property. Thus, $\mathbf{y}^{(1)}(t) = \mathbf{y}^{(2)}(t)$ with $\mathbf{y}_1(t) \neq \mathbf{y}_2(t)$ as soon as $t > t_1$. Thus, at the point $(t_1, \mathbf{y}^{(1)}(t_1))$ the Picard–Lindelöf theorem would be violated. In other words, if the assumptions of the Picard–Lindelöf theorem are satisfied everywhere, then any two solutions coinciding on some interval must coincide on the common interval of existence.

An important consequence of the above is that we can glue solutions together to obtain a solution defined on a possibly larger interval. For, if $\mathbf{y}(t)$ is a solution to (B.4.26) defined on an interval $[t_0-\alpha, t_0+\alpha]$ and $(t_0+\alpha, \mathbf{y}(t_0+\alpha))$ is a point around which the assumptions of the Picard–Lindelöf theorem are satisfied, then there is a solution passing through this point, defined on some interval $[t_0+\alpha-\alpha', t_0+\alpha+\alpha']$, $\alpha' > 0$. These two solutions coincide on $[t_0 + \alpha - \alpha', t_0 + \alpha]$ and therefore, by the comments above, they must coincide over the whole interval of their common existence, constituting, therefore, a solution of the original Cauchy problem defined at least on $[t_0 - \alpha, t_0 + \alpha + \alpha']$.

Following the discussion, we say that $[t_0, t_0+\alpha^*)$ is the maximal forward interval of existence for a solution $\mathbf{y}(t)$ to (B.4.26) if there is no solution $\mathbf{y}^{(1)}(t)$ defined on a longer time interval $[t_0, t_0 + \alpha^+)$, where $\alpha^+ > \alpha^*$, satisfying $\mathbf{y}(t) = \mathbf{y}^{(1)}(t)$ for $t \in [t_0, t_0 + \alpha^*)$. In other words, we cannot extend $\mathbf{y}(t)$ beyond $t_0 + \alpha^*$ so that it remains a solution of (B.4.26). The maximal backward interval of existence $(t_0 - \alpha^\#, t_0]$ is defined similarly.

We note that if the assumptions of the Picard–Lindelöf theorem are satisfied on $\mathbb{R} \times \mathbb{R}^d$, then the maximal forward interval of existence is open from the right—otherwise the Picard–Lindelöf theorem would give the solution to (B.4.26) with the initial data $(t_0 + \alpha^*, \mathbf{y}(t_0 + \alpha^*))$ on a bigger interval.

Theorem B.14 *Assume that* $\mathbf{f}$ *in* (B.4.26) *satisfies the assumptions of the Picard–Lindelöf theorem on* $\mathbb{R} \times \mathbb{R}^d$. *If the solution* $\mathbf{y}(t)$ *of* (B.4.26) *has a finite maximal forward interval of existence* $[t_0, t_0 + \alpha^*)$, *then*

$$\lim_{t\to t_0+\alpha^*} \|\mathbf{y}(t)\| = \infty. \tag{B.4.41}$$

Proof Assume that (B.4.41) does not hold. The meaning of (B.4.41) is that for any K there is t_K such that for any $t \in [t_K, t_0 + \alpha^*)$ we have $\|\mathbf{y}(t)\| \geq K$. Thus, by saying that (B.4.41) does not hold, we assert the existence of K such that for any $t < t_0 + \alpha^*$ there is $t < t' < t_0 + \alpha^*$ with $\|\mathbf{y}(t')\| < K$. In particular, there is a sequence $(t_n)_{n\in\mathbb{N}}$ such that $t_n \to t_0 + \alpha^*$ and $\|\mathbf{y}_n\| := \|\mathbf{y}(t_n)\| < K$. Consider the sequence of Cauchy problems

$$\mathbf{y}' = \mathbf{f}(\mathbf{y}), \quad \mathbf{y}(t_n) = \mathbf{y}_n. \tag{B.4.42}$$

Since all $\|\mathbf{y}_n\|$ are bounded by K and $\mathbf{f}$ satisfies the conditions of the Picard–Lindelöf theorem on $\mathbb{R} \times \mathbb{R}^d$, we can consider the above problems in sets $\Xi_n = \{(t, \mathbf{y}) : |t - t_n| < a, \|\mathbf{y} - \mathbf{y}_n\| < b\}$ for some fixed a, b. Moreover, all Ξ_ns are contained in the set $\Xi_K = \{(t, y) : t_0 - a < t < t_0+\alpha^*, -K - b < \|\mathbf{y}\| < K + b\}$ and the solutions of the problems (B.4.42) are defined on intervals $(t_n - \alpha, t_n + \alpha)$, where $\alpha = \min\left\{a, \frac{b}{M}\right\}$ and M can be taken as $\max_{(t,\mathbf{y})\in\Xi_K} \|\mathbf{f}(t, \mathbf{y})\|$ and is independent of n. If α^* was finite, then we could find t_n with $t_0 + \alpha^* - t_n < \alpha$

so that the solution could be continued beyond $t_0+\alpha^*$ contradicting the assumption that $[t_0, t_0+\alpha^*)$ is the maximal interval of existence. □

We often use Theorem B.14 as a contraposition. If we can prove that at no point does the solution $\boldsymbol{y}(t)$ blow up, that is, its norm never escapes to infinity, then $\boldsymbol{y}(t)$ is defined on $[t_0, \infty)$. While it seems that this statement is useless, as we do not know the solution, it turns out that we can use the properties of the equation to provide the required estimates, as illustrated in the following example.

Example B.15 Let us assume that $\boldsymbol{f}$ in (B.4.26) satisfies the assumptions of the Picard–Lindelöf theorem on $\mathbb{R}\times\mathbb{R}^d$ so that, in particular, it is globally Lipschitz, that is, there is $L > 0$ such that for $(t, \boldsymbol{y}_1), (t, \boldsymbol{y}_2) \in \mathbb{R}\times\mathbb{R}^d$ we have

$$\|\boldsymbol{f}(t, \boldsymbol{y}_1) - \boldsymbol{f}(t, \boldsymbol{y}_2)\| \le L\|\boldsymbol{y}_1 - \boldsymbol{y}_2\|. \tag{B.4.43}$$

Then, any solution to (B.4.26) is global in time. We prove it for $t > t_0$, the proof for $t < t_0$ being analogous. Using Gronwall's lemma, we obtain

$$\begin{aligned}
\|\boldsymbol{y}(t)\| &\le \|\mathring{\boldsymbol{y}}\| + \int_{t_0}^{t} \|\boldsymbol{f}(s, \boldsymbol{y}(s))\|ds \le \|\mathring{\boldsymbol{y}}\| + \int_{t_0}^{t} \|\boldsymbol{f}(s, \boldsymbol{y}(s)) - \boldsymbol{f}(s, \mathring{\boldsymbol{y}})\|ds \\
&\quad + \int_{t_0}^{t} \|\boldsymbol{f}(s, \mathring{\boldsymbol{y}})\|ds \\
&\le \|\mathring{\boldsymbol{y}}\| + \int_{t_0}^{t} \|\boldsymbol{f}(s, \mathring{\boldsymbol{y}})\|ds + L\int_{t_0}^{t} \|\boldsymbol{y}(s) - \mathring{\boldsymbol{y}}\|ds \le \|\mathring{\boldsymbol{y}}\| \\
&\quad + \int_{t_0}^{t} \|\boldsymbol{f}(s, \mathring{\boldsymbol{y}})\|ds + L\int_{t_0}^{t} \|\mathring{\boldsymbol{y}}\|ds + L\int_{t_0}^{t} \|\boldsymbol{y}(s)\|ds \\
&\le \|\mathring{\boldsymbol{y}}\| + \int_{t_0}^{t} \|\boldsymbol{f}(s, \mathring{\boldsymbol{y}})\|ds + L(t-t_0)\|\mathring{\boldsymbol{y}}\| + L\int_{t_0}^{t} \|\boldsymbol{y}(s)\|ds.
\end{aligned}$$

If $\boldsymbol{y}(t)$ is not defined for all t, then $\|\boldsymbol{y}(t)\|$ becomes unbounded as $t \to t_{max}$ for some t_{max}. Denoting

$$c = \|\mathring{\boldsymbol{y}}\| + \int_{t_0}^{t_{max}} \|\boldsymbol{f}(s, \mathring{\boldsymbol{y}})\|ds + L(t_{max} - t_0)\|\mathring{\boldsymbol{y}}\|,$$

which is finite as f is continuous for all t, we can write the above inequality as

$$\|\boldsymbol{y}(t)\| \leq c + L \int_{t_0}^{t} \|\boldsymbol{y}(s)\| ds$$

for any $t_0 \leq t \leq t_{max}$. Using Gronwall's lemma, we obtain

$$\|\boldsymbol{y}(t)\| \leq c \exp Lt \leq c \exp Lt_{max},$$

contradicting thus the definition of t_{max}.

One of the most pertinent aspects of applications of differential equations in modelling is that we rarely know precisely the data: the initial conditions and the parameters defining $\boldsymbol{f}$ are available only approximately. Thus, it is paramount to know that the solutions to two problems will be close to each other, provided that the equations and the input data do not differ much. Before we formulate and prove a relevant theorem, we need to make the concept of the Lipschitz continuity more precise.

First, let us note that the Lipschitz continuity was defined in (B.4.28) on a bounded closed subset Ξ. However, in applications, we often have to work with functions $\boldsymbol{f}$ defined sets of the form $\boldsymbol{\Omega} = [t_0, t_1] \times D \subset \mathbb{R} \times \mathbb{R}^d$, where D is not necessarily closed. In such a case, we say that $\boldsymbol{f}$ is locally Lipschitz on D (uniformly in $t \in [t_0, t_1]$) if for any $\boldsymbol{x} \in D$ there is $\rho_{\boldsymbol{x}} > 0$ and $L_{\boldsymbol{x}} = L(\boldsymbol{x}, \rho_{\boldsymbol{x}})$ such that for any $\boldsymbol{y} \in B_{\boldsymbol{x},\rho_x} = \{\boldsymbol{y} \in D : \ \|\boldsymbol{y} - \boldsymbol{x}\| < \rho_{\boldsymbol{x}}\}$

$$\|\boldsymbol{f}(t, \boldsymbol{y}) - \boldsymbol{f}(t, \boldsymbol{x})\| \leq L_{\boldsymbol{x}} \|\boldsymbol{y} - \boldsymbol{x}\|. \tag{B.4.44}$$

If D is compact (bounded and closed), then a locally Lipschitz function on D is globally Lipschitz there, that is, we can choose a universal, independent of $\boldsymbol{x}$, constant L, see [4, Proposition 6.4]. Indeed, the family $\left\{B_{\boldsymbol{x},\frac{1}{2}\rho_x}\right\}_{\boldsymbol{x}\in D}$ is a cover of D. Since D is compact, we have a finite cover $\left\{B_{\boldsymbol{x}_i,\frac{1}{2}\rho_{\boldsymbol{x}_i}}\right\}_{i=1,\dots,p}$. Define

$$\rho := \frac{1}{2} \min_{1\leq i\leq p} \rho_{\boldsymbol{x}_i} > 0.$$

We consider two cases. First, it may happen that $\|\boldsymbol{y} - \boldsymbol{x}\| < \rho$. Then, $\boldsymbol{x} \in B_{\boldsymbol{x}_i,\rho_{\boldsymbol{x}_i}}$ for some i and, since

$$\|\boldsymbol{y} - \boldsymbol{x}_i\| \leq \|\boldsymbol{y} - \boldsymbol{x}\| + \|\boldsymbol{x}_i - \boldsymbol{x}\| < \rho_i,$$

so that $\boldsymbol{x}, \boldsymbol{y} \in B_{\boldsymbol{x}_i,\rho_{\boldsymbol{x}_i}}$ for some i. Then

$$\|\boldsymbol{f}(t, \boldsymbol{y}) - \boldsymbol{f}(t, \boldsymbol{x})\| \leq L(\boldsymbol{x}_i, \rho_{\boldsymbol{x}_i})\|\boldsymbol{y} - \boldsymbol{x}\| \leq \max_{1 \leq i \leq p} L(\boldsymbol{x}_i, \rho_{\boldsymbol{x}_i})\|\boldsymbol{y} - \boldsymbol{x}\|.$$

If that is not the case, that is, $\|\boldsymbol{x} - \boldsymbol{y}\| \geq \rho$, then

$$\begin{aligned}\|\boldsymbol{f}(t, \boldsymbol{y}) - \boldsymbol{f}(t, \boldsymbol{x})\| &\leq 2 \max_{t\in[t_0,t_1],\boldsymbol{x}\in D} \|\boldsymbol{f}(t, \boldsymbol{x})\| \\ &= \frac{\max_{t\in[t_0,t_1],\boldsymbol{x}\in D} \|\boldsymbol{f}(t, \boldsymbol{x})\|}{\rho} \min_{1\leq i\leq p} \rho_{\boldsymbol{x}_i} \leq C\|\boldsymbol{x} - \boldsymbol{y}\|,\end{aligned}$$

where $C = \dfrac{\max_{t\in[t_0,t_1],\boldsymbol{x}\in D} \|\boldsymbol{f}(t,\boldsymbol{x})\|}{\rho}$.

Thus, $\boldsymbol{f}$ is globally Lipschitz on D with universal constant $L = \max\left\{\max_{1\leq i\leq p} L(\boldsymbol{x}_i, \rho_{\boldsymbol{x}_i}), C\right\}$.

Theorem B.16 *Let* $-\infty < t_0 < t_1 < \infty$, $D \subset \mathbb{R}^d$ *be an open and connected set,* $\lambda_0 \in \mathbb{R}^m$, $c > 0$, *and let* $[t_0, t_1] \times D \times \{\lambda : \|\lambda - \lambda_0\| \leq c\} \ni (t, \boldsymbol{x}, \lambda) \mapsto \boldsymbol{f}(t, \boldsymbol{x}, \lambda)$ *be a continuous function and locally Lipschitz continuous in* $\boldsymbol{x} \in D$, *uniformly in* t *and* λ. *Assume that the solution* $(t, \lambda_0) \mapsto \boldsymbol{y}(t, \lambda_0)$ *to*

$$\boldsymbol{x}' = \boldsymbol{f}(t, \boldsymbol{x}, \lambda_0), \qquad \boldsymbol{x}(t_0) = \boldsymbol{y}_0 \in D, \tag{B.4.45}$$

is defined on $[t_0, t_1]$ *and takes values in* D. *Then, for any* $\epsilon > 0$ *there is* $\delta > 0$ *such that if*

$$\|\boldsymbol{z}_0 - \boldsymbol{y}_0\| < \delta, \quad \|\lambda - \lambda_0\| < \delta,$$

then there is a unique solution $(t, \lambda) \mapsto \boldsymbol{z}(t, \lambda)$ *of*

$$\boldsymbol{x}' = \boldsymbol{f}(t, \boldsymbol{x}, \lambda), \qquad \boldsymbol{x}(t_0) = \boldsymbol{z}_0 \in D, \tag{B.4.46}$$

defined on $[t_0, t_1]$ *and satisfying*

$$\|\boldsymbol{z}(t, \lambda) - \boldsymbol{y}(t, \lambda_0)\| < \epsilon \tag{B.4.47}$$

for all $t \in [t_0, t_1]$.

Proof Since $\boldsymbol{y}(t, \lambda_0)$ is continuous and $[t_0, t_1]$ is compact, $\boldsymbol{y}(t, \lambda_0)$ is bounded on $[t_0, t_1]$. Let us define a tube U around $\boldsymbol{y}(t, \lambda_0)$ by

$$U := \{(t, \boldsymbol{x}) \in [t_0, t_1] \times \mathbb{R}^d : \|\boldsymbol{x} - \boldsymbol{y}(t, \lambda_0)\| \leq \epsilon\}.$$

We can assume that ϵ is small enough for $U \in [t_0, t_1] \times D$. Indeed, assume that this is not true. Then, for any n there is $(\tau_n, \boldsymbol{x}_n) \in U$ and $(\tau_n, \boldsymbol{x}_n) \notin [t_0, t_1] \times D$, that is, $\tau_n \in [t_0, t_1]$, $\|\boldsymbol{x}_n - \boldsymbol{y}(\tau_n, \lambda_0)\| \leq \frac{1}{n}$ and $\boldsymbol{x}_n \notin D$. By the compactness of $[t_0, t_1]$, we can assume that $\tau_n \to t' \in [t_0, t_1]$ and hence, by continuity, $\boldsymbol{y}(\tau_n, \lambda_0) \to \boldsymbol{y}(t', \lambda_0)$. Therefore, we have

$$\begin{aligned}\|\boldsymbol{x}_n - \boldsymbol{y}(t', \lambda_0)\| &\leq \|\boldsymbol{x}_n - \boldsymbol{y}(\tau_n, \lambda_0)\| + \|\boldsymbol{y}(\tau_n, \lambda_0) - \boldsymbol{y}(t', \lambda_0)\| \\ &\leq \frac{1}{n} + \|\boldsymbol{y}(\tau_n, \lambda_0) - \boldsymbol{y}(t', \lambda_0)\| \to 0,\end{aligned}$$

and thus $(\tau_n, \boldsymbol{x}_n) \to (t', \boldsymbol{y}(t', \lambda_0)) \in U$. On the other hand, since $\mathbb{R}^d \setminus D$ is closed, the limit of $\lim_{n\to\infty} \boldsymbol{x}_n \in \mathbb{R}^d \setminus D$, which is a contradiction.

Since U is compact, $\boldsymbol{f}$ is globally Lipschitz on U with respect to $\boldsymbol{x}$, uniformly in t with the Lipschitz constant, say, L. Using again the compactness of U, $\boldsymbol{f}$, being continuous on U, is uniformly continuous there, see [32, Chapter 1, Appendix A.2]. Hence, for any $\alpha > 0$ there is $\beta > 0$ (with $\beta < c$) such that for any $(t, \boldsymbol{x}) \in U$ and $\|\lambda - \lambda_0\| < \beta$ we have

$$\|\boldsymbol{f}(t, \boldsymbol{x}, \lambda) - \boldsymbol{f}(t, \boldsymbol{x}, \lambda_0)\| \leq \alpha.$$

Take $\alpha < \epsilon$ and $\|\mathring{z} - \mathring{\boldsymbol{y}}\| < \alpha$. By the Picard–Lindelöf theorem, there is a solution $\boldsymbol{z}(t, \lambda)$ on some interval $[t_0, t_0 + \Delta t]$. The solution starts inside U and can be extended in time as long as it stays in U. Indeed, assume that there is a first time, $\tau < t_1$, at which the solution leaves U. We write

$$\begin{aligned}\boldsymbol{z}'(t, \lambda) &= \boldsymbol{f}(t, \boldsymbol{z}(t, \lambda), \lambda) = \boldsymbol{f}(t, \boldsymbol{z}(t, \lambda), \lambda_0) + \boldsymbol{f}(t, \boldsymbol{z}(t, \lambda), \lambda) - \boldsymbol{f}(t, \boldsymbol{z}(t, \lambda), \lambda_0) \\ &=: \boldsymbol{f}(t, \boldsymbol{z}(t, \lambda), \lambda_0) + \boldsymbol{g}(t, \boldsymbol{z}(t, \lambda)),\end{aligned} \tag{B.4.48}$$

with $\|\boldsymbol{g}\| \leq \alpha$ on U. Hence, for any $t \in [t_0, \tau]$ we can integrate to get

$$\begin{aligned}\|\boldsymbol{y}(t, \lambda_0) - \boldsymbol{z}(t, \lambda)\| &\leq \|\mathring{\boldsymbol{y}} - \mathring{\boldsymbol{z}}\| + \left\| \int_{t_0}^{t} \boldsymbol{g}(t, \boldsymbol{z}(s, \lambda))ds \right\| \\ &\quad + \left\| \int_{t_0}^{t} \boldsymbol{f}(s, \boldsymbol{z}(s, \lambda), \lambda_0) - \boldsymbol{f}(s, \boldsymbol{y}(s, \lambda_0), \lambda_0)ds \right\| \\ &\leq \alpha + \alpha(t - t_0) + L \int_{t_0}^{t} \|\boldsymbol{y}(s, \lambda_0) - \boldsymbol{z}(s, \lambda)\|ds.\end{aligned} \tag{B.4.49}$$

Using Gronwall's lemma with $c = \alpha + \alpha(t_1 - t_0)$ and $g(s) = L$, we obtain

$$\|\boldsymbol{y}(t, \lambda_0) - \boldsymbol{z}(t, \lambda)\| \leq \alpha(1 + (t_1 - t_0))e^{L(t-t_0)}. \tag{B.4.50}$$

If we choose

$$\alpha < \epsilon \frac{e^{-L(t_1-t_0)}}{1 + (t_1 - t_0)},$$

we see that $\|\boldsymbol{y}(\tau, \lambda_0) - \boldsymbol{z}(\tau, \lambda)\| < \epsilon$ on $[t_0, \tau]$, which contradicts the assumption that at $t = \tau$ the solution leaves U. Thus $\boldsymbol{z}(t, \lambda)$ does not leave U at any $\tau < t_1$. To complete the proof, we select $\delta < \min\{\alpha, \beta\}$. □

Remark B.17 Applying a more refined version of Gronwall's inequality can improve estimate (B.4.50). Consider the inequality

$$0 \leq x(t) \leq g(t) + L \int_{t_0}^{t} x(s) ds, \quad t_0 \leq t \leq t_1, \tag{B.4.51}$$

where $L > 0$ is a constant and g a non-negative continuous function on $[t_0, t_1]$. If we let $X(t) = \int_{t_0}^{t} x(s) ds$, then (B.4.51) takes the form

$$X'(t) - LX(t) \leq g(t), \quad X(t_0) = 0.$$

Using the integrating factor e^{-Lt}, we get

$$(X(t)e^{-Lt})' \leq g(t)e^{-Lt}.$$

Integrating,

$$X(t) \leq \int_{t_0}^{t} g(s)e^{L(t-s)} ds,$$

which, substituted into (B.4.51), gives

$$0 \leq x(t) \leq g(t) + L \int_{t_0}^{t} g(s)e^{L(t-s)} ds. \tag{B.4.52}$$

Using (B.4.52) in (B.4.49) and integrating by parts, we obtain

$$\begin{aligned}\|\boldsymbol{y}(t,\lambda_0)-\boldsymbol{z}(t,\lambda)\| &\le \|\mathring{\boldsymbol{y}}-\mathring{\boldsymbol{z}}\|+\alpha(t-t_0)+L\int\limits_{t_0}^{t}(\|\mathring{\boldsymbol{y}}-\mathring{\boldsymbol{z}}\|+\alpha(s-t_0)e^{L(t-s)}ds\\ &= \|\mathring{\boldsymbol{y}}-\mathring{\boldsymbol{z}}\|e^{L(t-t_0)}+\alpha\int\limits_{t_0}^{t}e^{L(t-s)}ds\\ &= \|\mathring{\boldsymbol{y}}-\mathring{\boldsymbol{z}}\|e^{L(t-t_0)}+\frac{\alpha}{L}(e^{L(t-t_0)}-1),\end{aligned} \tag{B.4.53}$$

where α is the maximum of $\boldsymbol{g}$.

Theorem B.16 gives continuous dependence of the solution on the initial condition and a parameter. In some cases, we need a better regularity. In particular, we often need the differentiability of the flow with respect to all arguments. Here, we show the differentiability with respect to the initial condition under slightly stronger assumptions than necessary. A general version of this result can be found in [100, Theorem III §13].

Theorem B.18 ([33, pp. 15–16]) *Let $D \subset \mathbb{R}^d$ be an open set and suppose that $\boldsymbol{f}$ is defined and has continuous second-order partial derivatives with respect to all variables on $(-a, a) \times D$. Let further $\mathring{\boldsymbol{x}} \in D$. Then there are positive α and ρ such that for any $\boldsymbol{y}$ satisfying $\|\boldsymbol{y} - \mathring{\boldsymbol{x}}\| < \rho$, the problem*

$$\boldsymbol{x}' = \boldsymbol{f}(t, \boldsymbol{x}), \quad \boldsymbol{x}(0) = \boldsymbol{y}, \tag{B.4.54}$$

has a unique solution $\boldsymbol{x}(t) = \phi(t, \boldsymbol{y})$ on $(-\alpha, \alpha)$ such that at each point of the set

$$\Omega = \{(t, \boldsymbol{y}) : (t, \boldsymbol{y}) \in (-\alpha, \alpha) \times B_{\mathring{\boldsymbol{x}},\rho}\},$$

ϕ has continuous partial derivatives up to the third order with respect to t and of the first order with respect to $\boldsymbol{y}$.

Proof By Theorem B.10, there is a solution $\phi(t, \boldsymbol{y})$ to (B.4.54) which can be written as

$$\frac{\partial \phi(t, \boldsymbol{y})}{\partial t} = \boldsymbol{f}(t, \phi(t, \boldsymbol{y})). \tag{B.4.55}$$

Since $\boldsymbol{f}$ is twice differentiable with respect to all variables and $\phi(t, \boldsymbol{y})$ is differentiable with respect to t, the Chain Rule implies that (B.4.55) can be differentiated twice with respect to t, and hence $\phi(t, \boldsymbol{y})$ has third order partial derivatives with respect to this variable. Writing (B.4.55) as an integral equation, see (B.4.33), we get

$$\boldsymbol{\phi}(t, \boldsymbol{y}) = \boldsymbol{y} + \int_0^t \boldsymbol{f}(s, \boldsymbol{\phi}(s, \boldsymbol{y}))ds. \tag{B.4.56}$$

If we assume the differentiability of $\boldsymbol{\phi}$ with respect to $\boldsymbol{y}$, then, differentiating (B.4.56), we get

$$\frac{\partial \boldsymbol{\phi}(t, \boldsymbol{y})}{\partial y_i} = \boldsymbol{e}_i + \int_0^t \mathcal{J}_{\boldsymbol{f}}(s, \boldsymbol{\phi}(s, \boldsymbol{y}))\frac{\partial \boldsymbol{\phi}(s, \boldsymbol{y})}{\partial y_i}ds, \quad i = 1, \ldots, d, \tag{B.4.57}$$

where $\boldsymbol{e}_i$ is the ith versor of $\mathbb{R}^d$ and we used the abbreviation

$$\mathcal{J}_{\boldsymbol{f}}(s, \boldsymbol{\phi}(s, \boldsymbol{y})) = \nabla_{\boldsymbol{x}} \boldsymbol{f}(s, \boldsymbol{x})|_{\boldsymbol{x}=\boldsymbol{\phi}(s,\boldsymbol{y})}, \tag{B.4.58}$$

see (B.1.1). This suggests to consider the system of equations

$$\begin{aligned} \boldsymbol{\phi}(t, \boldsymbol{y}) &= \boldsymbol{y} + \int_0^t \boldsymbol{f}(s, \boldsymbol{\phi}(s, \boldsymbol{y}))ds, \\ \boldsymbol{\psi}^{(i)}(t) &= \boldsymbol{e}_i + \int_0^t \mathcal{J}_{\boldsymbol{f}}(s, \boldsymbol{\phi}(s, \boldsymbol{y}))\boldsymbol{\psi}^{(i)}(s)ds, \quad i = 1, \ldots, d. \end{aligned} \tag{B.4.59}$$

Since $\boldsymbol{f}$ is assumed to be twice differentiable, $\mathcal{J}_{\boldsymbol{f}}$ is differentiable and the system (B.4.59) satisfies assumptions of Theorem B.10 and therefore there exists a solution

$$(\boldsymbol{\phi}(t, \boldsymbol{y}), \boldsymbol{\psi}^{(1)}(t), \ldots, \boldsymbol{\psi}^{(d)}(t))$$

defined on some set Ω, where the first component is the solution of the original problem (B.4.54). By (B.4.34), the solution is given by the uniformly convergent sequence $(\boldsymbol{\phi}_n(t, \boldsymbol{y}), \boldsymbol{\psi}_n^{(1)}(t), \ldots, \boldsymbol{\psi}_{n+1}^{(d)}(t))$ defined by Picard iterates

$$\begin{aligned} \boldsymbol{\phi}_{n+1}(t, \boldsymbol{y}) &= \boldsymbol{y} + \int_0^t \boldsymbol{f}(s, \boldsymbol{\phi}_n(s, \boldsymbol{y}))ds, \\ \boldsymbol{\psi}_{n+1}^{(i)}(t) &= \boldsymbol{e}_i + \int_0^t \mathcal{J}_{\boldsymbol{f}}(s, \boldsymbol{\phi}_n(s, \boldsymbol{y}))\boldsymbol{\psi}_n^{(i)}(s)ds, \quad i = 1, \ldots, d, \end{aligned} \tag{B.4.60}$$

with the initial point $\phi_0(t, \mathbf{y}) = \mathbf{y}$ and $\psi_0^{(i)}(t, \mathbf{y}) = \mathbf{e}_i = \frac{\partial \phi_0(t,\mathbf{y})}{\partial y_i}, i = 1, \ldots, d$. Now, we make the induction assumption $\psi_n^{(i)}(t, \mathbf{y}) = \frac{\partial \phi_n(t,\mathbf{y})}{\partial y_i}$, which yields

$$\frac{\partial \phi_{n+1}(t, \mathbf{y})}{\partial y_i} = \mathbf{e}_i + \int_0^t \mathcal{J}_f(s, \phi_n(s, \mathbf{y}))\frac{\partial \phi_n(s, \mathbf{y})}{\partial y_i} ds$$

$$= \mathbf{e}_i + \int_0^t \mathcal{J}_f(s, \phi_n(s, \mathbf{y}))\psi_n^{(i)}(s)ds = \psi_{n+1}^{(i)}(t). \quad \text{(B.4.61)}$$

This shows that the sequences $(\phi_n(t, \mathbf{y}))_{n\in\mathbb{N}}$ and $\left(\frac{\partial \phi_n(t,\mathbf{y})}{\partial y_i}\right)_{n\in\mathbb{N}}, i = 1, \ldots, d$, are uniformly convergent on $(-\alpha, \alpha)$ and hence, by [31, Section 7.4], $\phi(t, \mathbf{y})$ is differentiable with respect to $\mathbf{y}$ and $\frac{\partial \phi(t,\mathbf{y})}{\partial y_i} = \psi^{(i)}(t)$. In particular, (B.4.57) is satisfied; in the differential form we have

$$\frac{\partial}{\partial t}\left(\frac{\partial \phi(t, \mathbf{y})}{\partial y_i}\right) = \mathcal{J}_f(t, \phi(t, \mathbf{y}))\frac{\partial \phi(t, \mathbf{y})}{\partial y_i}, \quad \frac{\partial \phi(0, \mathbf{y})}{\partial y_i} = \mathbf{e}_i, \quad i = 1, \ldots, d. \quad \text{(B.4.62)}$$

□

We observe that (B.4.62) is a linear equation in $\frac{\partial \phi(t,\mathbf{y})}{\partial y_i}$.

Remark B.19 As mentioned earlier, the assumption that $\mathbf{f}$ is twice-differentiable is excessive. We used it only to apply Theorem B.10 directly to (B.4.59). However, the first equation in (B.4.59) can be solved separately and, with given $\phi(t, \mathbf{y})$, the other d equations are linear in $\psi^{(1)}(t), \ldots, \psi^{(d)}(t)$, and for their solvability it is enough that $\mathcal{J}_f$ is continuous in s. With this approach, we cannot establish (B.4.61) in such a simple way. Nevertheless, it can be done by more careful estimates, [100, Theorems III §13. VI & XI], and hence, the second part of the theorem is valid if $\mathbf{f}$ is only once-differentiable.

B.4.2 Some Applications

The Barrier Lemma and Invariance

Consider for simplicity the autonomous system

$$\mathbf{x}' = \mathbf{f}(\mathbf{x}), \quad \mathbf{x}(0) = \mathring{\mathbf{x}}, \quad \text{(B.4.63)}$$

with (locally) Lipschitz function $\mathbf{f}$ in $\Omega \subset \mathbb{R}^d$. Recall that the invariant set for the flow generated by (B.4.63) was defined in Definition 6.21.

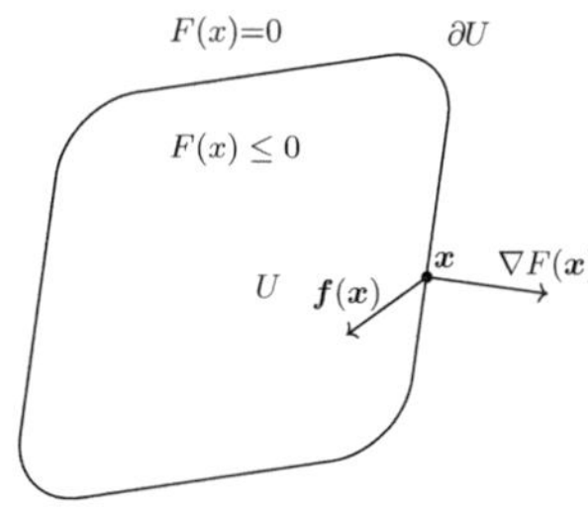

Fig. B.1 Configuration of $\boldsymbol{f}(\boldsymbol{x})$ and $\nabla F(\boldsymbol{x})$ in Lemma B.20

Let $F : \Omega \to \mathbb{R}$ be a C^1-function with Lipschitz continuous gradient (for less regular sets, see [100, Section III §10. XVI] or [19, Twierdzenie 3.12]), and let U be defined as the sublevel set

$$U = \{\boldsymbol{x} \in \Omega : \; F(\boldsymbol{x}) \leq 0\},$$

whose boundary is $\partial U := \{\boldsymbol{x} \in \Omega : \; F(\boldsymbol{x}) = 0\}$; we further assume that $\nabla F(\boldsymbol{x}) \neq 0$ for $\boldsymbol{x} \in \partial U$ (recall that $\nabla F(\boldsymbol{x})$ is a vector perpendicular to ∂U at $\boldsymbol{x}$ and the assumption ensures that the normal is well-defined).

Lemma B.20 *If for any $\boldsymbol{x} \in \partial U$ we have*

$$\langle \boldsymbol{f}(\boldsymbol{x}), \nabla F(\boldsymbol{x}) \rangle \leq 0, \tag{B.4.64}$$

then U is forward invariant under the flow defined by (B.4.63).

Geometrically, (B.4.64) means that the vector field of (B.4.63) never points outside U; see Fig. B.1.

Proof Consider a trajectory originating at $\mathring{\boldsymbol{x}} \in U$. For this trajectory to get outside U, it must cross the boundary ∂U at some time t_0 and point $\boldsymbol{x}_0$, where $F(\boldsymbol{x}_0) = 0$. First, let us assume that, at this point, we have

$$\langle \boldsymbol{f}(\boldsymbol{x}_0), \nabla F(\boldsymbol{x}_0) \rangle < 0.$$

By continuity, there is a neighbourhood V of $\boldsymbol{x}_0$ and $c > 0$ such that

$$\langle \boldsymbol{f}(\boldsymbol{x}), \nabla F(\boldsymbol{x}) \rangle \leq c < 0$$

for $\boldsymbol{x} \in V$. Then, again by continuity, for small $\tau > 0$ we have

$$\phi(t_0 + \tau, \mathring{\boldsymbol{x}}) = \phi(\boldsymbol{x}_0, \tau) \in V.$$

Now,

$$\begin{aligned} F(\boldsymbol{\phi}(\boldsymbol{x}_0,\tau)) &= \langle \nabla F(\boldsymbol{\phi}(\boldsymbol{x}_0,\tau)), \boldsymbol{\phi}'(\boldsymbol{x}_0,\tau)\rangle \\ &= \langle \nabla F(\boldsymbol{\phi}(\boldsymbol{x}_0,\tau)), \boldsymbol{f}(\boldsymbol{\phi}(\boldsymbol{x}_0,\tau))\rangle \le c < 0, \end{aligned}$$

hence $\tau \mapsto F(\boldsymbol{\phi}(\boldsymbol{x}_0,\tau))$ is strictly decreasing. Thus, $F(\boldsymbol{\phi}(\boldsymbol{x}_0,\tau)) < 0$ for small $\tau > 0$ which implies that $\boldsymbol{\phi}(\boldsymbol{x}_0,\tau)$ stays in the interior of U.

Next, assume that at $\boldsymbol{x}_0$ we have

$$\langle \boldsymbol{f}(\boldsymbol{x}_0), \nabla F(\boldsymbol{x}_0)\rangle = 0,$$

and consider the modified field

$$\boldsymbol{f}_\epsilon(\boldsymbol{x}) = \boldsymbol{f}(\boldsymbol{x}) - \epsilon\frac{\nabla F(\boldsymbol{x}_0)}{\|\nabla F(\boldsymbol{x}_0)\|^2}.$$

Since the perturbation is constant, the field $\boldsymbol{f}_\epsilon$ has the same regularity property as $\boldsymbol{f}$. Furthermore, we have

$$\langle \boldsymbol{f}_\epsilon(\boldsymbol{x}_0), \nabla F(\boldsymbol{x}_0)\rangle = -\epsilon < 0.$$

Let us fix some neighbourhood V of $\boldsymbol{x}_0$ in which $\|\boldsymbol{f}(\boldsymbol{x})\| \le c$ for some c, and let $\epsilon \le \epsilon_0$ for some ϵ_0. Then

$$\begin{aligned} \langle \boldsymbol{f}_\epsilon(\boldsymbol{x}), \nabla F(\boldsymbol{x})\rangle &= \langle \boldsymbol{f}_\epsilon(\boldsymbol{x}), \nabla F(\boldsymbol{x}) - \nabla F(\boldsymbol{x}_0)\rangle + \langle \boldsymbol{f}_\epsilon(\boldsymbol{x}) - \boldsymbol{f}_\epsilon(\boldsymbol{x}_0), \nabla F(\boldsymbol{x}_0)\rangle - \epsilon \\ &\le \|\boldsymbol{f}_\epsilon(\boldsymbol{x})\|\|\nabla F(\boldsymbol{x}) - \nabla F(\boldsymbol{x}_0)\| \\ &\quad + \|\boldsymbol{f}_\epsilon(\boldsymbol{x}) - \boldsymbol{f}_\epsilon(\boldsymbol{x}_0)\|\|\nabla F(\boldsymbol{x}_0)\| - \epsilon \\ &\le L_1\left(\|\boldsymbol{f}(\boldsymbol{x})\| + \epsilon\frac{1}{\|\nabla F(\boldsymbol{x}_0)\|}\right)\|\boldsymbol{x} - \boldsymbol{x}_0\| \\ &\quad + L_2\|\nabla F(\boldsymbol{x}_0)\|\|\boldsymbol{x} - \boldsymbol{x}_0\| - \epsilon \\ &\le L\|\boldsymbol{x} - \boldsymbol{x}_0\| - \epsilon, \end{aligned} \tag{B.4.65}$$

where L_1,L_2 are Lipschitz constants for $\boldsymbol{f}$ and ∇F, respectively, in V and

$$L = L_1\left(c + \frac{\epsilon_0}{\|\nabla F(\boldsymbol{x}_0)\|}\right) + L_2\|\nabla F(\boldsymbol{x}_0)\|$$

is independent of ϵ. Thus there is a ball $B\left(\boldsymbol{x}_0, \frac{\epsilon}{2L}\right)$ on which

$$\langle \boldsymbol{f}_\epsilon(\boldsymbol{x}), \nabla F(\boldsymbol{x})\rangle \le -\frac{\epsilon}{2} < 0.$$

Next, from the continuity, for sufficiently small $t_1 > 0$, we have $\phi(x_0, \tau) \in B\left(\boldsymbol{x}_0, \frac{\epsilon}{4L}\right)$ for $0 \leq \tau \leq t_1$. Let us denote by $\phi_\epsilon(\boldsymbol{x}_0, \tau)$ the flow generated by $\boldsymbol{f}_\epsilon$ passing through $\boldsymbol{x}_0$. Then, using (B.4.53), we have

$$\|\phi_\epsilon(\boldsymbol{x}_0, \tau) - \phi(\boldsymbol{x}_0, \tau)\| \leq \frac{\epsilon}{L_1}(e^{L_1 T} - 1)$$

(since the Lipschitz constant for $\boldsymbol{f}_\epsilon$ is the same as for $\boldsymbol{f}$), and we see that $\phi_\epsilon(\boldsymbol{x}_0, \tau) \in B\left(\boldsymbol{x}_0, \frac{\epsilon}{2L}\right)$ for $\tau \in [0, T]$ as long as

$$\frac{\epsilon}{L_1}(e^{L_1 T} - 1) \leq \frac{\epsilon}{4L},$$

that is, if

$$T \leq \frac{1}{L_1} \ln\left(1 + \frac{L_1}{4L}\right).$$

Hence, for such τ,

$$\langle \boldsymbol{f}_\epsilon(\phi_\epsilon(\boldsymbol{x}_0, \tau)), \nabla F(\phi_\epsilon(\boldsymbol{x}_0, \tau))\rangle \leq -\frac{\epsilon}{2} < 0$$

and, arguing as in the first part, $\phi_\epsilon(\boldsymbol{x}_0, \tau)$ is in the interior of U. Since T is independent of ϵ and $\phi_\epsilon(\boldsymbol{x}_0, \tau) \to \phi(\boldsymbol{x}_0, \tau)$ on $[0, T]$, we see that $\phi(\boldsymbol{x}_0, \tau) \in U$ as U is closed. □

Nonnegativity of Solutions

Let $\boldsymbol{f} : \mathbb{R} \times \Omega$, where $\Omega \subset \mathbb{R}^d$ is an open connected set, and $\boldsymbol{f} = (f_1, \ldots, f_d)$. Consider a system of differential equations

$$\boldsymbol{x}' = \boldsymbol{f}(t, \boldsymbol{x}), \quad \boldsymbol{x}(t_0) = \boldsymbol{x}_0. \tag{B.4.66}$$

We have the following theorem.

Theorem B.21 *Suppose $\boldsymbol{f}$ is continuous on $[t_0, t_1] \times \Omega$ for some $-\infty < t_0 < t_1 < \infty$ and locally Lipschitz with respect to $\boldsymbol{x}$ uniformly in t. Suppose further that for any $i = 1, \ldots, d$, if $\boldsymbol{x} \geq 0$ is of the form $\boldsymbol{x} = (x_1, \ldots, x_{i-1}, 0, x_{i+1}, \ldots, x_d)$, then $f_i(t, \boldsymbol{x}) \geq 0$. Then, the solution $\boldsymbol{x}(t)$ to* (B.4.66) *with $\boldsymbol{x}_0 \geq 0$ satisfies $\boldsymbol{x}(t) \geq 0$ for all $t \geq t_0$ for which it is defined.*

Proof First, we assume that $\boldsymbol{f}$ satisfies the stronger condition that for each i,

$$f_i(t, x_1, \ldots, x_{i-1}, 0, x_{i+1}, \ldots, x_d) > 0.$$

By the continuity of $\boldsymbol{x}(t)$, we have $\boldsymbol{x}(t) > 0$ on some interval $[0, \tau]$. Assume that the thesis does not hold, that is, there is k and t_k' for which $x_k(t_k') < 0$. Thus there exists $\tau_k := \inf\{t : x_k(t) < 0\}$, $\tau_k \geq 0$. We see that $\tau_k \notin \{t, x_k(t) < 0\}$ as then, by local preservation of sign, $x_k(t) < 0$ in some open neighbourhood of τ_k. So $x_k(\tau_k) \geq 0$. On the other hand, there is a sequence $\sigma_p \in \{t : x_k(t) < 0\}$ converging to τ_k so, by continuity, $0 > x_k(\sigma_p) \to x_k(\tau_k) \leq 0$. Thus, $x_k(\tau_k) = 0$. Let k_0 be the index for which τ_k is the smallest. Hence, $x_{k_0}(\tau_{k_0}) = 0$ and $x_k(\tau_{k_0}) \geq 0$ for $k \neq k_0$. By the (stronger) assumption, $x_{k_0}'(\tau_{k_0}) = f_{k_0}(\tau_{k_0}, \boldsymbol{x}(\tau_{k_0})) > 0$. Thus, $x_{k_0}(t) > 0$ on some interval (τ_{k_0}, τ'). This contradicts the earlier observation that $x_{k_0}(\sigma_p) < 0$ on some sequence converging to τ_{k_0} from above.

To extend the result to nonnegative $\boldsymbol{f}$, we consider

$$\boldsymbol{x}_\epsilon' = \boldsymbol{f}(t, \boldsymbol{x}_\epsilon) + \epsilon \boldsymbol{e}, \quad \boldsymbol{x}_\epsilon(t_0) = \boldsymbol{x}_0 \geq 0, \tag{B.4.67}$$

where $\boldsymbol{e} = (1, \ldots, 1)$ and $\epsilon > 0$. Then, $\boldsymbol{f}(t, \boldsymbol{x}_\epsilon) + \epsilon \boldsymbol{e}$ satisfies the stronger assumption introduced at the beginning of the proof. By Theorem B.16, for sufficiently small ϵ, the solutions $\boldsymbol{x}_\epsilon$ to (B.4.67) are defined on the same interval as $\boldsymbol{x}(t)$ and satisfy there

$$0 \leq \boldsymbol{x}_\epsilon(t) \to \boldsymbol{x}(t),$$

hence $\boldsymbol{x}(t) \geq 0$. □

Appendix C
Linear Algebra Tools

In this chapter, we continue our analysis of the structure of matrices, which is essential in solving systems of difference and differential equations discussed in Sect. 5.2.

C.1 Notation

Let $X = \mathbb{C}^d$ be a complex, d-dimensional vector space (we can also consider $X = \mathbb{R}^d$). Recall that by $\mathbb{M}_d$ we denoted the space of square $d \times d$ matrices on X. Matrix multiplications define linear operators on X. Still, it is important to note that the same operator on X may have different matrix representations depending on the choice of basis in X. Fortunately, there is an easy way to convert one matrix representation of an operator A to another. Let $\mathcal{v} = \{\boldsymbol{v}^{\mathbf{1}}, \ldots, \boldsymbol{v}^{\boldsymbol{d}}\}$ be a basis in X. For any vector $\boldsymbol{x}$ and operator A, we denote by $(\boldsymbol{x})_{\mathcal{v}}$ and $\mathcal{A} = (A)_{\mathcal{v}}$ the representation of, respectively, $\boldsymbol{x}$ and A in this basis. Thus, $(\boldsymbol{x})_{\mathcal{v}} = (c_1, \ldots, c_d) \in X$ means

$$\boldsymbol{x} = c_1\boldsymbol{v}^{\mathbf{1}} + \cdots + c_d\boldsymbol{v}^{\boldsymbol{d}}$$

and $(A)_{\mathcal{v}} = (\alpha_{ij})_{1 \le i,j \le d}$, hence

$$\begin{aligned} y_1\boldsymbol{v}^{\mathbf{1}} + \cdots + y_d\boldsymbol{v}^{\boldsymbol{d}} &= \boldsymbol{y} = A\boldsymbol{x} = x_1A\boldsymbol{v}^{\mathbf{1}} + \cdots + x_dA\boldsymbol{v}^{\boldsymbol{d}} \\ &= \left(\sum_{j=1}^{d}\alpha_{1j}x_j\right)\boldsymbol{v}^{\mathbf{1}} + \cdots + \left(\sum_{j=1}^{d}\alpha_{dj}x_j\right)\boldsymbol{v}^{\boldsymbol{d}}, \end{aligned}$$

J. Banasiak, *Introduction to Mathematical Methods in Population Theory*, Springer Undergraduate Mathematics Series,
https://doi.org/10.1007/978-3-031-65491-6

that is,

$$(\boldsymbol{y})_v = (A)_v(\boldsymbol{x})_v.$$

Consider another basis $w = \{\boldsymbol{w}^{\mathbf{1}}, \dots, \boldsymbol{w}^{\boldsymbol{d}}\}$ in X, that is, for a given $\boldsymbol{x} \in X$, we have

$$\boldsymbol{x} = w_1\boldsymbol{w}^1 + \cdots + w_d\boldsymbol{w}^d.$$

Let, for the basis $v = \{\boldsymbol{v}^{\mathbf{1}}, \dots, \boldsymbol{v}^{\boldsymbol{d}}\}$, $(\boldsymbol{v}^i)_w = (v_1^i, \dots, v_d^i)$ in the basis w, and for $\boldsymbol{x} \in \mathbb{C}^d$,

$$\boldsymbol{x} = c_1\boldsymbol{v}^{\mathbf{1}} + \cdots + c_d\boldsymbol{v}^{\boldsymbol{d}},$$

that is, $\boldsymbol{c} = (\boldsymbol{x})_v$. More precisely, let V be the matrix having vectors $(\boldsymbol{v}^i)_w$ as its columns,

$$V = \begin{pmatrix} | & \cdots & | \\ (\boldsymbol{v}^{\mathbf{1}})_w & \cdots & (\boldsymbol{v}^{\boldsymbol{d}})_w \\ | & \cdots & | \end{pmatrix}. \tag{C.1.1}$$

Note, that V is invertible as the vectors $\boldsymbol{v}^i$ are linearly independent, and hence we can write

$$(\boldsymbol{x})_v = V^{-1}(\boldsymbol{x})_w. \tag{C.1.2}$$

Since

$$(\boldsymbol{y})_w = (A)_w(\boldsymbol{x})_w$$

is equivalent to

$$V(\boldsymbol{y})_v = (A)_w V(\boldsymbol{x})_v,$$

we obtain

$$(\boldsymbol{y})_v = V^{-1}(A)_w V(\boldsymbol{x})_v,$$

that is,

$$(A)_v = V^{-1}(A)_w V, \tag{C.1.3}$$

where all matrices on the right-hand side are expressed in the basis w.

The special role is played by the canonical basis of X, $e = \{\boldsymbol{e}^{\mathbf{1}}, \dots, \boldsymbol{e}^{\boldsymbol{d}}\}$, where $\boldsymbol{e}^i = (0, \dots, 0, 1, 0, \dots, 0)$, with 1 at the i-th place, $1 \le i \le d$.

If in some considerations a basis in X is fixed and if it does not lead to any misunderstanding, we typically assume that it is the canonical basis, and we identify $\boldsymbol{x}$ with its coordinates, $\boldsymbol{x} = (\boldsymbol{x})_e = (x_1, \ldots, x_d)$ and similarly, we write $\mathcal{A} = (A)_e = (a_{ij})_{1 \le i,j \le d}$.

C.2 Jordan Normal Form

A crucial role in constructing a spacial basis of eigenvectors and associated eigenvectors described in Sect. 5.3.3, which was used to build the matrix exponential of $\mathcal{A}$, is played by the decomposition of X leading to the Jordan theorem, which is the subject of the present section.

Let us recall that the characteristic polynomial of a matrix $\mathcal{A} \in \mathbb{M}_d$ is defined as

$$p_{\mathcal{A}}(\lambda) = det(\mathcal{A} - \lambda\mathcal{I}) = \begin{vmatrix} a_{11} - \lambda \ldots & a_{1d} \\ \vdots & \vdots \\ a_{1d} & \ldots a_{nn} - \lambda \end{vmatrix} = p_0 + p_1\lambda + \ldots + (-1)^d\lambda^d, \tag{C.2.4}$$

for some constants $p_0, p_1, \ldots p_{d-1}$. The roots of this polynomial are called eigenvalues, and the set of all eigenvalues is denoted by $\sigma(\mathcal{A})$ and called the spectrum of $\mathcal{A}$, see (5.3.49) and the subsequent discussion. Since

$$det(A)_v = det\, \mathcal{V}^{-1}(A)_w\mathcal{V} = det\, \mathcal{V}^{-1}det(A)_w det\, \mathcal{V} = det(A)_w$$

the spectrum of a matrix does not change when the basis is changed. Thus, we can talk about the spectrum of the linear operator whose matrix representation in some basis is $\mathcal{A}$, and we often will use this terminology.

We recall that the spectral radius of $\mathcal{A}$ is defined by

$$r(\mathcal{A}) = \max\{|\lambda| : \lambda \in \sigma(\mathcal{A})\}. \tag{C.2.5}$$

The crucial role in the considerations is played by nilpotent matrices. We say that a matrix $\mathcal{T} \in \mathbb{M}_d$ is nilpotent of index q if q is the least natural number for which $\mathcal{T}^q = 0$, that is, for any $\boldsymbol{x} \in X$, $\mathcal{T}^q\boldsymbol{x} = \mathbf{0}$ and there is $\boldsymbol{x} \in X$ such that $\mathcal{T}^{q-1}\boldsymbol{x} \neq \mathbf{0}$. Since the powers of operators satisfy

$$(T)_v^n = \mathcal{V}^{-1}(T)_w^n\mathcal{V}, \quad n \ge 1,$$

we see that the property of being nilpotent is associated with the operator T and not with its matrix representation $\mathcal{T}$. Thus, hereafter, we will talk about nilpotent operators.

We shall need the following properties of nilpotent operators.

Proposition C.1 *Let T be a nilpotent operator of index q on $X = \mathbb{C}^d$. Then,*

1. $\sigma(T) = \{0\}$;
2. *if* $T^{q-1}\boldsymbol{x} \neq 0$, *then* $\{\boldsymbol{x}, T\boldsymbol{x}, \dots T^{q-1}\boldsymbol{x}\}$ *is linearly independent;*
3. $q \leq d$.

Proof

1. If for $\lambda \neq 0$ and $\boldsymbol{y} \neq 0$ we have $T\boldsymbol{y} = \lambda\boldsymbol{y}$, then, iterating, $T^q\boldsymbol{y} = \lambda^q\boldsymbol{y} \neq \boldsymbol{0}$ for any $q \in \mathbb{N}$, which is a contradiction.
2. Assume to the contrary, that for some $\{\alpha_0, \dots, \alpha_{q-1}\}$, with at least one nonzero α_j, we have

$$0 = \sum_{j=0}^{q-1} \alpha_j T^j \boldsymbol{x}.$$

Let j_0 be the smallest j for which $\alpha_j \neq 0$. Then, we can write

$$T^{j_0}\boldsymbol{x} = -\sum_{j=j_0+1}^{q-1} \frac{\alpha_j}{\alpha_{j_0}} T^j \boldsymbol{x}.$$

Hence,

$$T^{q-1}\boldsymbol{x} = T^{q-j_0-1}T^{j_0}\boldsymbol{x} = -T^q \sum_{j=j_0+1}^{q-1} \frac{\alpha_j}{\alpha_{j_0}} T^{j-j_0-1}\boldsymbol{x} = 0,$$

as $j - j_0 - 1 \geq 0$. This contradiction proves the claim.
3. If $q > d$, then there would be a set of q linearly independent vectors $\{\boldsymbol{x}, T\boldsymbol{x}, \dots T^{q-1}\boldsymbol{x}\}$ in a d-dimensional space X, which is a contradiction.

□

The main theorem of this section, the Jordan representation, will be preceded by a sequence of preliminary results. As we did above, if the discussed results are independent of the basis, we formulate them in terms of linear operators. Let us recall that if X and Y are vector spaces, their direct sum $X \oplus Y$ is defined as the vector space Z consisting of pairs $z = (\boldsymbol{x}, \boldsymbol{y})$, $\boldsymbol{x} \in X$, $\boldsymbol{y} \in Y$ with operation defined coordinate-wise. Elements $\boldsymbol{x}$ and $\boldsymbol{y}$ are identified with elements $(\boldsymbol{x}, 0)$ and $(0, \boldsymbol{y})$ of Z, respectively, and we will use this identification hereafter. We note that if X and Y are subspaces of Z, then $Z = X \oplus Y$ if for any $z \in Z$ there are $\boldsymbol{x} \in X$ and $\boldsymbol{y} \in Y$ such that $z = \boldsymbol{x} + \boldsymbol{y}$ and $X \cap Y = \{0\}$, so that the representation of z is unique. For linear operators A in X and B in Y, their direct sum $A \oplus B$ is a linear operator in $Z = X \oplus Y$ defined as

$$A \oplus B = \begin{pmatrix} A & 0 \\ 0 & B \end{pmatrix}. \tag{C.2.6}$$

Proposition C.2 *Any linear operator A on X is a direct sum of a nilpotent operator and an invertible operator. In other words, there are (unique) subspaces N and M such that $X = N \oplus M$ and which are invariant under A with $A|_N$ nilpotent and $A|_M$ invertible (on M).*

Proof Consider an operator $A : X \to X$ and, for $j \in \mathbb{N}_0$, define the subspaces

$$N_j = Ker\ A^j := \{\boldsymbol{x} \in X;\ A^j\boldsymbol{x} = 0\}, \quad N = \bigcup_{j \in \mathbb{N}_0} N_j,$$

$$M_j = A^j(X) := \{\boldsymbol{y} \in X;\ A^j\boldsymbol{x} = \boldsymbol{y} \text{ for some } \boldsymbol{x} \in X\}, \quad M = \bigcap_{j \in \mathbb{N}_0} M_j.$$

It is easy to see that both families of sets are nested:

$$0 = N_0 \subseteq N_1 \subseteq \ldots \subseteq N_j \subseteq \ldots \subseteq N,$$
$$X = M_0 \supseteq M_1 \supseteq \ldots \supseteq M_j \supseteq \ldots \supseteq M.$$

We observe that whenever $N_j \neq N_{j+1}$, then, since N_j is a linear space, there is an element $\boldsymbol{z} \in N_{j+1}$ which is linearly independent of N_j. Similarly, if $M_j \neq M_{j+1}$, then there is an element $\boldsymbol{y} \in M_j$ which is linearly independent of M_{j+1}. Since X is finite-dimensional, there are r and m such that

$$N_j = N_r, \quad j \geq r,$$
$$M_j = M_m, \quad j \geq m.$$

Hence, $N = N_r$ and $M = M_m$. We observe that N and M are invariant under A. Indeed, $\boldsymbol{x} \subset N$ if and only if $A^r\boldsymbol{x} = 0$. Then $A^rA\boldsymbol{x} = AA^r\boldsymbol{x} = A0 = 0$ and hence $A\boldsymbol{x} \in N$. Similarly, $\boldsymbol{y} \in M$ if and only if $\boldsymbol{y} = A^m\boldsymbol{x}$ for some $\boldsymbol{x} \in X$. Then $A\boldsymbol{y} = AA^m\boldsymbol{x} = A^mA\boldsymbol{x} = A^m\boldsymbol{z} \in M$, where $\boldsymbol{z} = A\boldsymbol{x}$.

Furthermore, it is clear that $A|_N$ is a nilpotent operator (of index r as $N_{r-1} \neq N_r$). On the other hand, we have $AM = M$. Indeed, since $M_{m+1} = M_m$, if $\boldsymbol{x} \in M_m = M_{m+1}$, then $\boldsymbol{x} = A^{m+1}\boldsymbol{y}$, $\boldsymbol{y} \in X$, and hence $\boldsymbol{x} = A\boldsymbol{z}$, where $\boldsymbol{z} = A^m\boldsymbol{y} \in M$. Thus, A is surjective onto M. It is a standard result in the theory of systems of linear equations that if a $k \times k$ equation always has a solution, then this solution is unique. This can be easily ascertained using a matrix representation of $A|_M$. Then, the fact that the system always has a solution means that the columns of the matrix span the whole space (and the solution is given by the coefficients of the expansion of the right-hand side in this basis). But since the columns form k vectors in a k-dimensional space, they must be a basis, so the solution is unique. Thus $A|_M$ is invertible.

Next, we show

$$X = N \oplus M.$$

Since $AM = M$, by induction we have $A^r M = M$, and hence $A^r|_M$ is also invertible and thus $A^r \boldsymbol{x} \neq 0$ for any $M \ni \boldsymbol{x} \neq 0$. Thus $M \cap N = \{0\}$. On the other hand, for any $\boldsymbol{v} \in X$, we have $A^m \boldsymbol{v} =: \boldsymbol{y} \in M$. Since also $A^m|_M$ is invertible, we have $A^m \boldsymbol{v} = A^m \boldsymbol{z}$ for some $\boldsymbol{z} \in M$, that is, $\boldsymbol{v} - \boldsymbol{z} \in N_m \subset N$. Writing $\boldsymbol{v} = \boldsymbol{z} + (\boldsymbol{v} - \boldsymbol{z})$ gives the desired decomposition of X.

To show that this decomposition is unique, we let $X = V \oplus W$ and $A|_V$ be nilpotent, and $A|_W$ be invertible. Hence, $(A|_V)^k = 0$ and $V \subset N_k \subset N$. On the other hand, if A is invertible on W, then $AW = W$, that is $W \subset M_k$ for any $k \geq 0$ and $W \subset M$. This shows $V = N$ and $W = M$. □

Remark C.3 We can show that $m = r$. Indeed, let us consider M_r instead of $M_m = M$. Then we also have

$$X = N \oplus M_r. \tag{C.2.7}$$

For, let $0 \neq x \in N \cap M_r$, that is, $A^r \boldsymbol{x} = 0$ and $\boldsymbol{x} = A^r \boldsymbol{y}$ for some $\boldsymbol{y} \in X$. Then, $A^{2r} \boldsymbol{y} = 0$ and $A^r \boldsymbol{y} = 0$ by the definition of r. Thus, $\boldsymbol{x} = A^r \boldsymbol{y} = 0$, contrary to the assumption. Therefore, the dimension of N is the nullity of A^r, by definition, and the dimension of M_r is the rank of A^r, and the sum of both is equal to the dimension of the space (see the Rank Theorem, [89, Section 2.4]). Hence, (C.2.7) is proved. We also have invertibility of A on M_r. Indeed, $AM_r \subset M_r$ and if $\boldsymbol{x} \in M_k$, that is, $\boldsymbol{x} = A^r \boldsymbol{y}$, $\boldsymbol{y} \in X$, then from $0 = A\boldsymbol{x} = A^{r+1} \boldsymbol{y}$ and the definition of r, it follows that $\boldsymbol{x} = A^r \boldsymbol{y} = 0$ hence, again, the Rank Theorem completes the proof of the invertibility. Thus, by the uniqueness, $M_r = M$.

Proposition C.4 *If A is a linear operator on X and $\sigma(A)$ consists of distinct eigenvalues $\lambda_1, \ldots, \lambda_p$ of algebraic multiplicities $m_1, \ldots, m_p$, $m_1 + \cdots + m_p = d$, then*

$$X = X_1 \oplus X_2 \oplus \ldots \oplus X_p, \tag{C.2.8}$$

where $X_1, \ldots, X_p$ are subspaces invariant under A and $A - \lambda_j I$ is nilpotent on X_j, $j = 1, \ldots, p$.

Here and below, by I we denote the identity operator in the respective space.

Proof Let us fix $j \in \{1, \ldots, p\}$ and apply Proposition C.2 to $A_j = A - \lambda_j I$ to obtain the subspaces $N(A_j) =: X_j$ and $M(A_j) =: Y_j$, which are invariant under A_j (and hence $A = A_j + \lambda_j I$), and such that A_j is nilpotent on X_j and invertible on Y_j.

Hence, for any λ, $A - \lambda I$ can be split into the direct sum of its restrictions to X_j and Y_j and hence

$$det\,(\mathcal{A} - \lambda I) = det\,(\mathcal{A} - \lambda I)|_{X_j} det\,(\mathcal{A} - \lambda I)|_{Y_j},$$

where $\mathcal{A}$ is the matrix representation of A written in an appropriate basis such that the direct sum takes the form of a block diagonal matrix, see (C.2.6). Since $A_j|_{Y_j}$ is invertible, $\lambda_j \notin \sigma(A|_{Y_j})$. Further, by Proposition C.1, 0 is the only eigenvalue of $A - \lambda_j I$ on X_j and hence λ_j is the only eigenvalue of $A|_{X_j}$. Therefore

$$det\,(\mathcal{A} - \lambda I)|_{X_j} = (\lambda - \lambda_j)^{m_j},$$

and thus $dim X_j = m_j$ as the degree of a characteristic polynomial of a matrix is equal to the dimension of the underlying space.

By Proposition C.1, the index of A_j is at most m_j, that is,

$$(A - \lambda_j I)^{q_j} X_j = 0 \tag{C.2.9}$$

for some $q_j \leq m_j$.

We are left with proving that the spaces X_j, $j = 1, \ldots, p$, form the decomposition of X into a direct sum. First, we show $X_j \cap X_k = \{0\}$ whenever $j \neq k$. Indeed, let $0 \neq \boldsymbol{x} \in X_j \cap X_k$. Then $(A - \lambda_k I)^{m_k}\boldsymbol{x} = \boldsymbol{0}$. Also $(A - \lambda_k I)^r \boldsymbol{x} \in X_j \cap X_k$ for any $0 \leq r < m_k$. Indeed, for any α, i, $(A - \alpha I)X_i \subset AX_i - \alpha X_i \subset X_i$ by Proposition C.2 and the linearity of X_i. So, if $\boldsymbol{x} \in X_j \cap X_k$, then $(A - \lambda_i I)^r \boldsymbol{x} \in X_j \cap X_k$ for $i = j, k$, and the statement for other r follows by induction.

Next, if we write $(A - \lambda_k I)^{m_k}\boldsymbol{x} = (A - \lambda_k I)(A - \lambda_k I)^{m_k - 1}\boldsymbol{x} = 0$, we see that we have two possibilities – either $(A - \lambda_k I)^{m_k-1}\boldsymbol{x} \neq 0$ or $(A - \lambda_k I)^{m_k-1}\boldsymbol{x} = 0$. In the first case, we have $0 \neq \boldsymbol{y} = (A - \lambda_k I)^{m_k-1}\boldsymbol{x} \in X_j$ and we can write

$$0 = (A - \lambda_k I)\boldsymbol{y} = (A - \lambda_j I)\boldsymbol{y} + (\lambda_j - \lambda_k)\boldsymbol{y},$$

that is,

$$(A - \lambda_j I)\boldsymbol{y} = (\lambda_k - \lambda_j)\boldsymbol{y}$$

and

$$0 = (A - \lambda_j I)^{m_j}\boldsymbol{y} = (\lambda_k - \lambda_j)^{m_j}\boldsymbol{y} \neq 0.$$

So, the first case is impossible unless $\lambda_k = \lambda_j \in \sigma(\mathcal{A}|_{X_j})$. Next, consider the second case and use induction. If, for some $2 \leq r \leq m_k - 1$, we have $(A - \lambda_k I)^{m_k - r + 1}\boldsymbol{x} = \boldsymbol{0}$ and $(A - \lambda_k I)^{m_k - r}\boldsymbol{x} \neq 0$, then the repetition of the above argument also gives a contradiction. This argument shows that any X_j is also disjoint (apart from 0) from the linear combination of all other X_k, $k \neq j$, as A restricted to such a combination cannot have λ_j as its eigenvalue. Indeed, let $\boldsymbol{x} \in X_j \cap (X_1 \oplus \cdots \oplus X_{j-1})$ for some $j \geq 2$ and then let

$$\boldsymbol{x} = \boldsymbol{x}_1 + \cdots + \boldsymbol{x}_{j-1}$$

with $\boldsymbol{x}_i \in X_i, i = 1, \ldots, j-1$, and

$$(\mathcal{A} - \lambda_1 \mathcal{I})^{m_1} \cdots (\mathcal{A} - \lambda_{j-1}\mathcal{I})^{m_{j-1}}\boldsymbol{x} = 0.$$

As before, there could be i such that

$$\boldsymbol{y} = (\mathcal{A} - \lambda_1 I)^{m_1} \cdots (\mathcal{A} - \lambda_i I)^{m_i - 1} \cdots (\mathcal{A} - \lambda_{j-1} I)^{m_{j-1}}\boldsymbol{x} \neq 0$$

and

$$(\mathcal{A} - \lambda_i I)\boldsymbol{y} = 0.$$

Since, by invariance, $\boldsymbol{y} \in X_j$, we can proceed as in the previous part of the proof to obtain a contradiction. If there is no such i, then we look for r for which

$$\boldsymbol{y} = (\mathcal{A} - \lambda_1 I)^{m_1} \cdots (\mathcal{A} - \lambda_r I)^{m_r - 2} \cdots (\mathcal{A} - \lambda_1 I)^{m_{j-1}}\boldsymbol{x} \neq 0,$$

and proceed as before. We continue until we exhaust all possibilities. This shows that $X = X_1 \oplus \cdots \oplus X_p$, see [71, Problem 5.9.14]. Hence, we have invariant subspaces X_j meeting only at $\{0\}$ and with dimensions summing up to d and (C.2.8) has been proved. □

The forthcoming discussion involves coordinates; hence, we return to the matrix notation. We recall that the default basis is the canonical one. Let for an operator A the canonical representation of $A - \lambda I$ be $\mathcal{A} - \lambda\mathcal{I}$.

Remark C.5 Proposition C.4 justifies the claim made in Sect. 5.2 that it is sufficient to solve

$$(\mathcal{A} - \lambda_j \mathcal{I})^k \boldsymbol{v} = \boldsymbol{0}$$

for $k \leq m_j$ to find all eigenvectors and associated eigenvectors belonging to λ_j. Thus $X_j = E_{\lambda_j}$, the generalized eigenspace belonging to λ_j.

If the index of $\mathcal{A}_j := \mathcal{A} - \lambda_j \mathcal{I}$ on E_{λ_j} is exactly m_j, then, according to Proposition C.1, the set $\mathcal{v} = \{\boldsymbol{v}^j_{\boldsymbol{1}}, \ldots, \boldsymbol{v}^j_{\boldsymbol{m_j}}\} = \{\mathcal{A}_j^{m_j - 1}\boldsymbol{x}, \ldots, \mathcal{A}_j\boldsymbol{x}, \boldsymbol{x}\}$ for appropriately chosen $\boldsymbol{x} \in E_{\lambda_j}$ is a basis of E_{λ_j}. Then, if we express $\boldsymbol{y} \in E_{\lambda_j}$ as

$$\boldsymbol{y} = \alpha_1 \boldsymbol{v}^j_{\boldsymbol{1}} + \cdots + \alpha_{m_j} \boldsymbol{v}^j_{\boldsymbol{m_j}},$$

then, using $\mathcal{A}_j^{m_j} E_{\lambda_j} = 0$, we obtain

$$\begin{aligned}\mathcal{A}_j \mathbf{y} &= \alpha_1 \mathcal{A}_j \mathbf{v}_1^j + \cdots + \alpha_{m_j} \mathcal{A}_j \mathbf{v}_{m_j}^j = \alpha_2 \mathcal{A}_j^{m_j-1} \mathbf{x} + \cdots + \alpha_{m_j} \mathcal{A}_j \mathbf{x} \\ &= \alpha_2 \mathbf{v}_1^j + \cdots + \alpha_{m_j} \mathbf{v}_{m_j-1}^j.\end{aligned}$$

This means that, in this basis, $A_j = A - \lambda_j \mathcal{I}$ has the matrix representation

$$(A_j)_v = \begin{pmatrix} 0 & 1 & 0 & \ldots & 0 \\ 0 & 0 & 1 & \ldots & \\ \vdots & \vdots & \ddots & \vdots & \vdots \\ 0 & 0 & 0 & \ldots & 1 \\ 0 & 0 & 0 & \ldots & 0 \end{pmatrix},$$

and hence

$$(A|_{E_{\lambda_i}})_v = \begin{pmatrix} \lambda_j & 1 & 0 & \ldots & 0 \\ 0 & \lambda_j & 1 & \ldots & \\ \vdots & \vdots & \ddots & \vdots & \vdots \\ 0 & 0 & 0 & \ldots & 1 \\ 0 & 0 & 0 & \ldots & \lambda_j \end{pmatrix}.$$

This is the so-called Jordan block of $\mathcal{A}$ corresponding to λ_j. The following theorem provides the corresponding construction in the general case. We formulate it for an arbitrary nilpotent operator T to shorten the notation.

Theorem C.6 *Let $T : X \to X$ be a linear operator satisfying $T^m = 0$ for some $m \geq 1$. Then there exist natural numbers $k, a_1, \ldots a_k$, and vectors $\mathbf{u}_1, \ldots \mathbf{u}_k$, such that the set*

$$u = \{\mathbf{u}_1, T\mathbf{u}_1, \ldots T^{a_1-1}\mathbf{u}_1, \ldots, \mathbf{u}_k, T\mathbf{u}_k, \ldots T^{a_k-1}\mathbf{u}_k\} \tag{C.2.10}$$

is a basis of X and, moreover, $T^{a_i}\mathbf{u}_i = \mathbf{0}$ for $1 \leq i \leq k$.

Proof We can take $d = dim\, X \geq 1$. We have $TX \subsetneq X$ as otherwise $T^m X = X$. Also, we can assume that $T \neq 0$, as, if this is the case, we can take any basis. So, henceforth we assume $\{0\} \subsetneq TX \subsetneq X$. We perform induction with respect to the dimension of X. If $dim\, X = 1$, then T is a scalar, and since T is nilpotent, it must be 0. Hence, the statement is obvious. We make the induction assumption that the theorem is valid for a nilpotent operator on any space of dimension smaller than d. In particular, we consider T on TX. This is possible as clearly, TX is invariant under T and $T|_{TX}$ is nilpotent. By the considerations at the beginning of the proof, $dim\, TX < d$. Hence, by the inductive assumption, there is a basis

$$\boldsymbol{v}_1, T\boldsymbol{v}_1, \ldots T^{b_1-1}\boldsymbol{v}_1, \ldots, \boldsymbol{v}_l, T\boldsymbol{v}_l, \ldots T^{b_l-1}\boldsymbol{v}_l, \tag{C.2.11}$$

of TX with $T^{b_1}\boldsymbol{v}_1 = 0, \ldots, T^{b_l}\boldsymbol{v}_l = 0$. Since, for any $i = 1, \ldots, l$, $\boldsymbol{v}_i \in TX$, there are $\boldsymbol{u}_i$ such that $\boldsymbol{v}_i = T\boldsymbol{u}_i$. The kernel of T contains linearly independent vectors $T^{b_1-1}\boldsymbol{v}_1, \ldots, T^{b_l-1}\boldsymbol{v}_l$ and we extend this set to a basis of the kernel by adding linearly independent vectors, say, $\boldsymbol{u}_{l+1}, \ldots, \boldsymbol{u}_k$. We claim that

$$\boldsymbol{u}_1, T\boldsymbol{u}_1, \ldots T^{b_1}\boldsymbol{u}_1, \ldots, \boldsymbol{u}_l, T\boldsymbol{u}_l, \ldots T^{b_l}\boldsymbol{u}_l, \boldsymbol{u}_{l+1}, \ldots, \boldsymbol{u}_k, \tag{C.2.12}$$

is a basis of X of the form (C.2.10). Indeed, $T^{b_i+1}\boldsymbol{u}_i = T^{b_i}\boldsymbol{v}_i = 0$, so $a_i = b_i + 1$ for $1 \leq i \leq l$ and $T\boldsymbol{u}_i = 0$ and hence $\alpha_1 = 1$ for $l + 1 \leq i \leq k$.

Assume there are not simultaneously vanishing scalars $\alpha_0^1, \ldots, \alpha_{b_l}^l, \gamma_{l+1}, \ldots, \gamma_k$, such that

$$\begin{aligned} 0 = \alpha_0^1\boldsymbol{u}_1 + \alpha_1^1 T\boldsymbol{u}_1 + \cdots + \alpha_{b_1}^1 T^{b_1}\boldsymbol{u}_1 + \cdots + \alpha_1^l\boldsymbol{u}_l + \cdots \\ + \alpha_{b_l}^l T^{b_l}\boldsymbol{u}_l + \gamma_{l+1}\boldsymbol{u}_{l+1} + \cdots + \gamma_k\boldsymbol{u}_k. \end{aligned}$$

Applying T to both sides, using $T\boldsymbol{u}_i = \boldsymbol{v}_i$ and $T^{b_i+1}\boldsymbol{u}_i = 0$, we obtain

$$0 = \alpha_0^1\boldsymbol{v}_1 + \alpha_1^1 T^2\boldsymbol{u}_1 + \cdots + \alpha_{b_1-1}^1 T^{b_1-1}\boldsymbol{v}_1 + \cdots + \alpha_1^l\boldsymbol{v}_l + \cdots + \alpha_{b_l-1}^l T^{b_l-1}\boldsymbol{v}_l.$$

Using the inductive assumptions, all scalars involved in the above equality must be zero. Thus, we obtain

$$0 = \alpha_{b_1}^1 T^{b_1}\boldsymbol{u}_1 + \cdots + \alpha_{b_l}^l T^{b_l}\boldsymbol{u}_l + \gamma_{l+1}\boldsymbol{u}_{l+1} + \cdots + \gamma_k\boldsymbol{u}_k.$$

Here, however, we have a linear combination of the basis of the kernel of T, which also shows that all coefficients present above must be zero. Hence, (C.2.12) is a linearly independent set. Further, by the Rank Theorem, $dim\, X$ is the sum of the dimensions of the kernel and range of T, that is, $dim X = l + k - l + b_1 + \cdots + b_l = k + b_1 + \cdots + b_l$. If we count the number of vectors in (C.2.12), we get $b_1 + 1 + \cdots + b_l + 1 + k - l = b_1 + \cdots + b_l + k$ and hence (C.2.12) is a basis in X. □

Thus, if

$$\begin{aligned} \boldsymbol{x} = \alpha_0^1 T^{a_1-1}\boldsymbol{u}_1 + \alpha_1^1 T^{a_1-2}\boldsymbol{u}_1 + \cdots + \alpha_{a_1-1}^1\boldsymbol{u}_1 + \cdots \\ + \alpha_0^k T^{a_k-1}\boldsymbol{u}_k + \alpha_1^k T^{a_k-2}\boldsymbol{u}_k + \cdots + \alpha_{a_k-1}^k\boldsymbol{u}_k, \end{aligned}$$

then

$$T\boldsymbol{x} = \alpha_1^1 T^{a_1-1}\boldsymbol{u}_1 + \cdots + \alpha_{a_1-1}^1 T\boldsymbol{u}_1 + \ldots + \alpha_1^k T^{a_k-1}\boldsymbol{u}_k + \ldots + \alpha_{a_k-1}^k T\boldsymbol{u}_k.$$

Hence, in this basis, we have the matrix representation

$$(T)_u = \begin{pmatrix} 0 & 1 & 0 & \ldots & 0 & 0 & 0 & 0 & \ldots & 0 \\ 0 & 0 & 1 & \ldots & 0 & 0 & 0 & 0 & \ldots & 0 \\ \vdots & \ddots & \ddots & \ddots & \vdots & \vdots & \vdots & \vdots & \vdots & \vdots \\ 0 & 0 & 0 & \ldots & 1 & 0 & 0 & 0 & \ldots & 0 \\ 0 & 0 & 0 & \ldots & 0 & 0 & 0 & 0 & \ldots & 0 \\ 0 & 0 & 0 & \ldots & 0 & 0 & 1 & 0 & \ldots & 0 \\ \vdots & \vdots & \vdots & \vdots & \vdots & \ddots & \ddots & \ddots & \vdots & \vdots \\ 0 & 0 & 0 & \ldots & 0 & 0 & 0 & 0 & \ldots & 1 \\ 0 & 0 & 0 & \ldots & 0 & 0 & 0 & 0 & \ldots & 0 \end{pmatrix},$$

that is, the matrix has zero entries everywhere apart from the super diagonal, and the super diagonal entries can be only zero or one. In the notation of Theorem C.2.10, the first diagonal block has the dimension $a_1 \times a_1$ and the last $a_k \times a_k$.

If we now use Proposition C.2.8, we see that there is a basis $q = \{u_1, \ldots, u_p\}$, where u_i is the basis of the form (C.2.10) with $T = A_i$, $1 \leq i \leq p$, in which any A_i has a matrix representation $\mathcal{J}_i$, called the Jordan normal form, given by

$$\mathcal{J}_i := (A_i)_q = \begin{pmatrix} \mathcal{J}_1 & 0 & 0 \\ \vdots & \ddots & \vdots \\ 0 & 0 & \mathcal{J}_p \end{pmatrix} = \mathcal{V}^{-1}\mathcal{A}_i\mathcal{V}, \tag{C.2.13}$$

where

$$\mathcal{J}_i = \begin{pmatrix} \lambda_i & 1 & 0 & \ldots & 0 & 0 & 0 & 0 & \ldots & 0 \\ 0 & \lambda_i & 1 & \ldots & 0 & 0 & 0 & 0 & \ldots & 0 \\ \vdots & \ddots & \ddots & \ddots & \vdots & \vdots & \vdots & \vdots & \vdots & \vdots \\ 0 & 0 & 0 & \ldots & 1 & 0 & 0 & 0 & \ldots & 0 \\ 0 & 0 & 0 & \ldots & \lambda_i & 0 & 0 & 0 & \ldots & 0 \\ 0 & 0 & 0 & \ldots & 0 & \lambda_i & 1 & 0 & \ldots & 0 \\ \vdots & \vdots & \vdots & \vdots & \vdots & \ddots & \ddots & \ddots & \vdots & \vdots \\ 0 & 0 & 0 & \ldots & 0 & 0 & 0 & 0 & \ldots & 1 \\ 0 & 0 & 0 & \ldots & 0 & 0 & 0 & 0 & \ldots & \lambda_i \end{pmatrix} \tag{C.2.14}$$

for $1 \leq i \leq p$ are called Jordan blocks and $\mathcal{V}$ is the transformation matrix between q and the canonical basis, as defined by (C.1.1).

Remark C.7 If we do not want to operate with different bases in $\mathbb{C}^d$, we can consider both sides of (C.2.13) in the canonical basis expressing the fact that A is similar to an operator J whose matrix representation in the canonical basis is $\mathcal{J}$.

The Jordan block representation provides an alternative proof of the fact that for any norm on X, $\lim\limits_{n\to\infty} \mathcal{A}^n = 0$ if and only if $r(\mathcal{A}) < 1$ and if $r(\mathcal{A}) > 1$, then $\lim_{n\to\infty} \|\mathcal{A}^n\| = \infty$. Indeed, if $\boldsymbol{v}$ is an eigenvector of $\mathcal{A}$, say $\mathcal{A}\boldsymbol{v} = \lambda_i \boldsymbol{v}$ for some $1 \leq i \leq p$, then

$$\mathcal{A}^n \boldsymbol{v} = \lambda_i^n \boldsymbol{v},$$

and, if the left-hand side converges to zero, then $|\lambda_i| < 1$. This also shows that if $|\lambda_i| > 1$ for some i, $\lim_{n\to\infty} \|\mathcal{A}^n \boldsymbol{v}\| = \infty$ and since $\boldsymbol{v} \neq 0$, we have the last statement. On the other hand, using (C.2.13) we have

$$\mathcal{A}^n = \mathcal{V}\mathcal{J}^n\mathcal{V}^{-1}$$

and, since $\mathcal{J}$ is block diagonal,

$$\mathcal{J}^n = \begin{pmatrix} \mathcal{J}_1^n & 0 & 0 \\ \vdots & \ddots & \vdots \\ 0 & 0 & \mathcal{J}_p^n \end{pmatrix}.$$

Let as fix $1 \leq i \leq p$. Then, using the fact that $\mathcal{J}_i = \lambda I_i + \mathcal{T}_i$, where $\mathcal{I}_i$ is the $m_i \times m_i$ diagonal matrix and $\mathcal{T}_i$ is a nilpotent matrix of index at most m_i, we have

$$\mathcal{J}_i^n = \lambda_i^n \sum_{j=0}^{m_i} \binom{n}{j} \lambda_i^{-j} \mathcal{T}_i^j.$$

We see that the sum in the formula above is a polynomial in n of degree at most m_j. Thus $\mathcal{J}_i^n \to 0$ provided $|\lambda_i| < 1$.

We can use this result to provide the Gelfand formula for $r(\mathcal{A})$.

Proposition C.8 *We have*

$$r(\mathcal{A}) = \lim_{n\to\infty} \sqrt[n]{\|\mathcal{A}^n\|}. \tag{C.2.15}$$

Proof We observe that if $r(\mathcal{A}) = 0$, then all eigenvalues of $\mathcal{A}$ are zero and thus $\mathcal{A}$ is nilpotent. Hence $\mathcal{A}^n = 0$ for $n > d$ and therefore

$$\lim_{n\to\infty} \sqrt[n]{\|\mathcal{A}^n\|} = 0 = r(\mathcal{A}).$$

Thus, we can assume that $r(\mathcal{A}) > 0$. Let us fix arbitrary $\epsilon \in (0, r(\mathcal{A}))$ and define

$$\mathcal{A}_{\pm\epsilon} = \frac{1}{r(\mathcal{A}) \pm \epsilon}\mathcal{A}.$$

Then $\lambda \in \sigma(\mathcal{A})$ if and only if $\frac{\lambda}{r(\mathcal{A})\pm\epsilon} \in \sigma(\mathcal{A}_{\pm\epsilon})$ and hence

$$r(\mathcal{A}_{\pm\epsilon}) = \frac{r(\mathcal{A})}{r(\mathcal{A}) \pm \epsilon}.$$

Since $r(\mathcal{A}_{+\epsilon}) < 1$, we have

$$\mathcal{A}^n_{+\epsilon} = \frac{\mathcal{A}^n}{(r(\mathcal{A}) + \epsilon)^n} \to 0 \quad \text{as } n \to \infty,$$

and hence, there is n_+ such that

$$\sqrt[n]{\|\mathcal{A}^n\|} \leq r(\mathcal{A}) + \epsilon, \quad n \geq n_+. \tag{C.2.16}$$

Further, since $r(\mathcal{A}_{-\epsilon}) > 1$, we have

$$\|\mathcal{A}^n_{-\epsilon}\| = \frac{1}{(r(\mathcal{A}) - \epsilon)^n}\|\mathcal{A}^n\| \to \infty \quad \text{as } n \to \infty,$$

and hence there is n_- such that

$$\sqrt[n]{\|\mathcal{A}^n\|} \geq r(\mathcal{A}) - \epsilon, \quad n \geq n_-. \tag{C.2.17}$$

Thus, for any sufficiently small ϵ there is $n_0 = \max\{n_+, n_-\}$ such that for any $n \geq n_0$

$$r(\mathcal{A}) - \epsilon \leq \sqrt[n]{\|\mathcal{A}^n\|} \leq r(\mathcal{A}) + \epsilon$$

and hence

$$r(\mathcal{A}) = \lim_{n\to\infty} \sqrt[n]{\|\mathcal{A}^n\|}.$$

□

C.3 The Perron–Frobenius Theorem

Since the solutions of equations describing the evolution of populations must necessarily be non-negative, it is unsurprising that the Perron–Frobenius theorem describing the spectral properties of matrices preserving nonnegativity was found extremely useful in many situations, both in discrete and continuous time cases. There are many proofs of this result, which were summarized in [68]. We have selected the one using complex analysis tools, which can be found in, e.g., [78, Chapter 10]. The reader preferring a more algebraic approach can find it in, e.g., [71, Chapter 8].

C.3.1 The Resolvent Operator

Let A be a linear operator on $X = \mathbb{C}^d$ and let $\mathcal{A} \in \mathbb{M}_d$ be its matrix in the canonical basis. In this section, it will be more convenient to use matrices. From the definition of the eigenvalues, the matrix $\lambda\mathcal{I} - \mathcal{A}$ is invertible for any $\lambda \in \mathbb{C} \setminus \sigma(\mathcal{A})$, that is, everywhere on the complex plane $\mathbb{C}$ except eigenvalues. Hence, as long as $\lambda \notin \sigma(\mathcal{A})$, there exists $(\lambda\mathcal{I} - \mathcal{A})^{-1}$ and is given by

$$(\lambda\mathcal{I} - \mathcal{A})^{-1} = \frac{1}{det\,(\lambda\mathcal{I} - \mathcal{A})}\mathcal{C}^T(\lambda), \tag{C.3.18}$$

where $\mathcal{C}(\lambda)^T$ is the transpose of $\mathcal{C}(\lambda) = (c_{ij}(\lambda))_{1\leq i,j\leq d}$, given by

$$c_{ij}(\lambda) = ((-1)^{i+j}m_{ij}(\lambda))_{1\leq i,j\leq d},$$

and $m_{ij}(\lambda)$ is the (i, j)-minor of $\lambda\mathcal{I} - \mathcal{A}$ defined as the determinant of the $(d-1)\times(d-1)$ matrix obtained by deleting the i-th row and j-th column from $\lambda\mathcal{I} - \mathcal{A}$. Since $det\,(\lambda\mathcal{I} - \mathcal{A})$ and $m_{ij}(\lambda)$, $1 \leq i, j \leq d$, are polynomials in λ, $(\lambda\mathcal{I} - \mathcal{A})^{-1}$ is an analytic function on $\mathbb{C}\backslash\sigma(\mathcal{A})$ which has a pole at each $\lambda \in \sigma(\mathcal{A})$. It is customary to denote

$$R(\lambda, \mathcal{A}) := (\lambda\mathcal{I} - \mathcal{A})^{-1}$$

and call it the *resolvent* of $\mathcal{A}$; the set $\rho(\mathcal{A}) := \mathbb{C} \setminus \sigma(\mathcal{A})$ is called the *resolvent set* of $\mathcal{A}$.

We observe that directly from the definition, we have

$$R(\lambda, \mathcal{A}) - R(\mu, \mathcal{A}) = (\mu - \lambda)R(\lambda, \mathcal{A})R(\mu, \mathcal{A}), \quad \lambda, \mu \in \rho(\mathcal{A}). \tag{C.3.19}$$

Indeed,

$$\begin{aligned}\mathcal{I} &= R(\lambda, \mathcal{A})(\lambda\mathcal{I} - \mathcal{A}) = R(\lambda, \mathcal{A})(\lambda\mathcal{I} - \mu\mathcal{I} + \mu\mathcal{I} - \mathcal{A}) \\ &= (\lambda - \mu)R(\lambda, \mathcal{A}) + R(\lambda, \mathcal{A})(\mu\mathcal{I} - \mathcal{A}),\end{aligned}$$

and (C.3.19) follows by multiplying the above equality by $R(\mu, \mathcal{A})$ from the right. Identity (C.3.19) implies that $R(\lambda, \mathcal{A})$ is differentiable, and hence analytic, in $\rho(\mathcal{A})$

and (C.3.18) shows that the eigenvalues of $\mathcal{A}$ are poles of $R(\lambda, \mathcal{A})$. Hence, for any $\rho \in \sigma(\mathcal{A})$, we can expand $R(\lambda, \mathcal{A})$ into the Laurent series

$$R(\lambda, \mathcal{A}) = \sum_{k=-h}^{\infty} (\lambda - \rho)^k \mathcal{A}_k, \tag{C.3.20}$$

where h, depending on ρ, is finite and $\mathcal{A}_{-h} \neq 0$.

We can provide an alternative representation of $R(\lambda, \mathcal{A})$ by expanding it in the Neumann series

$$(\lambda \mathcal{I} - \mathcal{A})^{-1} = \frac{1}{\lambda}\left(\mathcal{I} - \frac{1}{\lambda}\mathcal{A}\right)^{-1} = \frac{1}{\lambda}\sum_{k=1}^{\infty} \frac{\mathcal{A}^k}{\lambda^k}. \tag{C.3.21}$$

The classical result says that the above series converges for $|\lambda| > \|\mathcal{A}\|$. However, the Cauchy root test, see [78, Section 10.5], asserts that the series converges provided

$$\lim_{k\to\infty} \sqrt[k]{\|\mathcal{A}^k\|} < |\lambda|,$$

or, in other words, the series converges if $|\lambda| > r(\mathcal{A})$, see Proposition C.8.

Remark C.9 We observe that (C.3.20) and (C.3.21) can be viewed as Laurent expansions of the operator-valued function $\lambda \mapsto (\lambda \mathcal{I} - \mathcal{A})^{-1}$ (at respective singularities) which can be considered as an operator version of the function $\lambda \mapsto (\lambda - a)^{-1}$, $a \in \mathbb{C}$, constructed in the same way as the function $t \mapsto e^{t\mathcal{A}}$ which was used in Sect. 5.3.

Due to the coordinate-wise definition of the order in $\mathbb{R}^d$, the concept of a nonnegative operator refers to its matrix representation. Hence, in what follows, we shall be using the matrix representation of an operator in a fixed basis. We note that (C.3.21) implies $R(\lambda, \mathcal{A}) \geq 0$ for $\mathbb{R} \ni \lambda > r(\mathcal{A})$ provided $\mathcal{A} \geq 0$.

Proposition C.10 *If $\mathcal{A} \geq 0$, then $r(\mathcal{A}) \in \sigma(\mathcal{A})$.*

Proof Denote $r = r(\mathcal{A})$. The result is clear if $r = 0$, as in this case, all eigenvalues of $\mathcal{A}$ equal 0. Hence, assume that $r > 0$ is not an eigenvalue. Then, by (C.3.18),

$$\lim_{\lambda\to r} R(\lambda, \mathcal{A}) = R(r, \mathcal{A}),$$

that is, the limit is finite. Using the Neumann series expansion, for $\lambda \in \mathbb{R}$ we have

$$R(r, \mathcal{A}) = \lim_{\lambda\to r^+} \frac{1}{\lambda}\sum_{k=1}^{\infty} \frac{\mathcal{A}^k}{\lambda^k}.$$

For any n we have

$$\lim_{\lambda\to r^+} S_n(\lambda) := \lim_{\lambda\to r^+} \frac{1}{\lambda}\sum_{k=1}^{n}\frac{\mathcal{A}^k}{\lambda^k} = \frac{1}{r}\sum_{k=1}^{n}\frac{\mathcal{A}^k}{r^k} = S_n(r).$$

First, observe that the sequence $(S_n(r))_{n\in\mathbb{N}}$ is monotonically increasing with n (as $\mathcal{A} \geq 0$). Further, by (C.3.21), the function $\lambda \mapsto R(\lambda, \mathcal{A})$ is monotonically increasing as $\lambda \to r^+$, hence for any $\lambda > r$ and $n \in \mathbb{N}$,

$$S_n(\lambda) \leq R(\lambda, \mathcal{A}) \leq R(r, \mathcal{A}).$$

Letting $\lambda \to r^+$ on the right-hand side, we obtain

$$S_n(r) \leq R(r, \mathcal{A}).$$

Thus $(S_n(r))_{n\in\mathbb{N}}$ is also bounded, and hence it converges, that is,

$$R(r, \mathcal{A}) = \frac{1}{r}\sum_{k=1}^{\infty}\frac{\mathcal{A}^k}{r^k}.$$

This, however, shows that (C.3.21) converges for all λ with $|\lambda| = r$, contradicting the assumption that there is an eigenvalue on the circle $|\lambda| = r$. □

Next, we consider eigenvectors belonging to the eigenvalue r. We shall need the following lemma.

Lemma C.11 *Consider* (C.3.20). *We have*

$$\mathcal{A}_k\mathcal{A}_m = \begin{cases} -\mathcal{A}_{k+m+1} & \text{if } k, m \geq 0, \\ 0 & \text{if } k < 0 \text{ and } m \geq 0 \text{ or } k \geq 0 \text{ and } m < 0, \\ \mathcal{A}_{k+m+1} & \text{if } k, m < 0. \end{cases} \tag{C.3.22}$$

Proof By the Cauchy integral formula,

$$\mathcal{A}_k = \frac{1}{2\pi i}\int\limits_C \frac{R(\lambda, \mathcal{A})}{(\lambda-\rho)^{k+1}}d\lambda,$$

where C is a clockwise oriented circle centred at ρ and with radius sufficiently small not to enclose any other pole of $R(\lambda, \mathcal{A})$. To shorten the calculations, we shift ρ to zero by the substitution $\nu = \lambda - \rho$ and denote $R(\nu) = R(\nu + \rho, \mathcal{A})$. Then, taking circles C_1 and C_2 with centres at 0 and radii $r_1 < r_2$, we have, using (C.3.19),

$$
\begin{aligned}
\mathcal{A}_k\mathcal{A}_m &= -\frac{1}{4\pi^2 i^2}\int_{C_1}\int_{C_2}\frac{R(\nu)}{\nu^{k+1}}\frac{R(\mu)}{\mu^{m+1}}d\nu d\mu \\
&= -\frac{1}{4\pi^2 i^2}\int_{C_1}\int_{C_2}\frac{R(\nu)-R(\mu)}{\nu^{k+1}\mu^{m+1}(\nu-\mu)}d\nu d\mu \\
&= \frac{1}{2\pi i}\int_{C_1}\frac{R(\nu)}{\nu^{k+1}}\left(\frac{1}{2\pi i}\int_{C_2}\frac{1}{\mu^{m+1}(\mu-\nu)}d\mu\right)d\nu \\
&+ \frac{1}{2\pi i}\int_{C_2}\frac{R(\mu)}{\mu^{m+1}}\left(\frac{1}{2\pi i}\int_{C_1}\frac{1}{\nu^{k+1}(\nu-\mu)}d\nu\right)d\mu. \qquad \text{(C.3.23)}
\end{aligned}
$$

Now, the internal integrals are of the form

$$
\frac{1}{2\pi i}\int_C \frac{1}{z^n(z-w)}dz,
$$

where C is the circle centred at 0 with radius ρ and traversed counterclockwise, and w is either inside the circle (in the first term) or outside (in the second term). If $n \leq 0$ and $|w| > \rho$, then the integrand is analytic inside C; hence, the integral is zero. If $n \leq 0$ and $|w| < \rho$, then z^n is analytic inside C and the integral is the Cauchy integral giving w^n. On the other hand, if $n < 0$, then z^n has a pole at $z = 0$ and we can use the residue theorem to see that the integral is zero if $|w| < \rho$ as the integrals at the poles at $z = 0$ and $z = w$ are $\pm w^{-n}$ and they cancel each other, while if $|w| > \rho$, there is only one singularity inside C and the integral equals $-w^{-n}$. Summarizing,

$$
\frac{1}{2\pi i}\int_C \frac{1}{z^n(z-w)}dz = \begin{cases} 0 & \text{if } n \leq 0 \text{ and } |w| > \rho \;\text{ or } n > 0 \text{ and } |w| < \rho, \\ \frac{1}{w^n} & \text{if } n \leq 0 \text{ and } |w| < \rho, \\ -\frac{1}{w^n} & \text{if } n > 0 \text{ and } |w| > \rho, \end{cases}
$$

and (C.3.22) follows by substituting the appropriate formula in (C.3.23). □

Proposition C.12 *There are nonnegative eigenvectors belonging to the eigenvalue* $r = r(\mathcal{A})$.

Proof Since $\lambda = r$ is a pole of the resolvent, $R(\lambda, \mathcal{A})$ has the representation (C.3.20),

$$R(\lambda, \mathcal{A}) = \sum_{i=-h}^{\infty} (\lambda - r)^i \mathcal{A}_i, \tag{C.3.24}$$

for some finite h, where $\mathcal{A}_{-h} \neq 0$. First, observe that $R(\lambda, \mathcal{A}) \geq 0$ for $\lambda > r$ implies

$$\mathcal{A}_{-h} = \lim_{\lambda \to r^+} (\lambda - r)^h R(\lambda, \mathcal{A}) \geq 0. \tag{C.3.25}$$

Lemma C.11 implies, in particular, that

$$\mathcal{A}_i \mathcal{A}_{-h} = 0, \quad i \geq 0,$$

but also, since $i - h + 1 < -h$ whenever $i < -1$ and $\mathcal{A}_k = 0$ if $k < -h$,

$$\mathcal{A}_i \mathcal{A}_{-h} = 0, \quad i < -1.$$

Therefore,

$$R(\lambda, \mathcal{A})\mathcal{A}_{-h} = \sum_{i=-h}^{\infty} (\lambda - r)^i \mathcal{A}_i \mathcal{A}_{-h} = (\lambda - r)^{-1} \mathcal{A}_{-1} \mathcal{A}_{-h} = (\lambda - r)^{-1} \mathcal{A}_{-h}.$$

This shows that any column of $\mathcal{A}_{-h}$ is an eigenvector of $R(\lambda, \mathcal{A})$ and hence $\mathcal{A}$. By (C.3.25), each column of $\mathcal{A}_{-h}$ is nonnegative. □

Theorem C.13 *If $\mathcal{A} \gg 0$ (that is, recall, $a_{ij} > 0$ for all $1 \leq i, j \leq d$), then $r(\mathcal{A}) = r > 0$, it is a simple eigenvalue and the corresponding eigenvector $\boldsymbol{v}$ satisfies $\boldsymbol{v} \gg 0$. All other eigenvalues of $\mathcal{A}$ have absolute values strictly smaller than r.*

Proof Let $\boldsymbol{v} > \boldsymbol{0}$ be any eigenvector, whose existence was proved in Proposition C.12 ($\boldsymbol{v} \geq 0$ and $\boldsymbol{v} \neq 0$ by the definition of eigenvector). Since $\boldsymbol{v} > \boldsymbol{0}$ and $\mathcal{A} \gg 0$, we have

$$\boldsymbol{0} \ll \mathcal{A}\boldsymbol{v} = r\boldsymbol{v},$$

which shows that $r > 0$ and $\boldsymbol{v} \gg \boldsymbol{0}$. Next, we observe that

$$\mathcal{A}_{-h}\mathcal{A}_{-h} = \mathcal{A}_{-2h+1} = 0$$

for $h > 1$, as then $-2h + 1 < -h$. But columns of $\mathcal{A}_{-h}$ are strictly positive eigenvectors of $\mathcal{A}$, which contradicts the above equation. Thus $h = 1$ and $\lambda = r$

is a first-order pole of $R(\lambda, \mathcal{A})$. We shall prove that the algebraic and geometric multiplicities of r equal one. First, we shall prove that any eigenvector of $\mathcal{A}$ belonging to r is in the image of $\mathcal{A}_{-1}$, that is, it is a linear combination of columns of $\mathcal{A}_{-1}$. Indeed, if $r\boldsymbol{v} = \mathcal{A}\boldsymbol{v}$, then

$$R(\lambda, \mathcal{A})\boldsymbol{v} = \frac{1}{\lambda - r}\boldsymbol{v}$$

for any $\lambda \in \rho(\mathcal{A})$. Then, using (C.3.25),

$$\mathcal{A}_{-1}\boldsymbol{v} = \lim_{\lambda \to r} (\lambda - r)R(\lambda, \mathcal{A})\boldsymbol{v} = \boldsymbol{v}, \tag{C.3.26}$$

and hence $\boldsymbol{v}$ is in the range of $\mathcal{A}_{-1}$. Thus, all eigenvectors of $\mathcal{A}$ are in the linear span of columns of $\mathcal{A}_{-1}$. As we proved above, any nonnegative eigenvector of $\mathcal{A}$ is strictly positive. Let us take strictly positive eigenvectors $\boldsymbol{v}^{\mathbf{1}}$ and $\boldsymbol{v}^{\mathbf{2}}$. Then $\boldsymbol{v}^{\mathbf{1}} - \epsilon\boldsymbol{v}^{\mathbf{2}} \geq 0$ for sufficiently small $\epsilon > 0$, and it is still an eigenvector of $\mathcal{A}$. However, for

$$\epsilon_1 = \frac{v_j^1}{v_j^2} := \min_{1 \leq i \leq d} \frac{v_i^1}{v_i^2}$$

we have $v_i^1 - \epsilon_1 v_i^2 > 0, i \neq j$, and $v_j^1 - \epsilon_1 v_j^2 = 0$ and this contradicts the fact that all nonnegative eigenvectors are strictly positive. So, the geometric multiplicity of r is 1. To prove that the algebraic multiplicity of r is also one, let us assume that there is an associated eigenvector $\boldsymbol{u}$ satisfying $(r\mathcal{I} - \mathcal{A})^\nu \boldsymbol{u} = \boldsymbol{v} \neq \boldsymbol{0}$ for some $\nu > 0$, where $\boldsymbol{v}$ is an eigenvector. Then $\mathcal{A}_{-1}\boldsymbol{v} = \boldsymbol{v}$ by (C.3.26), and hence

$$\boldsymbol{v} = \mathcal{A}_{-1}\boldsymbol{v} = (r\mathcal{I} - \mathcal{A})^\nu \boldsymbol{u} = \mathcal{A}_{-1}(r\mathcal{I} - \mathcal{A})^\nu \boldsymbol{u} = (r\mathcal{I} - \mathcal{A})^\nu \mathcal{A}_{-1}\boldsymbol{u} = \boldsymbol{0},$$

as $\mathcal{A}_{-1}\boldsymbol{u}$ is in the range of $\mathcal{A}_{-1}$ and hence it is an eigenvector of $\mathcal{A}$ belonging to r (see the proof of Proposition C.12). This contradiction shows that there are no associated eigenvectors belonging to r.

To prove the last statement, consider $\epsilon > 0$ small enough for $\mathcal{A} - \epsilon\mathcal{I} \gg 0$. Then $r(\mathcal{A} - \epsilon\mathcal{I}) = r(\mathcal{A}) - \epsilon$ and thus all other eigenvalues λ_ϵ of $\mathcal{A} - \epsilon\mathcal{I}$ satisfy $|\lambda_\epsilon| \leq r(\mathcal{A}) - \epsilon$. On the other hand, $\lambda_\epsilon = \lambda - \epsilon$ for a corresponding $\lambda \in \sigma(\mathcal{A})$, that is, if $\lambda = x + iy$, then $\lambda_\epsilon = x - \epsilon + iy$ and $x < r(\mathcal{A}) - \epsilon$ unless $\lambda_\epsilon = r(\mathcal{A}) - \epsilon$. Hence

$$(r(\mathcal{A}) - \epsilon)^2 = (r(\mathcal{A}))^2 - 2\epsilon r(\mathcal{A}) + \epsilon^2 \geq x^2 - 2\epsilon x + \epsilon^2 + y^2$$

or

$$|\lambda|^2 = x^2 + y^2 \leq (r(\mathcal{A}))^2 - 2\epsilon(r(\mathcal{A}) - x) < (r(\mathcal{A}))^2$$

for any $\lambda \neq r(\mathcal{A})$. □

We have considered two extreme cases of nonnegative matrices—the case where, in principle, just one entry may be positive (ignoring the trivial case of the zero matrix) and where all entries must be positive. We can, however, identify intermediate cases which have well-defined spectral properties. We begin with irreducible matrices, introduced in Definition 5.32. To recall, using Remark 5.33, $\mathcal{A} = (a_{ij})_{1\le i,j\le d}$ is irreducible if for any $1 \le i, j \le d$, there is $0 \le k \le d-1$ such that

$$a_{ij}^{(k)} = \sum_{1\le i_r\le d, r=1,\ldots,k-1} a_{ii_1} \cdot a_{i_1 i_2} \cdot \ldots \cdot a_{i_{k-1} j} > 0,$$

where

$$\mathcal{A}^k = (a_{ij}^{(k)})_{1\le i,j\le d}.$$

Irreducible matrices are closely related to positive matrices. Namely, we have

Lemma C.14 *If $\mathcal{A}$ is irreducible, then $\mathcal{B} = (\mathcal{I} + \mathcal{A})^{d-1} \gg 0$.*

Proof We have

$$\mathcal{B} = (b_{ij})_{1\le i,j\le d} = \sum_{r=0}^{d-1} \binom{d-1}{r} \mathcal{A}^r$$

so that, for any $1 \le i, j \le d$,

$$b_{ij} = \sum_{r=0}^{d-1} \binom{d-1}{r} a_{ij}^{(r)},$$

where all terms of the sum are nonnegative and, since $\mathcal{A}$ is irreducible, one must be positive. □

Then we have

Theorem C.15 *Let (nonegative) $\mathcal{A}$ be irreducible. Then*

1. *$r(\mathcal{A}) > 0$ and $r = r(\mathcal{A}) \in \sigma(\mathcal{A})$;*
2. *r is a simple eigenvalue of $\mathcal{A}$;*
3. *there is positive eigenvector $\mathbf{v} \gg 0$ belonging to r;*
4. *there are no nonnegative eigenvectors of $\mathcal{A}$ except for positive multiples of $\mathbf{v}$.*

Proof Since $\mathcal{A}$ is nonnegative, Proposition C.10 gives $r \in \sigma(\mathcal{A})$. Let $\mathcal{J}_{\mathcal{A}}$ be the Jordan form of $\mathcal{A}$, that is,

$$\mathcal{J}_{\mathcal{A}} = \mathcal{V}^{-1}\mathcal{A}\mathcal{V}$$

for an appropriate matrix $\mathcal{V}$. Then

$$\tilde{\mathcal{J}} := \mathcal{J}_{\mathcal{I}+\mathcal{A}} = \mathcal{I} + \mathcal{J}_{\mathcal{A}} = \mathcal{V}^{-1}(\mathcal{I} + \mathcal{A})\mathcal{V},$$

that is, $\tilde{\mathcal{J}}$ consists of the same Jordan blocks with $1+\lambda$, $\lambda \in \sigma(\mathcal{A})$, on the diagonal. Using (C.2.13) and (C.2.14), we see that

$$\mathcal{V}^{-1}\mathcal{B}\mathcal{V} = \mathcal{V}^{-1}(\mathcal{I} + \mathcal{A})^{d-1}\mathcal{V} = (\mathcal{V}^{-1}(\mathcal{I} + \mathcal{A})\mathcal{V})^{d-1} = \mathcal{J}^{d-1}_{\mathcal{I}+\mathcal{A}}.$$

Hence, $\tilde{\mathcal{J}}^{d-1}$ is an upper triangular matrix with eigenvalues on the diagonal, which appear according to their algebraic multiplicity, and $(\mathcal{I} + \mathcal{A})^{d-1}$ has the same eigenvalues, with the same multiplicities. Then we have $\sigma(\mathcal{I} + \mathcal{A})^{d-1} = \{(1 + \lambda)^{d-1} : \lambda \in \sigma(\mathcal{A})\}$. Next, if $|\lambda| \leq r$, $\lambda = x + iy$, then $x \leq r$ and

$$|1 + \lambda| = \sqrt{1 + 2x + x^2 + y^2} \leq \sqrt{1 + 2r + r^2} = 1 + r$$

as $r \geq 0$. Hence, also $|(1+\lambda)^{d-1}| \leq (1+r)^{d-1}$ and thus $(1+r)^{d-1}$ is the dominant eigenvalue of $\mathcal{B}$. Since $\mathcal{B} \gg 0$, this eigenvalue is simple by Theorem C.13, that is, it appears only once on the diagonal of $\mathcal{V}^{-1}\mathcal{B}\mathcal{V}$. Thus, the Jordan block of $\mathcal{A}$ corresponding to r has dimension 1×1. Therefore, r is a simple eigenvalue of $\mathcal{A}$. Then, if $\boldsymbol{v}$ is a positive eigenvector of $\mathcal{B}$ belonging to $(1 + r)^{d-1}$, we have

$$\begin{aligned}(1 + r)^{d-1}\boldsymbol{v} &= \mathcal{B}\boldsymbol{v} = (\mathcal{I} + \mathcal{A})^{d-2}(\mathcal{I} + \mathcal{A})\boldsymbol{v} \\ &= (1 + r)(\mathcal{I} + \mathcal{A})^{d-2}\boldsymbol{v} = (1 + r)^{d-2}(\mathcal{I} + \mathcal{A})\boldsymbol{v},\end{aligned}$$

that is,

$$(1 + r)\boldsymbol{v} = (\mathcal{I} + \mathcal{A})\boldsymbol{v} = \boldsymbol{v} + \mathcal{A}\boldsymbol{v}.$$

Hence, $\boldsymbol{v}$ is the unique (up to the multiplication by a positive constant) positive eigenvector of $\mathcal{A}$ belonging to r. Then, as in the proof of Proposition C.13,

$$\mathbf{0} \ll \mathcal{A}\boldsymbol{v} = r\boldsymbol{v},$$

implying $r > 0$.

Finally, let $\boldsymbol{u} > 0$ be an eigenvector of $\mathcal{A}$ belonging to $\lambda \in \sigma(\mathcal{A}), \lambda \neq r$. Then, as above, it is also an eigenvector of $\mathcal{B} \gg 0$ belonging to $(1 + \lambda)^{d-1}$. Since the transpose $\mathcal{B}^T$ is also strictly positive with the same spectrum, there is $\boldsymbol{y} \gg 0$ satisfying $\mathcal{B}^T\boldsymbol{y} = r\boldsymbol{y}$. However, as we proved in Theorem 5.21,

$$\langle \boldsymbol{y}, \boldsymbol{u} \rangle = \sum_{i=1}^{d} y_i u_i = 0,$$

which is impossible as $y_i >$ and $u_i \geq 0$ for all i, with $u_i > 0$ for at least one i. □

We see that the only difference between Theorems C.15 and C.13 is the existence of eigenvalues on the spectral circle $C_{\mathcal{A}} := \{\lambda \in \mathbb{C} : |\lambda| = r(\mathcal{A})\}$. This leads to the important division of irreducible matrices into primitive and imprimitive matrices. In Sect. 5.4.3, we defined a primitive matrix by the condition

$$\mathcal{A}^m \gg 0 \tag{C.3.27}$$

for some $m \geq 1$. We have the following lemma.

Lemma C.16 *An irreducible matrix $\mathcal{A}$ satisfies* (C.3.27) *if and only if $\sigma(\mathcal{A}) \cap C_{\mathcal{A}} = r(\mathcal{A})$.*

Proof The spectral radius $r(\mathcal{A})$ is an eigenvalue of $\mathcal{A}$ on $C_{\mathcal{A}}$. If (C.3.27) is satisfied, we can apply Theorem C.13 to $\mathcal{A}^m$ to claim that $(r(\mathcal{A}))^m = r(\mathcal{A}^m)$ is the only eigenvalue of $\mathcal{A}^m$ on $C_{\mathcal{A}^m}$ and it is simple. If $r(\mathcal{A})$ was not simple, then, arguing as in Theorem C.15, $(r(\mathcal{A}))^m$ could not be simple. Similarly, if $\lambda = r(\mathcal{A})e^{i\theta}, \theta \neq 2\pi k$, was another eigenvalue of $\mathcal{A}$ on $C_{\mathcal{A}}$, then $\lambda^m = (r(\mathcal{A}))^m e^{im\theta}$ would be an eigenvalue of $\mathcal{A}^m$ on $C_{\mathcal{A}^m}$, contradicting either the simplicity of $r(\mathcal{A}^m)$ if $m\theta = 2q\pi$ for some integer q, or the fact that $r(\mathcal{A}^m)$ is the only eigenvalue on $C_{\mathcal{A}^m}$. Hence, $r(\mathcal{A})$ is the only eigenvalue of $\mathcal{A}$ on $C_{\mathcal{A}}$.

On the other hand, if $r(\mathcal{A}) > 0$ is simple and it is the only eigenvalue on $C_{\mathcal{A}}$, then 1 is the eigenvalue of $\mathcal{A}/r(\mathcal{A})$ with the same properties. Considering the Jordan form of $\mathcal{A}$, or the spectral decomposition (5.3.85), we see that

$$\lim_{k\to\infty} \frac{\mathcal{A}^k \boldsymbol{x}}{(r(\mathcal{A}))^k} = \langle \boldsymbol{f}, \boldsymbol{x}\rangle \boldsymbol{v} = \begin{pmatrix} f_1 v_1 & \dots & f_n v_1 \\ \vdots & \ddots & \vdots \\ f_1 v_n & \dots & f_n v_n \end{pmatrix} \begin{pmatrix} x_1 \\ \vdots \\ x_n \end{pmatrix},$$

where $\boldsymbol{f}$ and $\boldsymbol{v}$ are positive eigenvectors of, respectively, $\mathcal{A}^T$ and $\mathcal{A}$ belonging to $r(\mathcal{A})$, normalized so as $\langle \boldsymbol{f}, \boldsymbol{v}\rangle = 1$. Since $\boldsymbol{f} \gg 0, \boldsymbol{v} \gg 0$, the limit is strictly positive, thus $\frac{\mathcal{A}^m}{(r(\mathcal{A}))^m}$, and hence $\mathcal{A}^m$, must be positive for some m. □

Thus, a matrix $\mathcal{A}$ is imprimitive if and only if it has $h > 1$ eigenvalues on $C_{\mathcal{A}}$; h is called the imprimitivity index.

As we have seen, the Perron–Frobenius theorem for primitive matrices is exactly the same as for positive matrices. To provide a version of the Perron–Frobenius theorem for imprimitive matrices, we need the following result. Recall that for a matrix $\mathcal{B} = (b_{ij})_{1\leq i,j\leq d}$, $|\mathcal{B}| = (|b_{ij}|)_{1\leq i,j\leq d}$, and the same convention is applied to vectors.

Lemma C.17 *Let a matrix $\mathcal{B}$ satisfy $|\mathcal{B}| \leq \mathcal{A}$, where $\mathcal{A}$ is an irreducible matrix. Then $r(\mathcal{B}) \leq r(\mathcal{A})$. If $r(\mathcal{B}) = r(\mathcal{A})$, that is, there is, for some $\phi \in \mathbb{R}$, $\mu = r(\mathcal{A})e^{i\phi} \in \sigma(\mathcal{B})$, then there is a diagonal matrix*

$$\mathcal{D} = \begin{pmatrix} e^{i\theta_1} & 0 & \dots 0 & 0 \\ 0 & e^{i\theta_2} & \dots 0 & 0 \\ \vdots & \vdots & \ddots \vdots & \vdots \\ 0 & 0 & \dots 0 & e^{i\theta_d} \end{pmatrix}$$

for some $\theta_1, \dots, \theta_d$, such that

$$\mathcal{B} = e^{i\phi}\mathcal{D}\mathcal{A}\mathcal{D}^{-1}. \tag{C.3.28}$$

The converse statement is also valid.

Proof The formula $r(\mathcal{B}) \leq r(\mathcal{A})$ follows immediately by using the Gelfand formula, (C.2.15).

If $\boldsymbol{x}$ is an eigenvector of $\mathcal{B}$ belonging to μ satisfying $|\mu| = r = r(\mathcal{A})$, then

$$r|\boldsymbol{x}| = |\mu||\boldsymbol{x}| = |\mu\boldsymbol{x}| = |\mathcal{B}\boldsymbol{x}| \leq |\mathcal{B}||\boldsymbol{x}| \leq \mathcal{A}|\boldsymbol{x}|.$$

Assume $\mathcal{A}|\boldsymbol{x}| - r|\boldsymbol{x}| > 0$ and multiply from the left by the eigenvector $\boldsymbol{f} \gg 0$ of $\mathcal{A}^T$ belonging to r (that is, the left eigenvector of $\mathcal{A}$). Then

$$0 < \langle \boldsymbol{f}, \mathcal{A}|\boldsymbol{x}| - r|\boldsymbol{x}|\rangle = r\langle \boldsymbol{f}, |\boldsymbol{x}|\rangle - r\langle \boldsymbol{f}, |\boldsymbol{x}|\rangle = 0,$$

and we obtained a contradiction. Hence, $|\boldsymbol{x}| = \boldsymbol{v} \gg 0$ and therefore $\mathcal{A}|\boldsymbol{x}| - |\mathcal{B}||\boldsymbol{x}| = r|\boldsymbol{x}| - r|\boldsymbol{x}| = 0$. Since $\mathcal{A} \geq |\mathcal{B}| \geq 0$ and $|\boldsymbol{x}| \gg 0$, we obtain $\mathcal{A} = |\mathcal{B}|$.

Next, we observe that

$$\boldsymbol{x} = \mathcal{D}|\boldsymbol{x}|,$$

where $e^{i\theta_k} = \frac{x_k}{|x_k|}$ (well defined as $|\boldsymbol{x}| > 0$). Thus,

$$\mathcal{B}\mathcal{D}|\boldsymbol{x}| = \mathcal{B}\boldsymbol{x} = \mu\boldsymbol{x} = re^{i\phi}\boldsymbol{x} = e^{i\phi}r\mathcal{D}|\boldsymbol{x}| = e^{i\phi}\mathcal{D}\mathcal{A}|\boldsymbol{x}|.$$

Denote now $\mathcal{C} = e^{-i\phi}\mathcal{D}^{-1}\mathcal{B}\mathcal{D}$ and observe that $|\mathcal{C}| = |\mathcal{B}| = \mathcal{A}$. This allows for writing the above equality as

$$(|\mathcal{C}| - \mathcal{C})|\boldsymbol{x}| = 0.$$

In the above equation, only $\mathcal{C}$ can be complex and hence, in particular,

$$(|\mathcal{C}| - \Re\mathcal{C})|\boldsymbol{x}| = 0.$$

Since $\Re z < |z|$ for any complex z, we obtain $\Re\mathcal{C} \leq |\mathcal{C}|$ and, using $|\boldsymbol{x}| \gg 0$, we obtain

$$\Re C = |\mathcal{C}|.$$

Hence $\Im C = 0$, yielding $\mathcal{C} = |\mathcal{C}| = \mathcal{A}$. Using the definition of $\mathcal{C}$, we get

$$\mathcal{B} = e^{i\phi}\mathcal{D}\mathcal{A}\mathcal{D}^{-1}.$$

Conversely, (C.3.28) shows that $\mathcal{B}$ and $\mathcal{A}$ are similar and hence, have equal spectral radii. □

We are now ready to formulate and prove the final result in this section.

Theorem C.18 *If an irreducible matrix $\mathcal{A}$ with spectral radius $r = r(\mathcal{A})$ has h eigenvalues $\{\lambda_1, \ldots, \lambda_h\}$ on the spectral circle $C_{\mathcal{A}}$, then*

1. *the algebraic multiplicity of each λ_i, $1 \leq i \leq h$, is equal to 1,*
2. *$\{\lambda_1, \ldots, \lambda_h\} = \{r, r\omega, \ldots, \omega^{h-1}\}$, where $\omega = e^{i\frac{2\pi}{h}}$.*

Proof We apply Lemma C.3.28 to $\mathcal{B} = \mathcal{A}$ itself. We write $\lambda_j = re^{i\phi_j}$, $1 \leq j \leq h$, with $\lambda_1 = 1$. We can assume $0 < \phi_j < 2\pi$, $2 \leq j \leq h-1$. Consider some $\mu = \lambda_j = re^{i\phi_j}$, $\phi_j \neq 0$. Then there is a diagonal matrix $\mathcal{D}_{\phi_j}$ such that

$$\mathcal{A} = e^{i\phi_j}\mathcal{D}_{\phi_j}\mathcal{A}\mathcal{D}_{\phi_j}^{-1}. \tag{C.3.29}$$

Hence, $\mathcal{A}$ is similar to $e^{i\phi_j}\mathcal{A}$. Then, r is a simple eigenvalue of $\mathcal{A}$ if and only if $re^{i\phi_k}$ is a simple eigenvalue of $e^{i\phi_k}\mathcal{A}$. But, since the similarity transformation does not change the spectrum and its structure, $re^{i\phi_k}$ is a simple eigenvalue of $\mathcal{A}$ and the first statement is proved.

Next, repeating (C.3.29) for $re^{i\phi_k}$, we find

$$\mathcal{A} = e^{i(\phi_j+\phi_k)}\mathcal{D}_{\phi_j}\mathcal{D}_{\phi_k}\mathcal{A}\mathcal{D}_{\phi_k}^{-1}\mathcal{D}_{\phi_j}^{-1}, \tag{C.3.30}$$

and an argument as above shows that if $re^{i\phi_k}$ and $re^{i\phi_j}$ are eigenvalues on $C_{\mathcal{A}}$, then also $re^{i(\phi_k+\phi_j)}$ is an eigenvalue. Let $\phi = \min_{2\leq j\leq h-1}\phi_j$. Then $r\{1, e^{i\phi}, e^{i2\phi}, \ldots\}$ are eigenvalues of $\mathcal{A}$. First, let us assume that $\phi < 2\pi/h$. Then $h\phi < 2\pi$ and we would have more than h eigenvalues on $C_{\mathcal{A}}$ as all $e^{ik\phi}$, $0 \leq k \leq h$, are different. On the other hand, let $\phi > 2\pi/h$ and $k_0 < h$ be the largest index for which $k_0\phi < 2\pi$. Then, we have two cases. First, let $(k_0+1)\phi = 2\pi$. Then $e^{i(k_0+2)\phi} = e^{i\phi}$ and we start repeating the eigenvalues. However, in this case, we have $\phi = \frac{2\pi}{1+k_0}$ and, by assumption, $1 + k_0 < h$, which means we have at most $h-1$ eigenvalues. In the second case, $(k_0+1)\phi > 2\pi$ and $e^{i(k_0+1)\phi} = e^{i\varphi}$ for some $0 < \varphi < \phi$ and $re^{i\varphi} \in \sigma(\mathcal{A})$. But this contradicts the definition of ϕ. Therefore $\phi = 2\pi/h$ and the eigenvalues are $\{r, re^{2\pi i\frac{1}{h}}, re^{2\pi i\frac{2}{h}}, \ldots, re^{2\pi i\frac{h-1}{h}}\}$. □

C.3.2 Metzler Matrices

As we noted earlier, see Theorem 5.30 and Remark 5.31, in continuous dynamical systems, Metzler matrices play an analogous role to non-negative matrices in discrete systems. They are the only generators of nonnegative systems, whose dynamics is given then by matrix exponentials, and their long-term behaviour is, by (5.3.97), determined by the spectral bound, (5.3.51), in contrast to the spectral radius in discrete-time systems. A Metzler matrix $\mathcal{B}$ can always be written as

$$\mathcal{B} = \mathcal{A} - z\mathcal{I},$$

where $\mathcal{A} \geq 0$, and z is any real number not smaller than $\max_{1\leq i\leq n} |b_{ii}|$. Since the spectrum of $\mathcal{B}$ is the shift of $\mathcal{A}$ and $\rho(\mathcal{A}) = s(\mathcal{A})$, see Fig. C.1, Propositions C.10 and C.12 immediately imply

Theorem C.19 *Let $\mathcal{B}$ be a Metzler matrix. Then $s(\mathcal{B}) \in \sigma(\mathcal{B})$ and there are nonnegative left and right eigenvectors belonging to $s(\mathcal{B})$.*

Further results for Metzler matrices can be obtained if we impose the irreducibility assumption.

Definition C.20 A Metzler matrix $\mathcal{B}$ is irreducible if the non-negative matrix $\mathcal{A}$ is irreducible.

Using Definition 5.32, it is easy to see that the irreducibility of a matrix does not depend on the diagonal entries (as they do not provide a connection between different states). Hence, Definition C.20 does not depend on z. We also see no need to further subdivide the class of irreducible Metzler matrices into periodic (imprimitive) and primitive (aperiodic) matrices. It is clear from Fig. C.1, that if $\mathcal{B}$ is irreducible, then it has a simple eigenvalue $\lambda_{\max}$ and, even if $\mathcal{B}$ is imprimitive, $\lambda_{\max} - z > \Re\lambda$ for all other eigenvalues λ of $\mathcal{B}$. Precisely, we have, [85, Theorem 2.6],

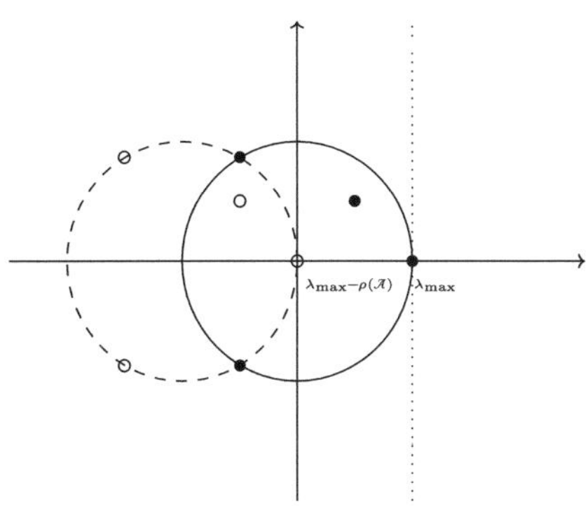

Fig. C.1 Spectra of $\mathcal{A}$ and $\mathcal{B}$. We see that even if the spectrum of $\mathcal{A}$ (full circles) has a periodic part on the spectral circle, the spectrum of its shift to the left by $\rho(\mathcal{A})$ (empty circles) $\mathcal{B}$ cannot have elements with real part $\lambda_{\max}$ other than $\lambda_{\max}$ and thus cannot give rise to periodic behaviour of the continuous-time system generated by $\mathcal{B}$

Theorem C.21 *Let* $\mathcal{B}$ *be an irreducible Metzler matrix. Then* $\lambda_{\max} = s(\mathcal{A})$ *is a simple eigenvalue of* $\mathcal{B}$ *to which there belong unique (up to scalar multipliers) strictly positive right and left eigenvectors.*

Many useful results of Metzler matrices can be proven using the theory of dynamical systems, and thus, they are considered in Sect. 8.1.

Appendix D
First Order Partial Differential Equations and the Method of Characteristics

We came across first-order partial differential equations in Sect. 5.5 and Chaps. 9 and 10, and hence they deserve a more detailed treatment.

D.1 One-Dimensional Problems

We start with the simplest transport equation

$$au_t + bu_x = 0, \quad t, x \in \mathbb{R}, \tag{D.1.1}$$

where a and b are constants. This equation can be written as

$$D_{\boldsymbol{v}} u = 0, \tag{D.1.2}$$

where $\boldsymbol{v} = a\boldsymbol{e}_1 + b\boldsymbol{e}_2$ (where here $\boldsymbol{e}_1$ and $\boldsymbol{e}_2$ are the unit vectors in, respectively, t and x directions), and $D_{\boldsymbol{v}} = \nabla u \cdot \boldsymbol{v}$ denotes the directional derivative in the direction of $\boldsymbol{v}$, compare Sect. 9.1. This means that the solution u is a constant function along each line having direction $\boldsymbol{v}$, that is, along each line of the equation $bt - ax = \xi$. Along each such line, the value of the parameter ξ remains constant. However, the solution can change from one line to another. Therefore, the solution is a function of ξ, that is, the solution to Eq. (D.1.1) is given by

$$u(t, x) = f(bt - ax), \tag{D.1.3}$$

where f is an arbitrary differentiable function. Such lines are called *characteristics* of the equation.

J. Banasiak, *Introduction to Mathematical Methods in Population Theory*, Springer Undergraduate Mathematics Series,
https://doi.org/10.1007/978-3-031-65491-6

Example D.1 Let us assume that $a \neq 0$, that is, there is a nontrivial temporal part of the problem. To obtain a unique solution, we must specify the initial value for u. Hence, let us consider the initial value problem for Eq. (D.1.1): find u satisfying

$$\begin{aligned} au_t + bu_x &= 0 \quad x \in \mathbb{R}, t > 0, \\ u(0, x) &= g(x), \quad x \in \mathbb{R}, \end{aligned} \tag{D.1.4}$$

where g is an arbitrary given function. From Eq. (D.1.3) we find that

$$u(t, x) = g\left(-\frac{bt - ax}{a}\right). \tag{D.1.5}$$

The initial shape propagates without any change along the characteristics, as seen below in Figs. D.1 and D.2 for the initial function $g = 1 - x^2$ for $|x| < 1$ and zero elsewhere. The speed $c = b/a$ is taken to be 1.

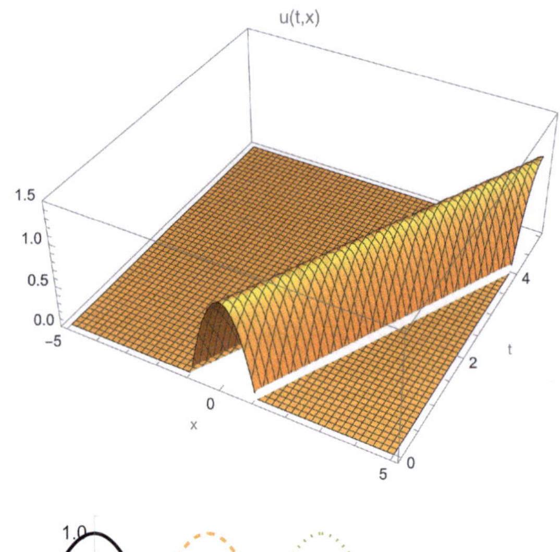

Fig. D.1 The graph of the solution in Example D.1

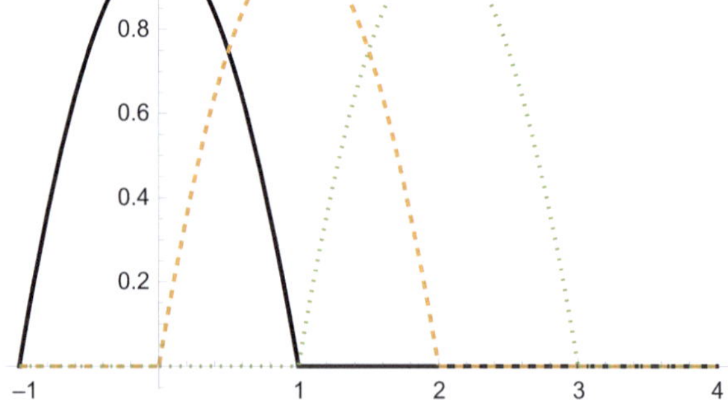

Fig. D.2 Travelling wave profile of the solution in Example D.1 for $t = 0$ (solid), $t = 1$ (dashed), $t = 2$ (dotted)

Example D.2 Let us consider a variation of this problem and solve the initial–boundary value problem

$$\begin{aligned} au_t + bu_x &= 0, \quad x, t > 0, \\ u(x, 0) &= g(x), \quad x > 0, \\ u(t, 0) &= h(t), \quad t > 0, \end{aligned} \tag{D.1.6}$$

for $a, b > 0$. From Example D.1, we have the general solution of the equation in the form

$$u(t, x) = f(bt - ax).$$

Putting $t = 0$ we get $f(-ax) = g(x)$ for $x > 0$, hence $f(x) = g\left(-\frac{x}{a}\right)$ for $x < 0$. Next, for $x = 0$ we obtain $f(bt) = h(t)$ for $t > 0$, hence $f(x) = h\left(\frac{x}{b}\right)$ for $x > 0$. Combining these two equations, we obtain

$$u(t, x) = \begin{cases} g(-\frac{bt-ax}{a}) \text{ for } x > \frac{bt}{a}, \\ h(\frac{bt-ax}{b}) \text{ for } x < \frac{bt}{a}, \end{cases}$$

as illustrated in Fig. D.3.

Example D.3 Assigning the boundary condition in transport problems requires some care. To explain, let us consider what happens if $a = 1 > 0$, $b = -1 < 0$. Then, the initial condition defines $f(x) = g(-x)$ for $x < 0$, and the boundary condition gives $f(x) = h(-x)$ also for $x < 0$! Hence, we cannot specify both initial and boundary conditions in an arbitrary way, as this could make the problem ill-posed.

Fig. D.3 The solution of the initial–boundary problem (D.1.6) with $a = b$, $g(x) = x$ and $h(t) = 1 - e^{-t}$ in Example D.2

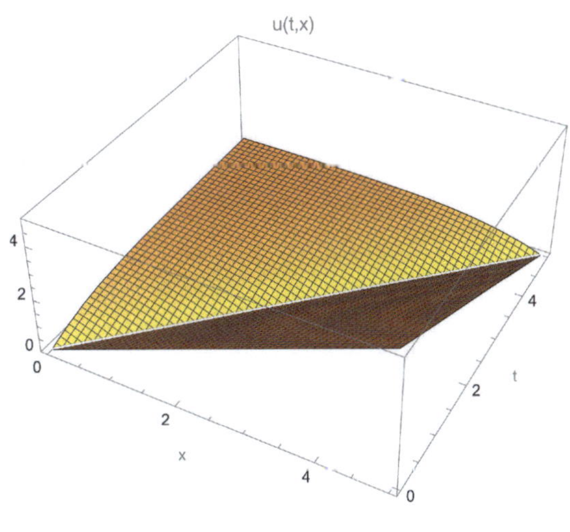

The physical explanation of this comes from the observation that since the characteristics are given by $\xi = x + t$, the flow occurs in the negative direction, and therefore, the values at $x = 0$ for any t are uniquely determined by the initial condition. Therefore, to have a well-posed problem, we must specify the boundary conditions at the point where the medium flows into the region.

The method we have just used is often called the geometric method. It is easy to understand but sometimes difficult to apply, especially for non-homogeneous problems. Fortunately, this method can be easily reformulated in a more analytic language. Let us introduce the change of variables according to $\xi = \xi(t, x), \ \eta = \eta(t, x)$, and define $v(\xi, \eta) = u(t, x)$. Then,

$$u_t = v_\xi \xi_t + v_\eta \eta_t, \quad u_x = v_\xi \xi_x + v_\eta \eta_x,$$

and the equation can be written as

$$a(v_\xi \xi_t + v_\eta \eta_t) + bv_\xi \xi_x + v_\eta \eta_x) = v_\xi(a\xi_t + b\xi_x) + v_\eta(a\eta_t + b\eta_x) = 0.$$

If we require the coefficient at v_η to be zero, the easiest way is to introduce $\eta_t = b, \ \eta_x = -a$, that is $\eta = bt - ax$. Note that this is exactly the characteristic direction! However, this is an incomplete change of variables as originally we had two independent variables t, x, but we have only one new variable η. That means that knowing η alone, we cannot determine the values of x and t. The trick is to introduce another variable, say, $\xi = \xi(x, t)$ in such a way that the system

$$\eta = bt - ax, \quad \xi = \xi(x, t),$$

is uniquely solvable. To keep things as simple as possible, we may take $\xi(x, t) = ct + dx$ with c, d selected so as $bd + ac \neq 0$. If a or b are not equal to zero, the easiest choice would be either $\xi = t$ or $\xi = x$, respectively. However, sometimes, it is more convenient to use the orthogonal lines given by $\xi = at + bx$.

We illustrate this approach in the following example.

Example D.4 Find the solution to the following equation

$$u_t + 2u_x - (x + t)u = -(x + t),$$

which satisfies the initial condition

$$u(0, x) = h(x), \quad x > 0,$$

and

$$u(t, 0) = g(t), \quad t > 0.$$

Introducing new variables according to $\xi = t, \ \eta = 2t - x$, we transform the equation into

$$v_\xi - (3\xi - \eta)v = -(3\xi - \eta).$$

This equation can be regarded as a linear first-order ordinary differential equation in ξ with parameter η.

To find the integrating factor μ, we solve the homogeneous equation

$$\mu_\xi = -(3\xi - \eta)\mu;$$

the integration gives

$$\mu(\xi, \eta) = e^{-\left(\frac{3}{2}\xi^2 - \eta\xi\right)}.$$

Multiplying both sides of the equation by μ and rearranging terms, we obtain

$$\left(e^{-\left(\frac{3}{2}\xi^2 - \eta\xi\right)} v(\xi, \eta)\right)_\xi = -(3\xi - \eta)e^{-\left(\frac{3}{2}\xi^2 - \eta\xi\right)}$$

Hence, the general solution is given by

$$v(\xi, \eta) = 1 + C(\eta)e^{\left(\frac{3}{2}\xi^2 - \eta\xi\right)},$$

where C is an arbitrary differentiable function of one variable.

In the original variables, we obtain

$$u(t, x) = 1 + C(2t - x)e^{\left(-\frac{t^2}{2} + tx\right)},$$

where C is an arbitrary differentiable function of one variable.

Now, we use the initial and boundary conditions. First,

$$h(x) = u(0, x) = 1 + C(-x)$$

for $x > 0$. To avoid misunderstanding, we introduce the variable s as the argument of C. Thus

$$C(s) = h(-s) - 1, \quad s < 0.$$

On the other hand,

$$g(t) = u(t,0) = 1 + C(2t)e^{-\frac{t^2}{2}},$$

thus

$$C(2t) = e^{\frac{t^2}{2}}(g(t) - 1).$$

Introducing $s = 2t$, $s > 0$ and

$$C(s) = e^{\frac{s^2}{8}}\left(g\left(\frac{s}{2}\right) - 1\right).$$

Thus, we have defined C for all values of the argument by

$$C(s) = \begin{cases} h(-s) - 1 & \text{for } s < 0, \\ e^{\frac{s^2}{8}}\left(g\left(\frac{s}{2}\right) - 1\right) & \text{for } s > 0. \end{cases}$$

In the solution to the given problem, the function C is evaluated at $s = 2t - x$, so C has different definitions for $2t - x > 0$ and $2t - x < 0$. Therefore, the solution to the given initial–boundary value problem is given by

$$u(t,x) = \begin{cases} 1 + (h(-2t + x) - 1)e^{\left(-\frac{t^2}{2}+tx\right)} & \text{for } x > 2t, \\ 1 + e^{\left(\frac{x^2+tx}{8}\right)}\left(g\left(t - \frac{x}{2}\right) - 1\right) & \text{for } x < 2t. \end{cases}$$

For example, if $h(x) = x$ and $g(t) = \sin t$, then the solution is given by

$$u(t,x) = \begin{cases} 1 + (-2t + x - 1)e^{\left(-\frac{t^2}{2}+tx\right)} & \text{for } x > 2t, \\ 1 + e^{\left(\frac{x^2+tx}{8}\right)}\left(\sin\left(t - \frac{x}{2}\right) - 1\right) & \text{for } x < 2t. \end{cases} \tag{D.1.7}$$

In the figures below, Figs. D.4 and D.5, note how the analytic description of the solution changes across the characteristic $x - 2t = 0$.

The same principle can be used to solve equations with variable coefficients. Consider the equation

$$a(t,x)u_t + b(t,x)u_x = 0. \tag{D.1.8}$$

This equation states that the derivative of u in the direction of the vector $(b(t,x), a(t,x))$ is equal to zero at each point (x,t), whenever this vector is not vanishing. We consider a family of curves $(t(s), x(s))$, $0 \le s < \infty$, which are

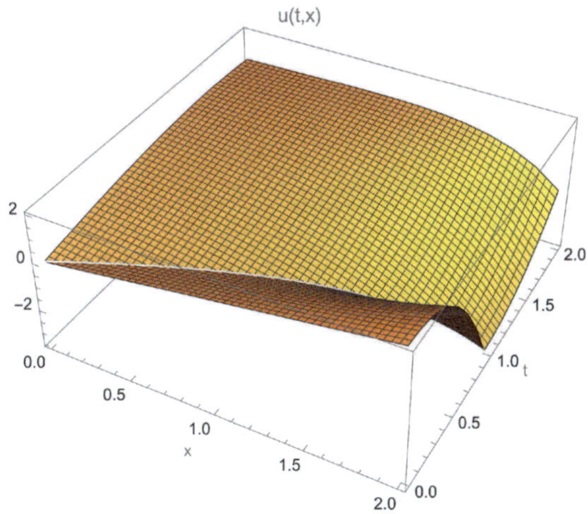

Fig. D.4 The graph of the solution (D.1.7) in Example D.4

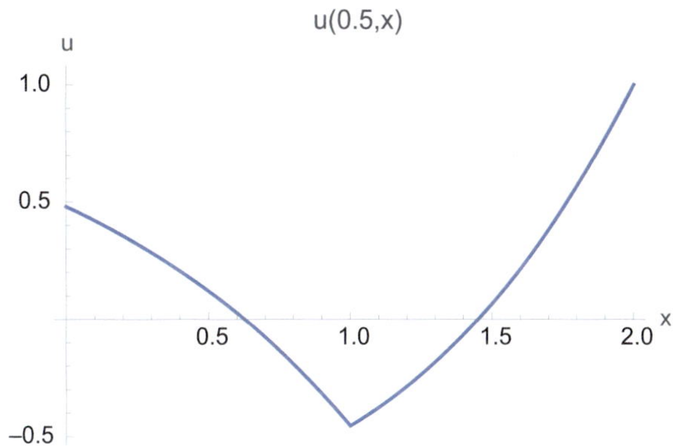

Fig. D.5 Snapshot of the solution (D.1.7) in Example D.4 at $t = 0.5$

tangent to these vectors at each point. In other words, these curves are solutions to the system

$$\begin{aligned} \frac{dx(s)}{ds} &= b(t(s), x(s)), \quad x(0) = \xi \\ \frac{dt(s)}{ds} &= a(t(s), x(s)), \quad t(0) = 0, \end{aligned} \tag{D.1.9}$$

where $\xi \in \mathbb{R}$. Then, for each ξ, we denote $v^\xi(s) := u(t(s), x(s))$ for $(x(s), t(s))$ satisfying (D.1.9). Differentiating, we obtain

$$\begin{aligned} \frac{dv^\xi(s)}{ds} &= u_t(t(s), x(s))t'(s) + u_x(t(s), x(s))x'(s) \\ &= u_t(t(s), x(s))a(t(s), x(s)) + u_x(t(s), x(s))b(t(s), x(s)) = 0. \end{aligned} \tag{D.1.10}$$

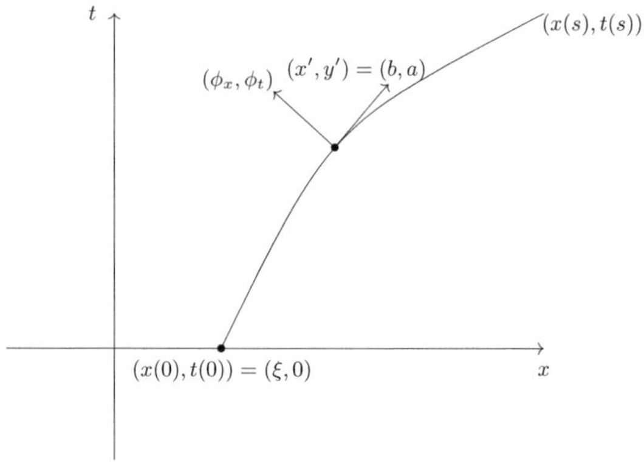

Fig. D.6 A characteristic line determined by the initial value $(\xi, 0)$

Thus, v^ξ is constant for each ξ; hence, $u(t, x)$ is constant along each trajectory defined by ξ. To express this fact in more mathematical terms, assume that the characteristics, defined by (D.1.9), can be written in an implicit form as $\phi(t, x) = \eta\ (= \phi(0, \xi))$, where η is an arbitrary constant. We know that the tangent vector at a point (x, t) of the characteristic is given by $(b(t, x), a(t, x))$ and the normal vector is given by the gradient of ϕ: $(\phi_x(t, x), \phi_t(t, x))$, see Fig. D.6. Then, for arbitrary differentiable f, the function

$$u(t, x) = f(\phi(t, x)), \tag{D.1.11}$$

solves (D.1.8). Indeed, inserting such u into (D.1.8) we obtain

$$au_t + bu_x = f' \cdot (a\phi_t + b\phi_x) = 0,$$

due to the orthogonality property of (ϕ_x, ϕ_t), which was mentioned above.

Extending the above considerations to the nonhomogeneous cases can be done along lines similar to the constant coefficient case. However, some difficulties can occur as $\eta = \phi(x, t)$ does not always give rise to a well-defined change of variables, for instance, if $a(t, x)$ vanishes for some values t or x. Thus, we shall present an alternative approach to solving inhomogeneous versions of (D.1.8) under the assumption $a \neq 0$, that is, that it can be written in the form

$$\begin{aligned} u_t + c(t, x)u_x &= f(t, x, u), \\ u(0, x) &= u_0(x). \end{aligned} \tag{D.1.12}$$

In this case, we have $s = t$, and then, as we explained above, the left-hand side of the equation is the derivative of u along characteristics defined by the equation $x' = c(x, t)$. Thus, we obtain the so-called characteristic system

$$\begin{aligned} v' &= f(t, x, v), \quad v(0) = u_0(\xi), \\ x' &= c(t, x), \quad x(0) = \xi, \end{aligned} \tag{D.1.13}$$

$\xi \in \mathbb{R}$. Here, as above, $v(t) = u(t, x(t))$, so $v(t)$ is $u(t, x)$ evaluated along a given characteristic. This system is a version of (D.1.9) and (D.1.10) for models with sources.

In principle, (D.1.13) can be solved. The second equation is independent of the first, and it determines the equation of characteristics $x = x(t, \xi)$. This solution can be substituted into the first equation, the solution of which gives the values of u as a function of t, a parameter along a characteristic, and the parameter ξ. Assuming that only one characteristic passes through each (x, t) and that it determines a unique $\xi = \xi(x, t)$ at which it crosses the x-axis, we can eliminate ξ and produce the solution u in terms of x and t only.

We emphasize that the above discussion presents a technique of obtaining a solution whose availability depends on being able to construct an invertible characteristic change of variables so that through any point (t, x), there passes a unique characteristic line allowing for the determination of ξ. This assumption is not always satisfied, especially if f depends also on u, see, e.g., [65, Chapter 2].

We illustrate these two methods in the example below.

Example D.5 Find the solution to the following initial value problem

$$u_t + xu_x + u = 0, \quad t > 0, x \in \mathbb{R},$$
$$u(0, x) = u_0(x),$$

where $u_0(x) = 1 - x^2$ for $|x| < 1$, and $u_0(x) = 0$ for $|x| \geq 1$.

The differential equation for characteristic curves is

$$x' = x,$$

which gives $x = \eta e^t$, thus $\eta = xe^{-t} = \phi(t, x)$, see Fig. D.7. If the equation did not contain the zero-order term u, that is, if it was of the form

$$u_t + xu_x = 0,$$

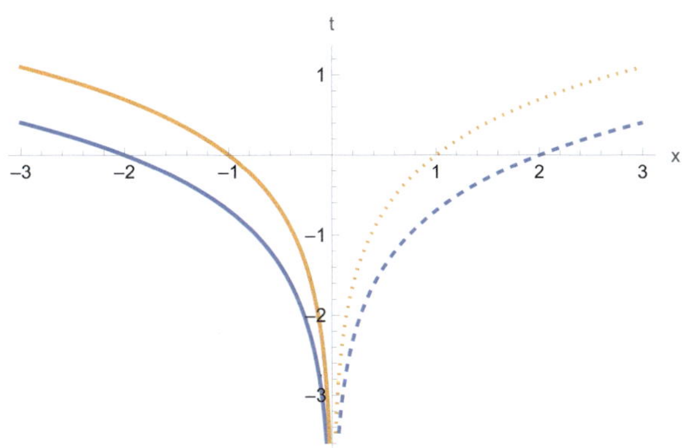

Fig. D.7 Characteristics of the equation in Example D.5

then the general solution would have had the form

$$u(t,x)=f(xe^{-t}),$$

for arbitrary function f.

However, since we have the additional term, we must perform the full change of variables. Fortunately, putting $\xi=t$ and $\eta=xe^{-t}$ produces an invertible change of variables, as $t=\xi$ and $x=e^{\xi}\eta$.

Defining $v(\xi,\eta)=u(x,t)$, we obtain that

$$0=u_t+xu_x+u=v_\xi\xi_t+v_\eta\eta_t+x(v_\xi\xi_x+v_\eta\eta_x)+v=v_\xi+v.$$

Therefore, we obtain the solution

$$v(\xi,\eta)=C(\eta)e^{-\xi},$$

that is,

$$u(t,x)=C(xe^{-t})e^{-t},$$

where C is an arbitrary function. Putting $t=0$, we see that $u(x,0)=C(x)=u_0(x)$, so that

$$u(x,t)=u_0(xe^{-t})e^{-t}.$$

For the chosen initial condition, we obtain

$$u(t,x)=\begin{cases}(1-x^2e^{-2t})e^{-t} & \text{for } |x|\le e^t,\\ 0 & \text{for } |x|>e^t,\end{cases} \tag{D.1.14}$$

see Figs. D.8 and D.9.

To use the characteristic system method, we write the characteristic system,

$$\begin{aligned}v'&=-v,\\ x'&=x,\end{aligned} \tag{D.1.15}$$

together with the initial conditions $v(0)=u_0(\xi)$, $x(0)=\xi$, where ξ is the intercept of the characteristic and the x-axis. The solution of (D.1.15) is

$$v(t)=ae^{-t},\qquad x(t)=be^t,$$

where a,b are integration constants. Using the initial conditions we get

$$v(0)=a=u_0(\xi),\qquad x(0)=b=\xi.$$

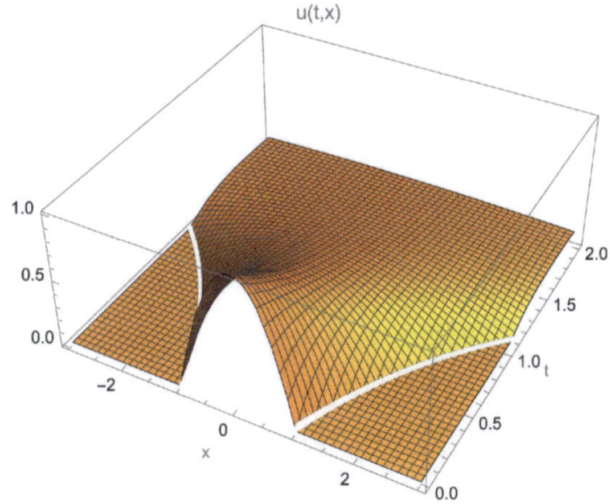

Fig. D.8 The graph of the solution (D.1.14)

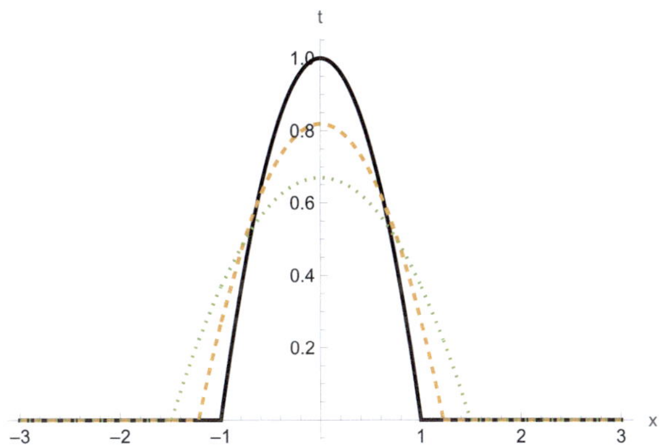

Fig. D.9 Snapshots of the solution (D.1.14) at times $t = 0$ (solid), $t = 0.2$ (dashed) and $t = 0.4$ (dotted)

Thus, $x(t) = \xi e^t$ and, eliminating ξ, we get

$$u(x, t) = u_0(xe^{-t})e^{-t},$$

in accordance with the formula derived by the other method.

D.2 Multidimensional Problems

The described procedure is not restricted to equations in two independent variables. Let us consider the equation

$$au_t + bu_x + cu_y = 0. \tag{D.2.16}$$

This equation expresses the fact that the directional derivative in the direction of the vector (a, b, c) is equal to zero, that is, that the solution does not change along any line with parametric equation $t = t_0 + as, x = x_0 + bs, y = y_0 + cs, s \in \mathbb{R}$. Such a line can also be written as a pair of equations $ax - bt = \xi, cy - bt = \eta$. For each pair ξ, η, the pair of equations describes a single line parallel to the vector (a, b, c), that is, the solution u can be a function of ξ and η only. Thus, we obtain the following general solution to (D.2.16) as

$$u(t, x, y) = f(ax - bt, ay - ct), \tag{D.2.17}$$

where f is an arbitrary differentiable function of two variables.

Example D.6 Find the solution to the following initial value problem:

$$\begin{aligned} u_t + 2u_x + 3u_y + u = 0, \quad (x, y) \in \mathbb{R}^2, t > 0, \\ u(0, x, y) = u_0(x, y) = e^{-x^2-y^2}. \end{aligned} \tag{D.2.18}$$

We use the change of coordinates suggested by the characteristics:

$$\xi = x - 2t, \quad \eta = y - 3t,$$

supplemented by

$$\tau = t,$$

so that the change of variables

$$t = \tau, \quad x = \xi + 2\tau, \quad y = \eta - 3\tau,$$

is invertible. Then, putting $u(t, x, y) = v(\tau, \xi, \eta)$, we find that

$$\begin{aligned} u_t &= -2v_\xi - 3v_\eta + v_\tau, \\ u_x &= v_\xi, \\ u_y &= v_\eta, \end{aligned}$$

so that

$$0 = u_t + 2u_x + 3u_y + u = v_\tau + v.$$

The last equation has the general solution

$$v(\tau, \xi, \eta) = C(\xi, \eta)e^{-\tau},$$

so that

$$u(t, x, y) = C(x - 2t, y - 3t)e^{-t}. \tag{D.2.19}$$

To solve the initial value problem, we put $t = 0$ to get

$$e^{-x^2-y^2} = C(x, y),$$

so that the solution of the initial value problem takes the form

$$u(x, y, t) = e^{-(x-2t)^2-(y-3t)^2}e^{-t},$$

see Fig. D.10.

Example D.7 The method of the characteristic system can be also easily adapted for equations in higher dimensions. In such a case, we obtain a system of $n + 1$

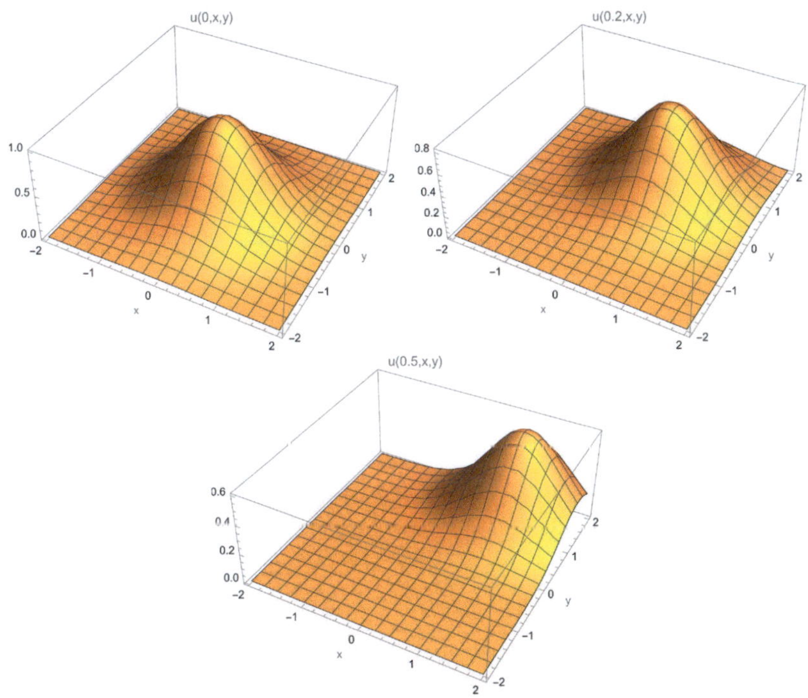

Fig. D.10 Snapshots of the solution in Example D.6 for $t = 0, 0.2$ and 0.5

equations, where n is the number of space variables. For instance, the characteristic system for the equation from the previous example is of the form

$$v' = -v,$$
$$x' = 2,$$
$$y' = 3,$$

with initial conditions $v(0) = u_0(\xi, \eta)$, $x(0) = \xi$, $y(0) = \eta$. Then, we obtain

$$v(t) = u_0(\xi, \eta)e^{-t}, \quad x = 2t + \xi, \quad y = 3t + \eta,$$

where ξ, η are constants. Eliminating ξ and η, we obtain

$$u(t, x, y) = u_0(x - 2t, y - 3t)e^{-t},$$

in agreement with (D.2.19).

References

1. Y.A. Abramovich, C.D. Aliprantis, *An Invitation to Operator Theory*. Graduate Studies in Mathematics, vol. 50 (American Mathematical Society, Providence, 2002)
2. R.A. Adams, J.J.F. Fournier, *Sobolev Spaces*. Pure and Applied Mathematics, vol. 140, 2nd edn. (Academic, New York, 2003)
3. E.S. Allman, J.A. Rhodes, *Mathematical Models in Biology: An Introduction* (Cambridge University Press, Cambridge, 2004)
4. H. Amann, *Ordinary Differential Equations*. de Gruyter Studies in Mathematics, vol. 13 (Walter de Gruyter & Co., Berlin, 1990)
5. H. Amann, J. Escher, *Analysis. I* (Birkhäuser Verlag, Basel, 2005)
6. H. Amann, J. Escher, *Analysis. II* (Birkhäuser Verlag, Basel, 2008)
7. R. Anguelov, Y. Dumont, J. Lubuma, Mathematical modeling of sterile insect technology for control of anopheles mosquito. Comput. Math. Appl. **64**(3), 374–389 (2012)
8. W. Arendt, A. Grabosch, G. Greiner, U. Groh, H.P. Lotz, U. Moustakas, R. Nagel, F. Neubrander, U. Schlotterbeck, *One-Parameter Semigroups of Positive Operators*. Lecture Notes in Mathematics, vol. 1184 (Springer, Berlin, 1986)
9. W. Arendt, C.J.K. Batty, M. Hieber, F. Neubrander, *Vector-Valued Laplace Transforms and Cauchy Problems*. Monographs in Mathematics, vol. 96, 2nd edn. (Birkhäuser/Springer Basel AG, Basel, 2011)
10. J. Banasiak, Kinetic models in natural sciences, in *Evolutionary Equations with Applications in Natural Sciences*. Lectures presented at the 2013 CIMPA-UNESCO-South Africa School, Cape Town, July 22–August 2, 2013 (Springer, Cham, 2015), pp. 133–198
11. J. Banasiak, Population models with projection matrix with some negative entries-a solution to the Natchez paradox. Bull. South Ural State Univ. Ser.: Math. Modell. Program. Comput. Softw. **11**(3), 18–28 (2018)
12. J. Banasiak, A note on the Tikhonov theorem on an infinite interval. Vietnam J. Math. **49**(1), 69–86 (2021)
13. J. Banasiak, L. Arlotti, *Perturbations of Positive Semigroups with Applications*. Springer Monographs in Mathematics (Springer-Verlag London, Ltd., London, 2006)
14. J. Banasiak, M. Lachowicz, *Methods of Small Parameter in Mathematical Biology*. Modeling and Simulation in Science, Engineering and Technology (Birkhäuser/Springer, Cham, 2014)
15. J. Banasiak, R.Y.M. Massoukou, Solvability of age-structured epidemiological models with intracohort transmission. Mediterr. J. Math. **12**(4), 1307–1321 (2015)

J. Banasiak, *Introduction to Mathematical Methods in Population Theory*, Springer Undergraduate Mathematics Series,
https://doi.org/10.1007/978-3-031-65491-6

16. J. Banasiak, S. Tchoumi, Multiscale malaria models and their uniform in-time asymptotic analysis. Math. Comput. Simul. **221**, 1–21 (2024)
17. J. Banasiak, R. Ouifki, W.A. Woldegerima, Some mathematical tools for modelling malaria: a subjective survey. BIOMATH **10**(2):Paper No. 2110029, 19 (2021)
18. J. Banasiak, Y. Dumont, I.V. Yatat Djeumen, Spreading speeds and traveling waves for monotone systems of impulsive reaction-diffusion equations: application to tree-grass interactions in fire-prone savannas. Differ. Equ. Dyn. Syst. **31**(3), 547–580 (2023)
19. J. Banasiak, K. Szymańska-Dębowska, *Układy dynamiczne w modelowaniu procesów przyrodniczych, społecznych i technologicznych* (PWN Polish Scientific Publisher, Warsaw, 2023)
20. A.T. Bharucha-Reid, *Elements of the Theory of Markov Processes and Their Applications*. McGraw-Hill Series in Probability and Statistics (McGraw-Hill Book Co., Inc., New York-Toronto-London, 1960)
21. A. Bobrowski, *Functional Analysis for Probability and Stochastic Processes* (Cambridge University Press, Cambridge, 2005)
22. Å. Brännström, D.J. Sumpter, The role of competition and clustering in population dynamics. Proc. R. Soc. B: Biol. Sci. **272**(1576), 2065–2072 (2005)
23. F. Brauer, C. Castillo-Chávez, *Mathematical Models in Population Biology and Epidemiology*. Texts in Applied Mathematics, vol. 40 (Springer, New York, 2001)
24. F. Brauer, P. Van den Driessche, J. Wu, L.J. Allen, *Mathematical Epidemiology*, vol. 1945 (Springer, Berlin, 2008)
25. M. Braun, *Differential Equations and Their Applications*. Texts in Applied Mathematics, vol. 11, 4th edn. (Springer, New York, 1993)
26. H. Brezis, *Functional Analysis, Sobolev Spaces and Partial Differential Equations*. Universitext (Springer, New York, 2011)
27. N.F. Britton, *Essential Mathematical Biology*. Springer Undergraduate Mathematics Series (Springer-Verlag London, Ltd., London, 2003)
28. K. Bryan, T. Leise, The $25,000,000,000$ eigenvector: the linear algebra behind google. SIAM Rev. **48**(3), 569–581 (2006)
29. N. Chitnis, J.M. Cushing, J.M. Hyman, Bifurcation analysis of a mathematical model for malaria transmission. SIAM J. Appl. Math. **67**(1), 24–45 (2006)
30. W.A. Coppel, *Stability and Asymptotic Behavior of Differential Equations* (D. C. Heath and Company, Boston, 1965)
31. R. Courant, F. John, *Introduction to Calculus and Analysis. Vol. I* (Springer, New York, 1989). Reprint of the 1965 edition
32. R. Courant, F. John, *Introduction to Calculus and Analysis. Vol. II* (Springer, New York, 1989)
33. J. Cronin, *Ordinary Differential Equations*. Pure and Applied Mathematics (Boca Raton), vol. 292, 3rd edn. (Chapman & Hall/CRC, Boca Raton, 2008)
34. P. Cull, M. Flahive, R. Robson, *Difference Equations*. Undergraduate Texts in Mathematics (Springer, New York, 2005)
35. J.M. Cushing, *An Introduction to Structured Population Dynamics*, CBMS-NSF Regional Conference Series in Applied Mathematics, vol. 71 (Society for Industrial and Applied Mathematics (SIAM), Philadelphia, 1998)
36. G. de Vries, T. Hillen, M. Lewis, J. Müller, B. Schönfisch, *A Course in Mathematical Biology*. Mathematical Modeling and Computation, vol. 12. (Society for Industrial and Applied Mathematics (SIAM), Philadelphia, 2006)
37. M.R. Dennis, P. Glendinning, P.A. Martin, F. Santosa, J. Tanner (eds.), *The Princeton Companion to Applied Mathematics* (Princeton University Press, Princeton, 2015)
38. O. Diekmann,J.A.P. Heesterbeek, *Mathematical Epidemiology of Infectious Diseases*. Wiley Series in Mathematical and Computational Biology (Wiley, Chichester, 2000)
39. O. Diekmann, J.A.P. Heesterbeek, J.A.J. Metz, On the definition and the computation of the basic reproduction ratio R_0 in models for infectious diseases in heterogeneous populations. J. Math. Biol. **28**(4), 365–382 (1990)

40. G. Doetsch, *Introduction to the Theory and Application of the Laplace Transformation* (Springer-Verlag, Berlin-Heidelberg-New York, 1974)
41. H. Dym, *Linear Algebra in Action*. Graduate Studies in Mathematics, vol. 78, 2nd edn. (American Mathematical Society, Providence, 2013)
42. S. Elaydi, *An Introduction to Difference Equations*. Undergraduate Texts in Mathematics, 3rd edn. (Springer, New York, 2005)
43. A. Fall, A. Iggidr, G. Sallet, J.J. Tewa, Epidemiological models and Lyapunov functions. Math. Model. Nat. Phenom. **2**(1), 55–73 (2007)
44. A.F. Filippov, An elementary proof of Jordan's theorem. Uspehi Matem. Nauk (N.S.) **5**(5(39)), 173–176 (1950)
45. R.A. Fisher, The wave of advance of advantageous genes. Ann. Eugen. **7**(4), 355–36 (1937)
46. F.R. Gantmacher, *Applications of the Theory of Matrices* (Interscience Publishers, Inc., New York; Interscience Publishers Ltd., London, 1959)
47. G.F. Gause, *The Struggle for Existence: A Classic of Mathematical Biology and Ecology* (Courier Dover Publications, Mineola, 2019)
48. P. Glendinning, *Stability, Instability and Chaos: An Introduction to the Theory of Nonlinear Differential Equations*. Cambridge Texts in Applied Mathematics (Cambridge University Press, Cambridge, 1994)
49. H.C.J. Godfray, *Parasitoids: Behavioral and Evolutionary Ecology*, vol. 67 (Princeton University Press, Princeton, 1994)
50. T. Gowers, J. Barrow-Green, I. Leader, *The Princeton Companion to Mathematics* (Princeton University Press, Princeton, 2010)
51. W.M. Haddad, V. Chellaboina, Q. Hui, *Nonnegative and Compartmental Dynamical Systems* (Princeton University Press, Princeton, 2010)
52. J.K. Hale, H. Koçak, *Dynamics and Bifurcations*. Texts in Applied Mathematics, vol. 3 (Springer, New York, 1991)
53. H.W. Hethcote, H.R. Thieme, Stability of the endemic equilibrium in epidemic models with subpopulations. Math. Biosci. **75**(2), 205–227 (1985)
54. M.W. Hirsch, S. Smale, *Differential Equations, Dynamical Systems, and Linear Algebra*, vol. 60 (Academic Press [Harcourt Brace Jovanovich, Publishers], New York-London, 1974)
55. M. Iannelli, *Mathematical Theory of Age-Structured Population Dynamics* (Giardini editori e stampatori in Pisa, 1995)
56. J.C. Kamgang, G. Sallet, Computation of threshold conditions for epidemiological models and global stability of the disease-free equilibrium (DFE). Math. Biosci. **213**(1), 1–12 (2008)
57. C.W. Kanyiri, M. Kimathi, L. Luboobi, Mathematical analysis of influenza A dynamics in the emergence of drug resistance. Comput. Math. Methods Med. Art. ID 2434560, 14 (2018)
58. H.K. Khalil, *Nonlinear Systems*, 2nd edn. (Prentice Hall, Upper Saddle River, 1996)
59. L. Kohaupt, Introduction to a Gram Schmidt type biorthogonalization method. Rocky Mt. J. Math. **44**(4), 1265–1279 (2014)
60. M. Kot, *Elements of Mathematical Ecology* (Cambridge University Press, Cambridge, 2001)
61. M.A. Krasnosel'skij, J.A. Lifshits, A.V. Sobolev, *Positive Linear Systems*. Sigma Series in Applied Mathematics, vol. 5 (Heldermann Verlag, Berlin, 1989)
62. A.K. Laird, Dynamics of tumour growth. Br. J. Cancer **18**(3), 490 (1964)
63. A. Lajmanovich, J.A. Yorke, A deterministic model for gonorrhoea in a nonhomogeneous population. Math. Biosci. **28**(3–4), 221–236 (1976)
64. J. LaSalle, S. Lefschetz, *Stability by Liapunov's Direct Method, with Applications*. Mathematics in Science and Engineering, vol. 4 (Academic Press, New York-London, 1961)
65. J.D. Logan, *An Introduction to Nonlinear Partial Differential Equations*. Pure and Applied Mathematics, 2nd edn. (Wiley-Interscience [John Wiley & Sons], Hoboken, 2008)
66. K.G. Lorenz, The Natchez of Southwest Mississippi, in *Indians of the Greater Southeast* (2000), pp. 142–177
67. D.G. Luenberger, *Dynamic Systems* (J. Wiley & Sons, Inc., New York-Chichester-Brisbane-Toronto, 1979)

68. C.R. MacCluer, The many proofs and applications of Perron's theorem. SIAM Rev. **42**(3), 487–498 (2000)
69. M. Martcheva, *An Introduction to Mathematical Epidemiology*. Texts in Applied Mathematics, vol. 61 (Springer, New York, 2015)
70. M. Mesterton-Gibbons, *A Concrete Approach to Mathematical Modelling*. A Wiley-Interscience Publication (John Wiley & Sons, Inc., New York, 1995)
71. P. Meyer-Nieberg, *Banach Lattices*. Universitext (Springer, Berlin, 1991)
72. R.E. Mickens, *Nonstandard Finite Difference Schemes—Methodology and Applications*, 2nd edn. (World Scientific Publishing Co. Pte. Ltd., Hackensack, 2021) ©2021.
73. J.D. Murray, *Mathematical Biology. I*. Interdisciplinary Applied Mathematics, vol. 17, 3rd edn. (Springer, New York, 2002)
74. J.M. Ortega, *Numerical Analysis*. Classics in Applied Mathematics, vol. 3, 2nd edn. (Society for Industrial and Applied Mathematics (SIAM), Philadelphia, 1990)
75. A. Pazy, *Semigroups of Linear Operators and Applications to Partial Differential Equations*. Applied Mathematical Sciences, vol. 44 (Springer, New York, 1983)
76. S. Portet, A primer on model selection using the Akaike Information Criterion. Infect. Dis. Model. **5**, 111–128 (2020)
77. H.L. Royden, *Real Analysis*, 3rd edn. (Macmillan Publishing Company, New York; Collier Macmillan Publishing, London, 1988)
78. W. Rudin, *Real and Complex Analysis*, 3rd edn. (McGraw-Hill Book Co., New York, 1987)
79. R. Rudnicki, *Modele i metody biologii matematycznej* (Instytut Matematyczny Polskiej Akademii Nauk, Warszawa, 2014)
80. D. Rumsfeld, There are known knowns (2002). https://en.wikipedia.org/wiki/Thereareunknownunknowns. Accessed 10 Oct 2023
81. T. Ruyama, *Analytical Population Dynamics* (Chapman & Hall, London, 1992)
82. G. Sallet, Mathematical epidemiology (2018). https://hal.science/hal-04688889
83. C.C. Sastri, Commentary-opinion: pure and applied mathematics. Not. Am. Math. Soc. **54**(4), 471 (2007)
84. J.L. Schiff, *The Laplace Transform: Theory and Applications*. Undergraduate Texts in Mathematics (Springer, New York, 1999)
85. E. Seneta, *Non-negative Matrices and Markov Chains*. Springer Series in Statistics (Springer, New York, 2006)
86. A.N. Sharkovsky, S.F. Kolyada, A.G. Sivak, V.V. Fedorenko, *Dynamics of One-Dimensional Maps*. Mathematics and Its Applications, vol. 407 (Kluwer Academic Publishers Group, Dordrecht, 1997)
87. C.P. Simon, J.A. Jacquez, Reproduction numbers and the stability of equilibria of SI models for heterogeneous populations. SIAM J. Appl. Math. **52**(2), 541–576 (1992)
88. H.L. Smith, P. Waltman, *The Theory of the Chemostat*. Cambridge Studies in Mathematical Biology, vol. 13 (Cambridge University Press, Cambridge, 1995)
89. G. Strang, *Introduction to Linear Algebra*, 3rd edn. (Wellesley-Cambridge Press, Wellesley, 2003)
90. S.H. Strogatz, *Nonlinear Dynamics and Chaos*, 2nd edn. (Westview Press, Boulder, 2015)
91. J.R. Swanton, *Indian Tribes of the Lower Mississippi Valley and Adjacent Coast of the Gulf of Mexico*, vol. 43 (US Government Printing Office, Washington, 1911)
92. A. Świerniak, A. Polański, J. Śmieja, M. Kimmel, J. Rzeszowska-Wolny, Asymptotic analysis of three random branching walk models arising in molecular biology. Control. Cybern. **32**, 147–161 (2003)
93. H.R. Thieme, *Mathematics in Population Biology*. Princeton Series in Theoretical and Computational Biology (Princeton University Press, Princeton, 2003)
94. H. Tverberg, A proof of the Jordan curve theorem. Bull. Lond. Math. Soc. **12**(1), 34–38 (1980)
95. G. Uribe, *On the Relationship Between Continuous and Discrete Models for Size-Structured Population Dynamics*. University of Arizona (1993)

96. P. van den Driessche, J. Watmough, Reproduction numbers and sub-threshold endemic equilibria for compartmental models of disease transmission. Math. Biosci. **180**, 29–48 (2002)
97. R.S. Varga, *Matrix Iterative Analysis*. Springer Ser. Comput. Math., vol. 27, 2nd revised and expanded edition (Springer, Berlin, 2000)
98. M. Vidyasagar, *Nonlinear Systems Analysis*. Classics in Applied Mathematics, vol. 42 (Society for Industrial and Applied Mathematics (SIAM), Philadelphia, 2002)
99. V.S. Vladimirov, *Generalized Functions in Mathematical Physics* (Mir Publishers, Moscow, 1979)
100. W. Walter, *Ordinary Differential Equations*. Graduate Texts in Mathematics, vol. 182 (Springer, New York, 1998)
101. Wikipedia, de Havilland Comet. https://en.wikipedia.org/wiki/De_Havilland_Comet.html. Accessed 26 Mar 2024

Index

J. Banasiak, *Introduction to Mathematical Methods in Population Theory*, Springer Undergraduate Mathematics Series,
https://doi.org/10.1007/978-3-031-65491-6

MIX
Papier aus verantwortungsvollen Quellen
Paper from responsible sources
FSC® C105338

If you have any concerns about our products,
you can contact us on
ProductSafety@springernature.com

In case Publisher is established outside the EU,
the EU authorized representative is:
Springer Nature Customer Service Center GmbH
Europaplatz 3, 69115 Heidelberg, Germany

Printed by Libri Plureos GmbH
in Hamburg, Germany